INTERNATIONAL SERIES OF MONOGRAPHS ON PHYSICS

SERIES EDITORS

J. BIRMAN	CITY UNIVERSITY OF NEW YORK
S. F. EDWARDS	UNIVERSITY OF CAMBRIDGE
R. FRIEND	UNIVERSITY OF CAMBRIDGE
M. REES	UNIVERSITY OF CAMBRIDGE
D. SHERRINGTON	UNIVERSITY OF OXFORD
G. VENEZIANO	CERN, GENEVA

AF895870

INTERNATIONAL SERIES OF MONOGRAPHS ON PHYSICS

163. B.J. Dalton, J. Jeffers, S.M. Barnett: *Phase space methods for degenerate quantum gases*
162. W.D. McComb: *Homogeneous, isotropic turbulence – phenomenology, renormalization and statistical closures*
161. V.Z. Kresin, H. Morawitz, S.A. Wolf: *Superconducting state – mechanisms and properties*
160. C. Barrabès, P.A. Hogan: *Advanced general relativity – gravity waves, spinning particles, and black holes*
159. W. Barford: *Electronic and optical properties of conjugated polymers, Second edition*
158. F. Strocchi: *An introduction to non-perturbative foundations of quantum field theory*
157. K.H. Bennemann, J.B. Ketterson: *Novel superfluids, Volume 2*
156. K.H. Bennemann, J.B. Ketterson: *Novel superfluids, Volume 1*
155. C. Kiefer: *Quantum gravity, Third edition*
154. L. Mestel: *Stellar magnetism, Second edition*
153. R.A. Klemm: *Layered superconductors, Volume 1*
152. E.L. Wolf: *Principles of electron tunneling spectroscopy, Second edition*
151. R. Blinc: *Advanced ferroelectricity*
150. L. Berthier, G. Biroli, J.-P. Bouchaud, W. van Saarloos, L. Cipelletti: *Dynamical heterogeneities in glasses, colloids, and granular media*
149. J. Wesson: *Tokamaks, Fourth edition*
148. H. Asada, T. Futamase, P. Hogan: *Equations of motion in general relativity*
147. A. Yaouanc, P. Dalmas de Réotier: *Muon spin rotation, relaxation, and resonance*
146. B. McCoy: *Advanced statistical mechanics*
145. M. Bordag, G.L. Klimchitskaya, U. Mohideen, V.M. Mostepanenko: *Advances in the Casimir effect*
144. T.R. Field: *Electromagnetic scattering from random media*
143. W. Götze: *Complex dynamics of glass-forming liquids – a mode-coupling theory*
142. V.M. Agranovich: *Excitations in organic solids*
141. W.T. Grandy: *Entropy and the time evolution of macroscopic systems*
140. M. Alcubierre: *Introduction to 3+1 numerical relativity*
139. A.L. Ivanov, S.G. Tikhodeev: *Problems of condensed matter physics – quantum coherence phenomena in electron-hole and coupled matter-light systems*
138. I.M. Vardavas, F.W. Taylor: *Radiation and climate*
137. A.F. Borghesani: *Ions and electrons in liquid helium*
135. V. Fortov, I. Iakubov, A. Khrapak: *Physics of strongly coupled plasma*
134. G. Fredrickson: *The equilibrium theory of inhomogeneous polymers*
133. H. Suhl: *Relaxation processes in micromagnetics*
132. J. Terning: *Modern supersymmetry*
131. M. Mariño: *Chern-Simons theory, matrix models, and topological strings*
130. V. Gantmakher: *Electrons and disorder in solids*
129. W. Barford: *Electronic and optical properties of conjugated polymers*
128. R.E. Raab, O.L. de Lange: *Multipole theory in electromagnetism*
127. A. Larkin, A. Varlamov: *Theory of fluctuations in superconductors*
126. P. Goldbart, N. Goldenfeld, D. Sherrington: *Stealing the gold*
125. S. Atzeni, J. Meyer-ter-Vehn: *The physics of inertial fusion*
123. T. Fujimoto: *Plasma spectroscopy*
122. K. Fujikawa, H. Suzuki: *Path integrals and quantum anomalies*
121. T. Giamarchi: *Quantum physics in one dimension*
120. M. Warner, E. Terentjev: *Liquid crystal elastomers*
119. L. Jacak, P. Sitko, K. Wieczorek, A. Wojs: *Quantum Hall systems*
117. G. Volovik: *The Universe in a helium droplet*
116. L. Pitaevskii, S. Stringari: *Bose-Einstein condensation*
115. G. Dissertori, I.G. Knowles, M. Schmelling: *Quantum chromodynamics*
114. B. DeWitt: *The global approach to quantum field theory*
113. J. Zinn-Justin: *Quantum field theory and critical phenomena, Fourth edition*
112. R.M. Mazo: *Brownian motion – fluctuations, dynamics, and applications*
111. H. Nishimori: *Statistical physics of spin glasses and information processing – an introduction*
110. N.B. Kopnin: *Theory of nonequilibrium superconductivity*

Advanced Statistical Mechanics

Barry M. McCoy
CN Yang Institute for Theoretical Physics
State University of New York
Stony Brook, NY

OXFORD
UNIVERSITY PRESS

OXFORD
UNIVERSITY PRESS

Great Clarendon Street, Oxford, OX2 6DP,
United Kingdom

Oxford University Press is a department of the University of Oxford.
It furthers the University's objective of excellence in research, scholarship,
and education by publishing worldwide. Oxford is a registered trade mark of
Oxford University Press in the UK and in certain other countries

© Barry McCoy 2010

The moral rights of the author have been asserted

First published 2010
First published in paperback 2015

All rights reserved. No part of this publication may be reproduced, stored in a retrieval system, or transmitted, in any form or by any means, without the prior permission in writing of Oxford University Press, or as expressly permitted by law, by licence or under terms agreed with the appropriate reprographics rights organization. Enquiries concerning reproduction outside the scope of the above should be sent to the Rights Department, Oxford University Press, at the address above

You must not circulate this work in any other form
and you must impose this same condition on any acquirer

Published in the United States of America by Oxford University Press
198 Madison Avenue, New York, NY 10016, United States of America

British Library Cataloguing in Publication Data
Data available

Library of Congress Cataloging in Publication Data
Data available

ISBN 978–0–19–955663–2 (Hbk.)
ISBN 978–0–19–874426–9 (Pbk.)

Links to third party websites are provided by Oxford in good faith and for information only. Oxford disclaims any responsibility for the materials contained in any third party website referenced in this work.

Preface

The best way to become acquainted with a subject is to write a book about it.

Benjamin Disraeli

The subject of statistical mechanics dates back to the 19th century. It has a rich history, and the basics of the subject are taught in all undergraduate and graduate physics programs. Consequently there is a wealth of books that explain the elementary aspects of the subject which form the foundation for all thermal properties of condensed matter systems. The content of these books is all rather similar in that they cover thermodynamics, ensemble theory, one-body problems and the perfect Bose and Fermi gases. These topics are all considered to be closed subjects which are thoroughly understood.

This book is an outgrowth of the author's teaching of advanced courses in statistical mechanics which go beyond the topics covered in elementary courses and is aimed at introducing the reader to topics in which there is ongoing research. In contrast to the material in an elementary course almost all topics lead to open questions, and the aim of this book is to present these topics of ongoing research to as wide an audience as possible. Consequently in almost all chapters there are sections on open questions and what I call missing theorems where one's physical intuition suggests that results should be true but for which no proof yet exists. It is hoped that, by highlighting the many places where there are unresolved questions, this book can stimulate progress in the field.

The selection of topics in any advanced treatment of a subject is affected by the tastes of the author and so several comments about my selection of topics are in order. I have chosen to divide the subject somewhat arbitrarily into three parts: exact general theorems; series expansions and numerical results; and solvable models. Each of these divisions has an immense literature and within the confines of one book it is not possible to state all results and prove all known theorems. I have therefore adopted the procedure of stating and explaining many results but have only given the proofs of a selection of the theorems stated. There is no other alternative since there are many important theorems whose proof in the literature requires papers of 40–50 pages. For example the proof of the stability of matter is the subject of a book in its own right by Elliot Lieb; Rodney Baxter devotes an entire book to the free energy and order parameters of the six-vertex, eight-vertex and hard hexagon models; T.T. Wu and the present author devote an entire book to the Ising model. In this sense this present book can be considered to be an introduction and guide to, but is hardly a substitute for, the literature of the past 50 years.

The reader will almost instantly note that there are several well-known topics in statistical physics which are not covered in this book: namely the renormalization group and mean field theory. This omission is deliberate since both of these topics are well covered in many books and it is, in my opinion, superfluous to give one more account of these methods.

Constant progress is being made in statistical mechanics and it is certain that even at the time of publication some topics will have advanced beyond what is presented here. In particular I draw the reader's attention to several examples of recent work: the proof discussed in chapter 4 of Kepler's conjecture that no packing of hard spheres in three dimensions can be more dense than the face centered cubic lattice, and the discovery also presented in chapter 4 that for ellipsoids there are packings denser than the fcc lattice. There are several computations presented which were initiated in part by the desire to clarify and better understand the existing literature, in particular the computation of the tenth order virial coefficients in chapter 7, the diagonal susceptibility and the evaluation of the form factor integrals of the Ising model in chapter 10 and 12 and the treatment of the TQ equation of the eight-vertex model in chapter 14.

There are also many interesting and important problems which are omitted merely for lack of space. In particular there are many solvable models which have not been mentioned. Furthermore there is no discussion of the methods of the coordinate and algebraic Bethe's ansatz and the mathematics of quantum groups and conformal field theory. These topics require much more space than this book allows and are treated extensively by other authors.

However, it is hoped that in spite of the many necessary omissions that this book covers a sufficiently large number of topics so that the reader will gain an appreciation of the great breadth of the subject; the many areas of progress which have been made in the past 40–50 years, and the places where future advances will be made.

I am fond of saying that "you cannot say that you understand a paper until you generalize it." This, of course, leads to the logical corollary that "no author can be said to understand his/her most recent paper." This book is an excellent demonstration of the truth and meaning of this corollary. In every chapter there are open questions and topics that need further research and explanation. Some of the more obvious and unavoidable of these questions have been singled out for discussion but many, if not most, are quietly hidden away waiting for the reader to discover them. There are many derivations and computations presented and, barring misprints, the proofs should be sufficient to prove the conclusions. But in no place is it ever shown that the given proof is actually necessary for the conclusion and that the steps exhibited actually reveal the mechanism for the phenomena being discussed. The most glaring example of this problem is the evaluation of integrals done in chapter 12 by the use of MAPLE for which, at the time of writing, no analytic derivation exists.

There are many places in this book where I need to thank collaborators and friends for their help and suggestions: Nathan Clisby for the evaluations of virial coefficients in chapter 7; Jean-Marie Maillard for teaching me how to do the symbolic evaluation of integrals on the computer in chapter 12; Klaus Fabricius for collaboration on the Q matrices of the eight-vertex model of chapter 14; and Jacques Perk and Helen Au-Yang for collaboration on chiral Potts models and for figures 4 and 5 of chapter 15. The method used in chapter 6 to prove the Mayer expansion comes from a set of lectures given by Hans Groeneveld in the late 1960s who has given me much valuable advice in the preparation of that chapter.

I am most grateful to the National Science Foundation for partial support during much of the time when I was writing this book and to the Rockefeller Foundation for a one-month residency at their Bellagio Conference and Study Center where several chapters were revised and perfected.

In conclusion I must thank and acknowledge two remarkable people to whom I am deeply indebted and without whose encouragement and inspiration this book would never have been completed. The first is my late wife, Tun-Hsu Martha McCoy , who helped me every step of the way and put up with the innumerable frustrations I have had during the far too many years I have spent in writing. The other is my classmate of 51 years ago from Catalina High School in Tucson, Arizona, Margaret Hagen Wright, who has given me profound friendship in a time of great need.

Stony Brook, New York
2009

Contents

PART II SERIES AND NUMERICAL METHODS

Part I

General Theory

A child of five could understand this. Fetch me a child of five.

Groucho Marx

1 Basic principles

This book is called *Advanced Statistical Mechanics* and as such it is assumed that the reader has studied thermodynamics and knows the ensemble formulation of statistical mechanics. The derivation of these principles from classical mechanics and the application of these laws to free and simple one-body systems is covered in previous courses at both the undergraduate and graduate level. This course is devoted to applying these principles to the study of interacting many-body systems to derive properties of macroscopic matter from microscopic interactions.

However, in order to have a common starting point for our investigations we will in this chapter summarize the laws and results of thermodynamics and present the formulation of ensemble theory in the form which it will be used in this course.

1.1 Thermodynamics

The study of large systems originates with the investigations of the 19th century into the subject of thermodynamics, a subject of great importance for the development of steam power which was at the heart of the industrial revolution. The laws of thermodynamics rule our daily lives and the thermodynamic notions of heat, temperature and efficiency are as common as the weather report and the kitchen stove. And yet for all of its ubiquitous truth thermodynamics is a strange subject. Its invention in the 19th century did not develop logically from Newton's laws of motion. Rather it seems to be a "final cause" in the sense of Aristotle. We are forced to study thermodynamics because nature empirically turns out to work this way. Because thermodynamic behavior exists it must follow from the microscopic laws of nature, regardless of how much we may not like it. Indeed, the late 19th century discussions of various "paradoxes" connected with Poincaré recurrence and irreversibility demonstrate that there were many who did not seem to like thermodynamics one bit.

In this section we will sketch these laws of thermodynamics. Their microscopic justification can only be said to have been completed in the late 20th century. The phenomena of thermodynamics form the simplest and best understood properties of large systems.

1.1.1 Macroscopic, extensive and intensive

Thermodynamics begins with a definition of the words macroscopic, extensive and intensive.

By macroscopic we mean that for a system of N particles in a volume V we are interested in the behavior when

$$N \to \infty \text{ and } V \to \infty \text{ with } V/N = v \text{ fixed} \tag{1.1}$$

This limit is called the thermodynamic limit, v is called the specific volume, and $1/v = \rho$ is called the (number) density.

The words extensive and intensive refer to the behavior of properties of this N-body system in the thermodynamic limit. A property is called intensive if it is independent of N in the limit (1.1) and is called extensive if it is linear in N. More precisely we have

$$\lim_{N\to\infty} \text{intensive} = \text{constant} \tag{1.2}$$

and

$$\lim_{N\to\infty} \text{extensive}/N = \text{constant} \tag{1.3}$$

and these limits are independent of the shape of the system. Thermodynamics assumes that there are no other ways that a property can depend on the number of particles. In particular, if thermodynamics is applicable to a system, its total energy must be extensive.

But this assertion that only intensive and extensive properties exist already imposes a limitation on the type of microscopic interactions we are allowed to deal with. In particular we must be able to guarantee that the system will neither collapse because of short distance attraction nor explode because of long distance repulsion.

The problem with short distance attraction is seen by considering the familiar case of a collection of N hard spheres of radius r and mass m where the interaction between two spheres of different masses separated by a distance d is the gravitational potential energy

$$V(d) = -\frac{Gm_1m_2}{d}. \tag{1.4}$$

To compute the potential energy of a spherical collection of these small spheres consider adding one sphere to a spherical collection of N spheres. If we let R denote the radius of the composite sphere, then the dependence of R on N is found by computing the volume of the collection of N spheres

$$(4/3)\pi R^3 = \text{constant } N(4/3)\pi r^3 \tag{1.5}$$

where the constant is geometrically determined by closest packing. Thus for large N

$$R \sim N^{1/3} \tag{1.6}$$

From (1.4) and (1.6) we see that the potential energy lost in adding one sphere to the collection of N spheres is

$$\frac{Gm(mN)}{R} \sim \frac{Gm^2N}{N^{1/3}} \sim N^{2/3} \tag{1.7}$$

and hence if we integrate from one to N we find that the average gravitational potential energy of the N closest packed gravitating hard spheres is proportional to $-N^{5/3}$ which is neither intensive nor extensive. Therefore we conclude that the early 19th century laws of thermodynamics will not apply to gravity which was the only force which had been mathematically investigated at the time that thermodynamics was first invented!

The possibility of a large system exploding in the thermodynamic limit is most dramatically seen in the construction of an atomic bomb. Here there is a critical mass beyond which an uncontrolled chain reaction sets in and there is no sense in which equilibrium describes the physics.

A slightly less drastic situation is the case of a system with a net charge. Here an electrostatic charge will reside on the surface of the system and the total energy will depend on the shape of the system.

The conditions on the potential which allow intensive and extensive behavior were not really understood until the mid 20th century, and it was not until the 1960s, when it was shown that the ground state energy of a system of electrons interacting with a neutralizing positive charge background is linear in N, that there was a proof that any realistic set of microscopic forces actually will have the dependence on the number of particles assumed by thermodynamics. We will study these questions in detail in chapter 3.

1.1.2 Equilibrium

The most "obvious" empirical property of macroscopic matter is that if we isolate it (for example by putting it in a Dewar) and wait long enough all changes in the system will die out. This is to be contrasted with a system of a few degrees of freedom such as two or three billiard balls on a frictionless table which will keep on bouncing forever.

Much of statistical mechanics deals with the computations of properties of this equilibrium state. However, the very statement of what is meant by equilibrium seems to fly in the face of the first law of motion. Consequently it is of great importance to understand where the concept comes from. To use the words proof or derivation is far too strong. The entire development of classical statistical mechanics can be said to be part of the proof of the existence of equilibrium.

The notion of equilibrium is in essence a statement that for a very large system there are several time scales. One of these time scales is finite when the number of particles goes to infinity. This scale may be on the order of microseconds or hours but once this time scale is reached the extensive and intensive properties of the large system cease to change. However, there may be other much longer time scales which become infinite as the size goes to infinity and on these extremely large scales change may again occur. Thermal equilibrium refers to the time after these finite time scales have been surpassed but the infinite ones have not yet been reached.

This statement of waiting long enough is admittedly vague because there are systems like glass which change slowly over a period of several hundreds of years. Consequently it is often useful to give an alternative definition and to say that something is in equilibrium if we can describe all of its properties that are subject to macroscopic measurements in terms of a few intensive and extensive variables (such as pressure and density) *which do not depend on the past history of the system.* These few variables needed to characterize the system are said to specify its state. Properties that depend only on the state of the system and not on the past history are called state functions. The independence of past history is clearly a genuine restriction on the types of phenomena that can be discussed. Thus the discussion of equilibrium properties marks only the beginning of the study of the properties of large systems.

The dramatic feature of equilibrium is that it is an irreversible concept which puts into physics a direction of time even if the underlying microscopic equations are time reversal invariant. One of the first tasks of microscopic statistical mechanics is to demonstrate the existence of these phenomena.

1.1.3 The four laws of thermodynamics

Thermodynamics is embodied in the following four laws which are abstracted from our observations of macroscopic matter in equilibrium:

Zeroth Law: If, of three bodies A, B and C, the bodies A and B are separately in equilibrium with C then A and B are in equilibrium with each other.

First Law: If the state of an otherwise isolated system is changed by the performance of work, the amount of work needed depends solely on the change accomplished and not on the means by which the work is performed, or on the intermediate stages through which the system passes between its initial and final states.

Second Law: (Clausius) It is impossible to devise an engine which, working in a cycle, shall produce no effect other than the transfer of heat from a colder to a hotter body;

(Kelvin) It is impossible to devise an engine which, working in a cycle, shall produce no effect other than the extraction of heat from a reservoir and the performance of an equal amount of work.

Third Law: (Nernst) The entropy change associated with any isothermal, reversible process approaches zero as $T \to 0$;

(Fowler and Guggenheim) It is impossible by any procedure to reduce any system to absolute zero in a finite number of operations.

The first three of these four laws have a purely classical origin and each leads to the existence of a new state function.

The zeroth law allows us to define the state function called empirical temperature (denoted by ϕ) which has the property that any two bodies in equilibrium have the same value of the empirical temperature. This empirical temperature defines a set of isotherms relative to one (arbitrarily) chosen standard system called a thermometer.

The first law lets us define the state function called internal energy. The change in this internal energy in going from one state to another is the work added to the system in thermal isolation in changing the state. The typical experimental demonstration of this is the Joule paddle wheel experiment in which the stirring of a mass of water raises the temperature. Once this state function is defined we can consider the more general situation where the change of state occurs without the system being thermally isolated. For example, by holding a Bunsen burner under a flask we can change the state without adding any mechanical work at all. In this general situation the change in internal energy consists of two terms; the work done on the system and the heat, ΔQ, transferred to the system. In symbols

$$\Delta U = \Delta W + \Delta Q. \tag{1.8}$$

Moreover, in connection with the first law we recognize that there are two distinct ways to change the state of a system. One way is to perform the change so slowly that equilibrium is maintained at all times. At any time we can reverse the direction of this change and thus such quasistatic changes are called reversible. The slow pulling out of a piston where

$$\Delta W = -PdV \tag{1.9}$$

exemplifies this type of change. On the other hand we could suddenly expand a gas into a free volume by breaking a membrane. Now only the initial and final states of the system are in equilibrium. Such changes cannot be reversed in time and are called irreversible.

The second law allows the definition of a state function, S, called entropy that describes the change in heat during a reversible change. In particular in a reversible change

$$\Delta Q = dQ = TdS, \tag{1.10}$$

and hence

$$dU = TdS - PdV. \tag{1.11}$$

Here T, called the absolute temperature, is a function of the empirical temperature alone and, if the thermometer is chosen as defined from the perfect gas law

$$Pv = k_B T \tag{1.12}$$

where $k_B = 1.38 \times 10^{-23}$Joules/degree Kelvin is called Boltzmann's constant, then the absolute and the empirical scales of temperature agree. This calibration of thermometers to agree with the perfect gas temperature is clearly a difficult task to do at very low temperatures where there are no good approximations to a perfect gas. Nevertheless great experimental ingenuity has gone into the design of thermometers that make experimental measurements in terms of absolute temperature with great accuracy and we will always unite the concepts of empirical and absolute temperature in this book and speak of temperature alone.

These state functions are measurable and of great importance in studying macroscopic systems. One of the most common of measurements is of the extensive quantity C called the heat capacity, defined as the change in the heat with respect to temperature at some fixed external conditions. For example, in a gas specified by the state variable P and V there are two commonly measured heat capacities:

$$C_v = \left(\frac{dQ}{dT}\right)_V \quad \text{and} \quad C_p = \left(\frac{dQ}{dT}\right)_P \tag{1.13}$$

It is also common to divide by the number of particle N and consider the specific heats

$$c_v = C_v/N \quad \text{and} \quad c_p = C_p/N. \tag{1.14}$$

The computation of specific heats will be one of the major topics of this book.

It remains to consider the third law. This was only discovered in the 20th century and stems from quantum instead of classical mechanics. Its major consequence for us can be seen if we combine the definition of the entropy state function (1.10) with the definition of heat capacity (1.13) to write

$$dS = C(T)dT/T. \tag{1.15}$$

Thus

$$S(T_2) - S(T_1) = \int_{T_1}^{T_2} dTC(T)/T. \tag{1.16}$$

The third law says that as $T_1 \to 0$ this entropy change goes to a constant and thus we see that the heat capacity of any system must vanish as $T \to 0$. The rate at which the heat capacity vanishes gives important information about the quantum interactions of the system.

The derivation of these state functions from the laws, and the relations between experimental quantities that can be derived from these laws, requires space and attention to detail. In particular we will find it useful at times to use the quantities

1. Helmholtz free energy

$$A = U - TS \quad \text{where } dA = -PdV - SdT \tag{1.17}$$

2. Gibbs function

$$G = U - TS + PV \quad \text{where } dG = VdP - SdT \tag{1.18}$$

3. Enthalpy

$$H = U + PV \quad \text{where } dH = TdS + VdP, \tag{1.19}$$

the four relations of Maxwell

$$\left(\frac{\partial T}{\partial V}\right)_S = -\left(\frac{\partial P}{\partial S}\right)_V$$
$$\left(\frac{\partial T}{\partial P}\right)_S = \left(\frac{\partial V}{\partial S}\right)_P$$
$$\left(\frac{\partial V}{\partial T}\right)_P = -\left(\frac{\partial S}{\partial P}\right)_T$$
$$\left(\frac{\partial P}{\partial T}\right)_V = \left(\frac{\partial S}{\partial V}\right)_T \tag{1.20}$$

and the internal energy of the perfect gas

$$U = \frac{3N}{2} k_B T. \tag{1.21}$$

An excellent set of derivations is given in the classic book *The Elements of Classical Thermodynamics* by A.B. Pippard [1].

1.2 Statistical mechanics

Classical statistical mechanics was invented at the end of the 19th century by Maxwell, Boltzmann and Gibbs as a way to compute the phenomena of thermal equilibrium from the microscopic laws of classical mechanics. However, if the study of thermal equilibrium required a detailed knowledge of the solution of Hamilton's equations even our knowledge in late 20th century would be insufficient. The reason that statistical mechanics could be founded long before such things as the KAM theorem were known is precisely because the relevant mechanics is statistical.

1.2.1 Statistical philosophy

From the beginning of philosophical inquiry 2400 years ago there has been a persistent dichotomy of thought as to what constitutes reality. On the one hand there are the empiricists who, starting with Aristotle, maintain that there exists an external reality which we learn of by perception. On the other hand there are the idealists who, starting with Plato, maintain that reality consists of mental ideas which we carry in our heads. An understanding of the relation between these two extreme points of view is necessary for an understanding of the relation of statistics to mechanics.

The fundamental point in all philosophic and scientific investigations is the distinction between the observer and observed, and the relationship they have with each other. This relationship is clarified if we distinguish three situations in terms of information content:

1. **The observer has infinitely more information capacity than the observed.**
2. **The observer and the observed have comparable information capacity.**
3. **The observed has infinitely more information capacity than the observer.**

It is the case 1), where the observer has infinitely more capacity to measure and store information than the observed object has degrees of freedom, that we most commonly think of when we consider Newtonian planetary motion or more generally any mechanical system described by a few degrees of freedom in the laboratory. In this case we have large and intricate devices for measuring positions and velocities and a great deal of capacity for storing the results of these measurements. Indeed the first major advances in astronomy were due to the ability to store and correlate data using a time period of over 1000 years starting with data taken in 720 B.C. in Mesopotamia. As our ability to make measurements increases in accuracy we are able to make use of the classical laws of motion to make predictions ever further into the future. This is the situation envisioned in the philosophy of empiricism. The limiting case of an infinite specification of the observed object is what the empirical philosophy defines as "physical reality.", Such infinite specification can only be done if the observer has infinite information capacity compared to the observed object.

Case 2) listed above is the situation which we commonly meet in ordinary human affairs. In our everyday dealings with people our knowledge is inevitably incomplete and nevertheless we are forced to act on the basis of what we actually know. While we can and often do use a fiction that there is an "objective reality" in practice we never know what it is. The manner in which we predict future human actions is dominated by the severe limitations on our observations of the external world.

The final case 3), where the observed object has infinitely more degrees of freedom than the observer has information capacity to measure, is the subject of statistical mechanics. In this case the idealist philosophy of Plato is indispensable because even if one wants to believe in an immutable external empirical world it is forever inaccessible. The inescapable Platonic idea that must be introduced into the study of infinite classical systems by finite means is the idea of a density of points in phase space $\rho(p_j, q_j, t)$. This density is a mandatory concept because by definition we can never specify an individual point.

To repeat:

The density of points in phase space is a pure Platonic idea. It exists in the mind only and cannot be measured empirically.

The object of statistical mechanics is to use these densities to make predictions of properties of infinite systems using only finite means.

1.2.2 The microcanonical ensemble

The initial question to be asked about the prediction of properties of infinite systems using only finite means is whether there are any finite means possible that will let us predict anything at all. This was the question asked by Maxwell, Boltzmann and Gibbs more than a century ago. They argued that, for a system whose energy is conserved, we at least know the value of the conserved energy. Then, if we knew absolutely nothing else about the system, we would be forced to say that we were studying a density which was constant on the surface of constant energy. This density function is called the

microcanonical ensemble:

$$\rho_{MC}(\{p_j.q_j\}) = \delta(H(\{p_j, q_j\}) - E)/\Omega(E) \tag{1.22}$$

where

$$\Omega(E) = \int d^{2N}\{p_j, q_j\}\delta(H(\{p_j, q_j\}) - E). \tag{1.23}$$

The normalizing factor $\Omega(E)$ is called the structure function.

However, it might be objected that not even the total energy can in principle be known exactly and hence we might consider as an alternative

$$\rho_{MC}(\{p_j, q_j\})' = 1/\Omega(E; \Delta) \text{ for } E \leq H \leq E + \Delta, \tag{1.24}$$

where

$$\Omega(E)' = \int_{E \leq H \leq E+\Delta} d^{2N}\{p_j.q_j\}. \tag{1.25}$$

We will refer to both of these as the microcanonical ensemble.

Maxwell, Boltzmann and Gibbs conjectured that the value of any macroscopic property A of a system with $2N$ degrees of freedom in thermal equilibrium is given in terms of this microcanonical density as

$$\langle A\rangle = \int d^{2N}\{p_j, q_j\}\rho A. \tag{1.26}$$

On the assumption that this ensemble and average formula do describe thermal equilibrium we will shortly find expressions for the thermodynamic state functions temperature, internal energy and entropy in terms of the structure function. This will demonstrate the existence of a formalism that allows the computations of nontrivial thermodynamic properties in terms of a microscopic Hamiltonian. This is done for a system of N identical particles in a volume V by identifying the thermodynamic entropy in terms of the structure function as

$$S(E,V) = k_B \ln\left[\Omega(E,V)/N!\right], \tag{1.27}$$

from which, by using(1.11), we identify the absolute temperature as

$$\frac{1}{T} = \frac{\partial S(E,V)}{\partial E}. \tag{1.28}$$

To prove that this identification of the entropy is correct we must establish two key properties:

1. The limit where $N \to \infty$, $V \to \infty$ with $V/N - v$ fixed

$$\lim_{N\to\infty} \frac{1}{N} S \text{ exists} \tag{1.29}$$

and

2. The temperatures of any two systems in thermal contact are equal.

The proof of these properties depends on the Hamiltonian of the system. The conditions on the potential which allow these properties will be discussed in detail in chapter 3.

1.2.3 The canonical ensemble

In the presentation of the microcanonical ensemble the typical experimental situation under consideration is one where a macroscopic system approaches equilibrium in thermal isolation from its surroundings. In practice, however, this is not the most typical experimental configuration. It is much more common to have the macroscopic system on which we are making measurements placed in thermal contact with an even larger macroscopic system (the heat bath) whose function is to keep the observed system in thermal equilibrium. In this experimental situation we have absolutely no interest in the dynamics of the heat bath whose only function is to define the temperature. We have no interest in discussing the question of how the heat bath itself approaches equilibrium. The only property of the heat bath that we will use is the empirical observation that heat baths exist.

We are only interested in the observed system and not the heat bath and thus we are free to describe the heat bath by any fiction we please just so long as thermal equilibrium is maintained. The most convenient way to do this is to let H be the Hamiltonian of the system under observation and to represent the heat bath by N_h systems each with the identical Hamiltonians H_j. Thus the total Hamiltonian of the system plus the heat bath is

$$H_{\text{tot}} = H + \sum_{j=1}^{N_h} H_j. \tag{1.30}$$

In the limit when $N_h \to \infty$ this represents the definition of a heat bath as having infinitely more degrees of freedom than the observed system. We represent the statement that the observed system is in thermal equilibrium with the heat bath by applying the microcanonical ensemble to the entire system (1.30) with the microcanonical density function

$$\rho_{MC} = \delta_{E_{tot},H_{tot}}/\Omega(E_{tot}). \tag{1.31}$$

To study the system H we need the density function with all the coordinates of the heat bath integrated out. Thus we consider

$$\rho(\vec{X}, E_{tot}) = \int d\vec{X}_{(1)} \cdots \int d\vec{X}_{(N_h)} \rho_{MC} \tag{1.32}$$

where $\vec{X}$ are the coordinates of H and $\vec{X}_{(j)}$ are the coordinates of H_j. In the limit $N_h \to \infty$ the ensemble represented by this density is called the canonical ensemble.

We evaluate the limit $N_h \to \infty$ by use of the method of steepest descents. To carry this out we will slightly simplify the formalism and assume that all energies are a common multiple of a unit ΔE. (Nothing will be lost in this argument since in the end nothing will depend on ΔE anyway.) We use this convention to write the Kronecker δ in (1.31) as

$$\delta_{j,k} = \frac{1}{2\pi} \int_{-\pi}^{\pi} d\theta \exp\left(i \frac{\theta(j-k)}{\Delta E} \right). \tag{1.33}$$

Thus, defining $\zeta = i\theta/\Delta E$, the denominator of (1.31) is written as

$$\begin{aligned}\Omega(E_{tot}) &= \int d\vec{X} d\vec{X}_{(1)} \cdots d\vec{X}_{(N)} \frac{1}{2\pi} \int_{-\pi}^{\pi} d\theta \exp\left(\frac{i\theta}{\Delta E} (E_{tot} - H - \sum_{j=1}^{N_h} H_j) \right) \\ &= \frac{\Delta E}{2\pi i} \int_{-i\pi/\Delta E}^{i\pi/\Delta E} d\zeta \; e^{(\zeta E_{tot} + \ln Z(\zeta) + N_h \ln Z_h(\zeta))}\end{aligned} \tag{1.34}$$

where we have defined

$$Z(\zeta) = \int d\vec{X} e^{-\zeta H} \text{ and } Z_h = \int d\vec{X}_{(j)} e^{-\zeta H_j}, \tag{1.35}$$

and the numerator of (1.32) is

$$\int d\vec{X}_{(1)} \cdots d\vec{X}_{(N_h)} \frac{1}{2\pi} \int_{-\pi}^{\pi} d\theta \exp\left(\frac{i\theta}{\Delta E}(E_{tot} - H - \sum_{j=1}^{N_h} H_j) \right)$$
$$= \frac{\Delta E}{2\pi i} \int_{-\pi i/\Delta E}^{\pi i/\Delta E} d\zeta e^{(\zeta(E_{tot}-H)+N_h \ln Z_h(\zeta))} \tag{1.36}$$

Consider first the structure function (1.34) in the limit $N_h \to \infty$, $E_{tot} \to \infty$ with $E_{tot}/N_h = \bar{E}$ fixed and write the integrand as

$$Z(\zeta) e^{N_h[\zeta \bar{E} + \ln Z_h(\zeta)]}. \tag{1.37}$$

The steepest descents point maximizes the value of the exponential as a function of ζ and is found as the solution of

$$\bar{E} + \frac{\partial}{\partial \zeta} \ln Z_h(\zeta) = 0. \tag{1.38}$$

We denote the solution of this equation as β which is easily shown to be real and positive. Then, deforming the contour of integration to the steepest descents path of constant phase which passes through β and on this path setting $\zeta = \beta + iy$ we find to leading order in N_h that

$$\Omega(E_{tot}) \sim Z(\beta) e^{N_h[\beta \bar{E} + \ln Z_h(\beta)]} \frac{\Delta E}{2\pi} \int_{-\infty}^{\infty} e^{-\frac{N_h}{2} y^2 \frac{\partial^2 \ln Z_h(\beta)}{\partial \beta^2}}$$
$$= Z(\beta) e^{N_h[\beta \bar{E} + \ln Z(\beta)]} \Delta E \left(2\pi N_h \frac{\partial^2 \ln Z_h(\beta)}{\partial \beta^2} \right)^{-1/2} \left[1 + O(\frac{1}{N_h}) \right]. \tag{1.39}$$

A similar steepest descents evaluation of the numerator (1.36) gives

$$e^{-\beta H} e^{N_h[\beta \bar{E} + \ln Z_h(\beta)]} \Delta E \left(2\pi N_h \frac{\partial^2 \ln Z_h(\beta)}{\partial \beta^2} \right)^{-1/2} \left[1 + O(\frac{1}{N_h}) \right]. \tag{1.40}$$

Therefore in the $N_h \to \infty$ limit all the dependence on N_h, Z_h and ΔE cancels out of (1.32) and we find that the resulting limiting density is

$$\rho_C = e^{-\beta H}/Z(\beta). \tag{1.41}$$

This is the density of the canonical ensemble and the normalizing factor $Z(\beta)$ is called the partition function. With this density the expectation value of any observable f is given as

$$\langle f \rangle = \int d\vec{X} f e^{-\beta H}/Z(\beta). \tag{1.42}$$

In particular, the internal energy is

$$U = \langle H \rangle = \int d\vec{X} H e^{-\beta H}/Z(\beta) = -\frac{\partial \ln Z(\beta)}{\partial \beta}. \tag{1.43}$$

It now remains to relate the canonical ensemble to thermodynamics. First note that, by construction, if H_1 and H_2 are two different systems in contact with the same heat bath they will both have a canonical density given by (1.41) with the same value of β. Therefore β is the same for any two systems in thermal equilibrium and thus, from the zeroth law, β must be a universal function of temperature. To find this dependence it thus suffices to consider the perfect gas

$$H = \frac{1}{2m}\sum_{j=1}^{N}\mathbf{p}_j^2 \tag{1.44}$$

confined in a volume V. For this system

$$Z(\beta) = \int d^{3N}x d^{3N}p e^{-\beta\frac{1}{2m}\sum_{j=1}^{N}\mathbf{p}_j^2} = V^N\left(\frac{2\pi m}{\beta}\right)^{3N/2} \tag{1.45}$$

and thus from (1.43) the internal energy is

$$U = \frac{3N}{2\beta}. \tag{1.46}$$

Thus comparing with the internal energy of the perfect gas obtained from thermodynamics (1.21)

$$U = \frac{3N}{2}k_BT \tag{1.47}$$

we obtain the general relation

$$\beta = \frac{1}{k_BT}. \tag{1.48}$$

To complete the relation with thermodynamics we need to express the Helmholtz free energy in terms of $Z(\beta)$. In particular we consider a system with N identical particles. In order to obtain a free energy which is extensive we set

$$Z_N(\beta) = Z(\beta)/N! \tag{1.49}$$

and define the Helmholtz free energy $A(V,T)$ from

$$Z_N(\beta) = e^{-\beta A(V,T)}. \tag{1.50}$$

Using Stirling's approximation

$$N! \sim N^{N+1/2}e^{-N}(2\pi)^{1/2} \tag{1.51}$$

we see from (1.45) that, for the perfect gas,

$$\lim_{\substack{V\to\infty,\ N\to\infty \\ v=V/N}} \frac{1}{V}\ln Z_N \tag{1.52}$$

with v fixed exists. We also find from (1.43) that

$$U = \frac{\partial}{\partial\beta}\beta A(V,T) = A(V,T) - T\frac{\partial}{\partial T}A(V,T) \tag{1.53}$$

from which, if we recall from thermodynamics (1.17) that

$$S = -\left(\frac{\partial A}{\partial T}\right)_V \text{ and } P = -\left(\frac{\partial A}{\partial V}\right)_T, \tag{1.54}$$

we conclude that the definition (1.50) does indeed satisfy all the required properties of the Helmholtz free energy.

In the canonical ensemble the energy is not precisely defined. Only the average energy is computable. But for a large system the probability that the true energy is far from the average energy is very small. As a measure of the fluctuation consider the variance $\langle H^2\rangle - \langle H\rangle^2$. On the one hand

$$\begin{aligned} \frac{\partial U}{\partial \beta} &= \frac{\partial}{\partial \beta}\int d\vec{X} H e^{-\beta H}/Z(\beta) \\ &= -\int d\vec{X} H^2 e^{-\beta H}/Z(\beta) + \left[\int \vec{X} H e^{-\beta H}/Z(\beta)\right]^2 \\ &= -\langle H^2\rangle + \langle H\rangle^2 \end{aligned} \tag{1.55}$$

and on the other hand from the thermodynamic definition of specific heat

$$-\frac{\partial U}{\partial \beta} = kT^2\left(\frac{\partial U}{\partial T}\right)_V = kT^2 N c_v \tag{1.56}$$

and thus

$$\frac{1}{N^2}\left[\langle H^2\rangle - \langle H\rangle^2\right] = kT^2 c_v/N. \tag{1.57}$$

The two terms on the left are each of order one but the right-hand side is of order $1/N$ as long at the specific heat c_v is finite in the thermodynamic limit. Thus we see that the fluctuations of the energy from the average value vanish as $N \to \infty$. This is an indication that large chaotic systems can have very predictable average quantities and is an illustration of the operation of the laws of large numbers in statistical mechanics.

1.2.4 The grand canonical ensemble

In order to use the canonical ensemble we must, in principle, compute the partition function $Z_N(\beta)$ for a system with a finite number of particles N in a finite volume V and take the limit

$$N \to \infty, \quad V \to \infty \text{ with } N/V = \frac{1}{v} = \rho \text{ fixed.} \tag{1.58}$$

However, there are situations where it is technically more convenient to consider, instead of the partition function $Z_N(\beta)$, the grand partition function

$$Q^{gr}(z) = \sum_{N=0}^{\infty} z^N Z_N(\beta). \tag{1.59}$$

In the grand canonical ensemble thermodynamic functions are expressed directly in terms of $Q^{gr}(z)$.

To express the thermodynamic functions in terms the grand partition function we use Cauchy's theorem to write

$$Z_N(\beta) = \frac{1}{2\pi i}\oint \frac{dz}{z^{N+1}} Q^{gr}(z), \tag{1.60}$$

where the contour of integration is around $z = 0$. We study $Z_N(\beta)$ in the limit (1.58) by first writing (1.60) as

$$Z_N(\beta) = \frac{1}{2\pi i}\oint \frac{dz}{z} e^{[\ln Q^{gr}(z) - N\ln z]} \tag{1.61}$$

and then using the method of steepest descents. The steepest descents point is at

$$\frac{\partial}{\partial z}[\ln Q^{gr}(z) - N\ln z] = 0 \tag{1.62}$$

which, under the assumption that $\ln Q^{gr}(z)$ is proportional to V, becomes in the limit (1.58)

$$\lim_{V\to\infty} \frac{1}{V} z\frac{\partial}{\partial z} \ln Q^{gr}(z) = \rho. \tag{1.63}$$

The value of z which solves this equation is often written as $z = e^{\beta\mu}$ and μ is called the chemical potential. We thus find

$$Z_N(\beta) \sim e^{[\ln Q^{gr}(z) - (N+1)\ln z]}\left(2\pi \frac{\partial^2 \ln Q^{gr}(z)}{\partial z^2}\right)^{-1/2} \tag{1.64}$$

and from (1.50) and (1.54)

$$P = -\frac{\partial}{\partial V}[-\beta^{-1}\ln Z_N(\beta)] \sim \beta^{-1}\frac{\partial}{\partial V}[\ln Q^{gr}(z) - (N+1)\ln z]. \tag{1.65}$$

Thus recalling that $\ln Q^{gr}(z)$ depends linearly on V we may use

$$\frac{\partial}{\partial V}\ln Q^{gr}(z) = \frac{1}{V}\ln Q^{gr}(z) \tag{1.66}$$

and the fact that z is independent of V to obtain the desired result

$$\frac{P}{k_B T} = \lim_{V\to\infty}\frac{1}{V}\ln Q^{gr}(z). \tag{1.67}$$

Finally we also obtain from (1.53)

$$\lim_{V\to\infty}\frac{U}{V} = -\frac{\partial}{\partial\beta}\lim_{V\to\infty}\frac{1}{V}\ln Q^{gr}(z). \tag{1.68}$$

1.2.5 Phases and ergodic components

The microcanonical ensemble as presented above assumes that the only conservation law possessed by the system is the conservation of energy and therefore the microcanonical ensemble averages over all states with a given energy. More precisely the microcanonical ensemble assumes that for almost all initial conditions the system dynamically evolves in time such that eventually the system will come arbitrarily close to any given point on the surface of constant energy. This is called the "ergodic hypothesis" and has been studied for well over a century.

The assumption that the energy is the only conserved quantity of a physical system in a large but finite box is technically possible because the boundary conditions of the box will break the translational invariance of the system. In the thermodynamic limit, however, boundary conditions should be irrelevant and thus we should be able to contemplate situations where the system is translationally and even rotationally invariant.

For such systems with translational and/or rotational invariance there are more absolute constants of the motion besides the energy, and consequently the phase space decomposes into ergodic components which are characterized by representations of these invariance groups. Because of the invariance symmetry group the dynamical evolution of the system is restricted to a single ergodic component.

These components which are characterized by different symmetry groups are said to be different pure phases of the system. If the representation is isotropic and homogeneous the system will be in a fluid phase but if the system has frozen into a crystal the allowed configurations will all have a crystalline symmetry and the only way that the system can pass from one phase into another is by means of interactions with the walls of the box. Therefore if we considered periodic instead of box boundary conditions each separate symmetry representation would have its own separate partition function and in order to find the "true" state of the system we would need to compute the partition function for all possible symmetry classes and then choose the free energy which was minimum. This partition function with the minimum free energy will on general depend on the temperature and a passage from say a face centered cubic to a body centered cubic phase will occur as a function of temperature when there is a crossing from one free energy to another. This is the phenomena of a first order phase transition. It cannot be seen by merely considering one phase in infinite space and can only be seen by comparing free energies of the different possible symmetry groups.

A discussion of the concept of different ergodic components in infinite space is given by Ruelle [2] and a detailed discussion of the relation of symmetry groups to phases and phase transitions is given by Landau and Lifshitz [3]. We will return to this topic in chapters 3 and 4 when we discuss theorems on the existence of thermodynamic limits and on crystalline order.

1.3 Quantum statistical mechanics

The classical statistical mechanics presented in the previous section is a powerful tool in the study of macroscopic phenomena starting from microscopic interactions. However, the ultimate laws of nature which describe microscopic interactions are not classical but are quantum mechanical. Therefore we must extend our statistical considerations to quantum statistics.

Quantum mechanics has many differences from classical mechanics. One of the most basic is that the operators of position q_j and of momentum p_j of a particle do not commute but instead satisfy the commutation relation

$$[q_j, p_k] = i\hbar\delta_{j,k}. \tag{1.69}$$

The consequence of this is that p_j and q_j cannot be simultaneously diagonalized and thus we cannot use the concept of orbit in phase space. But the concept of these orbits was basic to our classical understanding of the statistical philosophy that leads to the microcanonical ensemble.

A second fundamental difference between classical and quantum mechanics is that in a finite size system the quantum energy levels are discrete whereas the classical energy is continuous. This discreteness of energy levels is the very reason for the word "quantum." A major consequence of this discreteness is that the time behavior of any state or operator will be oscillatory and thus it is not to be expected that we can find ergodic behavior in the long time development of finite quantum systems. This is in great contrast to classical systems.

Indeed there is no logical derivation of quantum statistical mechanics from first principles. Thus instead of attempting to derive a quantum version of the microcanonical and canonical ensemble from still underived "first principles" we will here state the rule which is used for computing thermal averages in quantum statistical mechanics.

Quantum canonical ensemble:
The quantum thermal average of any operator A is given as

$$\langle A\rangle = \mathrm{Tr}A\rho \tag{1.70}$$

with

$$\rho = e^{-\beta H}/Z(\beta) \quad Z(\beta) = \mathrm{Tr}e^{-\beta H} \tag{1.71}$$

where $\beta = 1/k_BT$ and, by Tr, we mean the trace over all the states of the finite quantum system.

This formula is the most obvious quantum mechanical extension of the canonical ensemble of classical mechanics. Indeed it is the only such extension known. However, owing to our incomplete understanding of the energy levels and eigenfunctions of large nonintegrable quantum systems we confess that we do not have any microscopic quantum theory from which this rule can be derived in a logically compelling manner.

1.3.1 The relation of classical to quantum statistical mechanics

Bulk matter can be described by nuclei and electrons interacting by means of the Coulomb interaction and nonrelativistic quantum mechanics. The stability theory of this system will be presented in chapter 3 but in actual practice this theory is not very tractable for concrete computations. To obtain concrete results we are forced to resort to various idealizations and approximations and the most basic of these is to separate chemistry from statistical mechanics.

A fundamental tool in the application of quantum mechanics to chemistry is the Born–Oppenheimer approximation that the masses of the nuclei are infinitely heavier than the mass of the electron. This allows the separation of A) the electronic energy levels for which quantum mechanics is responsible for chemical binding into molecules from B) the interactions of the molecules which are responsible for the bulk phases of matter.

The Coulomb interactions of electrons and nuclei which cause the formation of molecules the electrons must be treated quantum mechanically, and the fact that the kinetic and potential energy do not commute must be taken into account. However, when we consider interactions between molecules with some effective potential, the quantum effects of the kinetic energy can usually be ignored and the problem becomes one of classical statistical mechanics. However, the price which must be paid for this reduction to classical mechanics is that the potential energy may no longer be considered to be the sum of pair potentials. For example a charged ion can induce a dipole moment in a neutral atom. For such systems three and higher body potentials are needed to describe the fact that molecules are not elementary but are composite. Furthermore, even though the Coulomb potential is spherically symmetric the effective molecular potentials very often have a complicated angular dependence of which the tetrahedral bonds of the carbon atom are probably the most familiar.

The study of bulk properties which depend on quantum mechanics such as superconductivity and the (fractional) quantum Hall effect are traditionally considered as part of condensed matter physics and not as part of statistical mechanics. This is in part because these inherently quantum mechanical phenomena are often treated at absolute zero. The physics that involves both quantum mechanics and phase transitions at nonzero temperatures is still poorly understood. For example there is as yet no generally accepted theory of the computation of the superconducting phase transition temperature of high T_c materials.

The reduction of quantum mechanics with Coulomb interactions to effective classical interactions between molecules is beyond the scope of this book. See [3] for further references. In fact, even the rough qualitative picture presented above is not established with any mathematical rigor. Nevertheless we will in this book make use of this qualitative intuition, and much of our considerations will be for classical systems. This will be seen to be completely adequate for dealing with the bulk phenomena of freezing, ferromagnetism and critical phenomena.

1.4 Quantum field theory

Statistical mechanics is commonly associated with condensed matter physics. However, perhaps of equal importance is the understanding which has developed over the past 40 years that statistical mechanics and in particular classical statistical mechanics, is exceptionally closely related to Euclidean quantum field theory.

The relation between these two apparently disconnected subjects is seen from the path integral formulation of quantum field theory where the expectation value of operators A is given as

$$\langle A \rangle = \frac{1}{Z}\int [d\phi] A e^{S/\hbar} \quad \text{with} \quad Z = \int [d\phi] e^{S/\hbar} \tag{1.72}$$

where the integral $\int[d\phi]$ is a functional integral over some appropriate space and S (which is called the action) is some functional of the fields ϕ. These formulas for Z and $\langle A\rangle$ are identical with the formulas for the partition function and for averages in the canonical ensemble and thus if we identity $\hbar$ in quantum field theory with the temperature T of statistical mechanics we see that the two subjects have a total formal similarity. In this interpretation quantum fluctuations caused by $\hbar$ are analogous to temperature fluctuations caused by T.

There are of course differences between the two subjects. First of all the functional integral $\int[d\phi]$ needs precise definition before calculations can be done. One such definition is to replace continuum space by a lattice of points and to start with finite volumes and take the thermodynamic limit.

A second difference is in the choice of the action S as opposed to the classical potential $U(r)$. Actions S tend to consist of a kinetic energy part and a single site potential energy while classical potentials tend to have a complicated two-body behavior. Nevertheless there are sufficient similarities that lattice gauge theories can be regarded as isomorphic to certain classical spin systems in statistical mechanics and the best known of the solvable lattice statistical models, the Ising model, has a precise quantum field theory analogue which will be discussed later in this book.

References

[1] A.B. Pippard, *The Elements of Classical Thermodynamics*, Cambridge Univ. Press (1961).
[2] D. Ruelle, *Statistical Mechanics* (Benjamin, New York 1969) chap. 6.
[3] L.D. Landau and E.M. Lifshitz, *Statistical Physics, third edition part 1* (Pergamon 1993). See especially p. 258, chap. XIII and chap XIV.

2

Reductionism, phenomena and models

In chapter 1 we presented the fundamental philosophy of statistical mechanics. In this chapter we present the experimental phenomena for which we will use statistical mechanics to study in the remainder of this book. In order to do this we need to begin with a review of the concept of reductionism. We then present the types of phenomena which will be studied and conclude with a discussion of the theoretical models used to describe these experimental phenomena.

2.1 Reductionism

The methods used in science to study systems depend crucially on their size. The smallest length scales are studied by physics; larger scales are studied by chemistry; still larger scales are studied by biology; larger still is the scale of economics and the largest scale of all is the the universe which is the province of astrophysics and cosmology. Examples of this hierarchy are shown in Table 2.1.

Table 2.1 Levels of reductionism with their degrees of freedom and the discipline which studies them.

Level	Degrees of freedom	Discipline
Universe	Stars	Cosmology
Societies	Stock prices	Economics
Living organisms	Humans	Biology
Cells	Amoeba, blood cells	Biology
Viruses	Polio, tobacco mosaic	Biochemistry
Polymers	Hemoglobin, cholesterol	Chemistry
Molecules	Water, carbon dioxide	Chemistry
Atoms	Hydrogen, helium, argon	Physics
Nuclei	Uranium and plutonium nuclei	Physics
Particle physics	Protons, neutrons, electrons	Physics
Quantum chromodynamics	Quarks, gluons	Physics
String theory	Strings, branes	Physics

Statistical mechanics can be used at any of these levels of description and indeed statistical methods and concepts are used from string theory to economics. But in this book we confine our attention to the level of atoms and molecules. Therefore we will consider nature to be described by

The Practical Person's Theory of Everything

The nonrelativistic quantum mechanics of nucleons and electrons interacting by means of Coulomb and magnetic interactions

This theory of everything excludes such things as nuclear fission, black holes and quantum gravity but in principle it is sufficiently general and powerful to explain most physical phenomena from the scale of atoms through the scale of societies. For the practical person this is more than enough.

Unfortunately our ability to use this theory for detailed calculations is extremely limited. Even the computation of the energy levels of helium must be done numerically, and computations of the ground state energy of the electron gas at finite density have only been carried to a few orders before unpleasant logarithms arise. Consequently what appear to be the simplest problems in atomic and condensed matter physics at zero temperature already involve serious computational complexity. How then is it possible to actually use the formalism of quantum statistical mechanics at positive temperatures and pressures which was presented theoretically in the preceding chapter?

The answer to this is an integral part of the philosophy of reductionism. Each scale is governed by its own set of laws, and reductionism is the attempt to derive the laws of the larger system from the laws of the smaller system. Thus string theory tries to explain quantum chromodynamics; quantum chromodynamics tries to explain nuclear physics; atoms are built from nuclei and electrons; molecules are built from atoms and so on. The levels of description are characterized by what are referred to as degrees of freedom: quarks and gluons for quantum chromodynamics, protons and neutrons for nuclear physics, the elements of the periodic table for atomic physics on up to human beings in societies. These degrees of freedom interact with each other. The behavior of a large number of degrees of freedom in interaction with each other is what is studied by statistical mechanics.

Therefore on the scale of atoms we will, for the purposes of statistical mechanics, not attempt to compute the properties of atoms from the theory of everything. This is left to the field of atomic physics. Similarly we will not attempt to compute the properties of molecules from the theory of everything. This is the province of chemistry. We will not attempt to compute the interaction potentials between these degrees of freedom. The laws of interaction will either be taken from experiment or, more commonly, simple force laws will be assumed which are hoped to incorporate the crucial features needed to explain the phenomena we are interested in.

Furthermore, since atoms such as argon and molecules such as water are heavy when compared to electrons and nuclei, quantum mechanics in terms of both the symmetry of the wave functions and the non-commutativity of kinetic and potential energy can often be safely ignored. Thus for many purposes we will describe the statistical mechanics of atoms and molecules in terms of classical mechanics where we specify the position and momentum of the degrees of freedom and characterize the interactions between them by a (possibly many-body) potential $U(\mathbf{r}_1, \ldots, \mathbf{r}_N)$. This description is only expected to be valid in some restricted range of temperature and pressure, and some care must be taken not to extend the statistical mechanics to regions of temperature and pressure where the underlying description of the degrees of freedom becomes invalid.

2.2 Phenomena

The phenomena of greatest interest in statistical mechanics are phase transitions and critical phenomena. We here survey some of these phenomena with an emphasis on their phase diagrams.

2.2.1 Monatomic insulators

The elements with the simplest experimentally observed phase diagram are the noble gases neon (Ne), argon (Ar), krypton (Kr) and xenon (Xe) in the region of pressure and temperature where they may be considered as monatomic insulators. This phase diagram is given schematically in Fig. 2.1 where we show the equation of state as a surface in the PTv space. In Fig. 2.2 we plot the projections of the phase boundaries in the PT and Pv planes The lines indicate the phase boundary where there is a **first order phase transition** between two coexisting phases which are either solid/liquid or liquid/gas. The curve in the Pv plane which separates the liquid and gas phases is known as the coexistence curve.

The **critical point** (P_c, T_c, v_c) is at the end of the first order line in the PT plane which separates the liquid/gas phases. It is possible to go from the gas to the liquid phase in a continuous path which does not cross the first order line and thus the liquid and gas phases are sometimes collectively referred to as the fluid phase.

The **triple point** (P_t, T_t) is a point in the PT plane where three first order lines meet and where the solid, liquid and gas phases coexist with specific volumes v_s, v_l and v_g.

We give the measured critical point and triple point data for neon, argon, krypton and xenon in Table 2.2.

The phase diagrams of these noble gases have been measured up to pressures of 6.0 GPa. These results are plotted in Fig. 2.3. On the scale of this figure the critical and triple point are not visible. The solid phase of all the noble gases is the fcc (face centered cubic) lattice. The definition of the fcc and all the other Bravais lattices is recalled in the appendix.

Table 2.2 Critical and triple point data for neon, argon, krypton and xenon taken from [2]. The unit for the pressure is 1 bar$= 10^5$N/m$^2 = 10^5$Pa, the unit for specific volume is cm^3/mole and we note that atmospheric pressure is 1.013 bar.

	T_c(K)	P_c(bar)	v_c	T_t(K)	P_t(bar)	v_g	v_l	v_s
Ne	44.4	26.5	41.8	24.55	0.4332	4600	16.18	14.06
Ar	150.7	48.6	74.9	83.80	0.6893	9867	28.24	24.6333
Kr	209.5	55.2	91.3	115.76	0.7298	12850	34.32	30.013
Xe	289.7	58.4	118.0	161.39	0.8160	16050	44.31	38.59

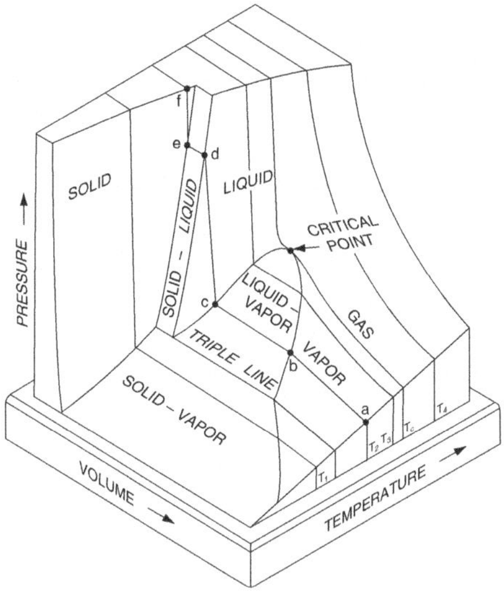

Fig. 2.1 The phase diagram in PTv space of a system which has a critical point and a triple point following [1].

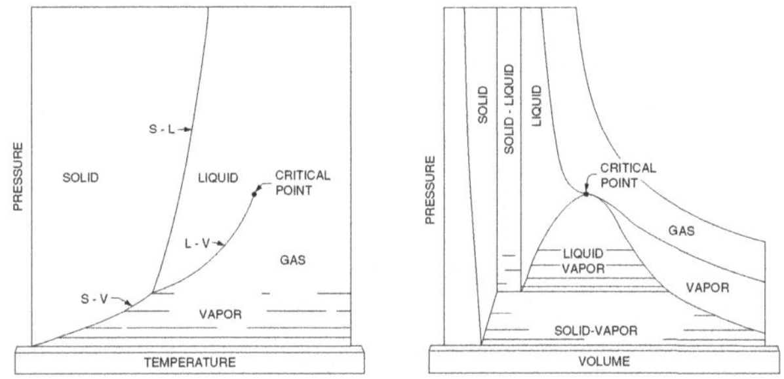

Fig. 2.2 (a)The projection on the PT plane of the PTv phase surface of Fig. 2.1. (b) The projection on the Pv plane of the PTv phase surface of Fig. 2.1 following [1]. The critical point is indicated by (P_c, T_c, v_c) and the triple point by $(P_t, T_t, v_s,\ v_l\ v_g)$.

2.2.2 Diatomic insulators

Many familiar elements such as nitrogen, oxygen and the halogens – fluorine, chlorine, iodine and bromine – exist at room temperature and pressure not as single atoms but as diatomic molecules. These diatomic molecules each have two additional rotational degrees of freedom in their kinetic energy which in the ideal gas approximation leads to a specific heat at constant volume of $5k_B/2$ instead of the $3k_B/2$ which characterizes the monatomic ideal gases. The interactions between these diatomic molecules now depends on the orientation of the molecules relative to each other.

Diatomic insulators have liquid–gas transitions and critical points just as do the monatomic insulators. We give the critical data for oxygen, nitrogen and the halogens in Table 2.3.

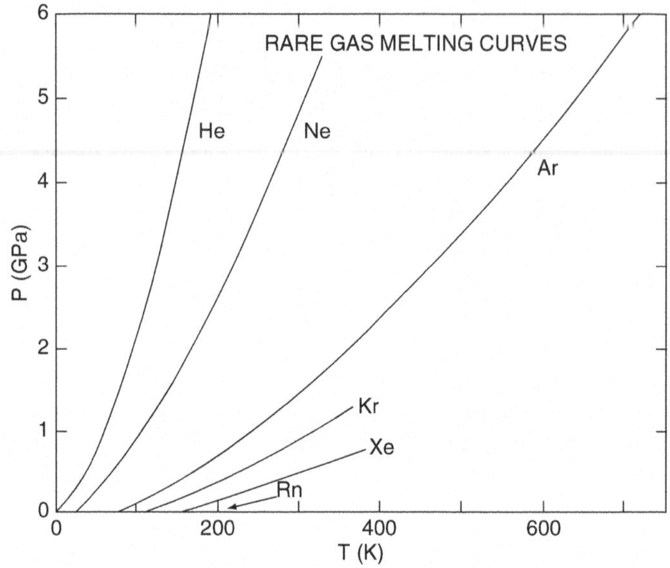

Fig. 2.3 Freezing/melting curves for the noble gases at high pressure following [3]. The unit of pressure is GPa= 10^9N/m^2. Note that 1 bar= 10^{-4}GPa and atmospheric pressure is 1.013×10^{-4}GPa so that the critical and triple points schematically shown in Figs. 2.1 and 2.2 will not be visible when plotted on this high pressure scale.

The phase diagrams of diatomic molecules are more complicated than the noble gases and show great variety which indicates that there are other properties beyond their diatomic nature which are important. We illustrate this diversity by giving in Figs. 2.4 and 2.5 the phase diagrams for fluorine, oxygen and nitrogen. The properties of the different phases of O_2 and N_2 are given in Tables 2.4.

Table 2.3 Critical point data for oxygen, nitrogen and the halogens taken from [3]. The unit for the pressure is 1 bar = 10^5N/m^2 = 10^5Pa and we note that atmospheric pressure is 1.013 bar.

	T_c(K)	P_c(bar)	v_c(cm^3/mole)
O_2	154.6	50.43	73.4
N_2	126.2	30.39	89.5
F_2	144.3	52.15	66.2
Cl_2	417	77.0	124.7
Br_2	584	103	127
I_2	819	103	155

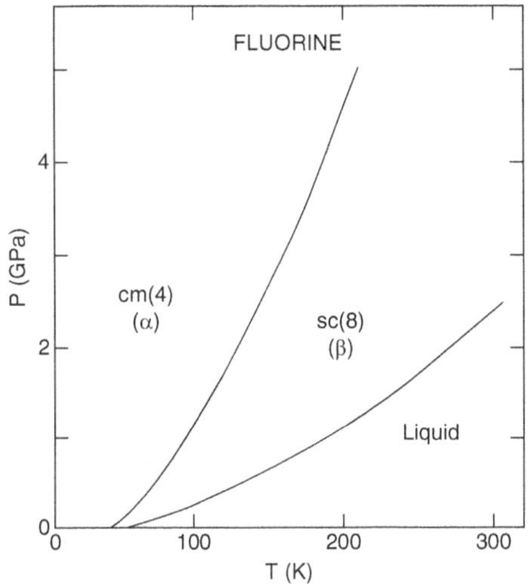

Fig. 2.4 The phase diagram of F_2 following [3]. The number in parenthesis indicates the number of molecules per unit cell. The symbols indicate the space group symmetry in the notation of appendix A.

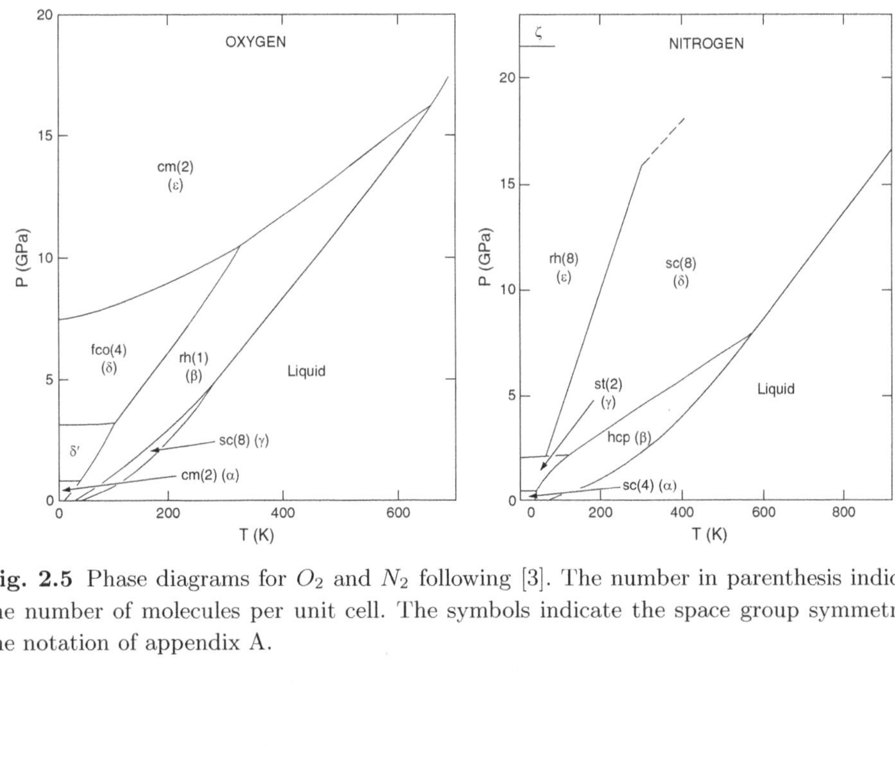

Fig. 2.5 Phase diagrams for O_2 and N_2 following [3]. The number in parenthesis indicates the number of molecules per unit cell. The symbols indicate the space group symmetry in the notation of appendix A.

Table 2.4 Properties of the phases of oxygen and nitrogen shown in Fig. 2.5 from [3].

Oxygen	
Phase	Properties
α	Monoclinic, antiferromagnetic
β	Rhombohedral, possible two-dimensional short range helicoidal order
γ	Paramagnetic
δ	Orthorhombic
δ'	Existence unclear
ϵ	Monoclinic
Nitrogen	
Phase	Properties
α	Orientationally ordered
β	Hex close packed, no orientational order
γ	Molecules in layers with common orientation
δ	Cubic
ϵ	Rhombohedral

2.2.3 Liquid crystals

Another very important class of materials where the molecular degrees of freedom behave as hard rods are organic liquid crystals. These materials are very different from the diatomic molecules in that the rods can have an orientational ordering even though the center of mass is still in a liquid state. There are three types of phases observed.

Nematic: The centers of mass have no order and form a liquid, while the rods have an orientational order.

Smectic: The centers of mass are ordered in layers and the rods are oriented perpendicular to the layers.

Cholesteric: The centers of mass are ordered in layers and the rods are ordered parallel to the layers where the direction of ordering smoothly rotates as one goes from one layer to the next.

2.2.4 Water

The most familiar of a polyatomic molecule is H_2O, water. The molecular structure of water is that the two hydrogen atoms bond to the oxygen atom with an angle of 105 degrees between the bonds. The triple and critical point data of water are given in Table 2.5. Water has the well-known but very atypical property that at atmospheric pressure it expands when it freezes. Furthermore there are many different phases of ice. The phase diagram is sketched in Fig. 2.6 in PTv space and in Fig. 2.7 in PT space.

Table 2.5 Critical and triple point data for water. The unit for the pressure is 1 bar= 10^5Newtons/m^2 = 10^5Pa and we note that atmospheric pressure is 1.013 bar.

	T_c(K)	P_c(bar)	T_t(K)	P_t(bar)
H_2O	647.30	2.2×10^3	273.16	6.0×10^{-2}

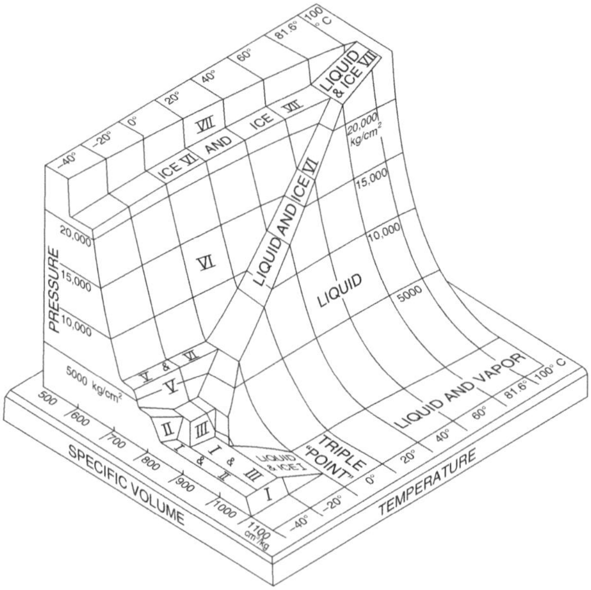

Fig. 2.6 The phase diagram for water and ice following [1].

2.2.5 Metals

Metals are characterized by the property of high electrical conductivity and therefore, unlike the monatomic and diatomic insulators, the degrees of freedom in the metallic phase must consist of ions and electrons instead of neutral molecules. Thus the phase diagrams for metals should be described by a theory much closer to the underlying Coulomb Hamiltonian than potentials used to describe neutral molecules. We show the phase diagrams of copper (Cu), gold (Ag) and silver (Au) in Fig. 2.8 and potassium (K) and sodium (Na) in Fig. 2.9. They illustrate the general rule that the phase diagrams of most metals are simpler than most diatomic and polyatomic insulators.

The critical point data of sodium and potassium are shown in Table 2.6.

Table 2.6 Critical point data for sodium and potassium from [3]. The unit for the pressure is 1 bar= 10^5Newtons/m^2 = 10^5Pa. The unit of specific volume is cm^3/mole.

	T_c(K)	P_c(bar)	v_c(cm^3/mole)
Na	2485	255	76.6
K	2198	150	

2.2.6 Helium

In the phase diagrams presented thus far the only property of the nucleus which played any role was the atomic number. However, in helium at low temperatures it makes a profound difference if the nucleus is He_3 or He_4 and this is one of the few elements where

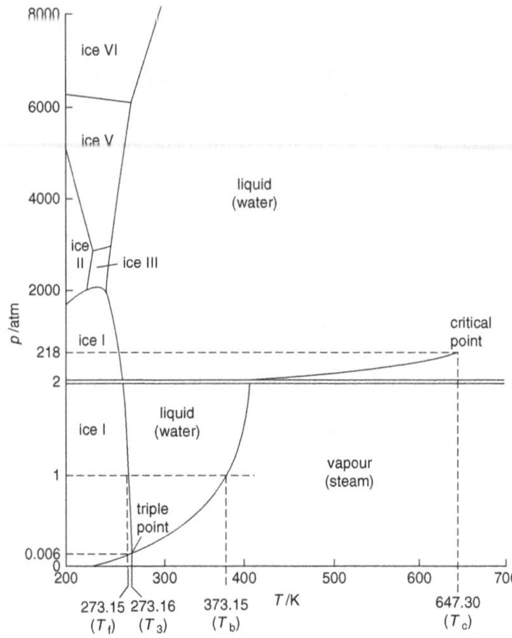

Fig. 2.7 Phase diagram in PT space for water and ice following [6]. The pressures are given in atmospheres, T_f, T_3 and T_b are respectively the temperatures of freezing, the triple point and boiling point at the pressure of one atmosphere.

the nonrelativistic Coulomb interaction must be supplemented by quantum mechanical considerations of symmetry which distinguish He_3, which obeys Fermi statistics, from He_4, which obeys Bose statistics. The low temperature phase diagrams for these two systems are shown in Fig. 2.10. For He_3 there is only one normal liquid phase in this temperature range whereas He_4 has a second order phase transition between a normal liquid I and a superfluid phase II. At much lower temperatures He_3 also shows a superfluid phase. The phase diagram at higher temperatures and pressures is shown in Fig. 2.11 where the difference between the two isotopes has become merely qualitative. The liquid/gas critical points of He_3 and He_4 are given in Table 2.7.

Table 2.7 Critical point data He_3 and He_4 taken from [3]. The unit for the pressure is 1 bar$=10^5 N/m^2 = 10^5 Pa$ and we note that atmospheric pressure is 1.013 bar.

	T_c(K)	P_c(bar)	v_c(cm^3/mole)
He_3	3.310	1.147	72.5
He_4	5.190	2.275	57.54

2.2.7 Magnetic transitions

There are two other types of very common phase transitions which need to be discussed: ferromagnetism and antiferromagnetism. For these phenomena spin degrees

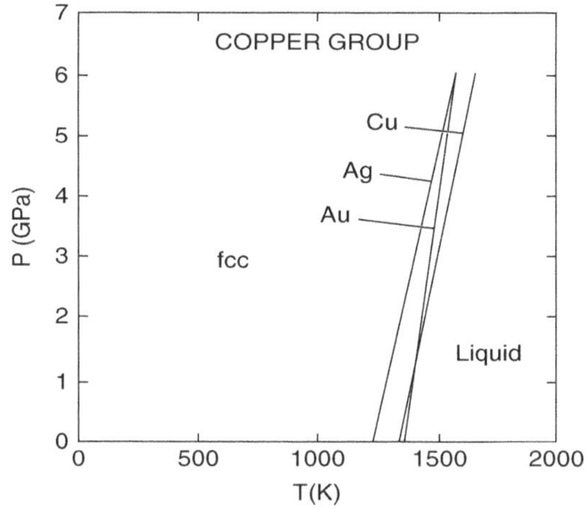

Fig. 2.8 The phase diagrams of copper, gold and silver following [3].

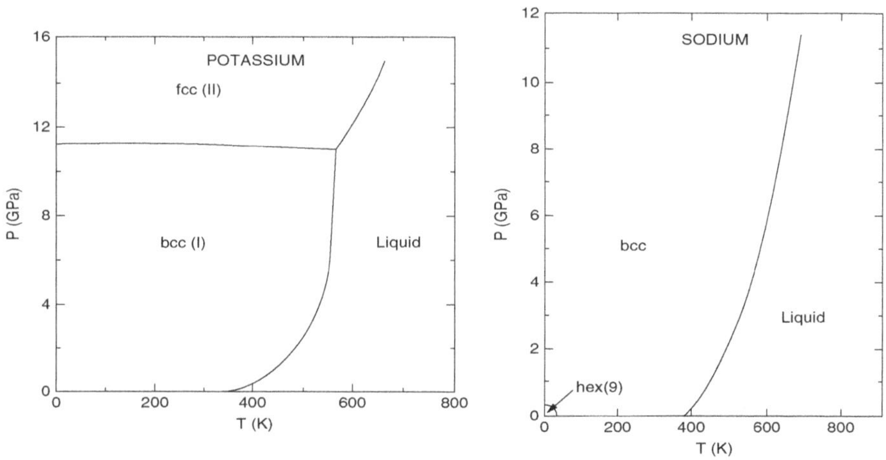

Fig. 2.9 The phase diagrams of potassium and sodium following [3].

of freedom must be included in the Hamiltonian and these degrees of freedom will interact with an external magnetic field. Phase transitions can take place in the spin degrees of freedom which are separate and distinct from the solid, liquid, gas transitions previously discussed.

Ferromagnetism

The ferromagnetic phase is characterized by the alignment of the spin degrees of freedom below a temperature T_c called the Curie temperature. Elements which have ferromagnetic phase transitions are shown in Table 2.8. Magnetism is, of course, an extremely important phenomenon and there are a very large number of alloys and compounds which are ferromagnetic. In Table 2.8 we compare the temperature T_c with the crystal phase structure at atmospheric pressure. This table reveals that

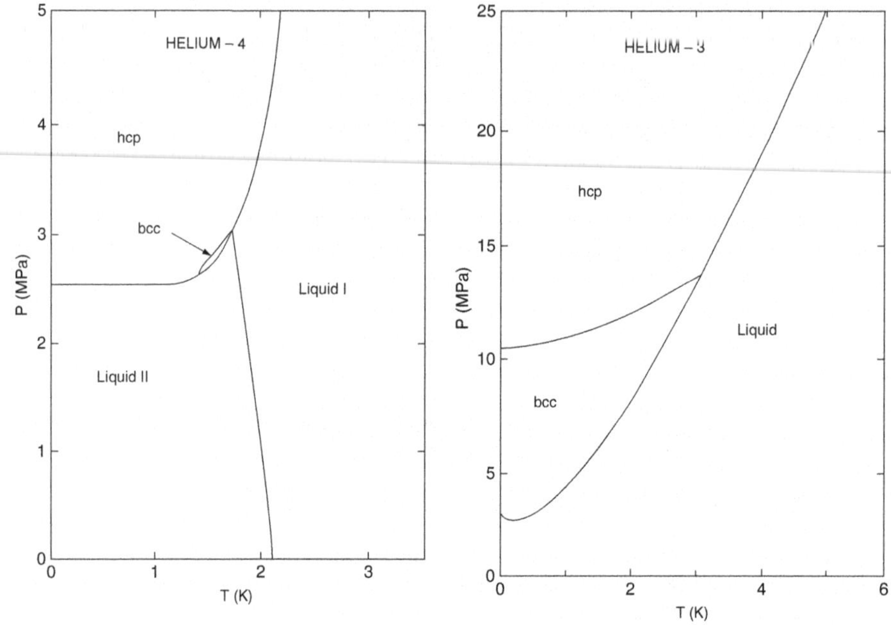

Fig. 2.10 The low temperature phase diagram of He_3 and He_4 following [3].

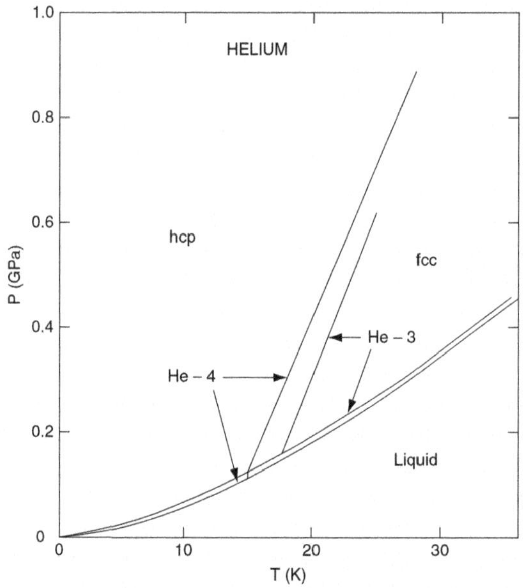

Fig. 2.11 The high pressure phase diagram of He_3 and He_4 from [3].

the ferromagnetic transition occurs substantially below the temperature at which the material has solidified into a lattice. It is therefore often possible to consider the spin degrees of freedom separately from the translational degrees of freedom.

Table 2.8 Critical (Curie) temperatures of elements which show ferromagnetism at atmospheric pressure. The phases of the lattice and the temperatures of transition are shown for comparison. The Curie temperatures are from [4] and the remaining properties from [3].

Material	$T_c(K)$	Phase boundaries
Fe	1043	bcc<1173<fcc<1660<bcc<1811(melting)
Co	1388	fcc<1768 (melting)
Ni	627	fcc<1728 (melting)
Gd	293	hcp<1520<bcc<1586
Dy	85	hcp<1655<bcc<1685

Antiferromagnetism

An antiferromagnet is characterized by the anti-alignment of nearest neighbor spin pairs below a temperature T_c which is called the Néel temperature. In Table 2.9 we give the Néel temperatures for two antiferromagnetic elements and several compounds. For the elements chromium and oxygen the Néel temperature coincides with the temperature at which there is a change in crystal structure even though in chromium the distortions in the tetragonal and orthorhombic phases from the bcc phase are small. In these cases it is much less apparent that a separation of spin degrees of freedom from translational degrees of freedom is an adequate representation of the system.

Table 2.9 Critical (Néel) temperatures of elements and compounds which show antiferromagnetism at atmospheric pressure [4] with some phase boundaries for comparison.

Material	$T_c(K)$	Phase boundaries
Cr	311	tetragonal<123<orthorhombic<311<bcc<2136 (melting)
O_2	23.7	monoclinic<23.7<rhombohedral<43.8<sc(8)<54.36 (melting)
FeO	198	
CoO	291	
NiO	600	

2.3 Models

Under the philosophy of reductionism all of the phenomena surveyed in the previous section should be explained by applying the methods of quantum statistical mechanics to the system of nonrelativistic nuclei of charges Z_i and masses m_i interacting with electrons by means of Coulomb interactions, possibly supplemented by the spins of the electron and the nuclei. This is very appealing because we are able to say that the degrees of freedom and their interactions are all known.

Unfortunately it has not proven possible to apply this method to explain or derive any of the phenomena of the preceding section. In order to make practical progress we must make some delineation of the boundaries between atomic physics and chemistry on the one hand, and statistical mechanics on the other.

One way to characterize the difference is to say that atomic physics and chemistry are concerned with the energy of the system while statistical mechanics is concerned with entropy.

In none of the phenomena of section 2.2 are the observed degrees of freedom nuclei and electrons. In the monatomic and diatomic insulators the observed degrees of freedom are molecules, and in metals they are ions and electrons. By molecules and ions we mean that the problem of solving the Schroedinger equation for the energy levels of the atoms and ions has already been solved for the individual molecules and ions in isolation.

We will apply statistical mechanics by considering these as our degrees of freedom instead of the nuclei and electrons but the price to pay for the use of these degrees of freedom is that we no longer have an a priori interaction between them. Reductionism says that in principle these interactions can be computed but in practice these computations are of limited use. Instead what is done in specific problems is to make a model that consists of a potential energy which is hoped contains the "essential" features of the interactions of the observed degrees of freedom and to use statistical mechanics to study this model.

It can be claimed that such computations only give information about the model and are not related to the genuine physical phenomena. This objection is in fact a very general philosophical problem which was extensively discussed by Kant 200 years ago in his famous statement that you can never understand the "thing in itself". No matter how much we try we can never study reality; what we do study is our mental model of reality. If our mental model of reality agrees with what we observe then the model is called a successful explanation. If our model makes predictions which disagree with observations we must discard the model no matter how elaborate the computations are or how esthetically appealing the model is.

We have indeed already adopted a model of reality when in the previous section we described phenomena by specifying nuclei such as argon, oxygen and potassium instead of specifying the number of neutrons and protons in the system and then contemplating a computation which would tell us what the various elements would be. This is manifestly beyond the scope of what the equilibrium statistical mechanics presented in the previous chapter can possibly deal with.

With this serious philosophic warning we will present some of the various models used to discuss the phenomena of the previous section. It must always be kept in mind that as the pressure and temperature change there may be need to change the model for the observed degrees of freedom. We are most interested in using statistical mechanics to study phase changes, and a model is only useful in explaining phase changes if it is the same on both sides of the phase boundary. Statistical mechanics has no predictive power if different models must be used in different phases.

2.3.1 Continuum models

We begin our theoretical discussion of the phenomena of section 2.2 by considering the phase diagram Fig. 2.2 and Fig. 2.3 of the noble gases. These elements are described as neutral spherically symmetric atoms. Because these atoms are all relatively heavy it is natural to treat their interactions classically.

The simplest model of the interaction of these atoms is to assume that there are only spherically symmetric two-body interactions to be considered and all models of this two-body interaction contain two principle features: 1) A short range repulsion which models the core of the atom and 2) a longer range attraction sometimes called a van der Waals force. The two most studied spherically symmetric two body potentials $U(r)$ are the

Lennard-Jones (m, n) potential

$$U(r) = (\sigma/r)^n - \epsilon(\sigma/r)^m \tag{2.1}$$

with $n > m > 3$ and the

Attractive square well

$$U(r) = \begin{cases} +\infty \text{ for } 0 \leq r \leq \sigma \\ -\epsilon \text{ for } \sigma \leq r \leq r_0 \\ 0 \text{ for } r_0 \leq r \end{cases} \tag{2.2}$$

For fixed values of (m, n) the Lennard-Jones potential has two parameters, whereas the attractive square well has three. The most popular values of (m, n) for the Lennard-Jones potential are $(6, 12)$. In the limit $\epsilon \to 0$ the Lennard-Jones potential (2.1) reduces to the

Inverse power law potential

$$U(r) = (\sigma/r)^n \tag{2.3}$$

and if further $n \to \infty$ the inverse power law potential reduces to the

Hard sphere potential

$$U(r) = \begin{cases} +\infty \text{ for } 0 \leq r \leq \sigma \\ 0 \text{ for } \sigma \leq r \end{cases} \tag{2.4}$$

The attractive square well also reduces to (2.4) when $\epsilon \to 0$.

To be considered as successful the model potentials (2.1) and (2.2) need to have a liquid–gas critical point, a first order freezing transition and a triple point. We will study these potentials in chapter 8. For the purely repulsive power law (2.3) and hard sphere potential (2.4) there is no expectation of a critical point but we intuitively might expect that the first order freezing transition will remain. This question will be extensively discussed in chapters 7–9. It is fair to say that none of these questions has been definitively answered and that even these very simple models have many interesting open questions.

The one feature which the phase diagrams of fluorine, oxygen and nitrogen of Fig. 2.4 and Fig. 2.5 have in common with the noble gases is the existence of a liquid–gas critical point. This leads to the suggestion that the physics of this critical point may not be very sensitive to the details of the potentials. This insensitivity of critical point phenomena to details of the potentials will be explored in chapter 5.

We next consider the diatomic insulators which we illustrated by the phase diagrams of fluorine, oxygen and nitrogen in Figs. 2.4 and 2.5. The observable degrees of freedom are diatomic molecules so the model for the interaction potential must depend on the angular orientation of the molecules. The fact that the interactions are no longer spherically symmetric is reflected in the fact that almost none of the solid phases have the high symmetry of the fcc lattice which was the solid phase of the noble gases but instead have lower symmetry of rhombic or monoclinic lattices. However, in spite of this lower symmetry of the solid phases these diatomic molecules have liquid–gas critical points just as do the spherically symmetric noble gases. This apparent insensitivity of the liquid–gas critical point is thought to indicate that the long range attractive part of the potential is similar to the long range part of the noble gas potential. This "universality" of the liquid–gas critical point will be explored in chapter 5.

The solid phases of the diatomic insulators, however, show great variety. This reflects the possibility that, in addition to angular dependent classical potentials which show the same type of short range repulsion and long range attraction that the noble gases have, new forces such as dipole interactions may also be present. In addition, for the phase boundaries of oxygen and nitrogen at low temperature and low pressure a classical approximation may not be adequate. This would seem particularly to be the case for oxygen where antiferromagnetism accompanies the phase change to the α phase in Fig. 2.5. We will briefly touch on calculations of classical potentials for hard convex molecules in chapter 7 but discussions of dipole or quantum effects are beyond the scope of our considerations.

The models for water are much more complicated than those for diatomic molecules. The nonconvex shape of the molecule with the 105 degree angle between the bonds to the hydrogen atoms plays a significant role in forming cage-like structures in the liquid phase. There is a large literature of explanations of the remarkable features of water such as its expansion on freezing but we will not discuss this most important of all transitions in detail.

The phase diagrams for metals, as illustrated by copper, gold, silver and potassium, in Fig. 2.6, however, are much simpler than either water or diatomic insulators. These phase diagrams share with the noble gases the property of having the fcc and bcc cubic symmetric lattices as the solid phases instead of the less symmetric rhomboid of monoclinic symmetry. This is a strong indication that the potentials are once again spherically symmetric. On the other hand metals must have free electrons as physical degrees of freedom because otherwise no conduction could occur. Therefore the degrees of freedom for which we build our statistical mechanics will have at least two species of particles: electrons and positively charged ions. However, the interactions cannot be pure classical Coulomb interactions. Not only because this is unstable but because the ions have an intrinsic size just as atoms do.

The elements that we have called metals in the previous section all have electrons and ions as their degrees of freedom at room temperature and pressure. But many other elements, even those that we called insulators have the potential of showing metallic behavior at elevated temperature or pressure. For example if the thermal energy in a gas is higher than the ionization energy of the atoms which constitute the gas then ionization will in fact occur and even a noble gas can in principle become a plasma. This is a change in the underlying atomic physics and will fundamentally change the properties of the system.

Similarly if any element is put under enough pressure so that the electrons of the individual atoms are forced to overlap, a classical description will cease to be valid and, at high enough pressure, a description in terms of a quantum mechanical electron gas will become appropriate. This system is a liquid, not a solid, at high pressure and zero temperature. This implies that at sufficiently high pressure the many melting curves seen in the phase diagrams of the previous section will eventually meet the axis of zero temperature. The pressures needed for this have never been reached in experimental practice but should happen in principle.

Finally the phase diagrams for He_3 and He_4 need to be mentioned. The high pressure phase diagram of Fig. 2.11 is quite normal and it can be expected that it can be explained by considerations of classical spherically symmetric potentials as are used to model the other noble gases. But the low temperature phase diagrams of Fig. 2.10 are drastically different, and quantum mechanics must be used. There is a wealth of literature on these two very important materials, and we will make no attempt to cover this topic here.

2.3.2 Lattice models

In the study of freezing transitions the underlying physics is described in the continuum, and the crystalline phases exist because of a phase transition. But we saw in the phase data for ferromagnetism in real materials in Table 2.7 that the Curie temperature, below which spontaneous magnetization occurs, is substantially below the freezing temperature. Therefore it is appropriate to discuss ferromagnetism in terms of a model of spins fixed to the sites of the underlying crystal lattice and to have the spins only interact with each other. The most important of these classical lattice spin models are the Ising model, the XY model and the Heisenberg model. These can be regarded as the cases $n = 1, 2, 3$ of the general n vector model with $O(n)$ symmetry..

The n vector model

The simplest classical model of a magnet has a spin variable of n components at each lattice site which has the Hamiltonian

$$\mathcal{H} = -\sum_{\mathbf{R}_1,\mathbf{R}_2} J(\mathbf{R}_1,\mathbf{R}_2)\mathbf{S}_{\mathbf{R}_1}\cdot\mathbf{S}_{\mathbf{R}_2} - H\sum_{\mathbf{R}_1} S^i_{\mathbf{R}_1} \tag{2.5}$$

where $\mathbf{S}_\mathbf{R}$ is a classical vector with n components at the site $\mathbf{R}$ which satisfies

$$\mathbf{S}^2 = 1, \tag{2.6}$$

and the magnetic field is chosen arbitrarily to point in the i direction. This model is variously called the n vector model, the $O(n)$ model, or the nonlinear sigma model. The special case $n = 3$ is called the Heisenberg model and the case $n = 2$ is called the XY model.

Ising model

The special case $n = 1$ of the n vector model (2.5) is called the Ising model and in this case the interaction energy is conventionally written as

$$\mathcal{E}_I = -\sum_{\mathbf{R},\mathbf{R}'} J(\mathbf{R},\mathbf{R}')\sigma_\mathbf{R}\sigma_{\mathbf{R}'} - H\sum_\mathbf{R}\sigma_\mathbf{R} \tag{2.7}$$

where $\sigma_\mathbf{R} = \pm 1$ and the sum is over nearest neighbor sites of the lattice. In two dimensions at $H = 0$ with the interactions restricted to nearest neighbors on the square, or triangular lattice, the free energy, spontaneous magnetization and correlation functions of this model can be exactly computed. We will study this model in detail in chapters 10–12.

The Quantum Heisenberg and XYZ model

The most general form of the quantum Heisenberg model is

$$\mathcal{H} = -\sum_{\mathbf{R}_1,\mathbf{R}_2} J(\mathbf{R}_1,\mathbf{R}_2)\mathbf{S}_{\mathbf{R}_1}\cdot\mathbf{S}_{\mathbf{R}_2} - H\sum_{\mathbf{R}_1} S^z_{\mathbf{R}_1} \tag{2.8}$$

where H is a magnetic field in the z direction and the sum is over all sites $\mathbf{R}_1, \mathbf{R}_2$ of the lattice. The spin operators on different sites commute with each other and, on the same site, the spin operators obey the commutation relations

$$[S^x_{\mathbf{R}}, S^y_{\mathbf{R}}] = iS^z_{\mathbf{R}}, \quad [S^y_{\mathbf{R}}, S^z_{\mathbf{R}}] = iS^x_{\mathbf{R}}, \quad [S^z_{\mathbf{R}}, S^x_{\mathbf{R}}] = iS^y_{\mathbf{R}} \tag{2.9}$$

with

$$\mathbf{S}^\dagger_{\mathbf{R}} = \mathbf{S}_{\mathbf{R}} \quad \text{and} \quad \mathbf{S}^2_{\mathbf{R}} = S(S+1) \tag{2.10}$$

where S is an integer or half integer and is the magnitude of the quantum spin. If $J(\mathbf{R}_1,\mathbf{R}_2) > 0$ the system is ferromagnetic. If $J(\mathbf{R}_1,\mathbf{R}_2) < 0$ the system is antiferromagnetic. In many applications only nearest neighbor interactions are allowed. The spin 1/2 case has been studied in great detail and for spin 1/2 with translationally invariant nearest neighbor interactions where

$$S^l_{\mathbf{R}} = \frac{1}{2}\sigma^l_{\mathbf{R}} \tag{2.11}$$

where $\sigma^j_{\mathbf{R}}$ are the Pauli spin matrices at site $\mathbf{R}$

$$\sigma^x = \begin{bmatrix} 0 & 1 \\ 1 & 0 \end{bmatrix}, \quad \sigma^y \begin{bmatrix} 0 & -i \\ i & 0 \end{bmatrix}, \quad \sigma^z = \begin{bmatrix} 1 & 0 \\ 0 & -1 \end{bmatrix}, \tag{2.12}$$

the Hamiltonian (2.8) reduces to

$$\mathcal{H}_{1/2} = -\frac{J}{4}\sum_{\mathbf{R},\mathbf{R}'}\{\sigma^x_{\mathbf{R}}\sigma^x_{\mathbf{R}'} + \sigma^y_{\mathbf{R}}\sigma^y_{\mathbf{R}'} + \sigma^z_{\mathbf{R}}\sigma^z_{\mathbf{R}'}\} - \frac{H}{2}\sum_{\mathbf{R}}\sigma^z_{\mathbf{R}}, \tag{2.13}$$

where $\mathbf{R}$ and $\mathbf{R}'$ are nearest neighbors.

The interaction (2.13) is isotropic in spin space. The generalization to the anisotropic case

$$\mathcal{H}_{\mathrm{XYZ}} = -\frac{1}{4}\sum_{\mathbf{R},\mathbf{R}'}\{J^x\sigma^x_{\mathbf{R}}\sigma^x_{\mathbf{R}'} + J^y\sigma^y_{\mathbf{R}}\sigma^y_{\mathbf{R}'} + J^z\sigma^z_{\mathbf{R}}\sigma^z_{\mathbf{R}'}\} - \frac{H}{2}\sum_{\mathbf{R}}\sigma^z_{\mathbf{R}} \tag{2.14}$$

is called the XYZ model and in one dimension this model is of immense importance. We will study it in detail in chapters 13 and 14.

In the special case $J^x = J^y = 0$ the quantum Hamiltonian (2.14) reduces to the classical Ising model (2.7).

Lattice gases

Even though the phase diagrams of the noble gases, diatomic molecules, water and the metals sodium and potassium are qualitatively very different they all share the common feature that they have a liquid–gas phase transition which ends in a critical point. Furthermore it is apparent in the data of section 2.2 that these critical points are all far removed in temperature and pressure from the triple point, melting curve and solid–solid transitions. Accordingly it is sensible to look for one common model which may explain critical points in isolation from all other features of a phase diagram. One such model is the lattice gas of Lee and Yang [5].

A lattice gas is a model for a gas or fluid in which continuum space is replaced by a lattice of points (or cells) and the particles are characterized by either being at the point (or cell) or not. The lattice may be in any dimension D and we denote by γ the number of nearest neighbors which any site has. Thus $\gamma = 4$ for the square lattice in $D = 2$, and $\gamma = 6$ for the cubic lattice in $D = 3$.

Any potential which will produce a critical point is expected to have a short range repulsive core and a longer range attraction. The lattice gas of Lee and Yang [5] models the short range repulsion by forbidding two or more particles to be at the same site, and models the attractive force by having a negative interaction energy $-\epsilon$ between particles which are nearest neighbors on the lattice. The potential is thus a lattice version of the attractive square well (2.2):

$$U(\mathbf{R}) = \begin{cases} +\infty \ \mathbf{R} = 0 \\ -\epsilon \ \mathbf{R} = \text{nearest neighbor} \\ 0 \ \text{otherwise} \end{cases} \tag{2.15}$$

Lattice gases are also used to gain insight into the phase transitions of hard core systems such as hard spheres and discs. One simple example is to have particle exclusion on the same and nearest neighbor sites so that

$$U(\mathbf{R}) = \begin{cases} +\infty \ \mathbf{R} = 0, \text{and nearest neighbor} \\ 0 \ \text{otherwise} \end{cases} \tag{2.16}$$

If the lattice is square (so that $\gamma = 4$) the model is called hard squares; if the lattice is triangular (so that $\gamma = 6$) the model is called hard hexagons. The hard hexagon model is particularly important because it is exactly solvable. It will be discussed in chapters 14 and 15.

The Ising/lattice gas correspondence

The Lee–Yang lattice gas and the Ising model, even though they are designed to model two apparently very different phenomena are in fact equivalent. To see this, consider first the lattice gas and define:

$$N = \text{the number of sites of the lattice} \tag{2.17}$$

$$N_a = \text{the number of atoms on the lattice} \tag{2.18}$$

$$N_{aa} = \text{the number of nearest neighbor pairs} \tag{2.19}$$

Then the interaction energy is

$$\mathcal{E}_G = -\epsilon N_{aa} \tag{2.20}$$

and thus the canonical partition function is

$$Q_G(N_a, T, N) = \frac{1}{N_a!} \sum e^{\beta\epsilon N_{aa}} = \sum_{N_{aa}=0}^{\infty} g(N_a, N_{aa}) e^{\beta\epsilon N_{aa}} \tag{2.21}$$

where the first sum is over all the ways of distributing N_a distinguishable atoms over the N lattice sites and $g(N_a, N_{aa})$ in the second sum is the number of ways to put N_a indistinguishable atoms on the lattice with N_{aa} nearest neighbor pairs. The grand partition function is thus

$$Q_G^{gr}(z, T, N) = \sum_{N_a=0}^{\infty} z^{N_a} Q_G(N_a, T) = \sum_{N_a=0}^{\infty} z^{N_a} \sum_{N_{aa}=0}^{\infty} g(N_a, N_{aa}) e^{\beta\epsilon N_{aa}} \tag{2.22}$$

and the equation of state is

$$\beta P_G = \frac{1}{N} \ln Q_G^{gr} \tag{2.23}$$

$$\frac{1}{v} = \frac{1}{N} z \frac{\partial}{\partial z} \ln Q_G^{gr}. \tag{2.24}$$

We consider the Ising model (2.7) by defining

$$\begin{aligned} N_+ &= \text{the number of } \sigma = +1 \\ N_- &= \text{the number of } \sigma = -1 \\ N_{++} &= \text{the number of } ++ \text{ nearest neighbor pairs} \\ N_{+-} &= \text{the number of } +- \text{ nearest neighbor pairs} \\ N_{--} &= \text{the number of } -- \text{ nearest neighbor pairs.} \end{aligned} \tag{2.25}$$

Then the Ising interaction energy (2.7) for the case $J(\mathbf{R}, \mathbf{R}') = E$ for $\mathbf{R}$ and $\mathbf{R}'$ nearest neighbors and zero otherwise is

$$\mathcal{E}_I = -E(N_{++} + N_{--} - N_{+-}) - H(N_+ - N_-). \tag{2.26}$$

The numbers N_+, N_-, N_{++}, N_{--}, N_{+-} and N are not independent. Clearly

$$N_- = N - N_+. \tag{2.27}$$

To find other relations consider drawing a line from each + spin to its γ nearest neighbors. There are γN_+ such lines. Between every ++ pair there are two lines, and between every +− pair there is one line. Thus

$$\gamma N_+ = 2N_{++} + N_{+-} \quad \text{and similarly} \quad \gamma N_- = 2N_{--} + N_{+-}. \tag{2.28}$$

Thus

$$N_{+-} = \gamma N_+ - 2N_{++} \quad \text{and} \quad N_{--} = \frac{\gamma}{2} N + N_{++} - \gamma N_+ \tag{2.29}$$

and therefore we may rewrite (2.26) as

$$\mathcal{E}_I = -4EN_{++} + 2(E\gamma - H)N_+ - (\frac{\gamma E}{2} - H)N. \tag{2.30}$$

Therefore letting $g(N_+, N_{++})$ be the number of configurations of spins with N_+ and N_{++} we find that the partition function $Z_I(H,T,N)$ of the Ising model is

$$Z_I(H,T,N) = e^{-N\beta(H-\gamma E/2)} \sum_{N_+=0}^{\infty} e^{-2\beta(E\gamma-H)N_+} \sum_{N_{++}=0}^{\infty} g(N_+, N_{++}) e^{4\beta N_{++}} \tag{2.31}$$

and the free energy F_I and magnetization M_I are given by

$$\beta F_I = -\frac{1}{N} \ln Z_I(H,T,N) \tag{2.32}$$

$$M_I = \frac{1}{N}(N_+ - N_-). \tag{2.33}$$

We now are able to compare the Ising model with the Lee–Yang lattice gas by noting that the definitions of $g(N_a, N_{aa})$ and $g(N_+, N_{++})$ are identical. Therefore if we identify N_a with N_+ and N_{aa} with N_{++} and compare (2.22)–(2.24) with (2.31)–(2.33) we obtain the correspondence of the Ising model with the lattice gas of Lee and Yang in Table 2.10.

Table 2.10 Correspondence of the Ising model with the Lee–Yang lattice gas.

Ising model	Lattice gas
N_+	N_a
N_-	$N - N_a$
$4E$	ϵ
$e^{2\beta(H-E\gamma)}$	z
$(M_I + 1)/2$	$1/v$
$-F_I - H + \gamma E/2$	P_G

The correspondence of the Ising model with the lattice gas of Lee and Yang suggests that the physics of the critical point of the liquid–gas transition and the Curie point of the ferromagnet are related. We will explore this correspondence in depth in chapter 5.

2.4 Discussion

The very simple models introduced in section 2.3 to explain the phase transitions presented in section 2.2 are all plausible representations of the interactions which in reality follow from the quantum mechanical treatment of nuclei and electrons. But just constructing a plausible model interaction does not in fact mean that the hoped for phase transitions actually do occur and that the properties of the model will give a qualitative agreement with observed properties of the system they purport to be a model of.

The remainder of this book is devoted to the study of the statistical mechanics of the models introduced in section 2.3. We will attempt to use analytic methods wherever possible and will in fact discover that there are a few model systems such as the Ising model in two dimensions with $H = 0$ and the XYZ model which can be exactly solved. We will also be able to prove rigorously the existence of phase transitions in some models for which we do not have exact solutions. However, there are many open unsolved problems even for these most simple of models where our best information comes from long series expansions and numerical simulations. None of the models have been completely solved and none of them totally reproduces the phase diagrams of section 2.2.

2.5 Appendix: Bravais lattices

In presenting phase diagrams the crystal structure of the solid phases has been specified by the abbreviation of their Bravais lattices. These 14 lattices are shown in Fig. 2.12 and their common nomenclature is given in Table 2.11. In the phase diagrams the number of atoms in the unit cell is given in parenthesis. The most common lattices of bcc(2), fcc(4) and hex(2) are commonly referred to as bcc (body centered cubic), fcc (face centered cubic) and hcp (hex close packed) respectively.

Table 2.11 Notation for the 14 Bravais lattices.

Bravais lattice	Abbreviation	Comments
simple cubic	sc	3 sides equal
body centered cubic	bcc	3 sides equal, 1 atom in center
face centered cubic	fcc	3 sides equal, 1 atom in each face
simple tetragonal	st	2 sides equal
centered tetragonal	ct	2 sides equal, 1 atom in center
simple orthorhombic	so	3 sides different
base centered orthorhombic	eco	3 sides different, atoms in 2 faces
body centered orthorhombic	bco	3 sides different, 1 atom in center
face centered orthorhombic	fco	3 sides different, 1 atom in each face
simple monoclinic	sm	3 sides different, 1 skew angle
centered monoclinic	cm	3 sides different, 1 skew angle, atoms in 2 faces
rhombohedral	rh	3 60 degree angles
hexagonal	hex	2 sides equal, 1 60 degree angle
triclinic	tr	3 sides different, 3 angles different

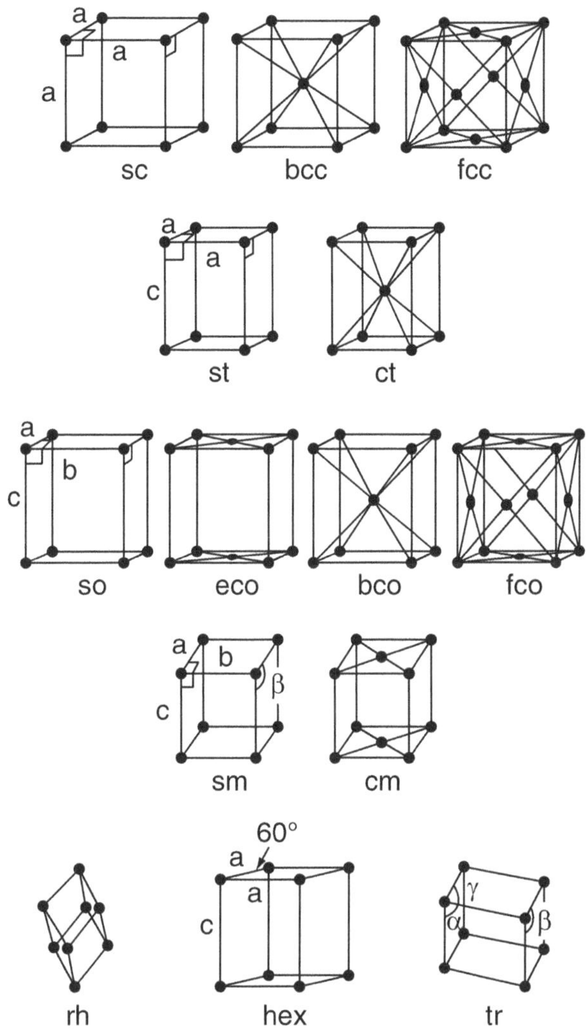

Fig. 2.12 The 14 Bravais lattices following [3]

References

[1] F.W. Sears, *An Introduction to Thermodynamics, the Kinetic Theory of Gases and Statistical Mechanics* (Addison-Wesley, Reading 1950)
[2] R.K. Crawford in *Rare Gas Solids*, M.L. Klein and J.A. Venables, eds. (Academic, London 1977) vol. 2, chap. 11.
[3] D. Young, *Phase Diagrams of the Elements* (University of California Press, Berkeley 1991)
[4] F. Keffer in *Handbuch der Physik*, vol. 18, part 2, (Springer, New York 1966).
[5] T.D. Lee and C.N. Yang, Statistical theory of equations of state and phase transitions. II. Lattice gas and Ising model, Phys. Rev. 87 (1952) 410–419.
[6] P.W. Atkins, *Physical Chemistry* (Oxford University Press 1982) edn. 2.

3

Stability, existence and uniqueness

In chapter 1 we saw that the macroscopic thermodynamic Helmholtz free energy $A(V,T)$ is computed classically from the partition function $Q_N(V,T)$ as

$$Q_N(V,T) = e^{-A(V,T)/k_BT}, \tag{3.1}$$

where

$$Q_N(V,T) = \frac{1}{N!}\int_V \prod_{j=1}^{N} d^D\mathbf{r}_j \int_{-\infty}^{\infty} \prod_{j=1}^{N} d^D\mathbf{p}_j e^{-H(\mathbf{r},\mathbf{p})/k_BT} \tag{3.2}$$

and $H(\mathbf{r},\mathbf{p})$ is the Hamiltonian of the system. For quantum statistical mechanics

$$Q_N(V,T) = \mathrm{Tr} e^{-H/k_BT}. \tag{3.3}$$

We will be most interested in the case of N identical particles of mass m interacting with a potential which depends only on position

$$H(\mathbf{r},\mathbf{p}) = \sum_{j=1}^{N} \frac{\mathbf{p}_j^2}{2m} + U^{(N)}(\mathbf{r}_1,\cdots,\mathbf{r}_N). \tag{3.4}$$

In this case the Gaussian integrals over the momentum $\mathbf{p}_j$ in the classical partition function (3.2) are easily done and we have

$$Q_N(V,T) = \frac{1}{N!\lambda^{DN}}\int_V d^D\mathbf{r}_1\cdots d^D\mathbf{r}_N e^{-U(\mathbf{r}_1,\cdots,\mathbf{r}_N)/k_BT} \tag{3.5}$$

where

$$\lambda = (2\pi m k_BT)^{-1/2}. \tag{3.6}$$

We defined the thermodynamic limit as the limit where the volume V and the number of particles N in the system both go to infinity such that the density

$$\rho = \lim_{\substack{N\to\infty \\ V\to\infty}} N/V \tag{3.7}$$

is fixed. In order for thermodynamics and statistical mechanics to apply, the Helmholtz free energy must be an extensive quantity. Therefore the limit

$$\lim_{\substack{V\to\infty,\ N\to\infty \\ v=V/N}} \frac{1}{N}\ln Q_N(V,T) \tag{3.8}$$

must 1) exist and 2) be independent of the shape of the volume V. We thus define the free energy per particle $F(v,T)$ as

$$\lim_{\substack{V\to\infty,\ N\to\infty \\ v=V/N}} \frac{1}{N}\ln Q_N(V,T) = -\frac{F(v,T)}{k_B T}, \tag{3.9}$$

and the free energy per unit volume $\tilde{F}(\rho,T)$ as

$$\lim_{\substack{V\to\infty,\ N\to\infty \\ v=V/N}} \frac{1}{V}\ln Q_N(V,T) = -\frac{\tilde{F}(\rho,T)}{k_B T}, \tag{3.10}$$

and note that

$$F(v,T) = \tilde{F}(\rho,T)/\rho. \tag{3.11}$$

This limit will not exist for all potentials and in this chapter we will study the restrictions needed on the potentials such that the thermodynamic limit can be taken. We will need restrictions on the potential such that 1) the system will not collapse and 2) the system will not explode.

The restrictions that prevent the system from collapsing are called stability conditions.

Classical stability
A classical potential for N particles in a fixed volume V is called stable if for all N

$$U^{(N)}(\mathbf{r}_1,\cdots,\mathbf{r}_N) \geq -NB \quad \text{with } B>0 \text{ and all } \mathbf{r}_i \in V. \tag{3.12}$$

From this condition we obtain the upper bound on the N-particle partition function in D dimensions

$$Q_N(V,T) = \frac{1}{N!\lambda^{DN}}\int_V d^D\mathbf{r}_1\cdots d^D\mathbf{r}_N e^{-U(\mathbf{r}_1,\cdots,\mathbf{r}_N)/k_BT} \leq \frac{1}{N!\lambda^{DN}} V^N e^{NB/k_BT} \tag{3.13}$$

which is a necessary upper bound for the existence of the thermodynamic limit. We are here interested in the case where the potential $U^{(N)}(\mathbf{r}_1,\cdots,\mathbf{r}_N)$ is a sum of two-body potentials

$$U^{(N)}(\mathbf{r}_1,\cdots,\mathbf{r}_N) = \sum_{1\leq i<j\leq N} U(\mathbf{r}_i-\mathbf{r}_j). \tag{3.14}$$

with

$$U(\mathbf{r}) = U(-\mathbf{r}). \tag{3.15}$$

In section 1 we will find sufficient conditions on the two-body pair potential $U(\mathbf{r})$ to ensure that the stability condition (3.12) holds.

Quantum Stability

A quantum mechanical system of N particles is called stable if

$$E_{gs}(N) \geq -NB \tag{3.16}$$

where $E_{gs}(N)$ is the ground state energy of the system. When this condition holds we have the bound

$$Q_N(V,T) = \text{Tr}e^{-H/k_BT} \leq e^{NB/k_BT} \times \text{number of states.} \tag{3.17}$$

As in the classical case this bound is necessary for the existence of the thermodynamic limit.

Systems that are classically stable are also stable quantum mechanically because the addition of the quantum mechanical kinetic energy will only serve to make the total energy more positive. Therefore the most interesting quantum mechanical potentials will be those which are classically unstable and the most important of these potentials is the Coulomb interaction of a collection of charged point particles which is macroscopically neutral. The most general version of this consists of N particles with masses m_j and charges e_j satisfying

$$0 \leq m_j \leq m \qquad -e \leq e_j \leq e \tag{3.18}$$

satisfying the neutrality condition

$$\sum_{i=1}^{N} e_i = 0 \tag{3.19}$$

interacting with the Hamiltonian

$$H = \sum_{i=1}^{N} \frac{\mathbf{p}_i^2}{2m_i} + \sum_{1\leq i<j\leq N} \frac{e_i e_j}{|\mathbf{r}_i - \mathbf{r}_j|} = -\sum_{i=1}^{N} \frac{\hbar^2 \nabla_i^2}{2m_i} + \sum_{1\leq i<j\leq N} \frac{e_i e_j}{|\mathbf{r}_i - \mathbf{r}_j|} \tag{3.20}$$

where

$$\nabla^2 = \sum_{j=1}^{3} \frac{\partial^2}{\partial r_j^2}. \tag{3.21}$$

The stability of this Hamiltonian (and its several special cases) is called the stability of matter, and the theory of this stability will be discussed in section 3.2.

To prove the existence of the free energy in the thermodynamic limit we need, in addition to the stability conditions of the previous sections, an additional restriction on the potential at large distances which will prevent the system from exploding. Two such general restrictions that will be sufficient for all cases except Coulomb and dipole interactions are weak and strong tempering.

Weak tempering

An N body potential $U^{(N)}(\mathbf{r}_1, \cdots \mathbf{r}_N)$ is said to be weakly tempered if, in dimension D, there exist two fixed positive bounds R_0 and u_B such that for $\epsilon > 0$

$$U^{(N_1+N_2)}(\mathbf{r}_1, \cdots, \mathbf{r}_{N_1}, \tilde{\mathbf{r}}_1, \cdots, \tilde{\mathbf{r}}_{N_2}) \\ -U^{(N_1)}(\mathbf{r}_1, \cdots \mathbf{r}_{N_1}) - U^{(N_2)}(\tilde{\mathbf{r}}_1, \cdots, \tilde{\mathbf{r}}_{N_2}) \leq \frac{N_1 N_2 u_B}{R^{D+\epsilon}} \tag{3.22}$$

if

$$|\mathbf{r}_i - \tilde{\mathbf{r}}_j| \geq R > R_0. \tag{3.23}$$

For a system with only a two-body potential (3.14) the bound (3.22) is assured if

$$U(\mathbf{r}) \leq \frac{A}{|\mathbf{r}|^{D+\epsilon}}, \qquad \text{with } A > 0 \text{ for } |\mathbf{r}| > R_0. \tag{3.24}$$

Strong tempering

A many-body potential is said to be strongly tempered if

$$U^{(N_1+N_2)}(\mathbf{r}_1, \cdots, \mathbf{r}_{N_1}, \tilde{\mathbf{r}}_1, \cdots, \tilde{\mathbf{r}}_{N_2}) \\ -U^{(N_1)}(\mathbf{r}_1, \cdots \mathbf{r}_{N_1}) - U^{(N_2)}(\tilde{\mathbf{r}}_1, \cdots, \tilde{\mathbf{r}}_{N_2}) \leq 0 \tag{3.25}$$

if

$$|\mathbf{r}_i - \tilde{\mathbf{r}}_j| \geq R > R_0. \tag{3.26}$$

For a system with only a two-body potential this bound is assured if

$$U(\mathbf{r}) \leq 0 \quad \text{for } |\mathbf{r}| \geq R_0. \tag{3.27}$$

We will study in section 3.3 the existence of the thermodynamic limit for stable systems that obey the tempering conditions (3.22) or (3.25) in both the canonical and grand canonical ensembles.

We will also study in section 3.3 the question of uniqueness of the free energy of the thermodynamic limit and its independence of boundary conditions. In particular we consider both "hard wall" boundary conditions where the particles are confined to a box with an infinite positive potential energy at the wall and periodic boundary conditions. Here we will see that the weakly tempered potentials lead to slightly more restrictive conditions on the shapes of the boundary boxes than to the strongly tempered potentials and that both give the same limiting free energy as do periodic boundary conditions. In 3.3.5 we will investigate the questions of continuity of the pressure. We will see by examples that the conditions of stability and temperedness are not sufficient to guarantee the pressure to be continuous everywhere and will find further sufficient conditions for the continuity to hold. In 3.4 we connect these theorems with the notion of ergodic components and first order phase transitions.

We conclude in section 3.5 with a discussion of extensions of these results to include surface and topological terms in the limiting behavior of the partition functions and present several open questions in section 3.6.

3.1 Classical stability

The subject of classical stability was studied in great detail by Ruelle (1963) [1], Fisher (1964) [2], Fisher and Ruelle (1966) [3], Ruelle (1969) [4] and Lenard and Sherman (1970) [5].

3.1.1 Catastrophic potentials

We first note that it is easy to give examples of a pair potential that does not satisfy the stability condition (3.12). Such potentials are referred to as catastrophic.

Example 1

$$U(\mathbf{r}) = -A < 0 \qquad \text{for } 0 \leq |\mathbf{r}| \leq a \tag{3.28}$$

Here we can have

$$|\mathbf{r}_j - \mathbf{r}_k| < a \quad \text{for} \quad 0 \leq j, k \leq N \tag{3.29}$$

so in this region, because each of the $N(N-1)/2$ pairs contributes $-A$ to the energy,

$$U^{(N)}(\mathbf{r}) = -AN(N-1)/2 \tag{3.30}$$

which certainly violates (3.12).

The argument of example 1 is valid for any pair potential such that $U(\mathbf{r}) < 0$ and thus we conclude that

$$U(0) \geq 0 \tag{3.31}$$

is a necessary condition for stability.

Example 2

Let the particles sit on the sites of a face centered cubic lattice. Each lattice site has 12 nearest neighbors at a distance R. Let the two body potential be

$$U(\mathbf{r}) = \begin{cases} a > 0 & \text{for } \mathbf{r} = 0 \\ -b < 0 & \text{for } |\mathbf{r}| = R \\ 0 & \text{otherwise.} \end{cases} \tag{3.32}$$

Consider a lattice of M sites and consider the configurations where there are n particles per site situated uniformly (with periodic boundary conditions). Then with $n = N/M$

$$U^{(N)}(\mathbf{r}_1, \cdots, \mathbf{r}_N) = M\frac{n(n-1)}{2}a - M\frac{n^2}{2}12b \tag{3.33}$$

which, using (3.32), will be negative and thus will violate (3.12) for fixed M and large N if

$$a - 12b < 0. \tag{3.34}$$

If we do not impose periodic boundary conditions then (3.33) must be modified by a surface term, which does not destroy the conclusion if M is chosen sufficiently large.

We will not attempt to make sense of the thermodynamics of catastrophic potentials in this book.

3.1.2 Conditions for stability

The following sufficient condition for classical stability (3.12) for a potential (3.14) which is the sum of two-body potentials is first given in [1] and [2].

Condition 1

If the pair potential $U(\mathbf{r})$ can be decomposed as

$$U(\mathbf{r}) = U_1(\mathbf{r}) + U_2(\mathbf{r}) \tag{3.35}$$

where $U_1(\mathbf{r})$ is nonnegative (but possibly discontinuous)

$$U_1(\mathbf{r}) \geq 0 \tag{3.36}$$

and $U_2(\mathbf{r})$ is continuous and whose the Fourier transform $\hat{U}_2(\mathbf{k})$, defined as

$$\hat{U}_2(\mathbf{k}) = \int d^D\mathbf{r} e^{i\mathbf{r}\cdot\mathbf{k}} U_2(\mathbf{r}), \tag{3.37}$$

exists for all k, is nonnegative

$$0 \leq \hat{U}_2(\mathbf{k}) < \infty \tag{3.38}$$

and is integrable

$$\int d^D\mathbf{k} \hat{U}_2(\mathbf{k}) < \infty, \tag{3.39}$$

then the condition for classical stability (3.12) holds. Functions $U_2(\mathbf{r})$ which are continuous and satisfy (3.38) and (3.39) are said to be of positive type. We prove several useful properties of positive type functions in appendix A.

The inverse of (3.37) is

$$U_2(\mathbf{r}) = \frac{1}{(2\pi)^D} \int d^D\mathbf{k} e^{-i\mathbf{r}\cdot\mathbf{k}} \hat{U}_2(\mathbf{k}) \tag{3.40}$$

and thus it follows from the positivity (3.38) and the integrability (3.39) of the Fourier transform $\hat{U}(\mathbf{k})$ that

$$|U_2(\mathbf{r})| \leq U_2(0) < \infty. \tag{3.41}$$

We note in particular that

$$0 < |U_2(\mathbf{r})| \leq U_2(0) \quad \text{if} \quad U_2(\mathbf{r}) \neq 0 \tag{3.42}$$

and that the equality in (3.42) will only hold for $\mathbf{r} \neq 0$ if $U_2(\mathbf{r})$ is periodic.

In (3.35) the nonnegative function $U_1(\mathbf{r})$ may diverge at $\mathbf{r} = 0$ or even have an infinite repulsive hard core of finite size. The function $U_2(\mathbf{r})$ contains all of the negative part of the potential.

Proof that condition 1 is sufficient for stability (3.12)

We use condition 1 to write the N-body potential (3.14) as

$$U^{(N)}(\mathbf{r}_1,\cdots,\mathbf{r}_N) = \sum_{1\le j<k\le N} (U_1(\mathbf{r}_j-\mathbf{r}_k)+U_2(\mathbf{r}_j-\mathbf{r}_k))$$

$$= \sum_{1\le j<k\le N} U_1(\mathbf{r}_j-\mathbf{r}_k) + \frac{1}{2}\left(\sum_{1\le j,k\le N} U_2(\mathbf{r}_j-\mathbf{r}_k) - NU_2(0)\right)$$

$$= \sum_{1\le j<k\le N} U_1(\mathbf{r}_j-\mathbf{r}_k) + \frac{1}{2(2\pi)^D}\int d^D\mathbf{k} \sum_{1\le j,k\le N} e^{-i\mathbf{k}\cdot(\mathbf{r}_j-\mathbf{r}_k)}\hat{U}_2(\mathbf{k}) - \frac{1}{2}NU_2(0)$$

$$= \sum_{1\le j<k\le N} U_1(\mathbf{r}_j-\mathbf{r}_k) + \frac{1}{2(2\pi)^D}\int d^D\mathbf{k} |\sum_{1\le j\le N} e^{i\mathbf{k}\cdot\mathbf{r}_j}|^2\hat{U}_2(\mathbf{k}) - \frac{1}{2}NU_2(0). \quad (3.43)$$

From (3.36) and (3.38) the first two terms are manifestly positive. Thus

$$U^{(N)}(\mathbf{r}_1,\cdots,\mathbf{r}_N) \ge -\frac{1}{2}NU_2(0) > -\infty \quad (3.44)$$

and using (3.41) the stability bound (3.12) follows.

Condition 1 is a very general condition for stability and for any given potential it requires a detailed computation to see if the condition holds. In practice, however, it is often sufficient to restrict our attention to potentials that obey the following condition, which is much easier to verify.

Condition 2

$$\text{for } |\mathbf{r}| < a_1, \quad U(\mathbf{r}) \ge C/|\mathbf{r}|^{D+\epsilon} \quad (3.45)$$

$$\text{for } a_1 \le |\mathbf{r}| \le a_2, \quad U(\mathbf{r}) \ge -w \quad (3.46)$$

$$\text{for } a_2 \le |\mathbf{r}|, \quad U(\mathbf{r}) \ge -C'/|\mathbf{r}|^{D+\epsilon'} \quad (3.47)$$

where $a_1, a_2, w, C, C', \epsilon$ and ϵ' are positive constants. The two-body potentials which satisfy this condition diverge to infinity and can never be considered to be a small perturbation of a free system. All potentials realistically encountered in practice have this divergent property.

Condition 2 is a slight specialization of the following more general condition.

Condition 3

$$\text{for } |\mathbf{r}| < a_1, \quad U(\mathbf{r}) \ge \xi(|\mathbf{r}|) \quad (3.48)$$

$$\text{for } a_1 \le |\mathbf{r}| \le a_2, \quad U(\mathbf{r}) \ge -w \quad (3.49)$$

$$\text{for } a_2 \le |\mathbf{r}|, \quad U(\mathbf{r}) \ge -\eta(|\mathbf{r}|) > -w \quad (3.50)$$

where $w > 0$ and $\xi(r)$ and $\eta(r)$ are monotonic decreasing functions of r satisfying

$$\int_0^{a_1} \xi(r) r^{D-1} dr = +\infty \quad \text{and} \quad \int_{a_2}^{\infty} \eta(r) r^{D-1} dr < \infty. \tag{3.51}$$

It is obvious that condition 2 follows from condition 3.

We will here prove that condition 1 follows from condition 3. We note that in condition 3 no assumptions have been made on the Fourier transforms of the bounding functions $\xi(r)$ and $\eta(r)$. The strategy of the proof will be to construct from $\xi(r)$ and $\eta(r)$ new bounding functions $\xi_1(r)$ and $\eta_3(r)$ whose Fourier transforms do have the desired properties.

Following the appendix of [3] it is sufficient to prove the following two lemmas:

Lemma 1

When $U(\mathbf{r})$ satisfies condition 3 there exists a nonnegative function $\eta_3(r)$ such that

$$U(\mathbf{r}) \geq -\eta_3(|\mathbf{r}|) \quad \text{for all } \mathbf{r} \tag{3.52}$$

and the Fourier transform $\hat{\eta}_3(p)$ of $\eta_3(|\mathbf{r}|)$ (which is not necessarily positive) satisfies

$$|\hat{\eta}_3(\mathbf{p})| \leq C_1(\mathbf{p}^2 + 1)^{-D} \tag{3.53}$$

for some positive constant $C_1 > 0$.

Lemma 2

When $U(\mathbf{r})$ satisfies condition 3 then there exists a nonnegative bounded function $\xi_1(\mathbf{r})$ such that

$$\xi_1(|\mathbf{r}|) \begin{cases} \leq U(\mathbf{r}) & \text{for } |\mathbf{r}| \leq a_1 \\ = 0 & \text{for } |\mathbf{r}| \geq a_1 \end{cases} \tag{3.54}$$

and the Fourier transform $\hat{\xi}_1(\mathbf{p})$ of $\xi_1(\mathbf{r})$ is integrable and has the lower bound

$$\hat{\xi}_1(\mathbf{p}) \geq C_2(\mathbf{p}^2 + 1)^{-D} \tag{3.55}$$

where $C_2 > C_1 > 0$.

Proof of lemma 1

From (3.49) and (3.50) of condition 3 we may define a nonnegative function $\eta_1(r)$ by

$$\eta_1(|\mathbf{r}|) = \begin{cases} w & \text{if } |\mathbf{r}| \leq a_2 \\ \eta(|\mathbf{r}|) & \text{if } |\mathbf{r}| \geq a_2 \end{cases} \tag{3.56}$$

such that

$$U(\mathbf{r}) \geq -\eta_1(|\mathbf{r}|). \tag{3.57}$$

Furthermore we introduce a quantity b such that $0 < b < a_2$ and define a second function $\eta_2(r)$ by

$$\eta_2(r) = \begin{cases} w & \text{if } r \leq b \\ \eta_1(r-b) & \text{if } b \leq r. \end{cases} \tag{3.58}$$

The functions $\eta_1(|\mathbf{r}|)$ and $\eta_2(|\mathbf{r}|)$ are obviously nonnegative and nonincreasing. The function $\eta_1(|\mathbf{r}|)$ is integrable in $\mathbf{r}$ because

$$\int_0^\infty \eta_1(r) r^{D-1} dr = w \int_0^{a_2} dr r^{D-1} + \int_{a_2}^\infty dr r^{D-1} \eta(r) < \infty \tag{3.59}$$

where in the last line we have used (3.51). The function $\eta_2(|\mathbf{r}|)$ is integrable because

$$\begin{aligned} \int_0^\infty dr r^{D-1} \eta_2(r) &= w \int_0^b dr r^{D-1} + \int_b^\infty dr r^{D-1} \eta_1(r-b) \\ &= w \int_0^b dr r^{D-1} + \int_0^\infty dr \eta_1(r)(r+b)^{D-1} < \infty \end{aligned} \tag{3.60}$$

where in the last line we have again used (3.51).

From the definition of $\eta_2(r)$ and the monotonicity of $\eta_1(r)$ and $\eta_2(r)$ it follows that

$$\eta_2(|\mathbf{r}'|) \geq \eta_1(|\mathbf{r}|) \quad \text{if} \quad |\mathbf{r}' - \mathbf{r}| \leq b. \tag{3.61}$$

Now let $\psi(\mathbf{r})$ be a nonnegative (smoothing) function of $\mathbf{r}$ with continuous derivatives of all orders which vanishes outside a sphere of radius b centered at the origin normalized such that

$$\int d^D \mathbf{r} \psi(\mathbf{r}) = 1. \tag{3.62}$$

Then if we define

$$\eta_3(|\mathbf{r}|) = \int d^D \mathbf{r}' \psi(\mathbf{r} - \mathbf{r}') \eta_2(|\mathbf{r}'|) \tag{3.63}$$

we have from (3.61)

$$\eta_3(|\mathbf{r}|) \geq \eta_1(|\mathbf{r}|). \tag{3.64}$$

Furthermore because $\eta_3(r)$ of (3.63) is defined as a convolution, the Fourier transform of $\eta_3(|\mathbf{r}|)$ is

$$\hat{\eta}_3(p) = \hat{\eta}_2(p) \hat{\psi}(\mathrm{p}). \tag{3.65}$$

We note that $\hat{\eta}(\mathbf{p})$ is continuous and bounded and that because $\psi(\mathbf{r})$ has been chosen to have continuous derivatives of all orders we see from theorem B1 proven in appendix B that $\hat{\psi}(\mathbf{p})$ is continuous and vanishes as $|\mathbf{p}| \to \infty$ faster at infinity than any inverse polynomial in $|\mathbf{p}|$. Therefore from (3.65) it follows that $\hat{\eta}_3(\mathbf{p})$ is continuous and decreases faster than any inverse polynomial at $|\mathbf{p}| \to \infty$. Lemma 1 follows from this fact and from the bounds (3.57) and (3.64) (where actually the inequality (3.53) can be made stronger by replacing the right-hand side by any inverse power of $\mathbf{p}^2$).

Proof of lemma 2

We begin the proof of lemma 2 by defining an auxiliary function $\chi(\mathbf{r})$ which is continuous and satisfies

$$\chi(\mathbf{r}) \geq 0 \quad \text{and} \quad \chi(\mathbf{r}) = 0 \text{ for } |\mathbf{r}| \geq 1/2. \tag{3.66}$$

From $\chi(\mathbf{r})$ define $\chi_1(\mathbf{r})$ by the convolution

$$\chi_1(\mathbf{r}) = \int d^D\mathbf{r}'\chi(\mathbf{r}-\mathbf{r}')\chi(\mathbf{r}'). \tag{3.67}$$

It is clear that $\chi_1(\mathbf{r})$ is continuous and, from (3.66), it follows that

$$\chi_1(\mathbf{r}) \geq 0 \quad \text{and} \quad \chi_1(\mathbf{r}) = 0 \text{ for } |\mathbf{r}| \geq 1. \tag{3.68}$$

Furthermore from (3.67) the Fourier transform of $\chi_1(\mathbf{r})$ is

$$\hat{\chi}_1(|\mathbf{p}|) = \hat{\chi}(\mathbf{p})^2 \geq 0 \tag{3.69}$$

and is continuous and nonzero in some neighborhood of the origin.

Next define the function $\chi_2(\mathbf{r})$ by

$$\chi_2(\mathbf{r}) = \chi_1(\mathbf{r}) \int d^D\mathbf{p}e^{i\mathbf{p}\cdot\mathbf{r}}(\mathbf{p}^2+1)^{-D}, \tag{3.70}$$

where we show in appendix B that

$$\int d^D\mathbf{p}e^{i\mathbf{p}\cdot\mathbf{r}}(\mathbf{p}^2+1)^{-D} = \frac{\pi^{1/2}\Gamma(\frac{D-1}{2})}{\Gamma(\frac{D}{2})}\left(\frac{r}{2}\right)^{D/2} K_{D/2}(r) \tag{3.71}$$

where $K_{D/2}(r)$ is the modified Bessel function of the third kind [6].

The function $\chi_2(\mathbf{r})$ also has the properties of $\chi_1(\mathbf{r})$ that it is continuous and

$$\chi_2(\mathbf{r}) \geq 0 \quad \text{and} \quad \chi_2(\mathbf{r}) = 0 \text{ for } |\mathbf{r}| \geq 1. \tag{3.72}$$

Finally define $\chi_3(\mathbf{r})$ by

$$\chi_3(\mathbf{r}) = \chi_2(\mathbf{r})/(\max \chi_2(\mathbf{r})). \tag{3.73}$$

The function $\chi_3(\mathbf{r})$ is continuous and from (3.72) satisfies

$$0 \leq \chi_3(\mathbf{r}) \leq 1 \quad \text{and} \quad \chi_3(\mathbf{r}) = 0 \text{ for } |\mathbf{r}| \geq 1. \tag{3.74}$$

Furthermore because (3.70) is a product of two functions of $\mathbf{r}$ the Fourier transform of $\chi_3(\mathbf{r})$ is given in terms of $\hat{\chi}_1(\mathbf{p})$ as the convolution

$$\hat{\chi}_3(\mathbf{p}) = \frac{1}{\max \chi_{2(\mathbf{r})}}(2\pi)^{-D}\int d^D\mathbf{p}'\hat{\chi}_1(\mathbf{p}')[(\mathbf{p}-\mathbf{p}')^2+1]^{-D}. \tag{3.75}$$

It follows from the nonnegativity of $\hat{\chi}_2(\mathbf{p})$ that

$$\hat{\chi}_3(\mathbf{p}) \geq 0 \tag{3.76}$$

and furthermore that if we restrict the region of integration in (3.75) to a small region around the origin and note that $\hat{\chi}_1(0) > 0$ we find the lower bound

$$\hat{\chi}_3(|\mathbf{p}|) \geq C'(|\mathbf{p}|^2 + 1)^{-D}. \tag{3.77}$$

We may now use $\chi_3(|\mathbf{r}|)$ to construct the function $\xi_1(|\mathbf{r}|)$ needed for lemma 2. The function $\xi(r)$ of (3.48) diverges to ∞ as $r \to 0$ and thus for sufficiently large positive integers n ($n \geq n_0$) we may define a set of $\alpha_n < 1$ as the solutions of

$$\xi(\alpha_n) = n. \tag{3.78}$$

Define the step function

$$\xi^*(r) = n \text{ for } \alpha_{n+1} < r < \alpha_n, \tag{3.79}$$

and note that

$$\xi^*(r) \leq \xi(r). \tag{3.80}$$

Furthermore define the unit step functions

$$\theta_n(r) = \begin{cases} 1 & \text{for } r \leq \alpha_n \\ 0 & \text{otherwise} \end{cases}. \tag{3.81}$$

Thus we have

$$\xi^*(r) = n_0 + \sum_{n'=n_0}^{n} \theta_{n'}(r). \tag{3.82}$$

But the properties (3.74) of $\chi_3(r)$ imply that

$$\chi_3(r/\alpha_n) \leq \theta_n(r) \tag{3.83}$$

where $\chi_3(r)$ has a positive Fourier transform. Thus we find from (3.80), (3.82) and (3.83) that

$$\sum_{n=n_0}^{\infty} \chi_3(r/\alpha_n) \leq \xi(r). \tag{3.84}$$

We may now define the function $\xi_1(r)$ of lemma 2 as

$$\xi_1(r) = \sum_{n=n_0}^{n_1} \chi_3(r/\alpha_n). \tag{3.85}$$

Thus since $\alpha_n < 1$ the lower bound (3.77) on $\hat{\chi}_3(\mathbf{p})$ yields the lower bound of the Fourier transform of $\xi_1(|\mathbf{r}|)$ of

$$\hat{\xi}_1(|\mathbf{p}|) = \sum_{n-n_0}^{n_1} \alpha_n^D \hat{\chi}_3(\alpha_n \mathbf{p}) \geq \left(\sum_{n=n_0}^{n_1} \alpha_n^D \right) C'(\mathbf{p}^2 + 1)^{-D}. \tag{3.86}$$

We further note from(3.51) that

$$\sum_{n=n_0}^{\infty} \alpha_n^D = +\infty \tag{3.87}$$

and thus by choosing n_1 sufficiently large in (3.86) we have proven the lower bound (3.55) with the constant C_2 greater than the constant C_1 in the upper bound of (3.53) of lemma 1. The boundedness of $\chi_1(r)$ follows from the boundedness of $\chi_3(r)$ and the integrability of $\hat{\chi}_1(|\mathbf{p}|)$ follows from the boundedness of $\xi_1(r)$. Thus lemma 2 has been established.

Proof of condition 1

We now may verify that condition 1 follows from lemmas 1 and 2 by choosing the function $U_2(\mathbf{r})$ in (3.35) as

$$U_2(\mathbf{r}) = \xi_1(|\mathbf{r}|) - \eta_3(|\mathbf{r}|). \tag{3.88}$$

The positivity of the Fourier transform $\hat{U}_2(\mathbf{p})$ follows from

$$\hat{U}_2(\mathbf{p}) = \xi_1(|\mathbf{p}|) - \eta_3(|\mathbf{p}|) \geq \xi_1(|\mathbf{p}|) - |\eta_3(|\mathbf{p}|)| \geq \xi_1(|\mathbf{p}|) - C_1(|\mathbf{p}|^2 + 1)^{-D} \geq 0 \tag{3.89}$$

where the second inequality follows from lemma 1 and the last inequality follows from lemma 2 with $C_2 > C_1$. The integrability of the Fourier transform follows from the integrability of the Fourier transforms of $\chi_3(|\mathbf{r})$ and $\xi_1(|\mathbf{r}|)$ separately. Thus we have established that all the requirements of condition 1 are satisfied.

We conclude this subsection by noting that condition 1 is not necessary for stability. This is demonstrated in one dimension by the potential exhibited in (1970) by Lenard and Sherman [5]

Example 3

In one dimension the pair potential

$$U(x) = \begin{cases} 2 & \text{if } \ 0 \leq |x| \leq 1 \\ -1 & \text{if } \ 1 < |x| < 2 \\ 0 & \text{if } \ 2 \leq |x| \end{cases} \tag{3.90}$$

is shown by Lenard and Sherman in [5] to be stable (3.12) but is not of the form (3.35). This potential is a one-dimensional version of the potential (3.32) considered in example 2 above where the inequality for catastrophic collapse (3.34) has been replaced by equality. This is made more precise in [5] where a family of potentials is constructed such that (3.90) is exactly at the boundary of stability. An example is also known in two dimensions [7] of a stable potential which does not satisfy (3.35).

It would be satisfying to be able to replace these examples by an example in three dimensions. The existence of such an example is conjectured in [5] but the author has been unable to find that such an example has actually ever been produced.

3.1.3 Superstability

A classical potential for N particles in a volume V is called superstable if for all N

$$U^{(N)}(\mathbf{r}_1,\cdots,\mathbf{r}_N) \geq N(A\frac{N}{V} - B) \text{ with } B, A > 0 \text{ and } \mathbf{r}_i \in V. \tag{3.91}$$

This condition is slightly stronger than stability and in particular says that, in the thermodynamic limit when $N \to \infty$ and $V \to \infty$ with the density $\rho = N/V$ fixed, if the density is sufficiently large then the N body potential is always positive.

A potential $U^{(N)}(\mathbf{r}_1,\cdots,\mathbf{r}_N)$ in the form (3.14) of the sum of two body potentials $U(\mathbf{r})$ is superstable (3.91) if

$$U(\mathbf{r}) \geq 0 \text{ and } U(0) > 0. \tag{3.92}$$

We note that if $U(\mathbf{r}) \geq 0$ and $U(0) = 0$ we can set $\mathbf{r}_j = \mathbf{r}_0$ for all $1 \leq j \leq N$ and for this configuration $U^{(N)}(\mathbf{r}_1,\cdots,\mathbf{r}_N) = 0$. Thus this potential is not superstable (3.91). This trivial example demonstrates that stability does not guarantee superstability.

However, it is true that if the pair potential satisfies condition 1 with $U_2(\mathbf{r}) \neq 0$ that the superstability bound (3.91) is satisfied. Moreover, if instead of condition 1 with $U_2(\mathbf{r}) \neq 0$ the pair potential satisfies

$$U(\mathbf{r}) \sim \frac{C}{|\mathbf{r}|^\lambda} \text{ with } \lambda > D \text{ as } |\mathbf{r}| \to 0 \tag{3.93}$$

then the superstable bound (3.91) is replaced by the stronger bound

$$U^{(N)}(\mathbf{r}_1,\cdots,\mathbf{r}_N) \geq N[A(\frac{N}{V})^{\lambda/D} - B] \text{ with } B, A > 0 \text{ and } \mathbf{r}_i \in V. \tag{3.94}$$

Proof of superstability from condition 1 with $U_2(\mathbf{r}) \neq 0$

To prove that (3.91) follows from condition 1 with $U_2(\mathbf{r}) \neq 0$ we follow [1]. We first write

$$\begin{aligned} U^{(N)}(\mathbf{r}_1,\cdots,\mathbf{r}_N) &= \sum_{1\leq j<k\leq N} (U_1(\mathbf{r}_j - \mathbf{r}_k) + U_2(\mathbf{r}_j - \mathbf{r}_k)) \\ &\geq \sum_{1\leq j<k\leq N} U_2(\mathbf{r}_j - \mathbf{r}_k) = \frac{1}{2}\left(\sum_{1\leq j,k\leq N} U_2(\mathbf{r}_j - \mathbf{r}_k) - NU_2(0)\right) \\ &= \frac{1}{2}\left(\int_V d^D\mathbf{x} d^D\mathbf{y} U_2(\mathbf{x}-\mathbf{y}) n(\mathbf{x};\{\mathbf{r}\}) n(\mathbf{y};\{\mathbf{r}\}) - NU_2(0)\right) \end{aligned} \tag{3.95}$$

where we have set

$$n(\mathbf{x};\{\mathbf{r}\}) = \sum_{j=1}^{N} \delta(\mathbf{x} - \mathbf{r}_j) \text{ with } \mathbf{r}_j \in V. \tag{3.96}$$

Then we use (3.40) to write

$$\int_V d^D\mathbf{x}d^D\mathbf{y}U_2(\mathbf{x}-\mathbf{y})n(\mathbf{x};\{\mathbf{r}\})n(\mathbf{y};\{\mathbf{r}\}) = \frac{1}{(2\pi)^D}\int d^D\mathbf{k}\hat{U}_2(\mathbf{k})\hat{n}(\mathbf{k};\{\mathbf{r}\})\hat{n}(-\mathbf{k};\{\mathbf{r}\}) \tag{3.97}$$

where

$$\hat{n}(\mathbf{k};\{\mathbf{r}\}) = \int_V d^D\mathbf{x}e^{i\mathbf{k}\cdot\mathbf{x}}n(\mathbf{x};\{\mathbf{r}\}). \tag{3.98}$$

In order to obtain a bound we first note that because $n(\mathbf{x};\{\mathbf{r}\})$ is real it follows from (3.98) that

$$\hat{n}(-\mathbf{k};\{\mathbf{r}\}) = \hat{n}^*(\mathbf{k};\{\mathbf{r}\}) \tag{3.99}$$

and thus

$$\hat{n}(-\mathbf{k};\{\mathbf{r}\})\hat{n}(\mathbf{k};\{\mathbf{r}\}) = |\hat{n}(-\mathbf{k};\{\mathbf{r}\})|^2 \geq 0. \tag{3.100}$$

Furthermore we write

$$\hat{n}(\mathbf{k};\{\mathbf{r}\})\hat{n}(-\mathbf{k};\{\mathbf{r}\}) = \int_V d^D\mathbf{x}d^D\mathbf{y}n(\mathbf{x};\{\mathbf{r}\})n(\mathbf{y};\{\mathbf{r}\})e^{i\mathbf{k}\cdot(\mathbf{x}-\mathbf{y})} \tag{3.101}$$

$$= \int_V d^D\mathbf{x}d^D\mathbf{y}n(\mathbf{x};\{\mathbf{r}\})n(\mathbf{y};\{\mathbf{r}\})\sum_{l=0}^{\infty}\frac{1}{l!}[i\mathbf{k}\cdot(\mathbf{x}-\mathbf{y})]^l, \tag{3.102}$$

and note (1) that the terms with l odd vanish under integration because of symmetry under the interchange of $\mathbf{x}$ and $\mathbf{y}$ and (2) that the terms with l a multiple of 4 will all be positive. Thus we find

$$\hat{n}(\mathbf{k};\{\mathbf{r}\})\hat{n}(-\mathbf{k};\{\mathbf{r}\}) \geq \int_V d^D\mathbf{x}d^D\mathbf{y}n(\mathbf{x};\{\mathbf{r}\})n(\mathbf{y};\{\mathbf{r}\})\left(1-\sum_{l=0}^{\infty}\frac{1}{(4l+2)!}[\mathbf{k}\cdot(\mathbf{x}-\mathbf{y})]^{4l+2}\right) \tag{3.103}$$

Then using

$$1-\sum_{l=0}^{\infty}\frac{x^{4l+2}}{(4l+2)!} = 1-\frac{1}{2}(\cosh x - \cos x), \tag{3.104}$$

and letting

$$f(x) = \max[0, 1-\frac{1}{2}(\cosh x - \cos x)] \tag{3.105}$$

we find from (3.100) and (3.103) that

$$\hat{n}(\mathbf{k};\{\mathbf{r}\})\hat{n}(-\mathbf{k};\{\mathbf{r}\}) \geq \int_V d^D\mathbf{x}d^D\mathbf{y}n(\mathbf{x};\{\mathbf{r}\})n(\mathbf{y};\{\mathbf{r}\})f(\mathbf{k}\cdot(\mathbf{x}-\mathbf{y})). \tag{3.106}$$

From (3.105) we see that $f(x)$ is a nonincreasing function of x and thus if we note that, in a D-dimensional cube of side L, the maximum of $|\mathbf{x}-\mathbf{y}|$ is $D^{1/2}L$ we find from (3.103) that

$$\hat{n}(\mathbf{k})\hat{n}(-\mathbf{k}) \geq \left(\int_V d^D\mathbf{x}n(\mathbf{x})\right)^2 f(D^{1/2}L|\mathbf{k}|)). \tag{3.107}$$

From (3.96) we have

$$\int_V d^D\mathbf{x}n(\mathbf{x};\{\mathbf{r}\}) = N \tag{3.108}$$

and thus for all $\mathbf{r_j}$

$$\hat{n}(\mathbf{k};\{\mathbf{r}\})\hat{n}(-\mathbf{k};\{\mathbf{r}\}) \geq N^2 f(D^{1/2}L|\mathbf{k}|) \tag{3.109}$$

which we may use in (3.97) to obtain

$$\int_V d^D\mathbf{x}d^D\mathbf{y}U^{(2)}(\mathbf{x}-\mathbf{y})n(\mathbf{x};\{\mathbf{r}\})n(\mathbf{y};\{\mathbf{r}\}) \geq \frac{N^2}{(2\pi)^D}\int d^D\mathbf{k}\hat{U}_2(\mathbf{k})f(D^{1/2}L|\mathbf{k}|). \tag{3.110}$$

Thus, replacing the nonnegative function $\hat{U}_2(\mathbf{k})$ (which is not identically zero) by its minimum positive value $U_{2,\min} > 0$ and letting $\zeta = D^{1/2}L\mathbf{k}$ we find

$$\int_V d^D\mathbf{x}d^D\mathbf{y}U_2(\mathbf{x}-\mathbf{y})n(\mathbf{x};\{\mathbf{r}\})n(\mathbf{y};\{\mathbf{r}\}) \geq \frac{N^2}{(2\pi)^D}\frac{U_{2,\min}}{V}\int d^D\zeta f(|\zeta|). \tag{3.111}$$

The integral over $d^D\zeta$ exists and is positive and thus there exists a nonnegative constant $2A$ such that

$$\int_V d^D\mathbf{x}d^D\mathbf{y}U^{(2)}(\mathbf{x}-\mathbf{y})n(\mathbf{x};\{\mathbf{r}\})n(\mathbf{y};\{\mathbf{r}\}) \geq 2A\frac{N^2}{DV}. \tag{3.112}$$

Using this bound in (3.95) the superstability bound (3.91) follows.

3.1.4 Multispecies interactions

Thus far we have considered the case where the system (3.4) contains only one species of particle. However, the considerations may be easily extended to a system with n types of particles where the number of particles of each species is N_α with

$$N = \sum_{\alpha=1}^{n} N_\alpha, \tag{3.113}$$

and the Hamiltonian is

$$H = \sum_{\alpha=1}^{n}\sum_{j(\alpha)=1}^{N_\alpha}\frac{\mathbf{p}_{j(\alpha)}^2}{2m_\alpha} + U^{(N)}(\mathbf{r}_{1(1)},\cdots,\mathbf{r}_{N_n(n)}) \tag{3.114}$$

with

$$U^{(N)}(\mathbf{r}_{1(1)},\cdots,\mathbf{r}_{N_n(n)}) = \sum_{\alpha=1}^{n}\sum_{j(\alpha)<k(\alpha)} U_{\alpha,\alpha}(\mathbf{r}_{j(\alpha)} - \mathbf{r}_{k(\alpha)}) + \sum_{\alpha<\beta}^{n}\sum_{j(\alpha)}\sum_{k(\beta)} U_{\alpha,\beta}(\mathbf{r}_{j(\beta)} - \mathbf{r}_{k(\alpha)}) \tag{3.115}$$

and

$$U_{\alpha,\beta}(\mathbf{r}) = U_{\beta,\alpha}(-\mathbf{r}). \tag{3.116}$$

In this case condition 1 is replaced by the

Multiparticle stability condition

Let the pair potential of (3.115) be of the form

$$U_{\alpha,\beta}(\mathbf{r}) = U_{1,\alpha,\beta}(\mathbf{r}) + U_{2,\alpha,\beta}(\mathbf{r}) \tag{3.117}$$

with

$$U_{1,\alpha,\beta}(\mathbf{r}) \geq 0 \tag{3.118}$$

$$U_{2,\alpha,\beta}(\mathbf{r}) = U_{2,\beta,\alpha}(-\mathbf{r}), \tag{3.119}$$

and let the Fourier transform of $U_{2,\alpha,\beta}(\mathbf{r})$

$$\hat{U}_{2,\alpha,\beta}(\mathbf{k}) = \int d^D\mathbf{r} e^{i\mathbf{k}\cdot\mathbf{r}} U_{2,\alpha,\beta}(\mathbf{r}) \tag{3.120}$$

have the properties that $\hat{U}_{2,\alpha,\beta}(\mathbf{k})$ is absolutely integrable

$$\int d^D\mathbf{k} |\hat{U}_{2,\alpha,\beta}(\mathbf{k})| < \infty \tag{3.121}$$

and that considered as an $n \times n$ matrix $\hat{U}_{2,\alpha,\beta}(\mathbf{k})$ has no negative eigenvalues for any value of $\mathbf{k}$. Then

$$U^{(N)}(\mathbf{r}_{1(1)}, \cdots, \mathbf{r}_{N_n(n)}) \geq -\frac{1}{2}\sum_{\alpha=1}^{n} N_\alpha U_{2,\alpha,\alpha}(0) \tag{3.122}$$

which is the multispecies generalization of (3.12).

We prove this by following [3] and first noting that from (3.118) and (3.119)

$$\hat{U}_{2,\alpha,\beta}(\mathbf{k}) = \int d^D\mathbf{r} U_{2,\beta,\alpha}(-\mathbf{r}) e^{i\mathbf{k}\cdot\mathbf{r}} = \int d^D\mathbf{r} U_{2,\beta,\alpha}(\mathbf{r}) e^{-i\mathbf{k}\cdot\mathbf{r}} = \hat{U}^*_{2,\beta,\alpha}(\mathbf{k}). \tag{3.123}$$

Thus $\hat{U}_{2,\alpha,\beta}(\mathbf{k})$ is Hermitian and hence its eigenvalues are real. Furthermore

$$U_{2,\alpha,\beta}(0) = \frac{1}{(2\pi)^D}\int d\mathbf{k}\hat{U}_{2,\alpha,\beta}(\mathbf{k}) \leq \frac{1}{(2\pi)^D}\int d\mathbf{k}|\hat{U}_{2,\alpha,\beta}(\mathbf{k})| < \infty. \tag{3.124}$$

From the definition (3.115) and the positivity (3.118) we have

$$\begin{aligned} &U^{(N)}(\mathbf{r}_{1(1)}, \cdots, \mathbf{r}_{N_n(n)}) \\ &\geq \sum_{\alpha=1}^{n}\sum_{j(\alpha)<k(\alpha)} U_{2,\alpha,\alpha}(\mathbf{r}_{j(\alpha)} - \mathbf{r}_{k(\alpha)}) + \sum_{\alpha<\beta}^{n}\sum_{j(\alpha)}\sum_{k(\beta)} U_{2,\alpha,\beta}(\mathbf{r}_{j(\alpha)} - \mathbf{r}_{k(\beta)}) \\ &= \frac{1}{2}W_N - \frac{1}{2}\sum_{\alpha=1}^{n} N_\alpha U_{2,\alpha,\alpha}(0) \end{aligned} \tag{3.125}$$

with

$$W_N = \sum_{\alpha=1}^{n}\sum_{\beta=1}^{n}\sum_{j(\alpha)}\sum_{k(\beta)} U_{2,\alpha,\beta}(\mathbf{r}_{j(\alpha)} - \mathbf{r}_{k(\beta)}) \tag{3.126}$$

which is written in terms of the Fourier transform as

$$\begin{aligned} W_N &= \sum_{\alpha=1}^{n}\sum_{\beta=1}^{n}\sum_{j(\alpha)}\sum_{k(\beta)} \int d^D\mathbf{k} e^{i\mathbf{k}\cdot(\mathbf{r}_{k(\beta)}-\mathbf{r}_{j(\alpha)})}\hat{U}_{2,\alpha,\beta}(\mathbf{k}) \\ &= \int d^D\mathbf{k} \sum_{\alpha=1}^{n}\sum_{\beta=1}^{n}\left(\sum_{j(\alpha)} e^{i\mathbf{k}\cdot\mathbf{r}_{j(\alpha)}}\right)^* \hat{U}_{2\alpha,\beta}(\mathbf{k})\left(\sum_{j(\beta)} e^{i\mathbf{k}\cdot\mathbf{r}_{j(\beta)}}\right). \end{aligned} \tag{3.127}$$

The integrand in (3.127) is of the form $\mathbf{z}^*\hat{\mathbf{U}}_2(\mathbf{k})\mathbf{z}$ with

$$\mathbf{z}|_\alpha = \sum_{j(\alpha)} e^{i\mathbf{k}\cdot\mathbf{r}_{j(\alpha)}} \quad \text{and} \quad \hat{\mathbf{U}}_2|_{\alpha,\beta} = \hat{U}_{2,\alpha,\beta}(\mathbf{k}) \tag{3.128}$$

and because the eigenvalues of $\hat{\mathbf{U}}_2(\mathbf{k})$ are assumed to be nonnegative for all $\mathbf{k}$, theorem 4 of appendix A guarantees that this integrand is nonnegative for all $\mathbf{r}_{j(\alpha)}$. Therefore $W_N \geq 0$ and thus the desired conclusion (3.122) follows from (3.125).

An important example of multispecies interactions is charged hard spheres which can be shown to satisfy this stability condition. However, the hard core is needed for the stability and the pure Coulomb interaction with no repulsive hard core is classically unstable and thus will not give thermodynamic behavior.

3.2 Quantum stability

As noted above, a potential which is classically stable is stable quantum mechanically so the only interesting new case to consider are potentials which are classically unstable but become stable in the quantum mechanical case. Of such classically unstable systems the one of by far the greatest importance is the Coulomb interaction (3.20). The stability of this system has been extensively studied by Ruelle [8], Dyson and Lenard [9–11], Lieb [12] and Conlon, Lieb and Yau [14]. These studies all concern themselves with properties of the ground state of the system and as such they may more properly be considered as a branch of atomic rather than statistical physics. Moreover, the proofs given in these papers are sufficiently complex that the reprints of the articles fill an entire book [15]. Therefore we will for the most part content ourselves with stating the results in subsection 3.2.1. Proofs of several of the simpler results will be given in subsection 3.2.2. For proofs of the more profound theorems we refer the reader to the original papers.

3.2.1 Stability of matter

In [3] and [9] the following bound is shown for the Hamiltonian (3.20) when there is no restriction (Bose or Fermi) on the symmetry of the wave function

Theorem 1

$$E_{gs}(N) \geq -\frac{1}{8}N^2(N-1)Ry \tag{3.129}$$

where $Ry = me^4/2\hbar^2$ and a better bound also given in ref. [9] is

Theorem 2

$$E_{gs}(N) \geq -N(N-1)2^{-1/2}Ry \tag{3.130}$$

The proofs of these will be given below.

A much better bound proven in [9] is
Theorem 3

$$E_{gs}(N) \geq -A_3 N^{5/3} Ry \tag{3.131}$$

where the best current value [12] of A_3 satisfies $A_3 \geq 14.01$. Furthermore in 1979 Lieb [13] proved

Theorem 4

If the positive charges have infinite mass and a charge $Ze \geq 0$ and if the negative particles have mass M and charge $-e$ then there is an upper bound

$$E_{gs}(N) \leq -\frac{Z^{4/3}}{108} N^{5/3} Ry. \tag{3.132}$$

On the other hand if the hypothesis (3.18) holds then Dyson [10] has proven the upper bound

Theorem 5

$$E_{gs}(N) < -A_5 N^{7/5} Ry \tag{3.133}$$

and finally in 1988 in [14] the companion lower bound for a neutral system of charged bosons under the hypothesis (3.18) was found:

Theorem 6

$$E_{gs}(N) \geq -0.31 N^{7/5} Ry. \tag{3.134}$$

It is clear from theorems 4-6 that for matter to be stable we must include the repulsion that comes from the Pauli exclusion principle for fermions. The first such theorem was proven in 1967 by Dyson and Lenard [9]:

Theorem 7

Suppose that the N particles whose masses and charges satisfy (3.18) belong to $q \geq 1$ distinct species of fermions. Then

$$E_{gs}(N) \geq -A_7 q^{2/3} N Ry \tag{3.135}$$

where $A_7 < 500$ is an absolute constant. Briefly, a system whose particles belong to a fixed number of fermion species is stable.

This theorem is a step in the right direction but is still not really sufficient to prove that real matter is stable. It ought not to require that all particles are fermions because the statistics of the nuclei should not matter. Thus in [9] it is stated and in [11] it is proven that

Theorem 8

Let N negatively charged particles belong to q different fermion species. Let their masses and charges be subject to (3.18). Let an arbitrary number of positively charged particles be subject only to the restriction (3.18) on the charges with their statistics and masses being arbitrary. Then we have

$$E_{gs}(N) \geq -A_8 q^{2/3} N Ry \tag{3.136}$$

where A_8 is an absolute constant.

In [11] the constant A_8 is about 10^{14}. This was greatly improved on by Lieb [12] in 1975 who obtained a value of about 23.

3.2.2 Proofs of theorems 1 and 2

We conclude this section with proofs of theorems 1 and 2 as given in [11].

Proof of Theorem 1

Write (3.20) in the form of a sum of two-body Hamiltonians as

$$H = \sum_{1 \leq i < j \leq N} H_{ij} \tag{3.137}$$

with

$$H_{ij} = -\frac{\hbar^2 \nabla_i^2}{2m_i(N-1)} - \frac{\hbar^2 \nabla_j^2}{2m_j(N-1)} + \frac{e_i e_j}{|\mathbf{r}_i - \mathbf{r}_j|}. \tag{3.138}$$

By definition the ground state energy is

$$E_{\min} = \text{Inf}\langle \Psi^*, H\Psi \rangle \tag{3.139}$$

where we use the notation

$$\langle f^*, g \rangle = \int f^*(x) g(x) dx \tag{3.140}$$

and the infimum is taken with respect to all N particle wavefunctions $\Psi(\mathbf{r}_1, \cdots, \mathbf{r}_N)$ normalized as $\langle \Psi^*, \Psi \rangle = 1$, over all values of the masses satisfying

$$0 < m_j \leq m \tag{3.141}$$

and all values of the charges satisfying

$$-e \leq e_j \leq e. \tag{3.142}$$

Therefore from (3.137) we have

$$E_{\min} = \text{Inf}\langle \Psi^*, H\psi \rangle \geq \sum_{1 \leq i < j \leq N} \text{Inf}\langle \Psi^*, H_{ij}\Psi \rangle. \tag{3.143}$$

The operators H_{ij} is the Hamiltonian of a two-body system with charges e_i and e_j and with masses $m_i(N-1)$ and $m_j(N-1)$. If $e_i e_j \geq 0$ this energy is positive. If $e_i e_j < 0$ the solution of the ground state energy problem is obtained from the solution of the ground state energy of the hydrogen atom. Therefore

$$\begin{aligned} \text{Inf}\langle \Psi^*, H_{ij}\Psi \rangle &= 0, \quad \text{for } \ e_i e_j \geq 0 \\ &= -\frac{(N-1)m_i m_j}{m_i + m_j}\frac{e_i^2 e_j^2}{2\hbar^2}, \quad \text{for } e_i e_j \leq 0. \end{aligned} \tag{3.144}$$

In the sum over i and j in (3.137) there are at most $N^2/4$ pairs with $e_i e_j < 0$ and for these pairs we see from the bounds (3.141) and (3.142) that

$$\frac{(N-1)m_i m_j}{m_i + m_j}\frac{e_i^2 e_j^2}{2\hbar^2} \leq \frac{(N-1)me^4}{4\hbar^2} = \frac{N-1}{2}Ry. \tag{3.145}$$

Therefore

$$E_{\min} \geq -\frac{N^2}{4}\frac{(N-1)}{2}Ry \tag{3.146}$$

and thus theorem 1 is proven.

To prove theorem 2 we need to use the

Cauchy–Schwarz inequality

$$|\int f^*(x)g(x)dx|^2 \leq \int |f(x)|^2 dx \int |g(x)|^2 dx. \tag{3.147}$$

Proof

Let $\Psi(x) = f(x) + \lambda g(x)$. Then, using the fact that $\langle \Psi^*, \Psi \rangle \geq 0$, we have

$$\langle \Psi^*, \Psi \rangle = \langle f^*, f \rangle + \lambda \langle f^*, g \rangle + \lambda^* \langle f, g^* \rangle + \lambda\lambda^* \langle g^*, g \rangle \geq 0. \tag{3.148}$$

Now in (3.148) set

$$\lambda = -\frac{\langle g^*, f \rangle}{\langle g^*, g \rangle} \tag{3.149}$$

and multiply by $\langle g^*, g \rangle$ to obtain

$$\langle f^*, f \rangle \langle g^*, g \rangle - \langle f^*, g \rangle \langle g^*, f \rangle - \langle f, g^* \rangle \langle g, f^* \rangle + \langle g^*, f \rangle \langle g, f^* \rangle \geq 0 \tag{3.150}$$

which is simplified to

$$\langle g^*, f \rangle \langle f^*, g \rangle \leq \langle f^*, f \rangle \langle g^*, g \rangle \tag{3.151}$$

from which (3.147) follows.

Proof of theorem 2

We prove theorem 2 by comparing the Coulomb interaction with the Yukawa interaction. This is done by first writing

$$H = \sum_{1 \le i < j \le N} H_{ij} + \sum_{1 \le i < j \le N} H'_{ij} \tag{3.152}$$

where now

$$H_{ij} = -\frac{\hbar^2}{2m_i(N-1)}\nabla_i^2 - \frac{\hbar^2}{2m_j(N-1)}\nabla_j^2 + \frac{e_i e_j}{|\mathbf{r}_i - \mathbf{r}_j|} e^{-\mu|\mathbf{r}_i - \mathbf{r}_j|} \tag{3.153}$$

and

$$H'_{ij} = \frac{e_i e_j}{|\mathbf{r}_i - \mathbf{r}_j|}(1 - e^{-\mu|\mathbf{r}_i - \mathbf{r}_j|}). \tag{3.154}$$

To proceed further we need the following:

Lemma

The Yukawa Hamiltonian

$$H_Y = -\frac{\hbar^2}{2m}\nabla^2 - \frac{e^2}{|\mathbf{r}|} e^{-\mu|\mathbf{r}|} \tag{3.155}$$

is positive definite (i.e. has no bound states) if

$$\mu \frac{\hbar^2}{me^2} \ge \sqrt{2}. \tag{3.156}$$

To prove the lemma we write the Hamiltonian in momentum space as

$$\langle \Psi^*, H_Y \Psi \rangle = \frac{\hbar^2}{2m} \int d^3\mathbf{k} |\mathbf{k}|^2 |\tilde{\Psi}(k)|^2 - \frac{e^2}{2\pi^2} \int d^3\mathbf{k} d^3\mathbf{k}' \frac{\tilde{\Psi}^*(\mathbf{k})\tilde{\Psi}(\mathbf{k}')}{\mu^2 + |\mathbf{k} - \mathbf{k}'|^2} \tag{3.157}$$

where

$$\tilde{\Psi}(\mathbf{k}) = \int d^3\mathbf{r} e^{i\mathbf{kr}} \Psi(\mathbf{r}) \tag{3.158}$$

is the Fourier transform of $\Psi(\mathbf{r})$. Thus if we use (the square root of) the Cauchy–Schwarz inequality (3.147) on the second term of (3.157) with

$$f^*(\mathbf{k}, \mathbf{k}') = \tilde{\Psi}^*(\mathbf{k})\tilde{\Psi}(\mathbf{k}')|\mathbf{k}||\mathbf{k}'| \tag{3.159}$$

$$g(\mathbf{k}, \mathbf{k}') = \frac{1}{|\mathbf{k}||\mathbf{k}'|(\mu^2 + |\mathbf{k} - \mathbf{k}'|^2)} \tag{3.160}$$

and

$$\int d^3\mathbf{k} d^3\mathbf{k}' |f(\mathbf{k}, \mathbf{k}')|^2 = \left(\int d^3\mathbf{k} \mathbf{k}^2 |\tilde{\psi}(\mathbf{k})|^2 \right)^2 \tag{3.161}$$

we obtain

$$\left| \int d^3\mathbf{k} d^3\mathbf{k}' \frac{\tilde{\Psi}^*(\mathbf{k})\tilde{\Psi}(\mathbf{k}')}{\mu^2 + |\mathbf{k} - \mathbf{k}'|^2} \right| \le J^{1/2} \int d^3\mathbf{k} |\mathbf{k}|^2 |\tilde{\Psi}(\mathbf{k})|^2 \tag{3.162}$$

with

$$J = \int d^3\mathbf{k} d^3\mathbf{k}' \frac{1}{|\mathbf{k}|^2|\mathbf{k}'|^2(\mu^2 + |\mathbf{k} - \mathbf{k}'|^2)^2} = \frac{2\pi^4}{\mu^2}. \tag{3.163}$$

Thus

$$\langle\Psi^*, H_Y\Psi\rangle \geq \left(\frac{\hbar^2}{2m} - \frac{e^2}{2\pi^2}J^{1/2}\right)\int d^3\mathbf{k}|\mathbf{k}|^2|\Psi(\mathbf{k})|^2 \tag{3.164}$$

which is positive if

$$\frac{\hbar^2}{2m} \geq \frac{e^2}{2\pi^2}J^{1/2} \tag{3.165}$$

which is the desired condition (3.156).

We now may prove theorem 2 by choosing in (3.156)

$$\mu = \frac{(N-1)me^2}{\hbar^2\sqrt{2}}. \tag{3.166}$$

With this choice of μ the lemma guarantees that the eigenvalues of H_{ij} are all positive and thus

$$E_{\min} \geq \sum_{1\leq i<j\leq N} H'_{ij} \tag{3.167}$$

which may be rewritten as

$$\sum_{1\leq i<j\leq N} H'_{ij} = \frac{1}{2}\sum_{i=1}^{N}\sum_{j=1}^{N}\frac{e_ie_j}{|\mathbf{r}_i - \mathbf{r}_j|}(1 - e^{-\mu|\mathbf{r}_i-\mathbf{r}_j|}) - \frac{1}{2}\mu\sum_{j=1}^{N}e_j^2. \tag{3.168}$$

The double sum in (3.168) is seen to be positive for any choice of e_j by writing it in terms of its Fourier transform as

$$\frac{1}{2}\sum_{i=1}^{N}\sum_{j=1}^{N}\frac{e_ie_j}{|\mathbf{r}_i - \mathbf{r}_j|}(1 - e^{-\mu|\mathbf{r}_i-\mathbf{r}_j|}) = \frac{1}{2\pi^2}\int d^3\mathbf{k}\left(\frac{1}{|\mathbf{k}|^2} - \frac{1}{|\mathbf{k}|^2 + \mu^2}\right)|\sum_{j=1}^{N}e_je^{i\mathbf{k}\mathbf{r}_j}|^2 \geq 0. \tag{3.169}$$

Therefore from (3.168) we find the desired result for theorem 2

$$E_{\min} > -\frac{1}{2}\mu Ne^2 = -\frac{N(N-1)}{\sqrt{2}}Ry \tag{3.170}$$

where in the last line we have used the value of μ defined by (3.166).

3.3 Existence and uniqueness of the thermodynamic limit

The existence of the thermodynamic limit must be unique and independent of boundary conditions in order for the laws of thermodynamics to follow from statistical mechanics. Thus in greatest generality we need to show that the specification of the potential used to confine a finite number of particles in a finite volume will be irrelevant in the thermodynamic limit.

In practice, however, this most general situation has never been addressed and attention has been restricted to box boundary conditions and periodic boundary conditions. We will see that both types lead to exactly the same limiting free energy which is independent of shape.

3.3.1 Box boundary conditions

One way to make concrete the boundary conditions used for taking the thermodynamic limit is to consider the particles to be confined to the interior of a domain $\mathcal{D}$ which has impenetrable walls (as specified for example by having an infinite repulsive potential on the exterior of the walls). Let the volume of $\mathcal{D}$ be denoted by $V(\mathcal{D})$ The domain will also have a shape and, in order for the thermodynamic limit to exist, the domain $\mathcal{D}$ will have to have the property that in some rough sense the ratio of surface area to volume goes to zero in the thermodynamic limit.

The most common shapes to consider are surely rectangular parallelepipeds with sides $L_j = a_j L$ with a_j fixed, and for this the ratio of surface to volume as $L \to \infty$ vanishes as $1/L$. However, for a domain like a snowflake where the boundary curve becomes fractal and all the volume lies close to the surface the very concept of bulk versus surface fails to apply and independence of shape should not be expected. It is thus physically necessary to put some restrictions on the types of shapes to be considered.

We will consider two different restrictions on the limiting shape in the thermodynamic limit first introduced by van Hove [16] and by Fisher [2].

Limit in the sense of van Hove

Denote by $V_h(\mathcal{D})$ the volume of $\mathcal{D}$ of all points whose distance from the boundary of $\mathcal{D}$ is less than or equal to h. We say that $V(\mathcal{D})$ goes to infinity in the sense of van Hove if

$$\lim_{V(\mathcal{D})\to\infty} V_h(\mathcal{D})/V(\mathcal{D}) = 0. \tag{3.171}$$

This restriction says that the fraction of the volume near the surface vanishes in the thermodynamic limit and will eliminate fractal domains.

If, in addition, the domain $\mathcal{D}$ were restricted to be convex (a natural condition associated with the word box) the restriction (3.171) is sufficient for the existence of the thermodynamic limit for strongly tempered potentials. However, if nonconvex domains are considered a further restriction is needed. To specify this restriction let $V_P(\mathcal{D})$ be the smallest parallelepiped which contains $\mathcal{D}$. We will then follow [2] and require that for the sequence of domains $\mathcal{D}_j$

$$\lim_{j\to\infty} V(\mathcal{D}_j)/V_P(\mathcal{D}_j) \geq \delta > 0 \tag{3.172}$$

as

$$\lim_{j\to\infty} V(\mathcal{D}_j) = \infty. \tag{3.173}$$

It is shown in [2] that for strongly tempered potentials the thermodynamic limit exists and is independent of shape when the restrictions (3.171) and (3.172) hold.

Even though the limit in the sense of van Hove (3.171) eliminates snowflake domains it is not sufficiently restrictive if the potential is weakly instead of strongly tempered. For these potentials we need to define

Limit in the sense of Fisher
We define the "shape function" $\sigma(\alpha;\mathcal{D})$ as

$$\sigma(h/V(\mathcal{D})^{1/D};\mathcal{D}) = V_h(\mathcal{D})/V(\mathcal{D}) \tag{3.174}$$

where $\alpha = h/V(\mathcal{D})^{1/\mathcal{D}}$ is fixed. This shape function is the fractional volume of points within a distance $h = \alpha V(\mathcal{D})^{1/D}$ of the surface. It should be noted that unlike the van Hove limit (3.171) where h is fixed that now h is some finite fraction of the total size of the system. We say that a shape function is regular if

$$\sigma(\alpha;\mathcal{D}) \to 0 \quad \text{as} \quad \alpha \to 0. \tag{3.175}$$

We say that a sequence of domains is regular if there is a fixed α' and some regular shape function $\sigma_0(\alpha)$ such that for all $j = 1, 2, \cdots$ and $\alpha < \alpha'$ we have

$$\sigma(\alpha, \mathcal{D}_j) < \sigma_0(\alpha). \tag{3.176}$$

When $V(\mathcal{D}_j) \to \infty$ through a sequence of regular domains we say that the thermodynamic limit is taken in the sense of Fisher. It is shown in [2] that for weakly tempered potential the limit in the sense of Fisher exists and is unique independent of shape.

Parallelepipeds where the sides are all proportional to L satisfy both the van Hove and the Fisher criteria but for parallelepipeds where the ratio of the sides vanishes in the thermodynamic limit van Hove and Fisher are different. To illustrate this, consider a two-dimensional rectangular domain $\mathcal{D}$ with

$$L_x = L^\gamma, \quad L_y = L^{2-\gamma}, \quad \text{with } \gamma < 1 \tag{3.177}$$

and hence

$$V(\mathcal{D}) = L^2. \tag{3.178}$$

To apply the van Hove condition we note that, for fixed h such that $2h < L_x$,

$$V_h(\mathcal{D}) = 2h(L_x + L_y) - 4h^2 \tag{3.179}$$

and thus

$$\lim_{L\to\infty} \frac{V_h(\mathcal{D})}{V(\mathcal{D})} = \frac{2h(L^\gamma + L^{2-\gamma} - 2h)}{L^2} = 0 \tag{3.180}$$

so the condition of van Hove (3.171) is satisfied. On the other hand, for the definition of the shape function $\sigma(\alpha;\mathcal{D})$ we have

$$h = \alpha L \tag{3.181}$$

whereas $L_x = L^\gamma$ with $\gamma < 1$. Therefore we have

$$h > L_x \quad \text{for fixed } \alpha > 0 \text{ and sufficiently large } L. \tag{3.182}$$

Thus it will be impossible to satisfy the conditions of the Fisher limit.

It is therefore an interesting question whether there are potentials that are weakly but not strongly tempered which lead to observable effects for domains which obey the van Hove but not the Fisher restrictions.

3.3.2 Periodic boundary conditions

To define what will be meant by periodic boundary conditions we let $\mathbf{e}_\alpha$ be D orthogonal unit vectors and then define lattice vectors

$$\mathbf{l}_\mathbf{t} = \sum_{\alpha=1}^{D} t_\alpha L_\alpha \mathbf{e}_\alpha \tag{3.183}$$

where L_α are the sides of our basic rectangular box (which will go to infinity in the thermodynamic limit) and t_α are positive or negative integers or zero.

We will consider for concreteness the pure pair potential (3.14) and extend it to a potential periodic on this lattice by defining

$$U_{\Pi}^{(N)}(\mathbf{L}, \mathbf{r}_1, \cdots \mathbf{r}_N) = \frac{1}{2}\sum_{j=1}^{N}\sum_{k=1}^{N}\sum_{\mathbf{t}}{}' U(\mathbf{r}_j - \mathbf{r}_k + \mathbf{l}_\mathbf{t}) \tag{3.184}$$

where the prime on the sum over $\mathbf{t}$ indicates that, when $j = k$, the term $\mathbf{t} = 0$ is omitted.

The partition function calculated for $U_{\Pi}^{(N)}(L, \mathbf{r}_1, \cdots, \mathbf{r}_N)$ with the particles confined to

$$0 \leq \mathbf{r}_\alpha \leq L_\alpha \tag{3.185}$$

is what we mean by periodic boundary conditions.

We note that a one-dimensional system with periodic boundary conditions may be thought of as a circle embedded in a two-dimensional space and a two-dimensional system with periodic boundary conditions may be thought of as a torus embedded in three-dimensional space. Both of these cases can be realized in the laboratory. However a three-dimensional system with periodic boundary conditions can only be embedded in four-dimensional space and therefore cannot be experimentally realized. This set of boundary conditions for the three-dimensional object is an example of what is known as a 3-manifold. Even though three dimensional periodic boundary conditions are unphysical in the sense that they cannot be embedded in three-dimensional space they are often used in practical computations because they are translationally invariant which is a property that the thermodynamic limit will also possess. These translationally invariant systems have a conserved total momentum which segments the phase space into different ergodic components.

3.3.3 Existence and uniqueness in the canonical ensemble

In the canonical ensemble when the classical (3.12) or quantum mechanical stability conditions (3.16) and one of the tempering conditions (3.22) or (3.25) are satisfied, the following properties hold:

1. **For finite N and V the function $\lambda^{DN}Q_N(V,T)$ is an entire function of T which is positive on the positive T axis.**

This property is obvious because the integrand in (3.5) is a positive entire function of T and the integration range is finite.

All entire functions may be expressed in terms of their zeros and growth order at infinity using the Weierstrass factorization theorem (also known as Hadamard's theorem [17, pages 18-22]). Thus (assuming the growth order is less than one) we have

$$\lambda^{DN} Q_N(V,T) = A_N(V) \prod_{j=1}^{\infty} [1 - (\beta/\beta_j)] \tag{3.186}$$

where $\beta = 1/k_B T$ and $A_N(V)$ is a normalizing function independent of β. For finite N and V the β_j will in general depend on V and cannot lie on the positive real axis but they can pinch the axis in the thermodynamic limit. The values of β where pinching occurs in the thermodynamic limit will be points of phase transition where the thermodynamic functions will have singularities. The following distributions of zeros are known to occur:

A. The limiting distribution is a finite collection of curves and hence pinching occurs only at points.

B. The limiting distribution in general fills an area but pinching still occurs at points.

C. The limiting distribution in general fills an area and pinching occurs at line segments.

2. **The free energy per unit volume $\tilde{F}(\rho,T)$ (and hence the free energy per particle $F(v)$) exists and is independent of shape in the sense of van Hove if the potential is strongly tempered (3.25).**

2′. **The free energy per unit volume $\tilde{F}(\rho,T)$ (and hence the free energy per particle) exists and is independent of shape in the sense of Fisher if the potential is weakly tempered (3.22).**

We discuss the proof of this basic theorem below.

3. **The free energy per unit volume $\tilde{F}(\rho,T)$ is a convex function of ρ**

$$\tilde{F}(\sum_j \omega_j \rho_j, T) \leq \sum_j \omega_j \tilde{F}(\rho_j, T) \text{ with } \sum_j \omega_j = 1 \tag{3.187}$$

and the free energy per particle $F(v,T)$ is a convex function of v.

We prove this below also.

From existence properties 2 and 2′ and the convexity property 3 the following further properties may be established:

4. **The free energy per unit volume $\tilde{F}(\rho,T)$ is a continuous function of ρ and the free energy per unit particle $F(v,T)$ is a continuous function of v. For $\rho \sim 0$ we have $\tilde{F}(\rho) \sim -k_B T \rho |\ln \rho|$ and as $v \to \infty$ we have $F(v) \to -k_B T \ln v$.**

5. **The free energy per particle $F(v,T)$ is a nonincreasing function of v.** We note that $\tilde{F}(\rho,T)$ does not need to be monotonic.

6. **The right- and left-hand derivatives of $F(v,T)$ with respect to v defined as**

$$\frac{\partial^{\pm}}{\partial v} F(v) = \lim_{\Delta v \to 0^{\pm}} \frac{F(v \pm \Delta) - F(v)}{\pm \Delta v} \tag{3.188}$$

exist (and are finite) everywhere and that

$$\frac{\partial^-}{\partial v}F(v,T) \leq \frac{\partial^+}{\partial v}F(v,T). \tag{3.189}$$

7. **The right- and left-derivatives of $F(v,T)$ with respect to v are equal almost everywhere (except at a countable number of points) and therefore the pressure exists, is positive everywhere**

$$P = -\frac{\partial F(v)}{\partial v} = -\left(\tilde{F}(\rho) - \rho\frac{\partial \tilde{F}(\rho)}{\partial \rho}\right) > 0 \tag{3.190}$$

and is continuous almost everywhere.

8. **The pressure is a nonincreasing function of v.**
9. **The inverse isothermal compressibility**

$$\frac{1}{\kappa_T} = -\frac{\partial P}{\partial v} > 0 \tag{3.191}$$

exists and is nonnegative almost everywhere because $P(v)$ is a monotonic nonincreasing function of v.

These features are illustrated in Figs.3.1–3.3 for the free energy per particle $F(v)$ and the pressure $P(v)$.

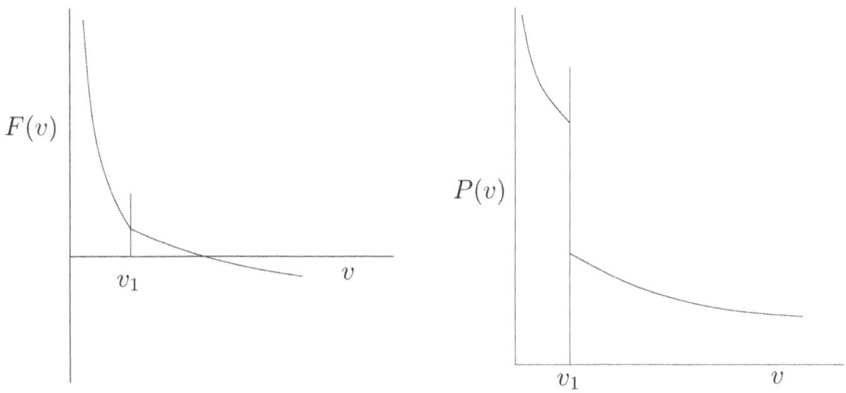

Fig. 3.1 A schematic plot of a free energy per particle $F(v)$ where, at $v = v_1$, the right- and left-derivatives are not equal, and of the corresponding pressure $P(v) = -\frac{\partial F(v)}{\partial v}$ which has a discontinuity as $v = v_1$.

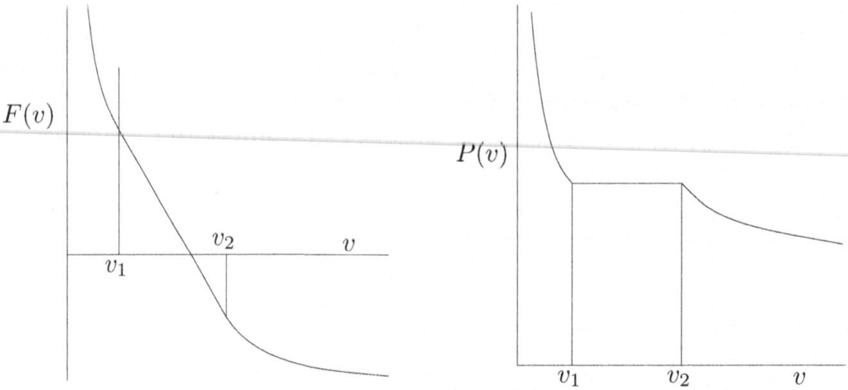

Fig. 3.2 A free energy per particle $F(v)$ with a continuous derivative but with a linear segment from $v_1 \le v \le v_2$, and the corresponding pressure $P(v) = -\frac{\partial F(v)}{\partial v}$. There is a first order phase transition at v_1 and v_2.

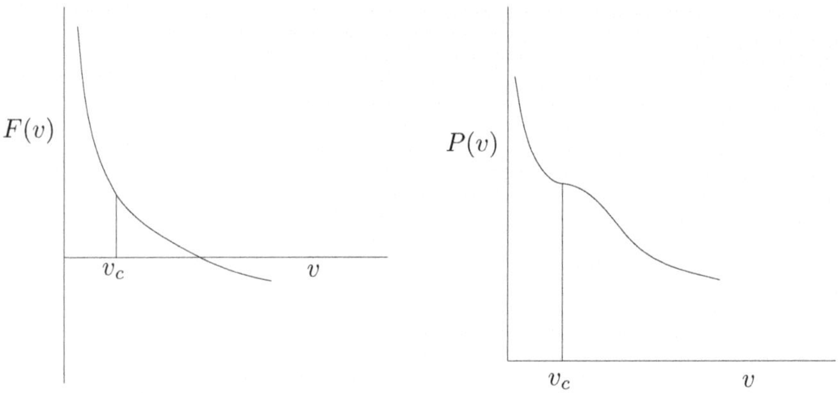

Fig. 3.3 A free energy per particle $F(v)$ with a continuous derivative but where the second derivative vanishes at $v = v_c$, and the corresponding pressure $P(v) = -\frac{\partial F(v)}{\partial v}$. This is the typical behavior at a critical point.

To prove the properties 2 and 2′ we consider two nonoverlapping subdomains $\mathcal{D}_1$ with N_1 particles and $\mathcal{D}_2$ with N_2 particles that are enclosed in the domain $\mathcal{D}$ and that are separated by at least a distance R. Then in the integrations in the definition (3.5), if we restrict the regions of integration to $\mathbf{r}_i \in \mathcal{D}_1$ for $1 \le i \le N_1$ and $\tilde{\mathbf{r}}_j \in \mathcal{D}_2$ for $1 \le j \le N_2$, we have

$$Q_N(\mathcal{D}) \ge \frac{1}{N!\lambda^{DN}} \sum_{N_1+N_2=N} \frac{N!}{N_1!N_2!} \int_{\mathcal{D}''} d\mathbf{r}_1 \cdots d\mathbf{r}_{N_1} \int_{\mathcal{D}_2} d\tilde{\mathbf{r}}_1 \cdots d\tilde{\mathbf{r}}_{N_2}$$
$$\times \exp[-(U^{(N_1)} + U^{(N_2)} + U^{(N_1,N_2)})/k_B T] \qquad (3.192)$$

with

$$U^{(N_1)} = U^{(N_1)}(\mathbf{r}_1, \cdots, \mathbf{r}_{N_1})$$

$$U^{(N_2)} = U^{(N_2)}(\tilde{\mathbf{r}}_1, \cdots, \tilde{\mathbf{r}}_{N_2})$$
$$U^{(N_1,N_2)} = U^N(\mathbf{r}_1, \cdots, \mathbf{r}_N) - U^{(N_1)}(\mathbf{r}_1, \cdots, \mathbf{r}_{N_1}) - U^{(N_2)}(\tilde{\mathbf{r}}_1, \cdots, \tilde{\mathbf{r}}_{N_2}). \quad (3.193)$$

Then since each term in (3.192) is positive we may drop all terms except one. If we use the weak tempering bound (3.22) the integrals factorize and we obtain the inequality

$$Q_{N_1+N_2}(\mathcal{D}) \geq Q_{N_1}(\mathcal{D}_1)Q_{N_2}(\mathcal{D}_2)\exp[-N_1N_2u_B/k_BTR^{D+\epsilon}] \quad (3.194)$$

which, if we define

$$\ln Q_N(\mathcal{D}) = -V\tilde{F}(\rho, \mathcal{D})/k_BT \quad (3.195)$$

and divide by V, gives

$$\tilde{F}(\rho, D) \leq \omega_1\tilde{F}(\rho_1, \mathcal{D}_1) + \omega_2\tilde{F}(\rho_2, \mathcal{D}_2) + \frac{N_1N_2u_B}{VR^{D+\epsilon}} \quad (3.196)$$

where

$$\omega_j = V_j/V, \quad \rho_j = N_j/V_j. \quad (3.197)$$

We may further subdivide $\mathcal{D}_1$ into subdomains $\mathcal{D}_j$ for $2 \leq j \leq n$. Each subdivision adds a repulsion term proportional to $(N - \sum_{j=2}^{n} N_j)N_n$ and thus using

$$\sum_{k=2}^{n}(N - \sum_{j=2}^{k} N_j)N_k \leq N\sum_{k=2}^{n} N_k \leq N^2 \quad (3.198)$$

we extend (3.196) to the general case

$$\tilde{F}(\rho, \mathcal{D}) \leq \sum_{m=1}^{n} \omega_m\tilde{F}(\rho_m, \mathcal{D}_m) + u_B\rho^2\xi \quad (3.199)$$

where

$$\xi = V/R^{D+\epsilon} \quad (3.200)$$

is called the repulsion parameter and

$$\rho = \sum_{m=1}^{n} \omega_m\rho_m. \quad (3.201)$$

We prove the existence of the limits in property 2 first for a set of "standard cubes" Γ_k defined as follows: Let the edge of Γ_k be

$$d_k = 2^k d_0 \quad (3.202)$$

so the volume of Γ_k is

$$V_k = V(\Gamma_k) = d_k^D = 2^{Dk}d_0^D. \quad (3.203)$$

We let the cube Γ_k have impenetrable walls of thickness

$$h_k = (2\theta_1)^k\alpha_0 d_0 \ \text{ with } \ 1/2 < \theta_1 < 1 \quad (3.204)$$

so that $h_{k+1} > h_k$. Put 2^D cubes Γ_k in one Γ_{k+1} as shown in Fig. 3.4 where the interiors of Γ_k are all in the interior of Γ_{k+1} and the interiors of Γ_k are as far apart

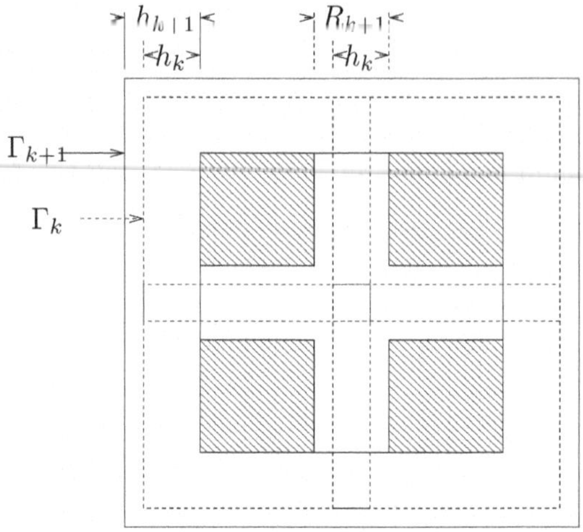

Fig. 3.4 The arrangement of domains Γ_k and Γ_{k+1} from [2]. The walls of Γ_{k_1} are drawn with solid lines and the walls of Γ_k are drawn with dotted lines. The interior of the four domains Γ_k are shaded.

from each other as possible. Then calling R_{k+1} the distance between the interiors of Γ_k we have

$$R_{k+1} = 2[h_k - (h_{k+1} - h_k)] = 4(1-\theta_1)\alpha_0 d_0 (2\theta_1)^k. \tag{3.205}$$

This will be greater than the constant R_0 in the weak tempering definition (3.22)–(3.23) if $\alpha_0 d_0$ is large enough.

The repulsion parameter ξ of (3.200) for the volume $V(\mathcal{D}_{k+1})$ is bounded above as

$$\xi_{k+1} = \frac{V(\mathcal{D}_{k+1})}{R_{k+1}^{D+\epsilon}} \leq \frac{V(\Gamma_{k+1})}{R_{k+1}^{D+\epsilon}} \tag{3.206}$$

which using (3.203) and (3.205) becomes

$$\xi_{k+1} \leq \xi_0 \theta_2^k \tag{3.207}$$

where

$$\xi_0 = \frac{(2d_0)^D}{[4(1-\theta)\alpha_0 d_0]^{D+\epsilon}} \tag{3.208}$$

and

$$\theta_2 = (2^\epsilon \theta_1^{D+\epsilon})^{-1}. \tag{3.209}$$

Choosing

$$\theta_1 = 2^{-\frac{\epsilon}{2(D+\epsilon)}} < 1 \tag{3.210}$$

we have

$$\theta_2 = 2^{-\epsilon/2} < 1 \tag{3.211}$$

and thus for this configuration $\xi_{k+1} \to 0$ as $k \to \infty$.

We now apply the inequality (3.199) to Γ_{k+1} where 2^{D-1} cubes have density ρ_1 and 2^{D-1} have density ρ_2 to obtain

$$\tilde{F}(\rho, \mathcal{D}_{k+1}) - u_B\rho^2\xi_0\theta_2^k \leq \frac{1}{2}\tilde{F}(\rho_1, \mathcal{D}_k) + \frac{1}{2}\tilde{F}(\rho_2, \mathcal{D}_k) \tag{3.212}$$

with

$$\rho = \frac{1}{2}\rho_1 + \frac{1}{2}\rho_2. \tag{3.213}$$

We further specialize (3.212) to $\rho_1 = \rho_2 = \rho$ to find

$$\tilde{F}(\rho, \mathcal{D}_{k+1}) - u_B\rho^2\xi_0\theta_2^k \leq \tilde{F}(\rho, \mathcal{D}_k). \tag{3.214}$$

Then setting

$$t_k = u_B\rho^2\xi_0\sum_{l=0}^{k-1}\theta_2^l \tag{3.215}$$

and subtracting t_k from both sides of (3.214) we find

$$\tilde{F}(\rho, \mathcal{D}_{k+1}) - t_{k+1} \leq \tilde{F}(\rho, \mathcal{D}_k) - t_k \tag{3.216}$$

which, with the definition

$$q_k = \tilde{F}(\rho, \mathcal{D}_k) - t_k \tag{3.217}$$

says that q_k is a monotonic nonincreasing function of k

$$q_{k+1} \leq q_k. \tag{3.218}$$

On the other hand from the classical stability bound (3.13)

$$Q_n(v.\mathcal{D}) \leq \frac{1}{N!\lambda^{DN}}V^N e^{NB/k_BT}. \tag{3.219}$$

Using the definition (3.195) we have

$$-V\tilde{F}(\rho, \mathcal{D})/k_BT \leq -\ln N! - N\ln\lambda^D + N\ln V + NB/k_BT \tag{3.220}$$

from which, when we use $\ln N! < N\ln N - N$, we obtain a lower bound on $\tilde{F}(\rho, \mathcal{D})$ of

$$\tilde{F}(\rho, \mathcal{D}) \geq \rho k_BT(\ln\rho\lambda^D - 1 - B/k_BT). \tag{3.221}$$

We also note from the definition (3.215) that t_k is a monotonic increasing function of k which is bounded above

$$t_k < t_{k+1} < \frac{u_b\rho^2\xi_0}{1-\theta_2} \tag{3.222}$$

and thus from (3.217), (3.221) and (3.222) we see that q_k is bounded below as

$$q_k \geq \rho k_BT(\ln\rho\lambda^D - 1 - B/k_BT) - \frac{u_B\rho^2\xi_0}{1-\theta_2}. \tag{3.223}$$

Consequently q_k is a monotonic nonincreasing function of k which is bounded below and therefore the limit of q_k exists as $k \to \infty$. Furthermore the limit of t_k as $k \to \infty$

is obviously $u_B\rho^{\Omega}\xi_0/(1-\theta_2)$ and therefore we have proven the desired result that the limit of $\tilde{F}(\rho, \mathcal{D}_k)$ exists as $k \to \infty$

$$\lim_{k\to\infty}\tilde{F}(\rho, \mathcal{D}_k) = \tilde{F}(\rho) \geq \rho k_B T(\ln\rho\lambda^D - 1 - B/k_B T). \tag{3.224}$$

Thus the existence theorem of properties 2 and 2' have been proven when the limit is taken through the set of domains cubical Γ_k.

To prove the independence of shape part of property 2 and 2′ we need to demonstrate that the shapes which satisfy the conditions of van Hove and Fisher can be suitably approximated by the cubical domains Γ_k and that the interactions of these domains can be controlled by the weak and strong tempering assumptions. This is done in [2] to which we refer the reader for details.

To prove the remaining properties 3–9 we need further convexity properties which follow from the above argument. In particular we take the thermodynamic limit in (3.199) and use the vanishing of the repulsion parameter ξ discussed above to find

$$\tilde{F}(\rho) \leq \sum_{m=1}^{n} \omega_m \tilde{F}(\rho_m) \text{ with } \rho = \sum_{m}^{n} \omega_m \rho_m \text{ and } \sum_{m=1}^{n} \omega_m = 1 \tag{3.225}$$

which is property 3 for $\tilde{F}(\rho)$.

The remaining properties 4–9 all follow from combining this basic existence argument with properties of convex functions. For example the continuity property 4 of $\tilde{F}(\rho)$ follows from the special case of (3.225)

$$\tilde{F}(\frac{\rho_1}{2} + \frac{\rho_2}{2}) \leq \frac{1}{2}\tilde{F}(\rho_1) + \tilde{F}(\rho_2) \tag{3.226}$$

and the limiting behavior at low densities $\rho \sim 0$

$$\tilde{F}(\rho) \sim -k_B T\rho|\ln\rho|. \tag{3.227}$$

Full details of the proof of all the properties are given in [1, 2, 4, 16] for box boundary conditions.

In [20] the case of periodic boundary conditions is studied and it is shown that box boundary conditions and periodic boundary conditions give the same unique free energy on the thermodynamic limit if a mild restriction is put on the pair potential which essentially eliminates certain potentials on the boundary of stability. Any potential satisfying conditions 2 or 3 will fulfill this extra stability requirement.

Just as for the case of stability, the Coulomb potential must be given a separate treatment because it does not satisfy the weak tempering assumption. Indeed, one of the basic features of a charged system is that the energy does indeed depend on the shape of the system. Therefore for the Coulomb case we must restrict our attention to neutral systems before we can expect properties 1–9 to hold. A proof is given in [18].

Properties 1–9 are all of a very general nature and give only qualitative features which any physics system must possess. They will prove to be useful because they must be obeyed by any approximate phenomenological description that we may want to use to describe an experimental system that is too complicated to deal with exactly. This will be seen in more detail when we discuss freezing and critical phenomena in later chapters.

3.3.4 Existence and uniqueness in the grand canonical ensemble

The grand partition function in a finite volume V is defined in terms of the canonical partition function as

$$Q^{gr}(z,T;V) = \sum_{N=0}^{\infty} (\lambda^D z)^N Q_N(V,T), \tag{3.228}$$

and from this the pressure and the density in the thermodynamic limit are expressed parametrically in terms of z (called the fugacity) as

$$\frac{p(z,T)}{k_BT} = \lim_{V\to\infty} \frac{1}{V} \ln Q^{gr}(z,T;V) \tag{3.229}$$

$$\rho(z,T) = \lim_{V\to\infty} \frac{1}{V} z \frac{\partial}{\partial z} \ln Q^{gr}(z,T;V) \tag{3.230}$$

where we denote the pressure as a function of the fugacity z by $p(z,T)$ to distinguish it from the pressure as a function of density ρ denoted by $P(\rho,T)$ and we will often suppress the T dependence in the notation. We note that z is a complex variable even though V and N are both real.

When the classical (3.12) or quantum mechanical stability conditions (3.16) hold we use the stability bound (3.13) on the canonical partition function to write

$$|Q^{gr}(z,T;V)| \leq \sum_{N=0}^{\infty} \frac{|z|^N}{N!} V^N e^{NB/k_BT} = e^{|z|Ve^{B/k_BT}}. \tag{3.231}$$

Moreover, when z is positive $Q^{gr}(z,T;V)$ is the sum of positive terms. Thus the following result holds:

1. When V is finite the grand partition function $Q^{gr}(z,T;V)$ is an entire function of z which is positive on the positive z axis.

Moreover, because the bound in (3.231) is exponential in $|z|$ we may use the Weierstrass (Hadamard) [17, pp. 18-22] factorization theorem to express Q^{gr} in terms of its zeros and thus

$$Q^{gr}(z,T;V) = e^{az} \prod_{j=1}^{\infty} [1-(z/z_j)]e^{z/z_j} \tag{3.232}$$

where in general the product contains an infinite number of zeros. Unlike (3.186) which in general has a normalizing factor independent of β which depends on N and V, the representation (3.232) is normalized to unity at $z=0$ because $Q_0(V,T)=1$. We also note that, for the ideal gas, there are no zeros and $a=V$.

When the potential is superstable, it was shown by Ruelle [1] that the growth order of $Q^{gr}(z,T:V)$ at $|z|\to\infty$ is less than one and thus the form (3.232) reduces to

$$Q^{gr}(z,T;V) = \prod_{j=1}^{\infty} [1-(z/z_j)]. \tag{3.233}$$

When the potential has a hard core, Yang and Lee [19] showed that the number of zeros in (3.233) is finite because any finite volume can contain only a finite number of particles and hence all $Q_N(V,T)$ must vanish for sufficiently large N (and thus $Q^{gr}(z,T;V)$ is a polynomial).

Furthermore in the thermodynamic limit for stable weakly tempered potentials we find from the existence of the limits in the canonical ensemble the following results:

2. $p(z,T)$ **is a convex function of** $\ln z$.
3. $p(z,T)$ **is a monotonic nondecreasing function of** $\ln z$.
4. **The derivative of** $p(z,T)$ **with respect to** $\ln z$ **exists almost everywhere and is equal to the average density. It is a nonnegative and nondecreasing function of** $\ln z$.
5. **The equation of state for the canonical ensemble is the same in the canonical ensemble as in the grand canonical ensemble.**

These results were established by Van Hove [16] and by Yang and Lee [19] for box boundary conditions and by Fisher and Lebowitz [20] for periodic boundary conditions.

Distribution of zeros

It is instructive to discuss the thermodynamic limit in terms of the behavior of the zeros of $Q^{gr}(z,T;V)$ of the product representations (3.232) and (3.233).

One possible behavior of the zeros z_j is that as $V \to \infty$ they will lie on curves in the complex z plane. If these limiting curves do not intersect the positive real z axis the pressure is analytic for all positive real z. But even though no zero can be on the positive real z axis for finite V the limiting distribution can pinch the axis.

The zeros will depend on temperature and it is possible that at high temperature the real axis will be zero free but that as the temperature is lowered a pinching will occur. For all temperatures below the temperature T_c where the pinch occurs the pressure will not be an analytic function of the density at the density corresponding to the value of z_c of the pinch. This behavior is illustrated in Fig.3.5. It was first shown to occur in the Ising model of a spin system in [21] and subsequently has been proven for many lattice models.

To the author's knowledge there are no theorems for continuum potentials which confine the zeros of the grand partition function to curves as $V \to \infty$ and thus we show in Fig. 3.6 two other possibilities: 1) the zeros can lie in areas of the z plane but can pinch the positive real axis only at points and 2) the zeros can pinch an entire segment of the positive real axis.

3.3.5 Continuity of the pressure

Property 8 of 3.3.3 leaves open the possibility that the pressure $P(v)$ can fail to be a continuous function of v at a countable number of points as illustrated in Fig.3.1. It is possible, however, to prove that if conditions stronger than stability and weak temperedness are imposed then $P(v)$ will in fact be continuous everywhere.

For classical continuum systems the continuity of the pressure is studied in [22] where it is shown for systems with pure pair interactions that if stability (3.12) is replaced by superstability (3.91) then the pressure is a continuous function of the density.

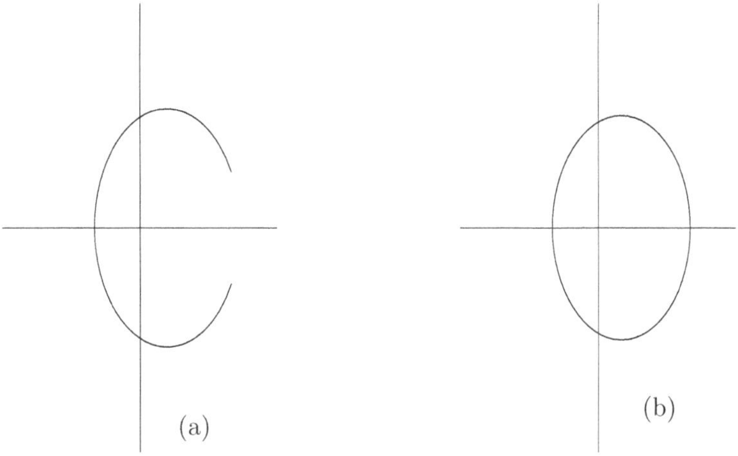

Fig. 3.5 A schematic distribution of zeros in the complex z plane of the grand partition function in a finite volume $Q^{gr}(z, T; V)$ where the limiting distribution is a set of curves. The solid lines represent the limiting curves of the location of zeros of the finite volume system. In Fig. 3.5a the zeros do not pinch the positive real axis. In Fig. 3.5b the limiting distribution does pinch the axis.

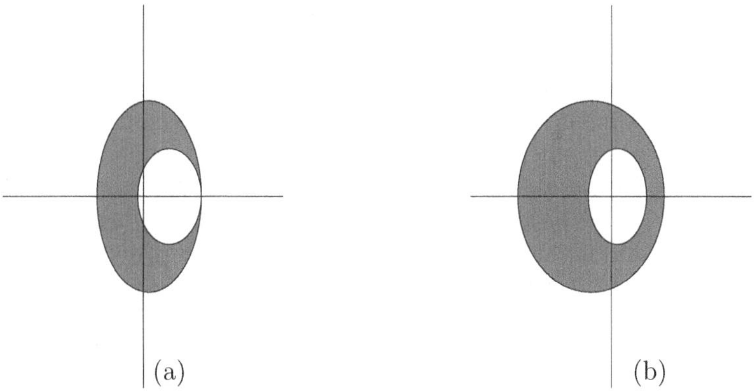

Fig. 3.6 A schematic distribution of zeros in the complex z plane of the grand partition function in a finite volume which fill up the area between the two bounding curves and pinch the positive real axis at a point. (b) A schematic distribution of zeros which fills up the area between the two bounding curves and pinches a line segment of the positive real axis.

For both classical and quantum systems on a lattice with many-body interactions of the form

$$U^{(N)}(\mathbf{r}_1, \cdots, \mathbf{r}_N) = \sum_{l=2}^{N} \sum_{1 \leq j_1 < j_2 < \cdots < j_l \leq N} \Phi^{(l)}(\mathbf{r}_{j_1}, \cdots, \mathbf{r}_{j_l}) \tag{3.234}$$

it was shown by Griffiths and Ruelle [24] in 1971 that the pressure is a continuous function of the density if for all $\mathbf{r}_j \in V$ and all N

$$\sum_{l=2}^{\infty} \sum_{2 \le j_2 < j_3 < \cdots < j_l \le \infty} |\Phi^{(l)}(\mathbf{r}_{j_1}, \mathbf{r}_{j_2} \cdots, \mathbf{r}_{j_l})| < \infty. \tag{3.235}$$

Fisher [25] in 1972 gave an example of a one-dimensional lattice many-body potential of the form (3.234) which does not satisfy (3.235) for which the pressure is a discontinuous function of the density. In 1983 Milton and Fisher [23] gave an example of a continuum system with many-body interactions for which the pressure is discontinuous. However, systems with only pair potentials which have a pressure which is a discontinuous function of the density do not seem to be known. A natural candidate for such an example would seen to be example 3 of section 3.1 but such a study does not in fact seem to have been made.

3.4 First order phase transitions, zeros and analyticity

The theorems of this chapter have made no reference to the phase of the system, and are valid for the various possible phases of crystalline order just as well as for isotropic fluids. Indeed for the box boundary conditions used in most of the proofs given above, the existence of the box breaks translational invariance and the exact classification of phases by means of the symmetry conditions discussed in chapter 1 is in the strictest sense impossible. Nevertheless in the thermodynamic limit the independence of boundary conditions implies that the symmetry of infinite space, which leads to the segmentation of the phase space into ergodic components that represent pure phases, should have important consequences for the free energy.

The theorems we have presented all assume that the variables such as the density and temperature are fixed before the thermodynamic limit is taken. The partition function averages over all states and thus averages over all ergodic components. Once the thermodynamic limit has been taken one (and in general only one) of these ergodic components will contribute. Therefore if after the thermodynamic limit has been taken we then consider varying the density we will only see properties of this one particular ergodic component even though once the parameters are changed the component may no longer dominate the free energy. In other words the process of changing the density will not need to commute with the process of taking the thermodynamic limit. This gives rise to the phenomena of first order phase transitions.

To examine first order phase transitions further consider the grand canonical partition function $Q^{gr}(z, T; V)$ as a function of z for finite volume V. From property 1 of 3.3.4 this grand partition function is an entire function of z and may be characterized by the zeros z_j in the product representations (3.232) and (3.233). Therefore as a function of z in finite volume the pressure,

$$\frac{p(z, T; V)}{k_B T} = \frac{1}{V} \ln Q^{gr}(z, T; V) \tag{3.236}$$

has logarithmic singularities, and the density

$$\rho(z,T;V) = \frac{1}{V} z \frac{\partial}{\partial z} \ln Q^{gr}(z,T;V). \tag{3.237}$$

has simple poles at $z = z_j$. These are the only singularities possible and none of them can lie on the positive z axis.

However, once the limit $V \to \infty$ is taken the situation is very different. Now the pressure and density are computed from (3.229) and (3.230)) and the singularities in $P(z,T)$ and $\rho(z,T)$ will in general differ from the singularities in $P(z,T;V)$ and $\rho(z,T;V)$ at V finite in several significant ways:

1. A theorem, proven by Yang and Lee [10] for potentials with a hard core and extended to general potentials by Ruelle [1], demonstrates that $p(z,T)$ and $\rho(z,T)$ are analytic (nonsingular) at values of z where $p(z,T;V)$ and $\rho(z,T;V)$ are free of zeros at finite V. However, it is perfectly possible that in the limit $V \to \infty$ that the limiting functions $p(z,T)$ and $\rho(z,T)$ may be analytically continued beyond some of the region occupied by zero in finite V.Thus, for example when the zeros fill up areas as in Fig.3.6b it is possible that $p(z,T)$ and $\rho(z,T)$ may be analytically continued from a zero free region into a region where for finite V there are zeros. In particular there is no a priori reason that $p(z,T)$ and $\rho(z,T)$ need to have singularities on the positive z axis at the values of z which bound the region of pinching in Fig.3.6b.
2. Where singularities do occur in the limiting functions $p(z,T)$ and $\rho(z,T)$ they are not restricted to be logarithms or simple poles.

There are very few (if any) examples known of continuum systems with a first order phase transition where the distribution of zeros for finite V is known. However, the following scenario is a plausible characterization of the relation of zeros to phases of the system.

1. When the zeros z_j of $Q^{gr}(z,T;V)$ divide the z plane into two or more disconnected zero free regions each region represents a phase of the system in the limit $V \to \infty$. Each phase represents a different ergodic component and each phase will have its own distinct symmetry (i.s. crystal structure).
2. The boundary of the zero free region represents the phase boundary in the $V \to \infty$ limit.
3. The segment of the z axis pinched by the zeros represents a region of two phase coexistence. In this region the pressure will be constant in the limit $V \to \infty$.
4. There is no reason to suppose that the pressure $p(z,T)$ and density $\rho(z,T)$ in a zero free region will fail to be analytic at the phase boundary.

When an analytic continuation of $p(z,T)$ and $\rho(z,T)$ beyond the phase boundary is possible the analytically continued phase is regarded as metastable. Such metastable situations are seen in the molecular dynamic computations presented in chapter 8.

On the other hand in the limiting case, where the pinching line segment reduces to a point, computations on the two-dimensional Ising model discussed in chapter 10 where the zeros are lie on a circle in the z plane, show that in that case there is an infinitely differentiable essential singularity at the phase boundary.

The possibility of analytic continuation through a first order phase transition is still an open question.

3.5 Discussion

We conclude this chapter with a few comments about the results we have presented.

First of all it cannot be emphasized too strongly that the stability conditions 2 and 3 of section 3.1 both have potentials which have non-integrable divergences at the origin. Therefore even though the function $U_2(\mathbf{r})$ in the stability condition 1 has a positive integrable Fourier transform and even though there are stable potentials which for which the Fourier transform does exist we are going to concentrate on those potentials for which the Fourier transform does not exist. We do this because potentials with nonintegrable divergences at the origin are what are seen to arise from atomic physics.

A very common two body potential is the Lennard-Jones (a, b) potential

$$U_{LJ}(r) = B(r_b/r)^b - A(r_a/r)^a \quad \text{with } A, B > 0. \tag{3.238}$$

When $A = 0$ and $b \to \infty$ this potential reduces to repulsive hard spheres and the case $a = 6,\ b = 12$ is commonly used to model nonpolar fluids such as carbon dioxide.

The nonintegrable divergence is an intimate part of the physics of fluids and this has the consequence that the interactions can never be treated as small perturbations of non-interacting systems. This is in great contrast to the perturbation theory of scattering where the Fourier transform of the potential is always assumed to exist.

A second point which is worth mentioning is connected with the shape independence of the free energy per particle $F(v, T)$. It is common to see the replacement of the limiting definition (3.9) of the free energy per particle by the more vague statement

$$Q_N(V, T) \sim e^{-NF(v,T)/k_BT} \tag{3.239}$$

without specifying what the symbol $\sim$ means. A much more precise expression for the large N behavior of $Q_N(V, T)$ for a system on a manifold with no boundary such as a surface of genus g embedded in three dimensions is

$$Q_N(V, T) = AN^{\gamma}e^{-NF(v,T)/k_BT}[1 + o(1)] \tag{3.240}$$

where $o(1)$ stands for terms which vanish as $N \to \infty$ and where it is important to note that even though $F(v, T)$ is independent of the shape of the system that both the exponent γ and the amplitude A can and often do depend on the system shape. In particular there are systems where at certain temperatures the exponent γ can depend on the topology of the genus g surface. These shape-dependent features of partition functions give information beyond the scope of thermodynamics which have become important in both mathematics and string theory in recent years.

If the system has a boundary of surface area $S \sim N^{(D-1)/D}$ the large N expansion becomes

$$Q_N(V, T) = AN^{\gamma}e^{-NF(v,T)/k_BT - SF_s(v,T)/k_BT}[1 + o(1)] \tag{3.241}$$

where $F_s(v, T)$ is the free energy of the free surface. Such surface free energies have been extensively studied and will be seen when we study the Ising model in chapter 10.

Finally it must be remarked that there are two classes of problems that occur commonly in practice and that can be treated in a very similar fashion to the continuum case treated in this chapter.

Rotational degrees of freedom

Most molecules are not spherically symmetric and in order to specify their phase space it is necessary to specify their orientation as well as the location of their center of mass. Therefore the kinetic energy will contain rotational terms and the potential energies will depend on the orientation of the molecules as well as on their relative positions. None of the existence theorems of this chapter are modified by the inclusion of these rotational degrees of freedom and all theorems can be taken over with no change. On the other hand the physics of these systems has significant effects which do not occur in spherically symmetric molecules. Very familiar examples are the several types of orientational ordering in liquid crystals and the phenomena of optical activity in organic molecules. Orientational effects are important in determining many crystal structures and they play an important role in the phenomena of the expansion of water under freezing.

Lattice models

We conclude by remarking that some of the most important problems in statistical mechanics are defined for cases where particles are confined to lie on the sites of a lattice (such as a square, triangular, or hexagonal lattice in two dimensions or a simple cubic, face centered cubic, or body centered cubic lattice in three dimensions). Often these problems either have an exclusion rule which forbids multiple occupancy of a site or have a variable on each site which takes on a discrete number of values. One such extremely important example which we will study extensively in later chapters is the Ising model where at each site $\mathbf{r}$ there is a variable $\sigma_{\mathbf{r}}$ which takes on the values ± 1 and the interaction energy is the sum over all nearest neighbor pair energies of the form

$$J(\mathbf{r}, \mathbf{r}')\sigma_{\mathbf{r}}\sigma_{\mathbf{r}'}. \tag{3.242}$$

As for rotational degrees of freedom the existence theorems of this chapter carry over to these lattice models and indeed in almost all cases the proofs are simpler than for the continuum case treated above. For this reason we do not give a separate treatment of these lattice cases.

However, there is one feature of lattice models which should be made explicit. No real lattice will be perfectly periodic due to the presence of impurities and in many cases these impurities are frozen in place and randomly distributed. Therefore a very realistic model of real dirty systems is to consider interaction constants such as $J(\mathbf{r}, \mathbf{r}')$ in (3.242) to be some sort of random variables. This leads to, if you will, a second level of statistics in the problem where the Hamiltonian itself has random elements. One striking feature of such dirty systems is that there are cases in which the temperature zeros of the partition function pinch the real axis in line segments. We will illustrate this for the Ising model in chapter 10.

3.6 Open questions

In this chapter we have proven that certain general properties of the potential (such as stability and temperedness) lead to a variety of general properties of the thermodynamic functions (such as existence and convexity of the free energy and the nondecreasing property of the pressure as a function of density). This has laid out a very general framework into which all of our subsequent work on specific potentials must fit. However, it is far from clear that the discovery of general as opposed to detailed specific properties of potentials is yet exhausted, and we wish here to make explicit some of the open questions which suggest themselves which have not been mentioned previously.

1. Possible bounds on derivatives

The properties of stability and weak temperedness were shown above to be sufficient to prove that the pressure was continuous and differentiable except at a countable number of points but that in order to have the pressure be continuous everywhere some additional property such as superstability needed to be imposed.

Superstability as defined by (3.91) says that at sufficiently high density the potential energy $U^{(N)}(\mathbf{r}_1, \cdots, \mathbf{r}_N)$ is positive for all configurations $\mathbf{r}_1, \cdots, \mathbf{r}_N$ and that this positive lower bound grows linearly in the density ρ as $\rho \to \infty$. But we noted in 3.1.3 potentials which are the sum of pair potentials $U(\mathbf{r})$ which diverge at $\mathbf{r} \to 0$ as $|\mathbf{r}|^{-\lambda}$ with $\lambda > D$ have a lower bound on $U^{(N)}(\mathbf{r}_1, \cdots \mathbf{r}_N)$ increases (3.94) as $\rho^{\lambda/D}$ as $\rho \to \infty$. The theorem on the continuity of the pressure in section 3.3 thus raises the question that if λ/D is made sufficiently large (but finite) then it might be possible to prove that, in addition to continuity, the pressure might also be a differentiable function of ρ for all $\rho > 0$. This could be considered to be plausible because there are no known examples where $P(\rho)$ is finite but $\partial P(\rho)/\partial\rho$ is infinite. If there is such a theorem then perhaps further increase in λ/D would make higher derivatives of $P(\rho)$ exist as well. On the other hand it is doubtful that there will be any theorems on the continuity of the derivative of the pressure because at any density where there is a first order transition (leading to a flat portion in the isotherm) the left and right derivatives are often seen experimentally to be discontinuous.

2. Upper bounds on the quantum mechanical Coulomb ground state energy

No real system has ever been seen where the pressure is a discontinuous function of the density even though, as discussed in section 3.5, this possibility is not excluded for classical systems with stable tempered potentials. However, in practice the only classical potentials we are physically interested in are those which arise from the Coulomb interactions of real matter. Therefore if we could prove that this quantum mechanical system satisfied the superstability requirement for the continuity of the pressure we would be assured that the classical many-body potentials used by Fisher [25] to produce a discontinuous pressure can never be realized in nature.

To the best of the author's knowledge there is no proof that real matter with Coulomb interactions does in fact satisfy the superstability bound, and in fact it might be argued that the reason that such a theorem has not been proven is that it is not true. It is thus a most interesting open question as to whether it can be proven that there is some upper bound on the Coulomb interaction energy which would demonstrate that superstability of this system is impossible.

3.7 Appendix A: Properties of functions of positive type

In the text we have made use of properties of continuous functions $f(\mathbf{r})$ of positive type which by definition is represented by its Fourier transform $\hat{f}(\mathbf{k})$ as

$$f(\mathbf{r}) = \frac{1}{(2\pi)^D} \int d^D\mathbf{k} \hat{f}(\mathbf{k}) e^{-i\mathbf{k}\cdot\mathbf{r}} \tag{3.243}$$

where $\hat{f}(\mathbf{k})$ is positive

$$\hat{f}(\mathbf{k}) \geq 0 \tag{3.244}$$

and integrable

$$\int d^D\mathbf{k} \hat{f}(\mathbf{k}) < \infty. \tag{3.245}$$

In this appendix we prove several properties of these functions used in the text.

Theorem A1
If a function $f(\mathbf{r})$ is of positive type then for all z_j and $\mathbf{r}_j$ and N we have

$$\sum_{1\leq j,k\leq N} z_j z_k^* f(\mathbf{r}_j - \mathbf{r}_k) \geq 0. \tag{3.246}$$

To prove this we note that because of (3.244) we have

$$\int d^D\mathbf{k} \hat{f}(\mathbf{k}) |\sum_{j=1}^{N} z_j e^{i\mathbf{k}\cdot\mathbf{r}_j}|^2 \geq 0. \tag{3.247}$$

Thus by rewriting the left-hand side of (3.247) as

$$\int d^D\mathbf{k} \hat{f}(\mathbf{k}) |\sum_{j=1}^{N} z_j e^{i\mathbf{k}\cdot\mathbf{r}_j}|^2 = \int d^D\mathbf{k} \hat{f}(\mathbf{k}) \sum_{1\leq j,k\leq N} z_j z_k^* e^{i\mathbf{k}\cdot(\mathbf{r}_j - \mathbf{r}_k)} = \sum_{1\leq j,k\leq N} z_j z_k^* f(\mathbf{r}_j - \mathbf{r}_k) \tag{3.248}$$

the inequality (3.246) follows.

Theorem A2
Any continuous function $f(\mathbf{r})$ which satisfies (3.246) is of positive type.

This theorem is the converse of theorem A1 and the proof, which is slightly more technical, can be found, for example, in [26].

Theorems A1 and A2 taken together state that the conditions of (3.246) and continuity of $f(\mathbf{r})$ can be (and often are) used as an alternative definition of functions of positive type.

We also note:

Theorem A3

If $f(\mathbf{r})$ is real and $f(\mathbf{r}) = f(-\mathbf{r})$ and (3.246) holds then all the eigenvalues of the $N \times N$ matrix $M_{jk} = f(\mathbf{r}_j - \mathbf{r}_k)$ are nonnegative.

To prove this let $\mathbf{z}$ be the N component vector with z_j as its components. We then may write

$$\sum_{1 \leq j,k \leq N} z_j z_k^* f(\mathbf{r}_j - \mathbf{r}_k) = \mathbf{z}^* M \mathbf{z}. \tag{3.249}$$

It follows from the fact that $f(\mathbf{r})$ is real and that $f(\mathbf{r}_j - \mathbf{r}_k) = f(\mathbf{r}_k - \mathbf{r}_j)$ that the matrix M is Hermitian and therefore can be diagonalized by a unitary transformation U (where $U^{-1} = U^{\dagger}$). Therefore

$$UMU^{\dagger} = \Lambda \tag{3.250}$$

where Λ is a diagonal matrix with diagonal entries λ_j. Thus we have, from (3.246),

$$0 \leq \mathbf{z}^* M \mathbf{z} = \mathbf{z}^* U^{\dagger} \Lambda U \mathbf{z} = \sum_{k=1}^{N} |(zU)_k|^2 \lambda_k. \tag{3.251}$$

and since (3.246) is assumed to hold for all z_j it follows that $\lambda_k \geq 0$ as desired.

Conversely we have:

Theorem A4

If $f(\mathbf{r})$ is real and $f(\mathbf{r})$ and the eigenvalues of $f(\mathbf{r}_j - \mathbf{r}_k)$ are nonnegative then (3.246) holds.

3.8 Appendix B: Fourier transforms

In the text, we have at times used various properties of a function $f(\mathbf{r})$ and its Fourier transform $\hat{f}(\mathbf{k})$. We will in this appendix prove several of the theorems we have used and give a few examples to illustrate them. Because of the reciprocity of the Fourier transform the role of $f(\mathbf{r})$ and $\hat{f}(\mathbf{k})$ can be interchanged in all theorems, and at times the theorem in the appendix may be interchanged from the way it is used in the text.

It is a general feature of Fourier transforms that if $f(\mathbf{r})$ is defined for $\mathbf{r}$ real and if $f(\mathbf{r})$ vanishes for $|\mathbf{r}| > R$ then the Fourier transform will in general oscillate as $|\mathbf{k}| \to \infty$ with $\mathbf{k}$ real and will grow as $e^{R|\mathrm{Im}\mathbf{k}|}$ when $|\mathbf{k}| \to \infty$ in the complex plane.

One example of this principle is seen in:

Theorem B1

Let $f(\mathbf{r})$ be a function which vanishes for $|\mathbf{r}| > R$ and has finite continuous derivatives of all orders. Then the Fourier transform $\hat{f}(\mathbf{k})$ is an entire function of $\mathbf{k}$ and has the property that, for all N and $|\mathbf{k}|$,

$$|\hat{f}(\mathbf{k})| < \frac{C_N e^{R|\mathrm{Im}\mathbf{k}|}}{(1 + |k|)^N} \tag{3.252}$$

where C_N is independent of $\mathbf{k}$. In particular it follows that $\hat{f}(\mathbf{k})$ is continuous and, when $\mathbf{k}$ is real, decreases as $|\mathbf{k}| \to \infty$ faster than any polynomial.

To prove theorem B1 we first note that, from the vanishing of $f(\mathbf{r})$ for $|\mathbf{r}| > R$,

$$\hat{f}(\mathbf{k}) = \int d^D\mathbf{r} e^{i\mathbf{k}\cdot\mathbf{r}} f(\mathbf{r}) = \int_{|\mathbf{r}|\leq R} d^D\mathbf{r} e^{i\mathbf{k}\cdot\mathbf{r}} f(\mathbf{r}). \tag{3.253}$$

The boundedness of the integration region ensures that all derivatives of $\hat{f}(\mathbf{k})$ exist for $|\mathbf{k}| < \infty$ and thus $\hat{f}(\mathbf{k})$ is an entire function of $\mathbf{k}$.

We next note that the existence of derivatives of all orders guarantees the existence of

$$\int_{|\mathbf{r}|<R} d^D\mathbf{r} e^{i\mathbf{k}\cdot\mathbf{r}} \frac{\partial^N}{\partial \mathbf{r}_{\alpha_1} \cdots \partial \mathbf{r}_{\alpha_N}} f(\mathbf{r}) \tag{3.254}$$

and furthermore we may integrate by parts N times to find

$$\prod_{j=1}^{N} (-ik_{\alpha_j}) \hat{f}(\mathbf{k}) = \int_{|\mathbf{r}|<R} d^D\mathbf{r} e^{i\mathbf{k}\cdot\mathbf{r}} \frac{\partial^N}{\partial \mathbf{r}_{\alpha_1} \cdots \partial \mathbf{r}_{\alpha_N}} f(\mathbf{r}) \tag{3.255}$$

where the boundary terms in the integration by parts all vanish because all derivatives are continuous. We may now take the absolute value of both sides to find

$$\prod_{j=1}^{N} |\mathbf{k}_{\alpha_j}||\hat{f}(\mathbf{k})| \leq \int_{|\mathbf{r}|<R} d^D\mathbf{r}|e^{i\mathbf{k}\cdot\mathbf{r}}||\frac{\partial^N}{\partial \mathbf{r}_{\alpha_1} \cdots \partial \mathbf{r}_{\alpha_N}} f(\mathbf{r})| \leq e^{R|\mathrm{Im}\mathbf{k}|}\tilde{C}_N \tag{3.256}$$

which follows from the boundedness of the region $|\mathbf{r}| < R$ and the existence of the derivatives. Therefore, for all $\mathbf{k}$,

$$|\hat{f}(\mathbf{k})| \leq \frac{\tilde{C}_N e^{R|\mathrm{Im}\mathbf{k}|}}{\prod_{j=1}^{N} |\mathbf{k}_{\alpha_j}|}. \tag{3.257}$$

This is sufficient to prove that, when $\mathbf{k}$ is real, $\hat{f}(\mathbf{k})$ vanishes more rapidly than a polynomial when $|\mathbf{k}| \to \infty$. If in addition we note that $\hat{f}(\mathbf{k})$ is bounded as $\mathbf{k} \to 0$ we may find a new constant $C_N > \tilde{C}_N$ such that the bound (3.252) holds.

Theorem B2
The converse of Theorem B1 also holds.

The proof is given in [26, pp.16,17]. Theorems B1 and B2 taken together are called the Paley–Wiener theorem.

When $f(\mathbf{r})$ vanishes for $|\mathbf{r}| > 0$ but some derivative is either discontinuous or diverges, the behavior of $\hat{f}(\mathbf{k})$ for real $\mathbf{k}$ as $|\mathbf{k}| \to \infty$ is determined by the boundary terms in the integration by parts which now no longer vanish. As an example of this we consider:

Example B1: The Fourier transform of a step function
Let

$$\begin{aligned} f(\mathbf{r}) = a \quad &\text{for } 0 \leq |\mathbf{r}| \leq R \\ 0 \quad &\text{for } R \leq |\mathbf{r}| \end{aligned} \tag{3.258}$$

To compute the Fourier transform we note that, for polar coordinates in D dimensions, if $\mathbf{r}$ makes an angle θ with the polar axis that the element of volume for functions which depends only on $|\mathbf{r}| = r$

$$d^D\mathbf{r} = r^{D-1}\Omega_{D-1}dr \tag{3.259}$$

with

$$\Omega_{D-1} = \frac{2\pi^{D/2}}{\Gamma(D/2)} \tag{3.260}$$

where $\Gamma(z)$ is the gamma function, and that for functions which depend on r and the angle θ between $\mathbf{r}$ and the axis defining the pole, the volume element is

$$d^D\mathbf{r} = \sin^{D-2}\theta r^{D-1}\Omega_{D-2}d\theta dr. \tag{3.261}$$

Thus the Fourier transform of (3.258) is

$$\hat{f}(\mathbf{k}) = \Omega_{D-2}a\int_0^R dRr^{D-1}\int_0^\pi d\theta \sin^{D-2}\theta e^{ikr\cos\theta}. \tag{3.262}$$

The integral over θ is expressible in terms of the Bessel function $J_\nu(z)$ using the Poisson integral representation [6, (9) on p.81] as.

$$\int_0^\pi d\theta \sin^{2\nu}\theta e^{iz\cos\theta} = \pi^{1/2}\Gamma(\nu+1/2)(2/z)^\nu J_\nu(z) \tag{3.263}$$

Thus using (3.263) with $\nu = (D-2)/2$ and setting $z = kr$ we have

$$\hat{f}(\mathbf{k}) = a\frac{(2\pi)^{D/2}}{k^D}\int_0^{kR} dz z^{D/2} J_{\frac{D}{2}-1}(z). \tag{3.264}$$

Then, using the property of Bessel functions [6, (50) on p.11] that

$$\frac{d}{dz}[z^\nu J_\nu(z)] = z^\nu J_{\nu-1}(z). \tag{3.265}$$

we obtain the result

$$\hat{f}(\mathbf{k}) = a\left(\frac{2\pi R}{k}\right)^{D/2} J_{D/2}(kR). \tag{3.266}$$

We note that $J_{D/2}(z)/z^{D/2}$ is an entire function which is bounded as $z \to 0$ and that, as $z \to \infty$,

$$J_{D/2}(z)/z^{D/2} \sim (\pi/2)^{1/2} r^{-(D+1)/2}\cos(z - \frac{D\pi}{4} - \frac{\pi}{2}). \tag{3.267}$$

Therefore the Fourier transform of the step function (3.258) is an entire function which falls off as $k^{-(D+1)/2}$ as $k \to \infty$.

If, however, instead of being a function of a real variable $\mathbf{r}$ which vanishes for $|\mathbf{r}| > R$ the function $F(\mathbf{r})$ is a function of a complex variable $\mathbf{r}$ which is analytic in a strip which includes the real $\mathbf{r}$ axis, then the Fourier transform will vanish exponentially as $|\mathbf{k}| \to \infty$ with $\mathbf{k}$ real. To illustrate this consider:

Example B2. Proof of (3.71)
Let

$$f(\mathbf{r}) = (\mathbf{r}^2 + 1)^{-D}. \tag{3.268}$$

Then

$$\begin{aligned}\hat{f}(\mathbf{k}) &= \int d\mathbf{r} e^{i\mathbf{r}\cdot\mathbf{k}} (\mathbf{r}^2 + 1)^{-D} = \int_0^\infty dr \frac{r^{D-1}}{(r^2+1)^D} \int_0^\pi d\theta \sin^{D-2}\theta e^{ikr\cos\theta} \\ &= \pi^{1/2}\Gamma(\frac{D-2}{2}) \left(\frac{2}{r}\right)^{\frac{D}{2}-1} \int_0^\infty dr \frac{r^{D/2}}{(r^2+1)^D} J_{\frac{D}{2}-1}(kr)\end{aligned} \tag{3.269}$$

where we have used (3.263). Using the Sonine-Gegenbauer integral [6, (51) on p.95] we find the desired result:

$$\int d\mathbf{r} e^{i\mathbf{r}\cdot\mathbf{k}} (\mathbf{r}^2 + 1)^{-D} = \frac{\pi^{1/2}\Gamma(\frac{D-1}{2})}{\Gamma(\frac{D}{2})} \left(\frac{k}{2}\right)^{D/2} K_{D/2}(k) > \tag{3.270}$$

We note that $k^{D/2}K_{D/2}(k)$ is bounded as $k \to 0$ and that as $k \to \infty$

$$k^{D/2}K_{D/2}(k) \sim (\pi/2)^{1/2} k^{(D-1)/2} e^{-k}. \tag{3.271}$$

References

[1] D. Ruelle, Classical statistical mechanics of a system of particles, Helv. Phys. Acta 36 (1963) 183.

[2] M.E. Fisher, Free energy of a macroscopic system, Arch. Rat. Mech. Anal. 17 (1964) 377–410.

[3] M.E. Fisher and D. Ruelle, The stability of many-particle systems, J. Math. Phys. 7 (1966) 260–270.

[4] D. Ruelle, *Statistical Mechanics*, (Benjamin 1969) chapter 3.

[5] A. Lenard and S. Sherman, Stable Potentials II, Comm. Math. Phys. 17 (1970), 91–97.

[6] A. Erdelyi, W. Magnus, F. Oberhettinger and F.G. Tricomi, *Higher Transcendental Functions* (McGraw-Hill New York 1953) vol. 2

[7] N. Angelescu, G. Nenciu and V. Protopopescu, On stable potentials, Comm. math. Phys. 22 (1971) 162–165.

[8] D. Ruelle, Statistical mechanics of quantum systems of particles Helv. Phys. Acta 36 (1963) 789.

[9] F.J. Dyson and A. Lenard, Stability of matter I, J. Math. Phys. 8 (1967) 423–434.

[10] F.J. Dyson, Ground–state energy of a finite System of charged Particles, J. Math. Phys. 8 (1967) 1538–1545.

[11] A. Lenard and F.J. Dyson, Stability of matter II, J. Math. Phys. 9 (1968) 698–711.

[12] E. H. Lieb, The stability of matter, Rev. Mod. Phys. 48 (1976) 553–569.

[13] E.H. Lieb, The $N^{5/3}$ law for bosons, Phys. Lett. 70A (1979) 71–73.

[14] J.G. Conlon, E.H. Lieb and H.–T. Yau, The $N^{7/5}$ law for charged bosons, Comm. Math. Phys. 116 (1988) 417–448.

[15] *The stability of matter: from atoms to stars; selecta of Elliott H. Lieb* ed. W. Thirring, (Springer-Verlag, Berlin and Heidelberg 1991).

[16] L. Van Hove, Quelque propriétés génerales de l'intégral de configuration d'un system de particules avec interaction, Physica 15 (1949) 951–961.

[17] R.P. Boas, *Entire Functions* (New York: Academic Press 1954).

[18] J. Lebowitz and F. Dyson, Existence of thermodynamics for real matter with Coulomb forces, Phys. Rev. Letts. 22 (1969) 631–634.

[19] C.N. Yang and T.D. Lee, Statistical theory of equations of state and phase transitions I. Theory of condensation, Phys. Rev. 87 (1952) 404–409.

[20] M.E. Fisher and J.L. Lebowitz, Asymptotic free energy of a system with periodic boundary conditions, Comm. Math. Phys. 19 (1970) 251–272.

[21] T.D. Lee and C.N.Yang, Statistical theory of equations of state and phase transitions II. Lattice gas and Ising model, Phys. Rev. 87 (1952) 410–419.

[22] D. Ruelle, Superstable interactions in classical statistical mechanics, Comm. Math. Phys. 18 (1970) 127–159

[23] G.W. Milton and M.E. Fisher, Continuum fluids with a discontinuity in the pressure, J. Stat. Phys. 32 (1983) 413–438.
[24] R.B. Griffiths and D. Ruelle, Strict convexity ("continuity") of the pressure in lattice systems, Comm. Math. Phys. 23 (1971) 169–175.
[25] M.E. Fisher, On the discontinuity of the pressure, Comm. Math. Phys. 26 (1972) 6–14.
[26] M. Reed and B. Simon, *Methods of Mathematical Physics II: Fourier analysis and self-adjointness* (Academic Press 1975), p. 13.

4
Theorems on order

In chapter 2 we presented several models which we want to use to describe freezing/melting transitions and critical phenomena in continuum systems and ferro- and antiferromagnetism in lattice magnetic systems. However, merely writing down a model does not demonstrate that it will actually exhibit the desired ordering, and in this chapter we turn our attention to a survey of the proofs of ordering for these potentials.

A chronology of some ordering theorems is presented in Table 4.1.

Studies of order date back to antiquity but in more modern times we may perhaps date the beginnings of the subject to the conjecture of Kepler [1] in 1611 that there is no denser packing of hard spheres than the face centered cubic lattice. Gauss proved [2] in 1831 that if we only consider periodic lattice packing that face centered cubic has maximum density but the demonstration that there is no nonperiodic structure that is denser than the fcc lattice was only given by Hales [12] in 2005. We will discuss these results in section 4.1.

An obvious extension of the Kepler problem is to hard ellipsoids of revolution such as we discussed in connection with models of liquid crystals. In 2004 an example was discovered by Donev, Stillinger, Chaikin and Torquato [11] of a configuration of hard ellipsoids which, for all aspect ratios, is denser than the density of fcc close packed hard spheres. This is also presented in section 4.1.

The modern studies of crystalline and magnetic order begin with the conjectures made on physical grounds by Peierls [13] and Landau [14] that in $D = 1, 2$ for $T > 0$ crystalline order does not occur and the proof by Peierls [3] in 1936 of the existence of spontaneous magnetization in the two-dimensional Ising model, The proof of this conjecture of nonexistence of crystalline order and the similar result of nonexistence of ferromagnetic and antiferromagnetic order in Heisenberg magnets was made in the mid 1960s and will be proven in detail in sections 4.2 and 4.3.

The proof of existence of spontaneous magnetization in the Ising model was first given by Peierls [3] in 1936. However, we will not give this proof because the spontaneous magnetization for the Ising model on the square lattice will be computed exactly in chapter 12. The existence of order in Heisenberg models is substantially more difficult to prove. For the classical case a proof of the existence of ferromagnetic and antiferromagnetic order in $D = 3$ at $T > 0$ for isotropic nearest neighbor cubic lattices was first given in 1976 [7]. We present the proof in section 4.4. The extension to the quantum case has only been done for antiferromagnets. The case $S \geq 1$ is done in [8] and $S = 1/2$ in [10]. At $T = 0$ antiferromagnetic order has been proven [9] in $D = 2$ for all $S \geq 1$. We survey these proofs in sections 4.5 and 4.6.

It is obvious that Table 4.1 does not include answers to many questions which are raised in the previous chapter. This is not because we have only made a limited selection of results but because there are indeed a large number of unanswered questions. We explicitly formulate a few of these important "missing theorems" in section 4.7.

Table 4.1 Chronology of selected theorems, conjectures and examples of order in $D = 1, 2, 3$.

Date	Authors	Theorem or conjecture
1611	Kepler [1]	Conjecture: fcc is densest structure for $D = 3$
1831	Gauss [2]	Proof: fcc is densest lattice for $D = 3$
1936	Peierls [3]	Proof: existence of ferromagnetic order in Ising model for $D = 2$
1940	Fejes Tóth [4]	Proof: hexagonal is densest structure for $D = 2$
1966	Mermin, Wagner [5]	Proof: lack of order in $D = 1, 2$ Heisenberg ferro- and antiferromagets for $T > 0$
1968	Mermin [6]	Proof: lack of crystalline order in $D = 1, 2$ for $T > 0$
1976	Froehlich, Simon, Spencer [7]	Proof: existence of order in $D \geq 3$ classical Heisenberg ferro- and antiferromagets for $T > 0$ for nearest neighbor isotropic cubic lattices
1978	Dyson, Lieb, Simon [8]	Proof: existence of antiferromagnetic order in quantum Heisenberg antiferromagnets for $S \geq 1$ at $T > 0$ for $D = 3$ nearest neighbor isotropic cubic lattices
1986	Neves, Perez [9]	Proof: existence of antiferromagnetic order in the $S \geq 1$ Heisenberg antiferromagnet at $T = 0$ for $D = 2$ nearest neighbor isotropic square lattices
1988	Kennedy, Lieb, Shastry [10]	Proof: existence of antiferromagnetic order in the $S = 1/2$ Heisenberg antiferromagnet at $T > 0$ for $D = 3$ nearest neighbor anisotropic cubic lattices
2004	Donev, Stillinger, Chaikin, Torquato [11]	Example: unusually dense packings of ellipsoids
2005	Hales [12]	Proof: Kepler's conjecture

4.1 Densest packing of hard spheres and ellipsoids

One of the oldest problems of order in condensed matter physics is the question of the closest packed arrangement of spheres of diameter σ in D dimensions.

In two dimensions this closest packed arrangement of hard discs is shown in Fig. 4.1. It is a translationally invariant lattice where each point has six nearest neighbors. This number of nearest neighbors is referred to by mathematicians as the kissing number. The fraction of space occupied by a configuration of spheres in dimension D is called the packing fraction. It is easy to compute that the packing fraction for the infinite hexagonal lattice in $D = 2$ of Fig. 4.1 is $\pi/\sqrt{12} = 0.9069\cdots$. The space in which the discs are situated can be separated into cells which surround the centers of the discs which contain the points which are closest to the center of the disc. This decomposition of space into hexagonal cells is shown in Fig. 4.2. These cells are called Voronoi cells by mathematicians and Wigner–Seitz cell by physicists.

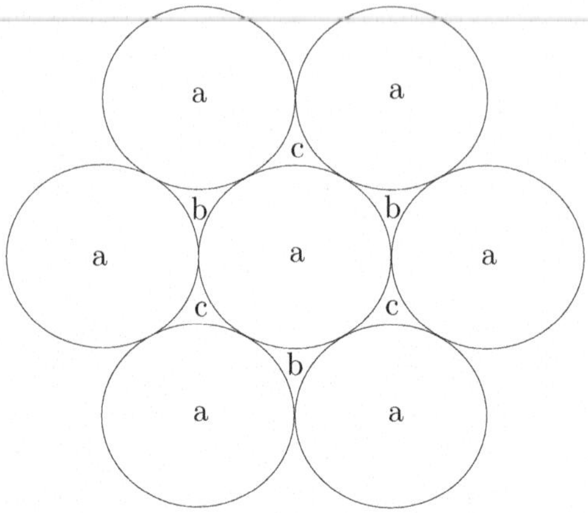

Fig. 4.1 The hexagonal lattice of closest packed discs in $D = 2$.

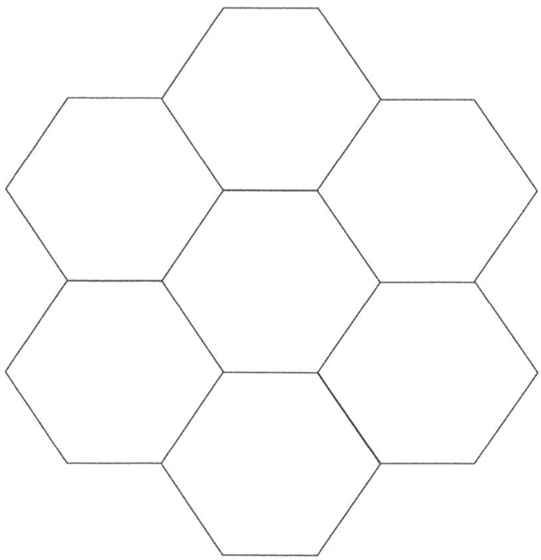

Fig. 4.2 The Voronoi or Wigner–Seitz cells for closest packed hard discs.

A three-dimensional lattice may be constructed by stacking these hexagonal lattices on top of each other. In Fig. 4.1, where we denote the location of the centers of the discs of the first lattice as a, the maximum density is achieved if the centers of the spheres of second layer are at either the positions b or c. The third layer may also be added in two possible ways in a close packed fashion. Thus every sequence of the letters a, b, c gives a lattice with the same density. The sequences of three positions

$$\cdots abcabcabc \cdots \quad \text{or} \quad \cdots acbacbacb \cdots \tag{4.1}$$

gives the face centered cubic lattice (fcc) and the sequence of two positions

$$\cdots ababab \cdots \tag{4.2}$$

gives the hex close packed lattice. The unit cells of these lattices are given in appendix A of chapter 2. Both of these lattices are translationally invariant, and each of them has 12 nearest neighbors and a packing fraction of $\pi/\sqrt{18} = 0.7405\cdots$. For the fcc lattice there is one site per unit cell which means that every lattice site is equivalent to every other lattice site. For the hcp lattice there are two sites per unit cell.

It was proven by Gauss [2] that there is no translationally invariant lattice in $D = 3$ with a packing fraction greater than the fcc lattice. The question of the densest translationally invariant packing of spheres in D dimensions has been extensively investigated and many results are tabulated in the book by Conway and Sloane [15].

Of course, most packings of spheres will not be translationally invariant. However in $D = 2$ it was proven by Fejes Tóth [4] in 1940 that the hexagonal lattice is indeed the densest packed structure.

In $D = 3$ the conjecture of Kepler [1] of 1611 asserts that none of these non translationally invariant lattices is more densely packed than the fcc lattice. The proof of this was an outstandingly difficult problem for almost 400 years and was only proven in 2005 by Hales [12] in a series of papers of great complexity. The reason for this difficulty is that it is possible to find non-lattice packings which are locally more dense than the fcc density of $\pi/\sqrt{18}$ and thus it is not obvious that a non-lattice packing must be less dense than a regular lattice packing. In fact for sufficiently large dimension D there are examples where non-lattice packings are more dense than regular lattice packings.

Hales first showed that it was sufficient to check the density of a finite (but large) number of configurations and then the properties of these configurations were painstakingly computed on a case by case basis with the assistance of a computer. Because of the computer-assisted nature of the proof there was not total agreement in the mathematics community [16] as to whether the proof could be considered logically complete and the controversy stands as a landmark in the philosophy of mathematical proof.

At high pressure, atoms of any material will be pressed together at high density and thus it may be expected that the repulsive core of the potential will play a dominant role in determining the crystal structure of the solid at high pressure. In chapter 2 we saw that many monotonic insulators and metals have fcc or hcp crystal structure at high pressure on the solid side of the melting curve. This is certainly consistent with a hard sphere model for the core of the potential of these systems.

A particularly simple model for the liquid crystals discussed in chapter 2 is the hard ellipsoid of revolution with aspect ratio $\alpha = a/b$ shown in Fig. 4.3, and thus it is interesting to find the densest packing of such molecules.

If we consider translationally invariant configurations with only one ellipsoid per unit cell then it can be shown, that for any aspect ratio, the maximum packing fraction is the same as the fcc lattice of hard spheres. However, if we allow several ellipsoids per unit cell with at least two inequivalent orientations it was found by Donev, Stillinger, Chaikin and Torquato [11] that more dense packings are possible. The densities they found depend on the aspect ratio α and are plotted in Fig. 4.4. For $\alpha \geq \sqrt{3}$ and $\alpha \leq 1/\sqrt{3}$ the packing fraction is 0.770732 and the ellipsoids each have 14 nearest neighbors which is to be compared with the 12 nearest neighbors of the fcc lattice. However, to quote [11] "There is nothing to suggest that the crystal packing we have presented here is indeed the densest for any aspect ratio other than the trivial case of spheres." This extension of Kepler's conjecture remains very much an open question and much more work need to be done to understand the ordering properties of even this most simple model of a liquid crystal.

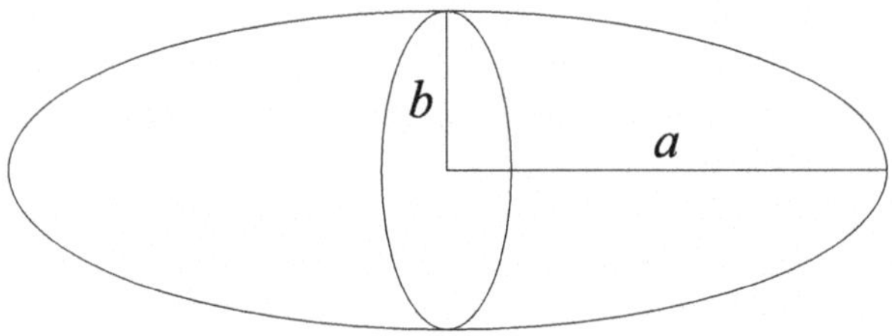

Fig. 4.3 An ellipsoid of revolution with aspect ratio $\alpha = a/b$.

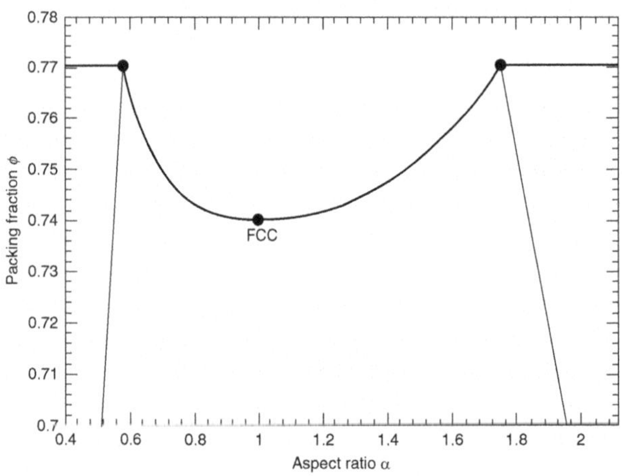

Fig. 4.4 The packing fraction as a function of the aspect ratio for the packings of ellipsoids of Donev, Stillinger, Chaikin and Torquato taken from [11].

4.2 Lack of order in the isotropic Heisenberg model in D = 1, 2

The discussion of densest packings of section 4.1 are all geometric properties which have nothing to do with temperature. However, we are principally interested in the dependence of order on temperature, and the possibility that long range order exists for $T > 0$ depends very much on the dimension D of the system.

We begin our considerations with the study of the isotropic quantum mechanical Heisenberg model of spin S introduced in chapter 2 which is defined by the Hamiltonian

$$\mathcal{H} = -\sum_{\mathbf{R},\,\mathbf{R}'} J(\mathbf{R}-\mathbf{R}')\mathbf{S}_{\mathbf{R}} \cdot \mathbf{S}_{\mathbf{R}'} - H\sum_{\mathbf{R}} S^z_{\mathbf{R}} \tag{4.3}$$

where the interaction constants are chosen without loss of generality to satisfy

$$J(\mathbf{R}) = J(-\mathbf{R}) \quad J(0) = 0. \tag{4.4}$$

The spin matrices at the site $\mathbf{R}$ are characterized by

$$[S^x_{\mathbf{R}}, S^y_{\mathbf{R}'}] = iS^z_{\mathbf{R}}\delta_{\mathbf{R},\mathbf{R}'}, \; [S^y_{\mathbf{R}}, S^z_{\mathbf{R}'}] = iS^x_{\mathbf{R}}\delta_{\mathbf{R},\mathbf{R}'}, \; [S^z_{\mathbf{R}}, S^x_{\mathbf{R}'}] = iS^y_{\mathbf{R}}\delta_{\mathbf{R},\mathbf{R}'} \tag{4.5}$$

$$S^{k\dagger}_{\mathbf{R}} = S^k_{\mathbf{R}} \tag{4.6}$$

and

$$\mathbf{S}^2 = S(S+1). \tag{4.7}$$

It is convenient to further introduce the notation

$$\mathbf{S}^{\pm}_{\mathbf{R}} = 2^{-1/2}(\mathbf{S}^x_{\mathbf{R}} \pm i\mathbf{S}^y_{\mathbf{R}}) \tag{4.8}$$

which from (4.5) satisfy the commutation relations

$$[\mathbf{S}^+_{\mathbf{R}}, \mathbf{S}^-_{\mathbf{R}'}] = \mathbf{S}^z_{\mathbf{R}}\delta_{\mathbf{R},\mathbf{R}'}, \qquad [\mathbf{S}^z_{\mathbf{R}}, \mathbf{S}^{\pm}_{\mathbf{R}'}] = \pm\mathbf{S}^{\pm}_{\mathbf{R}}\delta_{\mathbf{R},\mathbf{R}'} \tag{4.9}$$

For ease of notation we consider cubic lattices where $\mathbf{R} = \mathbf{N}$ where $\mathbf{N}$ is a D-dimensional vector with D components and $\mathbf{R}$ is restricted to the box Ω,

$$\mathbf{R} \in \Omega \;\text{ if }\; 0 \le N_j \le L_j - 1, \tag{4.10}$$

and we impose periodic boundary conditions on (4.3) by setting $N_j = L_j \equiv 0$. The total number of sites is $N = \prod_j L_j$. The extension to other lattices is merely a matter of extending the notation to other Bravais lattices.

The magnetization is defined as

$$M^z(H;N) = \frac{1}{N}\sum_{\mathbf{R}}\langle \mathbf{S}^z_{\mathbf{R}}\rangle \tag{4.11}$$

where we recall that the thermal average is given by

$$\langle X\rangle = \mathrm{Tr}Xe^{-\beta\mathcal{H}}/\mathrm{Tr}e^{-\beta\mathcal{H}}. \tag{4.12}$$

From (4.3) and (4.11) we see that

$$M^z(-H;N) = -M^z(H;N). \tag{4.13}$$

When the size of the system N is finite then $M^z(H;N)$ must be continuous (in fact analytic) at $H = 0$ and thus for finite N we must have $M^z(0;N) = 0$ in any dimension.

However, we are interested in the thermodynamic limit $N \to \infty$. In this limit continuity and analyticity at $H = 0$ are no longer guaranteed and thus we define the spontaneous magnetization (ferromagnetic order) as

$$M^z(0^+) = \lim_{H \to 0^+} \lim_{N \to \infty} M^z(H; N). \tag{4.14}$$

When $M^z(0^+)$ is positive we say that the magnet has ferromagnetic order.

The existence or nonexistence of ferromagnetic order in the isotropic Heisenberg magnet depends on the dimension D. The first theorem we prove for this Heisenberg magnet is the theorem of Mermin and Wagner [5].

Theorem 4.1: Mermin and Wagner
If we impose on the Heisenberg model of (4.3) the additional restriction

$$\sum_{\mathbf{R}} \mathbf{R}^2 |J(\mathbf{R})| < \infty \tag{4.15}$$

then the magnetization is bounded above for sufficiently small fields H by

$$|M^z| < \begin{cases} \text{const } T^{-2/3}|H|^{1/3} & \text{in } D = 1 \\ \text{const } T^{-1/2}|\ln|H||^{-1/2} & \text{in } D = 2. \end{cases} \tag{4.16}$$

Therefore, when (4.15) holds the isotropic Heisenberg magnet has no spontaneous magnetization.

We will prove theorem 4.1 by use of the following inequality of Bogoliubov [17].

Lemma 4.1: Bogoliubov's inequality
For any operators A, C and $\mathcal{H} = \mathcal{H}^\dagger$

$$\frac{1}{2}\langle AA^\dagger + A^\dagger A\rangle\langle[[C, \mathcal{H}], C^\dagger]\rangle \geq k_B T |\langle[C, A]\rangle|^2. \tag{4.17}$$

To prove the inequality (4.17) we first define a quantity which has come to be called the Duhamel two-point function [8].

Definition: Duhamel two-point function
The Duhamel two-point function (A, B) may be defined by

$$(A, B) = \sum_{i,j}{}' \langle i|A|j\rangle^* \langle i|B|j\rangle \frac{W_i - W_j}{E_j - E_i} \tag{4.18}$$

where E_i are the eigenvalues of $\mathcal{H}$, the sum is over all pairs of the eigenstates of $\mathcal{H}$ excluding pairs with equal energy, and

$$W_j = e^{-\beta E_j}/\text{Tr}e^{-\beta\mathcal{H}}. \tag{4.19}$$

We first prove that the Duhamel two-point function has the two properties needed for an inner product

$$(A, B) = (B, A)^* \tag{4.20}$$

$$0 < (A, A). \tag{4.21}$$

Property (4.20) is obvious from the definition (4.18). To prove property (4.21) we note that

$$\frac{e^{-\beta E_i} - e^{-\beta E_j}}{e^{-\beta E_i} + e^{-\beta E_j}} = \tanh \frac{\beta}{2}(E_j - E_i) \tag{4.22}$$

and

$$0 \leq \frac{\tanh \frac{\beta}{2}(E_j - E_i)}{E_j - E_i} \leq \beta/2. \tag{4.23}$$

It therefore follows that

$$0 \leq \frac{W_i - W_j}{E_j - E_i} \leq \frac{\beta}{2}(W_i + W_j) \tag{4.24}$$

and since

$$\sum_{i,j} \langle i|A|j\rangle^* \langle i|A|j\rangle W_i = \sum_i \langle i|AA^\dagger|i\rangle W_i = \langle AA^\dagger \rangle \tag{4.25}$$

we have the two inequalities

$$\begin{aligned} 0 &< {\sum_{i,j}}' \langle i|A|j\rangle^* \langle i|A|j\rangle \frac{W_i - W_j}{E_j - E_i} = (A, A) \\ &\leq \frac{\beta}{2} {\sum_{i,j}}' \langle i|A|j\rangle^* \langle i|A|j\rangle (W_i + W_j) = \frac{\beta}{2} \langle AA^\dagger + A^\dagger A \rangle. \end{aligned} \tag{4.26}$$

Property (4.21) is the inequality on the left-hand side of (4.26).

The properties (4.20) and (4.21) are the only properties used in the proof of the Cauchy–Schwarz inequality given in (3.147)–(3.151). Therefore the identical proof shows that

$$(A, A)(B, B) \geq |(A, B)|^2. \tag{4.27}$$

We may now derive the Bogoliubov inequality (4.17) by setting

$$B = [C^\dagger, \mathcal{H}]. \tag{4.28}$$

Then, using from the definition (4.18) of (A, B), we have

$$(A, B) = \langle [C^\dagger, A^\dagger] \rangle \tag{4.29}$$

$$0 < (B, B) = \langle [C^\dagger, [\mathcal{H}, C]] \rangle \tag{4.30}$$

and thus the inequality (4.27) becomes

$$(A, A)\langle [C^\dagger, [H, C]] \rangle \geq |\langle [C^\dagger, A^\dagger]] \rangle|^2. \tag{4.31}$$

The Bogoliubov inequality (4.17) now follows by using the inequality of the right-hand side of (4.26) in (4.31).

Proof of Theorem 4.1

In order to use the Bogoliubov inequality (4.17) to prove the bounds (4.16) we define the Fourier transform (and the inverse)

$$\tilde{\mathbf{S}}(\mathbf{k}) = \frac{1}{N^{1/2}} \sum_{\mathbf{R}} e^{-i\mathbf{k}\cdot\mathbf{R}} \mathbf{S}_{\mathbf{R}}, \quad \mathbf{S}_{\mathbf{R}} = \frac{1}{N^{1/2}} \sum_{\mathbf{k}} e^{i\mathbf{k}\cdot\mathbf{R}} \tilde{\mathbf{S}}(\mathbf{k}) \tag{4.32}$$

where the sum over $\mathbf{R}$ is over all space and in the sum over $\mathbf{k}$ we have $k_j = 2\pi n_j / L_j$ with $-L_j/2 < n_j \leq L_j/2$. Then defining

$$\tilde{\mathbf{S}}^{\pm}(\mathbf{k}) = 2^{-1/2} [\tilde{\mathbf{S}}^x(\mathbf{k}) \pm i \tilde{\mathbf{S}}^y(\mathbf{k})] \tag{4.33}$$

with

$$\tilde{\mathbf{S}}^{\pm\dagger}(\mathbf{k}) = \tilde{\mathbf{S}}^{\mp}(-\mathbf{k}) \tag{4.34}$$

we set

$$C = \tilde{\mathbf{S}}^{+}(\mathbf{k}), \quad A = \tilde{\mathbf{S}}^{-}(-\mathbf{k}). \tag{4.35}$$

We use (4.9) to compute

$$\langle [C, A] \rangle = \frac{1}{N} \sum_{\mathbf{R},\mathbf{R}'} \langle [S^+_{\mathbf{R}}, S^-_{\mathbf{R}'}] \rangle e^{i\mathbf{k}\cdot(\mathbf{R}-\mathbf{R}')} = \frac{1}{N} \sum_{\mathbf{R}} S^z_{\mathbf{R}} = M^z \tag{4.36}$$

and

$$\begin{aligned} &\langle [[C, \mathcal{H}], C^\dagger] \rangle \\ &= \frac{2}{N} \sum_{\mathbf{R},\mathbf{R}'} (1 - e^{-i\mathbf{k}\cdot(\mathbf{R}-\mathbf{R}')}) J(\mathbf{R}-\mathbf{R}') \langle S^-_{\mathbf{R}} S^+_{\mathbf{R}'} + S^z_{\mathbf{R}} S^z_{\mathbf{R}'} \rangle + HM^z \\ &= I_1(\mathbf{k}) + I_2(\mathbf{k}) + HM^z \end{aligned} \tag{4.37}$$

with

$$I_1(\mathbf{k}) = \frac{1}{N} \sum_{\mathbf{R},\mathbf{R}'} [1 - \cos \mathbf{k}\cdot(\mathbf{R}-\mathbf{R}')] J(\mathbf{R}-\mathbf{R}') \langle S^-_{\mathbf{R}} S^+_{\mathbf{R}'} + S^+_{\mathbf{R}} S^-_{\mathbf{R}'} + 2 S^z_{\mathbf{R}} S^z_{\mathbf{R}'} \rangle \tag{4.38}$$

$$I_2(\mathbf{k}) = \frac{1}{N} \sum_{\mathbf{R},\mathbf{R}'} i \sin \mathbf{k}\cdot(\mathbf{R}-\mathbf{R}') J(\mathbf{R}-\mathbf{R}') \langle S^-_{\mathbf{R}} S^+_{\mathbf{R}'} - S^+_{\mathbf{R}} S^-_{\mathbf{R}'} \rangle \tag{4.39}$$

where to obtain (4.37) we have averaged over the interchange of the summation variables $\mathbf{R}$ and $\mathbf{R}'$. We note that both $I_1(\mathbf{k})$ and $I_2(\mathbf{k})$ are real and that

$$I_1(\mathbf{k}) = I_1(-\mathbf{k}), \quad I_2(\mathbf{k}) = -I_2(-\mathbf{k}). \tag{4.40}$$

Furthermore from (4.30) we see that $\langle [[C, \mathcal{H}], C^\dagger] \rangle$ is positive and thus, using (4.40), we may add the same quantity with $\mathbf{k}$ replaced by $-\mathbf{k}$ to obtain

$$\langle [[C, \mathcal{H}], C^\dagger] \rangle = I_1(\mathbf{k}) + I_2(\mathbf{k}) + HM^z \leq 2 I_1(\mathbf{k}) + 2HM^z. \tag{4.41}$$

Therefore, using

$$1 - \cos \mathbf{k}\cdot(\mathbf{R}-\mathbf{R}') \le \frac{1}{2}[\mathbf{k}\cdot(\mathbf{R}-\mathbf{R}')]^2 \le \frac{1}{2}\mathbf{k}^2(\mathbf{R}-\mathbf{R}')^2 \tag{4.42}$$

we have

$$\begin{aligned}&\langle[[C,\mathcal{H}],C^\dagger]\rangle \\ &\le \frac{\mathbf{k}^2}{N}\sum_{\mathbf{R},\mathbf{R}'}(\mathbf{R}-\mathbf{R}')^2|J(\mathbf{R}-\mathbf{R}')||\langle S^-_{\mathbf{R}}S^+_{\mathbf{R}'} + S^+_{\mathbf{R}}S^-_{\mathbf{R}'} + 2S^z_{\mathbf{R}}S^z_{\mathbf{R}'}\rangle| + 2|HM^z|.\end{aligned} \tag{4.43}$$

We also note that from the Cauchy–Schwarz inequality and translational invariance

$$\begin{aligned}&|\langle S^-_{\mathbf{R}}S^+_{\mathbf{R}'} + S^+_{\mathbf{R}}S^-_{\mathbf{R}'} + 2S^z_{\mathbf{R}}S^z_{\mathbf{R}'}\rangle| \le |\langle S^-_0S^+_0 + S^+_0S^-_0 + 2S^{z2}_0\rangle| \\ &\le 2|\langle S^-_0S^+_0 + S^+_0S^-_0 + S^{z2}_0\rangle| = 2S(S+1)\end{aligned} \tag{4.44}$$

where in the last line we have used (4.7) and (4.33). Therefore (4.43) becomes

$$\langle[[C,\mathcal{H}],C^\dagger]\rangle \le \mathbf{k}^2 2S(S+1)\sum_{\mathbf{R}}\mathbf{R}^2|J(\mathbf{R})| + 2|HM^z|. \tag{4.45}$$

Thus, noting that

$$A^\dagger A + AA^\dagger = \tilde{\mathbf{S}}^+(\mathbf{k})\tilde{\mathbf{S}}^-(-\mathbf{k}) + \tilde{\mathbf{S}}^-(-\mathbf{k})\tilde{\mathbf{S}}^+(\mathbf{k}) \tag{4.46}$$

and using (4.36) and (4.45) in the Bogoliubov inequality (4.17) we obtain

$$\langle\tilde{\mathbf{S}}^+(\mathbf{k})\tilde{\mathbf{S}}^-(-\mathbf{k}) + \tilde{\mathbf{S}}^-(-\mathbf{k})\tilde{\mathbf{S}}^+(\mathbf{k})\rangle \ge \frac{k_BTM^{z\ 2}}{\mathbf{k}^2S(S+1)\sum_{\mathbf{R}}\mathbf{R}^2|J(\mathbf{R})| + |HM^z|}. \tag{4.47}$$

We now sum both sides of (4.47) over $\mathbf{k}$ and use

$$\begin{aligned}&\sum_{\mathbf{k}}\langle\tilde{\mathbf{S}}^+(\mathbf{k})\tilde{\mathbf{S}}^-(-\mathbf{k}) + \tilde{\mathbf{S}}^-(-\mathbf{k})\tilde{\mathbf{S}}^+(\mathbf{k})\rangle = \sum_{\mathbf{R}}\langle\mathbf{S}^+_{\mathbf{R}}S^-_{\mathbf{R}} + \mathbf{S}^-_{\mathbf{R}}\mathbf{S}^+_{\mathbf{R}}\rangle \\ &\le \sum_{\mathbf{R}}\langle\mathbf{S}^+_{\mathbf{R}}\mathbf{S}^-_{\mathbf{R}} + \mathbf{S}^-_{\mathbf{R}}\mathbf{S}^+_{\mathbf{R}} + \mathbf{S}^{z2}_{\mathbf{R}}\rangle = NS(S+1)\end{aligned} \tag{4.48}$$

to find

$$S(S+1) \ge k_BTM^{z\ 2}\frac{1}{N}\sum_{\mathbf{k}}\frac{1}{\mathbf{k}^2S(S+1)\sum_{\mathbf{R}}\mathbf{R}^2|J(\mathbf{R})| + |HM^z|}. \tag{4.49}$$

Then we obtain the thermodynamic limit $N \to \infty$ by replacing the sum by an integral to obtain

$$S(S+1) \ge k_BTM^{z\ 2}\frac{1}{(2\pi)^D}\int_{-\pi}^{\pi}\frac{d^D\mathbf{k}}{\mathbf{k}^2S(S+1)\sum_{\mathbf{R}}\mathbf{R}^2|J(\mathbf{R})| + |HM^z|}. \tag{4.50}$$

This inequality is only strengthened by restricting the integration to the largest sphere of radius π and thus, using

$$d^D\mathbf{k} = \Omega_D k^{D-1} dk \tag{4.51}$$

with

$$\Omega_D = \begin{cases} 1 & D = 1 \\ 2\pi & D = 2, \end{cases} \tag{4.52}$$

we have

$$S(S+1) \geq k_B T M^{z\ 2} \frac{\Omega_D}{(2\pi)^D} \int_0^{\pi} \frac{dk k^{D-1}}{\mathbf{k}^2 S(S+1) \sum_{\mathbf{R}} \mathbf{R}^2 |J(\mathbf{R})| + |HM^z|}. \tag{4.53}$$

The integral in (4.53) converges for $D = 3$ if $H = 0$ and therefore the magnetization at $H = 0$ is not mandated to vanish. However, for $D = 1$ and 2 the integral in (4.53) diverges if $H = 0$ and therefore the inequality (4.53) requires the magnetization M^z to vanish as $H \to 0$. In order to get an estimate of how rapidly M^z must vanish for $D = 1$ and 2 we evaluate the integral to obtain

$$S(S+1) \geq \frac{k_B T M^{z\ 2}}{2\pi(|HM^z| S(S+1)\omega)^{1/2}} \arctan\left(\pi^2 \frac{S(S+1)\omega}{|HM^z|}\right)^{1/2}, \quad D = 1 \tag{4.54}$$

and

$$S(S+1) \geq \frac{k_B T M^{z\ 2}}{4\pi S(S+1)\omega} \ln\left(1 + \frac{\pi^2 S(S+1)\omega}{|HM^z|}\right) \quad D = 2 \tag{4.55}$$

with

$$\omega = \sum_{\mathbf{R}} \mathbf{R}^2 |J(\mathbf{R})|. \tag{4.56}$$

Therefore

$$|M^z| \leq S(S+1)(|H|\omega)^{1/3} \left[\frac{k_B T}{2\pi} \arctan\left(\frac{\pi^2 S(S+1)\omega}{|HM^z|}\right)^{1/2}\right]^{-2/3} \quad \text{for } D = 1 \tag{4.57}$$

and

$$|M^z| \leq S(S+1) \left[\frac{k_B T}{4\pi\omega} \ln\left(1 + \frac{\pi^2 S(S+1)\omega}{|HM^z|}\right)\right]^{-1/2} \quad \text{for } D = 2 \tag{4.58}$$

from which the bounds (4.16) follow when H is small.

We have therefore proven theorem 4.1 that when (4.15) holds the spin S isotropic Heisenberg ferromagnet does not have spontaneous magnetization in dimensions $D = 1$ and 2.

In the classical limit $S \to \infty$ the Hamiltonian (4.3) will reduce to the Hamiltonian for the classical Heisenberg magnet

$$H_c = -\sum_{\mathbf{R},\mathbf{R}'} J_c(\mathbf{R}-\mathbf{R}')\mathbf{v}_{\mathbf{R}} \cdot \mathbf{v}_{\mathbf{R}'} - H_c \sum_{\mathbf{R}} v^z_{\mathbf{R}} \tag{4.59}$$

with $\mathbf{v}^2_R = 1$ if we set

$$J(\mathbf{R}) = J_c(\mathbf{R})/S^2, \quad H = H_c/S \quad \text{and} \quad \mathbf{v}_{\mathbf{R}} = \lim \mathbf{S}_{\mathbf{R}}/S. \tag{4.60}$$

Thus $\omega = O(S^{-2})$ and the inequalities continue to hold in the classical limit $S \to \infty$ if we consider the classical magnetization $v^z = M^z/S$. Therefore the lack of ferromagnetic order for $D = 1, 2$ holds for the classical case as well. An independent proof can be directly given.

A similar result holds for antiferromagnetic (staggered) order which is defined on the (bipartite) cubic lattice as

$$M^z_s(N) = \frac{1}{N}\sum_{\mathbf{R}} S^z_{\mathbf{R}} e^{i\mathbf{K}\cdot\mathbf{R}} \tag{4.61}$$

where the vector $\mathbf{K}$ is defined as

$$K_j = \pi \quad \text{for } 1 \le j \le D. \tag{4.62}$$

Then if we replace the term in (4.3) which depends on H by the "staggered" magnetic field

$$H_s \sum_{\mathbf{R}} S^z_{\mathbf{R}} e^{i\mathbf{K}\cdot\mathbf{R}} \tag{4.63}$$

antiferromagnetic order is said to exist if

$$\lim_{H_s \to 0^+} \lim_{N\to\infty} M^z_s(H_s; N) > 0. \tag{4.64}$$

The proof that there is no antiferromagnetic order in dimension $D = 1$ and 2 is proven by first setting setting

$$C = \tilde{\mathbf{S}}^+(\mathbf{k}), \quad A = \tilde{\mathbf{S}}^-(-\mathbf{k}-\mathbf{K}) \tag{4.65}$$

in the Bogoliubov inequality (4.17). The rest of the proof is identical with the proof given in the ferromagnetic case.

4.3 Lack of crystalline order in D = 1, 2

The principal model introduced in the previous chapter to study monotonic insulators is the classical N-body Hamiltonian

$$H = \frac{1}{2m}\sum_{j=1}^{N} \mathbf{p}^2_j + U^{(N)}(\mathbf{r}_1, \cdots \mathbf{r}_N) \tag{4.66}$$

where the potential $U^{(N)}(\mathbf{r}_1, \cdots \mathbf{r}_N)$ is the sum of two-body pair potentials

$$U^{(N)}(\mathbf{r}_1, \cdots \mathbf{r}_N) = \frac{1}{2}\sum_{i \neq j} U(\mathbf{r}_i - \mathbf{r}_j). \tag{4.67}$$

with $U(\mathbf{r}) = U(-\mathbf{r})$. The conditions on $U(\mathbf{r})$ needed for the existence of the thermodynamic limit have been extensively discussed in chapter 3.

The nonexistence of crystalline order for this classical system was investigated by Mermin [6] in 1968 by means of a classical analogue of the Bogoliubov inequality (4.17). The argument is more involved than what was done in the previous section for the Heisenberg magnet in two respects: 1) The restrictions on the potential are more subtle than (4.15) and 2) the characterization of crystallization is more involved than the definition of spontaneous magnetization (4.14).

Restrictions on the potential

We proved in chapter 3 that for the thermodynamic limit to exist it is sufficient that

$$U(\mathbf{r}) \geq \begin{cases} C_1/|\mathbf{r}|^{D+\epsilon} & \text{for } 0 \leq |\mathbf{r}| \leq a_1 \\ -C_2 & \text{for } a_1 < |\mathbf{r}| < a_2 \\ -C_3/|\mathbf{r}|^{D+\epsilon'} & \text{for } |\mathbf{r}| > a_2 \end{cases} \tag{4.68}$$

with the constants $C_j > 0$. However for the proof of nonexistence of crystalline order for $D = 1, 2$ of [6] to hold we need the further restrictions that:

1) $U(\mathbf{r})$ is twice differentiable for $\mathbf{r} \neq 0$. In particular, this excludes the case of any potential with a hard core.

2) The following bounds must hold with $C_j > 0$

$$\nabla^2 U(\mathbf{r}) = \frac{1}{r^{D-1}}\frac{\partial}{\partial r} r^{D-1}\frac{\partial}{\partial r}U(\mathbf{r}) \geq -C_4/r^{D+2+\epsilon} \text{ as } r \to \infty \tag{4.69}$$

$$U(\mathbf{r}) - \lambda r^2|\nabla^2 U(\mathbf{r})| = U(\mathbf{r}) - \lambda r^{3-D}|\frac{\partial}{\partial r} r^{D-1}\frac{\partial}{\partial r}U(\mathbf{r})| \geq C_5/r^{D+\epsilon}$$
$$\text{for some } \lambda > 0 \text{ as } r \to 0 \tag{4.70}$$

where $\nabla^2 = \sum_{i=1}^{D} \partial^2/\partial r_i^2$ and in the last term before the inequality sign we have specialized ∇^2 to the spherically symmetric case where $U(\mathbf{r}) = U(|\mathbf{r}|)$. The Lennard-Jones potential

$$U(\mathbf{r}) = (\sigma/r)^n - A(\sigma/r)^m \text{ for } n > m > D \tag{4.71}$$

and the oscillatory potential

$$U(\mathbf{r}) = \cos \mathbf{K} \cdot \mathbf{r}/r^n \text{ for } n > D + 2 \tag{4.72}$$

satisfy (4.69) and (4.70) while a potential which diverges at $r \to 0$ as

$$U(\mathbf{r}) = e^{B/r} \tag{4.73}$$

does not.

Definition of crystalline order

The physical concept of crystalline order is that the atoms are situated on the vertices of a periodic lattice such as fcc, bcc, or cubic in $D = 3$, or square or hexagonal in $D = 2$. The location and orientation of the crystal are obviously arbitrary, and in order to give a precise mathematical formulation of crystalline order we need to localize the position and orientation of the supposed lattice. We will do this by enclosing the N particles in a box with impenatrable walls of a shape consistent with the shape of the supposed crystalline order and take the limit as the size of the box goes to infinity. To be specific we concentrate on $D = 2$ and assume that the crystal has a Bravais lattice $n_1\mathbf{a}_1 + n_2\mathbf{a}_2$ (with n_i integers) specified by the two vectors $\mathbf{a}_1$ and $\mathbf{a}_2$ in a box commensurate with the Bravais lattice specified by the points

$$\mathbf{r} = x_1 N_1 \mathbf{a}_1 + x_2 N_2 \mathbf{a}_2 \in \Omega \quad \text{with } 0 \le x_1, x_2 \le 1 \tag{4.74}$$

where N_j are integers, the volume of Ω in $D = 2$ is $V = N_1 N_2 |\mathbf{a}_1 \times \mathbf{a}_2|$ and the total number of particles N is $N = nN_1N_2$ where n is the number of particles per unit cell.

The density at the point $\mathbf{r}$ when the N particles are at $\mathbf{r}_j$ is

$$\hat{\rho}(\mathbf{r}) = \frac{1}{V}\sum_{j=1}^{N}\delta(\mathbf{r} - \mathbf{r}_j) \tag{4.75}$$

and thus the integrated average density in $D = 2$ is $\rho_0 = n/|\mathbf{a}_1 \times \mathbf{a}_2|$. The Fourier transform of $\hat{\rho}(\mathbf{r})$ is

$$\tilde{\rho}(\mathbf{k}) = \frac{1}{V}\int_{\Omega} d\mathbf{r} e^{-i\mathbf{k}\cdot\mathbf{r}}\hat{\rho}(\mathbf{r}) = \frac{1}{V}\sum_{j=1}^{N} e^{-i\mathbf{k}\cdot\mathbf{r}_j}. \tag{4.76}$$

Thus we define the thermal average of the $\mathbf{k}$ component of the density as

$$\rho(\mathbf{k}) = \langle \tilde{\rho}(\mathbf{k}) \rangle \tag{4.77}$$

where, for the classical system (4.66), the thermal average of a quantity f is defined as

$$\langle f \rangle = \frac{1}{Q_N}\int_{\Omega} d\mathbf{r}_1 \cdots d\mathbf{r}_N e^{-\beta U^{(N)}(\mathbf{r}_1,\cdots\mathbf{r}_N)} f(\mathbf{r}_1 \cdots \mathbf{r}_N) \tag{4.78}$$

with

$$Q_N = \int_{\Omega} d\mathbf{r}_1 \cdots d\mathbf{r}_N e^{-\beta U^{(N)}(\mathbf{r}_1,\cdots\mathbf{r}_N)}. \tag{4.79}$$

The vector $\mathbf{K}$ is said to be in the reciprocal lattice of the Bravais lattice specified by $\mathbf{a}_i$ if with K_1 and K_2 integers

$$\mathbf{K} = K_1\mathbf{b}_1 + K_2\mathbf{b}_2 \quad \text{with } \mathbf{b}_j \cdot \mathbf{a}_k = 2\pi\delta_{j,k}. \tag{4.80}$$

If there is no crystalline order the density will be uniform in the thermodynamic limit, but if there is a periodic crystalline order there will be at least one vector in the reciprocal lattice for which $\rho(\mathbf{k})$ does not vanish. Thus we have the following criteria:

Crystalline order exists if and only if

$$A)\quad \lim_{N\to\infty} \rho(\mathbf{k}) = 0 \quad \mathbf{k} \text{ is not a reciprocal lattice vector}$$
$$B)\quad \lim_{N\to\infty} \rho(\mathbf{K}) \neq 0 \quad \text{for at least one nonzero reciprocal lattice vector} \tag{4.81}$$

where $\lim_{N\to\infty}$ denotes the thermodynamic limit.

We may now prove the theorem of Mermin [6]:

Theorem 4.2

Crystalline order in the sense of (4.81) does not exist for the classical Hamiltonian (4.66) in $D = 1, 2$ when the potential $U(\mathbf{r})$ satisfies restrictions 1 and 2 given above.

We prove the theorem by using a

Classical analogue of the Bogoliubov inequality

Let $\psi(\mathbf{r})$ be a differentiable function and $\phi(\mathbf{r})$ be a twice differentiable function which vanishes when $\mathbf{r}$ is on the surface of the box. Then we have

$$\langle |\sum_j \psi(\mathbf{r}_j)|^2\rangle \geq \frac{k_BT|\sum_j \langle \phi(\mathbf{r}_j)\nabla_j\psi(\mathbf{r}_j)\rangle|^2}{\langle \frac{1}{2}\sum_{i,j}|\phi(\mathbf{r}_i)-\phi(\mathbf{r}_j)|^2\nabla_j^2 U(\mathbf{r}_i-\mathbf{r}_j) + k_BT\sum_j|\nabla_j\phi(\mathbf{r}_j)|^2\rangle} \tag{4.82}$$

where we use the notation $\nabla_j = \partial/\partial\mathbf{r}_j$. To prove (4.82) we start from the obvious inequality valid for any scalar function $A(\mathbf{r}_1, \cdots, \mathbf{r}_N)$ and vector function $\mathbf{B}(\mathbf{r}_1, \cdots, \mathbf{r}_N)$

$$\langle \left|\mathbf{B} - \frac{A\langle A^*\mathbf{B}\rangle}{\langle |A|^2\rangle}\right|^2 \rangle \geq 0 \tag{4.83}$$

from which it follows that

$$\langle |A|^2\rangle \geq \frac{|\langle A^*\mathbf{B}\rangle|^2}{\langle |\mathbf{B}|^2\rangle}. \tag{4.84}$$

Choose A and $\mathbf{B}$ to be

$$A = \sum_{j=1}^{N} \psi(\mathbf{r}_j) \tag{4.85}$$

$$\mathbf{B} = -\frac{e^{\beta U^{(N)}}}{\beta}\sum_{j=1}^{N}\frac{\partial}{\partial\mathbf{r}_j}(\phi(\mathbf{r}_j)e^{-\beta U^{(N)}}) = \sum_{j=1}^{N}\left[\phi(\mathbf{r}_j)\frac{\partial U^{(N)}}{\partial\mathbf{r}_j} - \frac{1}{\beta}\frac{\partial\phi(\mathbf{r}_j)}{\partial\mathbf{r}_j}\right] \tag{4.86}$$

and note that using the first expression for $\mathbf{B}$ in (4.86) we have, for any differentiable function $X(\mathbf{r}_1, \cdots, \mathbf{r}_N)$,

$$\langle \mathbf{B}X\rangle = -\frac{1}{\beta Q}\int_\Omega d\mathbf{r}_1\cdots\mathbf{r}_N \sum_{j=1}^{N} X\frac{\partial}{\partial\mathbf{r}_j}\left(e^{-\beta U^{(N)}}\phi(\mathbf{r}_j)\right) = \frac{1}{\beta}\sum_{j=1}^{N}\langle \phi(\mathbf{r}_j)\frac{\partial X}{\partial\mathbf{r}_j}\rangle \tag{4.87}$$

where to obtain the last line we have integrated by parts and used the fact that $\phi(\mathbf{r}_j)$ vanishes on the boundary of the box. From (4.87) we find

$$\langle \mathbf{B}A^* \rangle = \frac{1}{\beta}\sum_{j=1}^{N}\langle \phi(\mathbf{r}_j)\frac{\partial \psi^*(\mathbf{r}_j)}{\partial \mathbf{r}_j}\rangle \tag{4.88}$$

and

$$\langle |\mathbf{B}|^2 \rangle = \langle \mathbf{B}\cdot\mathbf{B}^* \rangle = \frac{1}{\beta}\sum_{j=1}^{N}\langle \phi(\mathbf{r}_j)\frac{\partial}{\partial \mathbf{r}_j}\cdot\mathbf{B}^*\rangle. \tag{4.89}$$

Then, using the second expression for **B** of (4.86) in (4.89), we find

$$\begin{aligned}\langle |\mathbf{B}|^2 \rangle = &\frac{1}{\beta}\sum_{i,j=1}^{N}\langle \phi(\mathbf{r}_i)\phi^*(\mathbf{r}_j)\frac{\partial}{\partial \mathbf{r}_i}\cdot\frac{\partial}{\partial \mathbf{r}_j}U^{(N)}\rangle\\ &+\frac{1}{\beta}\sum_{j=1}^{N}\langle \phi(\mathbf{r}_j)\left[\frac{\partial \phi^*(\mathbf{r}_j)}{\partial \mathbf{r}_j}\cdot\frac{\partial U^{(N)}}{\partial \mathbf{r}_j}-\frac{1}{\beta}\nabla_j^2\phi^*(\mathbf{r}_j)\right]\rangle.\end{aligned} \tag{4.90}$$

The second term in (4.90) can be written as

$$-\frac{1}{\beta^2 Q}\int_{\Omega} d\mathbf{r}_1\cdots d\mathbf{r}_N\sum_{j=1}^{N}\phi(\mathbf{r}_j)\frac{\partial}{\partial \mathbf{r}_j}\cdot\left(e^{-\beta U^{(N)}}\frac{\partial \phi^*(\mathbf{r}_j)}{\partial \mathbf{r}_j}\right)=\frac{1}{\beta^2}\sum_{j=1}^{N}\langle|\frac{\partial \phi(\mathbf{r}_j)}{\mathbf{r}_j}|^2\rangle \tag{4.91}$$

where the last term is obtained by integrating by parts and noting that $\phi(\mathbf{r}_j)$ vanishes on the boundary. Thus

$$\langle |\mathbf{B}|^2 \rangle = \frac{1}{\beta}\sum_{i,j=1}^{N}\langle \phi(\mathbf{r}_i)\phi^*(\mathbf{r}_j)\frac{\partial}{\partial \mathbf{r}_i}\cdot\frac{\partial}{\partial \mathbf{r}_j}U^{(N)}\rangle+\frac{1}{\beta^2}\sum_{j=1}^{N}\langle|\frac{\partial \phi(\mathbf{r}_j)}{\partial \mathbf{r}_j}|^2\rangle \tag{4.92}$$

and therefore using

$$\sum_{i.j=1}^{N}\langle \phi(\mathbf{r}_i)\phi^*(\mathbf{r}_j)\frac{\partial}{\partial \mathbf{r}_i}\cdot\frac{\partial}{\partial \mathbf{r}_j}U^{(N)}\rangle=\frac{1}{2}\sum_{\substack{i.j=1\\ i\neq j}}^{N}|\phi(\mathbf{r}_i)-\phi(\mathbf{r}_j)|^2\frac{\partial}{\partial \mathbf{r}_i}\cdot\frac{\partial}{\partial \mathbf{r}_i}U(\mathbf{r}_i-\mathbf{r}_j) \tag{4.93}$$

we obtain the desired inequality (4.82) by using (4.88), (4.92) and (4.93) in (4.84).

Proof of theorem 4.2

We now prove theorem 4.2 by letting **K** be the vector in the reciprocal lattice where the density $\rho(\mathbf{K})$ is assumed not to vanish in the thermodynamic limit and specialize (4.82) by setting

$$\phi(\mathbf{r}) = \sin\mathbf{k}\cdot\mathbf{r} \quad \text{and} \quad \psi(\mathbf{r}) = e^{-i(\mathbf{k}+\mathbf{K})\cdot\mathbf{r}} \tag{4.94}$$

where

$$\mathbf{k} = \tilde{n}_1\mathbf{b}_1/N_1 + \tilde{n}_2\mathbf{b}_2/N_2 \tag{4.95}$$

with $-N_i/2 < \tilde{n}_i \le N_i/2$ for N_i even and $-(N_i-1)/2 \le \tilde{n}_i \le (N_i-1)/2$ for N_i odd. Thus we have

$$\langle|\sum_{j=1}^{N}\psi(\mathbf{r}_j)|^2\rangle = V^2\langle\tilde{\rho}(\mathbf{k}+\mathbf{K})\tilde{\rho}(-\mathbf{k}-\mathbf{K})\rangle \tag{4.96}$$

$$|\sum_{j=1}^{N}\langle\phi(\mathbf{r}_j)\nabla_j\psi(\mathbf{r}_j)\rangle|^2 = \frac{V^2}{4}(\mathbf{k}+\mathbf{K})^2|\langle\tilde{\rho}(\mathbf{K})-\tilde{\rho}(\mathbf{K}+2\mathbf{k})\rangle|^2 \tag{4.97}$$

and

$$\begin{aligned}
&\langle\frac{1}{2}\sum_{\substack{i,j=1\\i\neq j}}^{N}|\phi(\mathbf{r}_i)-\phi(\mathbf{r}_j)|^2\nabla_i^2U(\mathbf{r}_i-\mathbf{r}_j)+k_BT\sum_j|\nabla_j\phi(\mathbf{r}_j)|^2\rangle\\
&=\frac{1}{2}\sum_{\substack{i,j=1\\i\neq j}}^{N}\langle(\sin\mathbf{k}\cdot\mathbf{r}_i-\sin\mathbf{k}\cdot\mathbf{r}_j)^2\nabla_i^2U(\mathbf{r}_i-\mathbf{r}_j)\rangle+k_BT\mathbf{k}^2\langle\sum_{j=1}^{N}\cos^2\mathbf{k}\cdot\mathbf{r}_j\rangle\rangle\\
&\leq\left(\frac{1}{2}\sum_{\substack{i,j=1\\i\neq j}}^{N}(\mathbf{r}_i-\mathbf{r}_j)^2\nabla_i^2U(\mathbf{r}_i-\mathbf{r}_j)+Nk_BT\right)\mathbf{k}^2
\end{aligned} \tag{4.98}$$

where to obtain the last line of (4.98) we have used

$$(\sin\mathbf{k}\cdot\mathbf{r}_j-\sin\mathbf{k}\cdot\mathbf{r}_k)^2\leq\mathbf{k}^2(\mathbf{r}_j-\mathbf{r}_k)^2 \quad \text{and} \quad \cos^2\mathbf{k}\cdot\mathbf{r}\leq 1 \tag{4.99}$$

and hence using (4.96)–(4.98) in (4.82) we obtain

$$\frac{1}{N}\langle\tilde{\rho}(\mathbf{K}+\mathbf{k})\tilde{\rho}(-\mathbf{K}-\mathbf{k})\rangle\geq\frac{k_BT\langle\tilde{\rho}(\mathbf{K})-\tilde{\rho}(\mathbf{K}+2\mathbf{k})\rangle^2(\mathbf{k}+\mathbf{K})^2/4}{\mathbf{k}^2(k_BT+(1/2N)\sum_{i,j}\langle(\mathbf{r}_i-\mathbf{r}_j)^2\nabla_i^2U(\mathbf{r}_i-\mathbf{r}_j)\rangle)}. \tag{4.100}$$

Thus far the size of the system specified by N_k and the number of particles N have been finite. To complete the proof we need to take the thermodynamic limit $N_k, N\to\infty$.

We first prove that

$$\lim_{N_1,N_2,N\to\infty}\frac{1}{2N}\sum_{\substack{i,j=1\\i\neq j}}^{N}\langle(\mathbf{r}_i-\mathbf{r}_j)^2\nabla_i^2U(\mathbf{r}_i-\mathbf{r}_j)\rangle<\infty. \tag{4.101}$$

It is here that we will make use of the restrictions on the potential (4.69) and (4.70).

To prove (4.101) we define a new N-body potential

$$\begin{aligned}
U_\lambda^{(N)}(\mathbf{r}_1,\cdots,\mathbf{r}_N)&=U^{(N)}(\mathbf{r}_1,\cdots,\mathbf{r}_N)-\lambda\delta U^{(N)}(\mathbf{r}_1,\cdots,\mathbf{r}_N)\\
&=\frac{1}{2}\sum_{\substack{i,j\\i\neq j}}^{N}U_\lambda(\mathbf{r}_i-\mathbf{r}_j)
\end{aligned} \tag{4.102}$$

where

$$U_\lambda(\mathbf{r})=U(\mathbf{r})-\lambda\mathbf{r}^2\nabla^2U(\mathbf{r}). \tag{4.103}$$

For $U_\lambda^{(N)}$ we have the configurational contribution to the free energy per particle f_λ for the finite system

$$-\beta f_\lambda(N) = \frac{1}{N}\ln\int_\Omega d\mathbf{r}_1\cdots d\mathbf{r}_N e^{-\beta(U^{(N)}-\lambda\delta U^{(N)})}, \tag{4.104}$$

from which we find

$$-\frac{\partial f_\lambda(N)}{\partial\lambda} = \langle\delta U^{(N)}\rangle_\lambda \geq 0 \tag{4.105}$$

and

$$\frac{\partial}{\partial\lambda}\langle\delta U^{(N)}\rangle_\lambda = \beta(\langle\delta U^{(N)} - \langle\delta U^{(N)}\rangle_\lambda^2\rangle_\lambda \geq 0 \tag{4.106}$$

where $\langle\cdots\rangle_\lambda$ denotes a thermal average with respect to $U_\lambda^{(N)}$. From (4.105) and (4.106) we find

$$\begin{aligned} f_0(N) - f_\lambda(N) &= \frac{1}{N}\int_0^\lambda \langle\delta U^{(N)}\rangle_\mu d\mu \\ &\geq \lambda\langle\delta U^{(N)}\rangle_0 = \frac{\lambda}{2N}\sum_{\substack{i,j=1\\ i\neq j}}^{N}(\mathbf{r}_i-\mathbf{r}_j)^2|\nabla_j^2 U(\mathbf{r}_i-\mathbf{r}_j)| \geq 0. \end{aligned} \tag{4.107}$$

Therefore to prove (4.101) it is sufficient to prove the existence of the two thermodynamic limits

$$f_0 = \lim_{N\to\infty} f_0(N) < \infty \quad\text{and}\quad f_\lambda = \lim_{N\to\infty} f_\lambda(N) < \infty. \tag{4.108}$$

The proofs of chapter 3 demonstrate that the free energies per site f_0 and f_λ exist if the two body potentials $U(\mathbf{r})$ and $U_\lambda(\mathbf{r})$ each satisfy (4.68). Thus, noting that the restriction (4.68) for $U_\lambda(\mathbf{r})$ is the restrictions (4.69) and (4.70), the proof of (4.101) follows. Thus using (4.107) in (4.100) we obtain

$$\frac{1}{N}\langle\tilde\rho(\mathbf{K}+\mathbf{k})\tilde\rho(-\mathbf{K}-\mathbf{k})\rangle \geq \frac{k_BT\langle\tilde\rho(\mathbf{K})-\tilde\rho(\mathbf{K}+2\mathbf{k})\rangle^2(\mathbf{k}+\mathbf{K})^2/4}{\mathbf{k}^2(k_BT+(f_0-f_\lambda)/\lambda)}. \tag{4.109}$$

To complete the proof we define

$$g(\mathbf{q}) = e^{-\alpha\mathbf{q}^2} \quad\text{with}\quad \alpha > 0 \tag{4.110}$$

multiply both sides of (4.109) by $g(\mathbf{k}+\mathbf{K})$, divide by the volume V and sum over the vectors $\mathbf{k}\neq\mathbf{0}$ specified by (4.95). Each term in the sum on the right is positive and thus the right-hand side of the inequality is made smaller by restricting the sum of the vectors $\mathbf{k}\neq 0$ such that $2|\mathbf{k}|$ is less that the length K_0 of the shortest reciprocal lattice vector. For the vectors in this restricted sum the quantity $\langle\hat\rho(\mathbf{k}+\mathbf{K})\rangle$ vanishes in the thermodynamic limit by criterion A in the assumption of crystalline order. Therefore we find from (4.109) that

$$\frac{1}{V}\sum_{\mathbf{q}} g(\mathbf{q})\frac{\langle\tilde\rho(\mathbf{q})\tilde\rho(-\mathbf{q})\rangle}{N} \geq \frac{k_BT\mathbf{K}^2 g(\mathbf{K}_0/2)\rho(\mathbf{K})^2}{16[k_BT+(f_0-f_\lambda)/\lambda]}\frac{1}{V}\sum_{\substack{0\neq\mathbf{k}\\ |\mathbf{k}|<|\mathbf{K}_0|/2}}\frac{1}{\mathbf{k}^2}. \tag{4.111}$$

We now note that

$$\frac{1}{V}\sum_{\mathbf{q}} g(\mathbf{q})\frac{\langle\tilde{\rho}(\mathbf{q})\tilde{\rho}(-\mathbf{q})\rangle}{N} = \frac{1}{V}\sum_{\mathbf{q}} g(\mathbf{q}) + \frac{1}{N}\sum_{\substack{i,j=1\\ i\neq j}}^{N} \frac{1}{V}\sum_{\mathbf{q}}\langle g(\mathbf{q})e^{-i\mathbf{q}\cdot(\mathbf{r}_i-\mathbf{r}_j)}\rangle. \tag{4.112}$$

We show that the second term on (4.112) is bounded in the thermodynamic limit as we did above by considering the difference between the free energies of the pair potentials $U(\mathbf{r})$ and

$$U(\mathbf{r}) - \frac{1}{(2\pi)^2}\int d\mathbf{q} g(\mathbf{q})e^{i\mathbf{q}\cdot\mathbf{r}} \tag{4.113}$$

and noting that the free energy for (4.113) exists because the integral gives a Gaussian function of $\mathbf{r}$. Therefore the left-hand side of (4.111) is finite in the thermodynamic limit, and the sum on the right-hand side diverges as $\ln N$. Therefore in two dimensions

$$\lim_{N\to\infty} \rho(\mathbf{K}) = 0 \tag{4.114}$$

and hence crystalline order does not exist.

The similar argument shows that crystalline order does not exist in $D = 1$ but the argument fails in $D = 3$ because the limit of the sum on the right-hand side of (4.111) is finite.

4.4 Existence of ferromagnetic and antiferromagnetic order in the classical Heisenberg model (n vector model) in D = 3

The most extensively studied system for which order has been demonstrated to exist is the classical Heisenberg ferromagnet in three dimensions specified by the Hamiltonian

$$H = -\frac{1}{2}\sum_{\mathbf{R},\mathbf{R}'\in\Omega} J(\mathbf{R}-\mathbf{R}')\mathbf{S}_{\mathbf{R}}\cdot\mathbf{S}_{\mathbf{R}'} \tag{4.115}$$

where $J(\mathbf{R}) = J(-\mathbf{R})$, $J(0) = 0$, $\mathbf{R}$ a vector on a lattice which is confined to the box Ω with $N =$ sites, periodic boundary conditions imposed and $\mathbf{S}_{\mathbf{R}}$ is a classical vector of n components with fixed length

$$\mathbf{S}_{\mathbf{R}}{}^2 = 1. \tag{4.116}$$

Using (4.116) an alternative form for (4.115) is

$$H = -\frac{N}{2}\sum_{\mathbf{R}\in\Omega} J(\mathbf{R}) + \frac{1}{4}\sum_{\mathbf{R},\mathbf{R}'\in\Omega} J(\mathbf{R}-\mathbf{R}')(\mathbf{S}_{\mathbf{R}} - \mathbf{S}'_{\mathbf{R}})^2. \tag{4.117}$$

For this classical spin system thermal averages are defined as

$$\langle X\rangle = \frac{1}{Q}\int\prod_{\mathbf{R}} d\mathbf{S}_{\mathbf{R}}\delta(\mathbf{S}_{\mathbf{R}}^2 - 1)Xe^{-\beta H} \tag{4.118}$$

with

$$Q = \int \prod_{\mathbf{R}} d\mathbf{S}_{\mathbf{R}} \delta(\mathbf{S}_{\mathbf{R}}^2 - 1) e^{-\beta H}. \tag{4.119}$$

We say that a system has long range ferromagnetic order if

$$\lim_{\mathbf{r}\to\infty} \lim_{\mathbf{L}\to\infty} \langle S_0^j S_{\mathbf{r}}^j \rangle > 0. \tag{4.120}$$

This is equivalent to

$$\lim_{\mathbf{L}\to\infty} \langle \sum_{j=1}^{n} \left(\frac{1}{N} \sum_{\mathbf{R}\in\Omega} S_{\mathbf{R}}^j \right)^2 \rangle \neq 0 \tag{4.121}$$

which is written in terms of the Fourier transform (4.32) as

$$\lim_{\mathbf{L}\to\infty} \frac{1}{N} \langle \tilde{\mathbf{S}}^2(0) \rangle \neq 0. \tag{4.122}$$

The proof of the existence of long range ferromagnetic order (4.122) for the classical Heisenberg magnet (4.115) depends both on the properties of $J(\mathbf{R})$ and on the lattice on which $\mathbf{R}$ takes its values. However, the known cases where existence can be proven do not seem to exhaust the cases where it would seem that order should exist. Consequently we will first present the mechanism used for the proofs of long range order and after that we demonstrate that under certain restrictive assumptions the bounds used in the mechanism can be proven to hold.

4.4.1 The mechanism for ferromagnetic order

Long range order (4.122) will follow from the bound

$$\begin{aligned} &\frac{1}{2} \langle | \sum_{\mathbf{R},\mathbf{R}'\in\Omega} J(\mathbf{R}-\mathbf{R}')(\mathbf{h}_j(\mathbf{k}\cdot\mathbf{R}) - \mathbf{h}_j(\mathbf{k}\cdot\mathbf{R}') \cdot (\mathbf{S}_{\mathbf{R}} - \mathbf{S}_{\mathbf{R}'})|^2 \rangle \\ &\leq k_B T \sum_{\mathbf{R},\mathbf{R}'\in\Omega} |\mathbf{h}_j(\mathbf{k}\cdot\mathbf{R}) - \mathbf{h}_j(\mathbf{k}\cdot\mathbf{R}')|^2 J(\mathbf{R}-\mathbf{R}') \end{aligned} \tag{4.123}$$

where $\mathbf{h}_j(\mathbf{k}\cdot\mathbf{R})|_m = \delta_{jm} N^{-1/2} e^{i\mathbf{k}\cdot\mathbf{R}}$ are n component vectors

We use the bound (4.123) by fixing $\mathbf{k} \neq 0$ noting that

$$\sum_{\mathbf{R},\mathbf{R}'\in\Omega} |\mathbf{h}_j(\mathbf{k}\cdot\mathbf{R}) - \mathbf{h}_j(\mathbf{k}\cdot\mathbf{R}')|^2 J(\mathbf{R}-\mathbf{R}') = \sum_{\mathbf{R}\in\Omega} (1 - e^{i\mathbf{k}\cdot\mathbf{R}}) J(\mathbf{R}) = 2E_{\mathbf{k}} \tag{4.124}$$

and

$$\frac{1}{2}\sum_{\mathbf{R},\mathbf{R}'\in\Omega} J(\mathbf{R}-\mathbf{R}')(\mathbf{h}_j(\mathbf{k}\cdot\mathbf{R})-\mathbf{h}_j(\mathbf{k}\cdot\mathbf{R}'))\cdot(\mathbf{S}_\mathbf{R}-\mathbf{S}_{\mathbf{R}'})$$
$$=\sum_{\mathbf{R},\mathbf{R}'\in\Omega} J(\mathbf{R}-\mathbf{R}')(\mathbf{h}_j(\mathbf{k}\cdot\mathbf{R})-\mathbf{h}_j(\mathbf{k}\cdot\mathbf{R}'))\cdot\mathbf{S}_\mathbf{R}$$
$$=\sum_{\mathbf{R}\in\Omega}\frac{e^{i\mathbf{k}\cdot\mathbf{R}}}{N^{1/2}}\mathbf{S}^j_\mathbf{R}\sum_{\mathbf{R}'\in\Omega} J(\mathbf{R}')(1-e^{i\mathbf{k}\cdot\mathbf{R}'})=2E_\mathbf{k}\tilde{\mathbf{S}}^j(\mathbf{k}) \tag{4.125}$$

where we have defined

$$E_\mathbf{k}=\frac{1}{2}\sum_{\mathbf{R}\in\Omega}(1-e^{i\mathbf{k}\cdot\mathbf{R}})J(\mathbf{R})=\frac{1}{2}\sum_{\mathbf{R}\in\Omega}(1-\cos\mathbf{k}\cdot\mathbf{R})J(\mathbf{R}) \tag{4.126}$$

and used the definition (4.32) of the Fourier transform $\tilde{\mathbf{S}}(\mathbf{k})$. Thus we see that (4.123) specializes when $\mathbf{k}\neq 0$ to

$$0\le 4E^2_\mathbf{k}\langle\tilde{\mathbf{S}}^j(\mathbf{k})\tilde{\mathbf{S}}^j(-\mathbf{k})\rangle\le 2k_BTE_\mathbf{k} \tag{4.127}$$

or, in other words, for $\mathbf{k}\neq 0$

$$0\le\langle\tilde{\mathbf{S}}^j(\mathbf{k})\tilde{\mathbf{S}}^j(-\mathbf{k})\rangle\le\frac{k_BT}{2E_\mathbf{k}}. \tag{4.128}$$

We now note that the normalization condition (4.116) on $\mathbf{S}_\mathbf{R}$ may be written in terms of the Fourier transform $\tilde{\mathbf{S}}(\mathbf{k})$ as

$$1=\mathbf{S}^2_\mathbf{R}=\frac{1}{N}\sum_{\mathbf{R}\in\Omega}\mathbf{S}^2_\mathbf{R}=\frac{1}{N}\sum_\mathbf{k}\tilde{\mathbf{S}}(\mathbf{k})\cdot\tilde{\mathbf{S}}(-\mathbf{k}). \tag{4.129}$$

Then if we separate the sum over $\mathbf{k}$ in (4.129) into the terms $\mathbf{k}=0$ and $\mathbf{k}\neq 0$ and use the bound (4.128) for $\mathbf{k}\neq 0$ we obtain

$$1=\frac{1}{N}\langle\tilde{\mathbf{S}}(0)^2\rangle+\frac{1}{N}\sum_{\mathbf{k}\neq 0}\langle\tilde{\mathbf{S}}(\mathbf{k})\tilde{\mathbf{S}}(-\mathbf{k})\rangle$$
$$\le\frac{1}{N}\langle\tilde{\mathbf{S}}(0)^2\rangle+\frac{k_BTn}{2}\frac{1}{N}\sum_{\mathbf{k}\neq 0}\frac{1}{E_\mathbf{k}}. \tag{4.130}$$

We now take the thermodynamic limit where the sum in (4.130) may be replaced by an integral. Thus

$$1\le\lim_{\mathbf{L}\to\infty}\frac{1}{N}\langle\tilde{\mathbf{S}}(0)^2\rangle+\frac{k_BTn}{2}(2\pi)^{-3}\int\frac{d^3\mathbf{k}}{E_\mathbf{k}}. \tag{4.131}$$

But the integral in (4.131) is finite and thus for T sufficiently small we must have

$$\lim_{\mathbf{L}\to\infty}\frac{1}{N}\langle\tilde{\mathbf{S}}(0)^2\rangle>0 \tag{4.132}$$

which is the condition (4.122) for long range order.

In the special case of nearest neighbor isotropic interactions on the cubic lattice where $E_{\mathbf{k}} = J(3 - \cos k_1 - \cos k_2 - \cos k_3)$ the integral in (4.131) has been analytically evaluated by Watson [19] as

$$(2\pi)^{-3} \int_{-\pi}^{\pi} dk_1 \int_{-\pi}^{\pi} dk_2 \int_{-\pi}^{\pi} dk_3 \frac{1}{3 - \cos k_1 - \cos k_2 - \cos k_3} =$$

$$(18 + 12\sqrt{2} - 10\sqrt{3} - 7\sqrt{6})[\frac{2}{\pi} K((2 - \sqrt{3})(\sqrt{3} - \sqrt{2}))]^2 = 0.505462 \cdots \quad (4.133)$$

where $K(k)$ is the complete elliptic integral of the first kind

$$K(k) = \int_0^1 \frac{dx}{[(1 - x^2)(1 - k^2 x^2)]^{1/2}}. \quad (4.134)$$

Thus we find that long range order exists when

$$k_B T/J < \frac{3.95678}{n}. \quad (4.135)$$

In Table 4.2 we compare this bound with numerical values determined from high temperature series expansions (see chapter 9).

Table 4.2 Comparison of the lower bound on $k_B T_c/J$ computed from (4.135) and the values obtained from high temperature series expansion of chapter 9 for the Ising model $n = 1$ and the classical Heisenberg model $n = 3$.

n	Lower bound	T_c from series expansion
1	3.95678	4.5108
3	1.31893	1.44 ± 0.02

4.4.2 Proof of the bound (4.123)

It remains to formulate the set of restrictions on the interaction strengths $J(\mathbf{R})$ for which the bound (4.123) can be proven true. At the very least the desired inequality (4.128) puts the necessary condition on $J(\mathbf{R})$ that

$$E_{\mathbf{k}} = \frac{1}{2} \sum_{\mathbf{R} \in \Omega} (1 - \cos \mathbf{k} \cdot \mathbf{R}) J(\mathbf{R}) > 0 \quad \text{for all } \mathbf{k} \neq 0. \quad (4.136)$$

The most natural restriction on the $J(\mathbf{R})$ is the following

Conjecture

The necessary positivity condition (4.136) is also sufficient for the bound (4.123) to hold.

There is no general proof of this conjecture and we will content ourselves with the special case proven by Fröhlich, Simon and Spencer [7] in 1976 of nearest neighbors on the cubic lattice:

$$J(\mathbf{R}) = \begin{cases} J > 0 \text{ for } \mathbf{R} = (\pm 1, 0, 0),\ (0, \pm 1, 0),\ (0, 0, \pm 1) \\ 0 \ \text{ otherwise} \end{cases} \quad (4.137)$$

This case is extremely restrictive and it is impossible to believe that it exhausts the sets of $J(\mathbf{R})$ for which long range ferromagnetic order exists for the classical Heisenberg model. For example, it seems entirely reasonable to expect that the addition of positive bonds between any two sites can never decrease the long range order and thus, we would expect that long range order would exist on all cubic lattices with $J(\mathbf{R}) \geq 0$. This would be the case if, for the completely general lattice

$$H = -\frac{1}{2} \sum_{\mathbf{R},\mathbf{R}' \in \Omega} J(\mathbf{R}, \mathbf{R}')\mathbf{S}_{\mathbf{R}} \cdot \mathbf{S}_{\mathbf{R}'}, \tag{4.138}$$

we could prove that

$$\begin{aligned} &k_B T \frac{\partial}{\partial J(\mathbf{R}_3, \mathbf{R}_4)} \langle \mathbf{S}_{\mathbf{R}_1} \cdot \mathbf{S}_{\mathbf{R}_2} \rangle \\ &= \langle (\mathbf{S}_{\mathbf{R}_1} \cdot \mathbf{S}_{\mathbf{R}_2})(\mathbf{S}_{\mathbf{R}_3} \cdot \mathbf{S}_{\mathbf{R}_4}) \rangle - \langle \mathbf{S}_{\mathbf{R}_1} \cdot \mathbf{S}_{\mathbf{R}_2} \rangle \langle \mathbf{S}_{\mathbf{R}_3} \cdot \mathbf{S}_{\mathbf{R}_4} \rangle \geq 0. \end{aligned} \tag{4.139}$$

The inequality (4.139) has been proven true for $J(\mathbf{R}, \mathbf{R}') \geq 0$ for cases that $\mathbf{S}_{\mathbf{R}}$ has one component (the Ising model) by Griffiths [20] and for the two-component case (the rotator model) by Ginibre [21], but for three or more spin components, even though some inequalities for correlation functions have been obtained [22], the inequalities (4.139) have never been proven. On the other hand no counter example to (4.139) has been found for the case of three or more spin components and thus there is nothing to suggest that (4.137) is the only case with nonnegative interactions $J(\mathbf{R})$ where long range order exists. What seems to be lacking is the mathematical machinery to confirm our physical intuition.

We will here prove that the bound (4.123) holds for isotropic interactions on the nearest neighbor cubic lattice (4.137). We begin with the following elementary lemma:

Lemma 4.2
For any positive $F(x)$ and $G(x)$ and real h we have

$$0 \leq \int_{-\infty}^{\infty} dxdyF(x)e^{-\beta J(x-y+h)^2/4}G(y)| \leq ||F||_\beta ||G||_\beta \tag{4.140}$$

where

$$\int_{-\infty}^{\infty} dxdyF(x)e^{-\beta J(x-y)^2/4}F(x) \equiv ||F||_\beta^2. \tag{4.141}$$

To prove this lemma we define the Fourier and inverse Fourier transform

$$\tilde{F}(k) = \int_{-\infty}^{\infty} dxe^{ikx}F(x), \quad F(x) = \frac{1}{2\pi}\int_{-\infty}^{\infty} dke^{-ikx}\tilde{F}(x) \tag{4.142}$$

and write

$$\int_{-\infty}^{\infty} dxdyF^*(x)e^{-\beta J(x-y+h)^2/4}G(y) = \int_{-\infty}^{\infty} dk\tilde{F}^*(k)\tilde{G}(k)e^{-k^2/\beta J}e^{ihk} \tag{4.143}$$

where we have used

$$\int_{-\infty}^{\infty} dx e^{-\beta J(x-h)^2} e^{ikx} = e^{ikh} \int_{-\infty}^{\infty} dx e^{-\beta J x^2/4} e^{ikx} = e^{ikh} e^{-k^2/\beta J} \tag{4.144}$$

which holds for $\beta J > 0$. Furthermore if we note that $|e^{ihk}| = 1$ we find from (4.143) that

$$\int_{-\infty}^{\infty} dx dy F(x) e^{-\beta J(x-y+h)^2/4} G(y) \leq \int_{-\infty}^{\infty} dx dy F(x) e^{-\beta J(x-y)^2/4} G(y). \tag{4.145}$$

Then noting that the positivity of (4.141) is sufficient to prove the Schwarz inequality

$$\int_{-\infty}^{\infty} dx dy F(x) e^{-\beta J(x-y)^2/4} G(y) \leq ||F||_\beta ||G||_\beta. \tag{4.146}$$

the inequality (4.140) follows.

We now follow [7, 18, 23] and use lemma 4.2 to prove a bound called Gaussian domination.

Theorem 4.4. Gaussian domination
If we define

$$Z(\mathbf{h}_j(\mathbf{k}\cdot\mathbf{R})) = \int \prod_{\mathbf{R}} d\mathbf{S}_{\mathbf{R}} \delta(\mathbf{S}_{\mathbf{R}}^2 - 1) \exp\left(-\frac{\beta}{4}\sum_{\mathbf{R},\mathbf{R}'} J(\mathbf{R}-\mathbf{R}')[\mathbf{S}_{\mathbf{R}} - \mathbf{S}_{\mathbf{R}'} - \mathbf{h}_j(\mathbf{k}\cdot\mathbf{R}) + \mathbf{h}_j(\mathbf{k}\cdot\mathbf{R}')]^2\right) \tag{4.147}$$

and if $J(\mathbf{R})$ satisfies the restriction (4.137) of nearest neighbors on the cubic lattice with periodic boundary conditions then for all $\mathbf{h}_j(\mathbf{k}\cdot\mathbf{R})$

$$Z(\mathbf{h}_j(\mathbf{k}\cdot\mathbf{R})) \leq Z(0). \tag{4.148}$$

To prove this theorem we consider splitting the lattice into two pieces of equal size which transform into each other by reflection through a plane perpendicular to the 1 (x) axis which bisects the bonds between $R_1 = 0, 1$ and the bonds $R_1 = L, -L+1 \equiv L+1$ and split the Hamiltonian into three parts: those terms where $-L+1 \leq R_1, R_1' \leq 0$, those terms where $1 \leq R_1, R_1' \leq L$ and those terms $R_1 = 0, R_1' = 1$ and $R_1 = L, R' = -L+1$.

We introduce the notation

$$d\mu_{\mathbf{R}} = d\mathbf{S}_{\mathbf{R}} \delta(\mathbf{S}_{\mathbf{R}}^2 - 1) \tag{4.149}$$

and write (4.147) as

$$Z(\mathbf{h}_j(\mathbf{k}\cdot\mathbf{R})) =$$
$$\prod_{R_2,R_3}\int d\mu_{(0,R_2,R_3)}d\mu_{(1,R_2,R_3)}d\mu_{(L,R_2,R_3)}d\mu_{(-L+1,R_2,R_3)}$$
$$Z_-[\mathbf{S}_{(0,R_2,R_3)},\mathbf{S}_{(-L,R_2,R_3)}]$$
$$\exp\left(-\frac{\beta J}{4}[\mathbf{S}_{(0,R_2,R_3)}-\mathbf{S}_{(1,R_2,R_3)}-\mathbf{h}_j(\mathbf{k}\cdot(0,R_2,R_3)+\mathbf{h}_j(\mathbf{k}\cdot(1,R_2,R_3)]^2\right)$$
$$\exp\left(-\frac{\beta J}{4}[\mathbf{S}_{(L,R_2,R_3)}-\mathbf{S}_{(-L+1,R_2,R_3)}-\mathbf{h}_j(\mathbf{k}\cdot(L,R_2,R_3))+\mathbf{h}_j(\mathbf{k}\cdot(-L+1,R_2,R_3))]^2\right)$$
$$Z_+[\mathbf{S}_{(1,R_2,R_3)},\mathbf{S}_{(L,R_2,R_3)}] \tag{4.150}$$

where

$$Z_-[\mathbf{S}_{(0,R_2,R_3)},\mathbf{S}_{(-L,R_2,R_3)}] =$$
$$\int\prod_{\substack{-L+1\le R_1\le -1\\ -L_2\le R_2\le L_2,-L_3\le R_3\le L_3}} d\mu_{\mathbf{R}}\prod_{i=1}^{3}\exp\left(-\frac{\beta J}{4}[\mathbf{S}_{\mathbf{R}}-\mathbf{S}_{\mathbf{R}+\delta_i}-\mathbf{h}(\mathbf{k}\cdot\mathbf{R})+\mathbf{h}(\mathbf{k}\cdot(\mathbf{R}+\delta_{\mathbf{i}}))]^2\right) \tag{4.151}$$

and Z_+ is given by (4.151) with $1\le R_1\le L-1$. We now use the inequality (4.140) repeatedly with x,y the pairs of variables S^j_{0,R_2,R_3}, S^j_{0,R_2,R_3} and S^j_{L,R_2,R_3}, S^j_{-L+1,R_2,R_3} and thus we find

$$Z(\mathbf{h}_j(\mathbf{k}\cdot\mathbf{R})) = \left[\prod_{R_2,R_3}\int d\mu_{(0,R_2,R_3)}d\bar{\mu}_{(0,R_2,R_3)}d\bar{\mu}_{(-L+1,R_2,R_3)}d\mu_{(-L+1,R_2,R_3)}\right.$$
$$Z_-[\mathbf{S}_{(0,R_2,R_3)},\mathbf{S}_{(-L+1,R_2,R_3)}]\exp\left(-\frac{\beta J}{4}[\mathbf{S}_{(0,R_2,R_3)}-\mathbf{S}_{(1,R_2,R_3)}]^2\right)$$
$$\left.\exp\left(-\frac{\beta J}{4}[\mathbf{S}_{(L,R_2,R_3)}-\bar{\mathbf{S}}_{(0,R_2,R_3)}]^2\right)Z_+[\bar{\mathbf{S}}_{(0,R_2,R_3)},\bar{\mathbf{S}}_{(-L+1,R_2,R_3)}]\right]^{1/2}$$
$$\left[\prod_{R_2,R_3}\int d\bar{\mu}_{(1,R_2,R_3)}d\mu_{(1,R_2,R_3)}d\mu_{(L,R_2,R_3)}d\bar{\mu}_{(L,R_2,R_3)}\right.$$
$$Z_-[\bar{\mathbf{S}}_{(1,R_2,R_3)},\bar{\mathbf{S}}_{(L,R_2,R_3)}]\exp\left(-\frac{\beta J}{4}[\bar{\mathbf{S}}_{(1,R_2,R_3)}-\mathbf{S}_{(1,R_2,R_3)}]^2\right)$$
$$\left.\exp\left(-\frac{\beta J}{4}[\mathbf{S}_{(L,R_2,R_3)}-\bar{\mathbf{S}}_{(L,R_2,R_3)}]^2\right)Z_+[\mathbf{S}_{(1,R_2,R_3)},\mathbf{S}_{(L,R_2,R_3)}]\right]^{1/2} \tag{4.152}$$

where $\mathbf{S}_{\mathbf{R}}$ and $\bar{\mathbf{S}}_{\mathbf{R}}$ both indicate dummy variables of integration. It is easily recognized that the two factors in (4.152) are equal and are each equal to the original expression for $Z(\mathbf{h}_j(\mathbf{k}\cdot\mathbf{R}))$ with $\mathbf{h}_j(\mathbf{k}\cdot(0,R_1,R_2))-\mathbf{h}_j(\mathbf{k}\cdot(1,R_1,R_2))$ and $\mathbf{h}_j(\mathbf{k}\cdot(L,R_1,R_2))-\mathbf{h}_j(\mathbf{k}\cdot(-L+1,R_1,R_2))$ set equal to zero.

We now repeat this process for all other planes bisecting the lattice in the x, y and z directions. This sets all $\mathbf{h}_j(\mathbf{k}\cdot\mathbf{R})) = 0$ and thus the bound of Gaussian domination (4.148) is established.

To complete the deduction of (4.123) from (4.148), we see from the definition (4.147) of $Z(\mathbf{h}_j(\mathbf{k}\cdot\mathbf{R}))$ that

$$\begin{aligned} Z(\mathbf{h}_j(\mathbf{k}\cdot\mathbf{R})) &= \exp\left[-\frac{\beta}{4}\sum_{\mathbf{R},\mathbf{R}'} J(\mathbf{R}-\mathbf{R}')(\mathbf{h}_j(\mathbf{k}\cdot\mathbf{R}) - \mathbf{h}_j(\mathbf{k}\cdot\mathbf{R}'))^2\right] \\ &\times \int \prod_{\mathbf{R}} dS_{\mathbf{R}}\delta(\mathbf{S}_{\mathbf{R}}^2 - 1)\exp\left[-\frac{\beta}{4}\sum_{\mathbf{R},\mathbf{R}'} J(\mathbf{R}-\mathbf{R}')(\mathbf{S}_{\mathbf{R}} - \mathbf{S}_{\mathbf{R}'})^2\right] \\ &\times \exp\left[\frac{\beta}{2}\sum_{\mathbf{R},\mathbf{R}'} J(\mathbf{R}-\mathbf{R}')(\mathbf{S}_{\mathbf{R}} - \mathbf{S}_{\mathbf{R}'})\cdot(\mathbf{h}_j(\mathbf{k}\cdot\mathbf{R}) - \mathbf{h}_j(\mathbf{k}\cdot\mathbf{R}'))\right] \end{aligned} \tag{4.153}$$

and therefore noting that $Z(0)$ is the partition function of the system we find from (4.148) and (4.153) that

$$\begin{aligned} &\langle \exp\left[\frac{\beta}{2}\sum_{\mathbf{R},\mathbf{R}'} J(\mathbf{R}-\mathbf{R}')(\mathbf{S}_{\mathbf{R}} - \mathbf{S}_{\mathbf{R}'}\cdot(\mathbf{h}_j(\mathbf{k}\cdot\mathbf{R}) - \mathbf{h}_j(\mathbf{k}\cdot\mathbf{R}'))\right]\rangle \\ &\leq \exp\left[\frac{\beta}{4}\sum_{\mathbf{R},\mathbf{R}'} J(\mathbf{R}-\mathbf{R}')(\mathbf{h}_j(\mathbf{k}\cdot\mathbf{R}) - \mathbf{h}_j(\mathbf{k}\cdot\mathbf{R}'))^2\right]. \end{aligned} \tag{4.154}$$

Finally we replace $\mathbf{h}_j(\mathbf{k}\cdot\mathbf{R})$ by $\lambda\mathbf{h}_j(\mathbf{k}\cdot\mathbf{R})$ and expand (4.153) in λ. The term proportional to λ on the left-hand side vanishes because of translational invariance. Thus by equating the terms of λ^2 we obtain

$$\begin{aligned} &\beta^2/2\langle\left(\sum_{\mathbf{R},\mathbf{R}'} J(\mathbf{R}-\mathbf{R}')(\mathbf{S}_{\mathbf{R}} - \mathbf{S}_{\mathbf{R}'})\cdot(\mathbf{h}_j(\mathbf{k}\cdot\mathbf{R}) - \mathbf{h}_j(\mathbf{k}\cdot\mathbf{R}'))\right)^2\rangle \\ &\leq \beta\sum_{\mathbf{R},\mathbf{R}'} J(\mathbf{R}-\mathbf{R}')(\mathbf{h}_j(\mathbf{k}\cdot\mathbf{R}) - \mathbf{h}_j(\mathbf{k}\cdot\mathbf{R}'))^2 \end{aligned} \tag{4.155}$$

which holds when $J(\mathbf{r})$ is given by (4.137). In the proof of (4.155) we have assumed that $\mathbf{h}_j(\mathbf{k}\cdot\mathbf{R})$ is real but the final inequality extends to the case where $\mathbf{h}_j(\mathbf{k}\cdot\mathbf{R})$ is given by (4.137). This establishes (4.123) and thus the proof is complete that long range order exists at sufficiently low temperature for the classical Heisenberg magnet on the nearest neighbor cubic lattice.

4.4.3 Antiferromagnetism

For the classical Heisenberg magnet on the nearest neighbor isotropic cubic lattice, if we make the transformation

$$\mathbf{S}'_{\mathbf{R}} = (-1)^{R_1+R_2+R_3}\mathbf{S}_{\mathbf{R}} \tag{4.156}$$

we send the spontaneous magnetization into the staggered magnetization and the ferromagnetic Hamiltonian transforms into the antiferromagnetic Hamiltonian

$$H = -J\sum_{\mathbf{R}}\sum_{i=1}^{3}\mathbf{S}_{\mathbf{R}}\cdot\mathbf{S}_{\mathbf{R}+\delta_i} = J\sum_{\mathbf{R}}\sum_{i=1}^{3}\mathbf{S}'_{\mathbf{R}}\cdot\mathbf{S}'_{\mathbf{R}+\delta_i}. \tag{4.157}$$

Therefore if the ferromagnet has long range ferromagnetic order the antiferromagnet will also have antiferromagnetic order.

4.5 Existence of antiferromagnetic order in the quantum Heisenberg model for T > 0 and D = 3

The physics of the quantum Heisenberg magnet is substantially more involved than the classical case. For example we saw above that for the classical case the mapping (4.156) sends the ferromagnet into the antiferromagnet so the physics of these systems is essentially the same. However, in the quantum case the ferromagnet and the antiferromagnet are essentially different.This is easily seen by considering the very simple case of two spins of $S = 1/2$ with the interaction

$$H = -J\mathbf{S}_1\cdot\mathbf{S}_2. \tag{4.158}$$

Explicitly written out as a 4×4 matrix this Hamiltonian is

$$H = -\frac{J}{4}\begin{pmatrix} 1 & 0 & 0 & 0 \\ 0 & -1 & 2 & 0 \\ 0 & 2 & -1 & 0 \\ 0 & 0 & 0 & 1 \end{pmatrix}. \tag{4.159}$$

This matrix has a triply degenerate eigenvalue of $-J/4$ and a nondegenerate eigenvalue of $+3J/4$. Therefore the ground state for the ferromagnet $J > 0$ is triply degenerate while the ground state for the antiferromagnet $J < 0$ is nondegenerate.

The existence of order is much more difficult to prove in the quantum mechanical spin S Heisenberg model than for the classical case. For the quantum ferromagnet even nearest neighbor interactions on the cubic lattice where

$$H = -J\sum_{\mathbf{R}}\sum_{i=1}^{3}\mathbf{S}_{\mathbf{R}}\cdot\mathbf{S}_{\mathbf{R}+\delta_1} \tag{4.160}$$

with $J > 0$ and $\mathbf{S}_{\mathbf{R}}$ obeys the commutation relations (4.5), there is no proof of long range order for any spin $S < \infty$. For the quantum antiferromagnet on the nearest neighbor cubic lattice

$$H = J\sum_{\mathbf{R}}\sum_{i=1}^{3}\mathbf{S}_{\mathbf{R}}\cdot\mathbf{S}_{\mathbf{R}+\delta_1} \tag{4.161}$$

with $J > 0$, Dyson, Lieb and Simon [8] proved in 1978 that for the case $S \geq 1$ long range antiferromagnetic order exists for some $T > 0$. The extension to $S = 1/2$ was made by Kennedy, Lieb and Shastry [10] in 1988 who showed for a spatially anisotropic lattice

$$J(1,0,0) = J(0,1,0), \quad \text{and} \quad J(0,0,1) = rJ(1,0,0) \tag{4.162}$$

that long range antiferromagnetic order exists in the ground state for

$$0.16 \leq r \leq 1 \tag{4.163}$$

(and it is stated that the proof can be extended to some $T > 0$.)

The proof of [8] will only here be briefly discussed. The proof first obtains a bound on the Duhamel two-point function, defined here as

$$(A, B) = \int_0^1 \mathrm{Tr}\left(e^{-x\beta H} A e^{-(1-x)\beta H} B\right) / \mathrm{Tr} e^{-\beta H}, \tag{4.164}$$

of

$$\sum_{j=1}^{3} (\tilde{\mathbf{S}}^j(\mathbf{k}), \tilde{\mathbf{S}}^j(-\mathbf{k})) \leq \frac{3}{2E'_{\mathbf{k}}} \tag{4.165}$$

where in dimension D

$$E'_{\mathbf{k}} = D + \sum_{i=1}^{D} \cos k_i. \tag{4.166}$$

From this bound on $(\tilde{\mathbf{S}}^j(\mathbf{k}), \tilde{\mathbf{S}}^j(-\mathbf{k}))$ a bound is obtained on the thermal average $\langle \tilde{\mathbf{S}}^j(\mathbf{k}) \tilde{\mathbf{S}}^j(-\mathbf{k}) \rangle$ and then the mechanism of 4.4.1 is used to produce a lower bound T_L on the critical temperature T_c below which long range order occurs. This bound is obtained as the solution of

$$S(S+1) = (3|E_G|/2)(2\pi)^{-D} \int_{|k_i| \leq \pi} d^D\mathbf{k} (E_{\mathbf{k}}/E'_{\mathbf{k}})^{1/2} \coth[\beta_L (2|E_G| E_{\mathbf{k}} E'_{\mathbf{k}}/3D)^{1/2}] \tag{4.167}$$

where

$$E_{\mathbf{k}} = D - \sum_{i=1}^{D} \cos k_i \tag{4.168}$$

and E_G is the ground state energy per site of the antiferromagnetic chain.

The right-hand side of (4.167) is maximum at $T_L = 0$ ($\beta_L = \infty$). Therefore in order for antiferromagnetic order to exist the spin must satisfy the lower bound

$$S(S+1) \geq (3|E_G|/2)^{1/2} I_D \tag{4.169}$$

where

$$I_D = (2\pi)^{-D} \int_{|k_i| \leq \pi} d^D\mathbf{k} (E_{\mathbf{k}}/E'_{\mathbf{k}})^{1/2} \tag{4.170}$$

which has been numerically evaluated in $D = 3$ as $I_3 = 1.157$. Thus antiferromagnetic order will exist at finite temperature in $D = 3$ if

$$S(S+1) \geq (3|E_G|/2)^{1/2} 1.157. \tag{4.171}$$

It remains to find the ground state energy E_G. In [8] a bound is given (first proven by Anderson in 1951 [24]) that in D dimensions

$$|E_G| < DS(S+1/2) \tag{4.172}$$

and thus in three dimensions antiferromagnetism exists for

$$S(S+1) \geq \left(\frac{3S(S+1/2)}{2}\right)^{1/2} 1.157 \tag{4.173}$$

which is satisfied for all $S \geq 1$. Furthermore it was shown in [10] that the bounds can be improved to include the case $S = 1/2$ as well. Therefore it has been proven that, in $D = 3$ (and for all greater dimensions as well), for all spins S there is antiferromagnetic order in the quantum Heisenberg antiferromagnet in the nearest neighbor isotropic cubic lattice.

4.6 Existence of antiferromagnetic order in the quantum Heisenberg model for T = 0 and D = 2

We proved in section 4.2 that in two dimensions there is no long range order in either the spin S ferromagnet or antiferromagnet for $T > 0$. However, it is possible that there is long range order in the ground state even though order is impossible for $T > 0$. For the nearest neighbor Heisenberg antiferromagnet on the square lattice antiferromagnetic order has indeed been proven to exist for $S \geq 1$ by Neves and Perez [9] by a method which extends the work of [8]. The case of $D = 2$ and $S = 1/2$ remains an open question but numerical evidence [25] suggests that antiferromagnetism does exist in the ground state.

4.7 Missing theorems

A comparison of the many phase diagrams in chapter 2 which exhibit crystalline, ferromagnetic, or antiferromagnetic order with the few actual proofs of order for specific models given in this chapter demonstrates that we do not have proofs of the existence of order for many, if not most, cases of physical interest. Some of this lack of knowledge is presented in Table 4.3 where we list some of the "theorems" which we would like to prove (or disprove).

One of the glaring weaknesses in the proofs of order of sections 4.3–4.6 is that they crucially rely on the reflection properties of the lattice models. These restrictions were so severe that very few short range interactions are covered by the proofs and, for example, there is not even a proof of order for nearest neighbors on the bcc lattice. This restriction of the method to lattice systems makes it impossible to extend the proof of order in the one component classical magnet (Ising model) to the liquid–gas transition, crystallization or the existence of the solid–liquid–gas triple points of a genuine continuum fluid.

In addition the restrictions on the proof of order in the classical Heisenberg magnet discussed in section 4.4 seem very artificial, and for the quantum Heisenberg ferromagnet, even though [8] conjectures very plausible bounds which lead to bounds on critical temperatures for Heisenberg ferromagnets, none of these conjectures have been proven.

Orientational order for liquid crystals has been studied for fluids that are confined to move on lattices [26] but no rigorous theorems on liquid crystal ordering for particles moving in the continuum seem to exist.

It is very clear that the existence of order is a delicate and subtle phenomenon. This need for more exact knowledge of the general phenomena must be kept in mind in reading all the subsequent chapters on computations for specific models such as the high temperature series expansions for the Heisenberg model of chapter 9, which do indeed predict ferromagnetic order for the $S = 1/2$ model in three dimensions.

Table 4.3 Missing "theorems"

1	Existence of crystalline order for $D = 3$
2	(Non)existence of crystalline order for hard discs in $D = 2$
3	Existence of the liquid–gas transition in continuum fluids
4	Existence of the solid–liquid–gas triple point
5	Existence of order in the classical Heisenberg model for general lattices
6	Existence of ferromagnetic order in the quantum Heisenberg magnet for $D = 3$
7	Existence of antiferromagnetic order in the $S = 1/2$ Heisenberg magnet for $D = 2$ at $T = 0$
8	How does the spatial and orientational order of hard ellipsoids depend on D and on the aspect ratio α?

References

[1] J. Kepler, De Niue Sexangula, in the *Gesammelte Werke, Band IV, Kleinere Schriften 1602–1611* eds. M. Caspar and F. Hammer. C.H. Beck'sche Verlagsbuchhandlung (Munich 1901) 264–280.

[2] C.F. Gauss,Untersuchen über die Eigenschaften der positiven ternären quadratischen Formen von Ludwig August Seeber, Göttische gelehrte Anzeigen, July 9, 1831. Reprinted in *Werke*, Vol. 2 Königlichen Gesellschaft der Wissenschaften zu Göttingen, 1863, 188–196.

[3] R. Peierls, On Ising's model of ferromagnetism. Proc. Camb. Phil. Soc. 32 (1936) 477–481.

[4] L. Fejes Tóth, Uber einen geometrischen Satz, Math. Z. 46 (1940) 79–83.

[5] N.D. Mermin and H. Wagner, Absence of ferromagnetism or antiferromagnetism in one- or two- dimensional isotropic Heisenberg models, Phys. Rev. Letts. 17 (1966) 1133–1136.

[6] N.D. Mermin, Crystalline order in two dimensions, Phys. Rev. 176 (1968) 250–254.

[7] J. Fröhlich, B. Simon and T. Spencer, Infrared bounds, phase transitions and continuous symmetry breaking, Commun. Math. Phys. 50 (1976) 79–85.

[8] F.J. Dyson, E.H. Lieb and B. Simon, Phase transitions in quantum spin systems with isotropic and nonisotropic interactions, J. Stat. Phys. 18 (1978) 335–383.

[9] E.J. Neves and J. F. Perez, Long range order in the ground state of two-dimensional magnets, Phys. Letts. A114 (1986) 331–333.

[10] T. Kennedy, E.H. Lieb and B.S. Shastry, Existence of Néel order in some spin–1/2 Heisenberg antiferromagnets, J. Stat. Phys. 53 (1988) 1019–1030.

[11] A. Donev, F.H. Stillinger, P.M. Chaikin and S. Torquato, Unusually dense crystal packing of ellipsoids, Phys. Rev. Letts. 92 (2004) 255506–(1-4).

[12] T.C. Hales, A proof of the Kepler conjecture, Annals of Math. 162 (2005) 1065–1185; T.C. Hales with S.P. Fergeson, The Kepler conjecture, Discrete and Computational Geometry 36 (2006) 1–269.

[13] R.E. Peierls, On Ising's model of ferromagnetism, Proc. Camb.. Phil. Soc. 32 (1936) 477–481.

[14] L.D. Landau, Phys. Z. Sowjet. 11 (1937) 26.

[15] J. Conway and N.J.A. Sloane, *Sphere Packings, Lattices and Groups* (Springer Verlag 1999).

[16] Nature, Vol. 424 (3 July 2003).

[17] N.N. Bogoliubov, Physik. Abhandl. Sowjectunion 6 (1962) 1.113 and 229.

[18] J. Fröhlich, R. Israel, E.H. Lieb and B. Simon, Phase transitions and reflection positivity I. General theory and long range lattice models, Comm. Math. Phys. 62 (1978) 1–34.

[19] G.N. Watson, Three triple integrals, Quart. J. Math. 10 (1939) 266–276.
[20] R.B. Griffiths, Correlations in Ising ferromagnets, J. Math. Phys. 8 (1967) 478–483.
[21] J. Ginibre, General formulation of Griffiths' inequalities, Comm.Math. Phys. 16 (1970) 310–328.
[22] F. Dunlop, Correlation inequalities for multicomponent rotators, Comm. Math. Phys. 49 (1976) 247–256.
[23] J. Fröhlich and T. Spencer, Phase transitions in statistical mechanics and quantum field theory, *Cargese Lectures* (1976).
[24] P.W. Anderson, Limits on the energy of the antiferromagnetic ground state, Phys. Rev. 83 (1951) 1260.
[25] Shoudan Liang, Existence of Néel order at $T = 0$ in the spin-1/2 antiferromagnetic Heisenberg model on a square lattice, Phys. Rev. 42 (1990) 6555–6560.
[26] O.J. Heilmann and E.H. Lieb, Lattice models for liquid crystals, J. Stat. Phys. 20 (1979) 679–693.

5

Critical phenomena and scaling theory

A dominant feature of the phase diagrams of chapter 2 is that there are many different type of ordering which occur and most materials have several different crystalline phases. The study of these different phases must depend on the details of the interaction potentials. But all of the elements whose phase diagrams are presented in chapter 2 share one common feature in that they all have a gas and liquid phase separated by a first order line which terminates in a critical point. Furthermore the discussion of the relation of the lattice gas model to the Ising model given in chapter 2 gives a mapping between certain models of ferromagnets and the liquid–gas transition and critical point. We will devote this chapter to the discussion of these phenomena.

When we approach the critical point we find from experiment that thermodynamic properties such as the compressibility and the specific heat become infinite. This is taken to indicate that there are fluctuations in the system which extend over lengths which are much larger than the atomic size length scale and that at least some of the physics is insensitive to many of the details of the potential. There is thus the suggestion that there may be some properties of critical points which may in fact be independent of the material and might be predicted exactly by one of the simple models introduced in chapter 2. This is the origin of the concept of universality.

We know much more about critical phenomena than we do of crystalline ordering. We saw in chapter 4 that we do not have a proof that crystalline order exists for any potential, much less do we have a model for which freezing transitions can be analytically studied in microscopic detail. It is therefore of enormous benefit for the study of critical phenomena that in two dimensions there is a special case of the lattice-gas model introduced in chapter 2 which has a critical point and a coexistence curve that can be solved exactly, because it is identical to the Ising model at zero magnetic field.

The solvability of the two-dimensional Ising model in zero magnetic field is our most important tool in the study of critical phenomena. The calculations of exact properties of the Ising model will be presented in great detail in chapters 10–12 and from that study there emerges a very detailed microscopic picture of behavior near a critical point. From these results it is then possible to abstract a phenomenological picture which can be applied to other more realistic models such as the Ising model in three dimensions which can be hoped to be a realization of the hopes for universal properties of real three-dimensional fluids. The picture which emerges from this is referred to as scaling theory.

Scaling theory emerged in the 1960s from the independent work of several authors. Some of the principle papers are given in Table 5.1. It is one of the cornerstones of modern statistical mechanics and can be explained without reference to the Ising model. In that sense, scaling theory is a general principle. This is how it was developed, and this is how it will be presented here. Nevertheless it must be kept in mind that none of the considerations of the scaling theory of critical phenomena are rigorous in the sense of the proofs given in the last chapter. Scaling theory can perhaps best be described as a set of reasonable assumptions which have great plausibility because they are all proven to hold in the two-dimensional Ising model.

In section 5.1 we begin the discussion of scaling theory by presenting the concept of critical exponents for Ising-like systems and proving that they are constrained by certain thermodynamic inequalities. In section 5.2 we present scaling theory for Ising systems, which, among other things, predicts equalities between the critical exponents. In section 5.3 we will extend our considerations from Ising models to Heisenberg models and Lennard-Jones fluids. In section 5.4 we introduce the concept of universality and conclude in section 5.5 with a discussion of missing theorems.

Table 5.1 Chronology of the development of the scaling theory of critical phenomena.

Date	Reference	Development
1959	Fisher [1]	Critical exponent equality
1963	Rushbrooke [2]	Critical exponent inequality
1963	Essam, Fisher [3]	Critical exponent equality
1964	Widom [4]	Scaling theory
1964	Fisher [5]	Scaling theory
1965	Griffiths [6], [7]	Critical exponent inequalities
1965	Rushbrooke [8]	Critical exponent inequality
1966	Kadanoff [9]	Scaling theory
1967	Fisher [10]	Scaling theory
1967	Kadanoff et al. [11]	Scaling theory

5.1 Thermodynamic critical exponents and inequalities for Ising-like systems

We begin with the features of scaling theory which are obtained from the two-dimensional Ising model with nearest neighbor interactions

$$\mathcal{E} = -\sum_{j=1}^{L_v}\sum_{k=1}^{L_h}\{E^h\sigma_{j,k}\sigma_{j,k+1} + E^v\sigma_{j,k}\sigma_{j+1,k} + H\sigma_{j,k}\} \tag{5.1}$$

with $\sigma_{j,k} = \pm 1$.

The most basic feature of the scaling theory of the critical point as seen in the Ising model (5.1) is that in fact there is a unique isolated point $T = T_c, H = 0, M = 0$ in the ferromagnet or $T = T_c, P = P_c, v = v_c$ in the continuum fluid at which the thermodynamic properties have singularities. This is already a nontrivial assumption, because, for the inhomogeneous Ising model where interaction constants E^h and E^v are allowed to depend on the lattice position and vary randomly about some central value with some small standard deviation, it is known that the magnetization and susceptibility have singularities at different temperatures. (This will be discussed further in chapter 10.) Such random impurities simulate an impure (dirty) crystal, and thus it may be expected that for real materials the assumption that there is a single isolated T_c may fail.

For the Ising ferromagnet the behavior of the thermodynamic functions is conventionally parameterized as shown in Table 5.2.

Table 5.2 Critical exponent parameterization of the thermodynamic functions for the Ising–like magnets.

Property	Critical form	
specific heat	$c_H \sim A_c\|T - T_c\|^{-\alpha}$	$T \to T_c+$
	$c_H \sim A_c'\|T - T_c\|^{-\alpha'}$	$T \to T_c-$
spontaneous magnetization	$M \sim A_M\|T - T_c\|^{\beta}$	$T \to T_c-$
susceptibility at $H = 0$	$\chi \sim A_\chi\|T - T_c\|^{-\gamma}$	$T \to T_c+$
	$\chi \sim A_\chi'\|T - T_c\|^{-\gamma'}$	$T \to T_c-$
magnetization at T_c	$M \sim A_H H^{1/\delta}$	$H \to 0+$

There are many comments which must be made about Table 5.2. The notation for the exponents is standard in the literature and originates in [5]. The exponents α and γ refer to $T > T_c$ and α', γ' and β are for $T < T_c$ (and it is hoped that the exponent β will not be confused with $\beta = 1/k_BT$). There is no canonical notation for the amplitudes.

We also note that it is a pure assumption that the singular parts of the thermodynamic functions are parameterized by pure algebraic powers instead of more complicated forms such as

$$|T - T_c|^{-\alpha} \ln^p |T - T_c| \quad \text{or} \quad |T - T_c|^{-\alpha} \ln^p |T - T_c| \ln \ln |T - T_c|. \tag{5.2}$$

Indeed, the specific heat of the two-dimensional Ising model does have a logarithmic instead of an algebraic singularity. This logarithmic singularity is often "interpreted" as a limiting case of $\alpha = 0$, and $\delta = 1$ is often interpreted as $\ln H$.

But the most important point to be discussed about Table 5.2, for which there is no universal agreement in the literature, is the meaning of the symbol $\sim$. For the solvable case of the two-dimensional Ising model the free energy at $H = 0$, both above and below T_c, has the form

$$F(T) = F_1(T) + F_2(T)(T - T_c)^2 \ln |T - T_c| \tag{5.3}$$

where $F_1(T)$ and $F_2(T)$ are analytic at $T = T_c$. Similarly the magnetization has an isolated singularity at T_c and in many (if not most) papers it is tacitly assumed that the meaning of the critical exponent forms is that there is an isolated algebraic or logarithmic singularity at T_c. However, we will see in chapter 12 that for the two-dimensional Ising model the possibility exists that the singularity in the magnetic susceptibility at

T_c is not isolated but may be embedded in a natural boundary. Therefore any conclusions drawn from a tacit assumption that the singularities are isolated must be treated with caution.

These critical exponents are not independent but satisfy inequalities which follow from thermodynamics. Some of the most important of these thermodynamic inequalities are given in Table 5.3

Table 5.3 Thermodynamic inequalities for critical exponents in dimension D.

Date	Reference	Inequality
1963	Rushbrooke [2]	$\alpha' + 2\beta + \gamma' \geq 2$
1965	Griffiths [6] Rushbrooke [8]	$\alpha' + \beta(1+\delta) \geq 2$
1965	Griffiths [7]	$\gamma'(\delta+1) \geq (2-\alpha')(\delta-1)$ $\gamma' \geq \beta(\delta-1)$
1969	Gunton, Buckingham [12] Fisher [13]	$D(\delta-1)/(\delta+1) \geq (2-\eta)$

As an example of these inequalities, we prove the inequality of Rushbrooke [2]. To do this we start from the magnetic analogue of the relation between c_p and c_v for a fluid

$$\chi(c_H - c_M) = T\left(\frac{\partial M}{\partial T}\right)_H^2 \tag{5.4}$$

where the isothermal susceptibility χ is

$$\chi = \left(\frac{\partial M}{\partial H}\right)_T = -\left(\frac{\partial^2 F}{\partial H^2}\right)_T, \tag{5.5}$$

the two specific heats are

$$c_H = -T\left(\frac{\partial^2 F}{\partial T^2}\right)_H, \quad c_M = -T\left(\frac{\partial^2 \tilde{F}}{\partial T^2}\right)_M \tag{5.6}$$

and we use the notation that $F(T, H)$ denotes the free energy with T, H the independent variables and $\tilde{F}(T, M)$ denotes the free energy with T, M the independent variables.

There are now several cases. First consider

$$\lim_{T\to T_c} c_M/c_H = R < 1. \tag{5.7}$$

Then using the critical exponent forms of Table 5.2 in (5.4) we find that as $T \to T_c-$

$$A'_\chi A'_c(1-R)|T-T_c|^{-\gamma'-\alpha'} \sim T\left(A_M\beta|T-T_c|^{\beta-1}\right)^2 \tag{5.8}$$

and thus by matching the dependence on $T - T_c$ on both sides of the equation we have the equality

$$\alpha' + 2\beta + \gamma' = 2. \tag{5.9}$$

On the other hand, if $R = 1$ we expand with $x > 0$

$$c_M/c_H \sim 1 + r|T - T_c|^x \tag{5.10}$$

and we find

$$\alpha' + 2\beta + \gamma' = 2 + x > 2. \tag{5.11}$$

For the case of the Ising model where the specific heat has a logarithm but the spontaneous magnetization and susceptibility are pure power laws we write at $T \to T_c$

$$c_H \sim A'_c \ln|T - T_c| + B_H, \quad \text{and} \quad c_M \sim A'_c \ln|T - T_c| + B_M \tag{5.12}$$

where B_H and B_M may be different and we then find from (5.4) that

$$A'_\chi |T - T_c|^{-\gamma'}(B_H - B_M) = T\left(A_M \beta |T - T_c|^{\beta - 1}\right)^2 \tag{5.13}$$

from which we find

$$\gamma' = 2(1 - \beta) \tag{5.14}$$

and

$$B_H - B_M = T_c (A_M \beta)^2 / A'_\chi. \tag{5.15}$$

The remaining inequalities in Table 5.3 are proven in a similar fashion from convexity properties of thermodynamics in [6–8,12,13]. When scaling theory holds, all of these thermodynamic inequalities hold as equalities.

5.2 Scaling theory for Ising-like systems

Scaling theory extends the discussion of the description of critical phenomena from thermodynamic critical exponents to the position-dependent correlation functions. We begin by discussing the spin correlation function $\langle \sigma_0 \sigma_{\mathbf{R}} \rangle$ at zero magnetic field and consider the behavior for fixed T as $|\mathbf{R}| \to \infty$. For the Ising model in two dimensions we learn from the exact computations of chapters 11 and 12 that at $H = 0$ there are three separate cases which are summarized in Table 5.4

Table 5.4 Definition of the correlation exponents at $H = 0$ for Ising-like systems.

anomalous dimension	$T = T_c$	$\langle \sigma_0 \sigma_{\mathbf{R}} \rangle \sim C_{T_c}/R^{D-2+\eta}$
correlation length	$T > T_c$	$\langle \sigma_0 \sigma_{\mathbf{R}} \rangle \sim C_> e^{-R/\xi_>(T)}/R^p$
		$\xi_>(T) \sim A_\xi |T - T_c|^{-\nu}$
correlation length	$T < T_c$	$\langle \sigma_0 \sigma_{\mathbf{R}} \rangle \sim M^2 + C_< e^{-R/\xi_<(T)}/R^{p'}$
		$\xi_>(T) \sim A'_\xi |T - T_c|^{-\nu'}$

5.2.1 Scaling for $H = 0$

We begin by considering three separate cases of the $\mathbf{R} \to \infty$ behavior of the spin correlation function $\langle \sigma_0 \sigma_{\mathbf{R}} \rangle$.

Case 1. Asymptotic behavior of $\langle \sigma_0 \sigma_{\mathbf{R}} \rangle$ at $T = T_c$ and $H = 0$
On a lattice away from T_c the correlations tend to have the symmetry of the lattice; square, cubic, etc. But it is found in the two-dimensional Ising model on the isotropic square lattice with nearest neighbor interaction energies $E^v = E^h$ that as $T \to T_c$ for $H = 0$ the leading behavior of the correlations becomes rotationally invariant. More generally for the anisotropic lattice $E^v \neq E^h$ at $T = T_c$ the correlation has the leading behavior of

$$\langle \sigma_0 \sigma_{R_y, R_x} \rangle = \frac{C_{T_c}}{R^{1/4}}[1 + O(R^{-2})] \tag{5.16}$$

where

$$R^2 = (s_x R_x)^2 + (s_y R_y)^2 \tag{5.17}$$

where the correction term $O(R^{-2})$ depends on $s_x R_x / s_y R_y$ as well as on R.

From this model result we formulate a general form for the leading asymptotic behavior of the correlations at $T = T_c$ and $H = 0$ in dimension D of

$$\langle \sigma_0 \sigma_{\mathbf{R}} \rangle \sim \frac{C_{T_c}}{R^{D-2+\eta}}. \tag{5.18}$$

The exponent η is known as the anomalous dimension.

Case 2. Asymptotic behavior of $\langle \sigma_0 \sigma_{\mathbf{R}} \rangle$ at $T > T_c$ and $H = 0$
When $T > T_c$ and $H = 0$ the Ising spin–spin correlation function decays exponentially rapidly as $R \to \infty$ and for T fixed we have

$$\langle \sigma_0 \sigma_{\mathbf{R}} \rangle = \frac{C_>(T)}{R^{1/2}} e^{-R/\xi_>(T)}[1 + O(R^{-1})] \tag{5.19}$$

where as $T \to T_c+$

$$\xi_>(T) \sim A_\xi |T - T_c|^{-1} \quad \text{and} \quad C_>(T) \sim c_> |T - T_c|^{-1/4}. \tag{5.20}$$

The quantity $\xi_>(T)$ is called the correlation length. From this model result we formulate a general form for the leading asymptotic behavior of the correlations at $T > T_c$ in dimension D of

$$\langle \sigma_0 \sigma_{\mathbf{R}} \rangle \sim \frac{C_>(T)}{R^p} e^{-R/\xi_>(T)} \tag{5.21}$$

where

$$\xi_>(T) \sim A_\xi |T - T_c|^{-\nu}. \tag{5.22}$$

Case 3. Asymptotic behavior of $\langle\sigma_0\sigma_{\mathbf{R}}\rangle$ at $T < T_c$ and $H = 0$

When $T < T_c$ and $H = 0$ the Ising spin correlation $\langle\sigma_0\sigma_{\mathbf{R}}\rangle$ approaches the square of the spontaneous magnetization, the approach to this limit is exponential and, for T fixed, we have

$$\langle\sigma_0\sigma_{\mathbf{R}}\rangle = M^2\{1 + \frac{C_<(T)}{R^2} e^{-R/\xi_<(T)}[1 + O(R^{-1})]\} \tag{5.23}$$

where as $T \to T_c-$

$$\xi_<(T) \sim A'_\xi|T - T_c|^{-1} \quad \text{and} \quad C_<(T) \sim c_<|T - T_c|^{-2}. \tag{5.24}$$

From this model result we formulate a general form for the leading asymptotic behavior of the correlation function for $T < T_c$ in any dimension D of

$$\langle\sigma_0\sigma_{\mathbf{R}}\rangle \sim M^2\{1 + \frac{C_<(T)}{R^{p'}} e^{-R/\xi_<(T)}\} \tag{5.25}$$

where as $T \to T_c-$

$$\xi_<(T) \sim A'_\xi|T - T_c|^{-\nu'} \quad \text{and} \quad C_<(T) \sim c_<\xi_<(T)^{p'}. \tag{5.26}$$

A most important feature of these three cases of the $R \to \infty$ behavior of the correlation $\langle\sigma_0\sigma_{\mathbf{R}}\rangle$ is that they are not uniform, in the sense that if T is set equal to T_c, neither the $T > T_c$ form (5.19) nor the $T < T_c$ form (5.23) reduces to the $T = T_c$ result (5.16).

Scaling theory attempts to unite these three cases together by defining the scaling limit and the scaling function.

Scaling limit for $H = 0$

The scaling limit is defined at $H = 0$ as the limit $T \to T_c$ and $R \to \infty$ with

$$\lim_{\substack{T\to T_c-\\ R\to\infty}} R/\xi_<(T) = \lim_{\substack{T\to T_c-\\ R\to\infty}} A'^{-1}_\xi R|T - T_c|^{\nu'} = r \quad \text{for} \quad T < T_c \tag{5.27}$$

and

$$\lim_{\substack{T\to T_c+\\ R\to\infty}} R/\xi_>(T) = \lim_{\substack{T\to T_c+\\ R\to\infty}} A^{-1}_\xi R|T - T_c|^{\nu} = r \quad \text{for} \quad T > T_c \tag{5.28}$$

fixed.

Scaling function for $T < T_c$ and $H = 0$

The scaling two-point function $G_-(r)$ is defined for $T < T_c$ from the correlation function $\langle\sigma_0\sigma_{\mathbf{R}}\rangle$ as

$$G_-(r) = \lim_{\substack{T\to T_c-\\ \text{scaling}}} M(T)^{-2}\langle\sigma_0\sigma_{\mathbf{R}}\rangle \tag{5.29}$$

where $M(T)$ is the spontaneous magnetization which is parameterized as $T \to T_c-$ by Table 5.2. From (5.25) and (5.26) we must have

$$G_-(r) \sim 1 + \frac{c_< e^{-r}}{r^{p'}} \quad \text{for} \quad r \gg 1. \tag{5.30}$$

There is no a priori reason why the large $\mathbf{R}$ behavior of $< \sigma_0 \sigma_{\mathbf{R}} >$ at $T = T_c$ must agree with the small r behavior of $G_-(r)$. Scaling theory makes the assumption that these two behaviors do in fact agree. Thus scaling theory assumes that

$$G_-(r) \to \frac{\text{const}}{r^{D-2+\eta}} \quad \text{as} \quad r \to 0. \tag{5.31}$$

We then make contact with the large $\mathbf{R}$ behavior of $\langle \sigma_0 \sigma_{\mathbf{R}} \rangle$ at $T = T_c$ by extending (5.29) to

$$\langle \sigma_0 \sigma_{\mathbf{R}} \rangle \sim M(T)^2 G_-(R|T - T_c|^{\nu'}). \tag{5.32}$$

Then using the $T = T_c$ form (5.18) on the left-hand side and Table 5.2 for $M(T)$ and (5.31) for the small r behavior of $G_-(r)$ on the right-hand side we find

$$\frac{\text{const}}{R^{D-2+\eta}} \sim \frac{\text{const}'|T - T_c|^{2\beta}}{(R|T - T_c|^{\nu'})^{D-2+\eta}}. \tag{5.33}$$

Because the left-hand side is independent of $T - T_c$ the dependence of $T - T_c$ on the right-hand side must cancel. Thus we find the exponent relation

$$2\beta = \nu'(D - 2 + \eta) \tag{5.34}$$

which we will see in chapter 12 holds analytically for the two-dimensional Ising model.

Scaling function for $T > T_c$ and $H = 0$

The scaling two-point function $G_+(r)$ is defined for $T > T_c$ from the correlation function $\langle \sigma_0 \sigma_{\mathbf{R}} \rangle$ as

$$G_+(r) = \lim_{\substack{T \to T_c+ \\ \text{scaling}}} \text{const}|T - T_c|^{-\nu(D-2+\eta)} \langle \sigma_0 \sigma_{\mathbf{R}} \rangle \tag{5.35}$$

where const is some normalizing constant, and scaling theory assumes that $G_+(r)$ is finite and nonvanishing. To compare this with the $T = T_c$ form (5.18) of $\langle \sigma_0 \sigma_{\mathbf{R}} \rangle$ we extend (5.35) to

$$\langle \sigma_0 \sigma_{\mathbf{R}} \rangle \sim \text{const}'|T - T_c|^{\nu(D-2+\eta)} G_+(R|T - T_c|^{\nu}) \tag{5.36}$$

and we see that in order to regain (5.18) we need

$$G_+(r) \sim \frac{\text{const}}{r^{D-2+\eta}} \quad \text{as} \quad r \to 0. \tag{5.37}$$

Scaling functions for n-point correlations

We now may extend our definitions of scaling functions from the two to the n-spin correlations. For $T < T_c$ we define

$$G_-^{(n)}(\mathbf{r}_1, \cdots, \mathbf{r}_{n-1}) = \lim_{\text{scaling}} M(T)^{-n} \langle \sigma_0 \sigma_{\mathbf{R_1}} \cdots \sigma_{\mathbf{R_{n-1}}} \rangle \tag{5.38}$$

and similarly for $T > T_c$ we define

$$G_+^{(n)}(\mathbf{r}_1, \cdots, \mathbf{r}_{n-1}) = \lim_{\text{scaling}} |T - T_c|^{-n\nu(D-2+\eta)/2} \langle \sigma_0 \sigma_{\mathbf{R_1}} \cdots \sigma_{\mathbf{R_{n-1}}} \rangle \tag{5.39}$$

with the scaled lengths $\mathbf{r}_j$ defined as in (5.27) and (5.28).

Relation of correlation to thermodynamic exponents

We may now use these scaling functions to relate the thermodynamic exponents to the correlation exponents. For example consider the expansion of the magnetic susceptibility in terms of the two-point function

$$\chi(T) = \left(\frac{\partial M(H)}{\partial H}\right)_{H=0} = \frac{1}{k_B T}\sum_{\mathbf{R}}\{\langle\sigma_0\sigma_{\mathbf{R}}\rangle - M(T)^2\}. \quad (5.40)$$

As $T \to T_c-$ we use the scaling form (5.32) to write

$$\begin{aligned} k_B T\chi(T) &\sim \sum_{\mathbf{R}} M(T)^2\{G_-(R|T-T_c|^{\nu'}) - 1\} \\ &\sim M(T)^2|T-T_c|^{-D\nu'}\int d^D r\{G_-(r) - 1\} \\ &\sim \text{const}'\ |T-T_c|^{-(D\nu'-2\beta)}\int d^D r\{G_-(r) - 1\} \\ &\sim \text{const}'\ |T-T_c|^{-\nu'(2-\eta)}\int d^D r\{G_-(r) - 1\} \end{aligned} \quad (5.41)$$

where in the last line we used (5.34) and the constant depends on the details of the lattice interaction constants. Similarly for $T \to T_c+$

$$k_B T\chi(T) \sim \text{const}\ |T-T_c|^{-\nu(2-\eta)}\int d^D r\ G_+(r). \quad (5.42)$$

Thus we find the prediction of scaling theory for the Ising model that the susceptibility exponents γ and γ' satisfy

$$\gamma' = (2-\eta)\nu' \quad \text{and} \quad \gamma = (2-\eta)\nu. \quad (5.43)$$

This relation was first obtained in 1959 by Fisher [1] for the two-dimensional Ising model.

If we now use the exponent equality of Rushbrooke (5.9) with (5.34) and (5.43) we obtain the relation valid for $T < T_c$

$$D\nu' = 2 - \alpha'. \quad (5.44)$$

5.2.2 Scaling for $H \neq 0$

There are no exact computations known for the Ising model with $H \neq 0$ but we may consider expanding the correlations for $H \neq 0$ in a power series in $H/k_B T$ by writing the interaction energy as

$$\mathcal{E} = \mathcal{E}_0 - H\sum_{\mathbf{R}}\sigma_{\mathbf{R}}. \quad (5.45)$$

Thus

$$\langle\sigma_0\sigma_{\mathbf{R}}\rangle_H = Z(H)^{-1}\sum_{\sigma}\sigma_0\sigma_{\mathbf{R}}e^{-\mathcal{E}_0/k_B T + H/k_B T\sum_{\mathbf{R}}\sigma_{\mathbf{R}}} \quad (5.46)$$

and by expanding the exponential we obtain

$$\langle \sigma_0 \sigma_{\mathbf{R}} \rangle_H - M(H)^2 = \sum_{n=1}^{\infty} \frac{1}{n!} (H/k_B T)^n \sum_{\mathbf{R}_j} \langle \sigma_0 \sigma_{\mathbf{R}} \sigma_{\mathbf{R}_1} \cdots \sigma_{\mathbf{R}_n} \rangle_{H=0}^c \tag{5.47}$$

where the superscript c means that only the connected part of the correlation function is to be used. Then if we use the scaling form of the n-point function (5.38) and (5.39) we see that in order for a scaling function to exist we need the existence of the limit $T \to T_c$ and $H \to 0$ of

$$\begin{aligned} &\frac{1}{n!}\left(\frac{H}{k_B T}\right)^n M^{n+2} \sum_{\mathbf{R}_j} G_-(|T-T_c|^{\nu'}\mathbf{R}, |T-T_c|^{\nu'}\mathbf{R}_1, \cdots, |T-T_c|^{\nu'}\mathbf{R}_j) \\ &\sim \frac{\text{const}}{n!}\left(\frac{H}{k_B T}\right)^n |T-T_c|^{\beta(2+n)} |T-T_c|^{-D\nu' n} \int d^D\mathbf{r}_1 \cdots d\mathbf{r}_n^D G_-(\mathbf{r}, \mathbf{r}_1, \cdots \mathbf{r}_n) \end{aligned} \tag{5.48}$$

and hence we obtain for $T < T_c$ the definition of the scaled magnetic field

$$h = \lim_{\substack{H \to 0 \\ T \to T_c}} \frac{H}{|T-T_c|^{D\nu' - \beta}} = \lim_{\substack{H \to 0 \\ T \to T_c-}} \frac{H}{|T-T_c|^{\nu'(D+2-\eta)/2}}. \tag{5.49}$$

We have thus far treated T above and below T_c separately and have thereby introduced the distinction between the low temperature exponents α', γ' and ν' and their high temperature counterparts α, γ and ν. However, there are no singularities for $H \neq 0$ at $T = T_c$ so the scaling functions for $T < T_c$ and $T > T_c$ must smoothly and analytically match together at $T = T_c$ for $H \neq 0$. Furthermore from the exact calculations in $D = 2$ for the Ising model at $H = 0$ we find that in fact

$$\alpha' = \alpha, \quad \gamma' = \gamma \quad \nu' = \nu. \tag{5.50}$$

These facts will be incorporated in our scaling theory by dropping the distinction between high and low temperature exponents and by extending the definition of the scaled magnetic field to

$$h = \lim_{\substack{H \to 0 \\ T \to T_c}} \frac{H}{|T-T_c|^{\nu(D+2-\eta)/2}} \tag{5.51}$$

which is now taken to hold both above and below T_c. The distinction between $T > T_c$ and $T < T_c$ may be avoided if we define, in analogy to the scaled field h, a scaled temperature

$$\tau = \lim_{\substack{H \to 0 \\ T \to T_c}} (T - T_c) H^{-\frac{2}{\nu(D+2-\eta)}}. \tag{5.52}$$

We now may define

$$\kappa = |T - T_c|^{\nu} + H^{\frac{2}{D+2-\eta}} \tag{5.53}$$

and in terms of (5.53) we define a general scaling function

$$G(r,\tau) = A_G \lim_{\text{scaling}} \kappa^{-(D-2+\eta)} \langle \sigma_0 \sigma_{\mathbf{R}} \rangle \tag{5.54}$$

where A_G is independent of T and H but may depend on the lattice interaction constants, and where the scaling limit is defined as

$$T \to T_c, \quad H \to 0, \quad \mathbf{R} \to \infty \tag{5.55}$$

with

$$r = A_R \kappa R \text{ and } \tau \quad \text{both fixed} \tag{5.56}$$

where A_R depends on the lattice constants but is independent of T and H. Scaling theory assumes that the behavior of $\langle \sigma_0 \sigma_{\mathbf{R}} \rangle$ near $T = T_c$ and $H = 0$ is given in terms of the scaling function $G(r, \tau)$ defined by (5.54), as

$$\langle \sigma_0 \sigma_{\mathbf{R}} \rangle \sim \kappa^{D-2+\eta} A_G^{-1} G(\kappa R, t) \tag{5.57}$$

and that the lattice-dependent constants A_G and A_R can be chosen such that the scaling function $G(r, \tau)$ is independent of the lattice constants.

This definition of the scaling function will reduce to the previous definitions of $G_{\pm}(r)$ if we choose the lattice-dependent constants A_R and A_G such that as $T \to T_c-$

$$A_R \kappa \sim 1/\xi_< \text{ and } M^2(T) \sim A_G \kappa^{2\beta/\nu}. \tag{5.58}$$

Thus we find

$$G_{\pm}(r) = G(r, \pm\infty). \tag{5.59}$$

We may now make contact with the thermodynamic exponent δ by noting that from (5.57) the magnetization $M(T, H)$ is related to the scaling function $G(r, \tau)$ by

$$M^2(T,H) = \lim_{R\to\infty} \langle \sigma_0 \sigma_{\mathbf{R}} \rangle \sim \kappa^{D-2+\eta} A_G^{-1} G(\infty, \tau) \tag{5.60}$$

where $G(\infty, \tau)$ is assumed to be finite. At $T = T_c$ we have $\tau = 0$ and thus, with the assumption that $G(\infty, 0) \neq 0$, we compare the H dependence of $M^2(T_c, H)$ for $H \sim 0$ given by (5.60) with the definition of the exponent δ of Table 5.2 to find

$$\delta = \frac{D+2-\eta}{D-2+\eta} \tag{5.61}$$

which is also written as

$$2 - \eta = D\frac{\delta - 1}{\delta + 1}. \tag{5.62}$$

For the two-dimensional Ising model we have seen that $\eta = 1/4$, and thus (5.61) gives

$$\delta = 15. \tag{5.63}$$

Finally we note that the magnetization $M(T, H)$ and the susceptibility $\chi(T, H)$ are given in terms of the free energy as

$$M(T,H) = -\frac{\partial}{\partial H} F(T,H) \tag{5.64}$$

and

$$k_B T\chi(T,H) = k_B T\frac{\partial}{\partial H}M(T,H) = -k_B T\frac{\partial^2}{\partial H^2}F(T,H). \tag{5.65}$$

Near $T = T_c$ and $H = 0$ we write a scaling form for the free energy as

$$F(T,H) \sim F(T,H)_{\text{analytic}} + |T - T_c|^{2-\alpha} f(\tau) \tag{5.66}$$

and use

$$\frac{\partial}{\partial H} = \frac{\partial \tau}{\partial H}\frac{\partial}{\partial \tau} = -\frac{2}{\nu(D+2-\eta)}|T - T_c|^{-\nu(D+2-\eta)/2}|\tau|^{1+\nu(D+2-\eta)/2}\frac{\partial}{\partial \tau} \tag{5.67}$$

where we will define for convenience what is called the gap exponent

$$\Delta = \nu(D + 2 - \eta)/2. \tag{5.68}$$

Thus we obtain in the scaling limit

$$M(T,H) \sim \Delta^{-1}|T - T_c|^{2-\alpha-\Delta}|\tau|^{1+\Delta}\frac{\partial f(\tau)}{\partial \tau} \tag{5.69}$$

and

$$k_B T\chi(T,H) = \Delta^{-2}|T - T_c|^{2-2\Delta}|\tau|^{1+\Delta}\frac{\partial}{\partial \tau}|\tau|^{1+\Delta}\frac{\partial f(\tau)}{\partial \tau}. \tag{5.70}$$

Now if we set $H = 0$ and compare the temperature dependence with the parameterizations in Table 5.2 we find the exponent relations

$$\beta = 2 - \alpha - \Delta = 2 - \alpha - \nu(D + 2 - \eta)/2 \tag{5.71}$$

$$\gamma = \gamma' = 2 - \alpha - 2\Delta = 2 - \alpha - \nu(D + 2 - \eta). \tag{5.72}$$

The expression (5.71) for β will agree with the previously determined relation (5.34)

$$2\beta = D - 2 + \eta, \tag{5.73}$$

and the expression (5.72) will agree with (5.43) if

$$D\nu = 2 - \alpha \tag{5.74}$$

which is the relation (5.44) with the identification of high and low temperature exponents (5.50).

Unfortunately the Ising model has only been exactly solved at $H = 0$, and thus the only exact information we have comes from the free energy, spontaneous magnetization and susceptibility at $H = 0$, which is sufficient only to give the following results:

$$f(-\infty) = f(\infty) \neq 0 \tag{5.75}$$

$$\lim_{\tau\to\infty} \tau^{1+\Delta}\frac{\partial f(\tau)}{\partial \tau} = 0, \qquad \lim_{\tau\to-\infty} \tau^{1+\Delta}\frac{\partial f(\tau)}{\partial \tau} = \text{const} > 0 \tag{5.76}$$

$$\lim_{\tau\to\pm\infty} |\tau|^{1+\Delta}\frac{\partial}{\partial \tau}|\tau|^{1+\Delta}\frac{\partial f(\tau)}{\partial \tau} = \text{const}_\pm \neq 0 \tag{5.77}$$

where the constants $\text{const}_\pm$ in (5.77) are known to be quite different.

5.2.3 Summary of critical exponent equalities

The assumptions of scaling theory given above show that only two of the exponents are independent. These are usually taken to be the high temperature exponents for the specific heat α and for the susceptibility γ. All other exponents may be obtained by use of the scaling laws. The exponent β follows from (5.9) and (5.50) as

$$\beta = \frac{1}{2}(2 - \alpha - \gamma) \tag{5.78}$$

the exponent ν follows from (5.44) and (5.50) as

$$\nu = (2 - \alpha)/D \tag{5.79}$$

the exponent ν follows from (5.43) and (5.79) as

$$\eta = 2 - \frac{D\gamma}{2 - \alpha} \tag{5.80}$$

the exponent δ follows from (5.61) and (5.80) as

$$\delta = \frac{2 - \alpha + \gamma}{2 - \alpha - \gamma} \tag{5.81}$$

and the gap exponent Δ follows from (5.68) as (5.79) and (5.80)

$$\Delta = \frac{1}{2}(2 - \alpha + \gamma). \tag{5.82}$$

We also recall the assumed equality of high and low temperature exponents (5.50)

$$\alpha = \alpha' \quad \gamma = \gamma' \quad \nu = \nu'. \tag{5.83}$$

Thus, for Ising-like systems, the scaling laws compute all exponents, including the exponent of spontaneous magnetization which is only defined for $T < T_c$, in terms of the high temperature exponents α and γ.

5.3 Scaling for general systems

The scaling theory developed for the Ising model in the previous two sections began with critical exponents defined in the low temperature regime at $H = 0$ and, from the low temperature exponents and correlation functions, the definition of the low temperature scaling function for the two-point function (5.29) was constructed. This low temperature definition was then extended to temperatures above T_c and to include $H \neq 0$ as well. The final result was a theory which is valid in the scaling region $T \to T_c$ and $H \to 0$.

For more general systems, however, this construction which starts from the low temperature regime is not always either convenient or appropriate. Instead of attempting to construct a theory in the scaling region by beginning with a low temperature theory, we reverse the process and posit *ab initio* a scaling theory valid in the scaling region, and only afterwards will we attempt to make contact with results obtained for high and low temperatures.

We will illustrate this for the classical n vector model, the Heisenberg magnet and the Lennard-Jones fluid.

5.3.1 The classical n vector and quantum Heisenberg models

The classical n vector is defined by the Hamiltonian

$$\mathcal{H} = -J \sum_{\mathbf{R},\mathbf{R}'} \mathbf{S}_{\mathbf{R}} \cdot \mathbf{S}_{\mathbf{R}'} - H \sum_{\mathbf{R}} S^z_{\mathbf{R}} \tag{5.84}$$

where $\mathbf{S}_{\mathbf{R}}$ is a classical vector with $n \geq 2$ components at the site $\mathbf{R}$ which satisfies

$$\mathbf{S}^2 = 1 \tag{5.85}$$

and $\mathbf{R}$ and $\mathbf{R}'$ will be restricted to be nearest neighbors on some lattice. When $n = 3$ this is the classical Heisenberg magnet.

The Hamiltonian for the quantum mechanical Heisenberg model is the same as (5.84) where now $\mathbf{S}_{\mathbf{R}}$ are spin operators that commute with each other on different sites and on the same site obey the commutation relations

$$[S^x_{\mathbf{R}}, S^y_{\mathbf{R}}] = iS^z_{\mathbf{R}}, \quad [S^y_{\mathbf{R}}, S^z_{\mathbf{R}}] = iS^x_{\mathbf{R}}, \quad [S^z_{\mathbf{R}}, S^x_{\mathbf{R}}] = iS^y_{\mathbf{R}} \tag{5.86}$$

with

$$\mathbf{S}^\dagger_{\mathbf{R}} = \mathbf{S}_{\mathbf{R}} \quad \text{and} \quad \mathbf{S}^2_{\mathbf{R}} = S(S+1) \tag{5.87}$$

where S is an integer or half integer and is the magnitude of the quantum spin.

We will by convention assume that for $T < T_c$ the spontaneous magnetization is in the z direction. There will then be two types of correlations to consider: correlations of the spins $S^z_{\mathbf{R}}$ which we will call longitudinal correlations and correlations of spins $S^i_{\mathbf{R}}$ which we call transverse correlations. We note that, for $T > T_c$ and $H = 0$, it follows from rotational invariance in spin space that

$$\langle S^z_0 S^z_{\mathbf{R}} \rangle = \langle S^i_0 S^i_{\mathbf{R}} \rangle \quad \text{for all} \quad i. \tag{5.88}$$

However, for $H \neq 0$ and $H = 0$ with $T < T_c$ equality of the longitudinal and transverse correlations does not hold.

The longitudinal correlations are analogous to the Ising correlations whereas the transverse correlations are a new feature.

The magnetization of the n vector model is defined as

$$M(T,H) = \langle S^z \rangle = Z(T,H)^{-1} \int \prod_{\mathbf{R}} d\mathbf{S}_{\mathbf{R}} S^z_0 e^{-\mathcal{H}/k_B T} \tag{5.89}$$

where

$$Z(T,H) = \int \prod_{\mathbf{R}} d\mathbf{S}_{\mathbf{R}} e^{-\mathcal{H}/k_B T} \tag{5.90}$$

and for the quantum Heisenberg model the similar formula holds with the integral replaced by the trace over all states. Thus the magnetic susceptibility is given in terms of the longitudinal correlations as

$$\chi_{\|}(T,H) = \frac{\partial M(T,H)}{\partial H} = \sum_{\mathbf{R}} \{ \langle S^z_0 S^z_{\mathbf{R}} \rangle - M(T,H)^2 \} \tag{5.91}$$

and in an analogous fashion we define a transverse susceptibility as

$$\chi_\perp(T,H) = \sum_{\mathbf{R}} \langle S_0^x S_{\mathbf{R}}^x \rangle. \tag{5.92}$$

Scaling theory for the Ising model is based on the assumption that on the lattice there exists a diverging correlation length

$$\xi^{-1} = \kappa = |T - T_c|^\nu + H^{\frac{2}{D+2-\eta}} \tag{5.93}$$

and that this is the only length scale (in addition to the lattice length) in the system. If we make the identical assumption for the n vector and quantum Heisenberg model we define the longitudinal and transverse scaling functions in the scaling limit

$$T \to T_c, \ H \to 0, \ \mathbf{R} \to \infty \tag{5.94}$$

with

$$A_R \kappa \mathbf{R} = r \ \text{ and } \ \tau = (T - T_c) H^{-\frac{2}{\nu(D+2-\eta)}} \ \text{ both fixed} \tag{5.95}$$

as

$$G_\perp(r,\tau) = A_{G\perp} \lim_{\text{scaling}} \kappa^{-(D-2+\eta)} \langle S_0^x S_{\mathbf{R}}^x \rangle. \tag{5.96}$$

$$G_\parallel(r,\tau) = A_{G\parallel} \lim_{\text{scaling}} \kappa^{-(D-2+\eta)} \langle S_0^z S_{\mathbf{R}}^z \rangle \tag{5.97}$$

and

$$G_\parallel(\infty,\tau) = A_{G\parallel} \lim_{\text{scaling}} \kappa^{-(D-2+\eta)} M^2(T,H). \tag{5.98}$$

For the Ising model in $D = 2$ and $H = 0$ the scaling function $G(r, \pm\infty)$ in the limit $r \to \infty$ smoothly connects to the behavior of $\langle \sigma_0 \sigma_{\mathbf{R}} \rangle$ where $|T - T_c|$ is small but $R|T - T_c|$ is large. Thus the scaling of the two-dimensional Ising model is a means extending our considerations of the system away from the critical point into a region in the vicinity of $T = T_c$. Scaling theory for general systems assumes that this same connection principle holds in general and thus we formulate the following connection principles for the two correlation functions of the n vector model.

Connection of scaling functions at $r \to \infty$

1). The behavior of $G_\perp(r,\tau)$ when $r \to \infty$ smoothly connects to the behavior of $\langle S_0^x S_{\mathbf{R}}^x \rangle$ when R is first made large and then $T - T_c$ and H are allowed to become small.

2). The behavior of $G_\parallel(r,\tau)$ when $r \to \infty$ smoothly connects to the behavior of $\langle S_0^z S_{\mathbf{R}}^z \rangle$ when R is first large and then $T - T_c$ and H are allowed to become small.

We thus need to inquire about the behavior of the correlations $\langle S_0^x S_{\mathbf{R}}^x \rangle$ and $\langle S_0^z S_{\mathbf{R}}^z \rangle$ for large R on the lattice. The following summarizes what is known or assumed for these behaviors and the implications for the scaling functions which follow from the connection principle.

Connection at $r \to \infty$ for $H = 0$ and $T > T_c$ for $D = 3$
The large R behavior in $D = 3$ of the correlations is

$$\langle S_0^z S_{\mathbf{R}}^z \rangle = \langle S_0^x S_{\mathbf{R}}^x \rangle \sim \frac{C_>(T, \mathbf{\Omega}) e^{-R/\xi(T,\mathbf{\Omega})}}{R} \tag{5.99}$$

where $\mathbf{\Omega}$ specifies the direction of $\mathbf{R}$. Connection requires that for large r

$$G_{\|}(r, \infty) = G_{\perp}(r, \infty) \sim \frac{c_> \ e^{-r}}{r} \quad \text{as} \quad r \to \infty \tag{5.100}$$

where

$$0 < c_> < \infty \tag{5.101}$$

$$\lim_{T \to T_c} \xi(T, \mathbf{\Omega}) |T - T_c|^{\nu} = A_R^{-1} \tag{5.102}$$

and A_R is the lattice-dependent constant introduced in (5.95). From this connection we will be able to estimate the exponents γ and α from the high temperature series expansions we will develop in chapter 9.

Connection at $r \to \infty$ for $H \neq 0$ for $D = 3$
From low temperature spin wave computations for the quantum [14, 15] and classical [16, 17] Heisenberg magnet and from the inequality of Dunlop and Newman [18] for the classical $n = 2$ vector model

$$\langle S_0^x S_{\mathbf{R}}^x \rangle^2 \leq \{\langle S_0^z S_{\mathbf{R}}^z \rangle - M^2(T, H)\}\{\langle S_0^z S_{\mathbf{R}}^z \rangle + M^2(T, H)\} \tag{5.103}$$

it is inferred that as $R \to \infty$ for $H \neq 0$ and all $T \neq T_c$ that for $D = 3$

$$\langle S_0^z S_{\mathbf{R}}^z \rangle - M^2(T, H) \sim M^{-2}(T, H) \langle S_0^x S_{\mathbf{R}}^x \rangle^2. \tag{5.104}$$

For large $\mathbf{R}$ and all $T \neq T_c$ and H it is also inferred from low temperature spin wave computations that

$$\langle S_0^x S_{\mathbf{R}}^x \rangle \sim C_{\perp}(T, H, \mathbf{\Omega}) \frac{e^{-R/\xi_{\perp}(T,\mathbf{\Omega})}}{R} \tag{5.105}$$

and thus from (5.104) for large R

$$\langle S_0^z S_{\mathbf{R}}^z \rangle - M^2(T, H) \sim C_{\|}(T, H, \mathbf{\Omega}) \frac{e^{-2R/\xi_{\perp}(T,\mathbf{\Omega})}}{R^2} \tag{5.106}$$

where

$$C_{\|}(T, H, \mathbf{\Omega}) = M^{-2}(T, H) C_{\perp}^2(T, H, \mathbf{\Omega}). \tag{5.107}$$

Connection of these two expressions with the definitions of the scaling limit (5.96) and (5.97) requires that

$$C_{\perp}(T, H, \mathbf{\Omega}) \sim \kappa^{\eta} \tag{5.108}$$

$$C_{\parallel}(T,H,\boldsymbol{\Omega}) \sim \kappa^{-1+\eta} \tag{5.109}$$

and thus using (5.105), (5.106), (5.108) and (5.109) and the κ dependence of the scaling law for $M(T,H)$ (5.69) with (5.74)

$$M^2(T,H) \sim \kappa^{1+\eta} \tag{5.110}$$

in (5.103) we find

$$\kappa^{2\eta} \leq \kappa^{2\eta} \tag{5.111}$$

which is obviously satisfied. We thus see for τ finite and r large that the scaling functions behave as

$$G_{\perp}(r,\tau) \sim c_{\perp}(\tau)\,\frac{e^{-\kappa_{\perp}(\tau)r}}{r} \tag{5.112}$$

$$G_{\parallel}(r,\tau) \sim G_{\parallel}(\infty,\tau) + c_{\parallel}(\tau)\frac{e^{-2\kappa_{\perp}(\tau)r}}{r^2} \tag{5.113}$$

with

$$0 < c_{\perp}(\tau), G_{\parallel}(\infty,\tau), c_{\parallel}(\tau) < \infty \tag{5.114}$$

where $\kappa_{\perp}(\tau)$ is a new correlation length which satisfies

$$0 \leq \kappa_{\perp}(\tau) < \infty \tag{5.115}$$

for all τ including $\pm\infty$.

Connection at $r \to \infty$ for $H = 0$ and $T < T_c$ for $D = 3$
When $H \to 0$ it is known from low temperature spin wave computations [19, 20] that $\xi_{\perp}(T,H,\boldsymbol{\Omega}) \to \infty$ and thus the large R behavior of the correlations is

$$\langle S_0^x S_{\mathbf{R}}^x \rangle \sim \frac{C_{\perp}(T,0,\boldsymbol{\Omega})}{R} \tag{5.116}$$

$$\langle S_0^z S_{\mathbf{R}}^z \rangle - M^2(T,0) \sim \frac{C_{\parallel}(T,0,\boldsymbol{\Omega})}{R^2}. \tag{5.117}$$

Connection requires that for large r

$$G_{\perp}(r,-\infty) \sim \frac{c_{\perp}(-\infty)}{r} \tag{5.118}$$

$$G_{\parallel}(r,-\infty) \sim G_{\parallel}(\infty,-\infty) + \frac{c_{\parallel}(-\infty)}{r^2} \tag{5.119}$$

and thus we must have

$$\kappa_{\perp}(-\infty) = 0 \tag{5.120}$$

Connection of scaling functions at $T = T_c$ and $H = 0$
The scaling function $G(r,\tau)$ of the Ising model is also assumed to have the property that, when $r \to 0$, there is a smooth connection to the large R behavior of $\langle \sigma_0 \sigma_{\mathbf{R}} \rangle$

when $T = T_c$. For the n vector model this connection requires that at $T = T_c$ and $H = 0$ for $R \to \infty$

$$\langle S_0^z S_{\mathbf{R}}^z \rangle = \langle S_0^x S_{\mathbf{R}}^x \rangle \sim \frac{\text{const}}{R^{D-2+\eta}} \tag{5.121}$$

and that for $r \to 0$

$$G_\perp(r,\tau) = G_\parallel(r,\tau) \sim \frac{G}{r^{D-2+\eta}} \tag{5.122}$$

where G is a finite nonzero constant which is independent of τ. For the Ising model in $D = 2$ this is confirmed for $\tau = \pm\infty$ but no independent confirmation at finite τ yet exists.

Diverging susceptibility at $H = 0$ for $T < T_c$

The transverse and longitudinal susceptibilities are expressed in terms of the correlations by (5.91) and (5.92) which when we use forms (5.96) and (5.97) written as

$$\langle S_0^x S_{\mathbf{R}}^x \rangle \sim A_{G\perp}^{-1} \kappa^{D-2+\eta} G_\perp(\kappa r,\tau) \tag{5.123}$$
$$\langle S_0^z S_{\mathbf{R}}^z \rangle - M^2(T,H) \sim A_{G\parallel}^{-1} \kappa^{D-2+\eta} \{G_\parallel(\kappa r,t) - G_\parallel(\infty,\tau)\} \tag{5.124}$$

become in the scaling limit

$$\chi_\perp(T,H) \sim A_{G\perp}^{-1} \kappa^{-2+\eta} \int d^D r G_\perp(r,\tau) \tag{5.125}$$

$$\chi_\parallel(T,H) \sim A_{G\parallel}^{-1} \kappa^{-2+\eta} \int d^D r \{G_\parallel(r,t) - G(\infty,\tau)\}. \tag{5.126}$$

If for $H = 0$ and $T > T_c$ we use the asymptotic forms (5.100) and (5.122) we see that the integrals converge and thus find that the susceptibility exponent satisfies the exponent relation $\gamma = \nu(2-\eta)$. However, for $H = 0$ and $T < T_c$ we see from the asymptotic forms that the integrals in both (5.125) and (5.126) diverge. Thus the susceptibilities $\chi_\perp(T,H)$ and $\chi_\parallel(T,H)$ for $T < T_c$ both diverge as $H \to 0$ and thus the low temperature exponent γ' does not exist.

The divergence of $\chi_\parallel(T,H)$ was first seen by Dyson [19,20] in 1956 where he showed for the quantum Heisenberg magnet in three dimensions that for $T \sim 0$

$$\chi_\parallel(T,H) \sim \text{const}\ H^{-1/2}. \tag{5.127}$$

The computation is done in terms of "spin wave excitations" and is one of the classic computations of magnetism in condensed matter physics.

For the classical $n = 2$ vector model it was shown rigorously by Lebowitz and Penrose [21] in 1975 from the inequality (5.103) but without using scaling that

$$\chi_\parallel(T,H) \geq \begin{cases} \text{const}\ M(T,H)^{7/2} H^{-1/2} & \text{for } D = 3 \\ \text{const}\ M(T,H)^4 \ln \dfrac{\text{const}}{HM(T,H)} & \text{for } D = 4 \end{cases} \tag{5.128}$$

which diverge as $H \to 0$ for $T < T_c$ where by definition $M(T,0) > 0$.

These divergences for $H \to 0$ are obtained by using the forms (5.112) and (5.113) in the integrals of (5.125) and (5.126). Thus we find

$$\chi_\perp(T,H) \sim A_{G\perp}^{-1} c_\perp(-\infty)\kappa^{-2+\eta}\kappa_\perp^{-2} 4\pi \int_0^\infty dy\, y e^{-y} \tag{5.129}$$

$$\chi_\|(T,H) \sim A_{G\|}^{-1} c_\|(-\infty)\kappa^{-2+\eta}\kappa_\perp^{-1} 4\pi \int_0^\infty dy\, e^{-y}. \tag{5.130}$$

This will give a divergence of $H^{-1/2}$ in $D=3$ for $\chi_\|$ if as $t \to -\infty$

$$\kappa_\perp^{-1}(\tau) \sim H^{-1/2}. \tag{5.131}$$

Thus, since $\kappa_\perp(\tau)$ is a function of τ alone we find from (5.95) that as $\tau \to -\infty$ with $D=3$

$$\kappa_\perp^{-1}(\tau) \sim \tau^{\nu(5-\eta)/4} = H^{-1/2}(T_c - T)^{\nu(5-\eta)/4} \tag{5.132}$$

and hence we find the final results [17]

$$\chi_\perp(T,H) \sim A_{G\perp}^{-1} 4\pi c_\perp(-\infty)(T_c - T)^{\nu(1-\eta)/2} H^{-1} \tag{5.133}$$

$$\chi_\|(T,H) \sim A_G\|^{-1} 4\pi c_\|(-\infty)(T_c - T)^{-\nu 3(1-\eta)/4} H^{-1/2}. \tag{5.134}$$

5.3.2 Lennard-Jones fluids

The Lennard-Jones fluid introduced in chapter 2 is a classical continuum model with a two body interaction potential with $n > m$

$$U(\mathbf{R}) = (R/\sigma)^{-n} - (R/\sigma)^{-m}. \tag{5.135}$$

This potential decays as $(R/\sigma)^{-m}$ as $R \to \infty$, and consequently the two-particle correlation can decay as $R \to \infty$ no faster than R^{-m}. This algebraic decay means that we cannot define a correlation length $\xi(T)$ from the decay $e^{-R/\xi(T)}$ and hence we cannot define the exponent ν from the divergence of $\xi(T)$ as $T \to T_c$. Nevertheless we can still formally define a κ from the fluid analogue of (5.53) of the Ising model definition and scaling functions from the fluid analogue of (5.54). However, now the large r behavior of the scaling function will have to match for the power law behavior determined by the R^{-m}. The scaling laws for exponents will still formally hold but there are no high temperature expansions from which the exponents can be estimated.

5.4 Universality

We conclude this chapter with the introduction of the concept of universality. Loosely speaking this is the idea that near the critical point when the correlation length becomes large compared with the lengths over which the potential is nonzero that there should be properties of the behavior which do not depend on the details of the interaction but only on properties such as the dimensionality of space and the number of components in the n vector model.

The first place such a universality can be seen is in the comparison of the exact solution of the Ising model on the triangular and square lattice where all the dependence on the lattice constants is in the factors A_R and A_G of (5.95), (5.96), and (5.97). More generally, all the evidence is that, for any finite range interactions in an Ising system in either two or three dimensions, it is still the case that all dependence on the lattice interactions is contained in the A_R and A_G.

A less obvious statement of universality relates to the relation between the quantum and classical Heisenberg model. The weakest version of universality for these systems is the statement that the critical exponents do not depend on the quantum spin S of the system and thus that the classical and quantum Heisenberg magnets have the same exponents. Moreover for the classical Heisenberg magnet the ferromagnet and the antiferromagnet map onto each other if the lattice is bipartite, merely by changing the sign of the spin on one sublattice. Therefore universality will say that the critical exponents of the quantum ferromagnet and antiferromagnet are the same. This is a striking assertion because we saw in the previous chapter that while spontaneous antiferromagnetic order could be proven to exist in the three-dimensional quantum Heisenberg antiferromagnet there is no proof of order for the quantum ferromagnet. Clearly the concept of universality and scaling, if correct, is in advance of our ability to make rigorous proofs.

It can also be asked if, for the Heisenberg magnet, the complete scaling function is independent of the quantum spin. The answer to this seems not to be known.

Finally we would like to know if there is any universality between the three-dimensional lattice gas (which is isomorphic to the three-dimensional Ising model) and the Lennard-Jones fluid which is thought to describe real inert gases. It is obvious that the scaling functions are not the same, but it is commonly asserted that universality implies that the critical exponents of the three-dimensional Ising model are the same as the Lennard-Jones fluid and as real inert gases.

5.5 Missing theorems

When scaling theory is defined by (5.94)–(5.98) the predictive power of scaling comes from the assumption that the connection constants $c_>, c_\perp(\tau), c_\parallel(\infty, \tau)$ and $c_\parallel(\tau)$ of (5.100), (5.112), and (5.113) are finite and nonzero for all τ including $\pm\infty$, and G of (5.122) is finite, nonzero and independent of τ. The only case for which this has been proven true is the Ising model in two dimensions at $H = 0$.

These scaling assumptions allow us to predict the low temperature exponent β of the spontaneous magnetization and the exponent η of the correlations functions at $T = T_c$ from the high temperature exponents α and γ, and these two exponents can be estimated by means of high temperature series expansions.

From universality we predict that these exponents depend only on the dimension D of space and the number of components n of the spin variables (or order parameter) and are independent of all other details of the interaction energies if they are not too long range.

Scaling has powerful support from the fact that all the predictions are proven exactly true for the two-dimensional Ising model at $H = 0$. Universality is supported by the fact that the two-dimensional Ising model can be exactly solved on the triangular lattice with three different values of the interaction strengths on the three legs of the triangle and when the interactions are all positive the exponents and the scaling functions are independent of the interaction constants.

These observations were the starting point for the development of scaling and universality. However, these results were obtained by using the very strong integrability properties of the Ising model, and thus it may be asked whether for models which do not have these strong integrability properties all of the predictions of scaling and universality will continue to hold.

Because of the striking and powerful predictions of scaling and universality it would be most desirable to have some proofs of their validity which do not rely on exact integrability of the system. Unfortunately no such theorems exist at present. We list in Table 5.5 a few of these "missing theorems" which have arisen in the course of this chapter.

Table 5.5 Some "missing theorems" in the theory of scaling and universality.

1. Prove the assumed $T = T_c$ connection (5.122) for
A. The Ising model in $D = 2$ for $H \neq 0$
B. The Ising model in $D = 3$
C. The $n = 3$ classical Heisenberg model
D. The $n = 3$ spin S quantum Heisenberg ferromagnet.
2. Prove that the exponents of the spin S quantum Heisenberg ferromagnet are independent of S.
3. Prove that the exponents of the quantum Heisenberg antiferromagnet are the same as those of the ferromagnet
4. Prove that the exponents of the Ising and/or Heisenberg ferromagnet. are equal for the cubic, fcc and bcc lattices
5. Prove the inequality (5.103) for the $n = 3$ component classical Heisenberg ferromagnet.
6. Prove the inequality (5.103) for the spin S quantum Heisenberg ferromagnet.

References

[1] M.E. Fisher, The susceptibility of the plane Ising model, Physica 25 (1959) 521–524.

[2] G.S. Rushbrooke, On the thermodynamics of the critical region for the Ising problem, J. Chem. Phys. 39 (1963) 842–843.

[3] J.W. Essam and M.E. Fisher, Padé approximant studies of the lattice gas and Ising ferromagnet below the critical point, J. Chem. Phys. 38 (1963) 802–812.

[4] B. Widom, Equation of state in the neighborhood of the critical point, J. Chem. Phys.43 (1964) 3808–3905.

[5] M.E. Fisher, Correlation functions and the critical region of simple fluids, J. Math. Phys. 5 (1964) 944–962.

[6] R.B. Griffiths, Thermodynamic inequality near the critical point for ferromagnets and fluids, Phys. Rev. Letts. 14 (1965) 623–624.

[7] R.B. Griffiths, Ferromagnets and simple fluids near the critical point: some thermodynamic inequalities, J. Chem. Phys. 43 (1965) 1958–1968.

[8] G.S. Rushbrooke, On the Griffiths inequality at a critical point, J. Chem. Phys. 43 (1965) 3439–3441.

[9] L.P. Kadanoff, Scaling laws for Ising models near T_c, Physics 2 (1966) 263–272.

[10] M. E. Fisher, The theory of equilibrium critical phenomena, Reports in Progress in Physics, 36 (1967) 615–730.

[11] L.P. Kadanoff, W. Götze, D. Hamblen, R.Hecht, E.A.S. Lewis, V.V. Paliauskas, M. Rayl, J. Swift, D. Aspenes and J. Kane, Static phenomena near critical points: Theory and experiment, Rev. Mod. Phys. 39 (1967) 395–431.

[12] M.J. Buckingham and J.D. Gunton, Correlations at the critical point of the Ising model, Phys. Rev. 178 (1969) 848–853.

[13] M.E. Fisher, Rigorous inequalities for critical-point correlation exponents, Phys. Rev. 180 (1969) 594–600.

[14] R.B. Stinchcombe, G. Horwitz, F. Englert and R. Brout, Thermodynamic behavior of the Heisenberg ferromagnet, Phys. Rev. 130 (1963) 155–176.

[15] R. Silberglitt and A.B. Harris, Dynamics of the Heisenberg magnet at low temperatures, Phys. Rev 174 (1968) 640–658.

[16] A. Patashinskii and Z. Pokrovsky, Longitudinal susceptibility and correlations in degenerate systems, Sov. Phys. JETP 37 (1973) 733–736.

[17] M.E. Fisher, M.N. Barber and D. Jasnow, Helicity modulus, superfluidity and scaling in isotropic systems, Phys. Rev. A8 (1973) 1111–1124.

[18] F. Dunlop and C.M. Newman, Multicomponent field theories and classical rotators, Comm. Math. Phys. 44 (1975) 223–235.

[19] F.J. Dyson, General theory of spin wave interactions, Phys. Rev. 102 (1956) 1217–1230.

[20] F.J. Dyson, Thermodynamic behavior of an ideal ferromagnet, Phys. Rev. 102 (1956) 1230–1244.
[21] J.L. Lebowitz and O. Penrose, Divergent susceptibility of isotropic ferromagnets, Phys. Rev. Letts. 35 (1975) 549–552.

Part II

Series and Numerical Methods

...if one scheme of happiness fails, human nature turns to another, if the first calculation is wrong, we make a second better.

Jane Austen

6

Mayer virial expansions and Groeneveld's theorems

In this chapter we study the expansion of the equation of state when the density $\rho = 1/v = N/V$ is small for a system of N particles in a volume V of mass m interacting through the pair potential $U(\mathbf{r})$ with the Hamiltonian

$$H = \sum_{i=1}^{N} \frac{\mathbf{p}_i^2}{2m} + \sum_{1 \le i < j \le N} U(\mathbf{r}_i - \mathbf{r}_j). \tag{6.1}$$

These expansions were first derived by Mayer and Mayer [1].

The most elementary result is that as $v \to \infty$

$$\frac{Pv}{k_B T} = 1 - \frac{1}{2v} \int d^D \mathbf{r} (e^{-U(\mathbf{r})/k_B T} - 1) + O(v^{-2}). \tag{6.2}$$

The coefficient of $1/v$ is called the second virial coefficient. This will be derived in section 6.1.

In order to systematically extend (6.2) we introduce what is called the Mayer function

$$f(\mathbf{r}_{i,j}) \equiv f_{i,j} = e^{-U(\mathbf{r}_i - \mathbf{r}_j)/k_B T} - 1, \tag{6.3}$$

to write

$$\exp\{-\frac{1}{k_B T} \sum_{1 \le i < j \le N} U(\mathbf{r}_i - \mathbf{r}_j)\} = \prod_{1 \le i < j \le N} (1 + f_{i,j}) \tag{6.4}$$

and expand the product. The expansion (6.4) consists of 2^N terms, each of which can be represented in what is called a Mayer graph, by a set of points numbered from 1 to N and m lines representing the functions $f_{i,j}$ where m ranges from zero to $2^{N(N-1)/2}$ An example of a Mayer graph with ten points is given in Fig. 6.1. In general these graphs may consist of a number of disconnected pieces.

In the grand canonical ensemble in a finite volume V we will denote the pressure and density as a function of the activity (or fugacity) z by $p(z; V)$ and $\rho(z; V)$ respectively and refer to their expansions for small z as the activity (or cluster) expansion. It is given by Mayers' first theorem, which will be proven in section 6.2, as

$$\frac{p(z; V)}{k_B T} = \sum_{k=1}^{\infty} z^k b_k(V) \tag{6.5}$$

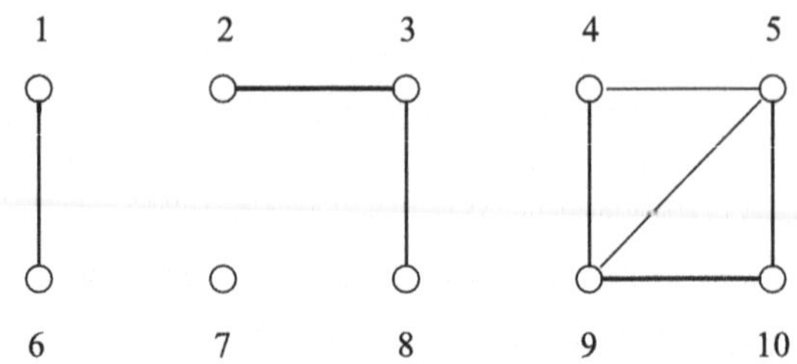

Fig. 6.1 An example of a Mayer graph with 10 points. This graph has four disconnected pieces and its value is $f_{1,6}f_{2,3}f_{3,8}f_{4,9}f_{5,9}f_{9,10}f_{5,10}f_{4,5}f_{5,10}$

$$\rho(z;V) = z\frac{d}{dz}(p(z;V)/k_BT) = \sum_{k=1}^{\infty} kz^k b_k(V) \tag{6.6}$$

where

$$b_k(V) = \frac{1}{k!V}\int_V U_k(\mathbf{r}_1, \cdots, \mathbf{r}_k) d^D\mathbf{r}_1 \cdots d\mathbf{r}_k^D \tag{6.7}$$

and $U_k(\mathbf{r}_1, \cdots, \mathbf{r}_k)$ is the sum of all connected numbered (labeled) Mayer graphs of k points. The quantities $b_k(V)$ are known as the cluster integrals and the functions $U_k(\mathbf{r}_1, \cdots, \mathbf{r}_k)$ are known as the Ursell functions.

In section 6.3 an expression in infinite volume is derived for the pressure $P(\rho)$ in terms of the (average) density $\rho = \lim_{V,N\to\infty} N/V$

$$\frac{P(\rho)}{k_BT} = \rho + \sum_{k=1} B_{k+1}\rho^{k+1} \tag{6.8}$$

with

$$B_{k+1} = -\frac{k}{k+1}\beta_k \tag{6.9}$$

where

$$\beta_k = \frac{1}{k!}\int V_{k+1}(\mathbf{r}_1, \cdots, \mathbf{r}_{k+1}) d^D\mathbf{r}_2 \cdots d^D\mathbf{r}_{k+1}, \tag{6.10}$$

the integrals are over all (infinite) space and $V_k(\mathbf{r}_1, \cdots, \mathbf{r}_k)$ are those connected numbered Mayer diagrams with k points which do not become disconnected by the removal of any one point. Such Mayer graphs are called either biconnected or irreducible and the functions $V_k(\mathbf{r}_1, \cdots, \mathbf{r}_k)$ are called Husimi functions. The expansion (6.8) is called the virial expansion, the β_k are known as the irreducible cluster integrals, the B_k are called the virial coefficients and the derivation is known as Mayer's second theorem. In the appendix we give the unlabeled irreducible (biconnected) graphs of four and five points and the number of labeled graphs to which they correspond. The six-point graphs are given in [2].

It is also possible to invert (6.6) to obtain $z(\rho;V)$ and use this in (6.5) to obtain a third expansion

$$\frac{P(\rho;V)}{k_BT} = \frac{p(z(\rho;V))}{k_BT} = \rho + \sum_{k=1} B_{k+1}(V)\rho^{k+1} \tag{6.11}$$

with

$$B_{k+1}(V) = -\frac{k}{k+1}\beta_k(V) \tag{6.12}$$

but unlike the $b_k(V)$ defined by (6.7), even though the $\beta_k(V)$ are given as polynomials in the $b_k(V)$, these $\beta_k(V)$ are *not* given by restricting the integrals in (6.10) to a finite volume. This expansion will be considered in section 6.5.

These expansions are useful whenever they converge, and we will consider four separate cases.

We call $R(V)$ the radius of convergence of (6.5) and (6.6) and thus the radius of convergence of the limiting case $V \to \infty$ of the series

$$\lim_{V\to\infty} \frac{p(z;V)}{k_BT} = \lim_{V\to\infty} \sum_{k=1}^{\infty} z^k b_k(V), \quad \lim_{V\to\infty} \rho(z;V) = \lim_{V\to\infty} \sum_{k=1}^{\infty} k z^k b_k(V) \tag{6.13}$$

is

$$\lim_{V\to\infty} R(V). \tag{6.14}$$

It follows from the discussion of first order phase transitions and grand partition function zeros in 3.4 that there are no phase transitions for

$$|z| \leq \lim_{V\to\infty} R(V). \tag{6.15}$$

We call R the radius of convergence of the series defined as

$$\sum_{k=1}^{\infty} z^k b_k = \frac{p(z)}{k_BT}, \quad \sum_{k=1}^{\infty} k z^k b_k = \rho(z) \tag{6.16}$$

with

$$b_k = \lim_{V\to\infty} b_k(V) = \frac{1}{k!}\int U_k(\mathbf{r}_1, \cdots, \mathbf{r}_k) d^D\mathbf{r}_2 \cdots d^D\mathbf{r}_k \tag{6.17}$$

where the integrals are over all space. The expansion (6.17) differs from (6.13) in that the $\lim_{V\to\infty}$ and the sum $\sum_{k=1}^{\infty}$ have been interchanged. In general

$$\lim_{V\to\infty} R(V) \leq R \tag{6.18}$$

which expresses the fact that the limiting position of the zero of $Q^{gr}(z,T;V)$ closest to the origin does not have to lead to a singularity of the pressure in the thermodynamic limit.

We will similarly define $\mathcal{R}$ to be the radius of convergence of (6.8) and $\mathcal{R}(V)$ to be the radius of convergence of (6.11). An argument similar to the argument given above shows that in general

$$\lim_{V\to\infty} \mathcal{R}(V) \leq \mathcal{R} \tag{6.19}$$

and that the system will not have any phase transitions for

$$|\rho| \leq \lim_{V\to\infty} \mathcal{R}(V). \tag{6.20}$$

In section 6.4 we prove the theorems of Groeneveld [3] on the radius of convergence R for non-negative potentials

$$U(\mathbf{r}) \geq 0. \tag{6.21}$$

In particular we will first prove that

$$0 \leq (-1)^{k-1} b_k(V) \leq (-1)^{k-1} b_k \tag{6.22}$$

and

$$\frac{1}{k} \leq \frac{b_k}{(2b_2)^{k-1}} \leq \frac{k^{k-2}}{k!}. \tag{6.23}$$

When $k = 2$ the upper and lower bounds are both equal to 1/2. From (6.23) it will follow that R

$$\frac{1}{2|b_2|e} < R < \frac{1}{2|b_2|} \tag{6.24}$$

and that the point $z = -R$ will be a singularity of $p(z)$. Furthermore, for nonnegative potentials, because the Ursell function U_k has been shown to have only one sign in the entire integration region, it follows that $b_k(V') \geq b_k(V)$ if V is contained in V'. Therefore it follows that

$$R \leq \lim_{V\to\infty} R(V). \tag{6.25}$$

Thus, combining (6.25) with the general result (6.18) we obtain the result of Lebowitz [9, (4.3)] that for nonnegative potentials

$$R = \lim_{V\to\infty} R(V). \tag{6.26}$$

The bounds may now be used to study the convergence of the virial expansion itself and in section 6.5 we prove a theorem of Lebowitz and Penrose [4] that

$$\frac{0.14476\cdots}{|2b_2|} \leq \lim_{V\to\infty} \mathcal{R}(V) \leq \mathcal{R} \tag{6.27}$$

and we will prove the bound

$$|\beta_k| \leq \frac{1}{k}\left(\frac{|\beta_1|}{0.14476\cdots}\right)^k = \frac{1}{k}\left(6.90799\cdots|\beta_1|\right)^k. \tag{6.28}$$

We note that $|\beta_1| = 2|b_2| = 2B_2$ so (6.27) is written as

$$\frac{0.0728}{B_2} \leq \lim_{V\to\infty} \mathcal{R}(V) \tag{6.29}$$

and (6.28) as

$$|\beta_k| \leq \frac{1}{k}(13.81598\cdots B_2)^k \tag{6.30}$$

from which using (6.9) we find

$$\frac{|B_k|}{B_2^{k-1}} \leq \frac{1}{k}(13.81598\cdots)^{k-1} \tag{6.31}$$

We also note the large k limiting form of the bound (6.24) on b_k obtained by using Stirling's approximation $k! \sim k^{k+1/2}e^{-k}(2\pi)^{1/2}$ in (6.23)

$$\frac{2^{k-1}}{k} \leq \frac{(-1)^{k-1}b_k}{B_2^{k-1}} \leq \frac{(2e)^k}{2(2\pi)^{1/2}k^{5/2}} = \frac{(5.43756\cdots)^k}{2(2\pi)^{1/2}k^{5/2}}. \tag{6.32}$$

It follows from (6.29) that there are no phase transitions for

$$0 \leq \rho \leq \frac{0.0728\cdots}{B_2} \tag{6.33}$$

The strongest known bound on the densities where there are no phase transitions for nonnegative potentials [4, footnote 18] is

$$0 \leq \rho \leq \frac{1}{(1+e)2B_2} = \frac{0.13447\cdots}{B_2} \tag{6.34}$$

which is proven by use of integral equations [5] and an inequality of Lieb [6]

The lower bound on $\mathcal{R}$ (6.27) is obtained from the bounds (6.23) on the cluster integrals b_k, and as such may be called an indirect method of studying the convergence of the virial series (6.8)–(6.10). However, direct methods that deal directly with the irreducible cluster integrals β_k have been developed by Groeneveld which improve the bound on β_k. For nonnegative potentials the best reported result [7, part IV eqn.(5.1)] is

$$|\beta_k| \leq \frac{1}{k}a_{k-1}(2B_2)^k \tag{6.35}$$

where $k \geq 2$ and the numbers a_k are defined by

$$a_k = \frac{1}{2\pi i}\oint \frac{d\xi}{\xi^{j+1}}e^{\xi}(1+e^{\xi})^{k-1} \tag{6.36}$$

for $k \geq 1$. For example [7, part IV eqn.(5.2)]

$$|\beta_2| \leq 2B_2^2, \quad |\beta_3| \leq \frac{20}{3}B_2^3 \text{ and } |\beta_4| \leq \frac{88}{3}B_2^4. \tag{6.37}$$

We remark that, unlike the cluster integrals b_k which alternate in sign for non-negative potentials, the signs of the β_k for non-negative potentials are only known for low orders by direct computation.

From (6.35) we find the simpler (but weaker) bound on the β_k

$$|\beta_k| \leq \frac{1}{k}\frac{(2B_2)^k}{(1+c_0)c_0^{k-1}} = \frac{0.21780\cdots}{k}(7.1823\cdots B_2)^k \tag{6.38}$$

where $c_0 = 0.27846\cdots$ is the positive root of the equation

$$c_0 e^{1+c_0} = 1 \tag{6.39}$$

and thus we have a bound of the virial coefficient [7, part IV eqn.(5.4)]

$$\frac{|B_k|}{B_2^{k-1}} \leq \frac{0.21780\cdots}{k}(7.1823\cdots)^{k-1} \tag{6.40}$$

which is a substantial improvement on (6.31).

From either(6.38) or (6.40) we derive the lower bound on the radius of convergence of the virial expansion for non negative potentials [7, part IV eqn.(5.5)]

$$\mathcal{R} \geq \frac{0.1392\cdots}{B_2} \tag{6.41}$$

which is a significant improvement on the bound (6.29) obtained from indirect methods. Unfortunately the details of the proof of (6.35) have not been published.

These bounds have been extended from non negative potentials to potentials which satisfy the stability condition (3.12)(with Φ replacing B)

$$\sum_{1\leq i<j\leq N} U(\mathbf{r}_i - \mathbf{r}_j) > -N\Phi \tag{6.42}$$

and for which the integral

$$C = \int_V d^D\mathbf{r}|e^{-U(\mathbf{r})/k_BT} - 1| \tag{6.43}$$

converges. For these potentials the radius of convergence of the cluster expansion in a finite volume satisfies

$$\frac{1}{Ce^{1+2\Phi/k_BT}} \leq R(V) \leq |e^{\Phi/k_BT}\frac{l}{l-1}b_l(V)|^{1/(l-1)} \tag{6.44}$$

for any l. The lower bound in (6.44) was first shown by Ruelle [8] and the upper bound was first shown by Penrose [9].

Lebowitz and Penrose [4] have shown that the the radius of convergence $\mathcal{R}$ of the virial expansion satisfies

$$\mathcal{R} \geq \lim_{V\to\infty} \mathcal{R}(V) \geq \frac{0.14476}{C}\frac{2}{1+u} \tag{6.45}$$

where

$$u = e^{2\Phi/k_BT}. \tag{6.46}$$

For non negative potentials $u = 1$ and $C = 2B_2$, and thus (6.45) reduces to (6.29).

The best results for β_k are those of Groeneveld [7, part IV eqns.(3.30),(3.31)]

$$|\beta_k| \leq \frac{C^k}{k}p_{k-1}(1+u) \tag{6.47}$$

where

$$p_j(y) = \frac{1}{2\pi i}\oint \frac{d\xi}{\xi^{j+1}}(ye^{\xi} - 1)^j \tag{6.48}$$

and the more refined bounds stated (but not proven) in [7, part IV eqns.(3.34),(3.35)]

$$\beta_k^- \leq \beta_k \leq \beta_k^+ \tag{6.49}$$

where for $k \geq 2$

$$\beta_k^\mu = \mu \frac{C^k}{2k}\{p_{k-1}(1+u)-1\} - \frac{(2B_2)^k}{2k}\{p_{k-1}(1-u)+1\}. \tag{6.50}$$

By use of the inequality which can be derived from (6.48)

$$p_k(1+u) \leq (1+u)^k \frac{k^k}{k!} \tag{6.51}$$

we obtain from (6.47) [7, part IV eqn.(3.39)]

$$|\beta_k| \leq C^k (1+u)^{k-1} \frac{(k-1)^{k-1}}{k!}. \tag{6.52}$$

An inequality stronger than (6.51) for large k is [7, part IV eqn.(3.41)]

$$p_k(1+u) \leq \frac{1}{c(u)^k} \tag{6.53}$$

where $c(u)$ is the smallest root of the equation [7, part IV eqn.(3.42)]

$$c(u)e^{-c(u)} = \frac{1}{(1+u)e}. \tag{6.54}$$

Thus using (6.53) in (6.47) we find [7, part IV eqn.(3.43)]

$$|\beta_k| \leq \frac{1}{k}\frac{C^k}{c(u)^{k-1}} \tag{6.55}$$

and thus the radius of convergence of the virial expansion $\mathcal{R}$ satisfies

$$\mathcal{R} \geq \frac{c(u)}{C}. \tag{6.56}$$

Finally, using

$$c(u) = \frac{e^{c(u)}}{(1+u)e} \geq \frac{1}{(1+u)e} \tag{6.57}$$

in (6.56) we obtain [7, part IV eqn.(5.46)]

$$\mathcal{R} \geq \frac{1}{2eC}\frac{2}{1+u} = \frac{0.18394\cdots}{C}\frac{2}{1+u} \tag{6.58}$$

which is stronger that (6.45). The convergence of the integral C (6.43) is guaranteed for weakly tempered potentials, and thus the lower bounds (6.45) and (6.58) prove that the virial expansion has a finite radius of convergence for all stable weakly tempered potentials.

We finally conclude in section 6.6 with theorems on the counting of Mayer diagrams.

6.1 The second virial coefficient

In the canonical ensemble the pressure is given as

$$P = -\left(\frac{\partial A}{\partial V}\right)_T \tag{6.59}$$

where A, the Helmholtz free energy, is defined from the partition function

$$Q_N(V,T) = \frac{1}{N!}\int d^D\mathbf{p}_1\cdots d^D\mathbf{p}_N \int_V d^D\mathbf{r}_1\cdots d^D\mathbf{r}_N e^{-H/k_BT} \tag{6.60}$$

as

$$A = -k_BT\,\ln Q_N(V,T). \tag{6.61}$$

For the class of Hamiltonians given by (6.1) the Gaussian integrals over the variable p_i are easily done, and thus (6.60) is more explicitly written as

$$Q_N(V,T) = \frac{1}{N!\lambda^{DN}}\int_V d^D\mathbf{r}_1\cdots d^D\mathbf{r}_N e^{-\frac{1}{k_BT}\sum_{i<j}U(\mathbf{r}_i-\mathbf{r}_j)} \tag{6.62}$$

where λ, which is called the thermal wavelength, is defined by

$$\lambda = \frac{1}{(2\pi m k_BT)^{1/2}}. \tag{6.63}$$

We are interested in the thermodynamic limit where

$$V\to\infty,\quad N\to\infty,\quad \text{with } \rho = \frac{1}{v} = \frac{N}{V} \text{ fixed.} \tag{6.64}$$

In the noninteracting case where $U(\mathbf{r}) = 0$ we trivially have

$$Q_N(V,T) = \frac{V^N}{N!\lambda^{DN}}. \tag{6.65}$$

Therefore from (6.61) we have

$$A = -k_BT(N\ln V - \ln(N!\lambda^{DN})) \tag{6.66}$$

and thus from (6.59)

$$P = \frac{k_BTN}{V} = \frac{k_BT}{v} \tag{6.67}$$

which is the equation of state for the ideal gas.

In general we write

$$Q_N(V,T) = \frac{V^N}{N!\lambda^{DN}}\tilde{Q}_N(V,T) \tag{6.68}$$

where

$$\tilde{Q}_N(V,T) = \frac{1}{V^N}\int_V d^D\mathbf{r}_1\cdots d^D\mathbf{r}_N e^{-\frac{1}{k_BT}\sum_{i<j}U(\mathbf{r}_i-\mathbf{r}_j)} \tag{6.69}$$

and define

$$\tilde{f}(v,T) = \lim_{N\to\infty,\ V\to\infty} \frac{1}{N}\ln\tilde{Q}_N(V,T) \tag{6.70}$$

under the assumption that, in the thermodynamic limit (6.64), the limit on the right-hand side exists. Then, using (6.70) in (6.68) and (6.61) we find from (6.59)

$$\frac{P}{k_B T} = \frac{1}{v} + \frac{\partial}{\partial(V/N)}\tilde{f}(v,T) \tag{6.71}$$

and thus

$$\frac{Pv}{k_B T} = 1 + v\frac{\partial}{\partial v}\tilde{f}(v,T). \tag{6.72}$$

To proceed further we use (6.4) in (6.69) to write

$$\tilde{Q}_N(V,T) = \frac{1}{V^N}\int_V d^D\mathbf{r}_1\cdots d^D\mathbf{r}_N \prod_{1\leq i<j\leq N}(1+f_{i,j}) \tag{6.73}$$

and expand the product. In order for the limit (6.70) to exist it is necessary that $\tilde{Q}_N(V,T)$ behave as an exponential in N as $N\to\infty$ and this is achieved in the lowest order of approximation by keeping only those terms in the expansion of the form

$$f_{j_1,j_2}f_{j_3,j_4}\cdots f_{j_{2n-1},j_{2n}} \tag{6.74}$$

where all of the j_i are distinct. For a given n the number of such terms is

$$\frac{N(N-1)(N-2)\cdots(N-2n+1)}{n!2^n} \tag{6.75}$$

and thus we have

$$\tilde{Q}_N(V,T) \sim 1 + \sum_{n=1}^{N/2}\frac{N(N-1)\cdots(N-2n+1)}{n!2^n}\left(\frac{1}{V}\int_V d^D\mathbf{r} f_{1,2}(\mathbf{r})\right)^n. \tag{6.76}$$

In the limit (6.64) this becomes

$$\tilde{Q}_N(V,T) \sim 1 + \sum_{n=1}\frac{N^n}{n!}\left(\frac{1}{2v}\int d^D\mathbf{r} f_{1,2}(\mathbf{r})\right)^n = \exp\left(N\frac{1}{2v}\int d^D\mathbf{r} f_{1,2}(\mathbf{r})\right). \tag{6.77}$$

Therefore from (6.70)

$$\tilde{f}(v,T) \sim \frac{1}{2v}\int d^D\mathbf{r} f_{1,2}(\mathbf{r}) \tag{6.78}$$

and from (6.72) we find

$$\frac{Pv}{k_B T} \sim 1 - \frac{1}{2v}\int d^D\mathbf{r} f_{1,2}(\mathbf{r}) \tag{6.79}$$

which is the desired result (6.2). Under the weakly tempered assumption of the pair potential (3.22) this integral converges at $\mathbf{r}\to\infty$. Comparing with the definition of the virial series (6.8) we see that the second virial coefficient is

$$B_2(T) = -\frac{1}{2}\int d^D\mathbf{r} f_{1,2}(\mathbf{r}) \tag{6.80}$$

and we note that $B_2(T)$ may be positive, negative or even zero. The temperature (if any) T_B at which $B_2(T_B) = 0$ is called the Boyle temperature.

6.2 Mayers' first theorem

In the derivation of the previous section it is not particularly clear that we are obtaining an expansion in terms of $1/v$. Equally unclear is how to systematically obtain the higher terms in this expansion. In this section we will begin to answer these questions by using the grand canonical ensemble to prove what is called Mayers' first theorem given by (6.5)–(6.7).

The grand partition function $Q^{gr}(z,T;V)$ is defined as

$$Q^{gr}(z,T;V) = \sum_{N=0}^{\infty}(\lambda^D z)^N Q_N(V,T) \tag{6.81}$$

and from this we have

$$\frac{p(z;V)}{k_B T} = \frac{1}{V}\ln Q^{gr}(z,T;V) \tag{6.82}$$

and

$$\rho(z;V) = \frac{1}{V}z\frac{\partial}{\partial z}\ln Q^{gr}(z,T;V). \tag{6.83}$$

Further we define $W_N(\mathbf{r}_1,\cdots,\mathbf{r}_N)$ to be the collection of all Mayer graphs of N numbered points and write (6.62) as

$$Q_N(V,T) = \frac{1}{N!\lambda^{DN}}\int_V d^D\mathbf{r}_1\cdots d^D\mathbf{r}_N W_N(\mathbf{r}_1\cdots\mathbf{r}_N). \tag{6.84}$$

Our first step in proving (6.5)–(6.7) is to express W_{N+1} in terms of U_{N+1}, and W_k and U_k with $k \le N$. To do this we single out the point $N+1$ for special attention and write $W_{N+1}(\mathbf{r}_1,\cdots,\mathbf{r}_{N+1})$ in terms of how many points the point $\mathbf{r}_{N+1}$ is connected to. For brevity we represent the coordinate $\mathbf{r}_k$ by the subscript k. Then we have the following recursion relation [10, (7.12)]

$$\begin{aligned} W_{N+1}(1\cdots N+1) &= U_{N+1}(1\cdots N+1) \\ &+ \sum_{j\le N} U_N(1\cdots\hat{j}\cdots N+1)W_1(j) \\ &+ \sum_{1\le j<k\le N} U_{N-1}(1\cdots\hat{j}\cdots\hat{k}\cdots N+1)W_2(j,k) \\ &+ \cdots \\ &+ U_1(N+1)W_N(1\cdots N) \end{aligned} \tag{6.85}$$

where the notation $\hat{j}$ means that the point j is omitted. In the term containing $U_{N-k+1}W_k$ the point $N+1$ is connected to $N-k$ other points and thus the k points of W_k are freely chosen from N possible points. Therefore there are

$$\binom{N}{k} = \frac{N!}{k!(N-k)!} \tag{6.86}$$

terms in the corresponding sum.

We now integrate all the coordinates $\mathbf{r}_k$ in the recursion relation (6.85) over the volume V by using the definitions (6.7) and (6.84) and the number of terms (6.86). The functions U_{N-k+1} and W_k contain no points in common. Thus the integral of $U_{N-k+1}W_k$ over V factorizes into the product of two separate integrals over V and thus we obtain

$$(N+1)!\lambda^{D(N+1)}Q_{N+1}(V,T) = \sum_{k=0}^{N} \binom{N}{k} V(k+1)!b_{k+1}(V)(N-k)!\lambda^{D(N-k)}Q_{N-k}(V,T) \tag{6.87}$$

which upon simplification becomes

$$(N+1)Q_{N+1}(V,T) = V\sum_{k=0}^{N}(k+1)b_{k+1}(V)\lambda^{-D(k+1)}Q_{N-k}(V,T). \tag{6.88}$$

We form a generating function by multiplying by $(z\lambda^D)^N$ and summing on N from 0 to ∞ and use the definition (6.81) of $Q^{gr}(z,T;V)$ to find

$$\sum_{N=0}^{\infty}(N+1)(\lambda^D z)^N Q_{N+1}(V,T) = \lambda^{-D}\frac{\partial}{\partial z}Q^{gr}(z,T;V)$$
$$= V\sum_{N=0}^{\infty}(\lambda^D z)^N \sum_{k=0}^{N}(k+1)b_{k+1}(V)\lambda^{-D(k+1)}Q_{N-k}(V,T). \tag{6.89}$$

We then interchange the order of summation on the right-hand side using

$$\sum_{N=0}^{\infty}\sum_{k=0}^{N} = \sum_{k=0}^{\infty}\sum_{N=k}^{\infty} \tag{6.90}$$

and replace the variable $N-k$ by a new variable m to obtain

$$\frac{\partial}{\partial z}Q^{gr}(z,T;V) = V\sum_{k=0}^{\infty}(k+1)z^k b_{k+1}(V)\sum_{m=0}^{\infty} z^m\lambda^{Dm}Q_m(V,T)$$
$$= VQ^{gr}(z,T;V)\sum_{k=0}^{\infty}(k+1)z^k b_{k+1}(V). \tag{6.91}$$

This differential equation is easily solved if we divide by $Q^{gr}(z,T;V)$ and use the initial condition which follows from the definition (6.81) of $Q^{gr}(z,T;V)$ that

$$Q^{gr}(0,T;V) = 1 \tag{6.92}$$

and we obtain

$$\frac{1}{V}\ln Q^{gr}(z,T;V) = \sum_{k=1}^{\infty} z^k b_k(V). \tag{6.93}$$

Thus, using the defining equations of the grand partition function (6.82) and (6.83), we obtain the desired results (6.5) and (6.6)

$$\frac{p(z;V)}{k_BT} = \sum_{k=1}^{\infty} z^k b_k(V) \tag{6.94}$$

and

$$\rho(z;V) = \sum_{k=1}^{\infty} k z^k b_k(V). \tag{6.95}$$

6.3 Mayers' second theorem

In this section we prove the virial expansion (6.8). This will be done in three separate steps.

6.3.1 Step 1

We begin by introducing the concept of an articulation point of a Mayer graph.

Definition

A point of a connected graph which has the property that if the point is removed the graph will separate into two or more disconnected pieces is called an articulation point.

Using this concept we now define a new class of graphs of k points [3]:

Definition of $T_{k-1}(1;2,\cdots,k)$

$$T_{k-1}(1;2,\cdots,k) = \text{those graphs in } U_k(1,\cdots,k) \text{ where the point 1 is not an articulation point.} \tag{6.96}$$

There is a recursion relation for $U_l(1,\cdots,l)$ in terms of T_{l-k} and U_k for $k \leq l-1$. This recursion relation is obtained by decomposing $U_l(1,2.\cdots,l)$ into the graphs where the point 1 is not an articulation point and all the ways in which the point 1 can be an articulation point which splits the graph is to two pieces; one of $l-k$ points where the point 1 is not an articulation point and which contains the point 2 and the remaining connected graph of k points where the point 1 may (but does not have to be) an articulation point. An example of this decomposition is given in Fig. 6.2.

From this decomposition we find [3, 7, 11]

$$\begin{aligned} U_l(1,2,\cdots,l) &= T_{l-1}(1;2,\cdots,l)U_1(1) \\ &+ \sum_{3\leq j\leq l} T_{l-2}(1;2,\cdots\hat{j}\cdots,l)U_2(1,j) \\ &+ \sum_{3\leq j_1<j_2\leq l} T_{l-3}(1;2\cdots\hat{j}_1\cdots\hat{j}_2\cdots l)U_3(1,j_1,j_2) \\ &+\cdots+T_1(1;2)U_l(1;3,\cdots,l). \end{aligned} \tag{6.97}$$

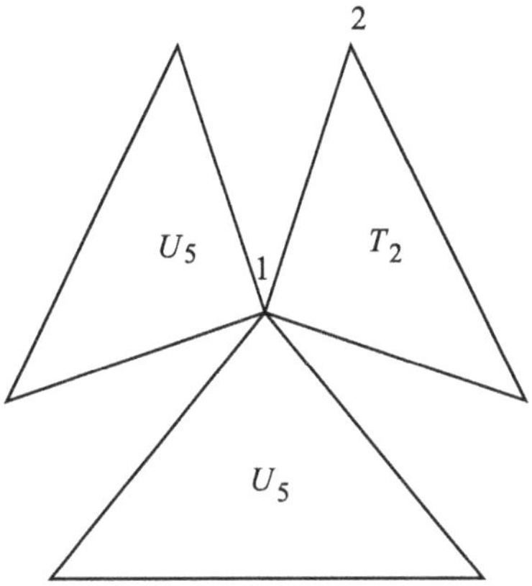

Fig. 6.2 An example of the decomposition of a connected graph with an articulation point at 1 into the product of a graph T_2 where 1 is not an articulation point and which contains the point 2 and a connected graph U_5 which does have 1 as an articulation point.

In this recursion relation the sum which contains $U_k(1, j_1, \ldots, j_{k-1})$ has $k-1$ variables freely chosen from the set $3, 4, \cdots, l$ and thus there are

$$\binom{l-2}{k-1} = \frac{(l-2)!}{(k-1)!(l-k-1)!} \tag{6.98}$$

terms in the sum.

We now define the integrated quantity

$$t_l = \frac{1}{l!} \int T_l(1; 2, \cdots, l+1) d^D\mathbf{r}_2 \cdots d^D\mathbf{r}_{l+1} \tag{6.99}$$

where the integrals are over all space and not just the finite volume V. Then we use (6.99) and the definition of the cluster integrals b_k in infinite volume (6.17) to integrate the variables $\mathbf{r}_2, \cdots, \mathbf{r}_l$ in (6.97) over all space. The only point that T_{l-k} and U_k have in common is the point $\mathbf{r}_1$ and therefore, because the variables $\mathbf{r}_2, \cdots, \mathbf{r}_l$ are integrated over all space, the integral of the product $T_{l-k}U_k$ factors into the product of T_{l-k} and U_k integrated separately and neither integral depends on the value of $\mathbf{r}_1$. (We note, however, that this factorization will *not* occur if the integrals are over a finite volume V). Thus we find

$$l!b_l = \sum_{k=1}^{l-1} k!b_k(l-k)!t_{l-k} \frac{(l-2)!}{(k-1)!(l-k-1)!} \tag{6.100}$$

and hence

$$l(l-1)b_l = \sum_{k=1}^{l-1} kb_k(l-k)t_{l-k}. \tag{6.101}$$

We now form a generating function by multiplying (6.101) by z^{l-1} and summing l from $2 \to \infty$. For the left-hand side we use (6.95) to write

$$\sum_{l=2}^{\infty} z^{l-1} l(l-1) b_l = z \frac{\partial}{\partial z} \frac{1}{z} \sum_{l=1}^{\infty} l z^l b_l = z \frac{\partial}{\partial z} \frac{\rho(z)}{z}. \tag{6.102}$$

For the right-hand side we interchange orders of summation using

$$\sum_{l=2}^{\infty} \sum_{k=1}^{l-1} = \sum_{k=1}^{\infty} \sum_{l=k+1}^{\infty} \tag{6.103}$$

and we introduce the generating function

$$T(z) = \sum_{n=1}^{\infty} t_n z^n \tag{6.104}$$

to find

$$\begin{aligned} \sum_{l=2}^{\infty} z^{l-1} \sum_{k=1}^{l-1} k b_k (l-k) t_{l-k} &= \sum_{k=1}^{\infty} k b_k z^k \sum_{l-k=1}^{\infty} z^{l-k-1} (l-k) t_{l-k} \\ &= \rho(z) \frac{\partial}{\partial z} T(z) \end{aligned} \tag{6.105}$$

and thus combining (6.102) and (6.105) we find

$$z \frac{\partial}{\partial z} \left(\frac{\rho(z)}{z} \right) = \rho(z) \frac{\partial}{\partial z} T(z). \tag{6.106}$$

This differential equation is easily solved and using the initial conditions

$$T(0) = 0 \text{ and } \lim_{z \to 0} \rho(z)/z = 1, \tag{6.107}$$

which follow from (6.104) and (6.95), we find

$$\frac{\rho(z)}{z} = e^{T(z)}. \tag{6.108}$$

6.3.2 Step 2

We begin the second step of the proof by defining another set of diagrams

Definition of $Y_l^{(s)}(\{\alpha\}_s; 1, 2, \cdots, l)$

$Y_l^{(s)}(\{\alpha\}_s; 1, 2, \cdots, l) =$
the subset of all diagrams with $l + s$ points which have the following two properties :
1. no path connects any pair of points in the set$\{\alpha\}_s$
2. the diagram becomes connected if all points of$\{\alpha\}$ are connected by lines. (6.109)

Some examples of these graphs are given in Fig. 6.3.

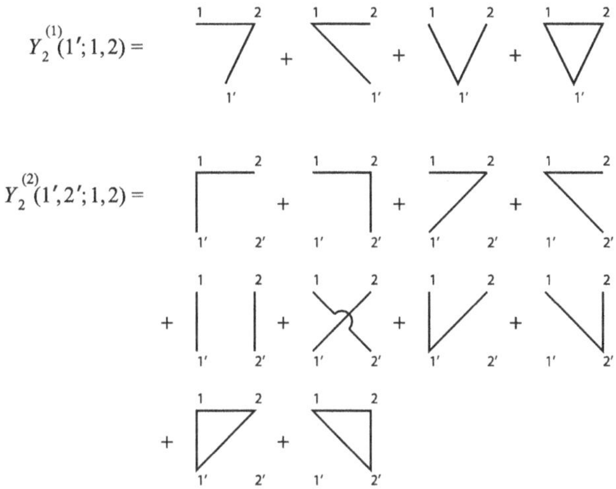

Fig. 6.3 Examples of the set of graphs $Y_l^{(s)}(\{\alpha\}_s;1,\cdots,l)$

There is a recursion relation which expresses the graphs $Y_l^{(s+1)}$ in terms of graphs in $Y_{l-k+1}^{(s)}$ and U_k for $k=1,\cdots,l+1$. This is obtained by adding the point $1'$ to the set of s points in $\{\alpha\}$ where this new added point is connected to $k-1$ of the points $1,2,\cdots,l$. Therefore

$$\begin{aligned}Y_l^{(s+1)}(\{1',\{\alpha\}_s\}_{s+1};1,2,\cdots,l) &= Y_l^{(s)}(\{\alpha\}_s;1,\cdots,l)U_1(1')\\ &+\sum_{1\leq j\leq l} Y_{l-1}^{(s)}(\{\alpha\}_s;1,\cdots\hat{j}\cdots l)U_2(1',j)\\ &+\sum_{1\leq j_1<j_2\leq l} Y_{l-2}^{(s)}(\{\alpha\}_s;1,\cdots\hat{j}_1\ldots\hat{j}_2\cdots l)U_2(1',j_1,j_2)\\ &+\cdots+Y_0^{(s)}(\{\alpha\}_s;)U_{l+1}(1',1,\cdots l).\end{aligned} \tag{6.110}$$

The sum which includes $U_k(1',1,\cdots,k-1)$ has $k-1$ points freely chosen from l possible points and thus the number of terms in the sum is

$$\binom{l}{k-1}=\frac{l!}{(k-1)!(l-k+1)!}. \tag{6.111}$$

We now define

$$y_l^{(s)}=\frac{1}{l!}\int Y_l^{(s)}(\{\alpha\}_s;1,\cdots l)d^D\mathbf{r}_1\cdots d^D\mathbf{r}_l \tag{6.112}$$

where the integrals are over all space and not just a finite volume V and integrate (6.110) over the coordinates $\mathbf{r}_1,\cdots\mathbf{r}_l$. The functions $Y_{l-k}^{(s)}$ and U_{k+1} have no points in common and thus the integral of the product $Y^{(s)}_{l-k}U_{k+1}$ factors into the product of the integrals of $Y^{(s)}_{l-k}$ and U_{k+1} separately. Therefore using the definitions of $y^{(s)}_l$ and b_l and (6.111) we find

$$l!y_l^{(s+1)} = \sum_{k=1}^{l+1} k!b_k(l-k+1)!y^{(s)}_{l-k+1}\frac{l!}{(k-1)!(l-k+1)!} \tag{6.113}$$

and hence

$$y_l^{(s+1)} = \sum_{k=1}^{l+1} kb_k y^{(s)}_{l-k+1}. \tag{6.114}$$

We then define a generating function

$$Y^{(s)}(z) = \sum_{l=0}^{\infty} y_l^{(s)} z^l \tag{6.115}$$

multiply (6.114) by z^l sum on l from $0 \to \infty$ and interchange orders of summation of the right-hand side to find

$$Y^{(s+1)}(z) = \rho(z)\frac{Y^{(s)}(z)}{z}. \tag{6.116}$$

We note further that

$$Y_l^{(1)}(\{1'\}_1; 1, \cdots, l) = U_{l+1}(1', 1, \cdots, l) \tag{6.117}$$

which when integrated over all coordinates gives

$$y_l^{(1)} = (l+1)b_{l+1}. \tag{6.118}$$

We then multiply (6.118) by z^l and sum on l from $0 \to \infty$ to obtain

$$Y^{(1)}(z) = \frac{\rho(z)}{z} \tag{6.119}$$

and using this as the initial condition for $s = 1$ we solve the recursion relation (6.116) and obtain

$$Y^{(s)}(z) = \left(\frac{\rho(z)}{z}\right)^s. \tag{6.120}$$

6.3.3 Step 3

For the final step in the proof we need to recall the definition of irreducible diagrams of l points given in the introduction

Definition of $V_l(1,\cdots,l)$

$$V_l(1,\cdots,l) = \text{the subset of diagrams in } U_l(1,\cdots,l) \text{ with no articulation points.} \tag{6.121}$$

We now construct a recursion relation that expresses $T_l(1';1,\cdots,l)$ in terms of V_{l+1-k} and $Y_k^{(l-k)}$ for $k = 0,1,\cdots,l-1$. We do this by considering $1'$ as a special point and dividing the graphs into the graphs that are biconnected to $1'$ and the rest, where the points which are biconnected to $1'$, are connected to the rest of the points through at least one articulation point. We count in terms of the number of such points that are connected to $1'$ though articulation points. Thus we find

$$\begin{aligned}
T_l(1';1, \quad &\cdots,l) = V_{l+1}(1',1,\cdots,l)Y_0^{(l)}(\{1,\cdots,l\}_l;) \\
&+ \sum_{1\le j\le l} V_l(1',1,\cdots\hat{j}\cdots l)Y_1^{(l-1)}(\{1.\cdots\hat{j}\cdots l\}_{l-1};j) \\
&+ \sum_{1\le j_1\le j_2\le l} V_{l-1}(1',1,\cdots\hat{j}_1\cdots\hat{j}_2\cdots l)Y_2^{(l-2)}(\{1,\cdots\hat{j}_1\cdots\hat{j}_2\cdots l\}_{l-2};j_1,j_2) \\
&+\cdots \sum_{1\le j\le l} V_2(1',j)Y_{l-1}^{(1)}(\{j\}_1;1,\cdots\hat{j}\cdots l).
\end{aligned} \tag{6.122}$$

For the sum containing $Y_{l-j}^{(j)}$ the j points in the set α are freely chosen from l possible points and thus the number of terms in the sum are

$$\binom{l}{j} = \frac{l!}{j!(l-j)!}. \tag{6.123}$$

We now integrate the coordinates $\mathbf{r}_1,\cdots,\mathbf{r}_l$ of (6.122) with $\mathbf{r}_{1'}$ fixed over all space and use the fact that the integral of the product $V_{l-k}Y_{k+1}^{(l-k-1)}$ is the product of the integral of V_{l-k} and the integral of $Y_{k+1}^{(l-k-1)}$ separately (and that they are independent of $\mathbf{r}_{1'}$) to find

$$l!t_l = \sum_{j=1}^{l} j!\beta_j(l-j)!y_{l-j}^{(j)}\frac{l!}{j!(l-j)!} \tag{6.124}$$

and thus

$$t_l = \sum_{j=1}^{l} \beta_j y_{l-j}^{(j)}. \tag{6.125}$$

Then multiply (6.125) by z^l, sum on l from $1 \to \infty$ and interchange orders of summation on the right-hand side to find

$$T(z) = \sum_{j=1}^{\infty} z^j\beta_j Y^{(j)}(z). \tag{6.126}$$

Thus, making use of the expression derived for $Y^{(s)}(z)$ in the previous step (6.120) we obtain

$$T(z) = \sum_{j=1}^{\infty} \rho^j \beta_j. \tag{6.127}$$

Finally, we may use the result of step 1 (6.108) to obtain the desired result

$$z(\rho) = \rho \exp(-\sum_{j=1}^{\infty} \beta_j \rho^j). \tag{6.128}$$

Thus we have succeeded in expressing z in terms of ρ which inverts the expression of ρ in terms of z (6.95).

It is now a simple matter to use the inversion formula (6.128) to obtain the virial expansion (6.8). We first recall (6.94) and (6.95) as

$$\frac{p(z)}{k_B T} = \sum_{l=1}^{\infty} b_l z^l \tag{6.129}$$

and

$$\frac{\partial}{\partial z}\left(\frac{p(z)}{k_B T}\right) = \sum_{l=1}^{\infty} l b_l z^{l-1} = \frac{\rho(z)}{z} \tag{6.130}$$

and hence we have

$$\frac{\partial}{\partial \rho}\left(\frac{P(\rho)}{k_B T}\right) = \frac{\partial z}{\partial \rho}\frac{\partial}{\partial z}\left(\frac{p(z)}{k_B T}\right) = \frac{\partial z}{\partial \rho}\frac{\rho}{z} = \frac{\partial \ln z}{\partial \ln \rho}. \tag{6.131}$$

But from (6.128) we have

$$\ln z = \ln \rho - \sum_{k=1}^{\infty} \beta_k \rho^k. \tag{6.132}$$

Thus we find

$$\begin{aligned} \frac{\partial \ln z}{\partial \ln \rho} &= 1 - \frac{\partial \rho}{\partial \ln \rho}\frac{\partial}{\partial \rho}\left(\sum_{k=1}^{\infty} \beta_k \rho^k\right) \\ &= 1 - \sum_{k=1}^{\infty} k \beta_k \rho^k \end{aligned} \tag{6.133}$$

and using this on the right-hand side of (6.131) we obtain

$$\frac{\partial}{\partial \rho}\left(\frac{P(\rho)}{k_B T}\right) = 1 - \sum_{k=1}^{\infty} k \beta_k \rho^k \tag{6.134}$$

which is easily integrated to obtain the desired result (6.8)

$$\frac{P(\rho)}{k_B T} = \rho - \sum_{k=1}^{\infty} \frac{k}{k+1} \beta_k \rho^{k+1}. \tag{6.135}$$

6.4 Non-negative potentials and Groeneveld's theorems

In this section we consider nonnegative potentials

$$U(\mathbf{r}) \geq 0 \tag{6.136}$$

and prove the three theorems of Groeneveld [3] given in the introduction (6.22)–(6.24).

A. The alternation in sign of b_l

We begin the proof of (6.22) by proving additional results for the function $T_{l-1}(1;2,\cdots,l)$ defined in (6.96). From this definition we see that the point 1 can be connected to k other points in a connected digram of $l-1$ points and therefore

$$\begin{aligned}
&T_{l-1}(1;2,\cdots,l)\\
&=\left(\sum_{2\leq j\leq l} f_{1,j}+\sum_{2\leq j_1<j_2\leq l} f_{1,j_1}f_{1,j_2}+\cdots+f_{1,2}f_{1,3}\cdots f_{1,l}\right)U_{l-1}(2,\cdots,l)\\
&=\left(\prod_{j=2}^{l}(1+f_{1,j})-1\right)U_{l-1}l(2,\cdots,l)\\
&=\left(\exp\{-\frac{1}{k_BT}\sum_{j=2}^{l}U(\mathbf{r}_1-\mathbf{r}_j)\}-1\right)U_{l-1}(2,\ldots,l).
\end{aligned} \tag{6.137}$$

For a nonnegative potential (6.136) we have

$$f_{i,j}=\exp\left(-\frac{U(\mathbf{r}_i-\mathbf{r}_j)}{k_BT}\right)-1\leq 0 \tag{6.138}$$

and therefore

$$\prod_{j=2}^{l}(1+f_{1,j})-1\leq 0 \tag{6.139}$$

so from (6.137) we conclude that T_l and U_l have opposite signs.

We now use this sign property along with the recursion relation (6.97) to prove the alternating sign property (6.22) by induction. First we note that

$$\begin{aligned}
U_1(1)&=1>0\\
U_2(1,2)&=f_{1,2}\leq 0\\
T_1(1;2)&=f_{1,2}\leq 0.
\end{aligned} \tag{6.140}$$

We thus may make the induction hypothesis that for integer L

$$\begin{aligned}
&\text{for } \ l\leq L-1\\
&(-1)^{l-1}U_l(1,\cdots,l)\geq 0\\
&(-1)^{l}T_l(1;1,\cdots,l)\geq 0
\end{aligned} \tag{6.141}$$

Therefore the sign of $T_{L-k}U_k$ is $(-1)^{L-1}$ and we find from (6.97) that the sign of U_L is $(-1)^{L-1}$ and thus the sign of T_L is $(-1)^L$ and hence the hypothesis (6.141) holds for $l = L$. From the definition (6.7) the sign of $b_k(V)$ is the same as the sign of $U_k(1, \cdots, k)$. Moreover the integrand in (6.7) never changes sign. Therefore the desired result (6.7) follows that

$$0 \leq (-1)^{k-1} b_k(V) \leq (-1)^{k-1} b_k. \tag{6.142}$$

B. The upper and lower bounds on b_l

We begin the proof of (6.23) by introducing the notation [3]

$$f(1; 2, \cdots, l) = \prod_{j=2}^{l} (1 + f_{1,j}) - 1 \tag{6.143}$$

and write (6.137) as

$$T_l(1; 2, \cdots, l+1) = f(1; 2, \cdots, l+1) U_l(2, \cdots, l+1). \tag{6.144}$$

We note that

$$\prod_{k=2}^{l} (1 + f_{1,k}) = (1 + f_{1,l}) \prod_{k=2}^{l-1} (1 + f_{1,k}) = f_{1,l} \prod_{k=2}^{l-1} (1 + f_{1,k}) + \prod_{k=2}^{l-1} (1 + f_{1,k}). \tag{6.145}$$

Therefore by induction

$$\prod_{k=2}^{l} (1 + f_{1,k}) = \sum_{k=3}^{l} f_{1,k} \prod_{j=2}^{k-1} (1 + f_{1,j}) + (1 + f_{1,2}) \tag{6.146}$$

and thus we have

$$f(1; 2, \cdots, l) = f_{1,2} + \sum_{k=3}^{l} f_{1,k} \prod_{j=2}^{k-1} (1 + f_{1,j}). \tag{6.147}$$

We now take the absolute value of this expression to obtain an upper bound

$$|f(1; 2, \cdots, l)| \leq |f_{1,2}| + \sum_{k=3}^{l} |f_{1,k}| \prod_{j=2}^{k-1} |1 + f_{i,j}| \tag{6.148}$$

from which, using the fact, that for nonnegative potentials (6.136), $f_{i,j} \leq 0$ and hence that $|1 + f_{i,j}| \leq 1$, we find

$$|f(1; 2, \cdots, l)| \leq \sum_{k=2}^{l} |f_{1,k}|. \tag{6.149}$$

Furthermore a lower bound on $|f(1; 2, \cdots, l)|$ is obtained if we retain only the first term in the sum (6.147). Thus we have

$$|f_{1,2}| \leq |f(1; 2, \cdots, l)| \leq \sum_{k=2}^{l} |f_{1,k}|. \tag{6.150}$$

We now integrate the variables $\mathbf{r}_2, \cdots, \mathbf{r}_{l+1}$ in (6.144) over all space and take the absolute value to write

$$\begin{aligned}|l!t_l| &= \left|\int f(1;2,\cdots,l+1)U_l(2,\cdots,l+1)d^D\mathbf{r}_2\cdots d^D\mathbf{r}_{l+1}\right| \\ &= \int |f(1;2,\cdots,l+1)||U_l(2,\cdots,l+1)|d^D\mathbf{r}_2\cdots d^D\mathbf{r}_{l+1}\end{aligned} \tag{6.151}$$

where the equality in the last line holds because of the condition of the signs of T_l and U_l previously established and now we may use the bounds of $|f(1;2,\cdots,l)|$ (6.150) to obtain

$$\begin{aligned}&\int |f_{1,2}||U_l(2,\cdots,l+1)|d^D\mathbf{r}_2\cdots d^D\mathbf{r}_{l+1} \\ &\leq |l!t_l| \leq \int \sum_{k=2}^{l+1} |f_{1,k}||U_l(2,\cdots,l+1)|d^D\mathbf{r}_2\cdots d^D\mathbf{r}_{l+1}.\end{aligned} \tag{6.152}$$

Now because the point 1 is not in common with any of the points in $U_l(2,\cdots,l+1)$ the integral over the product f_{1k} and $U_l(2,\cdots,l+1)$ factorizes into the product of the integrals and we obtain

$$\begin{aligned}&\int |f_{1,2}|d^D\mathbf{r}_2 \int |U_l(2,\cdots,l+1)|d^D\mathbf{r}_3\cdots d^D\mathbf{r}_{l+1} \\ &\leq |l!t_l| \leq \int l|f_{1,2}|d^D\mathbf{r}_2 \int |U_l(2,\cdots,l+1)|d^D\mathbf{r}_3\cdots d^D\mathbf{r}_{l+1}.\end{aligned} \tag{6.153}$$

and thus, using the definition of b_l (6.17), we have upper and lower bounds for $|t_l|$

$$2|b_2||b_l| \leq |t_l| \leq 2|b_2|l|b_l|. \tag{6.154}$$

In order to use the bounds (6.154) to find bounds on b_l we note that, because of the alternation of signs of t_l and b_l, we may rewrite (6.101) in terms of absolute values as

$$l(l-1)|b_l| = \sum_{k=1}^{l-1} k|b_k|(l-k)|t_{l-k}| \tag{6.155}$$

in which it is convenient to set for $l \geq 1$

$$l|b_l| = a_{l-1} \tag{6.156}$$

to get

$$la_l = \sum_{m=0}^{l-1} a_m(l-m)|t_{l-m}|. \tag{6.157}$$

We now define the quantities a_l^U and a_l^L as the solutions of the recursion relations identical in form to (6.157)

$$la_l^{U,L} = \sum_{m=0}^{l-1} a_m^{U,L}(l-m)t_{l-m}^{U,L} \tag{6.158}$$

with the initial condition

$$a_1^L = a_1^U = a_1 = B \tag{6.159}$$

where $t_l^{U,L}$ are the upper and lower bounds on the $|t_l|$ given by (6.154) as

$$t_l^L = B|b_l^L| = Ba_{l-1}^L/l \tag{6.160}$$

$$t_l^U = Bl|b_l^U| = Ba_{l-1}^U \tag{6.161}$$

and we have set

$$B = 2|b_2|. \tag{6.162}$$

We wish to prove that the $a_l^{U,L}$ defined by the recursion relations (6.158) actually have the properties of being upper and lower bounds

$$a_l^L \leq a_l \leq a_l^U. \tag{6.163}$$

This bounding property is satisfied for $l = 1$ by the initial conditions (6.159). For $l > 1$ we prove (6.163) by induction by noting that if (6.163) is assumed to hold up to $l - 1$ that

$$la_l = \sum_{m=0}^{l-1} a_m(l-m)|t_{l-m}| \leq \sum_{m=0}^{l-1} a_m^U(l-m)t_{l-m}^U = la_l^U \tag{6.164}$$

and

$$la_l = \sum_{m=0}^{l-1} a_m(l-m)|t_{l-m}| \geq \sum_{m=0}^{l-1} a_m^L(l-m)t_{l-m}^L = la_l^L \tag{6.165}$$

and thus (6.163) holds at l.

We now define the generating functions

$$A^{U,L}(z) = \sum_{l=0}^{\infty} a_l^{U,L} z^l, \tag{6.166}$$

$$T^{U,L}(z) = \sum_{l=0}^{\infty} t_l^{U,L} z^l. \tag{6.167}$$

Then the identical procedure which leads to (6.108) gives

$$A^{U,L}(z) = \exp(T^{U,L}(z)). \tag{6.168}$$

On the other hand we may use (6.160) and (6.161) in the definition (6.167) to find a second relation between $T^{U,L}(z)$ and $A^{U,L}(z)$. For the upper bound we directly substitute (6.161) into (6.167) to find

$$T^U(z) = BzA^U(z). \tag{6.169}$$

For the lower bound we substitute (6.160) into (6.167) to first obtain

$$T^L(z) = \sum_{l=1}^{\infty} z^l \frac{B}{l} a_{l-1}^L \tag{6.170}$$

from which we obtain the differential equation

$$\frac{\partial T^L(z)}{\partial z} = BA^L(z). \tag{6.171}$$

It remains to solve (6.169) and (6.171) with (6.168) to compute $a_l^{U,L}$.

The lower bound

We consider first the lower bound. Then from (6.168) we find

$$\frac{\partial A^L(z)}{\partial z} = A^L(z)\frac{\partial T^L(z)}{\partial z} \tag{6.172}$$

which, if we eliminate $\partial T^L(z)/\partial z$ using (6.171), gives the differential equation

$$\frac{\partial A^L(z)}{\partial z} = B[A^L(z)]^2. \tag{6.173}$$

Noting the initial condition

$$A^L(0) = 1 \tag{6.174}$$

which follows from $T^L(0) = 0$ we solve (6.173) to obtain

$$A^L(z) = \frac{1}{1 - Bz}. \tag{6.175}$$

Therefore recalling the definition of $A^L(z)$ (6.166) we find

$$a_l^L = B^l \tag{6.176}$$

and recalling the definition (6.156)

$$|b_l^L| = \frac{B^{l-1}}{l}. \tag{6.177}$$

Thus we have the lower bound

$$\frac{B^{l-1}}{l} \leq (-1)^{l-1} b_l \tag{6.178}$$

or recalling (6.162) and the sign alternation property (6.142) we obtain the final result

$$\frac{1}{l} \leq \frac{b_l}{(2b_2)^{l-1}}. \tag{6.179}$$

The upper bound

For the upper bound we substitute (6.169) into (6.168) to obtain

$$A^U(z) = e^{BzA^U(z)} \tag{6.180}$$

or, taking the logarithm,

$$z = \frac{\ln A^U(z)}{BA^U(z)}. \tag{6.181}$$

The desired coefficients a_l^U can be obtained from (6.181) by using the contour integral expression

$$a_l^U = \frac{1}{2\pi i}\int_C dz \frac{A^U(z)}{z^{l+1}} \tag{6.182}$$

where C is a closed contour enclosing $z = 0$ to recover the coefficient a_l^U from the generating function (6.166). From (6.181) we find

$$dz = (1 - \ln A^U)\frac{dA^U}{BA^{U2}}. \tag{6.183}$$

Then setting $A^U = e^\xi$ we have

$$\begin{aligned} z &= \frac{\xi}{Be^\xi} \\ dz &= \frac{1-\xi}{Be^\xi} d\xi \end{aligned} \tag{6.184}$$

and note that the closed contour C in the z plane will map into a closed contour C' enclosing $\xi = 0$ in the ξ plane. Thus we find

$$a_l^U = \frac{1}{2\pi i} B^l \int_{C'} d\xi \frac{1-\xi}{\xi^{l+1}} e^{\xi(l+1)} \tag{6.185}$$

which is readily computed by evaluating the residues at $\xi = 0$ to give

$$a_l^U = \frac{B^l(l+1)^{l-1}}{l!}. \tag{6.186}$$

Therefore, recalling the definitions of a_l^U (6.156) and B (6.162) and the sign alternation property (6.142), we obtain the desired upper bound

$$\frac{b_l}{(2b_2)^{l-1}} \leq \frac{l^{l-2}}{l!}. \tag{6.187}$$

C. The bounds on the radius of convergence

The radius of convergence R of the series (6.16) is obtained by applying the root test to the bounds of b_l. The upper bound R^U is found from the lower bound b_l^L as

$$R^U = \lim_{l\to\infty} (|b_l^L|)^{-1/l} = \lim_{l\to\infty} \left(\frac{|2b_2|^{l-1}}{l}\right)^{-1/l} = \frac{1}{|2b_2|} \tag{6.188}$$

where $\lim_{l\to\infty} l^{1/l} = 1$ has been used.

The lower bound R^L is found from the upper bound b_l^U as

$$R^L = \lim_{l\to\infty}(|b_l^U|)^{-1/l} = \frac{1}{2|b_2|}\lim_{l\to\infty}\left(\frac{l!}{l^{l-2}}\right)^{1/l}$$
$$= \frac{1}{2e|b_2|} \tag{6.189}$$

where, in the last line, Stirling's approximation

$$l! \sim l^{l+1/2}e^{-l}(2\pi)^{1/2} \quad \text{for} \quad l \to \infty \tag{6.190}$$

has been used. Thus the bounds (6.24) on the radius of convergence have been proven.

6.5 Convergence of virial expansions

In the introduction we introduced the expansion of the pressure $P(\rho;V)$ as a function of the density ρ in a finite volume V which is obtained by explicitly algebraically eliminating z between the equations (6.5) and (6.6) of the grand canonical ensemble in a finite volume. It is easy to see that this leads to algebraic relations between the finite volume $b_k(V)$ which have been shown to be given by the integrals of the Ursell functions U_N in a finite volume and the quantities $\beta_k(V)$ defined by (6.12) which are obtained from this algebraic elimination. Examples of these relations are

$$\beta_1(V) = 2b_2(V) \tag{6.191}$$
$$\beta_2(V) = 3b_3(V) - 6b_2(V)^2. \tag{6.192}$$

These algebraic relations are exactly the same as the relations which are obtained in infinite volume by using the formulas for β_k as integrals over the the Husimi functions V_k but, unlike the integral expressions for $b_k(V)$, the integral formulas for the β_k do *not* extend to the finite volume function $\beta_k(V)$ defined by these algebraic eliminations.

In this section we will follow [4] to find a lower bound on the radius of convergence $\mathcal{R}(V)$ of the finite volume density expansion (6.11) by use the results (6.23) and (6.24) of Groeneveld and from the general bound (6.19) will thus find the lower bound for the radius of convergence $\mathcal{R}$ of the virial expansion (6.8).

We recall that $p(z;V)$ and $\rho(z;V)$ are the pressure and density as a function of the fugacity z in finite volume. Thus we may obtain $P(\rho;V)$ from $p(z;V)$ and $\rho(z;V)$ by means of Cauchy's residue formula as

$$P(\rho;V) = \frac{1}{2\pi i}\int_C p(z;V)\frac{d\rho(z;V)}{dz}\frac{dz}{\rho(z;V)-\rho} \tag{6.193}$$

where C is a contour in the z plane which surrounds $z=0$ and on which the complex number ρ satisfies

$$|\rho| < \min|\rho(z;V)|. \tag{6.194}$$

When ρ satisfies (6.194) we may use the expansions of both

$$\frac{1}{\rho(z;V)-\rho} = \sum_{n=0}^{\infty}\frac{\rho^n}{\rho(z;V)^{n+1}} \tag{6.195}$$

and

$$P(\rho; V) = \sum_{n=1}^{\infty} c_n(V)\rho^n \tag{6.196}$$

in (6.193) to find

$$\begin{aligned} c_n(V) &= \frac{1}{2\pi i}\int_C dz \frac{d\rho(z;V)}{dz} p(z;V) \frac{1}{\rho(z;V)^{n+1}} \\ &= \frac{1}{2\pi i}\int_C dz p(z;V) \left(-\frac{d}{ndz}\rho(z;V)^{-n}\right) \end{aligned} \tag{6.197}$$

and integrate by parts to find

$$c_n(V) = \frac{1}{2\pi i}\int_C \frac{dp(z;V)}{dz} \frac{dz}{n(\rho(z;V))^n}. \tag{6.198}$$

The radius of convergence $\mathcal{R}(V)$ is thus determined by the minimum value of $\rho(z;V)$ on the contour C

$$\mathcal{R}(V) = \min_{z\in C}\rho(z;V). \tag{6.199}$$

Furthermore by comparison of (6.196) with (6.11) we see that

$$\frac{c_{l+1}(V)}{k_B T} = -\frac{l}{l+1}\beta_l(V) \tag{6.200}$$

and thus using the relation (6.6)

$$\rho(z;V) = \frac{z}{k_B T}\frac{dp(z;V)}{dz} \tag{6.201}$$

in (6.198) we find

$$l\beta_l(V) = -\frac{1}{2\pi i}\int_C \frac{dz}{z(\rho(z;V))^l}. \tag{6.202}$$

To study the minimum of $\rho(z;V)$ we recall that

$$\rho(z;V) = z + \sum_{l=2}^{\infty} l b_l(V) z^l \tag{6.203}$$

where the result (6.24) shows that the series converges if $|z| < 1/2eB_2$. where B_2 is the second virial coefficient in infinite volume. Therefore if in (6.203) we use the upper bound (6.23) on $b_l(V)$ we have

$$|\rho(z;V) - z| = |\sum_{l=2}^{\infty} l b_l(V) z^l| \leq \sum_{l=2}^{\infty} |l b_l(V)||z|^l \leq \frac{1}{2B_2}\sum_{l=2}^{\infty} l^{l-1}\frac{(2B_2|z|)^l}{l!}. \tag{6.204}$$

We now use a lovely formula of Euler which will be proven at the end of this section

$$w = \sum_{l=1}^{\infty} \frac{l^{l-1}(we^{-w})^l}{l!} \tag{6.205}$$

where the series converges for $0 \leq w$ and is unique for $0 \leq w \leq 1$.

Thus, assuming that $2B_2|z| \leq e^{-1}$, we define w as the smallest positive solution of

$$we^{-w} = 2B_2|z| \tag{6.206}$$

and write (6.204) as

$$|\rho(z;V) - z| \leq \frac{w}{2B_2} - |z|. \tag{6.207}$$

This upper bound in the distance of $\rho(z)$ from z leads to a lower bound on the distance of $\rho(z)$ from the origin by noting the triangle inequality

$$|\rho(z;V)| \geq |z| - |\rho(z;V) - z| \tag{6.208}$$

and thus using (6.207)

$$|\rho(z;V)| \geq |z| - \left(\frac{w}{2B_2} - |z|\right) = 2|z| - \frac{w}{2B_2} = (2e^{-w} - 1)\frac{w}{2B_2}. \tag{6.209}$$

The contour C can be chosen to be any circle with constant $|z|$ such that $|z| < 1/e2B_2$. This is satisfied for w such that $0 < w < 1$ and thus we find

$$|\rho(z;V)| \geq \max_{0\leq w\leq 1}(2e^{-w} - 1)\frac{w}{2B_2}. \tag{6.210}$$

This is maximized when

$$\frac{1}{2(1-w)} = e^{-w} \tag{6.211}$$

from which we find

$$w_{\max} = .31492\cdots \tag{6.212}$$

and thus

$$(2e^{-w_{\max}} - 1)w_{\max} = 0.14476\cdots \tag{6.213}$$

Thus we obtain from (6.199), (6.210) and (6.213) the desired result

$$\mathcal{R}(V) \geq \frac{0.14476\cdots}{2|B_2|}. \tag{6.214}$$

A bound on $|\beta_l(V)|$ is similarly obtained if in (6.202) we choose the contour C as $|z| =$ constant to obtain

$$l|\beta_l(V)| \leq \frac{1}{2\pi i}\oint_C dz\max_C\left|\frac{1}{z(\rho(z;V))^l}\right| = (\min_C|\rho(z;V)|)^l. \tag{6.215}$$

Therefore, choosing the radius of the circle C as before, we use the bound (6.213) to obtain

$$l|\beta_l(V)| \leq \left(\frac{2|B_2|}{0.14476}\right)^l. \tag{6.216}$$

It remains to prove the formula of Euler (6.205) which follows as an application of

Bürmann's theorem [12, pp. 128-131]

Let $\phi(z)$ be an analytic function of z in some neighborhood of zero and let $\phi'(0) \neq 0$. Then for z sufficiently close to zero, any function $f(z)$ analytic about zero may be expanded as

$$f(z) = f(0) + \sum_{m=1}^{\infty} \frac{\{\phi(z) - \phi(0)\}^m}{m!} \frac{d^{m-1}}{dz^{m-1}}[f'(z)\{\psi(z)\}^m]_{z=0} \tag{6.217}$$

where

$$\psi(z) = \frac{z}{\phi(z) - \phi(0)}. \tag{6.218}$$

To prove this we write

$$\begin{aligned} f(z) - f(0) &= \int_0^z f'(\zeta)d\zeta = \int_0^z d\zeta \frac{1}{2\pi i} \int_C dt \frac{f'(t)\phi'(\zeta)}{\phi(t) - \phi(\zeta)} \\ &= \frac{1}{2\pi i} \int_0^z d\zeta \int_C dt \frac{f'(t)\phi'(\zeta)}{\phi(t) - \phi(0)} \sum_{m=1}^{\infty} \left(\frac{\phi(\zeta) - \phi(0)}{\phi(t) - \phi(0)} \right)^{m-1} \\ &= \sum_{m=1}^{\infty} \int_0^z d\zeta \frac{1}{2\pi i} \phi'(\zeta)(\phi(\zeta) - \phi(0))^{m-1} \int_C dt \frac{f'(t)}{(\phi(t) - \phi(0))^m}. \end{aligned} \tag{6.219}$$

Then if we use (6.218) we obtain

$$\begin{aligned} f(z) &= f(0) + \sum_{m=1}^{\infty} \int_0^z d\zeta \phi'(\zeta)(\phi(\zeta) - \phi(0))^{m-1} \frac{1}{2\pi i} \int_C dt \frac{f'(t)\psi(t)^m}{t^m} \\ &= f(0) + \sum_{m=1}^{\infty} \int_0^z d\zeta \phi'(\zeta)(\phi(\zeta) - \phi(0))^{m-1} \frac{1}{(m-1)!} \frac{d^{m-1}}{dt^{m-1}} (f'(t)\psi(t)^m) \mid_{t=0} \end{aligned} \tag{6.220}$$

from which, when we do the integral over ζ, (6.217) follows.

The formula (6.205) now follows if in (6.217) we set

$$f(z) = z, \quad \phi(z) = ze^{-z} \tag{6.221}$$

along with

$$\psi(t) = e^t, \quad \frac{d^{m-1}}{dt^{m-1}} e^{mt}|_{t=0} = m^{m-1}. \tag{6.222}$$

6.6 Counting of Mayer graphs

We use the following notation:

$GL(k)$ is the number of all labeled Mayer graphs (both connected and disconnected) of order k.

$CL(k)$ is the number of labeled connected Mayer graphs of order k.

$IL(k)$ is the number of labeled irreducible Mayer graphs of order k.

$GU(k)$ is the number of all unlabeled Mayer graphs (both connected and disconnected) of order k.

$CU(k)$ is the number of unlabeled connected Mayer graphs of order k.

$IU(k)$ is the number of unlabeled irreducible Mayer graphs of order k.

By direct computation (either by hand or computer assisted) we have (from Table A3 of [13] and Table 1 of [14]) the results listed in Table 6.1.

Table 6.1 The number of unlabeled total, connected and irreducible (biconnected) Mayer graphs up to order 14.

k	$GU(k)$	$CU(k)$	$IU(k)$
1	1	1	0
2	2	1	1
3	4	2	1
4	11	6	3
5	34	21	10
6	156	112	56
7	1044	853	468
8	12,346	11,117	7,123
9	274,668	261,080	194,066
10	12,005,168	11,716,571	9,743,542
11	1,018,997,864	1,006,700,565	900,969,091
12	165,091,172,592	164,059,830,476	153,620,333,545
13	50,502,031,367,952	50,335,907,869,219	48,432,939,150,704
14	29,054,155,657,235,488	29,003,487,462,848,061	28,361,824,488,394,169

The number of all labeled Mayer graphs of k points is

$$GL(k) = 2^{k(k-1)/2}. \tag{6.223}$$

To prove this we let $g_l(k)$ be the number of labeled graphs with k points and l lines. We note that for any set of k points there are $k(k-1)/2$ distinct unordered pairs of points. In a graph of l lines these pairs are either connected by a line or they are not and therefore

$$g_l(k) = \binom{k(k-1)/2}{l}. \tag{6.224}$$

Thus we have

$$\begin{aligned} GL(k) = \sum_{l=0}^{k(k-1)/2} g_l(k) &= \sum_{l=0}^{k(k-1)/2} \binom{k(k-1)/2}{l} \\ &= 2^{k(k-1)/2} \end{aligned} \tag{6.225}$$

where in the last line we have used the binomial theorem. Thus (6.223) is established.

The number of all unlabeled Mayer graphs of k points behaves as $k \to \infty$ as [13, eqn.(9.1.25) p.199])

$$GU(k) = \frac{2^{k(k-1)/2}}{k!}\left(1 + \frac{k^2-k}{2^{k-1}} + \frac{(3k-7)k!}{2^{2k}(3k-9)(k-4)!} + O(\frac{k^5}{2^{5k/2}})\right). \tag{6.226}$$

The following results are also proven in [13, chapter 9].
Almost all unlabeled Mayer graphs are connected in the sense that

$$\lim_{k\to\infty} \frac{CU(k)}{GU(k)} = 1. \tag{6.227}$$

Almost all unlabeled Mayer graphs are irreducible in the sense that

$$\lim_{k\to\infty} \frac{IU(k)}{GU(k)} = 1. \tag{6.228}$$

The approach to these large k results is apparent in Table 6.1.

6.7 Appendix: The irreducible Mayer graphs of four and five points

We show in Fig. 6.4 the three irreducible unlabeled Mayer graphs of four points and the number of the corresponding labeled graphs. In Fig. 6.5 we show the irreducible Mayer graphs of five points and the number of the corresponding labeled graphs.

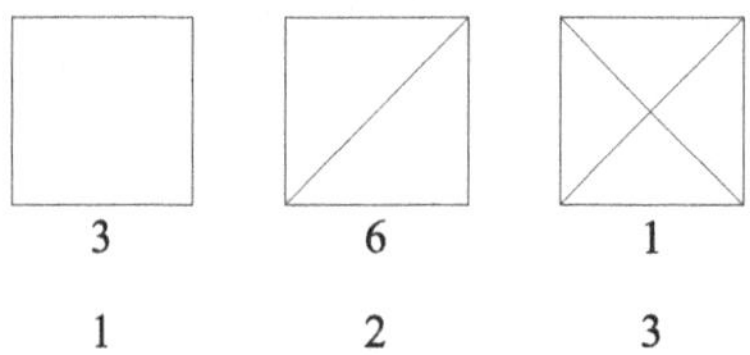

Fig. 6.4 The 3 unlabeled irreducible (biconnected) Mayer diagrams with four points. The first number under the graph is the number of labeled graphs corresponding to the unlabeled graph.

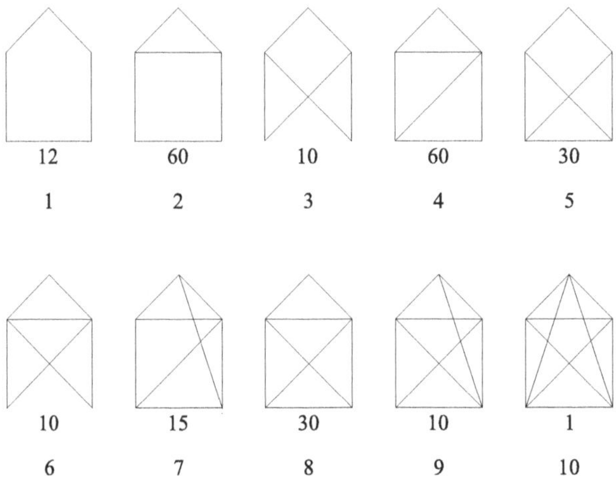

Fig. 6.5 The 10 unlabeled irreducible (biconnected) Mayer diagrams with five points. The first number under the graph is the number of labeled graphs corresponding to the unlabeled graph.

References

[1] J.E. Mayer and M.G. Mayer, *Statistical Mechanics* (Wiley 1940), chapter 13, 277–284.

[2] G.E. Uhlenbeck and G.W. Ford, "The theory of linear graphs with applications to the theory of the virial development of the properties of gases" in *Studies on Statistical Mechanics*, vol. 1, ed. J. de Boer and G.E. Uhlenbeck (North Holland 1962) 123–207.

[3] J. Groeneveld, Two theorems on classical many–particle systems, Phys. Letts. 3 (1962) 50–51.

[4] J.L. Lebowitz and O. Penrose, Convergence of virial expansions, J. Math. Phys. 5 (1964) 841–847.

[5] D. Ruelle, *Statistical Mechanics* (Benjamin, New York, 1969)

[6] E. Lieb, New method in the theory of imperfect gases and liquids, J.Math. Phys. 4 (1963) 671-678.

[7] J. Groeneveld, Estimation methods for Mayer graphical expansions, Doctor's thesis published in Proceedings of the Koninklijke Nederlandse Akademie van Wetenschappen, Series 70, Nrs. 4 and 5, 451–507 :(1967c) Part I Graphical Expansions; (1968a) Part II General Procedure; (1968b) Part III Estimation Methods of degrees 0 and 1; (1968c) Part IV Estimation methods of degree 2.

[8] D. Ruelle, Correlation functions of classical gases, Ann. Phys. 25 (1963) 109–120.

[9] O. Penrose, Convergence of fugacity expansion for fluids and lattice gases, J. Math. Phys. 4 (1963) 1312–1320.

[10] J. de Boer, Molecular distribution and the equation of state of gases, Reports on Progress in Physics, XII (1949) 305-374.

[11] J. Groeneveld, Estimation methods for Mayer's graphical expansions, in *Graph Theory and Theoretical Physics* ed. F. Harary (1967 Academic Press) 229-259.

[12] E.T, Whittaker and G.N. Watson, *A Course in Modern Analysis*, 4th edition (Cambridge University Press 1963).

[13] F. Harary and E.M. Palmer, *Graphical Enumeration* Academic Press (New York and London 1973)

[14] R.W. Robinson and T.R. Walsh, Inversion of cyclic index sum relations for 2- and 3- connected graphs, J. Comb. Theory, B57 (1993) 289–308.

7

Ree–Hoover virial expansion and hard particles

In this chapter we study what may be considered to be the simplest potential in classical statistical mechanics, hard spheres in D dimensions defined by the potential

$$U(\mathbf{r}) = \begin{cases} \infty & \text{if } |\mathbf{r}| \leq \sigma \\ 0 & \text{if } \sigma \leq |\mathbf{r}|. \end{cases} \tag{7.1}$$

Equation (7.1) says that the centers of the two spheres cannot be closer than σ which is traditionally called the hard core diameter of the sphere.

The potential (7.1) is particularly simple because the Boltzmann weights are independent of temperature. Thus P/k_BT is a function of density alone and the internal energy is the same as the perfect gas. There is no energy scale in the problem and all the physics is determined by entropy, geometry and combinatorics. What we learn from the study of hard spheres is basic to what has become known in recent years as "soft condensed matter physics".

The study of the virial coefficients of the hard sphere gas splits into two separate parts: the generation of the graphs and the evaluation of the integrals. The generation of the Mayer graphs is a serious problem because the number of these graphs grows very rapidly. The computation of integrals is difficult because there are strong cancellations between graphs, some of which are positive and some of which are negative.

A considerable simplification in the computation of virial coefficients was made in the mid 1960s by Ree and Hoover [1–3] who introduced a rearrangement of the Mayer graphs which decreases the number of graphs. We present this expansion in section 7.1. For the hard core gas (7.1) the Ree–Hoover expansion has the additional property that when the virial coefficient B_k is evaluated in D dimensions if $k-3 \geq D$ then some diagrams vanish for the hard core potential which do not vanish in general. diagrams This vanishing is vividly demonstrated in section 7.2 where we consider the very simple case of the Tonks gas [4], which is the one-dimensional case of the hard sphere gas (7.1), and show that there is exactly one nonvanishing Ree-Hoover graph for each virial coefficient.

In section 7.3 we analytically evaluate the virial coefficients for hard spheres B_2, B_3, and B_4 [5–13]. The results are given in Tables 7.5 and 7.6. The derivations for B_2 and B_3 are given in detail. The virial coefficients B_5–B_{10} have been evaluated numerically [1, 3, 14–17]. We discuss these computations in section 7.4 and tabulate the results in Table 7.8.

In section 7.5 we discuss the possible behavior of the hard sphere virial coefficients for $k \geq 11$ and examine the behavior of certain graphs for large values of k and in section 7.6 we use the results to estimate the radius of convergence of the virial series and discuss the location of the leading singularity in the complex density plane. We also present various of the approximate equations of state that have been proposed for hard spheres in three dimensions [18–26].

The hard sphere potential (7.1) is radially symmetric and the spatial variable $\mathbf{r}$ is in the continuum. However, the Mayer and Ree–Hoover expansions are equally valid for potentials which have an angular dependence and on a lattice the analogue of (7.1) is the pair potential $U(\mathbf{r}_k - \mathbf{r}_k)$ for a

lattice gas with nearest neighbor exclusion

$$\begin{aligned} U(0) &= +\infty \\ U(\mathbf{r}_j - \mathbf{r}_k) &= +\infty \text{ for } \mathbf{r}_j \text{ and } \mathbf{r}_k \text{ nearest neighbors on the lattice} \\ &\quad 0 \text{ otherwise} \end{aligned} \tag{7.2}$$

In section 7.7 we present the known results in the continuum for parallel hard squares and cubes [27] and the lattice gas results for hard squares on a square lattice [28, 29], hard hexagons on the triangular lattice [29–33] and hard particles with nearest neighbor exclusion on the cubic, fcc and bcc lattices in three dimensions [30].

In section 7.8 we consider nonspherical convex bodies which are allowed to rotate. In three dimensions the first computation of a second virial coefficient for nonspherical shapes was given by Onsager in 1949 [34], and in two dimensions B_2, B_3 and B_4 have been computed for ellipses, rectangles and needles [35, 36].

We conclude in section7.9 with a summary of some of the open questions in the study of virial expansions of hard particles.

7.1 The Ree–Hoover expansion

In the previous chapter we showed that the coefficients in the virial expansion of the pressure

$$\frac{Pv}{k_B T} = 1 + \sum_{k=2}^{\infty} B_k v^{1-k} \tag{7.3}$$

are

$$B_k = -\frac{k-1}{k!} \int V_k(\mathbf{r}_1, \cdots, \mathbf{r}_k) d^D\mathbf{r}_1 \cdots d^D\mathbf{r}_{k-1} \tag{7.4}$$

where $V_k(\mathbf{r}_1, \cdots, \mathbf{r}_k)$ is the sum of all numbered biconnected Mayer graphs with k points. In these Mayer graphs each line stands for the factor

$$f(\mathbf{r}_{i,j}) \equiv f_{i,j} = e^{-U(\mathbf{r}_i - \mathbf{r}_j)/k_B T} - 1. \tag{7.5}$$

The numbered Mayer graphs of a certain type are obtained by labeling the vertices of the basic unnumbered graph in all possible distinct ways. Calling $s_i[n]$ the number of labelings of the graph of type i with n points and calling $S_i[n]$ the integral of the unlabeled graph of type i and n points we have

$$B_n = -\frac{n-1}{n!}\sum_i s_i[n]S_i[n]. \tag{7.6}$$

In 1964 Ree and Hoover in [1–3] introduced a simple and very useful modification of this expansion by introducing in addition to (7.5) the function

$$\tilde{f} = e^{-U/k_BT} \tag{7.7}$$

with the property that

$$1 = \tilde{f}_{i,j} - f_{i,j}. \tag{7.8}$$

The basic idea of the re-expansion is to note that, in a general Mayer graph, points are either connected by a line which contributes a factor of f or they are not connected. Every pair of points which is not connected can be thought of as contributing a factor of 1 to the integral. The Ree–Hoover expansion is to replace this 1 by $\tilde{f} - f$ and to rewrite the expansion in terms of integrals where all points are connected by bonds which now are either f or $\tilde{f}$.

As a first example of the utility of this method, consider the fourth virial coefficient. There are three contributing graphs as shown in Fig. 7.1 and the symmetry numbers are

$$s_1[4] = 1, \quad s_2[4] = 6, \quad s_3[4] = 3. \tag{7.9}$$

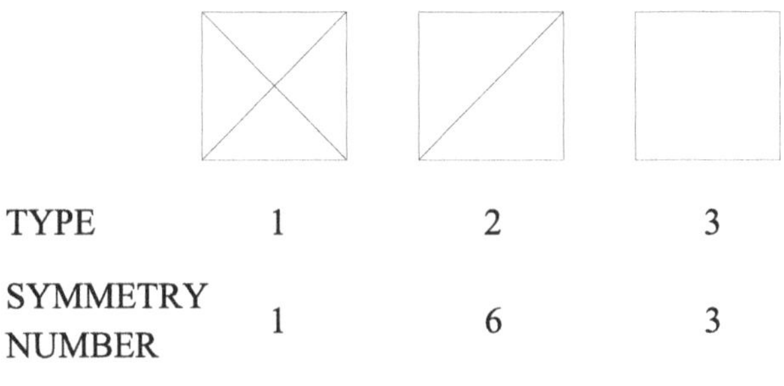

Fig. 7.1 The three unlabeled Mayer diagrams which contribute to B_4 and their symmetry numbers $s_i[4]$.

Thus

$$B_4 = -\frac{1}{8}\{S_1[4] + 6S_2[4] + 3S_3[4]\} \tag{7.10}$$

with

$$S_1[4] = \int f_{1,2}f_{2,3}f_{3,4}f_{4,1}f_{1,3}f_{2,4}d^D\mathbf{r}_1\cdots d^D\mathbf{r}_3 \tag{7.11}$$

$$S_2[4] = \int f_{1,2}f_{2,3}f_{3,4}f_{4,1}f_{1,3}d^D\mathbf{r}_1\cdots d^D\mathbf{r}_3 \tag{7.12}$$

$$S_3[4] = \int f_{1,2}f_{2,3}f_{3,4}f_{4,1}d^D\mathbf{r}_1\cdots d^D\mathbf{r}_3. \tag{7.13}$$

Now in $S_3[4]$ we insert the factors

$$1 = \tilde{f}_{1,3} - f_{1,3}, \quad \text{and} \quad 1 = \tilde{f}_{2,4} - f_{2,4} \tag{7.14}$$

and in $S_2[4]$ we insert the factor

$$1 = \tilde{f}_{2,4} - f_{2,4}, \tag{7.15}$$

and find that

$$S_1[4] + 6S_2[4] + 3S_3[4] = -2\tilde{S}_1[4] + 3\tilde{S}_3[4] \tag{7.16}$$

where

$$\tilde{S}_1[4] = \int f_{1,2}f_{2,3}f_{3,4}f_{4,1}f_{1,3}f_{2,4}d^D\mathbf{r}_1\cdots d^D\mathbf{r}_3 \tag{7.17}$$

$$\tilde{S}_3[4] = \int f_{1,2}f_{2,3}f_{3,4}\tilde{f}_{4,1}f_{1,3}\tilde{f}_{2,4}d^D\mathbf{r}_1\cdots d^D\mathbf{r}_3. \tag{7.18}$$

Hence we have reduced the number of diagrams to be considered from three to two. The computation is summarized graphically in Fig. 7.2 where the dotted lines represent the factor $\tilde{f}$.

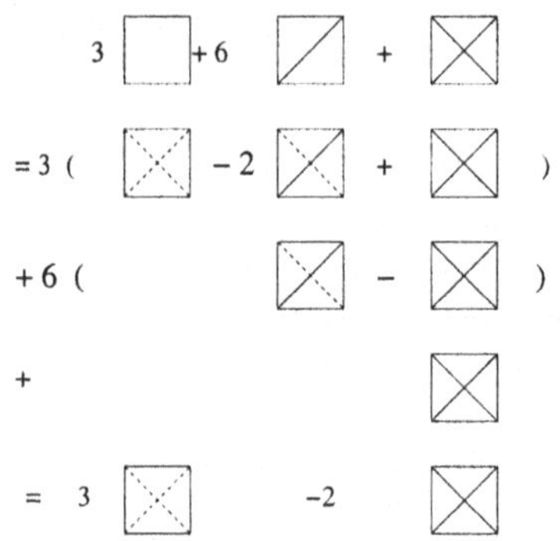

Fig. 7.2 The graphical representation of the reduction of the three Mayer diagrams for B_4 to the two Ree–Hoover diagrams. The dotted lines represent the factors $\tilde{f}$.

Thus we have the final result

$$B_4 = \frac{1}{4}\tilde{S}_1[4] - \frac{3}{8}\tilde{S}_2[4]. \tag{7.19}$$

To proceed further in a systematic fashion we follow [2] and denote the combinatorial factor in the Ree-Hoover expansion for the Ree–Hoover integral of k points

$\tilde{S}_i[k]$ as $\tilde{a}_i[k]$. We refer to $\tilde{a}_i[k]$ as the star content of the k point graph of type i. By definition

$$B_k = -\frac{k-1}{k!}\sum_i s_i(k)S_i(k) = -\frac{k-1}{k!}\sum_i \tilde{a}_i(k)s_i(k)\tilde{S}_i(k). \tag{7.20}$$

To compute $\tilde{a}_i[k]$, consider a Ree–Hoover diagram $\tilde{S}_j(k)$. This diagram is produced by expanding all those Mayer diagrams whose f functions are a subset of the f functions in $\tilde{S}_j(k)$. Denote those contributing diagrams as $S_l[j,k]$ and denote by Δf_l the number of f bonds in $\tilde{S}_j(k)$ that are not in the Mayer diagram $S_l[j,k]$. It is clear that the $S_l[j,k]$ are exactly those diagrams which can be formed by removing Δf_l f functions from the f functions in $\tilde{S}_j[k]$. Therefore we see that

$$\tilde{a}_j[k] = \sum_l (-1)^{\Delta f_l} \tag{7.21}$$

where the minus sign appears because the expansion of $(\tilde{f} - f)$ introduces a minus sign with each of the f functions.

The equation (7.21) can be expressed by the following rule [2, p.1637]:

Count the number of labeled Mayer graphs which can be formed by successively removing $0, 2, \cdots$ of the f functions from the Ree-Hoover diagram $\tilde{S}_j[k]$ and subtract from that the number of labeled Mayer graphs that can be formed from removing $1, 3, \cdots$ of the f functions from $\tilde{S}_j[k]$. The resulting number (which can be positive, negative or zero) is the star content $\tilde{a}_j[k]$.

There is a useful property of the star content $\tilde{a}_j[k]$ which helps in the computations. Namely if $\tilde{S}_j[k]$ and $\tilde{S}_j[k-1]$ have the same type of $\tilde{f}$ bonds and differ only in the f bonds then

$$\tilde{a}_j[k] = (-1)^{k-1}(k-2)\tilde{a}_j[k-1]. \tag{7.22}$$

From this we find recursively that

$$\tilde{a}_j[k] = (-1)^{k(k-1)/2-m(m-1)/2}\tilde{a}_j[m](k-2)!/(m-2)! \quad m < k \tag{7.23}$$

where m is the smallest total number of points possible for the given configuration of $\tilde{f}$ bonds.

From (7.23) we find that for the "complete Ree–Hoover star diagram" which contains no bonds $\tilde{f}$ we have $m = 2$ and $\tilde{a}_j[2] = 1$ and thus we find for the star content

$$a_{\text{star}}[k] = -(-1)^{k(k-1)/2}(k-2)! \tag{7.24}$$

For B_5 and B_6 we explicitly give all contributing Ree–Hoover graphs in Tables 7.1 and 7.2. Here we use the notation that the diagram specified by $B_k[m,i]$ has k points, and m points are connected by $\tilde{f}$ bonds. We list separately the factors $s_k[m,i]$. $\tilde{a}_k[m,i]$ and the product

$$C_k[m,i] = -\frac{k-1}{k!}s_k[m,i]\tilde{a}_k[m,i]. \tag{7.25}$$

We also specify the graphs in two different notations either by giving the $\tilde{f}$ or the f bonds.

Table 7.1 Ree–Hoover diagrams for B_5. For each diagram we give the values of the combinatorial factor $s_k[m,i]$, the star content $\tilde{a}_k[m,i]$ and the product $C_k[m,i] = -\frac{k-1}{k!}s_k[m,i]\tilde{a}_k[m,i]$ where m is the number of points conected by $\tilde{f}$ bonds.

Label	$s_k[m,i]$	$\tilde{a}_k[m,i]$	$C_k[m,i]$	$\tilde{f}$ notation	f notation
$B_5[0,1]$	1	-6	$6/30$	$\emptyset$	
$B_5[4,1]$	15	3	$-45/30$		
$B_5[5,1]$	30	-2	$60/30$		
$B_5[5,2]$	12	1	$-12/30$		
$B_5[5,3]$	10	1	$-10/30$		

The Ree–Hoover expansion has two definite advantages over the Mayer expansion. The first is that for any potential the number of Ree–Hoover graphs is smaller than the number of Mayer graphs. The second advantage is that for low dimensions certain diagrams vanish because of geometrical constraints. This effect is seen by examining $B_5[5,3]$ where in $D = 2$ (but not for $D \geq 3$) it is impossible to find any configuration which satisfies both the restrictions that $f(\mathbf{r}) = 0$ for $|\mathbf{r}| \geq \sigma$ and $\tilde{f}(\mathbf{r}) = 0$ for $|\mathbf{r}| < \sigma$.

The number of contributing diagrams for general potentials, hard disks and hard spheres is given in Table 7.3 up through B_{10}. The values of these virial coefficients will be computed in section 7.3 and 7.4.

7.2 The Tonks Gas

The Tonks gas [4] is the name given to the particularly simple case of the hard sphere gas (7.1) in one dimension. In this case the partition function is particularly easy to compute if we note that a collision of hard rods in one dimension in a volume L behaves kinematically in the same way as free particles which move in volume of $L - N\sigma$ where N is the number of rods. Thus if we replace V by $L - N\sigma$ in the equation of state of the free gas we have the equation of state of the Tonks gas

$$\frac{P(v-\sigma)}{k_B T} = 1. \tag{7.26}$$

When this is rewritten in the form of the virial expansion (7.3) we find

$$\frac{Pv}{k_B T} = \frac{1}{1-\sigma/v} = 1 + \sum_{k=1}^{\infty}(\sigma/v)^k \tag{7.27}$$

and thus

$$B_k = \sigma^{k-1} = B_2^{k-1} \tag{7.28}$$

which is to be compared with the bound (6.40)

$$|B_k|/B_2^{k-1} \leq \frac{0.21780\cdots}{k}(7.1823\cdots)^{k-1} \tag{7.29}$$

Table 7.2 Ree–Hoover diagrams for B_6. For each diagram we give the values of the combinatorial factor $s_k[m,i]$, the star content $\tilde{a}_k[m,i]$ and the product $C_k[m,i] = -\frac{k-1}{k!} s_k[m,i]\tilde{a}_k[m,i]$ where M is the number of points connected by $\tilde{f}$ bonds.

Label	$s_k[m,i]$	$\tilde{a}_k[m,i]$	$C_k[m,i]$	$\tilde{f}$ notation	f notation
$B_6[0,1]$	1	24	$-24/144$	∅	
$B_6[4,1]$	45	-12	$540/144$	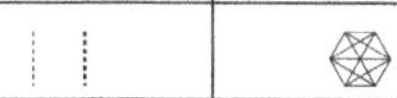	
$B_6[5,1]$	180	8	$-1440/144$	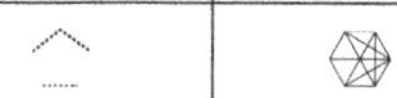	
$B_6[5,2]$	72	-4	$288/144$		
$B_6[5,3]$	60	-4	$240/144$		
$B_6[6,1]$	360	3	$-1080/144$	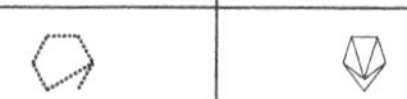	
$B_6[6,2]$	180	-2	$360/144$		
$B_6[6,3]$	60	1	$-60/144$		
$B_6[6,4]$	60	-6	$360/144$		
$B_6[6,5]$	180	-5	$900/144$		
$B_6[6,6]$	90	-4	$360/144$		
$B_6[6,7]$	45	4	$-180/144$		
$B_6[6,8]$	360	-1	$360/144$		
$B_6[6,9]$	360	-2	$720/144$		
$B_6[6,10]$	60	4	$-240/144$		
$B_6[6,11]$	15	16	$-240/144$		
$B_6[6,12]$	180	3	$-540/144$		
$B_6[6.13]$	360	1	$360/144$		
$B_6[6,14]$	90	-2	$180/144$		
$B_6[6,15]$	90	-1	$90/144$		
$B_6[6,16]$	180	1	$-180/144$		
$B_6[6,17]$	15	1	$-15/144$		
$B_6[6,18]$	10	4	$-40/144$	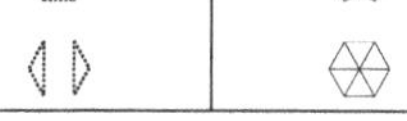	

Table 7.3 The number of Mayer and Ree–Hoover graphs which contribute to the virial coefficients up to order 10 from [17]. Some of the entries are only lower bounds because it is numerically difficult at times to distinguish between graphs which are very small and those which vanish identically.

	Mayer	R–H in general	R–H, $D=2$	R–H, $D=3$	R–H, $D=4$
B_2	1	1	1	1	1
B_3	1	1	1	1	1
B_4	3	2	2	2	2
B_5	10	5	4	5	5
B_6	56	23	15	22	23
B_7	468	171	73	161	169
B_8	7,123	2,606	$\gtrsim$ 647	>2334	>2556
B_9	194,066	81,564	$\gtrsim$ 8,417	$>60,902$	$>76,318$
B_{10}	9,743,542	4,980,756	$\gtrsim$ 110,529		

The Ree–Hoover expansion provides an exceptionally simple derivation of (7.27) because it is easily seen that in the evaluation of B_k all graphs vanish except the one in which each vertex is connected with every other vertex by the bond $f_{i,j}$. Therefore we find from (7.20) and (7.24)

$$\begin{aligned} B_k &= \frac{(-1)^{k(k-1)/2}}{k}\int dx_2\cdots dx_k \prod_{1\le i<j\le k} f_{i,j} \\ &= \frac{1}{k}\int dx_2\cdots dx_k \prod_{1\le i<j\le k} |f_{i,j}|. \end{aligned} \tag{7.30}$$

where, to obtain the last line, we have used the fact that the product contains $k(k-1)/2$ terms which are either -1 or zero. If we order the coordinates x_j as $x_1 < x_2 < \cdots < x_k$ and set $x_1 = 0$ by convention we see that (7.30) consists of $k!$ integration regions all of which contribute equally. Thus we obtain

$$\begin{aligned} B_k &= \frac{1}{k}k!\int_0^\sigma dx_2\int_{x_2}^\sigma dx_3\cdots\int_{x_{k-1}}^\sigma dx_k \\ &= \frac{1}{k}k!\frac{1}{(k-1)!}\int_0^\sigma dx_2\int_0^\sigma dx_3\cdots\int_0^\sigma dx_k = \sigma^{k-1} \end{aligned} \tag{7.31}$$

which agrees with (7.28).

It is also instructive to obtain the cluster integrals b_k. We first use (6.128) to find

$$z = \rho\exp\left(\sum_{k=1}^{\infty}\frac{k+1}{k}\sigma^k\rho^k\right). \tag{7.32}$$

Then noting that

$$\sum_{k=1}^{\infty}\frac{k+1}{k}\sigma^k\rho^k = \frac{\sigma\rho}{1-\sigma\rho} + \sum_{k=1}^{\infty}\frac{1}{k}\sigma^k\rho^k = \frac{\sigma\rho}{1-\sigma\rho} - \ln(1-\sigma\rho) \tag{7.33}$$

we have

$$z = \frac{P}{k_BT}\exp\frac{\sigma P}{k_BT}. \tag{7.34}$$

We now use Bürmann's theorem (6.217) with P as the variable z and set

$$f(P) = P, \ \ \phi(P) = \frac{P}{k_BT}\exp\frac{\sigma P}{k_BT} = z, \tag{7.35}$$

where

$$\psi(P) = k_BT\exp(-\frac{\sigma P}{k_BT}), \ \ \frac{d^{m-1}}{dP^{m-1}}\psi(P)^m|_{P=0} = k_BT(-m\sigma)^{m-1} \tag{7.36}$$

to obtain

$$\frac{P}{k_BT} = \sum_{m=1}^{\infty}\frac{z^m}{m!}(-m\sigma)^{m-1}. \tag{7.37}$$

Thus comparing with (6.5) we find

$$b_k = \frac{1}{k!}(-k\sigma)^{k-1}, \tag{7.38}$$

from which we find $b_2 = -\sigma$ and thus

$$\frac{b_k}{(2b_2)^{k-1}} = \frac{(k/2)^{k-1}}{k!} \tag{7.39}$$

which should be compared with Groeneveld's bounds (6.23).

7.3 Hard sphere virial coefficients B_2–B_4 in two and higher dimensions

We now turn to the evaluation of B_k for dimensions $D \geq 2$. The chronology of the computations is given in Table 7.4

The results for B_2, B_3 and B_4 are given in Tables 7.5 and 7.6. We will here derive the results for B_2 and B_3. The derivation of the results for B_4 is somewhat tedious and we refer the reader to the original papers.

7.3.1 Evaluation of B_2

The second virial coefficient for the hard sphere potential (7.1) is

$$B_2 = -\frac{1}{2}\int f(\mathbf{r})d^D\mathbf{r} \tag{7.40}$$

where

$$f(\mathbf{r}) = \begin{cases} -1 & \text{for } |\mathbf{r}| \leq \sigma \\ 0 & \text{otherwise} \end{cases} \tag{7.41}$$

and therefore

$$B_2 = \frac{1}{2}V_D(\sigma) \tag{7.42}$$

where $V_D(\sigma)$ is the volume of a sphere of radius σ in dimension D.

Table 7.4 Chronology of analytic computations of the hard sphere virial coefficients B_3 and B_4.

Date	Author(s)	Property
1899	Boltzmann [5]	B_3 for $D = 3$
1899	Boltzmann [5],van Laar [6]	B_4 for $D = 3$
1936	Tonks [4]	B_3 for $D = 2$
1951	Nijboer, van Hove [7]	Two center evaluation of B_4 in $D = 3$
1964	Rowlinson [8], Hemmer [9]	B_4 for $D = 2$
1982	Luban, Barum [10]	B_3 for arbitrary D
2003	Clisby, McCoy [11]	B_4 for $D = 4, 6, 8, 10, 12$
2005	Lyberg [12]	B_4 for $D = 5, 7, 9, 11$

One way to evaluate $V_D(r)$ is to note that (from dimensional considerations)

$$V_D(r) = C_D r^D \tag{7.43}$$

and that furthermore if we denote the surface area of the D-dimensional hypersphere by $\Omega_{D-1} r^{D-1}$ that

$$\Omega_{D-1} r^{D-1} = \frac{dV_D(r)}{dr} = D C_D r^{D-1}. \tag{7.44}$$

To evaluate C_D we consider the integral

$$\int_{-\infty}^{\infty} dx_1 \cdots \int_{-\infty}^{\infty} dx_D e^{-(x_1^2+\cdots+x_D^2)} = \left[\int_{-\infty}^{\infty} dx_1 e^{-x_1^2}\right]^D = \pi^{D/2}. \tag{7.45}$$

This integral is also expressed in terms of C_D as using

$$d^D\mathbf{r} = \Omega_{D-1} r^{D-1} dr \tag{7.46}$$

as

$$\int_{-\infty}^{\infty} dx_1 \cdots \int_{-\infty}^{\infty} dx_D e^{-(x_1^2+\cdots+x_D^2)} = \int_0^{\infty} dr \Omega_{D-1} r^{D-1} e^{-r^2}$$
$$= D C_D \int_0^{\infty} dr r^{D-1} e^{-r^2} = \frac{1}{2} D C_D \int_0^{\infty} dt t^{\frac{D}{2}-1} e^{-t} = C_D \Gamma(\frac{D}{2}+1). \tag{7.47}$$

Therefore

$$C_D = \frac{\pi^{D/2}}{\Gamma(\frac{D}{2}+1)} \tag{7.48}$$

and thus we have

$$V_D(r) = \frac{\pi^{D/2} r^D}{\Gamma(\frac{D}{2}+1)} \tag{7.49}$$

and

$$\Omega_{D-1} = \frac{2\pi^{D/2}}{\Gamma(\frac{D}{2})}. \tag{7.50}$$

Thus in any dimension D we have the desired answer

$$B_2 = \frac{\Omega_{D-1}a^D}{2D} = \frac{\sigma^D \pi^{D/2}}{2\Gamma(\frac{D}{2}+1)} = \begin{cases} \sigma^{2N}\pi^N/(2N!) \text{ if } D = 2N \\ \frac{\sigma^{2N+1}\pi^{N+1/2}}{2\prod_{l=0}^{N}(l+1/2)} \text{ if } D = 2N+1. \end{cases} \tag{7.51}$$

For large D we use Stirling's formula that for $z \to \infty$

$$\Gamma(z+a) \sim (2\pi)^{1/2} e^{(z+a-1/2)\ln z - z} \tag{7.52}$$

to obtain

$$B_2 \sim \sigma^D \pi^{(D-1)/2} 2^{-3/2} e^{1+D/2} e^{-\frac{1}{2}(D+1)\ln(1+D/2)}. \tag{7.53}$$

For low dimensions B_2 is explicitly given in Table 7.5.

Table 7.5 Exact and decimal results for B_2 and B_3 for $2 \le D \le 12$

D	B_2	B_3/B_2^2	B_3/B_2^2 decimal
2	$\pi\sigma^2/2$	$\frac{4}{3} - \frac{\sqrt{3}}{\pi}$	$0.78200443\cdots$
3	$2\pi\sigma^3/3$	$5/8$	0.625
4	$\pi^2\sigma^4/4$	$\frac{4}{3} - \frac{\sqrt{3}}{\pi}\frac{3}{2}$	$0.50633990\cdots$
5	$4\pi^2\sigma^5/15$	$53/2^7$	0.4140625
6	$\pi^3\sigma^6/12$	$\frac{4}{3} - \frac{\sqrt{3}}{\pi}\frac{9}{5}$	$0.34094132\cdots$
7	$8\pi^2\sigma^7/105$	$289/2^{10}$	$0.28222656\cdots$
8	$\pi^4\sigma^8/48$	$\frac{4}{3} - \frac{\sqrt{3}}{\pi}\frac{279}{140}$	$0.23461360\cdots$
9	$16\pi^4\sigma^9/48$	$6343/2^{15}$	$0.19357299\cdots$
10	$\pi^5\sigma^{10}/240$	$\frac{4}{3} - \frac{\sqrt{3}}{\pi}\frac{297}{140}$	$0.16372846\cdots$
11	$32\pi^5\sigma^{11}/10395$	$35995/2^{18}$	$0.13731002\cdots$
12	$\pi^6\sigma^{12}/1440$	$\frac{4}{3} - \frac{\sqrt{3}}{\pi}\frac{243}{110}$	$0.11539768\cdots$

7.3.2 Evaluation of B_3

The third virial coefficient is given by

$$\begin{aligned} B_3 &= -\frac{1}{3}\int d^D\mathbf{r}_1 d^D\mathbf{r}_2 f(\mathbf{r}_{1,2}) f(\mathbf{r}_2) f(\mathbf{r}_1) \\ &= -\frac{1}{3}\int d^D\mathbf{r}_1 f(\mathbf{r}_1) \int d^D\mathbf{r}_2 f(\mathbf{r}_{1,2}) f(\mathbf{r}_2) \end{aligned} \tag{7.54}$$

Since $f(\mathbf{r})$ is -1 for $|\mathbf{r}| \le a$ and 0 for $|\mathbf{r}| > a$,

$$B_3 = \frac{1}{3}\int_{|\mathbf{r}_1|<\sigma} d^D\mathbf{r}_1 \int d^D\mathbf{r}_2 f(\mathbf{r}_{1,2}) f(\mathbf{r}_2). \tag{7.55}$$

The above integrand either has the value 1 or 0, and $\mathbf{r}_1$ is now constrained to be within a ball of radius σ centered at the origin. Then $\mathbf{r}_2$ must be placed within distance a

from the origin, and also within unit distance from the point $\mathbf{r_1}$. The integral over all possible $\mathbf{r_2}$ thus corresponds to the volume of intersection of two D-dimensional spheres of radius a separated by a distance of $r_1 = |\mathbf{r_1}|$. Rescaling all distances by σ, replacing the integration variable r_1/σ by u, and denoting the volume of intersection of the two hyperspheres of unit radius by $V(u)$,

$$B_3 = \frac{\sigma^{2D}}{3}\int_{|\mathbf{u}|<1} d^D\mathbf{u}V(u) = \frac{\sigma^{2D}}{3}\Omega_{D-1}\int_0^1 du\, u^{D-1}V(u) \tag{7.56}$$

where Ω_D is given by (7.50). To calculate $V(u)$ integrate the one-dimensional overlap $L(h)$ of two line segments with radius 1 at a separation of h, over the $D-1$ dimensional hyperplane that is the perpendicular bisector of the line connecting the two hypersphere centers. This integration region is illustrated in Fig. 7.3

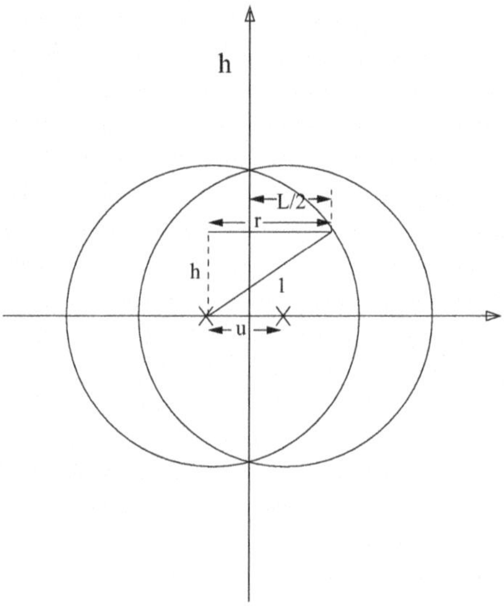

Fig. 7.3 Integration over h for B_3.

We see from Fig. 7.3 that

$$L(h) = 2r(h) - u = 2\sqrt{1-h^2} - u. \tag{7.57}$$

Therefore

$$\begin{aligned} V(u) &= \int_{|\mathbf{h}|<h_{max}} d^{D-1}\mathbf{h}L(h) = \Omega_{D-2}\int_0^{h_{max}} dh\, h^{D-2}L(h) \\ &= \Omega_{D-2}\int_0^{\sqrt{1-(\frac{u}{2})^2}} dh\, h^{D-2}(2\sqrt{1-h^2} - u) \end{aligned} \tag{7.58}$$

and hence

$$B_3 = \frac{\sigma^{2D}}{3}\Omega_{D-1}\Omega_{D-2}\int_0^1 du \int_0^{\sqrt{1-(\frac{u}{2})^2}} dh u^{D-1}h^{D-2}(2\sqrt{1-h^2}-u). \quad (7.59)$$

Substitute $\sin\phi = uh$ for h, and then reverse the order of integration.

$$\begin{aligned}
B_3 &= \frac{\sigma^{2D}}{3}\Omega_{D-1}\Omega_{D-2}\int_0^1 du \int_0^{2\arcsin(\frac{u}{2})} d\phi \cos\phi(\sin\phi)^{D-2}(\frac{2}{u}\sqrt{u^2-\sin^2\phi}-u)\\
&= \frac{\sigma^{2D}}{3}\Omega_{D-1}\Omega_{D-2}\int_0^{\frac{\pi}{3}} d\phi \cos\phi(\sin\phi)^{D-2}\int_{2\sin(\frac{\phi}{2})}^1 du(\frac{2}{u}\sqrt{u^2-\sin^2\phi}-u)\\
&= \frac{\sigma^{2D}}{3}\Omega_{D-1}\Omega_{D-2}\int_0^{\frac{\pi}{3}} d\phi \cos\phi(\sin\phi)^{D-2}\\
&\quad\times(-\frac{3}{2}-\pi\sin\phi+3\cos\phi+3\phi\sin\phi) \qquad (7.60)
\end{aligned}$$

Integrating directly the first and second terms, and integrating by parts the third and fourth terms gives:

$$\begin{aligned}
B_3 &= \frac{\sigma^{2D}}{3}\Omega_{D-1}\Omega_{D-2}\left[-\frac{3}{2}\frac{(\sin\phi)^{D-1}}{D-1}+\frac{3\cos\phi(\sin\phi)^{D-1}}{D-1}\right]_0^{\frac{\pi}{3}}\\
&\quad-\frac{\sigma^{2D}}{3}\Omega_{D-1}\Omega_{D-2}\int_0^{\frac{\pi}{3}} d\phi(\frac{-3(\sin\phi)^D}{D-1}+\frac{3(\sin\phi)^D}{D})\\
&= \frac{\sigma^{2D}\Omega_{D-1}\Omega_{D-2}}{D(D-1)}\int_0^{\frac{\pi}{3}} d\phi(\sin\phi)^D. \qquad (7.61)
\end{aligned}$$

Thus, using (7.51) we find the result of [10]:

$$B_3/B_2^2 = \frac{4\Gamma(\frac{D}{2}+1)}{\pi^{1/2}\Gamma(\frac{D}{2}+\frac{1}{2})}\int_0^{\frac{\pi}{3}} d\phi(\sin\phi)^D. \quad (7.62)$$

To complete the evaluation of the integral in (7.62) we use the identities

$$\begin{aligned}
\sin^{2N}\phi &= \frac{1}{2^{2N}}\left\{\binom{2N}{N}+2\sum_{m=1}^{N}(-1)^m\binom{2N}{N-m}\cos 2m\phi\right\} \qquad (7.63)\\
\sin^{2N+1}\phi &= \frac{1}{2^{2N}}\sum_{m=0}^{N}(-1)^m\binom{2N+1}{N-m}\sin(2m+1)\phi \qquad (7.64)
\end{aligned}$$

to find

$$\int_0^{\pi/3} d\phi \sin^{2N}\phi = \frac{1}{2^{2N}}\left\{\frac{\pi}{3}\binom{2N}{N}+2\sum_{m=1}^{N}\frac{(-1)^m}{2m}\binom{2N}{N-m}\sin\frac{2m\pi}{3}\right\} \quad (7.65)$$

$$\int_0^{\pi/3} d\phi \sin^{2N+1}\phi = \frac{1}{2^{2N}}\sum_{m=0}^{N}\frac{(-1)^m}{2m+1}\binom{2N+1}{N-m}(1-\cos\frac{2m+1}{3}\pi). \quad (7.66)$$

Thus for $D = 2N$ we use the duplication formula for the gamma function

$$\Gamma(N + \frac{1}{2}) = 2^{1-2N}\pi^{1/2}\Gamma(2N)/\Gamma(N) \tag{7.67}$$

and (7.65) in (7.62) to obtain

$$B_3/B_2^2 = \frac{4}{3} +$$
$$\frac{\sqrt{3}}{\pi}2(N!)^2\left(\sum_{\substack{m=1\\ m\equiv 1 \mod 3}}^{N}\frac{(-1)^m}{2m}\frac{1}{(N-m)!(N+m)!} - \sum_{\substack{m=1\\ m\equiv 2 \mod 3}}^{N}\frac{(-1)^m}{2m}\frac{1}{(N-m)!(N+m)!}\right) \tag{7.68}$$

and for $D = 2N + 1$ we use (7.66) and (7.67) in (7.62) to find

$$B_3/B_2^2 = \frac{[(2N+1)!]^2}{2^{4N}(N!)^2}\left\{\sum_{\substack{m=0\\ m\equiv 0,2 \mod 3}}^{N}\frac{(-1)^m}{2m+1}\frac{1}{(N-m)!(N+1+m)!}\right.$$
$$\left. +4\sum_{\substack{m=0\\ m\equiv 1 \mod 3}}^{N}\frac{(-1)^m}{2m+1}\frac{1}{(N-m)!(N+1+m)!}\right\} \tag{7.69}$$

For low dimensions B_3/B_2^2 is explicitly given in Table 7.5.

To obtain an expansion of B_3/B_2^2 for large D we expand the integrand in the integral of (7.62) about $\phi = \pi/3$ to find

$$\int_0^{\pi/3} d\phi(\sin\phi)^D \sim \int^{\pi/3} d\phi\left(\frac{\sqrt{3}}{2}\right)^D e^{-D(\frac{\pi}{3}-\phi)/\sqrt{3}} \sim \left(\frac{\sqrt{3}}{2}\right)^D\frac{\sqrt{3}}{D} \tag{7.70}$$

and thus, using the expansion as $z \to \infty$

$$\frac{\Gamma(z+\alpha)}{\Gamma(z+\beta)} \sim z^{\alpha-\beta}, \tag{7.71}$$

we obtain from (7.62) that as $D \to \infty$

$$B_3/B_2^2 \sim \frac{2\sqrt{6}}{\pi^{1/2}D^{1/2}}\left(\frac{\sqrt{3}}{2}\right)^D. \tag{7.72}$$

7.3.3 Evaluation of B_4

The evaluation of B_4 in even dimensions up through $D = 12$ was carried out in [11] using the Ree–Hoover formalism. The integrals are of the form of overlapping sphere volumes and generalize to B_4 the computations of the previous subsection for B_3. All integrals involved are elementary but their evaluation was sufficiently tedious that to obtain explicit results an algebraic computer program was used.

However, for odd dimensions the straightforward application of this method leads to elliptic integrals in intermediate steps. This is unfortunate because the final result of van Laar [6] for $D = 3$ makes no reference to elliptic integrals. For $D = 3$ the evaluation of B_4 was carried out by Nijber and van Hove [7] by means of the "two center" formalism which was invented by de Boer [39] in 1949. The evaluation of B_4 by means of this formalism has been extended by Lyberg [12] to odd dimensions up through $D = 11$. Once again only elementary integrals are encountered which, as for even dimensions, are sufficiently tedious that explicit evaluation was carried out using algebraic computer programs.

The final results for $2 \leq D \leq 12$ are given in Table 7.6. The most striking feature of B_4 is that it is positive for $2 \leq D \leq 7$ and is negative for $8 \leq D$. This feature was first seen in numerical computations by Ree and Hoover [13] in 1964.

Table 7.6 Exact and decimal results for B_4 for $2 \leq D \leq 12$. The first five digits of the decimal result are shown.

D	B_4/B_2^3 exact	decimal
2	$2 - \frac{9\sqrt{3}}{2\pi} + \frac{10}{\pi^2}$	0.53223
3	$\frac{2707}{4480} + \frac{219\sqrt{2}}{2240\pi} - \frac{4131}{4480}\frac{\arccos 1/3}{\pi}$	0.28694
4	$2 - \frac{27\sqrt{3}}{4\pi} + \frac{832}{45\pi^2}$	0.15184
5	$\frac{25315393}{32800768} + \frac{3888425\sqrt{2}}{16400384\pi} - \frac{67183425}{32800768}\frac{\arccos 1/3}{\pi}$	0.07597
6	$2 - \frac{81\sqrt{3}}{10\pi} + \frac{38848}{1575\pi^2}$	0.03336
7	$\frac{299189248759}{29059601184} + \frac{159966456685\sqrt{2}}{435894091776\pi} - \frac{292926667005}{96865353728}\frac{\arccos 1/3}{\pi}$	0.00986
8	$2 - \frac{2511\sqrt{3}}{280\pi} + \frac{17605024}{606375\pi^2}$	−0.00255
9	$\frac{2886207717678787}{2281372811001856} + \frac{2698457589952103\sqrt{2}}{7034320275046400\pi} - \frac{8656066770083523}{2281372811001856}\frac{\arccos 1/3}{\pi}$	−0.00858
10	$2 - \frac{2673\sqrt{3}}{280\pi} + \frac{49048616}{1528065\pi^2}$	−0.01096
11	$\frac{17357449486516274011}{11932824186709344256} + \frac{16554115383300832799\sqrt{2}}{29832060466773360640\pi} - \frac{52251492946866520923}{11932824186709344256}\frac{\arccos 1/3}{\pi}$	−0.01133
12	$2 - \frac{2187\sqrt{3}}{220\pi} + \frac{11565604768}{337702365\pi^2}$	−0.01067

7.4 Monte-Carlo evaluations of B_5–B_{10}

The preceding results on the second, third and fourth virial coefficients exhaust all known cases where hard sphere virial coefficients have been exactly computed. For B_5 exact analytic evaluation has only been done on selected diagrams of the Mayer type [37, 38].

To obtain further information we must turn to Monte-Carlo evaluations of the integrals on the computer. These studies were initiated by Ree and Hoover [1] in 1964 where the basic method is explained. The limits of the maximum order k for which B_k can be computed depend both on the speed and storage capacity of the computer used and on the efficiency of the algorithms for generating the diagrams and evaluating the integrals.

The chronology of numerical computations of virial coefficients is summarized in Table 7.7. The most extensive computations are those of Clisby and McCoy [17] which are given in Table 7.8.

Table 7.7 Chronology of numerical computation of the virial coefficients $B_5 - B_{10}$.

Date	Author(s)	Property
1964	Ree, Hoover [1]	B_5, B_6 for $D = 2, 3$
1967	Ree, Hoover [3]	B_7 for $D = 2, 3$
1993	van Rensburg [14]	B_8 for $D = 2, 3$
1999	Bishop, Masters, Clarke [15]	B_5, B_6 for $D = 4, 5$
2006	Clisby, McCoy [17]	B_k for $5 \leq k \leq 10$ and $2 \leq D \leq 12$

Table 7.8 Numerical results for B_k/B_2^{k-1} for $5 \leq k \leq 10$ taken from [17]. The underline indicates the position of the local minima and maxima. The error in the last digits is given in parenthesis.

D	B_5/B_2^4	B_6/B_2^5	B_7/B_2^6
2	0.33355604(1)	0.1988425(42)	0.11486728(43)
3	0.110252(1)	0.03888198(91)	0.01302354(91)
4	0.0357041(17)	0.0077359(16)	0.0014303(19)
5	0.0129551(13)	0.0009815(14)	0.0004162(19)
6	0.0075231(11)	−0.0017385(13)	0.0013066(18)
7	0.0070724(10)	−0.0035121(11)	0.0025386(16)
8	0.00743092(93)	−0.0045164(11)	0.0034149(18)
9	0.007438(6)	−0.00478(1)	
10	0.006969(5)	−0.00452(1)	
11	0.006176(4)	−0.00395(1)	
12	0.005244(4)	−0.003261(7)	

D	B_8/B_2^7	B_9/B_2^8	B_{10}/B_2^9
2	0.0649930(34)	0.0362193(35)	0.0199537(80)
3	0.0041832(11)	0.0013094(13)	0.0004035(15)
4	0.0002888(18)	0.0000441(22)	0.0000113(31)
5	−0.0001120(20)	0.0000747(26)	−0.0000492(48)
6	−0.0008950(30)	0.0006673(45)	−0.000525(16)
7	−0.0019937(28)	0.0016869(45)	−0.001514(14)
8	−0.0028624(26)	0.0025969(38)	−0.002511(13)

Further insight is obtained by the examination of the contribution of the individual diagrams. We list these in Tables 7.9 and 7.10 for B_5 and B_6 The results for the individual diagrams' contributions to B_7 in $D = 2, 3$ can be found in [3].

7.5 Hard sphere virial coefficients for $k \geq 11$

For virial coefficients B_k with $k \geq 11$ it has not yet been possible to evaluate all diagrams. We conclude our study here by examining, for $k \geq 11$, two particular diagrams

Table 7.9 The contributions of diagrams $B_5[m, i]$ given in Table 7.1 to the fifth virial coefficient for hard disks and spheres. The values given are the product of the integral and the combinatoric factor $C_k[m, i]$. The error in the last digits is given in parenthesis.

	disks	spheres	$D = 4$	$D = 5$	$D = 6$
B_5/B_2^4	0.3336	0.1103	0.03565(5)	0.01297(1)	0.007528(8)
$B_5[0,1]/B_2^4$	0.3618	0.1422	0.059015(9)	0.025442(1)	0.0112852(7)
$B_5[4,1]/B_2^4$	−0.0266	−0.0314	−0.02650(2)	−0.019184(5)	−0.012899(4)
$B_5[5,1]/B_2^4$	−0.0102	−0.0165	−0.01762(4)	−0.015511(4)	−0.012351(3)
$B_5[5,2]/B_2^4$	0.0086	0.0162	0.02131(2)	0.022980(7)	0.022332(6)
$B_5[5,3]/B_2^4$	0	−0.0002	−0.0005498(5)	−0.0007622(3)	−0.0008395(4)

Table 7.10 The contributions of the diagrams $B_6[m, i]$ given in Table 7.2 to the sixth virial coefficient for $D = 2, 3, 5, 6, 7$. The values given are the product of the integral and the combinatoric factor $C_k[m, i]$. The error in the last digits is given in parenthesis.

	disks	spheres	$D = 5$	$D = 6$	$D = 7$
B_6/B_2^5	0.1994	0.0386	0.00102(8)	−0.00176(2)	−0.00352(2)
$B_6[0,1]/B_2^5$	0.2292	0.0588	0.0048248(9)	0.0014771(1)	0.00046725(4)
$B_6[4,1]/B_2^5$	−0.0273	−0.0212	−0.00569(2)	−0.002600(3)	−0.001145(1)
$B_6[5,1]/B_2^5$	−0.0191	−0.0187	−0.00719(1)	−0.003800(4)	−0.001902(2)
$B_6[5,2]/B_2^5$	0.0090	0.0099	0.00498(2)	0.003038(4)	0.001752(2)
$B_6[5,3]/B_2^5$	0	−0.0002	−0.000244(1)	−0.0001759(3)	−0.0001124(2)
$B_6[6,1]/B_2^5$	0.0088	0.0132	0.01029(4)	0.00735(1)	0.004866(6)
$B_6[6,2]/B_2^5$	0.0077	0.0121	0.01064(4)	0.008204(8)	0.005887(5)
$B_6[6,3]/B_2^5$	−0.0051	−0.0109	−0.01702(5)	−0.01693(2)	−0.01540(2)
$B_6[6,4]/B_2^5$	−0.0019	−0.0027	−0.001520(4)	−0.0009332(8)	−0.0005349(5)
$B_6[6,5]/B_2^5$	−0.0010	−0.0022	−0.001559(9)	−0.001001(2)	−0.000589(1)
$B_6[6,6]/B_2^5$	−0.0009	−0.0011	−0.000553(1)	−0.0003335(3)	−0.0001883(1)
$B_6[6,7]/B_2^5$	−0.0005	−0.0007	−0.000437(1)	−0.0002948(3)	−0.0001858(2)
$B_6[6,8]/B_2^5$	0.0004	0.0008	0.000712(2)	0.0004958(8)	0.0003158(4)
$B_6[6,9]/B_2^5$	0.0001	0.0012	0.00256(2)	0.002356(7)	0.001912(4)
$B_6[6,10]/B_2^5$	0.0000	0.0002	0.0003086	0.0002481(2)	0.0001778(1)
$B_6[6,11]/B_2^5$	−0.0000	−0.0003	−0.0002596(7)	−0.0001587(1)	−0.00008782(7)
$B_6[6,12]/B_2^5$	0	−0.0002	−0.000472(2)	−0.0003931(6)	−0.0002848(4)
$B_6[6,13]/B_2^5$	0	0.0002	0.000318(4)	0.000285(1)	0.0002231(6)
$B_6[6,14]/B_2^5$	0	−0.0000	−0.0000991(5)	−0.0001007(2)	−0.0000868(2)
$B_6[6,15]/B_2^5$	0	0.0000	0.000046(2)	0.0000486(9)	0.0000447(4)
$B_6[6,16]/B_2^5$	0	0.0004	0.00138(1)	0.001463(2)	0.001358(2)
$B_6[6,17]/B_2^5$	0	−0?	$-2.00(1) \times 10^{-6}$	$-3.176(6) \times 10^{-6}$	$-3.774(8) \times 10^{-6}$
$B_6[6,18]/B_2^5$	0	0	2.59×10^{-7}	$3.696(8) \times 10^{-7}$	$3.944(9) \times 10^{-7}$

which in a sense represent two extreme cases.

1) For $D = 2$, for both B_5 and B_6, the largest diagram (by almost a factor of 10) is the complete star $B_5[0, 1]$ and $B_6[0, 1]$ where each point is connected to every other point by a bond f. Because the bonds f vanish when $|\mathbf{r}| > \sigma$ the points of this diagram are as close together as possible. We call such a diagram "close packed." The contribution of this close packed diagram to B_k is positive for for all k.

2) For large D and any k the largest diagram is the "ring diagram" where there are only k bonds of type f which join the k together in a ring where each particle is connected to only two other particles and all other bonds are of type $\tilde{f}$. For B_5 and B_6 these diagrams are given in Tables 7.1 and 7.2 as $B_5[5, 2]$ and $B_6[6, 3]$, and their

Table 7.11 The ratio of the ring to the star diagrams $\mathsf{R}/\emptyset$. The error in the last digits is given in parenthesis.

Order	$D=2$	$D=3$	$D=4$	$D=5$
4	−0.02986(16)	−0.09417(35)	−0.19636(79)	−0.34048(67)
5	0.02375(16)	0.11433(66)	0.3618(32)	0.9111(55)
6	−0.02296(37)	−0.1844(20)	−0.912(10)	−3.471(48)
7	0.02373(46)	0.3368(69)	2.915(56)	$1.877(52) \times 10^1$
8	−0.02773(56)	−0.732(20)	$-1.041(29) \times 10^1$	$-1.000(28) \times 10^2$
9	0.03496(76)	1.673(47)	$4.36(12) \times 10^1$	$7.06(32) \times 10^2$
10	−0.0428(10)	−4.07(11)	$-1.767(67) \times 10^2$	$-5.14(78) \times 10^3$
11	0.0559(15)	$1.030(29) \times 10^1$	$8.63(83) \times 10^2$	
12	−0.0759(21)	$-2.598(73) \times 10^1$	$-3.04(76) \times 10^3$	
13	0.1024(29)	$6.01(49) \times 10^1$		
14	−0.1330(49)	$-1.60(30) \times 10^2$		
15	0.1819(98)			
16	−0.237(20)			
17	0.387(48)			
18	−0.53(11)			

values are given in Tables 7.9 and 7.10. In these diagrams the particles are as far apart as they can possibly be. We call such a diagram "loosely packed." The contribution of these diagrams to B_k has the sign $(-1)^{k-1}$.

Some insight into the sign of B_k may be gained by studying the ratio of the ring to the star diagram as a function of k and D. This has been done numerically [16] and the results for the ratio are given in Table 7.11. There we see that for $D = 2$ the star dominates the ring for k as large as 18. This is some indication that for $D = 2$ the virial coefficient B_k can be expected to be positive for at least $k \leq 18$. However, for $D = 3$ the ring is slightly larger than the star for $k = 9$ and at $D = 4$ the ring dominates for $k = 8$. The dominance of the ring over the star diagram seems to be a (weak) necessary condition for B_k to have a possible negative sign. Using this as a criterion it may be possible to see a negative sign for B_k in $D = 3, 4$ for k not much larger than 9 but there is no evidence that, for $D = 2$, B_k can be negative for $k \leq 18$.

7.6 Radius of convergence and approximate equations of state

The most important property of the virial coefficients B_k is not their actual values for k less than some finite number but rather their asymptotic behavior as $k \to \infty$ because it is this asymptotic behavior which determines the radius of convergence of the series. In the preceding sections we have examined two properties of the large k behavior of B_k of hard spheres in D dimensions: the identical vanishing of Ree–Hoover diagrams due to geometric reasons seen in Table 7.3 and the loose packed dominance seen in Table 7.11. This therefore suggests the existence of two criteria which need to be fulfilled in order for B_k to be in the asymptotic large k regime:

Criterion 1
The number of nonzero Ree–Hoover diagrams has approached its large k behavior.
Criterion 2
The loose packed diagrams with the number of $\tilde{f}$ bonds near their maximum value numerically dominate B_k.

We see from Table 7.3 for $k = 10$ that criterion 1 is only completely fulfilled for $D = 2$ and is not fulfilled at all for $D \geq 5$.

We see from Table 7.11 that for $D = 3$ and $k \geq 12$ that criterion 2 is well satisfied and that as D increases the criterion is satisfied for smaller values of k. However, for $D = 2$ the criterion is not satisfied even for k as large as 18.

We thus conclude that there is no dimension in which both of these criteria are fully satisfied although for $D = 3$ and $D = 4$ it is possible that both criteria could hold for some moderate value of k of the order of 12 to 14. It is thus probable that the true radius of convergence cannot be determined from the virial coefficients up to B_{10} given in Tables 7.5, 7.6 and 7.8.

Nevertheless it is of interest to see what results if we attempt to study the radius of convergence by means of the ratio method, and accordingly the ratios $B_k\rho_{cp}/B_{k-1}$ are plotted in Figs. 7.4–7.5 where the ratios have been normalized to the closest packed densities given in Table 7.12 instead of to the second virial coefficient B_2. In this table we also give the lower bound $[(1+e)2B_2]^{-1}$ (6.34) on the density of the fluid phase and the fluid and solid ends of the first order phase transition which are seen in the computer experiments to be presented in chapter 8.

Table 7.12 Values for hard spheres of diameter σ in dimensions $D = 2, \cdots, 8$ of B_2, the density ρ_{cp} of the known densest packed lattices from [40, table 1.2], the fluid ρ_f and solid ρ_s ends of the computer-determined first order transition and the lower bound $\rho_{LB} = [(1+e)2B_2]^{-1}$ (6.34) on the density where melting can occur.

D	B_2	ρ_{cp}	$B_2\rho_{cp} = 2^{D-1}\eta_{cp}$	ρ_f/ρ_{cp}	ρ_s/ρ_{cp}	ρ_{LB}/ρ_{cp}
2	$\frac{\pi\sigma^2}{2}$	$\frac{2}{\sqrt{3}\sigma^2}$	$1.8137\cdots$	0.78	0.81	$0.0744\cdots$
3	$\frac{2\pi\sigma^3}{3}$	$\frac{\sqrt{2}}{\sigma^3}$	$2.9619\cdots$	0.66	0.75	$0.0454\cdots$
4	$\frac{\pi^2\sigma^4}{4}$	$\frac{2}{\sigma^4}$	$4.9348\cdots$	0.50	0.68	$0.0272\cdots$
5	$\frac{4\pi^2\sigma^5}{15}$	$\frac{2\sqrt{2}}{\sigma^5}$	$7.4441\cdots$	0.41	0.62	$0.0180\cdots$
6	$\frac{\pi^3\sigma^6}{12}$	$\frac{8}{\sqrt{3}\sigma^6}$	$11.9343\cdots$			$0.0112\cdots$
7	$\frac{8\pi^3\sigma^7}{105}$	$\frac{8}{\sigma^7}$	$18.8990\cdots$			$0.0071\cdots$
8	$\frac{\pi^4\sigma^8}{48}$	$\frac{16}{\sigma^8}$	$32.4696\cdots$			$0.0041\cdots$

In Fig. 7.4 we plot the ratios for $D = 2, 3$ and see that the ratios appear to smoothly extrapolate to a radius of convergence at a value of the density which is greater than the close packed density ρ_{cp}.

In Fig. 7.5 we plot the ratios for $D = 4$. Now the plots are not smooth but have oscillations which become increasingly strong as k increases. No useful extrapolation of these ratios is possible and it may be that negative virial coefficients will appear for higher k.

For $D \geq 5$ there are oscillations of sign of the virial coefficients which, if this occurs in the true asymptotic behavior, indicates that the radius of convergence is not on the real axis.

There is no dimension for which a study of the ratio plots gives evidence that

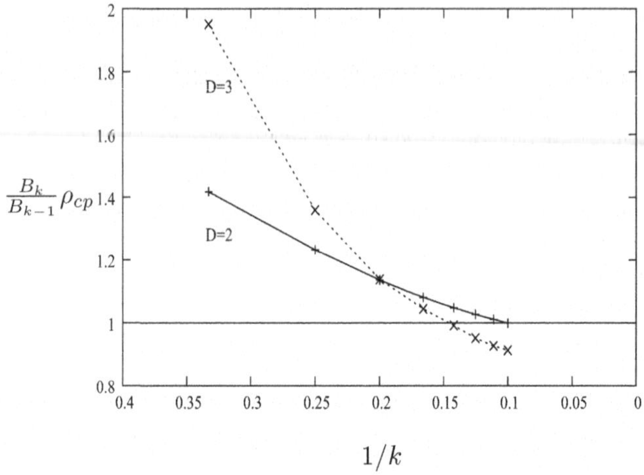

Fig. 7.4 Ratio plot for hard sphere virial coefficients of Tables 7.5, 7.6, 7.8 in dimensions $D = 2, 3$.

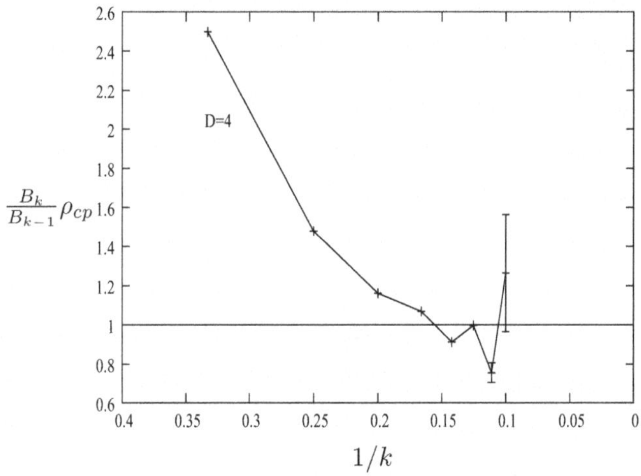

Fig. 7.5 Ratio plot for hard sphere virial coefficients of Tables 7.5, 7.6, 7.8 in dimension $D = 4$.

the asymptotic regime of large k has been reached and thus there is little point in attempting to use the first 10 terms in the virial expansion to study the analytic properties of the equation of state. Nevertheless there have been many attempts to fit the first eight terms of the virial expansion to an approximate equation of state. It is conventional to express these approximate equations of state in terms of the packing fraction

$$\eta = \frac{B_2}{4}\rho = \frac{\pi\sigma^3}{6}\rho. \tag{7.73}$$

The packing fraction equals one at the density that would obtain if the hard spheres fill all space which is an unphysical density above ρ_{cp}.

Some of the approximate equations of state for hard spheres have singularities at $\eta = 1$:

$$\frac{Pv}{k_BT} = \frac{1+\eta+\eta^2}{(1-\eta)^3} \qquad [18\text{–}20] \tag{7.74}$$

$$= \frac{1+2\eta+3\eta^3}{(1-\eta)^2} \quad [18\text{–}20] \tag{7.75}$$

$$= \frac{1}{(1-\eta)^4} \qquad [21] \tag{7.76}$$

$$= \frac{1+\eta+\eta^2-\eta^3}{(1-\eta)^3} \quad [22], \tag{7.77}$$

and some have multiple poles at η_{cp}

$$\frac{Pv}{k_BT} = 1 + \frac{3\eta}{\eta_{cp}-\eta} + \frac{5.372804\eta^2}{(1-1.041709\eta)^2} + \sum_{n=1}^{6} b_n\eta^n \quad [23] \tag{7.78}$$

where the b_n are given in Table 4 of [23].

Some have simple poles at complex values of η determined by Padé approximates determined from the eight virial coefficients of [14] such as

$$\frac{Pv}{k_BT} = 1 + 4\eta\left(\frac{1+0.30184\eta+0.27726\eta^2}{1-2.198\eta+1.214\eta^2}\right) \tag{7.79}$$

which has poles at

$$\eta = 0.9052 \pm 0.0648i = \eta_{cp}(1.222 \pm 0.0875i). \tag{7.80}$$

Some have branch point singularities at what is called the "random close packed density" η_{rcp} determined from a D-log Padé approximate such as [24]:

$$\frac{Pv}{k_BT} = 1 + 4\eta\frac{1+2.5\eta+4.5904\eta^2+4.515439\eta^3}{[1-(\eta/\eta_{rcp})^3]^{.67802}} \tag{7.81}$$

with $\eta_{rcp} = 0.6435$, or the form

$$\frac{Pv}{k_BT} = 1 + 4\eta\frac{\sum_n c_n\eta^n}{(1-\eta/\eta_{rcp})^s} \tag{7.82}$$

where in [25] $\eta_{rcp} = 0.6435$, $s = .84$ and the c_n obtained from [25, Table II] and where in [26] a wide range of values for η_{rcp} and s are obtained depending on which form of the D-log Padé is used.

It is clear from these many forms that the first eight virial coefficients are far from estimating either the true radius of convergence or the nature of the leading singularity of the virial series for hard spheres.

7.7 Parallel hard squares, parallel hard cubes and hard hexagons on a lattice

The derivation of the Mayer expansion of the virial coefficients presented in chapter 6 is valid not only for spherically symmetric potentials but also for potentials with a directional dependence as long as the orientation of the molecules is fixed and the integrations are carried out only over the center of mass coordinates. One such example is hard squares or cubes whose edges are all fixed to be parallel. The integrals for this problem are dramatically simpler than for hard spheres. The first seven virial coefficients were computed analytically in 1962 [27] (five years before B_7 was numerically evaluated for hard spheres [3]). We give the results in Table 7.13.

Table 7.13 Virial coefficients B_k for for parallel hard squares and cubes (with $\sigma = 1$) from [27].

k	$B_k$for squares	$B_k$for cubes
1	1	1
2	2	4
3	3	9
4	$\frac{11}{3}$	$\frac{34}{3}$
5	$\frac{65}{18}$	$\frac{455}{144}$
6	$\frac{121}{40}$	$-\frac{2039}{108}$
7	$\frac{17,827}{10,800}$	$-\frac{167,149,119}{3,888,000}$

Table 7.14 Virial coefficients kB_k for the hard square, simple cubic, fcc and bcc lattice gas with nearest neighbor exclusion from [28,30].

k	kB_k squares	kB_k cubic	kB_k fcc	kB_k bcc
1	1	1	1	1
2	5	7	13	9
3	13	19	85	25
4	17	7	385	−87
5	−19	−149	1,261	−1,070
6	−175	−833	1,633	−3,910
7	−503	−3,569	−167,154	3,613
8	−695	−14,553		100,977
9	373	−53,405		308,041
10	633	−165,413		−1,828,761
11	−2,007	−4,400,021		−21,205,645
12	−58,207			
13	−237,691			

Furthermore the parallel hard squares (cubes) can be restricted to lie on a square (cubical lattice). In this case the evaluation of the integrals reduces to a problem of pure combinatorics, and substantially more terms can be obtained.

The case of the lattice gas of hard squares was studied in [28] and in three dimensions the simple cubic, fcc and bcc lattice were studied in [30]. The results are given in Table 7.14. Note that for hard squares, terms up to order 42 can be obtained from the work of [29]. Even more can be computed for hard hexagons on the triangular lattice

Table 7.15 Virial coefficients B_k for the hard hexagon lattice gas from table 11 of [33] where $kB_k = -(k-1)\beta_{k-1}$.

k	kB_k
1	1
2	7
3	31
4	115
5	391
6	1237
7	3529
8	8155
9	8311
10	−6, 2543
11	−612, 809
12	−3, 759, 551
13	−19, 472, 387
14	−91, 607, 873
15	−402, 535, 529
16	−1, 671, 753, 125
17	−6, 585, 730, 265
18	−24, 544, 637, 087
19	−85, 671, 502, 739
20	−273, 505, 952, 615
21	−753, 160, 139, 729
22	−1, 456, 884, 883, 535
23	860, 351, 408, 035
24	30, 699, 547, 973, 425
25	288, 155, 349, 143, 341

which has the remarkable property discovered by Baxter [31,32] that it can be solved exactly for all densities. This exact solution will be discussed in chapter 15 but it is useful here to note that the radius of convergence of the low density virial expansion is determined by a singularity in the complex ρ plane at

$$\begin{aligned}\rho &= \frac{1}{2} - \frac{1}{20}\sqrt{10}[(4\sqrt{10} - 5\sqrt{5} - 4\sqrt{2} + 7)^{1/2} \pm i(4\sqrt{10} - 5\sqrt{5} + 4\sqrt{2} - 7)^{1/2}] \\ &= 0.234862\cdots \pm i0.0560413\cdots \end{aligned} \tag{7.83}$$

The first 25 virial coefficients have been determined by Joyce [33] and are given in Table 7.15.

There are distinct differences in the patterns of signs of the B_k for the hard spheres of Table 7.8 and the hard squares, cubes and hexagons of Tables 7.13–7.15.

For the hard hexagons of Table 7.15 the signs oscillate with a rather large period and because the location of the leading singularity is known exactly (7.83) to lie in the complex plane near the real axis this oscillation will continue indefinitely as $k \to \infty$.

For hard squares the signs of the B_k in Table 7.14 also oscillate and because of the similarity with the hard hexagons of Table 7.15 it is very natural to conjecture that this oscillation also contributes for all k and that the leading singularity is in the complex plane. The cubic and bcc lattice also have this property.

For the parallel hard cubes of Table 7.13 it is tempting to interpret the negative signs of B_6 and B_7 as the beginning of an oscillation that will give a leading singularity in the complex plane.

However, for the parallel hard square results in Table 7.13 and the hard sphere results in Tables 7.5 and 7.6 the interpretation is less clear. The signs of the parallel hard square virial coefficients and the hard spheres in $D = 2, 3, 4$ are always positive which is consistent with a leading singularity on positive real axis while the $(-1)^{k-1}$ oscillation of sign for hard spheres with $D \geq 5$, if carried out to infinity, would put the leading singularity on the negative real axis.

7.8 Convex nonspherical hard particles

We finally note that for potentials that are not radially symmetric the more physically relevant problem will integrate over the orientation of the molecules as well as their center of mass.

The first computation of this kind was done by Onsager in 1949 [34] who found that the second virial coefficient for a cylinder of diameter d and length l which is capped on both ends by a hemisphere of diameter d is

$$B_2 = \frac{2\pi}{2}d^3 + \frac{5\pi}{4}d^2 l. \tag{7.84}$$

Boublik [35] showed in two dimensions that in general for hard bodies with proper area A and perimeter s the second virial coefficient is

$$B_2 = A + \frac{s^2}{\pi}. \tag{7.85}$$

The cases of ellipses with semi-major axes $l_1^e \geq l_2^e$, rectangles with sides l_1^r and l_2^r and needles (the limiting case of rectangles where $l_2^r \to 0$) of length l have been studied in [36]. The second virial coefficients are (with $A = 1$ for ellipses and rectangles):

ellipses

$$B_2^e = 1 + \frac{4k}{\pi^2}\{E[(1-k^{-2})^{1/2}]\}^2 \text{ with } k = l_1^e/l_2^e \geq 1 \tag{7.86}$$

where

$$E(k) = \int_0^{\pi/2} d\theta (1 - k^2 \sin^2\theta)^{1/2} \tag{7.87}$$

rectangles

$$B_2^r = 1 + \frac{(k+1)^2}{\pi k} \text{ with } k = l_1^r / l_2^r \tag{7.88}$$

needles

$$B_2^n = \frac{1}{\pi}. \tag{7.89}$$

For needles the third virial coefficient has been analytically evaluated as [36]:

$$\begin{aligned} B_3/B_2^2 &= \frac{1}{648}\left[-366\ln 2 + 96\ln(1+\sqrt{2}/2) - 9\pi^2 - 66\pi - 160\sqrt{2} + 1058\right] \\ &= 0.51420248\cdots \end{aligned} \tag{7.90}$$

For ellipses and rectangles the virial coefficients B_3 and B_4 have been evaluated by Monte-Carlo [36] and are given in Table 7.16. However, it should be noted, that many, if not most, molecules do not have a convex shape and that there is no example of a nonconvex body for which even the second virial coefficient has been computed.

Table 7.16 Virial coefficients B_3/B_2^2 and B_4/B_2^3 for hard ellipses and rectangles as a function of the aspect ratio from [36].

	B_3/B_2^2		B_4/B_2^3	
k	ellipses	rectangles	ellipses	rectangles
1	0.7820···	0.770(2)	0.5322···	0.506(6)
2	0.750(1)	0.750(1)	0.471(2)	0.467(5)
4	0.677(1)	0.696(3)	0.323(2)	0.363(15)
5	0.651(1)	0.675(3)	0.272(2)	0.320(20)
6	0.631(1)	0.658(4)	0.230(2)	0.290(30)
15		0.588(9)		0.112(80)
∞		0.5142···		−0.031(7)

7.9 Open questions

We concluded the preceding chapters on order and scaling theory by presenting a selection of results which physical intuition suggests should hold but for which no proof has yet been found. We referred to these as "missing theorems." The study of hard particles in this present chapter also has revealed a large number of places where desired results are missing. However, in contrast to the cases of order and scaling, here we have little or no intuition as to what results are to be expected and consequently we here discuss "open questions" in place of "missing theorems".

The most important property of the virial series we would like to know is the location of the leading singularity and the radius of convergence. If the virial coefficients are all positive the leading singularity will be on the positive density axis. Unfortunately we found in section 7.6 that while the the virial coefficients for hard spheres in $D = 2, 3$ up through B_{10} are all positive they extrapolate to give a radius of convergence which is greater than the close packed density. Thus it is plausible that B_k is not in an asymptotic large k regime for $k = 10$.

Table 7.17 Open questions for virial expansions of hard particles.

1.	Are there negative B_k for hard discs and spheres?
2.	What fraction of Ree–Hoover diagrams for B_k vanish identically for large k?
3.	What is the analytic expression for B_4 for arbitrary D?
4.	Can B_k be evaluated analytically for hard discs and spheres for $k \geq 5$?
5.	What is the true radius of convergence for hard spheres in D dimensions?

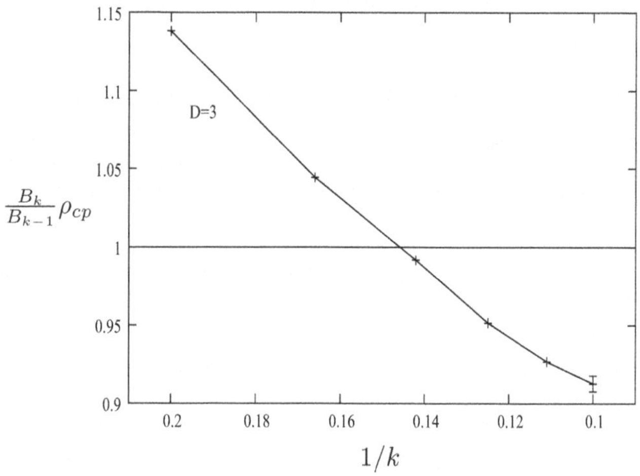

Fig. 7.6 Ratio plot on an expanded scale for hard sphere virial coefficients in dimensions $D = 3$. The point at $k = 7$ is very slightly above the line joining $k = 6$ and $k = 8$.

Indeed, because negative B_k occur for hard spheres for $D \geq 5$, for parallel hard cubes and for the lattice gases with nearest neighbor exclusion in both $D = 2$ and $D = 3$ it seems fair to suggest that negative virial coefficients generically occur for hard particles and that it will require some special property in order to make the coefficients all positive. However, we see in Table 7.15 that, for hard hexagons, the first negative coefficient occurred only at B_{10} so that a large value of k may be necessary to see a negative B_k in $D = 2$. However, for $D = 3$ a close inspection of the ratio plot in Fig 7.6 shows that there is a slight deviation from monotonicity in the slopes and this may indicate the beginning of oscillations such as are seen in the ratios for $D = 4$ of Fig. 7.5 which can build up to eventually give changes in sign of B_k.

It is furthermore to be noted that it is not particularly satisfactory that for the analytic evaluation of B_4 we needed to resort to the use of algebraic computer programs. This prevents us from determining an analytic formula for arbitrary D.

Finally the question can be raised as to whether it is possible in principle to analytically evaluate all B_k for hard spheres in D dimensions in terms of simple numbers such as $\sqrt{3}/\pi$ and $\pi^{-1}\arccos(1/3)$ as was the case for B_4. For B_5 seven of the ten Mayer diagrams given in appendix A of chapter 6 have been evaluated for $D = 3$ in [37] and [38] and are given in Table 7.18. Because hard sphere virial coefficients are geometrical objects related to the overlap of spheres it is appealing to conjecture

Table 7.18 The known evaluations of the 10 Mayer diagrams (in the notation appendix A of chapter 6) which contribute to the fifth virial coefficient for hard spheres. The results for diagrams 1–5 are from [37]. The results for diagrams 6 and 7 are from [38].

diagram	contribution
$B_{5;1}/B_2^4$	$-\frac{40949}{10752}$
$B_{5;2}/B_2^4$	$\frac{68419}{26880}$
$B_{5;3}/B_2^4$	$\frac{82}{35}$
$B_{5;4}/B_2^4$	$-\frac{34133}{17920}$
$B_{5;5}/B_2^4$	$-\frac{73491}{35840}$
$B_{5;6}/B_2^4$	$-\frac{18583}{5376}+\frac{33291}{9800}\frac{\sqrt{3}}{\pi}$
$B_{5;7}/B_2^4$	$\frac{35117}{6720}+\frac{1458339}{627200}\frac{\sqrt{2}}{\pi}-\frac{114201}{35840\pi}\arccos(1/3)$ $+\frac{3609}{280\pi}\arcsin(1/\sqrt{3})-\frac{369}{35\pi}\arcsin(5\sqrt{3}/9)$ $-\frac{33291}{9800}\frac{\sqrt{3}}{\pi}$
$B_{5;8}/B_2^4$	unknown
$B_{5;9}/B_2^4$	unknown
$B_{5;10}/B_2^4$	unknown

that such analytic evaluations are possible for all diagrams. In particular it should be remarked that a crucial step in the Hales proof of the fcc densest packing of hard spheres discussed in chapter 4 is the reduction of the problem to a *finite* number of configurations. It would be of great importance if this finitization has relevance to the computation of the virial coefficients.

References

[1] F.H. Ree and W.G. Hoover, Fifth and sixth virial coefficients for hard spheres and hard disks, J. Chem. Phys. 40 (1964) 939–950.

[2] F.H. Ree and W.G. Hoover, Reformulation of the virial series for classical fluids, J. Chem. Phys. 41 (1964) 1635–1645.

[3] F.H. Ree and W.G. Hoover, Seventh virial coefficients for hard spheres and hard disks, J. Chem. Phys. 46 (1967) 4181–4196.

[4] L. Tonks, The complete equation of state of one, two and three dimensional gases of hard elastic spheres, Phys. Rev. 50 (1936) 955-963.

[5] L. Boltzmann, Verslag. Gewonee Vergadering Afd. Natuurk. Nederlandse Akad. Wtensch. 7 (1899) 484.

[6] J.J. van Laar, Berekening der tweede correctie op de grootheid b der toestandsvergliijking vab der Waals, Amsterdam Akad. Versl. 7 (1899) 350–364.

[7] B.R.A. Nijboer and L. van Hove, Radial distribution function of a gas of hard spheres and the superposition approximation, Phys. Rev. 85 (1952) 777–783.

[8] J.S. Rowlinson, The virial expansion in two dimensions, Mol. Phys. 7 (1964) 593–594.

[9] P.C. Hemmer, Virial coefficients for the hard-core gas in two dimensions, J. Chem. Phys. 42 (1964) 1116–1118.

[10] M. Luban and A. Barum, Third and fourth virial coefficients of hard hyperspheres in arbitrary dimensionality, J. Chem. Phys. 76 (1092) 3233–3241.

[11] N. Clisby and B.M. McCoy, Analytic calculation of B_4 for hard spheres in even dimensions, J. Stat. Phys. 114 (2004) 1343–1360.

[12] I. Lyberg, The fourth virial coefficient of a fluid of hard spheres in odd dimensions, J. Stat. Phys. 119 (2005) 747–764.

[13] F.R. Ree and W.G. Hoover, On the signs of the hard sphere virial coefficients, J. Chem. Phys. 41 (1964) 1635–1636.

[14] E.J. Janse van Rensburg, Virial coefficients for hard discs and hard spheres, J. Phys. A 26 (1993) 4805–4818.

[15] M. Bishop, A. Masters and J.H.R. Clarke, Equation of state of hard and Weeks–Chandler–Anderson hyperspheres in four and five dimensions, J. Chem. Phys. 110 (1999) 11449–11453.

[16] N. Clisby and B.M. McCoy, Negative virial coefficients and the dominance of loosely packed diagrams for D-dimensional hard spheres, J. Stat. Phys. 114 (2004) 1361–1393.

[17] N. Clisby and B.M. McCoy, Ninth and tenth order virial virial coefficients for hard spheres in D dimensions, J. Stat. Phys. 122 (2006) 15–57.

[18] E. Thiele, Equation of state for hard spheres, J. Chem. Phys. 39 (1963) 474–479.

[19] M.S. Wertheim, Exact solution of the Percus–Yevick equation for hard spheres, Phys. Rev. Letts. 10 (1963) 321–323.
[20] M.S. Wertheim, Analytic solution of the Percus-Yevick equation, J. Math. Phys. 5 (1964) 643–651.
[21] E.A. Guggenheim. Variations on van der Waal equation of state for high densities, Mol. Phys. 9 (1965) 199–200.
[22] N.F. Carnahan and K.F. Starling, Equation of state for nonattracting rigid spheres, J. Chem. Phys. 51 (1969) 635–636.
[23] R. Hoeste and W.D. Dael, Equation of state for hard sphere and hard disc systems, J. Chem. Soc. Faraday Trans. 80 (1984) 477–488.
[24] D. Ma and G. Ahmadi, An equation of state for dense rigid sphere gases, J. Chem. Phys. 84 (1986) 3449–3450.
[25] Y. Song, R.M. Stratt and A.E. Mason, The equation of state of hard spheres and the approach to random closest packing, J. Chem. Phys. 88 (1988) 1126–1133.
[26] S. Jasty, M. Al-Naghy and M. de Llano, Critical exponent for glassy packing of rigid hard spheres and disks, Phys. Rev. A35 (1987) 1376–1381.
[27] W.G. Hoover and A.G. De Rocco, Sixth and seventh virial coefficients for the parallel hard cube model, J. Chem. Phys. 36 (1962) 3141–3162.
[28] D.S. Gaunt and M.E. Fisher, Hard-sphere lattice gases I. Plane-square lattice, J. Chem. Phys. 43 (1965) 2840–2863.
[29] R.J. Baxter, I.G. Enting and S.K. Tsang, Hard square lattice gas, J. Stat. Phys. 22 (1980) 465–489.
[30] D.S. Gaunt, Hard-sphere lattice gases II. Plane triangular and three-dimensional lattices, J. Chem. Phys. 46 (1967) 3237–3259.
[31] R.J. Baxter, Hard hexagons: exact solution, J. Phys. A13 (1980) L61–L70.
[32] R.J. Baxter, *Exactly solved models in statistical mechanics* London (Academic Press 1982).
[33] G.S. Joyce, On the hard-hexagon model and the theory of modular functions, Phil. Trans. R. Soc. Lond. A325 (1988) 643–702.
[34] L. Onsager, The effects of shape on the interaction of colloidal particles, Ann. N.Y. Acad. Sci. 51 (1949) 627–659.
[35] T. Boublik, Two dimensional particle liquid, Mol. Phys. 29 (1975) 421–428.
[36] G. Tarjus, P. Voit, S.M. Ricci, and J. Talbot, New analytical and numerical results on virial coefficients for 2-D convex bodies, Mol. Phys. 73 (1991) 773–787.
[37] J.S. Rowlinson, The fifth virial coefficient of a fluid of hard spheres, Proc. Roy. Soc. London A279 (1964) 147–160.
[38] S. Kim and D. Henderson, Exact values of two cluster integrals in the fifth virial coefficient for hard spheres, Phys. Letts. 27A (1968) 378–397.
[39] J. de Boer, Reports on Progress in Physics, Phys. Soc. (London), 12 (1949) 305.
[40] J.H. Conway and N.J.A. Sloane, *Sphere packings, lattices and groups* (Springer–Verlag 1988).

8

High density expansions

The virial expansion of the previous chapter can be applied to any stable and weakly tempered classical potential at low density. However, this method is not applicable to any phase other than fluid and, in particular, first order phase transitions which change the symmetry of the order parameter will not be detected by this method. Therefore new methods must be used to compute the phase diagrams in high density regions which are inaccessible to the virial expansion.

Unfortunately there are very few analytic results available for high density systems beyond the theorems on order presented in chapter 4. Instead, for the past 50 years the study of high density systems has been done numerically on the computer either by means of Monte-Carlo estimates of the free energy or by molecular dynamics simulations of the dynamic evolution of the motion of a large number of particles. We very briefly discuss what is meant by a molecular dynamics computation in section 8.1. However, in each result in the literature there are detailed discussions of method given which affect the range of reliability of the results. Each separate result in the end must be separately evaluated for its own reliability. As an example there are at least two competing phase diagrams for the Lennard-Jones potential in the literature. These numerical studies are ongoing and in all cases a value judgment is needed to assess the reliability of the results.

In the remainder of this chapter we present results of these studies to explore the phase diagrams of several model systems.

In section 8.2 we study hard spheres in two and three dimensions and will see that there are phase transitions in both two and three dimensions when the density is sufficiently large. In section 8.3 we extend these considerations to the repulsive inverse power law potential

$$U(\mathbf{r}) = \epsilon \left(\frac{\sigma}{|\mathbf{r}|} \right)^n \tag{8.1}$$

and will see that there is a range of n for which the phase diagram has both an fcc and a bcc phase. In section 8.4 we consider the hard sphere potential with an additional square well

$$\begin{aligned} U(\mathbf{r}) &= +\infty \quad \text{for} \quad 0 \leq |\mathbf{r}| < \sigma & (8.2)\\ &= \pm\epsilon \quad \text{for} \quad \sigma \leq |\mathbf{r}| \leq c\sigma & (8.3)\\ &= 0 \quad \text{for} \quad c\sigma < |\mathbf{r}|. & (8.4) \end{aligned}$$

When the + sign is used the potential is referred to as the step potential. It has first order transitions and a triple point. When the negative sign is used the potential is called the attractive square well and in addition to first order lines the phase diagram may have critical points. Finally in section 8.5 we consider the Lennard-Jones (6,12) potential

$$U(\mathbf{r}) = A/|\mathbf{r}|^{12} - B/|\mathbf{r}|^{6}. \tag{8.5}$$

8.1 Molecular dynamics

The basic method of molecular dynamics for hard spheres is presented in [1–4]. The basic principle is to consider a finite number N of particles in a finite volume (usually with periodic boundary conditions). An initial condition is assumed and the system is allowed to evolve in time by following the exact trajectories of all the N particles. The pressure is calculated from the trajectories of the particles by use of the virial theorem in the form [3] (which is slightly different from [2]):

$$\frac{Pv}{k_B T} = 1 + \lim_{\tau \to \infty} \frac{1}{N \langle v^2 \rangle \tau} \sum_{c} (\mathbf{r}_i - \mathbf{r}_j)_c \cdot (\mathbf{v}_i - \mathbf{v}_j)_c \tag{8.6}$$

where $\langle v^2 \rangle$ is the mean square velocity, the sum is over all collisions which occur in the time τ and $\mathbf{r}_i, \mathbf{v}_i$, and $\mathbf{r}_j, \mathbf{v}_j$ are the position and velocity of the two particles involved in the collision. By the conservation of energy and momentum either the initial or final velocities may be used.

In the thermodynamic limit the equilibrium pressure will be a unique well-defined function of the volume. However, for finite size systems with a finite number of particles N and a finite collision time (or number of collisions) the pressure computed from (8.6) may not always be in equilibrium and may depend on the history of the preparation of the state.

To determine the high density phase we set the initial positions of the N particles to be on an ordered lattice which, at the highest density, will be the the closest packed lattice (hexagonal in two dimensions and face centered cubic in three dimensions). Once the system is equilibrated we may then decrease the density by increasing the size of the confining periodic box and again run the program until the system equilibrates.

To determine the low density phase of the system we set the initial conditions in a disordered configuration. Once the system is equilibrated we may slowly change the size of the system and obtain higher densities.

However, only the cubic and square lattices are compatible with periodic boundary conditions. Therefore when studying closest packed configurations in a finite volume there will be a slight mismatch caused by the boundary conditions. In the thermodynamic limit the ordered fcc has a symmetry which the disordered fluid does not have and therefore in the thermodynamic limit these two phases will be in different representations of the space group and hence will lie in different ergodic components of the phase space. This ergodic decomposition is broken only by the periodic boundary conditions on the finite lattice. This dependence on the initial and boundary conditions is of great importance in the study of first order transitions.

Even in this brief summary it is obvious that there are many assumptions that have the potentiality of giving misleading results. Therefore for a critical assessment of the reliability of any of the results presented in this chapter it is ultimately necessary to read the original publications.

8.2 Hard spheres and discs

Hard spheres and discs have the property that there is a maximum density which results when the discs or spheres are closest packed. The fraction of the volume which is occupied by the discs or spheres is called the packing fraction. We denote the packing fraction in D dimensions of the closest packed configuration as η_{cpD}.

We considered this problem of closest packing in chapter 4 and saw that in two dimensions the closest packed configuration was the hexagonal lattice of Fig. 4.1. From that figure we see that η_{cp2} is the area of the circle with radius $\sigma/2$

$$A_o = \pi(\sigma/2)^2 \tag{8.7}$$

divided by the area of the circumscribing hexagon (called the Voronoi or Wigner–Seitz cell). This area is 12 times the area of the right triangle with base $\sigma/2$ and altitude $(\sigma/2)/\sqrt{3}$. Thus

$$A_{hex} = \frac{6}{\sqrt{3}}(\sigma/2)^2 \tag{8.8}$$

and thus the packing fraction for closest packed hard discs is

$$\eta_{cp2} = A_o/A_{hex} = \frac{\pi}{2\sqrt{3}} \sim 0.90690\cdots \tag{8.9}$$

Similarly, in three dimensions, we saw that there are no packings more dense than the fcc lattice which has the packing fraction of closest packed hard spheres:

$$\eta_{cp3} = \frac{\pi\sqrt{2}}{6} \sim 0.74048\cdots \tag{8.10}$$

In two dimensions the total occupied volume at close packing is

$$V_{cp2} = \eta_{cp2}L^2. \tag{8.11}$$

Alternatively if we call N_{cp2} the number of discs in the close packed configuration then because the volume of a single disc is $\pi(\sigma/2)^2$ we have the alternative expression

$$V_{cp2} = N_{cp2}\pi(\sigma/2)^2 \tag{8.12}$$

and thus by equating (8.11) and (8.12) we have

$$N_{cp2} = \frac{\eta_{cp2}L^2}{\pi(\sigma/2)^2}. \tag{8.13}$$

Therefore the closest packed number density for hard discs ρ_{cp2} is

$$\rho_{cp2} = N_{cp}/L^2 = \frac{2}{\sigma^2\sqrt{3}} \tag{8.14}$$

and the closest packed specific volume is

$$v_{cp2} = 1/\rho_{cp2} = \sigma^2\sqrt{3}/2. \tag{8.15}$$

Similarly in three dimensions

$$\rho_{cp3} = \frac{\sqrt{2}}{\sigma^3} \tag{8.16}$$

and

$$v_{cp3} = 1/\rho_{cp3} = \sigma^3/\sqrt{2}. \tag{8.17}$$

The limiting behavior of the pressure of the hard sphere system in D dimensions as $\rho \to \rho_{cp}$ is argued in the "free volume theory" to be [5]

$$\frac{P}{k_B T} \sim \frac{D}{v - v_{cp}} = \frac{D\rho}{1 - \rho/\rho_{cp}}. \tag{8.18}$$

This is believed to be the first term in a systematic expansion:

$$\frac{P v_{cp}}{k_B T} = \frac{D}{v/v_{cp} - 1} + \sum_{n=0}^{\infty} C_n (v/v_{cp} - 1)^n. \tag{8.19}$$

It is commonly accepted that (8.18) is indeed the correct equation of state near close packing but in fact the best rigorous result known is the 1965 result of Fisher [6]:

$$\eta_1 \ln(v/v_{cp} - 1)^{-1} \leq P v_{cp}/k_B T \leq \frac{\eta_2 D \ln(v/v_{cp} - 1)^{-1}}{v/v_{cp} - 1} \tag{8.20}$$

with $\eta_1 < 1 < \eta_2$.

There are no further exact results known for the high density properties of hard spheres and hard discs. Thus for further information it has been necessary to rely on numerical computations.

8.2.1 Behavior near close packing

A molecular dynamics study of the equation of state for hard spheres and discs near close packing was done in 1968 by Alder, Hoover and Young [4]. These computations are done by the method of molecular dynamics where the calculation begins in the closest packed fcc lattice with periodic boundary conditions. The density is then reduced and the program is run for a sufficiently long time so that the system reaches an equilibrium at the lower density. Thus there is a history in the computation which connects the state with the initial high density state which is fcc in three dimensions and hex close packed in two dimensions.

We reproduce the results of their paper for Pv/k_BT as a function of v/v_{cp} in Table 8.1 for $D = 3$ and in Table 8.2 for $D = 2$. For comparison with (8.19) and later work we also give $Pv_{cp}/k_BT = P/(kT\rho_{cp})$.

In order to compare with the assumed high density form (8.19) we plot $Pv/k_BT - D/(v/v_{cp} - 1)$ versus $\alpha = v/v_{cp} - 1$ in Fig. 8.1 for $D = 3$ and Fig. 8.2 for $D = 2$.

In both cases we see that $Pv/k_BT - D/(v/v_{cp} - 1)$ approaches a finite value at $\alpha = 0$ with a finite slope. Thus we may write

$$Pv/k_BT - D/\alpha = \tilde{C}_0 + \tilde{C}_1\alpha + \cdots \tag{8.21}$$

Table 8.1 Molecular dynamical determination of the pressure for hard spheres $D = 3$ in the high density regime. The data for Pv/k_BT is taken from table II of the paper of Alder, Hoover and Young [4]. The number of particles is denoted by N.

v/v_{cp}	ρ/ρ_{cp}	η	Pv/k_BT	$P/(kT\rho_{cp})$	N
1.42	0.742	0.549	10.170(10)	7.546	4000
1.3448	0.744	0.551	11.542(06)	8.587	500
1.25	0.800	0.592	14.720(09)	11.776	500
1.20	0.833	0.616	17.680(10)	14.727	500
1.15	0.869	0.643	22.638(10)	19.677	500
1.05	0.952	0.704	62.582(10)	59.580	500
1.02	0.980	0.725	152.573(50)	149.58	500
1.01	0.990	0.733	302.560(40)	299.53	500
1.005	0.995	0.736	604.067(50)	601.64	4000

Table 8.2 Molecular dynamical determination of the pressure for hard discs $D = 2$ at high densities. The data for Pv/k_BT is taken from Table I of the paper of [4]. The number of discs is denoted by N.

v/v_{cp}	ρ/ρ_{cp}	η	Pv/k_BT	$P/(kT\rho_{cp})$	N
1.25	0.800	0.725	10.170(13)	8.136	870
1.20	0.833	0.755	12.069(08)	10.054	7968
1.15	0.869	0.788	15.340(14)	13.330	870
1.10	0.909	0.824	21.965(06)	19.953	870
1.07	0.934	0.847	30.515(04)	28.501	870
1.03	0.971	0.881	68.586(12)	66.597	870
1.01	0.990	0.898	201.911(06)	199.971	7968

which gives C_0 and C_1 of (8.19)

$$C_0 = D + \tilde{C}_0, \quad C_1 = \tilde{C}_0 + \tilde{C}_1. \tag{8.22}$$

The values of C_0 and C_1 as obtained in [4] from these plots are for hard spheres $D = 3$

$$\tilde{C}_0 = 2.555 \pm 0.02 \qquad \tilde{C}_1 = 0.56 \pm 0.08 \tag{8.23}$$

and for hard discs $D = 2$

$$\tilde{C}_0 = 1.898 \pm 0.01 \qquad \tilde{C}_1 = 0.67 \pm 0.07. \tag{8.24}$$

We thus conclude that the molecular dynamics computations support the form of the high density expansion (8.19).

8.2.2 Freezing of hard spheres

To proceed further we examine molecular dynamics data where, at low density, the state is in the disordered fluid phase. The density is then increased in steps and at each step the program is run for a sufficiently long time to reach an equilibrium. The

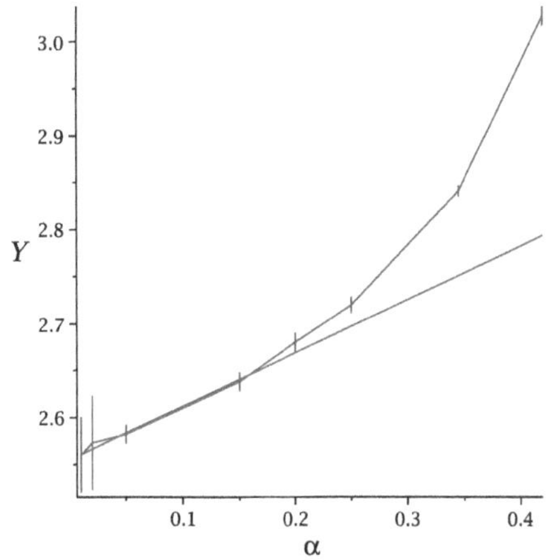

Fig. 8.1 A plot for hard spheres $D = 3$ of $Y = Pv/k_BT - 3/(v/v_{cp} - 1)$ as a function of $\alpha = v/v_{cp} - 1$ from Table 8.1 and the straight line fit (8.21) with $\tilde{C}_0 = 2.56 \pm 0.02$ and $\tilde{C}_1 = 0.55 \pm 0.08$.

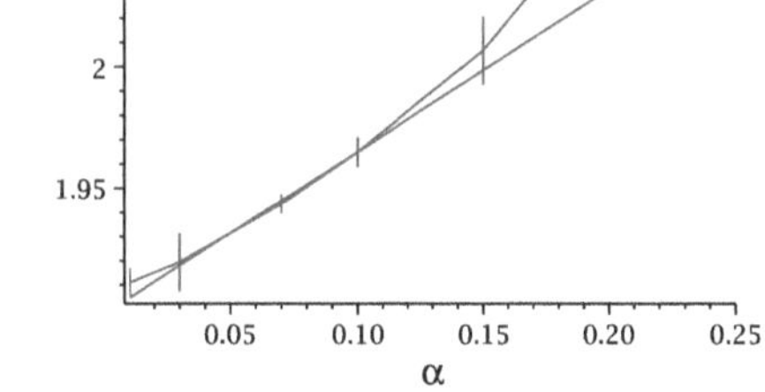

Fig. 8.2 A plot for hard discs $D = 2$ of $Y = Pv_{cp}/k_BT - 2/(v/v_{cp} - 1)$ as a function of $\alpha = v/v_{cp} - 1$ from Table 8.2 and the straight line fit (8.21) with $\tilde{C}_0 = 1.898 \pm 0.01$ and $\tilde{C}_1 = 0.67 \pm 0.07$.

first such studies for the equation of state for hard spheres is the 1957 papers of Alder and Wainwright [7] and Wood and Jacobson [8]. Since these early papers there have been many increasingly refined numerical studies and the best low density values of the pressure have been determined by [9] in 1984 for $0 \leq \rho/\rho_{cp} \leq 0.588$ and by [10] in 1997 for $0.625 \leq \rho/\rho_{cp} \leq 0.770$ and are given in Table 8.3.

We plot in Fig. 8.3 the high density of Table 8.1 and the low density data of Table 8.3. This plot reveals the striking feature that the low and high density branches of the equation of state do not smoothly join one another but that there is a sizable region of density from $\rho/\rho_{cp} = 0.742$ to $\rho/\rho_{cp} = 0.770$ where they overlap and for one density the branch which connects to the fcc solid phase lies below the phase which connects to the low density fluid phase. This phenomenon was first seen in 1957 in [7,8].

However, by the very definition an equilibrium state cannot depend on the past history of how the state was obtained. Therefore in the region of overlap at least one of the branches must be in a metastable as opposed to a stable state. Furthermore, we proved in some detail in chapter 3 that the pressure at equilibrium must be a monotonic function of the density. Therefore the simplest interpretation of this molecular dynamics data is that in the overlap there is a first order phase transition where the system is in a two-phase region in which the pressure is constant as the density varies from the density of the pure fluid ρ_f to the density of the pure solid ρ_s. The flat portion of the isotherm which connects ρ_f and ρ_s is called the tie line. When the density is between ρ_f and ρ_s both branches are metastable and, given sufficient, time the system will fluctuate between the two. Thus a metastable "crystal" can melt into a fluid if the density is below ρ_s and a metastable "fluid" can freeze into a solid if the density is above ρ_f.

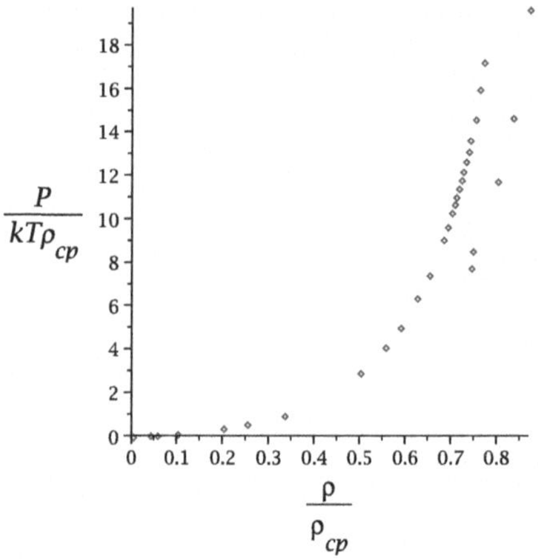

Fig. 8.3 A plot of the equation of state of hard spheres as determined from the molecular dynamics data of [4] as given in Tables 8.1 and 8.3.

The question now arises as to how the pressure and the endpoints of the tie line can be computed. In principle, this question was answered in some detail in chapter 3 where we proved that the free energy must be a convex function of the density. Therefore in principle we should calculate the free energy of the low density fluid and the high density fcc solid which is illustrated in Fig. 8.4 where we note that because in the thermodynamic limit (the only place where the free energy is in fact defined) the fluid and the fcc solid lie in different ergodic components of the phase space there is no a priori reason that the free energy of the two phases should be analytic continuations of each other. The true free energy of the system is then determined by constructing

Table 8.3 Molecular dynamical determination of the pressure for the hard sphere fluid in the low density regime. For $0 \leq \rho/\rho_{cp} \leq 0.588$ the data is from Table 1 of [9] and for $0.625 \leq \rho/\rho_{cp} \leq 0.770$ the data is from Table 1 of [10]. The locations of ρ_f/ρ_{cp} and ρ_s/ρ_{cp} are marked by horizontal lines. The number of spheres used in the computation is denoted by N. Where the pressure was calculated for several different values of N we have listed only the largest one used in the table.

ρ/ρ_{cp}	η	Pv/k_BT	$P/(k_BT\rho_{cp})$	N
0.040	0.030	1.12777(03)	0.045S1	4000
0.055	0.041	1.18282(05)	0.0650	4000
0.100	0.074	1.35939(07)	0.1359	4000
0.200	0.148	1.88839(22)	0.3776	4000
0.250	0.185	2.24356(36)	0.560	4000
0.333	0.246	3.03162(28)	1.009	4000
0.500	0.370	5.85016(85)	2.9250	4000
0.555	0.411	7.4304(13)	4.1238	4000
0.588	0.435	8.6003(12)	5.0589	4000
0.625	0.462	10.203(03)	6.3768	1372
0.650	0.481	11.498(04)	7.4737	1372
0.680	0.503	13.341(04)	9.0718	1372
0.690	0.511	14.049(05)	9.6938	1372
0.700	0.518	14.787(04)	10.350	4000
0.705	0.521	15.187(08)	10.706	1372
0.710	0.525	15.587(10)	11.066	1372
0.715	0.529	16.020(10)	11.454	1372
0.720	0.532	16.457(08)	11.849	4000
0.725	0.536	16.904(11)	12.255	1372
0.730	0.540	17.351(19)	12.666	4000
0.735	0.544	17.905(40)	13.160	1372
0.740	0.547	18.42(05)	13.636	1372
0.750	0.555	19.55(10)	14.662	1372
0.760	0.562	21.05(10)	15.998	1372
0.770	0.569	22.40(15)	17.248	1372

the tangent to the two individual free energies as is also indicated in Fig. 8.4.

Unfortunately, the free energy is not a dynamical quantity which can be directly obtained from a molecular dynamics computation. Instead it must be deduced by integrating the equation of state with respect to the density. This integration will in general involve an arbitrary constant which cannot be determined from the equation of state and which can be different in the two phases. The difference between these two constants of integration will influence the position of the tie line. Consequently we need some further assumptions to predict where the tie line will lie.

In the past 50 years several scenarios have been advanced to resolve the question of how to compute freezing and melting densities.

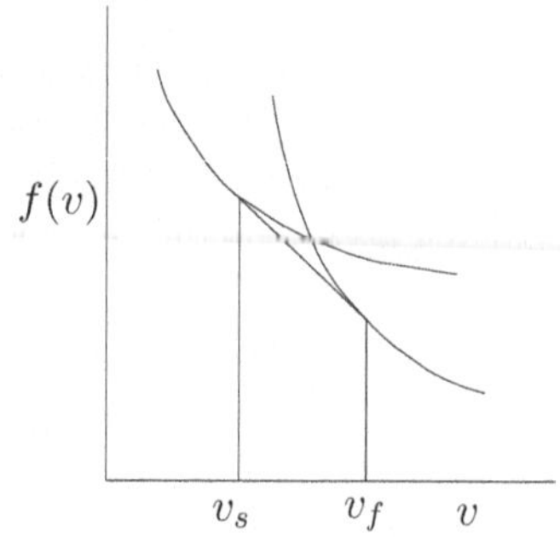

Fig. 8.4 A schematic plot of the free energy of the fluid and solid branches of the hard sphere system. The tie line is determined from the tangent which connects the two branches and we have assumed in this schematic that the two branches are not analytic continuations of each other.

One such theoretical proposal comes from the work of Fisher [11] and Langer [12] who argued that for the first order phase transition associated with liquid–gas condensation that there is an essential singularity at the liquid–gas coexistence curve. This essential singularity can indeed be proven to exist in the two-dimensional Ising lattice gas. However, in condensation the liquid and the gas have the same symmetry and thus lie in the same ergodic component of the phase space. This is not the situation faced in a fluid–solid transition and thus the analogy is in doubt. Furthermore even if this scenario does exist in principle, the effect of the essential singularity is so small that it has not been seen in numerical computations [10].

A second method of determining the tie line relies on assumptions of an analytic continuation of the fluid to the solid phase [13]. This has the advantage that it can actually be applied to the numerical equations of state. The original determination in [13] leads to

$$\rho_f/\rho_{cp} = 0.667 \pm 0.003 \qquad \eta_f = 0.491 \pm 0.002 \tag{8.25}$$

$$\rho_s/\rho_{cp} = 0.736 \pm 0.003 \qquad \eta_s = 0.542 \pm 0.002 \tag{8.26}$$

at a pressure of

$$\frac{P}{k_B T} = (8.27 \pm 0.13)\rho_{cp} \tag{8.27}$$

and the most recent determination is given in [10] as

$$\rho_f/\rho_{cp} = 0.663 \pm 0.002 \qquad \eta_f = 0.491 \pm 0.002 \tag{8.28}$$

$$\rho_s/\rho_{cp} = 0.733 \pm 0.002 \qquad \eta_s = 0.542 \pm 0.002. \tag{8.29}$$

These values give rise to the following observation of [10] ... *the equilibrium freezing density of fluids coincides with the lowest density where the crystal survives without melting in simulation studies. Hard-sphere crystals can be simulated for a few million collisions at* $\rho/\rho_{cp} = 0.67$ *before melting but they melt quickly at* $\rho/\rho_{cp} = 0.66$. *It is interesting that the highest density where the fluid can be simulated without freezing,* $\rho/\rho_{cp} = 0.73$, *is also close to the equilibrium melting density of the crystal. These 'rules' probably stem from the difficulty of fitting coexisting phases in a small cell with periodic boundaries.*

8.2.3 The phase transition for hard discs

The physics of hard discs is far more obscure and controversial than the physics of hard spheres just presented. To begin with we proved in chapter 4 that the phenomenon of crystalline order does not exist in two dimensions for a large class of potentials which are singular only at the origin. Hard discs, of course, have a potential which is singular at $|\mathbf{r}| = \sigma$ and thus the proven theorem does not apply. Nevertheless it is commonly believed that in fact there is no long range order in the hard disc systems.

This (assumed) lack of long range crystalline order in two dimensions might suggest that there will be no phase transition in the hard disc system. However, in 1962 Adler and Wainwright [14] found by means of a molecular dynamics computation that there is a region of density where a phase transition does indeed appear to take place, which they described in terms of a first order transition with a van der Waals loop. They estimated that

$$\rho_f/\rho_{cp} = 0.762 \quad \rho_s/\rho_{cp} = 0.789 \tag{8.30}$$

$$\frac{P}{k_B T \rho_{cp}} = 7.72. \tag{8.31}$$

This critical region can be seen by plotting in Fig. 8.5 the high density molecular dynamics results [4] given in Table 8.2 with the lower density results of [3] given in Table 8.4.

To further study this apparent phase transition, further numerical studies have been done by [16, 18]. Their results are given in Table 8.5 and plotted in Fig. 8.6.

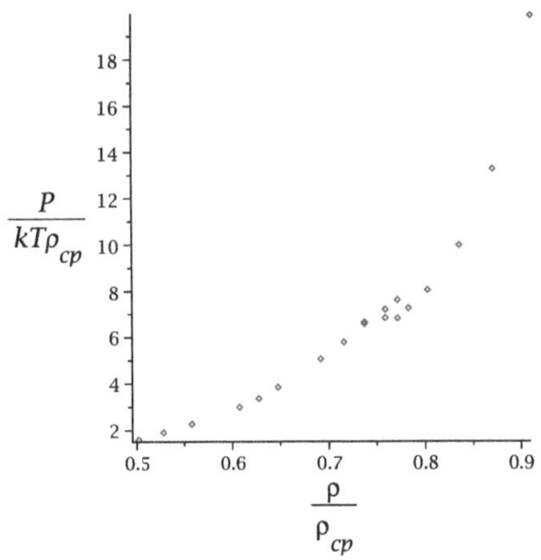

Fig. 8.5 A plot of the equation of state of hard discs as determined from the molecular dynamics data of [3] as given in Tables 8.2 and 8.4.

In both Figs. 8.5 and 8.6 there is certainly an anomalous behavior in the density region (8.30). The behavior shown in Fig. 8.5 outside of this density region can certainly be interpreted as belonging to two different branches of the free energy representing two different phases just as was seen in the hard sphere system. However, the uncertainty in the data in the transition region precludes any observation of metastable states as seen in Fig. 8.3 for hard spheres.

The data in Fig. 8.6 are more disturbing because the reported pressure is not monotonic in the density as required by thermodynamics and can indeed be taken as an indication that thermal equilibrium has not been achieved.

However, what does seem incontrovertible is that the behavior in the two sides of the transition region does not smoothly connect together.

In order to understand this phenomenon in more detail Kosterlitz and Thouless [19] and Halperin and Nelson [20] introduced a new "topological" order parameter for the two-dimensional system which can be nonzero even when there is no long range crystalline order.

To define this new order parameter we take a configuration of hard discs and make the Voronoi (or Wigner–Seitz) cell construction as done in chapter 4. This gives an unambiguous definition of nearest neighbors and we consider connecting the centers of the nearest neighbor discs by straight lines (which we call bonds). If we then take the direction of a bond connecting a particular particle k and one of its neighbors as a reference direction, we can then define the angle between the bond connecting

Table 8.4 Molecular dynamical determination of the pressure for the hard disc system. The data for Pv/k_BT is taken from the 1967 paper of [3]. The number of discs is 72.

ρ/ρ_{cp}	η	Pv/k_BT	$P/(kT_B\rho_{cp})$
0.500	0.453	3.39	1.69
0.526	0.477	3.78	1.98
0.555	0.503	4.24	2.35
0.588	0.533	4.76	2.79
0.606	0.549	5.13	3.10
0.625	0.567	5.56	3.47
0.645	0.585	6.08	3.92
0.689	0.625	7.47	5.14
0.714	0.648	8.25	5.89
0.735	0.667	9.20 ± 0.07	6.7 ± 0.05
0.746	0.677	9.4 ± 0.3	7.0 ± 0.2
0.756	0.686	9.5 ± 0.3	7.1 ± 0.2
0.763	0.692	9.2 ± 0.3	7.0 ± 0.2
0.769	0.697	9.6 ± 0.5	7.3 ± 0.4
0.775	0.703	9.16 ± 0.03	7.10 ± 0.02
0.781	0.708	9.38 ± 0.01	7.32 ± 0.01
0.800	0.726	10.06	8.04
0.909	0.824	21.2	19.3

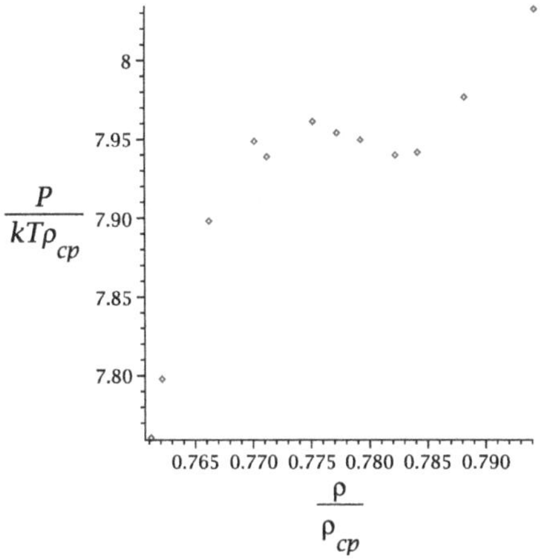

Fig. 8.6 A plot of the equation of state of hard discs in the transition region as determined from the Monte-Carlo data of [16] and [18] as given in Table 8.5.

particle k and a nearest neighbor j with respect to this reference direction as ϕ_{kj}. For a perfect hexagonal structure this angle will be a multiple of $2\pi/6$. We then define for each particle k the quantity

$$\Psi_k = \frac{1}{6}\sum_j e^{6i\phi_{kj}} \tag{8.32}$$

Table 8.5 Monte-Carlo determination of the pressure for the hard disc system near the region of the possible transition as taken from Table 1 of [16] and Table II of [18]. The ρ^* of [16] and the ρ of [18] is $(2/\sqrt{3})(\rho/\rho_{cp})$. In both cases we have used the data with $N = 128^2 = 16384$ discs.

ρ/ρ_{cp}	η	$P/(k_B T \rho_{cp})$ [16]	$P/(k_B T \rho_{cp})$ [18]
0.761	0.690	7.762(03)	
0.762	0.691		7.799(07)
0.766	0.695	7.890(04)	7.899(08)
0.770	0.698		7.950(09)
0.771	0.699	7.940(08)	
0.775	0.703		7.963(09)
0.777	0.705	7.952(08)	7.955(05)
0.779	0.706		7.951(07)
0.782	0.709	7.941(05)	
0.784	0.711		7.943(08)
0.788	0.715	7.978(07)	7.928(05)
0.794	0.720	8.034(05)	

where the sum is over the nearest neighbors of k, and from this we define the order parameter

$$\Psi = \lim_{N\to\infty} \frac{1}{N} |\sum_k \Psi_k| \tag{8.33}$$

where the sum over k is over all N discs in the system.

This order parameter is unity for the closest packed configuration of hard discs. The system is said to be ordered when $\Psi > 0$ and to be disordered when $\Psi = 0$. When $\Psi = 0$ it is possible to consider the correlation

$$g_6(r) = |\langle \Psi_k \Psi_j \rangle| \quad \text{where} \quad r = |\mathbf{r}_k - \mathbf{r}_j|. \tag{8.34}$$

In terms of this order parameter a first order transition as originally suggested by Alder and Wainwright will have $\Psi > 0$ in the high density phase, $\Psi = 0$ in the low density phase and in the transition region the low and high density phases are connected by a tie line just as for hard spheres.

However, a second alternative exists [19–22] because in a disordered region where $\Psi = 0$ there are two possible behaviors of $g_6(r)$: for $r \to \infty$ either the correlation $g_6(r)$ decays exponentially

$$g_6(r) \sim e^{-r/\xi} \tag{8.35}$$

or it decays algebraically

$$g_6(r) \sim r^{-\eta}. \tag{8.36}$$

We thus have a scenario where for $\rho < \rho_f$ there is exponential decay of $g_6(r)$ while for $\rho_f < \rho < \rho_s$ the correlation $g_6(r)$ decays algebraically. Such a phase, if it exists, is called the *hexatic* phase and in this region of density the pressure will not be constant but will be monotonic increasing.

Since the inception of this hexatic phase there has been much effort to determine which of the two possible scenarios is correct for hard discs. To date there is no definitive answer to the question. As an example we quote the conclusion of the 2002 paper of Binder, Sengupta and Nielaba [23]:

It has been shown that the currently available simulation data are compatible with a continuous transition from the fluid to the hexatic phase (with divergent bond orientation susceptibility) at $\rho_f = 0.899$ *and with a hexatic to crystal transition at* $\rho \sim 0.914 \pm 0.002$. *However, no simulations that reach full thermal equilibrium in the density range* $0.90 \leq \rho \leq 0.915$ *and show directly the existence of the hexatic phase are available so far. Without such direct evidence, the possibility of a (very weak) first order transition from the fluid to the crystal cannot yet be firmly ruled out, although so far clear signals of two phase coexistence are also lacking.*

8.3 The inverse power law potential

The next simplest potential after the hard sphere potential is the inverse power law potential (8.1) which is the only potential that has the property that the free energy depends on one combination of the variable T and ρ and does not depend on T and ρ separately. Thus, while first order freezing transitions are allowed, triple points are forbidden. In the limit that $n \to \infty$ the inverse power law potential (8.1) reduces to the hard sphere potential and thus it is expected that for sufficiently large n there will be a freezing transition to an fcc solid. We will see, however, that this phase diagram is valid only for $n > 7$. For $3 \leq n \leq 7$ the system develops a second phase transition with a bcc phase lying between the fluid and the fcc phase.

We first present the scaling argument which reduces the thermodynamic functions to depend on one variable rather than T and ρ separately and will then discuss the numerical computations of the phase diagram and the evidence for a second bcc phase.

8.3.1 Scaling behavior

For the power law potential (8.1) the Mayer function is

$$f_{ij}(\mathbf{r}_i - \mathbf{r}_j) = \exp\left(-\frac{\epsilon}{k_BT}(\frac{\sigma}{|\mathbf{r}_i - \mathbf{r}_j|})^n\right) - 1 \tag{8.37}$$

and therefore every integral in the Mayer or the Ree-Hoover expansion for the virial coefficient B_k is of the form

$$I_k(T) = \int d^D\mathbf{r}_2 \cdots d^D\mathbf{r}_k F_k\left[\frac{\epsilon}{k_BT}\left(\frac{\sigma}{|\mathbf{r}_i - \mathbf{r}_j|}\right)^n\right]. \tag{8.38}$$

If we now make the substitution

$$\mathbf{r}_j = (k_BT/\epsilon)^{-\frac{1}{n}}\sigma\mathbf{x}_j \tag{8.39}$$

we find

$$I_k(T) = (k_BT/\epsilon)^{-D(k-1)/n}\sigma^{D(k-1)}\int d^D\mathbf{x}_2 \cdots d^D\mathbf{x}_k F_k\left[\left(\frac{1}{|\mathbf{x}_i - \mathbf{x}_j|}\right)^n\right] \tag{8.40}$$

and therefore

$$B_k(T) = \left(\frac{\epsilon}{k_BT}\right)^{\frac{D}{n}(k-1)}\sigma^{D(k-1)}\tilde{B}_k \tag{8.41}$$

where $\tilde{B}_k$ is independent of T, ϵ and σ. Thus we have

$$\frac{Pv}{k_BT} = 1 + \sum_{k=2}^{\infty}\tilde{B}_k\sigma^{D(k-1)}\left(\frac{\epsilon}{k_BT}\right)^{D(k-1)/n}\rho^{k-1}. \tag{8.42}$$

Therefore if we define

$$\tilde{\rho} = \sigma^D\left(\frac{\epsilon}{k_BT}\right)^{D/n}\rho \tag{8.43}$$

we see that

$$P = (k_BT)^{1+D/n}(\sigma\epsilon^{1/n})^{-D}\left(\tilde{\rho} + \sum_{k=2}^{\infty}\tilde{B}_k\tilde{\rho}^k\right). \tag{8.44}$$

This is valid in the low density regime where the virial expansion is valid. However, an identical argument can be made on the partition function itself which is valid for all densities. Therefore we conclude for the potential (8.1) that $P/(k_BT)^{1+D/n}$ is a function of the single variable $v(k_BT)^{D/n}$ and is not a function of v and T separately. In the limit of hard spheres this reduces to the statement that P/k_BT is a function only of the density ρ and is independent of the temperature. The repulsive power law potential (8.1) is the only potential where the equation of state effectively depends on only one instead of two variables.

For hard spheres and power law potentials the special feature that $P/(k_BT)^{1+D/n}$ depends only on the single variable $v(kT)^{D/n}$ means that if there is a flat portion for one isotherm then there will be a flat portion on all isotherms. In Fig. 8.7 we schematically plot the (P, v) diagram for a power law potential with some finite value of n for both the case of one and two phase transitions.

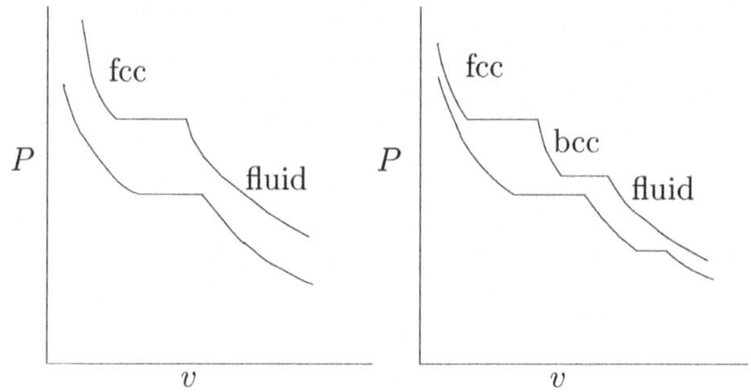

Fig. 8.7 A schematic plot of isotherms for a power law potential with some finite value of n. The plot on the left has single fcc crystalline. The plot on the right has both an fcc and a bcc phase.The endpoints of the phases are proportional to $(k_BT)^{-3/n}$.

8.3.2 Numerical computations

The molecular dynamics computations for inverse power law potentials are substantially more involved than for hard spheres for several reasons. We will only sketch here the highlights of the procedure. A full explanation is to be found in the original papers [24–28].

First of all there is now no well-defined notion of collision, and the equations of motion for all N particles have to be integrated for (small) discrete time steps.

Secondly it must be realized that there is no temperature variable per se in a molecular dynamics computation. In these dynamical computations it is the energy which is kept constant and not the temperature (which does not appear). The temperature is kept constant by rescaling the velocities after every time step. However, because of the scaling property of this potential only one isotherm needs to be computed.

With these two modifications the calculations in the fluid phase may now be done in a manner identical to the corresponding fluid phase computations done for hard spheres.

The calculations in the bcc and fcc phases are more difficult than for hard spheres for two reasons. Firstly there is the need to keep the bcc and fcc phases in their proper ergodic component, and secondly the need to find a method of computing the integration constant in the free energy for the (P, v) data computed by molecular dynamics which is compatible with the integration constant in the fluid phase free energy. For example in [27, page 75] the following procedure is used:

Here the system is transformed continuously through the use of a coupling constant denoted here by λ, from the fully interacting N particle system ($\lambda = 0$) to a collection of N independent (Einstein) harmonic oscillators centered at the lattice sites of the crystal, for which the free energy can be calculated analytically. The coupling to the lattice sites prevents the system from melting as the interparticle pair potentials are scaled to zero. Note that we have assumed that the crystal has zero concentration of vacancies, since density functional methods show that the small equilibrium concentration of vacancies makes a negligible contribution to the bulk free energy.

The lattice coupling potential energy used in the simulation of [27] is

$$\Phi(\lambda) = \Phi_0 + (1-\lambda)^2 \left(\sum_{k<j} (r_0/|r_k - r_j|)^n - \Phi_0 \right) + \lambda k_{\max} \sum_i (\mathbf{R}_i - \mathbf{R}_i^0)^2 \tag{8.45}$$

where Φ_0 is the ($N = \infty$) static lattice energy, $\{\mathbf{R}_i^0\}$ is the set of crystal lattice sites, and $k_{\max}$ is chosen such that the mean squared lattice displacement of the Einstein oscillators is approximately that of the actual uncoupled system.

Furthermore [27, page 76] has the proviso

Since the system becomes increasingly non-ergodic as the Einstein crystal limit $\lambda \to 1$ is approached, the molecular dynamics procedure was modified so as to reset periodically the particle velocities from a Boltzmann distribution defined by the temperature, allowing a more complete sampling of the phase space.

Finally once the free energies for the separate phases are computed the tie line between the phases is computed using a double tangent construction.

There are thus several assumptions made in these computations which are not present in the previous computations for hard spheres.

Over the years there have been a variety of numerical simulations of the inverse power law potential: the fluid and fcc phases for $n = 12$ in [24]; the fluid and fcc phases for $n = 4, 6, 9$ in [25]; the fluid, bcc and fcc phases for $n = 4, 6$ in [26]; the fluid, bcc and fcc phases for $n = 6$ in [27]; the fluid, bcc and fcc phases for $n = 4, 6, 9, 12$ in [28]. Each paper uses somewhat different assumptions and numerical methods but even though there are some quantitative differences in the various results they all confirm the qualitative three-phase picture of Fig. 8.7 for $3 \leq n \leq 7$.

8.4 Hard spheres with an additional square well

The next potential to consider is the square well potential (8.4) which may be either repulsive or attractive depending on the sign in front of ϵ. Molecular dynamics studies of these potentials were initiated by Young and Alder for the repulsive step potential [29] in 1979 and for the attractive square well [30] in 1980. In contrast to the hard sphere and inverse power law potentials these potentials depend on density and temperature separately. It is thus possible to obtain triple points and critical points in addition to lines of first order transitions.

We restrict ourselves here to the case of the attractive square well (8.4) with the negative sign, which has become an important model for the study of real liquids. We present here the results of three selected studies of this system [30–32] which reveal a rich variety of phenomena that change as the width of the attractive region is varied. As in the case of the potentials previously studied the computations have various approximations and caveats which must be taken into consideration when evaluating these results.

The pioneering paper of [30] made a molecular dynamics study of the particular case $c = 1.5$. The results of this study are reproduced in Figs. 8.8 and 8.9 .These results have the very striking feature that, in addition to the expected liquid–vapor critical point and the liquid–vapor–solid triple point, there is a first order transition to a hex close packed phase and, even more remarkable, there is an isostructural transition in the fcc phase that ends in a critical point.

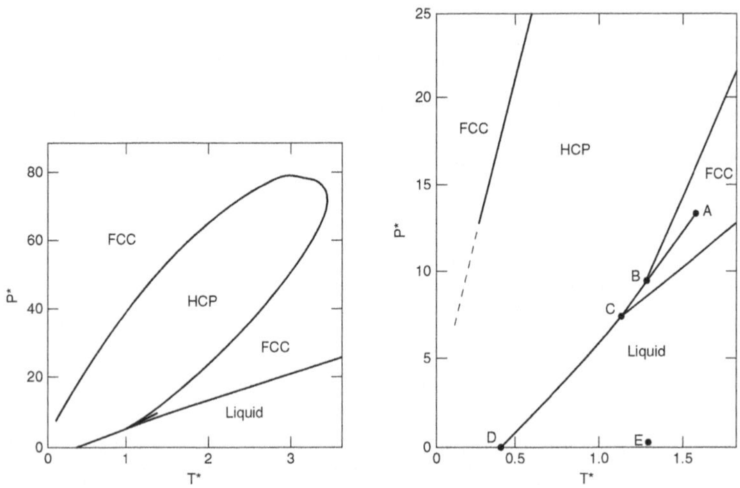

Fig. 8.8 The phase diagram of the attractive square well potential with $c = 1.5$ taken from [30]. The figure on the right is an enlarged detail of the figure on the left where the labeled points are (A) the fcc–fcc critical point; (B) hcp–fcc–fcc triple point; (C) hcp–fcc–liquid triple point; (D) hcp-liquid–vapor triple point;(E) liquid–vapor critical point. In both figures the horizontal axis is $T* = k_B T/\epsilon$ and the vertical axis is $P* = P/\rho_{cp}\epsilon$.

Further study of this isostructural transition was made in [31] by varying the parameter c in the range $1.01 \leq c \leq 1.06$ and it was found that as c in increased the distance between the hcp–fcc–fcc triple point and the fcc–fcc critical point decreases. Quantitatively, however, the fact that the isostructural transition disappeared at $c = 1.06$ is at variance with the observation of the transition in [30] at $c = 1.5$.

Finally a more recent numerical study in conjunction with additional theoretical considerations [32] has reported a much more elaborate structure of phases than re-

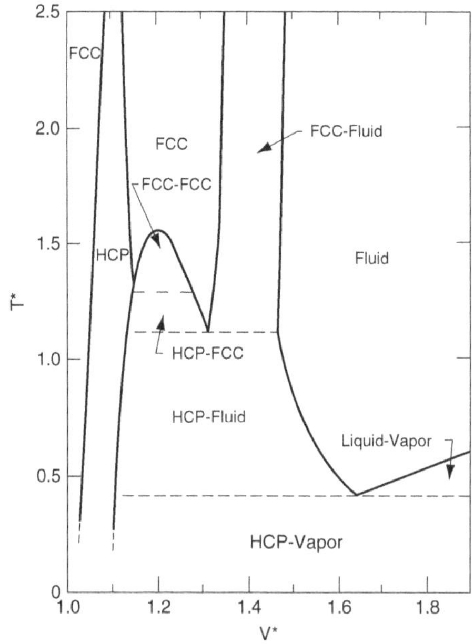

Fig. 8.9 Phase boundaries of the $c = 1.5$ square-well potential in the T–V plane from [30]. The horizontal axis is v/v_{cp} and the vertical axis is $T* = k_BT/\epsilon$.

ported in either [30] or [31] which we reproduce in Fig. 8.10 where the numbers in parenthesis indicate the number of atoms in a unit cell.

8.5 Lennard-Jones potentials

The final potential we discuss is the famous Lennard-Jones (6, 12) potential (8.5) which was first introduced in 1924 [33] and has been used many times to model the behavior of the noble gases which have the simple phase diagrams given in chapter 2 with one solid fcc phase, an fcc–fluid–vapor triple point and a fluid–vapor critical point. Many numerical computations have been done on this system beginning with the pioneering computation in 1969 of Hansen and Verlet [35] which yielded a phase diagram in qualitative agreement with the noble gases.

However, instead of being an example of the successes of numerical computations this agreement of numerical computations with experiment is actually, in the words of Choi, Ree and Ree [36] a "continuing scandal" because it has been known ever since the work of Kihara and Koba [34] in 1952 that, at low temperature and low pressure, the crystalline structure of the Lennard-Jones potential is not the face centered cubic lattice with one atom per unit cell observed in the noble gases but is rather the hex close packed lattice with two atoms per unit cell. As shown by Stillinger [37] an fcc phase at $T = 0$ can only be produced by putting the system under sufficiently high pressure. The phase boundary between the fcc and hcp phases has been studied in some detail in [36] and the phase diagram looks qualitatively like the phase diagram

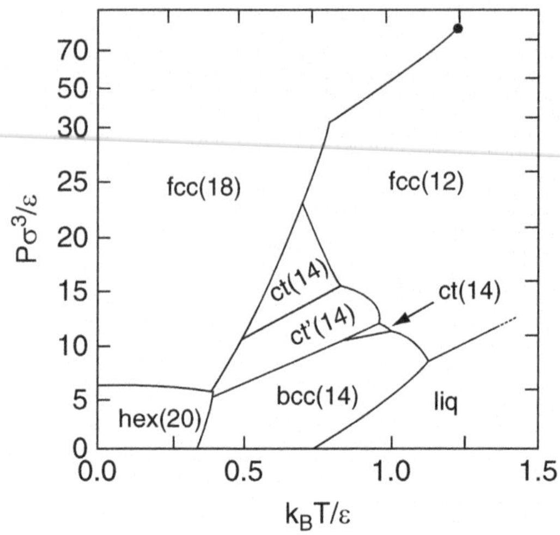

Fig. 8.10 Phase boundaries of the attractive square well potential for $c = 1.43$ as given in [32]

of the attractive square well of Fig. 8.8 with the exception that for the Lennard-Jones potential no isostructural transition in the fcc phase has been observed.

Because of the disagreement of many numerical simulations with the results of Kihara and Koba it is perhaps appropriate to conclude with some discussion of how such discrepancies can arise in addition to the caveats already discussed.

First of all there is the almost irresistible tendency to perform simulations with an hcp incompatible cubic box with periodic boundary conditions, and then to choose an integer number N of particles which is a "magic number" for the fcc structure. This alone will strongly predispose towards the formation of fcc versus hcp in finite a N simulation. Furthermore some simulations will cut off the Lennard-Jones potential at some finite distance for numerical simplicity and thus the potential is actually not Lennard-Jones at all. It is possible that such a truncated potential could have an fcc structure and low temperature and pressure but such a potential is in fact not Lennard-Jones.

8.6 Conclusions

Statistical mechanics as presented in chapter 1 of this book holds out the promise of explaining the macroscopic properties of bulk matter in equilibrium from the knowledge of the underlying microscopic interactions. One of the most fundamental properties of a macroscopic system that we would like to know is the phase diagram, and many such examples of phase diagrams were given in chapter 2. The most prominent feature of these diagrams is that there are lines of first order transitions which separate phases with different order parameters. It is these first order transitions that we have investigated with the model potentials in this chapter and because of the ubiquity of these transitions it is disappointing that even for these simple potentials there are so few exact results known and that there are so many unresolved controversies in the numerical results. To quote again from [36] there is *a general lack of reliable theoretical techniques to predict the theoretical structure of the simplest crystalline solids from a knowledge of their chemical composition.* If we do not even have a satisfactory explanation of why the heavy noble gas solids exist in an fcc rather than a hcp phase we clearly have much to learn about how to use the machinery of statistical mechanics to actually predict the properties of bulk matter.

References

[1] B.J. Alder and T.E. Wainwright, Studies in molecular dynamics I. General method, J. Chem. Phys. 31 (1959) 459–466.

[2] B.J. Alder and T.E Wainwright, Studies in molecular dynamics II: Behavior of small numbers of elastic hard spheres, J. Chem. Phys. 33 (1960) 1439–1451.

[3] W.G. Hoover and B.J. Alder, Studies in molecular dynamics IV. The pressure, collision rate and their number dependence for hard disks, J. Chem. Phys. 46 (1967) 686–691.

[4] B.J. Alder, W.G. Hoover and D.A. Young, Studies in molecular dynamics. V. High-density equation of state and entropy for hard discs and sphere, J. Chem. Phys. 49 (1968) 3688–3696.

[5] T.L. Hill, *Statistical Mechanics* (McGraw-Hill, NY 1956)

[6] M.E. Fisher, Bounds for the derivatives of the free energy and pressure of a hard-core system near close packing, J. Chem. Phys. 42 (1965) 3852–3856.

[7] B.J. Alder and T.E. Wainwright, Phase transition for a hard sphere system, J. Chem. Phys. 27 (1957) 1208–1209.

[8] W.W. Wood and J.D. Jacobson, Preliminary results from a recalculation of the Monte-Carlo equation of state of hard spheres, J. Chem. Phys. 27 (1957) 1207–1208,

[9] J.R. Erpenback and W.W. Wood, Molecular dynamics calculations of the hard-sphere equation of state, J. Stat. Phys. 35 (1984) 321–340.

[10] R.J. Speedy, Pressure of the metastable hard-sphere fluid, J. Phys.: Cond. Matt. 9 (1997) 8591–8599.

[11] M.E. Fisher, Theory of condensation and the critical point, Physics 3 (1967) 255–283.

[12] J.S. Langer, Theory of the condensation point, Ann. Phys. 41 (1967) 108–157.

[13] W. G. Hoover and F.H. Ree, Melting transition and communal entropy for hard spheres, J. Chem. Phys. 49 (1968) 3609–3617

[14] B.J. Alder and T.E. Wainwright, Phase transitions in elastic disks, Phys. Rev. 127 (1962) 359–361.

[15] J.A. Zollweg, G.V. Chester and P.W. Leung, Size dependent properties of two-dimensional solids, Phys. Rev. B39 (1989) 9518–9530.

[16] J.A. Zollweg and G.V. Chester, Melting in two dimensions, Phys. Rev. 46 (1992) 11186–11189.

[17] J. Lee and K.J. Strandburg, First-order melting transition of the hard disk system, Phys. Rev. B46 (1992) 11190–11193.

[18] A. Jaster, Computer simulations of the two–dimensional melting transition using hard discs, Phys. Rev. E 59 (1999) 2594–2602.

[19] J. M. Kosterlitz and D.J. Thouless, Ordering, metastability and phase transitions in two dimensional systems, J. Phys. C6 (1973) 1181–1203.
[20] B.I. Halperin and D.R. Nelson, Theory of two-dimensional melting, Phys. Rev. Lett. 41 (1978) 121–124.
[21] D.R. Nelson and B.I. Halperin, Dislocation mediated melting in 2 dimensions, Phys. Rev. B 19 (1979) 2457–2484.
[22] A.P. Young, Melting and the vector Coulomb gas in 2 dimensions Phys. Rev. B 19 (1979) 1855–1866.
[23] K. Binder, S. Sangupta and P. Nielaba, The liquid–solid transition of hard discs: first order transition or Kosterlitz–Thouless–Halperin–Nelson–Young scenario. J. Phys. Cond. Matt. 14 (2002) 2323–2333.
[24] W.G. Hoover, M. Ross, K.W. Johnson, D. Henderson, J.A. Barker and B.C. Brown, Soft-sphere equation of state, J. Chem. Phys. 52 (1970) 4931–4941.
[25] W.G. Hoover, S.G. Gray, and K.W. Johnson, Thermodynamic properties of the fluid and solid phases for inverse power potentials, J.Chem. Phys. 55 (1971) 1128–1136.
[26] W.G. Hoover, D.A. Young and R. Grover, Statistical mechanics of phase diagrams. I. Inverse power potentials and the close-packed to body centered cubic transition, J. Chem. Phys. 56 (1972) 2207–2210.
[27] B.B. Laird and A.D.J. Haymet, Phase diagram for the inverse sixth power potential system from molecular dynamics computer simulation, Mol. Phys. 75 (1992) 71–80.
[28] R. Agrawal and D.A. Kofke, Solid–fluid coexistence for inverse-power potentials, Phys. Rev. letts. 74 (1995) 122–125.
[29] D.A. Young and B.J. Alder, Studies in molecular dynamics XVII. Phase diagrams for "step" potentials in two and three dimensions, J. Chem. Phys. 70 (1979) 473–481.
[30] D.A. Young and B. J. Alder, Studies in molecular dynamics XVIII, Square well phase diagram, J.Chem Phys. 73 (1980) 2430–2434.
[31] P. Bolhuis, M. Hagen, D. Frenkel, Isostructural solid–solid transition in crystalline systems with short-ranged interactions, Phys. Rev. E 50 (1994) 4880–4890.
[32] J. Serrano-Illan, G. Navascues and E. Velasco, Noncompact crystalline solids in the square-well potential, Phys. Rev. E73 (2006) 01110–(1–11).
[33] J.E. Lennard-Jones, Proc. R. Soc. London, ser. A 106 (1924),463
[34] T. Kihara and S. Koba, Crystal structures and intermolecular forces of rare gases, J. Phys. Soc. Jpn. 7 (1952) 348–354.
[35] J-P Hansen and L. Verlet, Phase transitions of the Lennard-Jones system, Phys. Rev. 184 (1969) 151–161.
[36] Y. Choi, T. Ree and F.H. Ree, Phase diagram of a Lennard-Jones solid, J. Chem. Phys. 99 (1993) 9917–9919.
[37] F.R. Stillinger, Lattice sums and their phase diagram implications for the classical Lennard-Jones model, J. Chem. Phys. 115 (2001) 5208–5212.

9

High temperature expansions for magnets at H = 0

In this chapter we will develop the theory of high temperature series expansions for the classical n vector and the quantum spin S Heisenberg model . Our goal is to estimate the critical exponents α for the specific heat and γ for the zero field susceptibility introduced in the general theory of critical phenomena and to test the conjecture of universality presented in chapter 5.

We write the Hamiltonian for the classical n vector model on a lattice of N sites as

$$H = -Jn \sum_{\mathbf{R},\mathbf{R}'} \mathbf{n}_{\mathbf{R}} \cdot \mathbf{n}_{\mathbf{R}'} - H \sum_{\mathbf{R}} n^z_{\mathbf{R}} - H_s \sum_{\mathbf{R}} \eta_{\mathbf{R}} n^z_{\mathbf{R}} \tag{9.1}$$

where $\mathbf{n}^2_{\mathbf{R}} = 1$. The sum over $\mathbf{R}$ and $\mathbf{R}'$ is over nearest neighbor pairs (counted once) of an appropriate lattice. The term with $\eta_{\mathbf{R}}$ is to be included for the square, cubic and bcc (bipartite) lattices and $\eta_{\mathbf{R}} = \pm 1$ on the respective sublattices.

When $J > 0$ the model is ferromagnetic, and H couples to the magnetization

$$M(H,T) = \frac{1}{N} \sum_{\mathbf{R}} n^z_{\mathbf{R}}. \tag{9.2}$$

When $J < 0$ the model is antiferromagnetic and on the cubic and bcc (bipartite) lattices H_s couples to the staggered magnetization

$$M_s(H_s,T) = \frac{1}{N} \sum_{\mathbf{R}} \eta_{\mathbf{R}} n^z_{\mathbf{R}}. \tag{9.3}$$

We will concentrate on the particular cases of $n = 1, 2, 3$ which are called Ising, XY and Heisenberg models respectively.

We write the Hamiltonian of the spin S quantum Heisenberg model as

$$H = -\tilde{J} \sum_{\mathbf{R},\mathbf{R}'} \mathbf{S}_{\mathbf{R}} \cdot \mathbf{S}_{\mathbf{R}'} - \tilde{H} \sum_{\mathbf{R}} S^z_{\mathbf{R}} - \tilde{H}_s \sum_{\mathbf{R}} \eta_{\mathbf{R}} S^z_{\mathbf{R}} \tag{9.4}$$

where, again, the sum is over nearest neighbor pairs counted once, the quantum spin operators satisfy the commutation relations

$$[S^x_{\mathbf{R}}, S^y_{\mathbf{R}'}] = iS^z_{\mathbf{R}} \delta_{\mathbf{R},\mathbf{R}'}, \quad [S^y_{\mathbf{R}}, S^z_{\mathbf{R}'}] = iS^x_{\mathbf{R}} \delta_{\mathbf{R},\mathbf{R}'}, \quad [S^z_{\mathbf{R}}, S^x_{\mathbf{R}'}] = iS^y_{\mathbf{R}} \delta_{\mathbf{R},\mathbf{R}'} \tag{9.5}$$

and

$$\mathbf{S}^2 = S(S+1). \tag{9.6}$$

The classical n vector model with $n = 3$ is the limit $S \to \infty$ of the spin S Heisenberg model if we set

$$\tilde{J}S(S+1) = 3J, \quad \tilde{H}[S(S+1)]^{1/2} = H, \quad \tilde{H}_s[S(S+1)]^{1/2} = H_s \tag{9.7}$$

and

$$\mathbf{n_R} = \lim_{S\to\infty} \frac{\mathbf{S_R}}{[S(S+1)]^{1/2}}. \tag{9.8}$$

We will thus normalize the magnetizations for the quantum case as

$$M(H,T) = \frac{1}{N}\sum_{\mathbf{R}} \frac{S^z_{\mathbf{R}}}{[S(S+1)]^{1/2}}, \quad M_s(H_s,T) = \frac{1}{N}\sum_{\mathbf{R}} \eta_{\mathbf{R}} \frac{S^z_{\mathbf{R}}}{[S(S+1)]^{1/2}} \tag{9.9}$$

and define

$$k_BT\chi(T) = \frac{\partial M(T,H)}{\partial H}|_{H=0}, \quad k_BT\chi_s(T) = \frac{\partial M_s(T,H_s)}{\partial H_s}|_{H_s=0}. \tag{9.10}$$

The Hamiltonians for the classical n vector model on the bipartite lattices is invariant if $J \to -J$ and $\mathbf{n_R} \to -\mathbf{n_R}$ on one of the sublattices. Therefore for the n vector model on a bipartite lattice the specific heat at $H = H_s = 0$ has the symmetry in J/k_BT

$$c(-J/k_BT) = c(J/k_BT) \tag{9.11}$$

and the two susceptibilities satisfy

$$\chi_s(-J/k_BT) = \chi(J/k_BT) \tag{9.12}$$

and thus the ferromagnetic and antiferromagnetic cases are essentially identical. For the quantum case the symmetries (9.11) and (9.12) do not hold and we will need to study the ferromagnetic and antiferromagnetic cases separately by giving series both for $\chi(T)$ and $\chi_s(T)$.

High temperature series expansions for these lattice models were developed in the 1950s and 1960s. They are all combinatorial graph enumeration problems and do not involve any integrations as were needed for virial coefficients of the continuum fluids.

We study the n vector model in section 9.1. We restrict our attention to the expansions for $D = 2$ and $D = 3$ where the expansions have been carried out to high order. From these high-order expansions we see that in $D = 2$ there is a striking difference between $n \leq 2$ where all known coefficients of the susceptibility are positive and hence the leading singularity is on the real axis, and $n > 2$ where the evidence is that negative coefficients occur and that the leading singularity is in the complex plane. We estimate these radii of convergence first by means of a ratio analysis and then by the methods of Padé and differential approximates. Wherever possible we use these series to estimate the critical exponents γ and α.

The spin S quantum Heisenberg model is studied in section 9.2. The series for these quantum models are shorter than for the classical n vector model. We investigate in detail the evidence for the prediction of universality that the critical exponents for both the ferromagnetic and antiferromagnetic models will be independent of the quantum spin S. In section 9.3 we discuss the significance of these computations for magnets and conclude in section 9.4 with the interpretation of the classical n vector model as a quantum field theory and in particular how the classical $n = 3$ Heisenberg magnet in two dimensions illustrates the concept of asymptotic freedom.

9.1 Classical n vector model for D = 2, 3

Partition functions for continuum fluids interacting via two-body potentials $U(\mathbf{r})$ are integrals over the Boltzmann weights

$$e^{-U(\mathbf{r})/k_BT} \tag{9.13}$$

and if the two-body potential $U(\mathbf{r})$ were finite for all $\mathbf{r}$ then the limit $T \to \infty$ would be trivial and an expansion about $T = \infty$ could be developed. Unfortunately as we have seen in previous chapters all realistic continuum potentials are infinite at $\mathbf{r} = 0$ and hence expansions about $T \to \infty$ are not possible. This was illustrated most vividly for the hard sphere gas where there is only one isotherm.

However the Hamiltonian for the classical n vector model (9.1) does not contain any infinite terms and thus in the partition function

$$Z = \int \prod_{\mathbf{R}} d\mathbf{n}_{\mathbf{R}} e^{-H/k_BT} \tag{9.14}$$

the limit $T \to \infty$ is trivially given by

$$Z = S_n^N \tag{9.15}$$

where N is the number of sites of the lattice and S_n is the surface area of the n-dimensional sphere. Thus we may obtain a high temperature series expansion by expanding the exponential in (9.14) as

$$Z = S_n^N \sum_{j=0}^{\infty} \frac{\mu_n}{(k_BT)^j j!} \tag{9.16}$$

with

$$\mu_j = \frac{1}{S_n^N} \int \prod_{\mathbf{R}} d\mathbf{n}_{\mathbf{R}} (-H)^j. \tag{9.17}$$

The partition function Z is of course exponentially large in N and we are interested in the thermodynamic limit of the free energy

$$F = -k_BT \lim_{N \to \infty} \frac{1}{N} \ln Z. \tag{9.18}$$

Therefore we need

$$\ln Z = NS_n + \ln \sum_{j=0}^{\infty} \frac{\mu_j}{(k_BT)^j j!} = NS_n + \sum_{j=1}^{\infty} \frac{\lambda_j}{(k_BT)^j j!} \tag{9.19}$$

which implicitly defines the λ_j in terms of μ_j. Explicitly we have for the first few terms

$$\lambda_1 = \mu_1 \tag{9.20}$$
$$\lambda_2 = \mu_2 - \mu_1^2 \tag{9.21}$$
$$\lambda_3 = \mu_3 - 3\mu_1\mu_2 + 2\mu_1^3 \tag{9.22}$$
$$\lambda_4 = \mu_4 - 4\mu_3\mu_1 - 3\mu_2^2 + 12\mu_2\mu_1^2 - 6\mu_1^4. \tag{9.23}$$

The μ_n will contain all powers of N up to N^n. However, the λ_n must contain only terms linear in N because the free energy is extensive. Thus we must have

$$\lambda_n = N \times \text{coefficient of } N \text{ in } \mu_n. \tag{9.24}$$

This is completely analogous to the step in the derivation of the cluster expansion in chapter 6 of the passage from all diagrams to all connected Mayer diagrams.

The computation of λ_n using (9.24) now involves two steps: the integrals over $d\mathbf{n}_\mathbf{R}$ which are easily done, and a combinatorial sum which may be expressed as a graph counting problem. In this chapter we will not explore further details of the relevant graph counting problem and in any event the counting of the graphs must be done by computer. Instead we refer the reader to the original papers where the results are derived and focus instead on the presentation and interpretation of the results of five decades of work.

High temperature series expansions for the n vector model are generally written for the specific heat in the form

$$c_H = \frac{q}{2n}(J/k_BT)^2 \sum_{k=0}^{\infty} e_k (J/k_BT)^k \tag{9.25}$$

and for the susceptibility

$$k_BT\chi(T) = \frac{1}{n}\sum_{k=0}^{\infty} a_k (J/k_BT)^k \tag{9.26}$$

where q is the number of nearest neighbors and the normalization is chosen such that

$$e_0 = a_0 = 1. \tag{9.27}$$

The only exception is the Ising model ($n = 1$) where instead the variable

$$v = \tanh J/k_BT \tag{9.28}$$

is often used as the expansion variable and the expansions are of the form

$$c_H = \frac{q}{2}(J/k_BT)^2 \sum_{k=0}^{\infty} e_k v^k \tag{9.29}$$

where q is the coordination number of the lattice and

$$k_BT\chi(T) = \sum_{k=0}^{\infty} a_k v^k. \tag{9.30}$$

In practice we are only able to compute a finite number of terms in a high temperature series expansion and thus in order to use these expansions to study critical phenomena some method extrapolation will be needed. All methods of extrapolation make assumptions which may be open to criticism and thus it is extremely useful that there are two special cases for which exact computations may be carried out: the case $n = 1$ (the Ising model) in $D = 2$ and the limit $n \to \infty$ in arbitrary D.

The Ising model in $D = 2$ will be treated in detail in chapters 10–12. The internal energy of the Ising model on the isotropic square lattice is

$$u = -J \coth 2J/k_BT[1 + 2\pi^{-1}(2\tanh^2 2J/k_BT - 1)K(k)] \tag{9.31}$$

with

$$k = 2\frac{\sinh 2J/k_BT}{(\cosh 2J/k_BT)^2} \tag{9.32}$$

where $K(k)$, the complete elliptic integral of the first kind is defined as

$$K(k) = \int_0^{\pi/2} \frac{d\phi}{(1 - k^2 \sin^2 \phi)^{1/2}}. \tag{9.33}$$

The specific heat is given in terms of the internal energy u as

$$c = \frac{\partial u}{\partial T} = -\frac{1}{k_BT^2}\frac{\partial u}{\partial \beta} \tag{9.34}$$

and thus, noting that the complete elliptic integral has the expansion for small k

$$K(k) = \frac{\pi}{2}\sum_{n=0}^{\infty}\left(\frac{(1/2)(1+1/2)\cdots(n+1/2)}{n!}\right)^2 k^{2n} \tag{9.35}$$

an exact expression for the high temperature series expansion of the specific heat can be obtained.

From the exact expressions (9.31) and (9.34) we find several important properties of the specific heat. We first note that the complete elliptic integral $K(k)$ has a logarithmic singularity at $k^2 = 1$ and is analytic everywhere else. It thus follows from (9.34) that the specific heat has a logarithmic singularity at $k^2 = 1$ which corresponds to

$$\sinh 2J/k_BT = \pm 1. \tag{9.36}$$

Any method of using a finite number of terms in a high temperature series expansion to extract a critical temperature and critical exponent must be tested against this exact result.

We also note that the internal energy depends on J/k_BT only through the variable k and thus is a periodic function of J/k_BT with period $i\pi$. This feature of periodicity is only a feature of the case $n = 1$ and does not hold for any other value of n. Consequently when we use the $n = 1$ model to compare with general values of n we will often expand in terms of J/k_BT and not a variable such as $v = \tanh J/k_BT$ which is only appropriate for $n = 1$.

The $n \to \infty$ case is called the spherical model and is exactly solvable [1–3]. For $D > 2$ the model has a positive critical point with exponents

$$\alpha = (D-4)/(D-2) \quad \gamma = 2/(D-2) \quad \text{for } 2 < D < 4 \tag{9.37}$$

$$\alpha = 0, \quad \gamma = 1 \quad D > 4. \tag{9.38}$$

The case of $D = 2$ is particularly interesting. Here if we define the variable k from

$$\frac{J}{k_B T} = \frac{k}{2\pi} K(k) \tag{9.39}$$

the susceptibility is [4]

$$k_B T \chi = \frac{k_B T}{J} \frac{k}{4(1-k)} = \frac{\pi}{2(1-k)K(k)}. \tag{9.40}$$

This susceptibility is analytic for all positive T and diverges exponentially as $T \to 0$.

For $D = 2$ and $n = 1$ the susceptibility of the square lattice was expanded in 2001 to order 330 by Orrick, Nickel, Guttmann and Perk [5]. For arbitrary n in $D = 2$ the longest series for the susceptibility on the square lattice is in the 1996 paper of Butera and Comi [6]. On the triangular lattice the longest series for the susceptibility for $n = 2, 3$ is from the 1996 articles of Campostrini, Pelissetto, Rossi and Vicari [7,8].

For $D = 2$ and $n = 1$ the specific heat of the square lattice was exactly computed in 1944 by Onsager [9] and for the triangular lattice in 1950 by Houtapple [10]. For $D = 2$ and $n = 2, 3$ the longest series for the specific heat on the square and triangular lattices is from the 1996 articles of Campostrini, Pelissetto, Rossi and Vicari [7,8].

For $D = 3$ the longest series for the susceptibility for arbitrary n for cubic and bcc lattices are the 1999 results of Butera and Comi [11]. On the fcc lattice the longest series for the susceptibility in $n = 3$ and $D = 3$ are the 1974 results to 10th order of Rushbrooke, Baker and Wood [12] and the 11 and 12 order results done in 1982 of McKenzie, Domb and Hunter [13]. For $n = 2$ on the fcc lattice the longest series is the 1967 work of Bowers and Joyce [15]. For $n = 1$ on the bcc lattice the longest series (to order $k = 21$) was first obtained by Nickel [14] in 1980 and on the fcc lattice the longest series are from the 1975 work of McKenzie [18].

For $D = 3$ the longest series for specific heats for arbitrary n for cubic and bcc latices are the 1999 results of Butera and Comi [19]. For $n = 1$ the longest series for the cubic lattice (to order $k = 24$) was obtained Guttmann and Enting [21,22] in 1994. For the fcc lattice the longest series for the specific heat for $n = 1$ are the 1972 results of Sykes, Hunter, McKenzie and Heap [20] and for arbitrary n the longest series are the 1979 results of English, Hunter and Domb [23].

These references are summarized in Table 9.1.

9.1.1 Results for D = 2

In chapter 4 we saw that in two dimensions the n vector model with $n \geq 2$ has no spontaneous order while in three dimensions there is a positive temperature below which spontaneous order exists. Consequently the physics in two and three dimensions

Table 9.1 Summary of references for high temperature series expansion for the classical n vector model.

D	n	susceptibility	specific heat
2 sq	1	Orrick, Nickel Guttmann, Perk 2001 [5]	Onsager (1944 exact) [9]
2 tri	1		Houtapple (1950 exact) [10]
2 sq	arb	Butera, Comi 1996 [6]	
	2	Campostrini et al 1996 [8]	Campostrini et al 1996 [8]
	3	Butera, Comi 1996 [6] Campostrini et al 1996 [7]	Campostrini et al 1996 [7]
2 tri	2	Campostrini et al 1996 [8]	Campostrini et al 1996 [8]
	3	Campostrini et al 1996 [7]	Campostrini et al 1996 [7]
3 cub bcc	arb	Butera,Comi 1999 [11]	Butera,Comi 1999 [19]
3 cub	1	Nickel 1980 [14]	Guttmann, Enting 1993-94 [21,22]
3 fcc	arb		English, Hunter Domb 1979 [23]
	1	S. McKenzie 1975 [18]	Sykes,Hunter McKenzie, Heap 1972 [20]
	2	Bowers, Joyce 1967 [15]	
	3	Rushbrooke, Baker, Wood 1973 [12] McKenzie, Domb, Hunter 1982 [13]	

will be quite different and this difference is very apparent in the high temperature expansions.

The specific heat for $D = 2$ has been expanded to order 21 on the square lattice for $n = 2, 3, 4, 8$ and on the triangular lattice to order 14 by Campostrini, Pelissetto, Rossi and Vicari [7,8]. The results for $n = 2, 3$ are given in Table 9.2 and indeed there is a dramatic difference between the case $n = 2$ where on both the triangular and square lattice all coefficients beginning with e_6 are negative and the case $n = 3$ where on both the square and triangular lattice there is oscillation of signs as k increases. This oscillation of signs is also seen in the results of [8] for $n = 4$ and 8. The difference is made even more striking when compared with the exact result for the Ising model $n = 1$ where all coefficients are known to be positive.

The susceptibility for $D = 2$ has been expanded to order 21 on the square lattice for arbitrary n by Butera and Comi [6] and for $n = 2, 3, 4, 8$ by Campostrini, Pelissetto, Rossi and Vicari [7], [8], and on the triangular lattice to order 15 [7], [8]. The results are given in Table 9.3.

In contrast with the specific heat coefficients of Table 9.2 all entries in Table 9.3 are positive. Nevertheless it is expected, that for sufficiently large order, the coefficients for $n = 3$ will oscillate just as did the coefficients for the specific heat. These oscillations can be inferred by an examination of the coefficients for arbitrary n of Butera and Comi [6] which we reproduce in the appendix and are shown in Figures 9.1 and 9.2 where we plot $a_k/2^k$ versus k. These results for the n vector model with finite n are to be compared with the exact result (9.40), (9.39) for the susceptibility of the spherical

Table 9.2 Normalized high temperature expansion coefficients e_k for the specific heat of n vector model in $D = 2$. The data for $n = 2$ are from [8]. The data for $n = 3$ are from [7].

	$n = 3$ Heisenberg		$n = 2$ XY	
k	square	triangular	square	triangular
0	1	1	1	1
1	0	4	0	4
2	4.2	10.2	4.50	10.5
3	0	16	0	20
4	−3.42857	3.77143	1.66666	29.1666
5	0	−84.48	0	28
6	−27.512	−430.664	−4.52083	−35.4375
7	0	−1607.00	0	−418.777
8	−102.259	−5193.59	−54.825	−40.4705
9	0	−14622.4	0	−7821.
10	−206.407	−34794.1	−223.354	−25667.1
11	0	−65031.6	0	−77734.3
12	140.219	−67930.6	−684.634	−225152.
13	0	137022.	0	−633142.
14	2293.62	1.2417×10^6	−1748.06	-1.7347×10^6
15	0		0	
16	7755.67		−4233.76	
17	0		0	
18	17992.4		−8017.15	
19	0			
20	16097.7		−210209.4	

model which is expanded to order 62 in Table 9.4. In Fig. 9.3 we plot for the spherical model $a_k(45.484177)^k$ where the normalizing factor is the exact radius of convergence.

In Fig. 9.1 we see that for $n = 1, 2$ there is no indication that a_k ever vanishes.We also see that a_k is positive for all values of n for $k \leq 10$. However, we also see in Fig. 9.2 that for $n \geq 5$ negative values of a_k occur for $k \leq 21$ and it is extremely plausible that negative values of a_k also occur for $n = 3, 4$ at values of k which are not unreasonably large. This behavior of a_k as a function of n is very reminiscent of the behavior of the hard sphere virial coefficients as a function of dimension D.

From Figs. 9.1–9.3 we infer that for all $n > 2$ the coefficients a_k oscillate in sign as $k \to \infty$ and that as a function of k the location of the first minimum decreases monotonically as a function of n as n increases. For the $n \to \infty$ spherical model limit the period of oscillation in k is 8 which locates the leading singularity of the susceptibility in the complex plane at $Re^{\pi i/4}$ where R is the radius of convergence. Unfortunately, the dependence on n (if any) of the period of oscillation cannot be obtained from the data for $k \leq 21$.

Table 9.3 Normalized high temperature expansion coefficients a_k for the susceptibility of the $n = 1, 2, 3$ vector model square and triangular lattices in $D = 2$. The data for the square lattice are from [6–8]. The data for the triangular lattice are from [7, 8]. The entry in the table is to be multiplied by the power of 10 given in parentheses.

	$n = 3$ Heisenberg		$n = 2$ XY		n=1 Ising
k	square	triangular	square	triangular	square
0	1	1	1	1	1
1	4	6	4	6	4
2	12	30	12	30	12
3	33.6	134.4	34	135	34.6666
4	85.6	560.4	88	570	92
5	2.06857(2)	2.21974(3)	2.19333(2)	2306	2.40533(2)
6	4.80823(2)	8.45020(3)	529	9.04147(3)	6.11200(2)
7	1.07756(3)	3.11315(4)	1.24442(3)	3.45821(4)	1.53818(3)
8	2.34924(3)	1.11528(5)	2.86868(3)	1.29634(5)	3.80964(3)
9	4.97829(3)	3.89999(5)	6.48988(3)	4.77988(5)	9.36466(3)
10	1.03249(4)	1.33503(6)	1.44917(4)	1.73825(6)	2.28208(4)
11	2.09755(4)	4.48333(6)	3.19527(4)	6.24694(6)	5.53176(4)
12	4.18002(4)	1.47960(7)	6.97111(4)	2.22202(7)	1.33267(5)
13	8.18813(4)	4.80442(7)	1.50671(5)	7.83250(7)	3.19862(5)
14	1.57742(5)	1.53661(8)	3.23002(5)	2.73888(8)	7.64296(5)
15	2.99110(5)	4.84492(8)	6.87193(5)	9.50902(8)	1.82100(6)
16	5.58647(5)		1.45252(6)		4.32421(6)
17	1.02851(6)		3.05148(6)		1.02449(7)
18	1.86601(6)		6.37548(6)		2.42092(7)
19	3.33902(6)		1.32535(7)		5.71013(7)
20	5.89034(6)		2.74238(7)		1.34401(8)
21	1.02425(7)		5.65014(7)		3.15860(8)

9.1.2 A qualitative interpretation of the D = 2 data

The three different types of behavior of the high temperature series for $n = 1, 2, 3$ vividly illustrates both the utility and the limitations of high temperature series in the study of phase transitions and critical phenomena.

Consider first the Ising model $n = 1$. Here the coefficients of both the specific heat and the susceptibility do not oscillate in sign as $k \to \infty$ which means that the leading singularity occurs at a positive value of T. This value may be estimated from the series expansions and is found to be the same for both the specific heat and the susceptibility. Moreover because the coefficients, in addition to having no oscillations in sign are all positive the specific heat and the susceptibility both monotonically increase as T decreases and, indeed will diverge at T_c. The nature of these divergences will be studied in detail in chapters 10 and 12.

In contrast when $n \geq 3$ the coefficients of both the specific heat and the susceptibility oscillate in sign as $k \to \infty$ and therefore the leading singularity occurs not on the positive T axis but in the complex plane. Indeed we see in the spherical model

Table 9.4 High temperature expansion coefficients of the susceptibility a_k in $D = 2$ for the $n \to \infty$ (spherical model) in terms of the variable Jn/k_BT. The data are from [4].

k	a_k	k	a_k	k	a_k
0	1	21	−47360	42	−1873359399232
1	4	22	677568	43	−5306076848128
2	12	23	2097152	44	−6720867874112
3	32	24	2918416	45	1226303567872
4	76	25	407296	46	382358039040
5	160	26	−12607296	47	107615266537472
6	304	27	−37715968	48	135063348070288
7	512	28	−50921792	49	−29380330962176
8	748	29	−2226176	50	−788936558830912
9	928	30	240405248	51	−2209168717692928
10	880	31	703987712	52	−2751199251809600
11	512	32	929619628	53	68130811657216
12	80	33	−29107808	54	16430657778461440
13	256	34	−4684667536	55	45816112359145472
14	2752	35	−13519134208	56	56679265503639232
15	8192	36	−17550216752	57	−155517312752318464
16	12332	37	1607717632	58	−344954419919964416
17	5536	38	92968930496	59	−958513093597462528
18	−37008	39	265456566272	60	−11789689975952248323
19	−122368	40	339970952848	61	349764119023599616
20	−178096	41	−47971688192	62	7293149768561830912

that there are an infinite number of singularities in the complex T plane and that on the real T axis the free energy and susceptibility only fail to be analytic at $T = 0$. It is believed that this is the case for all $n \geq 3$. It is not possible, at least without strong additional assumptions, to study the behavior of such systems as $T \to 0$ from the high temperature expansions.

The feature of having the critical temperature at $T = 0$ is referred to as being "asymptotically free" and is a property which the $n = 3$ model in $D = 2$ shares with gauge theories of strong interactions in four dimensions.

We finally consider the case of $n = 2$. Once again there is no oscillation in the coefficients of either the specific heat or the susceptibility and thus the leading singularity is on the positive real axis. But now the coefficients in the specific heat are negative as $k \to \infty$ so the specific heat, which must be positive, is not monotonic and cannot diverge to infinity at $T \to \infty$. Indeed, further study shows that the singularity is not a power law singularity but instead has a weak essential singularity of exponential form.

Furthermore below T_c it follows from the theorems in chapter 4 that unlike the Ising model $n = 1$ there can be no spontaneous magnetization below T_c and thus the transition at T_c for $n = 2$ in $D = 2$ is very different from the typical magnetic transition of the Ising model in $D = 2, 3$ or the Heisenberg model in $D = 3$. This new type of transition was first discovered by Kosterlitz and Thouless [24] in 1973.

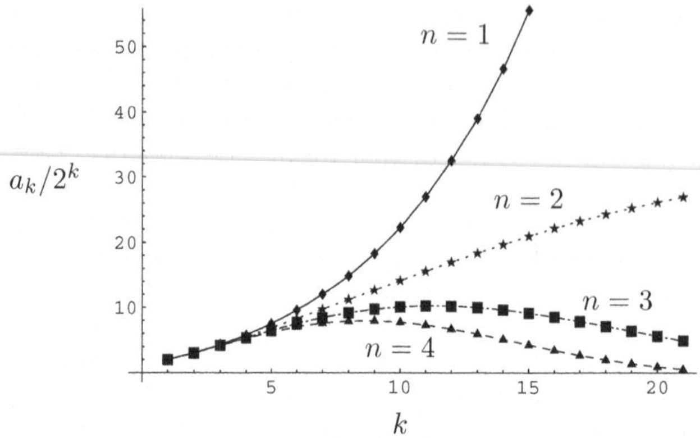

Fig. 9.1 A plot of the normalized coefficients $a_k/2^k$ of [6] of the susceptibility of the n vector model for $n = 1, 2, 3, 4$ in $D = 2$ for $1 \leq k \leq 21$. The coefficients are reproduced in the appendix.

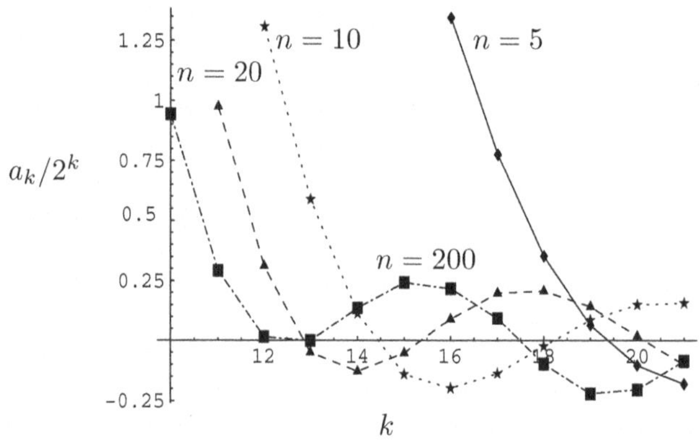

Fig. 9.2 A plot of the normalized coefficients $a_k/2^k$ of [6] of the susceptibility of the n vector model for $n = 5, 10, 20, 200$ in $D = 2$ for $1 \leq k \leq 21$. The coefficients are reproduced in the appendix.

9.1.3 Results for D = 3

The series expansions for $D = 3$ are much more uniform than those of $D = 2$ because the signs of all the coefficients for both the specific heat and the susceptibility are positive for all n. Thus the leading singularity for all cases is on the positive real T axis. Furthermore we saw in chapter 4 that for all n for $D = 3$ there is spontaneous magnetization at sufficiently low temperatures. Therefore the n vector model for arbitrary n in $D = 3$ will behave qualitatively as does the $n = 1$ Ising model in $D = 2$ and the phenomena of the Kosterlitz–Thouless transition and asymptotic freedom do not exist in $D = 3$.

The results of the series expansions are presented below. The series for the susceptibility of the Ising model is particularly elegant in terms of the variable v because all the expansion coefficients are integers. The results of this expansion are given in

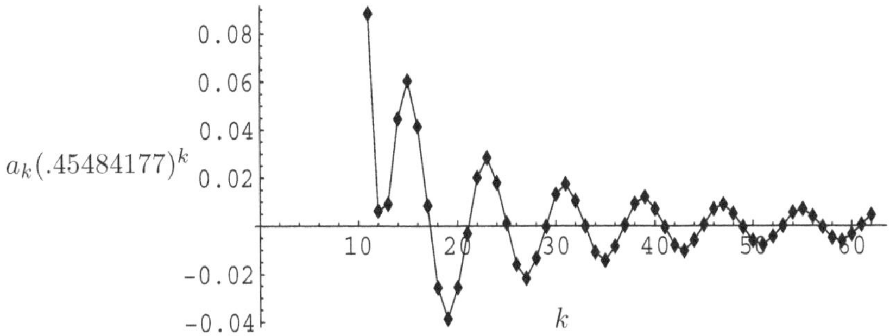

Fig. 9.3 A plot of the normalized coefficients $a_k(.45484177)^k$ of the susceptibility of the spherical ($n \to \infty$) model for $11 \leq k \leq 62$.

Table 9.5. For comparison with $n \geq 2$ where the variable v is not useful we give in Table 9.6 the expansion coefficients in terms of the expansion variable J/k_BT. The expansion of the specific heat in terms of v and J/k_BT are both used in the literature but unlike the susceptibility expansion the coefficients are not in general integers and because our focus is on comparison with $n \geq 2$ the expansion in Table 9.7 is in terms of the variable J/k_BT.

For $n = 2, 3$ the numerical values for the susceptibility coefficients are given in Table 9.8. For the cubic and bcc lattices the data are from [11], the fcc data for $n = 2$ is from [15] and the fcc data for $n = 3$ is from [12] and [13]. The numerical values for the specific heat coefficients are given in Table 9.9. For the cubic and bcc lattices the data are from [19] and for the fcc lattice the data is from [23]. We note that the Hamiltonian used in the papers of Butera and Comi [11] and [19] is normalized to unity whereas in (9.1) there is a factor of Jn. Thus the critical values β_c^{BC} of [11] and [19] are related to our critical variable J/k_BT_c by

$$\beta_c^{BC} = n\frac{J}{k_BT_c}. \tag{9.41}$$

9.1.4 Critical exponents

A major reason for the computation of high temperature series expansions is the desire to use them to study the phase transitions and critical exponents of the system. Logically, of course, this can never be done with only a finite number of terms. However, for systems that have not been solved exactly the extrapolation of high temperature expansions is one of the very few ways in which critical phenomena can be studied. All of the methods by very definition of an extrapolation scheme must make assumptions about the behavior of the function being extrapolated which ultimately cannot be verified with only a finite number of terms.

To see the need for making assumptions, consider the magnetic susceptibility at $H = 0$ which for $T > T_c$ is given in terms of the spin correlation functions as

$$k_BT\chi(T) = k_BT\frac{\partial M(T,H)}{\partial H}|_{H=0} = \sum_{\mathbf{R}}\langle n_0^z n_{\mathbf{R}}^z\rangle \tag{9.42}$$

Table 9.5 Normalized high temperature expansion coefficients a_k using v as the expansion variable for the susceptibility of the $n = 1$ (Ising) model on the simple cubic, fcc and bcc lattices in $D = 3$. The data for cubic is from [16, 17], the data for bcc is from [14] and the data for fcc is from [18].

k	cubic	bcc	fcc
0	1	1	1
1	6	8	12
2	30	56	132
3	150	392	1404
4	726	2648	14652
5	3510	17864	151116
6	16710	118760	1546332
7	79494	789032	15734460
8	375174	5201048	159425580
9	1769686	34268104	1609987708
10	8306862	224679864	16215457188
11	38975286	1472595144	162961837500
12	182265822	9619740648	1634743178420
13	852063558	62823141192	16373484437340
14	3973784886	409297617672	163778159931180
15	18527532310	2665987056200	1636328839130860
16	86228667894	17333875251192	
17	401225368086	112680746646856	
18	1864308847838	731466943653464	
19	8660961643254	4747546469665832	
20		30779106675700312	
21		199518218638233896	

and the internal energy which is given in terms of the nearest neighbor two-spin correlation as

$$U = -Jnq\langle \mathbf{n}_0 \cdot \mathbf{n}_{\mathbf{R}_{nn}} \rangle = -Jn^2 q \langle n^z{}_0 \cdot n^z{}_{\mathbf{R}_{nn}} \rangle \qquad (9.43)$$

where $\mathbf{R}_{nn}$ is one of the q nearest neighbors of the origin.

The susceptibility $\chi(T)$ diverges as $T \to T_c$ and in chapter 5 we defined the critical exponent γ as

$$k_B T \chi(T) \sim A_\chi (T - T_c)^{-\gamma} \qquad (9.44)$$

as $T \to T_c$. But from the expansion of the susceptibility in terms of spin correlation (9.42) we see that (9.44) is not the only singularity at T_c present in $\chi(T)$ because, from (9.43), the nearest neighbor correlation will have a singularity, coming from the singularity in the specific heat, of

$$(T - T_c)^{1-\alpha}. \qquad (9.45)$$

Indeed, from the study of the Ising model in chapters 10–12 we not only expect singularities of the form (9.45) in further neighbor correlations but higher order singularities such as

$$(T - T_c)^{m(1-\alpha)} \qquad (9.46)$$

as well.

Table 9.6 Normalized high temperature expansion coefficients a_k using J/k_BT as the expansion variable for the susceptibility of the $n = 1$ vector (Ising) model for simple cubic, fcc and bcc lattices in $D = 3$ as obtained from Table 9.5. The entry in the table is to be multiplied by the power of 10 given in parentheses.

k	cubic	bcc	fcc
0	1	1	1
1	6	8	12
2	30	56	132
3	148	3.89333(2)	1400
4	706	2.61057(3)	14564
5	333600(3)	1.74731(4)	1.49714(5)
6	1.57533(4)	1.15250(5)	1.52685(6)
7	7.37537(4)	7.59546(5)	1.54836(7)
8	3.42619(5)	4.96669(6)	1.56350(8)
9	1.59037(6)	3.24586(7)	1.57354(9)
10	7.34697(6)	2.11101(8)	1.58301(10)
11	3.39206(7)	1.37234(9)	1.58183(11)
12	1.56104(8)	8.89225(9)	1.58123(12)
13	7.18079(8)	5.75990(10)	1.57843(13)
14	3.29549(9)	3.72216(11)	1.57341(14)
15	1.51188(10)	2.40468(12)	1.56501(15)
16	6.92398(10)	1.55078(13)	
17	3.17010(11)	9.99874(13)	
18	1.44943(12)	6.43783(14)	
19	6.62559(12)	4.14433(15)	
20	3.02535(13)	2.66495(16)	
21	1.38116(14)	1.71338(17)	

It is not unreasonable to expect that similar problems will in general also be present in the specific heat. Thus we expect that as more and more terms are added to a high temperature series expansion eventually more and more of this complicated singularity structure will be needed to explain the data and that any conclusions drawn from a finite number of terms in a high temperature series expansion run the risk of being misleading.

Nevertheless many schemes have been used in the past 50 years to estimate critical exponents from high temperature series expansions and the subject is exhaustively and extensively discussed by Guttmann [25–27] with many examples. The most general of these methods is the method of differential approximates which approximates the susceptibility and specific heat by a function $F(x)$ which satisfies a linear inhomogeneous finite order differential equation

$$\sum_{k=0}^{K} Q_k(x)\frac{d^kF(x)}{dx^k} = P(x) \tag{9.47}$$

where x is some suitable variable such as J/k_BT or $v = \tanh J/k_BT$ and $Q_k(x)$ and

Table 9.7 Normalized high temperature expansion coefficients e_k in terms of the variable J/k_BT for the specific heat of the $n = 1$ (Ising) model on the simple cubic, fcc and bcc lattices in $D = 3$. The data for cubic and bcc are from [11, 19]. The fcc data are from [20]. The entry in the table is to be multiplied by the power of 10 given in parentheses.

k	cubic	bcc	fcc
0	1	1	1
1	0	0	8
2	11	35	65
3	0	0	5.33333(2)
4	1.80600(2)	9.90667(2)	4.43067(3)
5	0	0	3.77291(4)
6	2.74549(3)	3.10012(4)	3.28515(5)
7	0	0	2.90479(6)
8	4.59474(4)	1.03199(6)	2.59697(7)
9	0	0	2.34183(8)
10	7.99761(5)	3.55827(7)	2.12669(9)
11	0	0	1.94281(10)
12	1.42349(7)	1.25638(9)	1.78387(11)
13	0	0	
14	2.57857(8)	4.51468(10)	
15	0	0	
16	4.73436(9)	1.64433(12)	
17	0	0	
18	8.78548(10)	6.05317(13)	
19	0	0	
20	1.64447(12)	2.24770(15)	
21	0	0	

$P(x)$ are polynomials. Several special cases of this most general form are extensively used and must be noted:

1) **Padé approximants**
Here $K = 0$ and the approximating function is

$$F(x) = P(x)/Q_0(x) \tag{9.48}$$

with $P(x)$ and $Q(x)$ polynomials. This will be useful if the function being approximated has poles as its only singularities.

2) **dlog-Padé Approximants**
Here $K = 1$ and $P(x) = 0$ so the approximating function satisfies

$$\frac{dF(x)}{dx} = -\frac{Q_0(x)}{Q_1(x)}F(x) \tag{9.49}$$

This form is useful if the function being approximated is of the form

$$B(x)(1 - x/x_0)^{-p} \tag{9.50}$$

with $B(x)$ regular at x_0.

Table 9.8 Normalized high temperature expansion coefficients a_k for the susceptibility of the $n = 2, 3$ vector model simple cubic, fcc and bcc lattices in $D = 3$. The entry in the table is to be multiplied by the power of 10 given in parentheses. The data for cubic and bcc are from [11], for fcc with $n = 2$ from [15], and for fcc with $n = 3$ from [12] and [13]. Note that the coefficients here for $n = 3$ and for $n = 2$ on the cubic and bcc lattices are n^k times the coefficients given in the references because of the factor of n in our Hamiltonian (9.1).

	$n = 3$ Heisenberg			$n = 2$ XY		
k	cubic	bcc	fcc	cubic	bcc	fcc
0	1	1	1	1	1	1
1	6	8	12	6	8	12
2	30	56	132	30	56	132
3	146.40	387.20	1396.8	147	388	1398
4	690.2	2580.8	14455.2	696	2592	14496
5	3.22388(3)	1.70856(4)	1.47439(5)	3.27498(3)	1.72307(4)	1.48294(5)
6	1.48249(4)	1.11400(5)	1.48869(6)	1.51715(4)	1.12839(5)	1.50307(6)
7	6.77947(4)	7.23335(5)	1.49196(7)	7.00091(4)	7.36260(5)	1.51324(7)
8	3.07496(5)	4.65826(6)	1.48668(8)	3.20512(5)	4.77381(6)	1.51568(8)
9	1.38975(6)	2.99144(7)	1.47468(9)	1.46371(6)	3.08660(7)	
10	6.24927(6)	1.91123(8)	1.45734(10)	6.65206(6)	1.98584(8)	
11	2.80303(7)	1.21858(9)	1.43578(11)	3.01779(7)	1.27584(9)	
12	1.25292(8)	7.74262(9)	1.41087(12)	1.36457(8)	8.16951(9)	
13	5.59000(8)	4.91184(10)		6.16220(8)	5.22557(10)	
14	2.48777(9)	3.10821(11)		2.776167(9)	3.33444(11)	
15	1.10557(10)	1.96376(12)		1.24946(10)	2.12595(12)	
16	4.90388(10)	1.23925(13)		5.61341(10)	1.35332(13)	
17	2.17245(11)	7.80998(13)		2.51990(11)	8.60485(13)	
18	9.61217(11)	4.91466(14)		1.12964(12)	5.46486(14)	
19	4.24853(12)	3.09023(15)		5.06076(12)	3.46878(15)	
20	1.87561(13)	1.94068(16)		2.26469(13)	2.19929(16)	
21	8.27185(13)	1.21803(17)		1.01291(14)	13.9377(17)	

3) **First order approximants**

Here $K = 1$, and $P(x)$ does not vanish identically. These functions are useful if the function being approximated is of the form

$$A(x) + B(x)(1 - x/x_0)^{-p} \tag{9.51}$$

with $A(x)$ and $B(x)$ regular at $x = x_0$.

4) **Second order approximants**

Here $K = 2$. These functions are useful if the function being approximated is of the form

$$A(x) + B_1(x)(1 - x/x_0)^{-p_1} + B_2(x)(1 - x/x_0)^{-p_2} \tag{9.52}$$

with $A(x)$ and $B_k(x)$ regular at $x = x_0$. Approximates with $K > 2$ will allow there to be K singularities.

To use these approximants the orders of the polynomials are chosen such that the coefficients in the polynomials are determined by having the approximating function agree with the finite number of terms computed in the high temperature expansion.

Table 9.9 Normalized high temperature expansion coefficients e_k for the specific heat of the $n = 2, 3$ vector model simple cubic, fcc and bcc lattices in $D = 3$. The data for cubic and bcc lattices are from [19] and the data for the fcc lattice are from [23]. The entries in the table are to be multiplied by the power of 10 given in parentheses.

	$n = 3$ Heisenberg			$n = 2$ XY		
k	cubic	bcc	fcc	cubic	bcc	fcc
0	1	1	1	1	1	1
1	0	0	8	0	0	8
2	10.2	34.2	62.4	10.4	34.5	64.5
3	0	0	511.998	0	0	520
4	1.50571(2)	8.96573(2)	4.09426(3)	1.61667(2)	9.31667(2)	4.21667(3)
5	0	0	3.32025(4)	0	0	3.49162(4)
6	1.94176(3)	2.51956(4)	2.75047(5)	2.23256(3)	2.73649(4)	2.95292(5)
7	0	0	2.31400(6)	0	0	2.53723(6)
8	2.81999(4)	7.52999(5)	1.97004(7)	3.44788(4)	8.56307(5)	2.20588(7)
9	0	0	1.69327(8)	0	0	1.93569(8)
10	4.30072(5)	2.34097(7)	1.46705(9)	5.57100(5)	2.8417(7)	1.71171(9)
11	0	0	1.27987(10)	0	0	
12	6.74276(6)	7.48567(8)		9.19269(6)	9.29620(8)	
13	0	0		0	0	
14	1.08142(8)	2.44625(10)		1.56205(8)	3.1677(10)	
15	0	0		0	0	
16	1.76635(9)	8.13204(11)		2.68515(9)	1.09565(12)	
17	0	0		0	0	
18	2.92754(10)	2.74057(13)		4.67471(10)	3.83781(13)	
19	0	0		0	0	
20	4.90992(11)	9.33970(14)		8.22264(11)	1.35783(15)	

The critical point is then determined by the smallest value of $1/T$ at which the approximating function is singular. This singularity is at the zeros of the leading coefficient $Q_K(x)$. At this singularity the approximating function in general has power law singularities which are easily determined from the indicial equation of the linear differential equation (9.47).

We thus see that in principle the susceptibility needs to be approximated by at least a second order approximate in order to encompass the singularities γ and $1 - \alpha$.

High temperature series are one of the very few tools available to study critical phenomena for systems which are not exactly solvable and thus they are widely used to gain insight into critical phenomena. Nevertheless, the validity of the exponents obtained is in the last analysis a subjective decision to be made by the reader.

The exponents γ and α will be estimated below in section 1.6 using the various cases of differential approximates. However, to gain a qualitative insight into the analysis we begin with a discussion of the most primitive of all methods; the ratio method.

9.1.5 The ratio method

The most elementary way to study the radius of convergence and leading singularity in the high temperature series expansion is by analysis of the ratios of coefficients.

The method begins with the elementary expansion valid for $p > 0$

$$\sum_{k=0}^{\infty}(x/x_0)^k\frac{\Gamma(k+p)}{k!\Gamma(p)} = [1-(x/x_0)]^{-p}. \tag{9.53}$$

Thus calling c_k the coefficient of x^k we see that as $k \to \infty$

$$c_k \to x_0^{-k}k^{p-1}/\Gamma(p)] \tag{9.54}$$

and thus

$$r_k = c_{k+1}/c_k = x_0^{-1}(1+1/k)^{(p-1)} \sim x_0^{-1}[1+(p-1)/k + O(k^{-2})]. \tag{9.55}$$

The simplest version of the ratio test uses this simple relation to estimate the critical value x_0 as

$$kr_k - (k-1)r_{k-1} = x_0^{-1}[1+O(1/k^2)] \tag{9.56}$$

or, more generally in a form which reveals some of the arbitrariness inherent in the method

$$(k+\epsilon)r_k - (k+\epsilon-1)r_{k-1} = x_0^{-1}[1+O(1/k^2)]. \tag{9.57}$$

The exponent p is then estimated (with $\epsilon = 0$) as

$$p = \frac{k(2-k)r_k - (k-1)^2 r_{k-1}}{kr_k - (k-1)r_{k+1}}[1+O(1/k)]. \tag{9.58}$$

The estimates (9.56)–(9.58) are known as unbiased estimates. If the critical value x_0 is known the exponent p may be determined from

$$p_k = k\{x_0(c_{k+1}/c_k) - 1\} + 1 + O(1/k). \tag{9.59}$$

This estimation method is referred to as a biased estimate.

Ratios for D = 3

The qualitative behavior of the critical phenomena for $D = 3$ is the same for all values of n in the sense that for all n the critical temperature is greater than zero. Consequently, even though in the presentation of the results for the Ising model ($n = 1$) it was most natural to use the variable $v = \tanh J/k_BT$ for which the expansion coefficients of the susceptibility (as seen in Table 9.5) were positive integers we will here use the variable J/k_BT as the expansion variable for $n = 1$ as well as for $n \geq 2$

In Table 9.10 we give the ratio $a_{k+1}/a_k q$ for the coefficients of the susceptibility expansion in terms of the variable J/k_BT normalized by the number of nearest neighbors q for the $n = 1, 2, 3$. For convenience we restrict our analysis to the bipartite cubic and bcc lattices for which the series are substantially longer than the closest packed fcc lattice.

The ratios scaled by q are plotted in Fig. 9.4 for $n = 1, 2, 3$ for the cubic lattice. We see here that there is a small odd-even oscillation which is due to the fact that for the bipartite lattices that in addition to the singularity at $T = T_c$, there is an additional (antiferromagnetic) singularity at $T = -T_c$ due to the symmetry (9.12). This oscillation rapidly decreases as k increases and the points rapidly approach a straight line.

Table 9.10 The ratios $r_k = a_{k+1}/a_k q$ for the susceptibility $k_B T\chi(T)$ in terms of the expansion variable $J/k_B T$ of the $n = 1, 2, 3$ vector model for $D = 3$ for the cubic and bcc lattices. The values for $r_\infty^{e,o}$ are obtained by a linear extrapolation (9.60).

	$a_{k+1}/a_k q$ for $n=1$		$a_{k+1}/a_k q$ for $n=2$		$a_{k+1}/a_k q$ for $n=3$	
k	cubic	bcc	cubic	bcc	cubic	bcc
2	.822222	.869047	.816667	.866071	.813333	.864286
3	.795045	.838155	.789116	.835052	.785747	.833161
4	.787535	.836652	.784239	.830956	.778489	.827534
5	.787035	.824482	.772091	.818590	.766411	.815014
6	.780299	.823803	.769087	.815609	.762172	.811642
7	.774241	.817378	.763024	.810483	.755949	.804997
8	.773634	.816907	.761131	.808212	.753262	.802725
9	.769943	.812963	.757443	.804218	.749447	.798625
10	.769492	.812609	.756104	.803086	.747536	.796948
11	.767007	.809953	.753625	.800405	.744980	.794226
12	.766667	.809680	.752642	.799554	.743596	.792987
13	.764886	.807776	.750859	.797626	.741732	.790999
14	.764621	.807554	.749991	.796967	.740804	.789747
15	.763286	.806126	.748898	.795715	.739135	.788825
16	.763073	.805944	.748179	.794791	.738344	.787773
17	.762032	.804830	.747146	.793863	.737429	.786599
18	.761862	.804683	.746663	.793428	.736658	.785972
19	.761027	.803794	.745833	.792530	.735789	.785006
20	.760882	.803664	.745437	.792170	.735036	.784538
r_∞^e	.75207	.79450	.73440	.78084	.72043	.77163
r_∞^o	.75248	.79599	.73467	.78119	.72184	.77146

The limiting value for $k \to \infty$ is estimated by linearly extrapolating the even and odd values of k using

$$r_\infty = \frac{1}{2}\{(k+2)r_{k+2} - kr_k\} \tag{9.60}$$

and the exponent γ is estimated from (9.59) using the estimator

$$\gamma_k = 1 + k(\frac{r_k}{r_\infty} - 1) + O(1/k) \text{ with } r_k = a_{k+1}/a_k q. \tag{9.61}$$

In Table 9.10 in the entries marked $r_\infty^{e,o}$ we give the average of the extrapolations obtained from (9.60) using the values at $k = 20, 18$ and $k = 19, 17$. In Table 9.11 we give the estimators (9.61) for the exponent γ. There we see that there is serious discrepancy between the odd and even estimators and that the cubic and bcc exponents are not in accord with the expectations of universality. The conclusion is that, particularly with the odd-even oscillation, the ratio method is far too crude to be used for quantitative results for critical exponents.

The ratio method is applied to the specific heat for the cubic and bcc lattices by computing in Table 9.12, for even k, the ratios scaled by $r_k^\alpha = e_{k+2}/e_k q^2$ from the

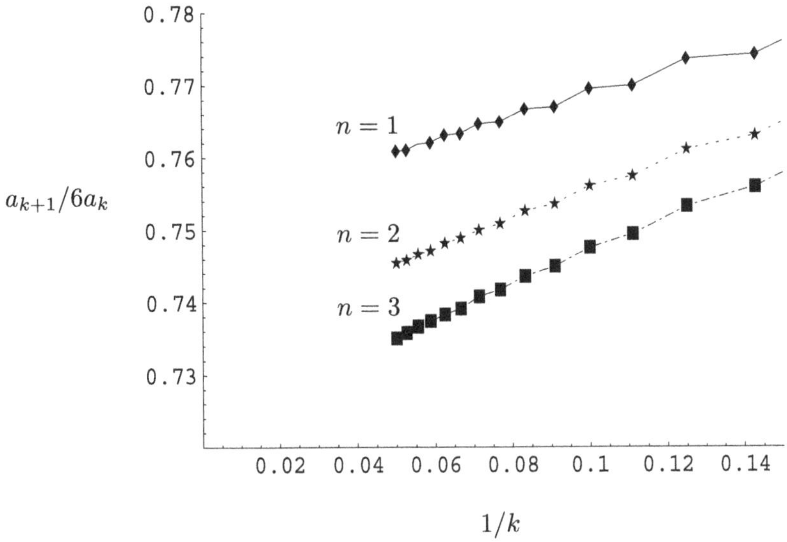

Fig. 9.4 A plot for the cubic lattice with $n = 1, 2, 3$ of the ratios of the susceptibility coefficients normalized by q the number of nearest neighbors $a_{k+1}/6a_k$ versus $1/k$.

Table 9.11 The estimators $\gamma_k = 1 + k(\frac{r_k}{r_\infty} - 1)$ for the susceptibility exponent γ of the $n = 1, 2, 3$ vector model for $D = 3$ for the cubic and bcc lattices.

	γ_k for $n = 1$		γ_k for $n = 2$		γ_k for $n = 3$	
k	cubic	bcc	cubic	bcc	cubic	bcc
20	1.2343	1.2305	1.3003	1.2860	1.4054	1.3345
19	1.2271	1.1862	1.2886	1.2758	1.3671	1.3220
18	1.2343	1.2307	1.3004	1.2862	1.4054	1.3345
17	1.2157	1.1887	1.2886	1.2757	1.3671	1.3336

coefficients of Table 9.9. These ratios are plotted in Fig. 9.5. When the values of $\sqrt{r_\infty^\alpha}$ in the last line are compared to the values of $r_\infty^{e,o}$ of Table 9.10 we see that the agreement is not better than 1 or 2 percent even though they should be the same. This is another indication of the crudity of the ratio method.

The difficulties and the arbitrariness in the use of the ratio method are well illustrated by an attempt to estimate the exponent α from the ratios of Table 9.12 by use of an estimator analagous to (9.61) used for the susceptibility

$$\alpha_k^{(1)} = 1 + \frac{k}{2}\left(\frac{r_k^\alpha}{r_\infty^\alpha} - 1\right) \tag{9.62}$$

or

$$\alpha_k^{(2)} = 1 - \frac{k}{2}\left(\frac{r_\infty^\alpha}{r_k^\alpha} - 1\right) \tag{9.63}$$

We first note that these two estimators, which must give identical results in the limit $k \to \infty$, are substantially different for the data in Table 9.12. Furthermore the esti-

Table 9.12 The ratio $r_k^\alpha = e_{k+2}/e_k q^2$ as obtained from Tables 9.7 and 9.9 for the specific heat of the $n = 1, 2, 3$ vector model for $D = 3$ for the cubic and bcc lattices. The values for r_∞^α are obtained by a linear extrapolation from $k = 16, 18$. The values of $\sqrt{r_\infty^\alpha}$ are to be compared with the corresponding values of $r_\infty^{e,o}$ in Table 9.10.

	$n=1$		$n=2$		$n=3$	
k	cubic	bcc	cubic	bcc	cubic	bcc
2	.45606	.44226	.43180	.41006	.41005	.40961
4	.42227	.48895	.38360	.45893	.35822	.43909
6	.46487	.52013	.43464	.48894	.40341	.46697
8	.48350	.53874	.44882	.51852	.42383	.48576
10	.49441	.55169	.45836	.51114	.43550	.50640
12	.50317	.56146	.47200	.53242	.44550	.51060
14	.51001	.56909	.47749	.54044	.45371	.51942
16	.51546	.57560	.48256	.54730	.46038	.52657
18	.51994	.58019	.48860	.55281	.46587	.53249
r_∞^α	.5557	.6169	.5369	.5968	.5097	.5798
$\sqrt{r_\infty^\alpha}$	.7454	.7854	.7328	.7725	.7139	.7614

mators are quite sensitive to the value of r_∞^α, and there are significant differences, if instead of the value obtained from the specific heat ratios in Table 9.12, we used the value obtained from the susceptibility ratios of Table 9.10. If we compute the estimator $\alpha^{(2)}$ using $r_\infty^\alpha = r_\infty^{o2}$ with r_∞^{o2} obtained from Table 9.12 we obtain the estimates in Table 9.13.

Table 9.13 The estimators $\alpha_k^{(2)}$ (9.63) for $k = 16, 18$ and the extrapolant $9\alpha_{18} - 8\alpha_{16}$ for the cubic lattice using $r_\infty^\alpha = r_\infty^{o2}$ where r_∞^o is obtained from Table 9.12.

	$n=1$	$n=2$	$n=3$
$\alpha_{18}^{(2)}$	0.1988	0.0580	−0.0660
$\alpha_{16}^{(2)}$	0.2121	0.0586	−0.0543
$9\alpha_{18}^{(2)} - 8\alpha_{16}^{(2)}$	0.0923	0.0529	−0.1600

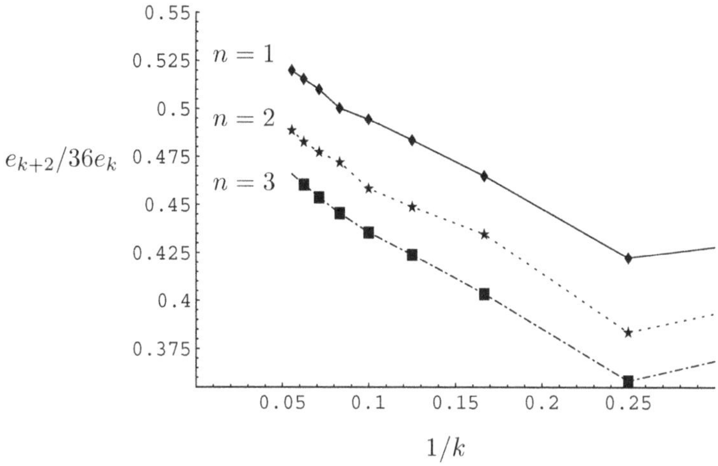

Fig. 9.5 A plot for the cubic lattice with $n = 1, 2, 3$ of the ratios of the specific heat coefficients normalized by q^2 the number of nearest neighbors $e_{k+2}/e_k 36$ versus $1/k$.

Ratios for D = 2

The ratios for the susceptibility for $D = 2$ are given in Table 9.14 and plotted in Fig 9.6 for $n = 1, 2, 3$. We see in the figure that for $n = 3$ there is no linearity to be observed in the plot. This is to expected from the observation that the coefficients are expected to oscillate in sign for sufficiently large k and thus the true large k behavior of the ratios is not expected to be seen in the behavior of the first 20 ratios.

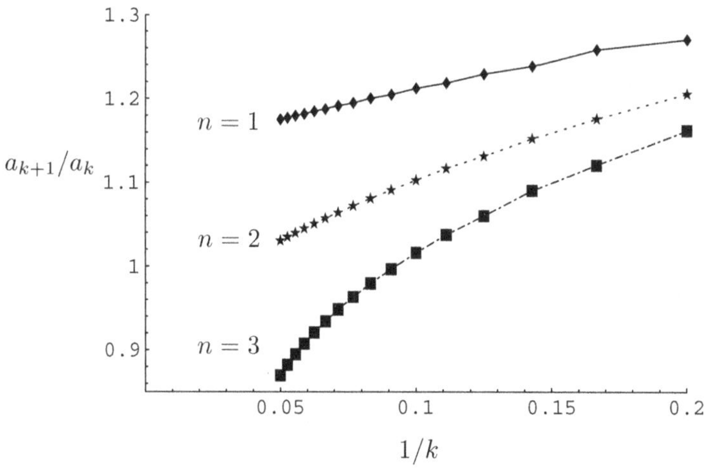

Fig. 9.6 A plot for the square lattice with $n = 1, 2, 3$ of the ratios of the susceptibility coefficients a_{k+1}/a_k versus $1/k$.

Table 9.14 The ratios a_{k+1}/a_k for the susceptibilities $k_B T\chi(T)$ of the $n = 1, 2, 3$ vector model for the $D = 2$ square and triangular lattices computed from the data of Table 9.3.

	$n = 1$	$n = 2$		$n = 3$	
k	square	square	triangular	square	triangular
2	1.44444	1.4166	2.2500	0.9333	1.4333
3	1.32692	1.2941	2.1111	0.8492	1.3898
4	1.30725	1.2462	2.0226	0.8055	1.3203
5	1.27051	1.2059	1.9604	0.7748	1.2689
6	1.25833	1.1762	1.9124	0.7470	1.2280
7	1.23836	1.1525	1.8742	0.7267	1.1941
8	1.22907	1.1312	1.8436	0.7063	1.1656
9	1.21845	1.1164	1.8183	0.6913	1.1410
10	1.21200	1.1024		0.6771	
11	1.20456	1.0908		0.6642	
12	1.20008	1.0806		0.6529	
13	1.19473	1.0718		0.6421	
14	1.19129	1.0637		0.6323	
15	1.18732	1.0568		0.6225	
16	1.18460	1.0504		0.6139	
17	1.18152	1.0446		0.6047	
18	1.17933	1.0394		0.5964	
19	1.17686	1.0340		0.5880	
20	1.17507	1.0301		0.5795	

9.1.6 Estimates from differential approximates

The best estimates of the exponents γ and α come from the analysis of the high temperature series expansions by means of the differential approximants discussed in section 9.1.4 and many such studies have been made. Early studies used Padé approximants (9.48) and dlog-Padé approximants (9.49). However, as the length of the series was increased and the necessity of accounting for confluent singularities became apparent, it became clear that these extrapolation techniques are not sufficient.

We present in Table 9.15 the results of the analysis of [11, 19] of the series for the cubic and bcc lattices which uses first and second order approximants and is thus capable of including one confluent singularity. Even with this tool there are many variations. In particular the orders of the polynomials $Q_k(x)$ and $P(x)$ can be freely chosen and each choice will give a different exponent at the critical point. No one choice is preferred and in practice many different choices are made and the spread of the resulting exponents is taken as a measure of the accuracy of the estimate.

Furthermore one may compute T_c from the specific heat and susceptibility series separately or the assumption can be made that the T_c for both must be the same and use the value which is considered the most accurate. Such extrapolations which use extra assumptions beyond the series themselves are said to be biased. In [11,19] results

Table 9.15 The estimates of [11, 19] for the critical temperatures, susceptibility exponent γ and the specific heat exponent α for the n vector model with $n = 1, 2, 3$. The error in the last digit of the estimate is given in the parenthesis.

n	1	2	3
β_c cub	0.221663(9)	0.227095(2)	0.23101(1)
β_c bcc	0.157379(2)	0.160214(2)	0.162268(1)
γ cub	1.244(3)	1.327(4)	1.404(4)
γ bcc	1.243(2)	1.322(3)	1.396(3)
α cub	0.103(8)	−0.014(9)	−0.11(2)
α bcc	0.105(9)	−0.019(8)	−0.13(2)

of several such additional assumptions are presented along with unbiased estimates and all results are in good agreement.

The estimates presented in Table 9.15 are the unbiased estimates reported in [11, 19]. The number in parentheses represents a subjective estimate of the uncertainty in the result for the exponent. We have defined $\beta_c = J/k_B T_c$ and note that because the Hamiltonian (9.1) has a different normalization from [11,19] that

$$\frac{1}{qr_\infty} = \beta_c = (\beta_c)_{BC}/n \tag{9.64}$$

where $(\beta_c)_{BC}$ is the value of β_c in Table II of [11] and

$$qr_\infty = \lim_{k\to\infty} \frac{a_{k+1}}{a_k} \tag{9.65}$$

We will discuss the significance of these estimates of the critical exponents in section 9.3 after the corresponding results for the quantum case have been derived.

9.2 Quantum Heisenberg model

High temperature expansions for the Heisenberg model may be developed using the same techniques as were used for the classical n vector model with the replacement

$$\int \prod_{\mathbf{R}} d\mathbf{n}_{\mathbf{R}} \to \mathrm{Tr} \tag{9.66}$$

where the trace is over all states of the system and we note that at infinite temperature the partition function reduces to

$$Z = \mathrm{Tr}1 = (2S+1)^N. \tag{9.67}$$

The details of the resulting graphical expansion are given in the article of Baker, Gilbert, Eve and Rushbrooke [28].

We will here present the results of the computations using the notation

$$k_B T \chi(T) = \sum_{k=0}^{\infty} a_k (J/k_B T)^n \tag{9.68}$$

$$k_B T \chi_s(T) = \sum_{k=1}^{\infty} a_k^s (-J/k_B T)^n \tag{9.69}$$

$$c_H/k_B = \frac{3q}{2}(J/k_B T)^2 \sum_{k=0}^{\infty} e_k (J/k_B T)^n \tag{9.70}$$

where q is the coordination number (number of nearest neighbors) of the lattice.

For $D = 2$ on both the square and triangular lattices the most extensive results known for arbitrary S for both the susceptibility and specific heat are given in the 1958 work of Rushbrooke and Wood [29] as extended in 1968 by Stephenson, Pirne, Wood and Eve [30]. For $S = 1/2$ the most extensive data for the specific heat and susceptibility are in the 1996 work of Oitmaa and Bonilla [33]. For $S = 1/2$ the most extensive data for the staggered susceptibility on square lattice are in the 2000 work of Pan [31].

For $D = 3$ on the cubic, bcc and fcc lattice the most extensive results known for arbitrary S for both the susceptibility and specific heat are given in the 1958 work of Rushbrooke and Wood [29] as extended in 1968 by Stephenson, Pirne, Wood and Eve [30], and the most extensive data for the staggered susceptibility on the cubic and bcc lattice in in the 1963 work of Rushbrooke and Wood [32]. For $S = 1/2, 1, 3/2$ on the cubic and bcc lattices the most extensive data for the susceptibility and staggered susceptibility is the 2004 work of Oitmaa and Zheng [34].

These references are summarized in Table 9.16

Table 9.16 Summary of references for high temperature series expansion for the quantum spin S Heisenberg model.

D	S	χ	χ^s	specific heat
2 sq tri	arb.	Rushbrook, Wood 1958 [29] Stephenson, Pirne Wood, Eve 1968 [30]		Rushbrooke, Wood 1958 [29] Stephenson, Pirne Wood, Eve 1968 [30]
2 sq tri	1/2	Oitmaa, Bonilla 1996 [33]	Pan 2000 [31]	Oitmaa, Bonilla 1996 [33]
3 cub bcc fcc	arb	Rushbrooke, Wood 1958 [29] Stephenson, Pirne Wood, Eve 1968 [30]		Rushbrooke, Wood 1958 [29] Stephenson, Pirne Wood, Eve 1968 [30]
3 cub bcc	arb		Rushbrooke, Wood 1963 [32]	
3 cub bcc fcc	1/2	Oitmaa, Bonilla 1996 [33]		Oitmaa Bonilla 1996 [33]
3 cub bcc	1/2,1,3/2	Oitmaa, Zheng 2004 [34]	Oitmaa, Zheng 2004 [34]	

9.2.1 Results for D = 2

For $S = 1/2$ the most extensive results known are the 14 term expansions of Oitmaa and Bornilla [33] for the square (sq) and triangular (tri) lattices in two dimensions. The results are given in Table 9.17 where we use the notation

$$e_n = \frac{\tilde{e}_n(1/2)}{4^n n!} \quad \text{and} \quad a_n(1/2) = \frac{\tilde{a}_n(1/2)}{4^n n!}. \tag{9.71}$$

It seems that series for the staggered susceptibility and for $S \geq 1$ are not known.

Table 9.17 High temperature expansion coefficients $\tilde{e}_n(1/2)$ and $\tilde{a}_n(1/2)$ for the specific heat and susceptibility of the spin 1/2 quantum Heisenberg model for the square and triangular lattices in $D = 2$ from Table 1 of [33]. We have set $e_n(1/2) = \tilde{e}_n(1/2)/(4^n n!)$ and $a_n(1/2) = \tilde{a}_n(1/2)/(4^n n!)$. Note that the coupling J of [33] is one half of the $\tilde{J}$ of (9.4) and e_k is e_{k+2} of [33].

	Specific heat	
n	$\tilde{e}_n(1/2)$ square	$\tilde{e}_n(1/2)$ triangular
0	1	1
1	−2	2
2	−14	−34
3	20	−360
4	520	5464
5	−41,552	162,960
6	153,488	−1,514,000
7	1,437,9136	−130,296,448
8	−205,339,264	8,361,856
9	−6,828,335,360	154,693,752,576
10	231,955,086,080	2,047,410,296,064
11	349,717,905,356	
12	−299,728,629,046,272	
	Susceptibility	
n	$\tilde{a}_n(1/2)$ square	$\tilde{a}_n(1/2)$ triangular
0	1	1
1	4	6
2	16	48
3	64	408
4	416	3,600
5	4,544	42,336
6	23,488	781,728
7	−207,616	13,646,016
8	4,205,056	90,893,568
9	198,295,552	−1,798,204,416
10	−2,574,439,424	70,794,720,768
11	−112,886,362,112	7,538,546,211,840
12	3,567,419,838,464	6,813,109,782,528
13	94,446,596,145,152	
14	−1,798,371,774,277,632	

9.2.2 Results for D = 3

For the specific heat with $S = 1/2$ the most extensive results known are the 14-term expansions of Oitmaa and Bornilla [33] for the simple cubic (sc), bcc and fcc lattices. The results for e_n are given in Table 9.18 where we use the notation of (9.71). There seem to be no results known for $S \geq 1$.

Table 9.18 Normalized High temperature expansion coefficients $\tilde{e}_n(1/2)$ for the specific heat of the spin 1/2 quantum Heisenberg model for simple cubic, face centered cubic and body centered cubic lattices in $D = 3$ from table 1 of [33] We have set $e_n(1/2) = \tilde{e}_n(1/2)/(4^n n!)$.Note that the coupling J of [33] is one half of the $\tilde{J}$ of (9.4) and e_k is e_{k+2} of [33].

n	$\tilde{e}_n(1/2)$ for simple cubic	$\tilde{e}_n(1/2)$ for bcc
0	1	1
1	−2	−2
2	−18	14
3	280	95
4	3,688	2,040
5	−113, 232	−24, 752
6	−867, 216	2,334,768
7	80,440,192	−44, 473, 472
8	288, 502, 656	3, 429, 683, 056
9	−95, 126, 989, 944	−41, 940, 628, 224
10	709,294,331,648	4,416,784,659,712
11	150,744,103,377,920	−153, 284, 724, 083, 712
12	−3, 074, 209, 362, 326, 528	20, 508, 575, 418, 559, 488
n	$\tilde{e}_n(1/2)$ for fcc	
0	1	
1	6	
2	10	
3	−280	
4	9,000	
5	809,200	
6	31, 291, 856	
7	974, 702, 208	
8	4,168,957,2736	
9	3,147,043,161,856	
10	276,332,034,732,800	

For the susceptibilities $\chi(T)$ and $\chi_s(T)$ with $S = 1/2, 1, 3/2$ the best results for cubic, bcc, and fcc are those of Oitmaa and Zheng [34]. We present these in Tables 9.19–9.22 where we use the notation

$$a_n(1/2) = \frac{\tilde{a}_n(1/2)}{4^{n+1}n!}, \quad a_n(1) = \frac{2\tilde{a}_n(1)}{3^{n+1}n!}, \quad a_n(3/2) = \frac{5\tilde{a}_n(3/2)}{2^{n+2}n!} \tag{9.72}$$

$$a^s_n(1/2) = \frac{\tilde{a}^s_n(1/2)}{4^{n+1}(n+1)!}, \quad a^s_n(1) = \frac{2\tilde{a}^s_n(1)}{3^{n+1}(n+1)!}, \quad a^s_n(3/2) = \frac{5\tilde{a}^s_n(3/2)}{2^{n+3}(n+1)!}. \tag{9.73}$$

Table 9.19 High temperature expansion coefficients for the susceptibilities $\chi(T)$ and $\chi_s(T)$ of the spin 1/2 quantum Heisenberg model for cubic and bcc lattice from Table 1 of [34]. We have set $a_n(1/2) = \tilde{a}_n(1/2)/(4^{n+1}n!)$ and $a^s_n(1/2) = \tilde{a}^s_n(1/2)/(4^{n+1}(n+1)!$

n	$\tilde{a}_n(1/2)$ for the cubic lattice	$\tilde{a}^s_n(1/2)$ for the cubic lattice
0	1	1
1	6	12
2	48	168
3	528	2,880
4	7,920	59,376
5	149,856	1,478,592
6	3,169,248	42,537,024
7	77,046,528	1,353,271,296
8	2,231,209,728	48,089,027,328
9	71,938,507,776	1,908,863,705,088
10	2, 446, 325, 534, 208	83,357,870,602,752
11	92, 886, 269, 386, 752	3,926,123,179,720,704
12	3,995,799,894,239,232	198,436,560,561,973,248
13	180,512,165,153,832,960	10,823,888,709,015,846,912
14	8, 443, 006, 907, 441, 565, 696	635,114,442,481,347,244,032
n	$\tilde{a}_n(1/2)$ for the bcc lattice	$\tilde{a}^s_n(1/2)$ for the bcc lattice
0	1	1
1	8	16
2	96	320
3	1664	8,192
4	36,800	248,768
5	1,008,768	8,919,296
6	32,626,560	367,854,720
7	1,221,399,040	17,216,475,136
8	51,734,584,320	899,434,884,096
9	2, 459, 086, 364, 672	51,925,815,320,576
10	129,082,499,311,616	3,280,345,760,086,016
11	7,432,690,738,003,068	225,270,705,859,919,872
12	464,885,622,793,134,080	16,704,037,174,526,894,080
13	31,456,185,663,820,136,448	1,330,557,135,528,577,925,120
14	2,284,815,238,218,471,260,160	113,282,648,639,921,512,955,904

9.2.3 Analysis of results

The first observation to be made about the high temperature series expansion of the quantum Heisenberg model for both the specific heat and the susceptibility is that the signs of the coefficients appear to be significantly more irregular than in the classical case. In the case of $D = 2$, Table 9.17, the series for the specific heat shows the same sort of oscillations in sign as seen in the classical case and the susceptibility for the square and triangular lattices have minus signs at orders 7 and 9 respectively whereas in the classical case minus signs were not seen in the susceptibility up to order 21 and were inferred at higher order only by an analysis of n vector models with $n \geq 3$. It is believed that the quantum Heisenberg model in $D = 2$ shares with the classical model in $D = 2$ the property of having a phase transition only at $T = 0$ which is consistent with the leading singularities of the specific heat and susceptibility having a complex value. It is thus not to be expected that information about the phase transition can be

Table 9.20 High temperature expansion coefficients for the susceptibility $\chi(T)$ of the spin 1/2 quantum Heisenberg model for the fcc lattice from Table 2 of [34]. We have set $a_n(1/2) = \tilde{a}_n(1/2)/(4^{n+1}n!)$

n	$\tilde{a}_n(1/2)$ for fcc lattice
1	12
2	240
3	6,624
4	234,720
5	10,208,832
6	526,810,176
7	31,434,585,600
8	2,127,785,02,024
9	161, 064, 469, 168, 128
10	13,483,480,670,745,600
11	1,237,073,710,591,635,456
12	123,437,675,536,945,410,048

obtained from the series expansion without further assumptions. In this respect the classical and quantum spin 1/2 Heisenberg models are similar in $D = 2$.

For $D = 3$, however, there is much less similarity between the classical and the quantum case. In the classical case the coefficients in the specific heat expansion are all positive while from Table 9.18 we see that there are many minus signs for the cubic and bcc lattices and there is even one isolated minus sign in the fcc lattice. The oscillations of sign for the cubic and bcc lattices make it fruitless to attempt an analysis of the series with the assumption that the leading singularity is on the positive temperature axis.

On the other hand Tables 9.19–9.22 for the direct and staggered susceptibility have only positive signs so an assumption of a leading singularity at a positive temperature is not precluded even though such analysis could not be done for the specific heat.

The simplest such analysis is the ratio method. This has been done by Oitmaa and Zheng [34] and we plot these ratios for both the cubic and the bcc lattice for $S = 1/2, 1$ and ∞ in Fig. 9.7. From this figure we see that aside from the small oscillation between even and odd ratios which was seen even in the classical case that, in all cases except $S = 1/2$ on the cubic lattice, the ratios seem to be linear in $1/n$ for large n. The intercept of this plot at $1/n = 0$ estimates $k_BT_c/(JS(S+1))$ and clearly shows that the critical temperature of the staggered (antiferromagnetic) susceptibility is larger than the critical temperature of the direct (ferromagnetic) susceptibility. The slope gives the critical exponent γ appears to be independent of S and is the same for both the direct and staggered susceptibility.

A more quantitative estimate of the critical temperature and exponents γ is made in [34] by use of a dlog-Padé analysis. The resulting estimates for the scaled critical temperature $k_BT_c/(JS(S+1))$ are given in Table 9.23. The estimates of the exponent γ are given in Table 9.24 where we give the range of values obtained by the use of different degrees of the polynomials $Q_1(x)$ and $P(x)$ used in the dlog-Padé analysis.

Table 9.21 High temperature expansion coefficients for the susceptibilities $\chi(T)$ and $\chi_s(T)$ of the spin 1 quantum Heisenberg model for cubic and bcc lattice from Table 3 of [34]. We have set $a_n(1) = 2\tilde{a}_n(1)/(3^{n+1}n!)$ and $a_n^s(1) = 2\tilde{a}_n^s/(3^{n+1}(n+1)!)$

n	$\tilde{a}_n(1)$ for the cubic lattice	$\tilde{a}_n^s(1)$ for the cubic lattice
0	1	1
1	12	24
2	222	702
3	5,904	26,280
4	201,870	1,184,526
5	8,556,912	63,357,984
6	426,905,802	3,887,604,666
7	2,467,414,724	270,348,199,128
8	1,616,505,223,518	20,988,390,679,758
9	118,701,556,096,392	1,802,403,961,243,776
10	9,628,527,879,611,262	169,418,364,565,523,958
11	856,813,238,084,411,136	17,314,303,199,655,636,792
12	82,856,991,914,713,902,402	
n	$\tilde{a}_n(1)$ for the bcc lattice	$\tilde{a}_n^s(1)$ for the bcc lattice
0	1	1
1	16	32
2	424	1320
3	16,512	71136
4	819,240	4,588,968
5	50,363,136	351,263,232
6	3,652,143,480	30,873,601,080
7	307,454,670,000	3,082,065,903,648
8	29,310,549,057,000	343,320,789,071,016
9	3,133,368,921,937,824	42,320,100,429,654,912
10	370,060,173,560,963,304	5,709,664,512,091,086,984
11	47,968,071,364,509,850,944	837,942,419,330,764,322,976
12	6,756,542,767,252,059,234,840	

9.3 Discussion

For the classical n vector model in three dimensions with $n = 1$ (Ising), $n = 2$ (XY) and $n = 3$ (Heisenberg) the estimates of Table 9.15 obtained from the 31-term series expansions as extrapolated by differential approximants are the most unbiased and systematic computations of the critical exponents α and γ which exist. In particular the agreement within the error bars of the exponents for the cubic and bcc lattices may be taken as good evidence for the hypothesis of universality.

Similarly the agreement of the exponents γ for the spin S Heisenberg magnet of Table 9.24 with the classical value of Table 9.15 may be taken as good evidence of the extension of universality from the classical to the quantum system.

Using the assumptions of scaling presented in chapter 5 we may compute all other critical exponents in terms of the high temperature exponents α and β estimated in Table 9.15 for the classical model, by use of equations (5.78)–(5.82) of chapter 5. These estimates of the critical exponents are summarized in Table 9.25. Therefore from one point of view it can be argued that we understand a great deal of the physics of the quantum magnet as well.

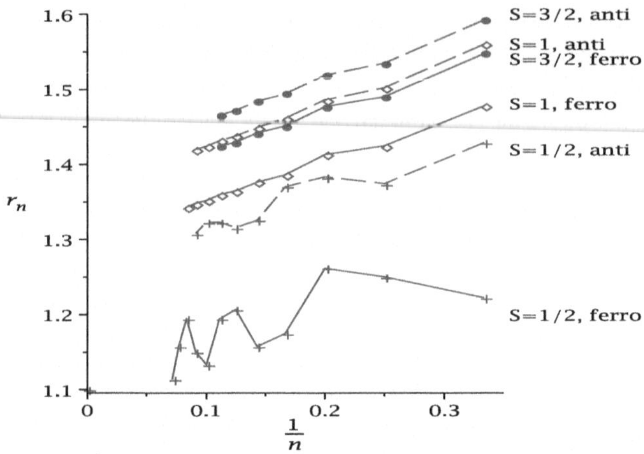

Fig. 9.7 Ratio plots from [34] for the ferromagnetic and antiferromagnetic susceptibilities for the cubic lattices for $S = 1/2, 1, 3/2$. The ratios plotted are $r_n = a_n/(S(S+1)a_{n-1})$ where the a_n are given in Tables (9.19), (9.21) for $S = 1/2$ and 1 and from the classical Heisenberg model for $S = \infty$. The ferromagnetic cases are given by the solid lines and the antiferromagnetic cases by the dashed lines. The spin $S = 1/2$ is given by crosses, $S = 1$ by diamonds and $S = 3/2$ by solid circles.

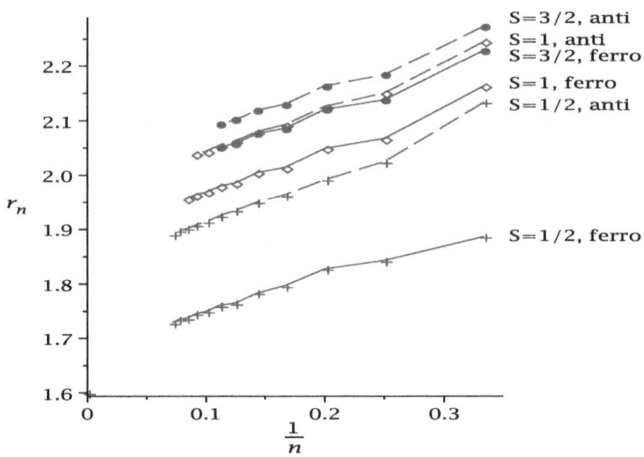

Fig. 9.8 Ratio plots from [34] for the ferromagnetic and antiferromagnetic susceptibilities for the bcc lattices for $S = 1/2, 1, 3/2$. The ratios plotted are $r_n = a_n/(S(S+1)a_{n-1})$ where the a_n are given in Tables (9.19), (9.21) for $S = 1/2$ and 1 and from the classical Heisenberg model for $S = \infty$. The ferromagnetic cases are given by the solid lines and the antiferromagnetic cases by the dashed lines. The spin $S = 1/2$ is given by crosses, $S = 1$ by diamonds and $S = 3/2$ by solid circles.

Table 9.22 High temperature expansion coefficients for the susceptibilities $\chi(T)$ and $\chi_s(T)$ of the spin 3/2 quantum Heisenberg model for cubic and bcc lattice from [34]. We have set $a_n(3/2) = 5\tilde{a}_n(3/2)/(2^{n+2}n!)$ and $a_n^s(3/2) = 5\tilde{a}_n^s(3/2)/(2^{n+3}(n+1)!)$.

n	$\tilde{a}_n(3/2)$ for the cubic lattice	$\tilde{a}_n^s(3/2)$ for the cubic lattice
0	1	2
1	60	60
2	1440	2220
3	50,136	106,032
4	2,241,660	6,103,230
5	124,125,372	417,121,164
6	8,102,868,414	32,715,943,017
7	613,292,153,184	2,911,926,450,048
8	52,599,376,466,556	2,289,263,779,556,198
9	50,561,988,998,505,288	31,792,485,934,519,488
n	$\tilde{a}_n(3/2)$ for the bcc lattice	$\tilde{a}_n^s(3/2)$ for the bcc lattice
0	1	2
1	80	80
2	2,720	4160
3	136,448	283,776
4	8,751,600	23,240,440
5	696,028,496	2,263,139,152
6	65,331,028,472	253,095,247,076
7	7,121,212,898,544	32,175,304,799,424
8	879,298,191,968,624	4,563,926,306,507,096
9	121, 768, 840, 349, 153, 216	716,734,730,963,510,496

Table 9.23 Estimate of $k_BT_c/(JS(S+1))$ by use of a dlog-Padé analysis from [34].

cubic						
$S = 1/2$		$S = 1$		$S = 3/2$		$S = \infty$
χ	χ^s	χ	χ^s	χ	χ^s	χ
1.119(2)	1.259(2)	1.2994(5)	1.3676(7)	1.37(2)	1.404(7)	1.4429
bcc						
$S = 1/2$		$S = 1$		$S = 3/2$		$S = \infty$
χ	χ^s	χ	χ^s	χ	χ^s	χ
1.6803(6)	1.8350(5)	1.894(1)	1.967(1)	1.97(1)	2.009(6)	2.0542

On the other hand this analysis has made many assumptions which are untested and in some cases contradictory.

In particular consider the result shown in Table 9.23 that for the quantum spin S Heisenberg model the critical temperature of the ferromagnetic susceptibility lies below the critical temperature of the antiferromagnetic susceptibility. This is troubling because these singularities are not independent of each other and, for example, the oscillation between even and odd coefficients of the direct susceptibility is caused by

Table 9.24 Estimate of γ by use of a dlog-Padé analysis from [34]. We give here the range of estimates of γ obtained by use of different degrees of the approximating polynomials $Q_1(x)$ and $P(x)$.

χ	χ^s
$S = 1/2$ cubic	
1.411–1.421	1.440–1.425
$S = 1/2$ bcc	
1.416–1.423	1.431–1.436
$S = 1$ cubic	
1.406–1.411	1.409–1.417
$S = 1$ bcc	
1.398–1.404	1.390–1.405

Table 9.25 Estimates of the critical exponents in $D = 3$ for the classical n vector model and the quantum Heisenberg model from Table 9.15 and the exponent equalities (5.78)–(5.82) of chapter 5. The values of α and γ for $n = \infty$ are the spherical model values from (9.37).

	n			
exponent	1	2	3	∞
α	0.10	−0.02	−0.12	−1
γ	1.24	1.32	1.40	2
β	0.33	0.35	0.36	1/2
δ	4.6	4.8	4.9	5
ν	0.63	0.67	0.71	1
η	0.042	0.039	0.019	0
Δ	1.57	1.67	1.76	5/2

the existence of a second singularity at $-T_c$ which, by sending $\mathbf{n_R} \to -\mathbf{n_R}$ on one of the sublattices of the cubic or bcc lattices, is equivalent to the singularity in the antiferromagnetic susceptibility.

For the classical case the existence of this second singularity causes no harm, but for the quantum case where the critical temperature of the ferromagnet is less than the critical temperature of the antiferromagnet this means that ultimately for sufficiently large order the coefficients of the ferromagnetic susceptibility must oscillate between positive and negative values as the order of the coefficient varies from even to odd. There is no trace seen in the data of Tables 9.19, 9.21, and 9.22 of this necessary oscillation in the coefficients of the ferromagnetic susceptibility. Doubt can therefore be cast on the assumption that the series for the quantum Heisenberg magnet are sufficiently long for the true asymptotic behavior to be seen. This is indeed in agreement with the observation already made that the alteration in signs of the specific heat coefficients for the cubic and bcc lattices of Table 9.18 indicates that these series are not long enough to be in the asymptotic regime where the critical exponent α can be obtained.

Worse still is the fact that since the physical singularity for the ferromagnet is not leading many more terms in the series expansion for the ferromagnetic case should be needed than for the antiferromagnetic case. Furthermore when we recall that in Fig. 9.7 the coefficients for the $S = 1/2$ Heisenberg magnet on the cubic lattice were certainly not smooth and from chapter 4 that for the quantum Heisenberg magnet the proof of spontaneous order exists for the antiferromagnet but not for the ferromagnetic case it may be suggested that perhaps the high temperature expansions have not actually given us much useful information about the Heisenberg ferromagnet after all.

These problems should be resolved before we consider the phase transition in the quantum Heisenberg model as fully understood.

9.4 Statistical mechanics versus quantum field theory

At the end of chapter 1 we briefly commented that the path integral formulation of quantum field theory where averages of operators A are calculated as

$$\langle A \rangle = \frac{1}{Z} \int [d\phi] A e^{S/\hbar} \tag{9.74}$$

with

$$Z = \int [d\phi] e^{S/\hbar} \tag{9.75}$$

where S is the action looks "formally" equivalent to the statistical mechanical formulas where averages of operators O are calculated as

$$\langle O \rangle = \frac{1}{Z} \sum_{\text{all states}} O e^{-\mathcal{E}/k_B T} \tag{9.76}$$

where the partition function Z is

$$Z = \sum_{\text{all states}} e^{-\mathcal{E}/kRT} \tag{9.77}$$

where $\mathcal{E}$ is the interaction energy of the system. This formal equivalence is fully developed in the discipline of lattice gauge theory and it is out of place and beyond the scope of this book to develop this in detail. Nevertheless it is perhaps not out of place to conclude this chapter by giving what can be called a dictionary which translates the language from one field to the other.

Consider the Euclidean field theory of the n component nonlinear sigma model in two dimensions where the action is

$$S = \sum_{j=1}^{n} \sum_{\mu=1}^{2} \left(\frac{\partial \phi^j(\mathbf{x})}{\partial x_\mu} \right)^2 \tag{9.78}$$

and the n component field $\phi^j(\mathbf{x})$ satisfies the constraint

$$\sum_{j=1}^{n} (\phi^j(\mathbf{x}))^2 = 1. \tag{9.79}$$

For this field theory the path integral may be precisely defined by "imposing a lattice cut-off" and writing derivatives as differences. When this is done we obtain the interaction energy of the n vector model which for $n = 1$ is the Ising model and for $n \geq 2$

is studied by high temperature expansions. The field theory of the nonlinear sigma model in continuum space is then obtained by taking the scaling limit as discussed in chapter 5. With this as the definition of the Euclidean quantum field theory of the nonlinear sigma model we may now make contact between the language of statistical mechanics and the language of quantum field theory. Some of the important elements in the translation of the two languages are presented in Table 9.26.

Table 9.26 A dictionary of translation of classical statistical mechanics of the n vector model and the quantum field theory of the nonlinear sigma model.

Statistical mechanics	Quantum field theory
Classical interaction energy	Lagrangian
Sum over all states	Lattice regularized path integral
Scaling limit	Renormalization procedure
Temperature T	$\hbar$
$T_c > 0$	Nontrivial fixed point
$T_c = 0$	Asymptotically free
Kosterlitz–Thouless phase	Quantum electrodynamics

The first important translation in Table 9.26 is that $\hbar$ in quantum field theory is equivalent to $k_B T$ in statistical mechanics. This may at first sight seem very strange because in quantum field theory units are often said to be chosen such that $\hbar$ is set equal to unity and this merely sets a scale whereas the major goal in the study of statistical mechanics is to determine the properties of the system as a function of T It is thus most important to recognize that in fact for the nonlinear sigma model $\hbar$ must not be treated as a scale setting number but should be thought of as a "coupling constant" which can in principle be varied.

For the n vector model in two dimensions there are three distinct cases for $n = 1$, $n = 2$, and $n \geq 3$.

Consider first $n = 1$. This is the Ising model where there is a critical point at $T = T_c > 0$ and in the vicinity of that critical point the scaling limit exists. In the corresponding field theory language the value of $\hbar$ or if you will of the coupling constant g which corresponds to T_c is called a nontrivial fixed point. We have seen that there are two distinct scaling limits which may be constructed by approaching T_c either from above or below. This same two-phase structure must therefore exist for the $n = 1$ component field theory.

The second case to consider is $n = 2$. In this case the n vector model has a Kosterlitz–Thouless temperature $T_{KT} > 0$ such that for $T < T_{KT}$ the system has no mass gap. Thus for this system there are also two ways to compute a scaling limit. Either $T \to T_c+$, in which case there will be massive excitations in the system, or T can be allowed to have any value below T_{KT}, in which case the excitations are massless. The nonlinear sigma model will therefore also have these two different phases and we note that the low temperature phase is in the same spirit as quantum electrodynamics where the photon in massless but the coupling constant can be varied.

The final case is $n \geq 3$. In this case the n vector model becomes massless only at

$T = 0$ and the only phase which can be constructed is the high temperature phase $T \to 0+$. In quantum field theory this is called "asymptotic freedom". We saw that this high temperature phase had the unfortunate feature that it could not be studied in the limit $T \to 0$ by means of high temperature series expansions because the leading singularities were not at $T = 0$ but were in the complex plane with a positive real part. We now come to what is quite possibly a controversial aspect to the comparison of quantum field theory and statistical mechanics because quantum field theories are very often defined by "dimensional regularization" which in the present case amount to allowing the dimension D of the system to be $D > 2$. At least in some formal sense this can be done and in the same formal sense the theorem proven in chapter 5 on lack of order for the $n \geq 2$ vector model breaks down and there will be a nontrivial fixed point for $T_c > 0$. Thus for $D > 2$ a low temperature phase of the nonlinear sigma model can be constructed and then the dimension D may be allowed to return to two. However, any such results constructed in this manner will have no analogue or relevance to the n vector model constructed on the lattice where only a high temperature phase exists.

9.5 Appendix: The expansion coefficients for the susceptibility on the square lattice

We list here the 21 coefficients $a_k(n)$ for the susceptibility on the square lattice as given in [6, pages 15838–15840]

$$a_1(n) = 4 \tag{9.80}$$
$$a_2(n) = 12 \tag{9.81}$$
$$a_3(n) = (72 + 32n)/(n + 2) \tag{9.82}$$
$$a_4(n) = (200 + 76n)/(n + 2) \tag{9.83}$$
$$a_5(n) = 8(284 + 147n + 20n^2)/((n + 2)(n + 4)) \tag{9.84}$$
$$a_6(n) = 16(780 + 719n + 201n^2 + 19n^3)/((n + 2)^2(n + 4)) \tag{9.85}$$

For the remaining coefficients we write

$$a_k(n) = P_k(n)/Q_k(n) \tag{9.86}$$

and list $P_k(n)$ and $Q_k(n)$ separately.

$$P_7(n) = 16(26064 + 38076n + 20742n^2 + 5280n^3 + 655n^4 + 32n^5) \tag{9.87}$$
$$Q_7(n) = (n + 2)^3(n + 4)(n + 6) \tag{9.88}$$

$$P_8(n) = 4(283968 + 383568n + 186912n^2 + 41000n^3 + 4392n^4 + 187n^5) \tag{9.89}$$
$$Q_8(n) = (n + 2)^3(n + 4)(n + 4)(n + 6) \tag{9.90}$$

$$P_9(n) = 8(3123456 + 4186336n + 2087128n^2 + 492220n^3 + 62386n^4$$
$$+4161n^5 + 116n^6) \tag{9.91}$$
$$Q_9(n) = (n+2)^3(n+4)(n+6)(n+8) \tag{9.92}$$

$$P_{10}(n) = 16(33868800 + 66758016n + 53214272n^2 + 22126648n^3$$
$$+5211372n^4 + 719330n^5 + 58789n^6 + 2684n^7 + 55n^8) \tag{9.93}$$
$$Q_{10}(n) = (n+2)^4(n+4)^2(n+6)(n+8) \tag{9.94}$$

$$P_{11}(n) = 32(3695370240 + 9913385984n + 11437289216n^2 + 7427564992n^3$$
$$+2989987696n^4 + 776848144n^5 + 132130072n^6 + 14693596n^7 + 1052911n^8$$
$$+46923n^9 + 1225n^{10} + 16n^{11}) \tag{9.95}$$
$$Q_{11}(n) = (n+2)^5(n+4)^3(n+6)(n+8)(n+10) \tag{9.96}$$

$$P_{12}(n) = 16(4990955520 + 11511967232n + 10992991488n^2 + 5609888352n^3$$
$$+1649559472n^4 + 281912408n^5 + 27080244n^6 + 1334568n^7 + 22368n^8$$
$$-199n^9 + 5n^{10}) \tag{9.97}$$
$$Q_{12}(n) = (n+2)^5(n+4)^2(n+6)(n+8)(n+10)$$

$$P_{13}(n) = 64(162478080000 + 406158981120n + 431982472192n^2$$
$$+25491324928n^3 + 90288340864n^4 + 19721001832n^5 + 2561904944n^6$$
$$+170376718n^7 + 1211742n^8 - 616479n^9 - 37625n^{10} - 635n^{11} + 4n^{12}) \tag{9.98}$$
$$Q_{13}(n) = (n+2)^5(n+4)^3(n+6)(n+8)(n+10)(n+12) \tag{9.99}$$

$$P_{14}(n) = 16(21007560867840 + 63770201063424n + 84400316350464n^2$$
$$63787725946880n^3 + 30245054013440n^4 + 9275137432448n^5$$
$$1810742519232n^6 + 204284290016n^7 + 7651128688n^8 - 1135009056n^9$$
$$-177417600n^{10} - 9861216n^{11} - 161554n^{12} + 5277n^{13} + 172n^{14}) \tag{9.100}$$
$$Q_{14}(n) = (n+2)^6(n+4)^3(n+6)^2(n+8)(n+10)(n+12) \tag{9.101}$$

$$P_{15}(n) = 16(9537349044142080 + 3464137448513888n$$
$$+5616903063129292 8n^2 + 5353702843652505 6n^3$$
$$+3321683038173593 6n^4 + 1400654267503513 6n^5$$
$$+777531907925504n^7 + 87227510881024n^8 + 1944102682560n^9$$
$$-1061882170400n^{10} - 183809104832n^{11} - 14563826832n^{12}$$
$$-515944376n^{13} + 583019n^{14} + 1259012n^{15} + 43647n^{16} + 512n^{17}) \quad (9.102)$$
$$Q_{15}(n) = (n+2)^7(n+4)^3(n+6)^3(n+8)(n+10)(n+12)(n+14) \quad (9.103)$$

$$P_{16}(n) = 4(410123375487221760 + 153712994478037401 6n + 258398341169047142n^2$$
$$+256697559569557094 4n^3 + 166908428335133491 2n^4 + 741114014711103488n^5$$
$$+225948162044579840n^6 + 45385102417264640n^7 + 4996850176026624n^8$$
$$-6480424496896n^9 - 122658733213440n^{10} - 20909429640960n^{11}$$
$$-1752208241536n^{12} - 56642417728n^{13} + 3062606512n^{14}$$
$$+412508368n^{15} + 18713696n^{16} + 395328n^{17} + 3083n^{18}) \quad (9.104)$$
$$Q_{16}(n) = (n+2)^7(n+4)^4(n+6)^3(n+8)(n+10)(n+12)(n+14) \quad (9.105)$$

$$P_{17}(n) = 8(3536181502805016576 0 + 1385396668872586690 56n$$
$$+245436909326998437888n^2 + 259375081142913859584n^3$$
$$+181396134616565809152n^4 + 87793764370648399872n^5$$
$$+29673202500166647808n^6 + 6770762601709142016n^7 + 893508862130341888n^8$$
$$+52293940767672 32n^9 - 24934992828139008n^{10} - 5547918408527104n^{11}$$
$$-625740097598720n^{12} - 33521607263744n^{13} + 738107699392n^{14}$$
$$+272358030048n^{15} + 21867513640n^{16} + 937447020n^{17} + 22261658n^{18}$$
$$+250495n^{19} + 692n^{20}) \quad (9.106)$$
$$Q_{17}(n) = (n+2)^7(n+2)^5(n+6)^3(n+8)(n+10)(n+12)(n+14)(n+16) \quad (9.107)$$

$$P_{18}(n) = 16(7589368385404246425 60 + 3343934774878956158976n$$
$$+6720650800795024883712n^2 + 8141819950133007089664n^3$$
$$+6611686534391523180544n^4 + 3777832155850593533952n^5$$
$$+1543669324445646061568n^6 + 443919373830353158144n^7$$
$$+82653304109539049472n^8 + 6333077582288386048n^9$$
$$-1419388196952978432n^{10} - 578178922758906368n^{11} - 99290612095487744n^{12}$$
$$-9300735775467264n^{13} - 256677768200576n^{14} + 53852331942080n^{15}$$
$$+8516631212960n^{16} + 629479458104n^{17} + 26896421724n^{18}$$
$$+607483694n^{19} + 2912825n^{20} - 154080n^{21} - 2313n^{22}) \quad (9.108)$$
$$Q_{18}(n) = (n+2)^8(n+4)^5(n+6)^3(n+8)^2(n+10)(n+12)(n+14)(n+16) \quad (9.109)$$

$$
\begin{aligned}
P_{19}(n) &= 32(29379296213298566922240\,0 + 1452203904509587992084480n \\
&+3305764874023051160715264n^2 + 4587498272216547279765504n^3 \\
&+4326547244550747303444480n^4 + 2922303204727243671601152n^5 \\
&+1446836388063996470624256n^6 + 524786164553898279829504n^7 \\
&+134502578329442459254784n^8 + 21104878727348218859520n^9 \\
&+4119916010010724884480n^{10} - 794587338452494176256n^{11} \\
&-2421412948365837516800n^{12} - 393733079929782138880n^{13} \\
&-35838546659172825600n^{14} - 518790729415521280n^{15} \\
&+36069375006840576n^{16} + 5868286096676352n^{17} \\
&+508264525824336n^{18} + 27200961065872n^{19} + 811699909040n^{20} \\
&+3015005636n^{21} - 793163459n^{22} - 32254806n^{23} - 562185n^{24} - 3824n^{25})
\end{aligned} \tag{9.110}
$$

$$
\begin{aligned}
Q_{19}(n) &= (n+2)^9(n+4)^5(n+6)^3(n+8)^3(n+10)(n+12) \\
&(n+14)(n+16)(n+18)
\end{aligned} \tag{9.111}
$$

$$
\begin{aligned}
P_{20}(n) &= 2518403141305883317764096\,0 + 1209918483517384823679221\,76n \\
&+266735758564462825159262208n^2 + 356790558744797070187560960n^3 \\
&+322230995604339689974136832n^4 + 206410863077042429645291520n^5 \\
&+95396352174319203631759360n^6 + 31338576528595789665009664n^7 \\
&+6741729364322236678275072n^8 + 606442638553174662709248n^9 \\
&-158967889827748034248704n^{10} - 78160816104265974611968n^{11} \\
&-16419585036974248984576n^{12} - 1908913019215816540160n^{13} \\
&-62769211853172834304n^{14} + 19736004882625224704n^{15} \\
&+4130688305419677696n^{16} + 421867767284303872n^{17} + 25241612021992960n^{18} \\
&+693193236915968n^{19} - 18814089206912n^{20} - 2684704080320n^{21} \\
&-120121949760n^{22} - 2872757568n^{23} - 35919232n^{24} - 178096n^{25}
\end{aligned} \tag{9.112}
$$

$$
\begin{aligned}
Q_{20}(n) &= (n+2)^9(n+4)^5(n+6)^3(n+8)^3(n+10)(n+12)(n+14) \\
&\times(n+16)(n+18)
\end{aligned} \tag{9.113}
$$

$$
\begin{aligned}
P_{21}(n) &= 1351534773860942603511398400 + 6365460757030282723495772160n \\
&+13733120620155454487896522752n^2 + 17929816694573858749176348672n^3 \\
&+15741441712440400461107822592n^4 + 9736866800557416285986619392n^5 \\
&+4291917210346656516848746496n^6 + 1307842644179764469724872704n^7 \\
&+238056916142751738992001024n^8 + 3543432139088928241090560n^9 \\
&-12741108714260109576372224n^{10} - 4295859766813135292465152n^{11} \\
&-755125851031606609051648n^{12} - 65023497347980126945280n^{13} \\
&+2821567036764628910080n^{14} + 1727752072630205923328n^{15} \\
&+264572837767576051712n^{16} + 22840865368902557696n^{17} \\
&+1035703381014509568n^{18} - 6131507476388352n^{19} \\
&-4493143518510080n^{20} - 338088589058432n^{21} \\
&-14045533700352n^{22} - 35536140880n^{23} \\
&-5200818400n^{24} - 35916176n^{25} - 47360n^{26}
\end{aligned} \tag{9.114}
$$

$$
\begin{aligned}
Q_{21}(n) &= (n+2)^9(n+4)^5(n+6)^3(n+8)^3(n+10)(n+12)(n+14) \\
&(n+16)(n+18)(n+20)
\end{aligned} \tag{9.115}
$$

References

[1] Th. Berlin and M. Kac, The spherical model of a ferromagnet, Phys. Rev. 86 (1952) 821–835.

[2] H.E. Stanley, Spherical model as the limit of infinite spin dimensionality, Phys. Rev. 176 (1968) 718–722.

[3] R.J. Baxter, *Exactly Solved Models in Statistical Mechanics*, (Academic Press, London 1982).

[4] P. Butera, M. Comi, G. Marchesini and E. Onofri, Complex temperature singularities for the two-dimensional Heisenberg $O(\infty)$ model, Nucl. Phys. B326 (1989) 758–774.

[5] W.P. Orrick, B.G. Nickel, A.J.Guttmann and J.H.H. Perk, The susceptibility of the square lattice Ising model: new developments, J. Stat. Phys. 102 (2001) 795–841.

[6] P. Butera and M. Comi, Perturbative renormalization group, exact results and high-temperature series to order 21 for the N-vector spin models on the square lattice, Phys. Rev. B 54 (1996) 15828–15848.

[7] M. Campostrini, A. Pelissetto, P. Rossi and E. Vicari, Strong coupling analysis of two-dimensional $O(N)$ σ models with $N \geq 2$ on square, triangular and honeycomb lattices, Phys. Rev. D54 (1996) 1782–1808.

[8] M. Campostrini, A. Pelissetto, P. Rossi and E. Vicari, Strong coupling analysis of two-dimensional $O(N)$ σ models with $N \leq 2$ on square, triangular , and honeycomb lattices, Phys. Rev. B54 (1996) 7301–7317.

[9] L. Onsager, Crystal statistics I. A two dimensional model with an order disorder transition, Phys. Rev. 65 (1944) 117–149.

[10] R.M.F. Houtapple, Order–disorder in hexagonal lattices, Physica 16 (1950) 425–455.

[11] P. Butera and M. Comi, N vector spin models on the simple-cubic and body-centered-cubic lattices: a study of the critical behavior of the susceptibility and of the correlation length by high-temperature series extended to order β^{21}.Phys. Rev. B 56 (1997) 8212–8240.

[12] G.S. Rushbrooke, G.A. Baker and P.J. Wood, The Heisenberg model, in *Phase transitions and critical phenomena* vol. 3, ed. C. Domb and M.S. Green, chap 4. (Academic Press 1974)

[13] S. McKenzie, C. Domb and D.L. Hunter, Extended high-temperature series for the classical Heisenberg model in three dimensions, J. Phys. A15 (1982) 3899–3907.

[14] B.G. Nickel in *Phase Transitions: Cargese 1980*, M. Levy, J.C. Le Guillou and J. Zinn-Justin, eds. (Plenum Press, New York,1982) p. 217.

[15] R.G. Bowers and G.S. Joyce, Lattice model for the λ transition in a Bose fluid, Phys. Rev. Letts. 19 (1967) 630–632.

[16] M.F. Sykes, D.S. Gaunt, J.D. Roberts and J.A. Wyles, High temperature series for the susceptibility of the Ising model II. Three dimensional lattices, J. Phys. A 5 (1972) 640–652.
[17] D.S. Gaunt and M.F. Sykes, The critical exponent γ for the three dimensional Ising model, J. Phys. A12 (1979) L25–L28
[18] S. McKenzie, High-temperature reduced susceptibility of the Ising model, J. Phys. A 8 (1975) L102–L105
[19] P. Butera and M. Comi, Critical specific heats of the N vector spin models on the simple cubic and bcc lattices, Phys. Rev. B 60 (1999) 6749–6760.
[20] M.F. Sykes, D.L. Hunter, D.S. McKenzie and R. Heap, J. Phys. A 5 (1972) 667–673.
[21] A.J. Guttmann and I.G. Enting, Series studies of the Potts model:I. the simple cubic Ising model, J. Phys. A 26 (1993) 807–821.
[22] A.J. Guttmann and I.G. Enting, The high–temperature specific heat exponent of the 3D Ising model, J. Phys. A 27 (1994) 8007–8010.
[23] P.S. English, D.L. Hunter and C. Domb, Extension of the high-temperature, free energy series for the classical vector model of ferromagnetism in general spin dimensionality, J. Phys. A 12 (1979) 2111–2130.
[24] M. Kosterlitz and D.J. Thouless, Ordering, metastability and phase transitions in two-dimensional systems, J. Phys. C6 (1973) 1181–1203.
[25] A.J. Guttmann, On the critical behavior of self-avoiding walks, J. Math. Phys. A 20 (1987) 1839–1854.
[26] A.J. Guttmann, The high-temperature susceptibility and spin-spin correlation function of the three dimensional Ising model, J.Phys. A20 (1987u) 1855–1863.
[27] A.J. Guttmann, Asymptotic analysis of power-series expansions in *Phase Transition and Critical Phenomena*, (Academic Press, San Diego 1989) ed. C. Domb and J.L.lebowitz, vol. 13. pp. 1–234.
[28] G.A. Baker, Jr., H.E. Gilbert, J. Eve and G.S. Rushbrooke, Phys. Rev. 164 (1967) 800–817.
[29] G.S. Rushbrooke and P.J. Wood, On the Curie points and high temperature susceptibilities of Heisenberg model ferromagnetics, Mol. Phys. 1 (1958) 257–283.
[30] R.L. Stephenson, K. Pirnie, P.J. Wood and J. Eve, On the high temperature susceptibility and specific heat of the Heisenberg magnet for a general spin, Phys. Letts. A27 (1968) 2–3.
[31] K–K. Pan, Néel temperature of quantum quasi-two-dimensional Heisenberg antiferromagnets, Phys. Letts. A 271 (2000) 291–295.
[32] G.S. Rushbrooke and P.J. Wood, On the high temperature staggered susceptibility of Heisenberg model antiferromagnetics, Mol. Phys. 6 (1963) 409–421.
[33] J. Oitmaa and E. Bornilla, High temperature study of the spin 1/2 Heisenberg ferromagnet, Phys. Rev. B 53 (1996) 14228–14235.
[34] J. Oitmaa and W-H. Zheng, Curie and Néel temperatures of quantum magnets, J. Phys. Cond. Mat. 16 (2004) 8653–8660.

Part III

Exactly Solvable Models

A thing of beauty is a joy forever.
John Keats

10
The Ising model in two dimensions: summary of results

The remainder of this book will be devoted to statistical models for which exact computations may be carried out. These are called integrable or solvable models. By far the most important and most extensively studied of these systems is the Ising model in two dimensions defined by the interaction energy

$$\mathcal{E} = -\sum_{j=1}^{L_v}\sum_{k=1}^{L_h}\{E^h\sigma_{j,k}\sigma_{j,k+1} + E^v\sigma_{j,k}\sigma_{j+1,k} + H\sigma_{j,k}\} \tag{10.1}$$

where the variables $\sigma_{j,k} = \pm 1$ are located at the j row and k column of a square lattice with L_v rows and L_h columns, and either free, cylindrical or periodic (toroidal) boundary conditions may be imposed. This interaction energy (10.1) was first considered by Lenz [1] in 1920 and the free energy in one dimension was computed by Ising [2] in 1925. In two dimensions the existence of long range order was proven by Peierls [3] in 1936 and the critical temperature was located by Kramers and Wannier [4,5] in 1941 by means of a duality transformation.

The discovery that exact computations can be done for the two-dimensional Ising model (10.1) at $H = 0$ was made in 1944 by Lars Onsager [6] who exactly computed the free energy. Since then many properties including the spontaneous magnetization and correlation functions have been exactly computed and the magnetic susceptibility has been studied in substantial detail. An historical overview of some of the major developments is given in Table 10.1.

The great importance of the Ising model in two dimensions is that at $H = 0$ it exhibits a phase transition at a critical temperature $T = T_c$. The model exhibits all the features of critical phenomena presented in chapter 5 and in particular the critical exponents at $H = 0$ which are given in Table 10.2 have been computed exactly and are seen to obey the scaling laws of chapter 5. Indeed many of the scaling laws of chapter 5 were first seen in the Ising model and it is no exaggeration to say that the general theory of critical phenomena and scaling theory presented in the previous chapters is a generalization of results first obtained for the Ising model. This general theory is universally used to describe critical phenomena in three dimensions.

The first purpose of this chapter is to summarize the many exact results of the Ising model at $H = 0$ on which the general theory of critical phenomena is erected. We will also present the exact results for surface phenomena which also exhibit critical exponents and scaling behavior. In addition we will give the results for layered systems

Table 10.1 Historical overview of major developments in the study of the two dimensional Ising model at $H = 0$.

Date	Author(s)	Property
1920	Lenz [1]	Ising model introduced
1925	Ising [2]	On dimensional model solved
1936	Peierls [3]	Existence of spontaneous magnetization
1941	Kramers, Wannier [4, 5]	Duality and T_c
1944	Onsager [6]	Free energy
1949	Onsager [7]	Spontaneous magnetization
1949	Kaufman [8]	Partition function on torus
1949	Kaufman, Onsager [9]	Short range correlations two-spin correlation at T_c
1952	Yang [10]	Spontaneous magnetization
1963	Kastelyn [11]	Ising model reduced to dimers
1963	Montroll, Potts, Ward [12]	Correlations as determinants
1966-67	Cheng, Wu [13, 14]	Large separation behavior of the two-point function
1967	McCoy, Wu [15]	Boundary critical behavior
1968-69	McCoy, Wu [16–18]	Random layered lattices
1969	Griffiths [19]	Random lattice singularities
1973-1976	Barouch, McCoy, Tracy, Wu [20–22]	Painlevé III representation for the scaled two point function
1976-1980	Sato, Miwa, Jimbo [23–27]	Holonomic field theory
1978	McCoy , Tracy, Wu [28–30]	n point functions
1980-1981	McCoy, Perk, Wu [31–34]	Partial difference equations for correlation functions
1981	Jimbo,Miwa [35]	Painlevé VI for diagonal correlations
1999-2001	Orrick, Nickel, Guttmann, Perk [36–39]	Bulk susceptibility
2004	Zenine, Boukraa, Hassani, Maillard [41–43]	Fuchsian equation for 3 and 4 particle contributions to bulk susceptibility
2007	Boukraa, Hassani Maillard, McCoy Orrick, Zenine [44]	Factorization of form factors
2007	Boukraa, Hassani Maillard, McCoy Weil, Zenine [45]	Diagonal Ising susceptibility

where the interaction energies E^v and E^h vary randomly from row to row and show that, for these random systems, the critical exponent description does not hold.

However, all of the results at $H = 0$ depend on very special properties which allow the exact solvability and these remarkable properties can not possibly be shared by any real system in the laboratory. It may therefore be asked why the phenomenology derived from this Ising model seems to be so very accurate and powerful when applied

Table 10.2 Critical exponents for the two-dimensional Ising model.

Property	Critical exponent
Bulk properties	
Specific heat at	$\alpha = 0 \quad \ln\lvert T - T_c\rvert$
Spontaneous magnetization	$\beta = 1/8$
Magnetic susceptibility	$\gamma = 7/4$
Magnetization at $T = T_c$	$\delta = 1/15$
Correlation length	$\nu = 1$
Anomalous dimension	$\eta = 1/4$
Boundary properties	
Specific heat	$\alpha_b = 1$
Spontaneous magnetization	$\beta_b = 1/2$
Magnetic susceptibility	$\gamma_b = 0 \quad \ln\lvert T - T_c\rvert$
Magnetization at T_c	$\delta_b = 1 \quad H_b \ln\lvert H_b\rvert$
Correlation length	$\nu_b = 1$
Anomalous dimension at $H_b = 0$	$\eta_b = 1$
Anomalous dimension at $H_b \neq 0$	$\eta_b = 4$

to real systems in three dimensions and what, if anything, is missing in the intuition derived from the exact results obtained for the Ising model. In particular since the general theory of critical phenomena was erected as a generalization of the Ising model and confirmed by all other models which have special properties (to be discussed in later chapters) which allow them to be exactly solved, we may ask if it could happen that this general theory of critical phenomena is only applicable to these special integrable systems.

We attempt to give some insight into this most important question by considering the Ising model at $H \neq 0$. An historical overview of some of the major developments is given in Table 10.3. For small H a perturbation technique [48] shows that in the scaling limit there is a sort of "confinement phenomenon" which takes place and a very remarkable computation done in 1989 by Zamolodchikov [50] shows that, in the scaling limit at $T = T_c$, the model is again integrable.

In the present chapter we will summarize and discuss these results. In the following chapter we will present in detail one of the methods of computing the partition function

Table 10.3 Historical overview of major developments in the study of the two dimensional Ising model at $H \neq 0$.

Date	Author(s)	Property
1952	Lee, Yang [46]	Circle theorem
1952,1967	Lee, Yang [46] McCoy, Wu [47]	Solution at $H/k_BT = i\pi/2$
1978	McCoy, Wu [48]	Confinement for $T < T_c$
1984	Isakov [49]	Nonanalyticity at $H = 0$ for $T < T_c$
1988	Zamolodchikov [50]	E_8 scaled Ising model at $T = T_c$
2003	Fonseca, Zamolodchikov [51]	Extended analyticity for free energy

on the finite lattice and derive determinental expressions for the correlation functions. In chapter 12 we derive the spontaneous magnetization and the form factor expansion of the correlation functions from the determinants. For all other derivations we refer the reader to the original derivations in the literature or for some of the older results to the book *The Two Dimensional Ising Model* by T.T. Wu and the present author [52].

10.1 The homogeneous lattice at $H = 0$

We begin by presenting the results for the homogeneous Ising model (10.1) on a lattice with L_v rows and L_h columns.

10.1.1 Partition function on the torus

For periodic boundary conditions, the partition function was found by Kaufman [8] to be given by the sum of four terms

$$Z(T, L_v, L_h) = \frac{1}{2}\{-Z_{ee}(T, L_v, L_h) + Z_{eo}(T, L_v, L_h) + Z_{oe}(T, L_v, L_h) + Z_{oo}(T, L_v, L_h)\} \quad (10.2)$$

where

$$Z_j(T, L_v, L_h) = 2^{L_v L_h} \prod_{\theta_1} \prod_{\theta_2} [\cosh 2K^h \cosh 2K^v - \sinh 2K^h \cos\theta_1 - \sinh 2K^v \cos\theta_2]^{1/2} \quad (10.3)$$

where

$$K^h = E^h/k_B T \quad \text{and} \quad K^v = E^v/k_B T \quad (10.4)$$

and θ_1 and θ_2 are chosen as in Table 10.4 with $n_1 = 1, 2, \cdots, L_h$ and $n_2 = 1, 2, \cdots, L_v$.

Table 10.4 Allowed values of θ_i

j	θ_1	θ_2
ee	$2\pi n_1/L_h$	$2\pi n_2/L_v$
eo	$2\pi n_1/L_h$	$\pi(2n_2 - 1)/L_v$
oe	$\pi(2n_1 - 1)/L_h$	$2\pi n_2/L_v$
oo	$\pi(2n_1 - 1)/L_h$	$\pi(2n_2 - 1)/L_v$

The square root in (10.3) is only apparent because all terms appear in pairs. These apparent square roots are defined to be positive when $T > T_c$. When L_v and L_h are both even we explicitly have

$$Z_{oo} = 2^{L_v L_h} \prod_{n_1=1}^{L_h/2} \prod_{n_2=1}^{L_v/2} \{\cosh 2K^v \cosh 2K^h - \sinh 2K^h \cos\pi(2n_1 - 1)/L_h - \sinh 2K^v \cos\pi(2n_2 - 1)/L_v\}^2, \quad (10.5)$$

$$Z_{eo} = 2^{L_v L_h} \prod_{n_2=1}^{L_v/2} \{([\cosh 2K^v \cosh 2K^h - \sinh 2K^v \cos \pi(2n_2 - 1)/L_v]^2 - \sinh^2 2K^h)$$
$$\times \prod_{n_1=1}^{L_h/2-1} [\cosh 2K^v \cosh 2K^h - \sinh 2K^h \cos 2\pi n_1/L_h - \sinh 2K^v \cos \pi(2n_1 - 1)/L^v]^2\} \tag{10.6}$$

$$Z_{oe} = 2^{L_v L_h} \prod_{n_1=1}^{L_h/2} \{([\cosh 2K^v \cosh 2K^h - \sinh 2K^h \cos \pi(2n_1 - 1)/L_h]^2 - \sinh^2 2K^v)$$
$$\times \prod_{n_2=1}^{L_h/2-1} [\cosh 2K^v \cosh 2K^h - \sinh 2K^v \cos 2\pi n_2/L_v - \sinh 2K^h \cos \pi(2n_2 - 1)/L^h]^2\}, \tag{10.7}$$

and

$$Z_{ee} = 2^{L_v L_h}(1 - \sinh^2 2K^h \sinh^2 2K^v)$$
$$\times \prod_{n_1=1}^{L_h/2-1} ([\cosh 2K^v \cosh 2K^h - \sinh 2K^h \cos \pi 2n_1/L_h]^2 - \sinh^2 2K^v)$$
$$\times \prod_{n_2=1}^{L_v/2-1} ([\cosh 2K^v \cosh 2K^h - \sinh 2K^v \cos \pi 2n_2/L_v]^2 - \sinh^2 2K^h)$$
$$\times \prod_{n_1=1}^{L_h/2-1} \prod_{n_2=1}^{L_v/2-1} \{\cosh 2K^h \cosh 2K^v - \sinh 2K^h \cos 2\pi n_1/L_h - \sinh 2K^v \cos 2\pi n_2/L_v\}^2, \tag{10.8}$$

10.1.2 Zeros of the partition function

The zeroes in the complex T plane of the four $Z_j(T, L_v, L_h)$ are obvious from the factored form (10.3) but the zeros of the partition function itself are tedious to locate. However, for large L_v and L_h the limiting distributions of the zeroes of the partition function and of the four $Z_j(T, L_v, L_h)$ are identical and are obtained from the equation

$$\cosh 2K^h \cosh 2K^v - \sinh 2K^h \cos \theta_1 - \sinh 2K^v \cos \theta_2 = 0 \tag{10.9}$$

with $0 \leq \theta_1, \theta_2 \leq 2\pi$.

In the isotropic case when $E^v = E^h = E$ the formula for the location of the zeros (10.9) reduces to

$$1 + \sinh^2 2K - \sinh 2K(\cos \theta_1 + \cos \theta_2) = 0 \tag{10.10}$$

from which we see that

$$\sinh 2K = \frac{1}{2}\{\cos\theta_1 + \cos\theta_2 \pm i[4 - (\cos\theta_1 + \cos\theta_2)^2]^{1/2}]\} \tag{10.11}$$

and thus the zeros lie on a circle

$$|\sinh 2K|^2 = 1. \tag{10.12}$$

For the general anisotropic case the zeros obtained from (10.9) fill up areas. However for real E^v and E^h these regions of zeros pinch the real axis only at points T_c determined from

$$\sinh 2E^v/k_BT_c \sinh 2E^h/k_BT_c = \pm 1. \tag{10.13}$$

Consequently these zeros satisfy the condition discussed in chapter 5 which is required for a second order phase transition.

It is convenient in the isotropic case to study the zeros of the partition function by introducing the variable

$$x = e^{-2K} \tag{10.14}$$

and write

$$Z(T, L_v, L_h) = x^{L_vL_h}\tilde{Z}(x) \tag{10.15}$$

where $\tilde{Z}(x)$ is a polynomial in x of degree L_vL_h given by (10.2) with $Z_j(T, L_v, L_h) \to \tilde{Z}_j(x)$ where

$$\tilde{Z}_j(x) = \prod_{\theta_1}\prod_{\theta_2}\{(1+x^2)^2 + 2x(1-x^2)(\cos\theta_1 + \cos\theta_2)\}^{1/2} \tag{10.16}$$

which is a polynomial in

$$u = x^2. \tag{10.17}$$

The zeros of the polynomials $\tilde{Z}_j(x)$ lie on the two circles

$$|x \pm 1| = \sqrt{2} \tag{10.18}$$

but the zeros of the full partition function lies on these circles only in the thermodynamic limit.

It was discovered in 1974 by Brascamp and Kunz [53] that for the isotropic lattice the zeros of the partition function will lie exactly on the circles (10.18) if L_h is even and the boundary conditions are chosen as follows:

1) There are periodic boundary conditions in the horizontal direction.

2) The spins in the upper boundary row interact with a row of spins all fixed to be +.

3) The spins in the lower boundary row interact with a row of spins which are fixed and alternate + and −.

For this choice of boundary conditions the polynomial partition function normalized to unity at $x = 0$ is

$$Z_{BK}(x) = \prod_{j=1}^{L_h/2}\prod_{k=1}^{L_v}\{(1+x^2)^2 + 2x(1-x^2)[\cos((2j-1)\pi/L_h) + \cos(k\pi/(L_v+1))]\}. \tag{10.19}$$

We compare the zero distributions for the toroidal and the Brascamp–Kunz boundary conditions by plotting the zeros for an 18×18 lattice in the complex u plane in

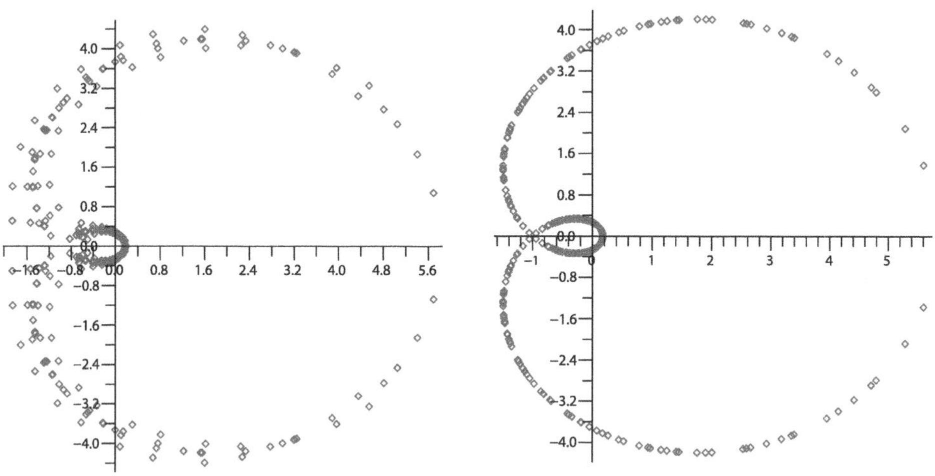

Fig. 10.1 Comparison of the zeros in the plane $u = x^2$ of the Ising partition functions with periodic boundary conditions on the left and Brascamp–Kunz boundary conditions on the right for an 18×18 lattice at $H = 0$.

Fig. 10.1. Zeros for various finite size lattices with other boundary conditions have been studied by Matveev and Shrock [54].

10.1.3 Bulk free energy per site

The free energy per site is computed from the partition function (10.2) as

$$-F/k_BT = \lim_{L_v, L_h \to \infty} \frac{1}{L_v L_h} \ln Z(T, L_v, L_h) \tag{10.20}$$

and when T is in a zero-free region of the plane we obtain the famous result of Onsager [6]:

$$\begin{aligned} F/k_BT &= -\ln 2 \\ &- \frac{1}{8\pi^2} \int_0^{2\pi} d\theta_1 \int_0^{2\pi} d\theta_2 \ln[\cosh 2K^h \cosh 2K^v - \sinh 2K^h \cos\theta_1 - \sinh 2K^v \cos\theta_2] \end{aligned} \tag{10.21}$$

This free energy is independent of the signs of E^v and E^h and we will henceforth take these two interaction energies to be real and positive.

The argument of the logarithm in (10.21) is nonnegative for $0 \leq \theta_1, \theta_2 < 2\pi$. The argument only vanishes at $\theta_1, \theta_2 = 0, \pi$ and this vanishing imposes the restriction $T = T_c$

$$\cosh 2E^h/k_BT_c \cosh 2E^v/k_BT_c \pm \sinh 2E^h/k_BT_c \pm \sinh 2E^v/k_BT_c = 0 \tag{10.22}$$

which is equivalent to (10.13) which is the location in the complex T plane where the zeros of the partition function pinch the real axis. We will by convention define T_c to be the solution of (10.13) with the positive sign on the right-hand side. We also note that, if we solve (10.13) (with the positive sign) for e^{2E^v/k_BT_c}, we obtain an equivalent condition for T_c of

$$e^{2E^v/k_BT_c} = \frac{e^{2E^h/k_BT_c}+1}{e^{2E^h/k_BT_c}-1} \tag{10.23}$$

and the companion equation with $E^v \leftrightarrow E^h$.

The free energy (10.21) is singular at $T = T_c$. However, this is the only feature of the distribution of the zeros of the partition function that remains in the free energy. In particular the free energy may be analytically continued through and beyond the regions in the complex plane where the finite size partition function has zeros.

The internal energy is obtained from the free energy (10.21) as

$$u = \frac{\partial \beta F}{\partial \beta} = -E^h\langle\sigma_{0,0}\sigma_{0,1}\rangle - E^v\langle\sigma_{0,0}\sigma_{1,0}\rangle \tag{10.24}$$

where $\beta = 1/k_BT$ and

$$\langle\sigma_{0,0}\sigma_{0,1}\rangle = \frac{1}{2\pi}\int_0^{2\pi} d\theta \left[\frac{(1-\alpha_1 e^{i\theta})(1-\alpha_2 e^{-i\theta})}{(1-\alpha_1 e^{-i\theta})(1-\alpha_2 e^{i\theta})}\right]^{1/2} \tag{10.25}$$

with

$$\alpha_1 = e^{-2K^v}\tanh K^h \quad\text{and}\quad \alpha_2 = e^{-2K^v}\coth K^h \tag{10.26}$$

and $\langle\sigma_{0,0}\sigma_{0,1}\rangle = \langle\sigma_{0,0}\sigma_{1,0}\rangle$ with K^h and K^v interchanged. The square root is defined positive at $\theta = \pi$.

In the isotropic case $E^v = E^h = E$ the internal energy may be written in terms of the complete elliptic integral

$$K(k) = \int_0^{\pi/2} \frac{d\phi}{(1-k^2\sin^2\phi)^{1/2}} = \int_0^1 \frac{dx}{2[x(1-x)(1-k^2x)]^{1/2}} \tag{10.27}$$

as

$$\begin{aligned} u &= -E\coth 2\beta E[1 + 2\pi^{-1}(2-\tanh^2 2\beta E)K(k)] \\ &= -E\{2[1+(1-k^2)^{1/2}]\}^{1/2}\{1 \mp \pi^{-1}(1-k^2)^{1/2}K(k)\} \end{aligned} \tag{10.28}$$

where

$$k = 2\frac{\sinh 2\beta E}{(\cosh 2\beta E)^2} = \frac{2}{\sinh 2\beta E + \sinh^{-1} 2\beta E} \tag{10.29}$$

and in the second line of (10.28) the minus (plus) sign is chosen for $T > T_c$ ($T < T_c$).

We note for real T that $|\alpha_1| < 1$ but that α_2 has no such restriction. In particular we may have $\alpha_2 = 1$ and at this point the integral in (10.25) fails to be analytic because the square root branch cuts at $\alpha_2^{\pm 1}$ pinch the contour of integration. From (10.26) we see that the condition $\alpha_2 = 1$ is identical to the form of the T_c condition (10.23). We also note that near T_c we have

$$\begin{aligned}\alpha_2 &\sim 1 - (\beta - \beta_c) 2\{E^v + \frac{E^h}{\sinh 2E^h/k_B T_c}\} \\ &= 1 - (\beta - \beta_c) 2\{E^v + E^h \sinh 2E^v/k_B T_c\} \\ &= 1 - (\beta - \beta_c) 2 \tanh 2E^v/k_B T_c \{E^v \coth 2E^v/k_B T_c + E^h \coth 2E^h/k_B T_c\}\end{aligned} \tag{10.30}$$

where in the last line we have used the identity

$$\frac{\coth 2E^h/k_B T_c}{\coth 2E^v/k_B T_c} = \left(\frac{\sinh 2E^v/k_B T_c}{\sinh 2E^h/k_B T_c}\right)^{1/2} = \frac{\cosh 2E^v/k_B T_c}{\cosh 2E^h/k_B T_c} \tag{10.31}$$

which follows from the condition for T_c (10.13) or (10.23). Near T_c we find

$$\langle \sigma_{0,0}\sigma_{0,1}\rangle \sim 2\pi \coth 2K_c^h \mathrm{gd} 2K_c^h - \frac{2}{\pi}(\beta - \beta_c)(E^v + E^h \sinh 2E^v/k_B T_c) \ln|\beta - \beta_c| \tag{10.32}$$

where $\mathrm{gd}x = \arctan \sinh x$ is the Gudermannian of x.

From the internal energy (10.24) we obtain the specific heat and find that near T_c

$$c = \frac{\partial u}{\partial T} \sim -\frac{2k_B}{\pi}(K_c^{h2} \sinh 2K_c^v + 2K_c^v K_c^h + K_c^{v2} \sinh 2K_c^h) \ln|1 - T/T_c| + O(1) \tag{10.33}$$

where the term $O(1)$ is the same for T above and below T_c. Thus the specific heat has a logarithmic divergence at T_c which corresponds to a critical index of $\alpha = 0$. The specific heat is plotted in Fig. 10.2 as a function of temperature for various value of the anisotropy.

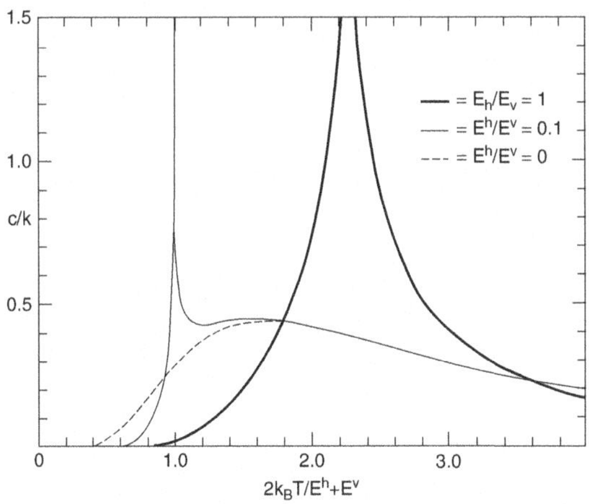

Fig. 10.2 Specific heat of the Ising model for $E^v/E^h = 1$, $E^v/E^h = 0.1$ and $E^v/E^h = 0$.

10.1.4 Partition function at $T = T_c$

If $T \neq T_c$ we see from (10.2) and (10.21) that for $L_v, L_h \gg 1$

$$Z(T, L_v, L_h) = e^{-L_v L_h F/k_B T}\{1 + \text{terms exponentially small in } L_v L_h\}\}. \qquad (10.34)$$

However, if $T = T_c$ (10.34) does not hold and instead we have the result of Ferdinand and Fisher [60] that if $L_v, L_h \gg 1$ then

$$Z(T_c, L_v, L_h) \sim \frac{1}{2}\left[|\frac{\theta_2(0;\tau)}{\eta(\tau)}| + |\frac{\theta_3(0;\tau)}{\eta(\tau)}| + |\frac{\theta_4(0;\tau)}{\eta(\tau)}|\right] \exp(-L_v L_h F/k_B T_c) \qquad (10.35)$$

where

$$\eta(\tau) = e^{i\tau/24}\prod_{j=1}^{\infty}(1 - e^{ij\tau}) \qquad (10.36)$$

and the theta functions are $\theta_j(v;\tau)$

$$\theta_1(v;\tau) = \sum_{n=-\infty}^{\infty}(-1)^n e^{i\tau(n+1/2)^2} e^{2\pi i v(n+1/2)/K} \qquad (10.37)$$

$$\theta_2(v;\tau) = \sum_{n=-\infty}^{\infty} e^{i\tau(n+1/2)^2} e^{2\pi i v(n+1/2)/K} \qquad (10.38)$$

$$\theta_3(v;\tau) = \sum_{n=-\infty}^{\infty} e^{i\tau n^2} e^{2\pi i v n/K} \qquad (10.39)$$

$$\theta_4(v;\tau) = \sum_{n=-\infty}^{\infty}(-1)^n e^{i\tau n^2} e^{2\pi i v n/K} \qquad (10.40)$$

where

$$\tau = i\frac{L_v \cosh 2K_c^h}{L_h \cosh 2K_c^v}. \qquad (10.41)$$

10.1.5 Spontaneous magnetization

The spontaneous magnetization $M_-(T)$ was announced by Onsager [6] in 1949 and a derivation was given by Yang [10] in 1952. The remarkably simple result is

$$\begin{aligned} M_-(T) &= [1 - (\sinh 2E^h/k_B T \sinh 2E^v/k_B T)^{-2}]^{1/8} \quad \text{for} \quad 0 < T \leq T_c \\ &= 0 \quad \text{for} \quad T > T_c. \end{aligned} \qquad (10.42)$$

This shares the feature with the free energy that it is singular only at the points where the T_c condition (10.13) holds and that there are no further singularities at the location of the zeros of the partition function. Near T_c we have

$$\begin{aligned} &\sinh 2E^h/k_B T \sinh 2E^v/k_B T \\ &\sim 1 + (\beta - \beta_c)\left(2E^h \coth 2E^h/k_B T_c + 2E^v \coth 2E^v/k_B T_c\right) \end{aligned} \qquad (10.43)$$

and thus from (10.42) we have as $T \to T_c-$

$$M_-(T) \sim \left[(\beta - \beta_c)4(E^h \coth 2E^h/k_B T_c + E^v \coth 2E^v/k_B T_c)\right]^{1/8} \qquad (10.44)$$

from which we see that the critical exponent $\beta = 1/8$.

The spontaneous magnetization is plotted in Fig. 10.3 for the isotropic case $E^v = E^h$ along with the magnetization in the boundary row in the half plane lattice given in (10.259).

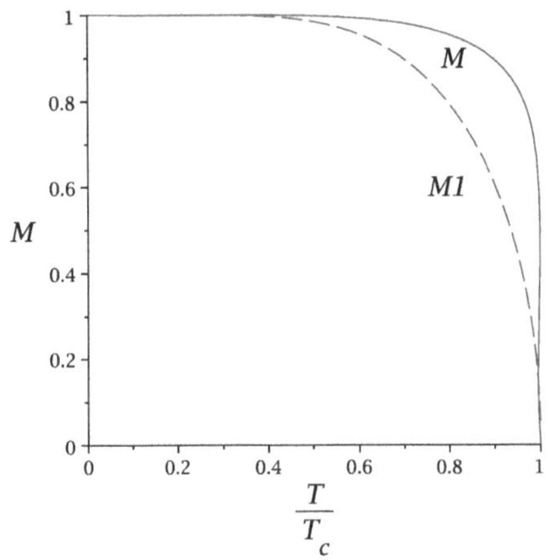

Fig. 10.3 Comparison of the spontaneous magnetization M (10.42) with the magnetization M_1 (10.259) in the boundary row of the half plane lattice for the isotropic lattice $E^v = E^h$ as a function of temperature.

10.1.6 Row and diagonal spin correlation functions

It was shown by Montroll, Potts and Ward [12] that all spin correlation functions $C(M, N) = \langle \sigma_{0,0}\sigma_{M,N} \rangle$ may be expressed as determinants. Indeed, every correlation function may be expressed as a determinant in an infinite number of ways in the thermodynamic limit. However, more is known about the diagonal and row correlation than is known for the case of general $C(M, N)$ and therefore we will treat these cases separately.

The diagonal correlation $C(N, N) = \langle \sigma_{0,0}\sigma_{N,N} \rangle$ and the row $C(0, N) = \langle \sigma_{0,0}\sigma_{0,N} \rangle$ correlation functions can both be written as $N \times N$ Toeplitz determinants

$$D_N = \begin{vmatrix} a_0 & a_{-1} & \cdots & a_{-N+1} \\ a_1 & a_0 & \cdots & a_{-N+2} \\ \vdots & \vdots & & \vdots \\ a_{N-1} & a_{N-2} & \cdots & a_0 \end{vmatrix} \tag{10.45}$$

with

$$a_n = \frac{1}{2\pi} \int_0^{2\pi} d\theta e^{-in\theta} \left[\frac{(1 - \alpha_1 e^{i\theta})(1 - \alpha_2 e^{-i\theta})}{(1 - \alpha_1 e^{-i\theta})(1 - \alpha_2 e^{i\theta})} \right]^{1/2} \tag{10.46}$$

where for $C(N, N)$

$$\alpha_1 = 0, \quad \alpha_2 = (\sinh 2K^v \sinh 2K^h)^{-1} \tag{10.47}$$

and for $C(0, N)$

$$\alpha_1 = e^{-2K^v} \tanh K^h, \quad \alpha_2 = e^{-2K^v} \coth K^h \tag{10.48}$$

and the square roots are defined to be positive at $\theta = \pi$. These determinants are very efficient for the calculation of the correlations when N is small.

For the diagonal correlation function all matrix elements may be expressed in terms of the integral representation the hypergeometric function [55, chapter 2] valid for $\mathrm{Re}\, c > \mathrm{Re}\, b > 0$:

$$F(a,b;c;t) = \frac{\Gamma(c)}{\Gamma(c-b)\Gamma(b)} \int_0^1 dx x^{b-1}(1-x)^{c-b-1}(1-tx)^{-a}. \tag{10.49}$$

For $T < T_c$ with

$$t = (\sinh 2K^v \sinh 2K^h)^{-2} = \alpha_2^2 < 1 \tag{10.50}$$

we have for $n \geq 0$

$$a_n = t^{n/2} \frac{\Gamma(n+1/2)}{\pi^{1/2} n!} F(-1/2, n+1/2; n+1; t) \tag{10.51}$$

and for $n \leq -1$

$$-t^{|n|/2} \frac{\Gamma(|n|-1/2)}{2\pi^{1/2}(|n|)!} F(1/2, |n|-1/2; |n|+1; t) \tag{10.52}$$

For $T > T_c$ with

$$t = (\sinh 2K^v \sinh 2K^h)^{2} = \alpha_2^{-2} < 1 \tag{10.53}$$

we have for $n \geq 0$

$$a_n = \frac{t^{(n+1)/2}\Gamma(n+1/2)}{2\pi^{1/2}(n+1)!} F(1/2, n+1/2; n+2; t) \tag{10.54}$$

and for $n \leq -1$

$$a_n = -\frac{t^{(|n|-1)/2}\Gamma(|n|-1/2)}{\pi^{1/2}(|n|-1)!} F(-1/2, |n|-1/2; |n|; t) \tag{10.55}$$

By use of the contiguous relations of hypergeometric functions such as [55, chapter 2],

$$F(a,b;c;t) = \frac{(1-c)_m t^{-m}}{(b-c+1)_m} \sum_{k=0}^{m} \binom{m}{k} (t-1)^{m-k} F(a-k, b; c-m; t), \tag{10.56}$$

where

$$(a)_m = \Gamma(a+m)/\Gamma(a), \tag{10.57}$$

these may be rewritten in terms of the complete elliptic integrals of the first and second kind

$$K(t^{1/2}) = \int_0^{\pi/2} \frac{d\phi}{(1-t\sin^2\phi)^{1/2}} = \frac{\pi}{2} F(1/2, 1/2; 1; t) \tag{10.58}$$

$$E(t^{1/2}) = \int_0^{\pi/2} d\phi (1 - t\sin^2\phi)^{1/2} = \frac{\pi}{2} F(-1/2, 1/2; 1; t). \tag{10.59}$$

Using the abbreviated notations

$$\tilde{K} = \frac{2}{\pi} K(t^{1/2}) \quad \text{and} \quad \tilde{E} = \frac{2}{\pi} E(t^{1/2}) \tag{10.60}$$

we have, for example when $T < T_c$

$$\langle \sigma_{0,0}\sigma_{1,1} \rangle = \tilde{E} \tag{10.61}$$

$$\langle \sigma_{0,0}\sigma_{2,2} \rangle = \frac{1}{3t}\{(5t-1)\tilde{E}^2 + 2(t-1)^2\tilde{E}\tilde{K} - (t-1)^2\tilde{K}^2\} \tag{10.62}$$

and for $T > T_c$

$$\langle \sigma_{0,0}\sigma_{1,1} \rangle = t^{-1/2}\{\tilde{E} - (1-t)\tilde{K}\} \tag{10.63}$$

$$\langle \sigma_{0,0}\sigma_{2,2} \rangle = \frac{1}{3t}\{(5-t)\tilde{E}^2 + 8(t-1)\tilde{K}\tilde{E} + 3(t-1)^2\tilde{K}^2\}. \tag{10.64}$$

Furthermore as $T \to T_c$ each element a_n has an expansion

$$a_n \sim \sum_{k=0}^{\infty} a_{n,k}^{\pm} |T - T_c|^k + \ln|T - T_c| \sum_{k=1}^{\infty} b_{n,k}^{\pm} |T - T_c|^k \tag{10.65}$$

and using (10.65) in (10.45) we find that cancellations take place and that as $T \to T_c\pm$ the determinants D_N have the form for both the row and diagonal correlations:

$$D_N = \sum_{k=0}^{N} |T - T_c|^{k^2} \ln^k |T - T_c| \sum_{j=0}^{\infty} d_{k,j}^{\pm} |T - T_c|^j \tag{10.66}$$

Form factor expansion

For $T < T_c$ the determinant D_N may also be expressed in an exponential form [56]

$$D_N = (1-t)^{1/4} e^{F_N} \tag{10.67}$$

where

$$F_N = \sum_{n=1}^{\infty} \lambda^{2n} F_N^{(2n)} \tag{10.68}$$

with t given by (10.50),

$$F_N^{(2n)} = \frac{(-1)^{n+1}}{n 2^{2n}} \oint \prod_{j=1}^{2n} \frac{dz_j \, z_j^N}{1 - z_j z_{j+1}} \prod_{j=1}^{n} Q(z_{2j-1}) Q(z_{2j-1}^{-1}) P(z_{2j}) P(z_{2j}^{-1}) \tag{10.69}$$

where

$$\lambda = 1/\pi \tag{10.70}$$

the contours of integration are $|z_j| = 1 - \epsilon,\ z_{2n+1} \equiv z_1$ and

$$P(z) = 1/Q(z) = \left(\frac{1-\alpha_2 z}{1-\alpha_1 z}\right)^{1/2}. \tag{10.71}$$

For $T > T_c$ the corresponding result is

$$D_N = (1-t)^{1/4} X_N e^{\hat{F}_{N+1}} \tag{10.72}$$

where t is given by (10.53), the function $\hat{F}_N$ is given by (10.68) and (10.69) with $P(z)$ and $Q(z)$ replaced by

$$\hat{P}(z) = 1/\hat{Q}z = [(1-\alpha_1 z)(1-\alpha_2^{-1}z)]^{-1/2} \tag{10.73}$$

and

$$X_N = \sum_{n=0}^{\infty} \lambda^{2n+1} X_N^{(2n+1)} \tag{10.74}$$

with

$$\begin{aligned} X_N^{(2n+1)} &= \frac{1}{(2i)^{2n+1}} \oint \prod_{j=1}^{n+1} (z_j^{N+1} dz_j) \frac{1}{z_1 z_{2n+1}} \prod_{j=1}^{2n} \frac{1}{1 - z_j z_{j+1}} \\ &\times \prod_{j=1}^{n+1} \hat{P}(z_{2j-1}) \hat{P}(z_{2j-1}^{-1}) \prod_{j=1}^{n} \hat{Q}(z_{2j}) \hat{Q}(z_{2j}^{-1}). \end{aligned} \tag{10.75}$$

The exponentials in (10.67) and (10.72) may be expanded and thus we obtain the form factor expressions for the correlation functions. For $T < T_c$ the form factor expansion is

$$D_N = (1-t)^{1/4}\{1 + \sum_{n=1}^{\infty} \lambda^{2n} f_N^{(2n)}\} \tag{10.76}$$

with

$$\begin{aligned} f_N^{(2n)} &= \frac{1}{(n!)^2} \frac{1}{(2i)^{2n}} \oint \prod_{j=1}^{2n} dz_j z_j^N Q(z_{2j-1}) Q(z_{2j-1}^{-1}) P(z_{2j}) P(z_{2j}^{-1}) \\ &\times \prod_{1\le j\le n} \prod_{1\le k\le n} \left(\frac{1}{1 - z_{2j-1} z_{2k}}\right)^2 \prod_{1\le j<k\le n} (z_{2j-1} - z_{2k-1})^2 (z_{2j} - z_{2k})^2 \end{aligned} \tag{10.77}$$

and for $T > T_c$ the form factor expansion is

$$D_N = (1-t)^{1/4} \sum_{n=0}^{\infty} \lambda^{2n+1} f_N^{(2n+1)} \tag{10.78}$$

with

$$f_N^{(2n+1)}$$
$$\frac{1}{n!(n+1)!(2i)^{2n+1}} \oint \prod_{j=1}^{2n+1} (dz_j z_j^N) \prod_{j=1}^{n+1} z_{2j-1}^{-1} \hat{P}(z_{2j-1}) \hat{P}(z_{2j-1}^{-1}) \prod_{j=1}^{n} z_{2j} \hat{Q}(z_{2j}) \hat{Q}(z_{2j}^{-1})$$
$$\times \prod_{1 \le j \le n+1} \prod_{1 \le k \le n} \left(\frac{1}{1 - z_{2j-1} z_{2k}} \right)^2 \prod_{1 \le j < k \le n+1} (z_{2j-1} - z_{2k-1})^2 \prod_{1 \le j < k \le n} (z_{2j} - z_{2k})^2 \tag{10.79}$$

It was discovered in [44] by means of computer calculation that the form factors of the diagonal correlation has the property that the multiple integrals in (10.77) and (10.79) all decompose into sums of products of the complete elliptic integrals $K(t^{1/2})$ and $E(t^{1/2})$. Using the notation of (10.60) a few examples of this factorization are:

$$2f_0^{(2)} = (\tilde{K} - \tilde{E})\tilde{K} \tag{10.80}$$
$$2f_1^{(2)} = 1 - 3\tilde{K}\tilde{E} - (t-2)\tilde{K}^2 \tag{10.81}$$
$$6tf_2^{(2)} = 6t - (6t^2 - 11t + 2)\tilde{K}^2 - (15t - 4)\tilde{K}\tilde{E} - 2(t+1)\tilde{E}^2 \tag{10.82}$$

$$6f_0^{(3)} = \tilde{K} - (t-2)\tilde{K}^3 - 3\tilde{K}^2\tilde{E} \tag{10.83}$$
$$6t^{1/2}f_1^{(3)} = 4(\tilde{K} - \tilde{E}) - 6\tilde{K}^2\tilde{E} - (2t-3)\tilde{K}^3 + 3\tilde{K}\tilde{E}^2 \tag{10.84}$$

$$24f_0^{(4)} = 4(\tilde{K} - \tilde{E})\tilde{K} - (2t-3)\tilde{K}^4 - 6\tilde{K}^3\tilde{E} + 3\tilde{K}^2\tilde{E}^2 \tag{10.85}$$
$$24f_1^{(4)} = 9 - 30\tilde{K}\tilde{E} - 10(t-2)\tilde{K}^2 + (t^2 - 6t + 6)\tilde{K}^4 + 15\tilde{K}^2\tilde{E}^2 + 10(t-2)\tilde{K}^3\tilde{E} \tag{10.86}$$

A Painlevé VI representation for $C(N, N)$

The form factor representations for the diagonal correlation (10.76) for $T < T_c$ and (10.78) for $T > T_c$ may be used for all values of λ to define functions $C_\pm(N, N; \lambda)$ where the true diagonal correlation is

$$C_\pm(N, N) = C_\pm(N, N; 1/\pi). \tag{10.87}$$

This generalized function has the striking property first found by Jimbo and Miwa [35] that if we define a function σ for $T < T_c$ by

$$\sigma_N(t) = t(t-1)\frac{d \ln C_-(N, N; \lambda)}{dt} - \frac{t}{4} \tag{10.88}$$

with t defined by (10.50), and for $T > T_c$ by

$$\sigma_N(t) = t(t-1)\frac{d \ln C_+(N, N)}{dt} - \frac{1}{4} \tag{10.89}$$

where t is defined by (10.53), then for both $T < T_c$ and $T > T_c$ and all values of λ the function σ_N satisfies

$$\left(t(t-1)\frac{d^2\sigma}{dt^2}\right)^2 = N^2\left((t-1)\frac{d\sigma}{dt}-\sigma\right)^2 - 4\frac{d\sigma}{dt}\left((t-1)\frac{d\sigma}{dt}-\sigma-\frac{1}{4}\right)\left(t\frac{d\sigma}{dt}-\sigma\right). \tag{10.90}$$

We note that, for the Ising case $\lambda = 1/\pi$, $C(N,N;1/\pi)$ must satisfy the normalization condition for $T < T_c$

$$C_-(N,N;1/\pi) = 1 + O(t) \quad \text{for} \quad t \to 0 \tag{10.91}$$

and for $T > T_c$

$$C_+(N,N;1/\pi) = \frac{(1/2)_N}{N!}t^{N/2}(1+O(t)) \quad \text{for} \quad t \to 0. \tag{10.92}$$

Equation (10.90) is a special case of the more general equation

$$\begin{aligned}\frac{dh}{dt}\left(t(1-t)\frac{d^2h}{dt^2}\right)^2 + \left[\frac{dh}{dt}\left(2h-(2t-1)\frac{dh}{dt}\right)+b_1b_2b_3b_4\right]^2 \\ = \left(\frac{dh}{dt}+b_1^2\right)\left(\frac{dh}{dt}+b_2^2\right)\left(\frac{dh}{dt}+b_3^2\right)\left(\frac{dh}{dt}+b_4^2\right)\end{aligned} \tag{10.93}$$

with, for example

$$b_1 = b_4 = N/2, \; b_2 = (1-N)/2, \; b_3 = (1+N)/2. \tag{10.94}$$

The equation (10.93) has been shown [57, 58] to be equivalent to the Painlevé VI equation

$$\begin{aligned}\frac{d^2q}{dt^2} &= \frac{1}{2}\left(\frac{1}{q}+\frac{1}{1-q}+\frac{1}{q-t}\right)\left(\frac{dq}{dt}\right)^2 - \left(\frac{1}{t}+\frac{1}{1-t}+\frac{1}{q-t}\right)\frac{dq}{dt} \\ &+ \frac{q(q-1)(q-t)}{t^2(t-1)^2}\left(\alpha+\beta\frac{t}{q^2}+\gamma\frac{t-1}{(q-1)^2}+\delta\frac{t(t-1)}{(q-t)^2}\right)\end{aligned} \tag{10.95}$$

by the following set of birational transformations [58, (2.5)–(2.7)]:

$$q = \frac{1}{2A}\left[(b_3+b_4)B + \left(\frac{dh}{dt}-b_3b_4\right)C\right] \tag{10.96}$$

where

$$A = \left(\frac{dh}{dt}+b_3^2\right)\left(\frac{dh}{dt}+b_4^2\right) \tag{10.97}$$

$$B = t(t-1)\frac{d^2h}{dt^2} + (b_1+b_2+b_3+b_4)\frac{dh}{dt} - (b_1b_2b_3+b_1b_2b_4+b_1b_3b_4+b_2b_3b_4) \tag{10.98}$$

and

$$C = 2\left(t\frac{dh}{dt}-h\right) - (b_1b_2+b_1b_3+b_1b_4+b_2b_3+b_2b_4+b_3b_4) \tag{10.99}$$

where the relation between b_j and α, β, γ, δ is given by [58, (1.2)] as

$$b_1 + b_2 = (-2\beta)^{1/2} \tag{10.100}$$

$$b_1 - b_2 = (2\gamma)^{1/2} \tag{10.101}$$
$$b_3 + b_4 + 1 = (1 - 2\delta)^{1/2} \tag{10.102}$$
$$b_3 - b_4 = (2\alpha)^{1/2}. \tag{10.103}$$

Conversely h is given in terms of q by (2.1) of [58]

$$h = q(q-1)(q-t)p^2 - \{b_1(2q-1)(q-t) - b_2(q-t) + (b_3+b_4)q(q-1)\}p$$
$$+ (b_1+b_2)(b_1+b_2)q - b_1^2 t - \frac{1}{2}(b_1b_2 + b_1b_3 + b_1b_4 + b_2b_3 + b_2b_4 + b_3b_4) \tag{10.104}$$

where p is determined in terms of q from (0.6) of [58]

$$t(t-1)\frac{dq}{dt} = \frac{\partial h}{\partial p} = 2q(q-1)(q-t)p - \{b_1(2q-1)(q-t) - b_2(q-t) + (b_3+b_4)q(q-1)\}. \tag{10.105}$$

The equation for h is invariant under permutations of b_j and the change of any two signs of b_j but the transformation equations clearly do not have this symmetry. Thus there are several Painlevé VI equations (10.95) that lead to the same equation for σ.

Asymptotic behavior as $N \to \infty$

There are three distinct behaviors of $C(N,N)$ and $C(0,N)$ as $N \to \infty$: $T > T_c$, $T = T_c$ and $T < T_c$ which need to be treated separately.

The simplest case to consider is $C(N,N)$ at $T = T_c$. In this case the determinental representation (10.45) reduces to an $N \times N$ Cauchy determinant which reduces to the remarkably simple form

$$\langle \sigma_{0,0}\sigma_{N,N}\rangle = \left(\frac{2}{\pi}\right)^N \prod_{l=1}^{N-1} [1 - \frac{1}{4l^2}]^{l-N} \tag{10.106}$$

and for $N \to \infty$ this behaves as

$$\langle \sigma_{0,0}\sigma_{N,N}\rangle \sim AN^{-1/4}\left(1 - \frac{1}{64N^2} + O(N^{-4})\right) \tag{10.107}$$

where the transcendental constant A is expressed in terms of the zeta function $\zeta(z)$ as

$$A = 2^{1/12}\exp[3\zeta'(-1)] \sim 0.6450\cdots. \tag{10.108}$$

and we note that (10.107) is independent of the ratio E^v/E^h.

For correlations in the same row (column) a more refined computation [13] gives

$$\langle \sigma_{0,0}\sigma_{0,N}\rangle = A\left(\frac{\cosh 2K_c^h}{N}\right)^{1/4}\left[1 + \frac{1}{64N^2}\left(\frac{2}{\sinh^2(2K^v)} - 1\right) + O(N^{-4})\right]$$
$$\langle \sigma_{0,0}\sigma_{M,0}\rangle = A\left(\frac{\cosh 2K_c^v}{M}\right)^{1/4}\left[1 + \frac{1}{64M^2}\left(\frac{2}{\sinh^2(2K^h)} - 1\right) + O(M^{-4})\right]. \tag{10.109}$$

From both (10.107) and (10.109) we see that the anomalous dimension is $\eta = 1/4$.

For both $T < T_c$ and $T > T_c$ the large N behavior is given by the leading terms of the form factor expansions (10.76) and (10.78). Thus on the diagonal we find:
for $T < T_c$

$$\langle \sigma_{0,0}\sigma_{N,N}\rangle \sim (1-t)^{1/4}\{1+f_N^{(2)}+\cdots\} = (1-t)^{1/4}\{1+\frac{2t^{N+1}}{2\pi N^2(t-1)^2}+\cdots\} \quad (10.110)$$

and for $T > T_c$

$$\langle \sigma_{0,0}\sigma_{N,N}\rangle \sim (1-t)^{1/4} f_N^{(1)} = \frac{1}{(\pi N)^{1/2}}\frac{t^{N/2}}{(1-t)^{1/4}}+\cdots \quad (10.111)$$

For correlations in the same row (column) we have for $T < T_c$ ($\alpha_2 < 1$)

$$\langle \sigma_{0,0}\sigma_{0,N}\rangle \sim (1-t)^{1/4}\{1+f_N^{(2)}+\cdots\} = (1-t)^{1/4}\{1+\frac{1}{2\pi N^2}\frac{\alpha_2^{2N}}{(\alpha_2^{-1}-\alpha_2)^2}+\cdots\} \quad (10.112)$$

and for $T > T_c$ ($\alpha_2 > 1$)

$$\langle \sigma_{0,0}\sigma_{0,N}\rangle \sim (1-t)^{1/4} f_N^{(1)} = \frac{(1-t)^{1/4}}{(\pi N)^{1/2}}\frac{\alpha_2^{-N}}{(\alpha_2-\alpha_2^{-1})^{1/2}}+\cdots \quad (10.113)$$

These correlation functions for $T \neq T_c$ approach their values at $N, M \to \infty$ exponentially rapidly and thus we may define the correlation length on the diagonal ξ_d by

$$t^N = e^{-\sqrt{2}N/\xi_{d\pm}} \quad (10.114)$$

and the correlation length in a row ξ_h by

$$\alpha_2^{\pm N} = e^{-N/\xi_{h\pm}}. \quad (10.115)$$

The correlation in a column ξ_v is obtained from ξ_h by interchanging $E^v \leftrightarrow E^h$.

As $T \to T_c$ we find from (10.43) and (10.114) that

$$\begin{aligned}\xi_{d\pm}^{-1} &= 2^{-1/2}|\ln t| \sim 2^{-1/2}|1-k_\pm| \\ &\sim \sqrt{2}|\beta-\beta_c|(E^v\coth 2E^v/k_BT_c + E^h\coth 2E^h/k_BT_c)\end{aligned} \quad (10.116)$$

and from (10.30) and (10.115) that

$$\begin{aligned}\xi_{h\pm}^{-1} &= |\ln\alpha_2| \sim |1-\alpha_2| \\ &\sim |\beta-\beta_c| 2\tanh 2E^v/k_BT_c(E^v\coth 2E^v/k_BT_c + E^h\coth 2E^h/k_BT_c)\end{aligned} \quad (10.117)$$

and

$$\xi_{v\pm}^{-1} \sim |\beta-\beta_c| 2\tanh 2E^h/k_BT_c(E^v\coth 2E^v/k_BT_c + E^h\coth 2E^h/k_BT_c) \quad (10.118)$$

These correlation lengths diverge linearly as $T \to T_c$ and thus the critical exponent $\nu = 1$.

We see that in general

$$\lim_{T\to T_c} \xi_{h\pm}/\xi_{d\pm} = 2^{-1/2}\coth 2E^v/k_BT_c$$
$$\lim_{T\to T_c} \xi_{v\pm}/\xi_{d\pm} = 2^{-1/2}\coth 2E^h/k_BT_c \tag{10.119}$$

and that, for $E^v = E^h = E$ where $\coth 2E/k_BT_c = \sqrt{2}$,

$$\lim_{T\to T_c} |\beta - \beta_c|\xi_h = \lim_{T\to T_c} |\beta - \beta_c|\xi_v = \lim_{T\to T_c} |\beta - \beta_c|\xi_d \tag{10.120}$$

which is consistent with the expected rotational invariance of the correlation functions of the isotropic lattice near T_c.

10.1.7 The correlation $C(M, N)$ for general M, N

For general values of M and N it is more efficient to express the correlations as infinite (Fredholm) determinants and from these expressions the following results are obtained [22]:

For $T < T_c$ the generalization of $\langle\sigma_{0,0}\sigma_{0,N}\rangle$ and $\langle\sigma_{0,0}\sigma_{N,N}\rangle$ given by (10.67)–(10.71) is

$$\langle\sigma_{0,0}\sigma_{M,N}\rangle = (1-t)^{1/4}\exp\left(-F_{M,N}\right) \tag{10.121}$$

where

$$F_{M,N} = \sum_{n=1}^{\infty} F_{M,N}^{(2n)} \tag{10.122}$$

and

$$F_{M,N}^{(2n)} = (-1)^n[2z_v(1-z_h^2)]^{2n}(2n)^{-1}(2\pi)^{-4n}$$
$$\int_{-\pi}^{\pi} d\phi_1 \cdots \int_{-\pi}^{\pi} d\phi_{4n} \prod_{j=1}^{2n}\left(\frac{e^{-iM\phi_{j-1}-iN\phi_{2j}}\sin\frac{1}{2}(\phi_{2j-1}-\phi_{2j+1})}{\Delta(\phi_{2j-1},\phi_{2j})\sin\frac{1}{2}(\phi_{2j}+\phi_{2j+2})}\right) \tag{10.123}$$

with $\phi_{4n+1} = \phi_1, \phi_{4n+2} = \phi_2$, $\mathrm{Im}\phi_j < 0$ and

$$\Delta(\phi_{2j-1},\phi_{2j}) = (1+z_h^2)(1+z_v^2) - 2z_v(1-z_h^2)\cos\phi_{2j-1} - 2z_h(1-z_v^2)\cos\phi_{2j} \tag{10.124}$$

and

$$z_v = \tanh E^v/k_BT, \quad z_h = \tanh E^h/k_BT. \tag{10.125}$$

For $T > T_c$ the generalization of $\langle\sigma_{0,0}\sigma_{0,N}\rangle$ and $\langle\sigma_{0,0}\sigma_{N,N}\rangle$ given by (10.72)–(10.75) is

$$\langle\sigma_{0,0}\sigma_{M,N}\rangle = (1-t)^{1/4}X_{M,N}\exp\left(-F_{M,N}\right) \tag{10.126}$$

where

$$X_{M,N} = \sum_{n=0}^{\infty} X_{M,N}^{(2n+1)} \tag{10.127}$$

with

$$X_{M,N}^{(2n+1)} = (2\pi)^{-4n-2}[2iz_v(1-z_h^2)]^{2n}\int_{-\pi}^{\pi} d\phi_1 \cdots \int_{-\pi}^{\pi} d\phi_{4n+2}$$
$$e^{-i\phi_{4n+1}-i\phi_{4n+2}}\prod_{j=1}^{2n+1}\frac{e^{-i(M-1)\phi_{2j-1}-i(N-1)\phi_{2j}}}{\Delta(\phi_{2j-1},\phi_{2j})}\prod_{j=1}^{2n}\frac{e^{-i\phi_{2j-1}}-e^{-i\phi_{2j+1}}}{1-e^{i\phi_{2j}+i\phi_{2j+2}}}. \tag{10.128}$$

As with the row and diagonal correlations these exponentials can be expanded to obtain form factor expansions [22, 36–38] similar to (10.76) and (10.78).

Nonlinear difference equations

The spin correlations on the lattice also satisfy nonlinear partial difference equations with respect to the locations of the spins and (systems of) differential equations in the temperature. To give these equations we define new constants for what is called the dual lattice by K^{v*} and K^{h*} by

$$\sinh 2K^{v*}\sinh 2K^h = 1 \quad \text{and} \quad \sinh 2K^{h*}\sinh 2K^v = 1 \tag{10.129}$$

and let $\langle\sigma_{0,0}\sigma_{M,N}\rangle^*$ be the correlation functions with $K^v \to K^{h*}$ and $K^h \to K^{v*}$. Then we have [33]

$$\frac{\langle\sigma_{0,0}\sigma_{M,N}\rangle^*}{\langle\sigma_{0,0}\sigma_{M,N}\rangle} = (\sinh 2K^v \sinh 2K^h)^{1/2}X_{M,N} \tag{10.130}$$

where $X_{M,N}$ given by (10.127) satisfies the partial difference equation for all M,N except $M=N=0$

$$\begin{aligned}
&[\nabla_L^2 - 2(a-\gamma_1-\gamma_2)]X_{M,N}\\
&= -\frac{X_{M+1,N}+X_{M-1,N}}{1-X_{M+1,N}X_{M-1,N+1}X_{M,N+1}X_{M,N-1}}\\
&\times\{\gamma_2[X_{M,N}^2 - X_{M,N+1}X_{M,N-1}]\\
&-\gamma_1X_{M,N+1}X_{M,N-1}[X_{M,N}^2 - X_{M+1,N}X_{M-1,N}]\}\\
&+ M\leftrightarrow N, \gamma_1\leftrightarrow\gamma_2
\end{aligned} \tag{10.131}$$

where the lattice Laplacian is defined as

$$\begin{aligned}
\nabla_L X_{M,N} &= \gamma_1[X_{M+1,N}+X_{M-1,N}-2X_{M,N}]\\
&+\gamma_2[X_{M,N+1}+X_{M,N-1}-2X_{M,N}].
\end{aligned} \tag{10.132}$$

The correlations are given in terms of $X_{M,N}$ for $T<T_c$ by (10.121) and for $T>T_c$ by (10.126) where

$$F_{M,N} = \sum_{k=1}^{\infty}\ln f_{M,k} \tag{10.133}$$

with

$$f_{M,N} = \prod_{j=0}^{\infty} \left[\frac{1 - X_{M+j,N} X_{M+1+j,N+1}}{1 - X_{M+1+j} X_{M+j,N+1}} \right]. \tag{10.134}$$

The correlations themselves satisfy the two nonlinear partial difference equations

$$\begin{aligned} &\langle \sigma_{0,0} \sigma_{M,N} \rangle^2 - \langle \sigma_{0,0} \sigma_{M-1,N} \rangle \langle \sigma_{0,0} \sigma_{M+1,N} \rangle \\ &= -\sinh^2 2K^v \left(\langle \sigma_{0,0} \sigma_{M,N} \rangle^{*2} - \langle \sigma_{0,0} \sigma_{M,N-1} \rangle^* \langle \sigma_{0,0} \sigma_{M,N+1} \rangle^* \right) \end{aligned} \tag{10.135}$$

and

$$\begin{aligned} &\langle \sigma_{0,0} \sigma_{M,N} \rangle^2 - \langle \sigma_{0,0} \sigma_{M,N-1} \rangle \langle \sigma_{0,0} \sigma_{M,N+1} \rangle \\ &= -\sinh^2 2K^h \left(\langle \sigma_{0,0} \sigma_{M,N} \rangle^{*2} - \langle \sigma_{0,0} \sigma_{M-1,N} \rangle^* \langle \sigma_{0,0} \sigma_{M+1,N} \rangle^* \right). \end{aligned} \tag{10.136}$$

At $T = T_c$ we have $\langle \sigma_{0,0} \sigma_{M,N} \rangle = \langle \sigma_{0,0} \sigma_{M,N} \rangle^*$ and these two equations reduce to the remarkably simple result valid except at $M = N = 0$

$$\begin{aligned} &\sinh^2 2K^v \left(\langle \sigma_{0,0} \sigma_{M,N} \rangle^2 - \langle \sigma_{0,0} \sigma_{M-1,N} \rangle \langle \sigma_{0,0} \sigma_{M+1,N} \rangle \right) \\ &= \sinh^2 2K^h \left(\langle \sigma_{0,0} \sigma_{M,N} \rangle^{*2} - \langle \sigma_{0,0} \sigma_{M,N-1} \rangle^* \langle \sigma_{0,0} \sigma_{M,N+1} \rangle^* \right). \end{aligned} \tag{10.137}$$

Asymptotic behavior for $M, N \to \infty$

For $T \neq T_c$ there are two cases. Asymptotically as M and/or $N \to \infty$ we have for $T < T_c$

$$\langle \sigma_{0,0} \sigma_{N,M} \rangle \sim M_-^2 \{1 + \frac{\exp(-2M\theta_1 - 2N\theta_2)}{8\pi(M \sinh\theta_1 \cosh\theta_2 + N \cosh\theta_1 \sinh\theta_2)^2} + \cdots\} \tag{10.138}$$

and for $T > T_c$

$$\langle \sigma_{0,0} \sigma_{M,N} \rangle \sim M_+^2 \frac{\exp(-M\theta_1 - N\theta_2)}{[2\pi(M \sinh\theta_1 \cosh\theta_2 + N \cosh\theta_1 \sinh\theta_2)]^{1/2}} + \cdots \tag{10.139}$$

where θ_1 is defined by

$$\cosh\theta_1 = \frac{(M^2/\gamma_1)(a^2 - \gamma_2^2) + N^2\gamma_1}{aM^2 + [M^2N^2a^2 + (M^2 - N^2)(M^2\gamma_2^2 - N^2\gamma_1^2)]^{1/2}} \tag{10.140}$$

with

$$a = (1 + z_h^2)(1 + z_v^2), \quad \gamma_1 = 2z_v(1 - z_h^2) \quad \gamma_2 = 2z_h(1 - z_v^2) \tag{10.141}$$

and θ_2 is obtained from θ_1 by the interchange $M \leftrightarrow N,\ E^v \leftrightarrow E^h$.

10.1.8 Scaling limit

The asymptotic behavior for $N, M \gg 1$ of the correlation functions derived in the previous subsection is not uniform and the asymptotic expansion at $T = T_c$ cannot be obtained by letting $T \to T_c$ in the expansions for $T \neq T_c$. In order to smoothly connect these three different behaviors together we must study the scaling limit introduced in chapter 5.

The concept of the scaling limit has been presented in detail in chapter 5. To apply this definition to the Ising model, define scaled coordinates n and m by

$$n = N/\xi_h, \quad m = M/\xi_v \tag{10.142}$$

with

$$r = \sqrt{n^2 + m^2}. \tag{10.143}$$

The scaling limit is then defined as the limit $T \to T_c$, $N \to \infty$, $M \to \infty$ with n and m, fixed. In this limit the correlation functions are constant on the ellipses of fixed r which, using (10.31), is equivalent to

$$\left(\frac{\sinh 2K_c^h}{\sinh 2K_c^v}\right)^{1/2} M^2 + \left(\frac{\sinh 2K_c^v}{\sinh 2K_c^h}\right)^{1/2} N^2 = \text{const.} \tag{10.144}$$

The scaled two point-function is thus defined as

$$G_{\pm}(r) = \lim_{\text{scaling}} M_{\pm}^{-2} \langle \sigma_{0,0}\sigma_{M,N}\rangle \tag{10.145}$$

where $M_{\mp} = (1-\alpha_2^{\pm 2})^{1/8}$. This function is a function of r alone and thus it is sufficient to consider the scaling function on the diagonal

$$G_{\pm}(r) = \lim_{\text{scaling}} M_{\pm}^{-2} \langle \sigma_{0,0}\sigma_{N,N}\rangle \tag{10.146}$$

where $\langle \sigma_{0,0}\sigma_{N,N}\rangle$ is given by the various expansions of section 10.1.6 with $\alpha_1 = 0$. The scaling limit is obtained by letting $\alpha_2 \to 1$ and $N \to \infty$ where for $T < T_c$ we define

$$r = N(1-\alpha_2^2)/2 = N(1-t)/2 \tag{10.147}$$

and for $T > T_c$

$$r = N(1-\alpha_2^{-2})/2 = N(1-t)/2 \tag{10.148}$$

where on the right-hand side we have used the definitions of t of (10.50) and (10.53).

Scaled form factor expansion

For $T < T_c$ the exponential form of the scaling function is obtained by first deforming the contour in (10.69) from $|z_j| = 1$ to the branch cut which runs from 0 to α_2, and setting $z_j = \alpha_2 x_j$ to obtain

$$F_N^{(2n)} = \frac{(-1)^{n+1}\alpha_2^{2n(N+1)}}{n} \int_0^1 \prod_{j=1}^{2n} \frac{dx_j x_j^N}{1-\alpha_2^2 x_j x_{j+1}} \prod_{j=1}^{n} \left[\frac{(1-\alpha_2^2 x_{2j})(x_{2j}^{-1}-1)}{(1-\alpha_2^2 x_{2j-1})(x_{2j-1}^{-1}-1)}\right]^{1/2}. \tag{10.149}$$

The scaling limit is then obtained by setting

$$x_j = 1 - y_j(1-\alpha_2^2) \tag{10.150}$$

and using the following approximations which are valid in the scaling limit

$$\alpha_2^{2n(N+1)} = e^{n(N+1)\ln\alpha_2^2} \to e^{-n(N+1)(1-\alpha_2^2)} \to e^{-2nr} \tag{10.151}$$

$$x_k^N = e^{N\ln[1-y_k(1-\alpha_2^2)]} \to e^{-N(1-\alpha_2^2)y_k} = e^{-2ry_k} \tag{10.152}$$

$$x_k^{-1} - 1 = \frac{1}{1 - y_k(1-\alpha_2^2)} - 1 \to (1-\alpha_2^2)y_k \tag{10.153}$$

$$1 - \alpha_2^2 x_k = 1 - \alpha_2^2[1 - y_k(1-\alpha_2^2)] \to (1-\alpha_2^2)(1+y_k) \tag{10.154}$$

$$\begin{aligned} 1 - \alpha_2^2 x_j x_k &= 1 - \alpha_2^2[1-(1-\alpha_2^2)y_j][1-(1-\alpha_2^2)j_k] \\ &\to (1-\alpha_2^2)[1+y_j+y_k]. \end{aligned} \tag{10.155}$$

We thus find from (10.67)–(10.69) and (10.146) that

$$G_-(r) = \exp\sum_{n=1}^{\infty} \lambda^{2n}\tilde{F}^{(2n)}(r) \tag{10.156}$$

where

$$\tilde{F}^{(2n)}(r) = \frac{(-1)^{n+1}e^{-2nr}}{n}\int_0^\infty \prod_{j=1}^{2n} \frac{dy_k e^{-2ry_j}}{1+y_j+y_{j+1}} \prod_{j=1}^{n}\left[\frac{y_{2j}(1+y_{2j})}{y_{2j-1}(1+y_{2j-1})}\right]^{1/2} \tag{10.157}$$

with $y_{2n+1} \equiv y_1$ and

$$\lambda = 1/\pi \tag{10.158}$$

which is displayed in a more symmetric form by setting

$$y_k = (\tilde{y}_k - 1)/2 \tag{10.159}$$

to find

$$\tilde{F}_{2n}(r) = \frac{(-1)^{n+1}}{n}\int_1^\infty \prod_{j=1}^{2n} \frac{dy_j e^{-ry_j}}{y_j + y_{j+1}} \prod_{j=1}^{n}\left[\frac{y_{2j}^2 - 1}{y_{2j-1}^2 - 1}\right]^{1/2} \tag{10.160}$$

(where the tilde on the y has been suppressed).

In a similar fashion we find from (10.76) and (10.77) the scaled form factor expansion for $T < T_c$

$$G_-(r) = 1 + \sum_{n=1}^{\infty}\lambda^{2n}\tilde{f}^{(2n)}(r) \tag{10.161}$$

with

$$\tilde{f}^{(2n)}(r) = \frac{1}{(n!)^2}\int_1^\infty \prod_{j=1}^{2n} dy_j e^{-ry_j} \prod_{j=1}^{n} dy_j \left[\frac{y_{2j}^2-1}{y_{2j-1}^2-1}\right]^{1/2}$$
$$\times \prod_{1\le j\le n}\prod_{1\le k\le n}\frac{1}{(y_{2j-1}+y_{2k})^2}\prod_{1\le j<k\le n}(y_{2j-1}-y_{2k-1})^2(y_{2j}-y_{2k})^2 \quad (10.162)$$

For $T > T_c$ we find from (10.72)–(10.75)

$$G_+(r) = \tilde{X}(r)G_-(r) \quad (10.163)$$

where from (10.127) and (10.128)

$$\tilde{X}(r) = \sum_{n=0} \lambda^{2n+1}\tilde{X}_{2n+1}(r) \quad (10.164)$$

with

$$\tilde{X}_{2n+1}(r) = (-1)^n \int_1^\infty \prod_{j=1}^{2n+1}\frac{dy_j e^{-ry_j}}{(y_j^2-1)^{1/2}}\prod_{j=1}^{2n}\frac{1}{y_j+y_{j+1}}\prod_{j=1}^{n}(y_{2j}^2-1) \quad (10.165)$$

where again $\lambda = 1/\pi$.

From (10.78) and (10.79) we obtain for $T > T_c$ the scaled form factor expansion

$$G_+(r) = \sum_{n=0}^{\infty} \lambda^{2n+1}\tilde{f}^{(2n+1)}(r) \quad (10.166)$$

with

$$f^{(2n+1)}(r) = \frac{1}{n!(n+1)!}\int_1^\infty \prod_{j=1}^{2n+1} dy_j e^{-ry_j}\prod_{j=1}^{n}(y_{2j}^2-1)^{1/2}\prod_{j=1}^{n+1}(y_{2j-1}^2-1)^{-1/2}$$
$$\times \prod_{1\le j\le n+1}\prod_{1\le k\le n}\frac{1}{(y_{2j-1}+y_{2k})^2}\prod_{1\le j<k\le n+1}(y_{2j-1}-y_{2k-1})^2\prod_{1\le j<k\le n}(y_{2j}-y_{2k})^2.$$
$$(10.167)$$

Expansion for $r \gg 1$

For $T > T_c$ the asymptotic behavior for $r \to \infty$ is obtained from (10.163) with $n = 0$ in $X(r)$ of (10.164) and $G_-(r) \sim 1$ as

$$G_+(r) \sim \frac{1}{\pi}K_0(r) \quad (10.168)$$

where $K_n(r)$ is the modified Bessel function of order n.

For $T < T_c$ the asymptotic behavior for $r \to \infty$ is obtained from $G_-(r)$ (10.161) with $n = 1$ as

$$G_-(r) \sim 1 + \pi^{-2}\{r^2[K_1^2(r) - K_0^2(r)] - rK_0(r)K_1(r) + \frac{1}{2}K_0^2(r)\}. \quad (10.169)$$

Painlevé representations

All further terms in $G_{\pm}(r)$ are exponentially smaller, as $r \to \infty$, than the terms of (10.168) and (10.169). However, in the opposite limit $r \to 0$ all terms diverge as powers of $\ln t$ and contribute to the result. To effectively study this limit we need the scaling limit of the nonlinear equation (10.90) which is obtained by the use of (10.88), (10.89),(10.146), (10.147) and (10.148) for both $G_{\pm}(r)$ as

$$(r\zeta'')^2 = 4(r\zeta' - \zeta)^2 - 4(\zeta')^2(r\zeta' - \zeta - 1/4) \tag{10.170}$$

where

$$\zeta = r\frac{d \ln G_{\pm}}{dr}. \tag{10.171}$$

Equation (10.170) has an intimate relation with the Painlevé V equation [57].

The first nonlinear equation for the scaling function $G_{\pm}(r)$ which was announced in 1973 [20] and for which the derivation was published in 1976 [22] is

$$G_{\pm}(r) = \frac{1}{2}[1 \mp \eta(r/2)]\eta(r/2)^{-1/2} \exp \int_{r/2}^{\infty} d\theta \frac{1}{4}\theta\eta^{-2}\left[(1-\eta^2)^2 - (\eta')^2\right] \tag{10.172}$$

where $\eta(\theta)$ satisfies the Painlevé III equation

$$\frac{d^2\eta}{d\theta^2} = \frac{1}{\eta}\left(\frac{d\eta}{d\theta}\right)^2 - \frac{1}{\theta}\frac{d\eta}{d\theta} + \eta^3 - \eta^{-1} \tag{10.173}$$

with the boundary condition

$$\eta(\theta) \sim 1 - 2\lambda K_0(2\theta) \text{ as } \theta \to \infty \text{ where } \lambda = 1/\pi. \tag{10.174}$$

If we set

$$\eta(\theta) = e^{-\psi(r)} \text{ with } r = 2\theta \tag{10.175}$$

the Painlevé III equation (10.173) reduces to the radial sinh-Gordon equation

$$\frac{d^2\psi}{dr^2} + \frac{1}{r}\frac{d\psi}{dr} = \frac{1}{2}\sinh 2\psi \tag{10.176}$$

and the correlation function becomes

$$G_{\pm}(r) = \begin{pmatrix}\sinh \frac{1}{2}\psi \\ \cosh \frac{1}{2}\psi\end{pmatrix} \exp\{\frac{1}{4}\int_r^{\infty} dss[-\left(\frac{d\psi}{ds}\right)^2 + \sinh^2\psi]\}. \tag{10.177}$$

The equivalence between (10.170) and (10.177) was directly proven in [59]. An alternative derivation is given in [40].

Expansion for $r \sim 0$

We may now obtain the expansion of $G_{\pm}(r)$ for $r \sim 0$ by studying the Painlevé III differential equation (10.173). This is done in two steps: first find a local expansion of $\eta(r)$ for $r \sim 0$ and then connect this local expansion with the coefficient $1/\pi$ in the large r expansion (10.168).

The local expansion of $\eta(r)$ near $r \sim 0$ is easily obtained as

$$\eta(r/2) = Br^{\sigma}\{1 - \frac{1}{16}B^{-2}(1-\sigma)^2 r^{2-\sigma} + O(r^2)\}. \tag{10.178}$$

However, to obtain the relation between σ and B in terms of λ we must solve a connection problem and in general analytic solutions to connection problems do not exist. However, the six Painlevé equations are very special, and for these equations the connection problems can be solved. The solution of the connection problems for the Painlevé III equation (10.173) was first obtained by Tracy, Wu and the present author [61]:

$$\sigma(\lambda) = \frac{2}{\pi}\arcsin(\pi\lambda), \quad B = B(\sigma) = 2^{-3\sigma}\frac{\Gamma((1-\sigma)/2)}{\Gamma((1+\sigma)/2)} \tag{10.179}$$

When $\lambda \to 1/\pi$ then $\sigma \to 1$ and the correlation behaves as

$$G_{\pm}(r) = \text{const } r^{-1/4}\left(1 \pm \frac{1}{2}r[\ln(r/8) + \gamma_E] + \frac{1}{16}r^2 + \cdots\right) \tag{10.180}$$

where γ_E is Euler's constant.

To complete the verification of scaling theory we must show that the constant in (10.180) is the same as the one which is obtained from the transcendental constant A in (10.107). This requires a separate computation and was first done by Tracy [62] in 1991.

We have thus demonstrated that all the features of the scaling theory of the two-point function presented in chapter 5 hold. From the initial results on correlation functions in 1949 of Kaufman and Onsager [9] to the final computation in 1991 of the constant in (10.180) took 42 years. The Ising model in two dimensions is the only system for which all the properties of the two-point function required by scaling theory have ever been verified.

10.1.9 Magnetic susceptibility of the bulk

At $H \neq 0$ the magnetization per site is

$$M(H) = Z(H)^{-1} \sum_{\sigma=\pm 1} \sigma_{0,0} e^{-\mathcal{E}/k_B T} \tag{10.181}$$

where $\mathcal{E}$ is the interaction energy (10.1) and $Z(H)$ is the magnetic field dependent partition function. Therefore the zero field magnetic susceptibility

$$\chi(T) = \frac{\partial M(H)}{\partial H}|_{H=0} \tag{10.182}$$

is given in terms of the two spin correlation function as

$$k_B T\chi(T) = \sum_{M=-\infty}^{\infty}\sum_{N=-\infty}^{\infty}\{\langle\sigma_{0,0}\sigma_{M,N}\rangle - M_-^2(T)\} \tag{10.183}$$

where $M_-(T)$ is the spontaneous magnetization given by (10.42) (which vanishes for $T > T_c$).

By expanding the exponential in the expansions for the correlation functions on the lattice (10.121) and (10.126), we can write explicit expressions for (10.183) in terms of integrals

$$k_B T\chi_+(T) = (1-t)^{1/4} t^{-1/4} \sum_{j=0}^{\infty} \hat{\chi}^{(2j+1)}(T) \quad \text{for } T > T_c \tag{10.184}$$

$$k_B T\chi_-(T) = (1-t)^{1/4} \sum_{j=1}^{\infty} \hat{\chi}^{(2j)}(T) \quad \text{for } T < T_c \tag{10.185}$$

where

$$\hat{\chi}^{(j)}(T) = \frac{\cot^j \alpha}{j!} \int_{-\pi}^{\pi} \frac{d\omega_1}{2\pi} \cdots \int_{-\pi}^{\pi} \frac{d\omega_{j-1}}{2\pi} \left(\prod_{n=1}^{j} \frac{1}{\sinh \gamma_n} \right) H^{(j)} \frac{1 + \prod_{n=1}^{j} x_n}{1 - \prod_{n=1}^{j} x_n} \tag{10.186}$$

with

$$x_n = \cot^2 \alpha \left[\xi - \cos\omega_n - \sqrt{(\xi - \cos\omega_n)^2 - (\cot\alpha)^{-4}} \right] \tag{10.187}$$

$$\sinh \gamma_n = \cot^2 \alpha \sqrt{(\xi - \cos\omega_n)^2 - (\cot\alpha)^{-4}} \tag{10.188}$$

where

$$\cot\alpha = \sqrt{s_h/s_v} \tag{10.189}$$

$$\xi = (1 + s_h^{-2})^{1/2} (1 + s_v^2)^{1/2} \tag{10.190}$$

$$s_v = \sinh 2E^v/k_B T \qquad s_h = \sinh 2E^h/k_B T \tag{10.191}$$

and ω_j is defined in terms of the remaining ω_i from $\omega_1 + \cdots \omega_j = 0 \bmod 2\pi$.

There are many equivalent expressions $H^{(j)}$. The original expression [22] comes directly from the expansion of the exponential. Subsequent developments [63–66, 36] have discovered more compact explicit expressions. Reference [63] is for the isotropic lattice $E^v = E^h$. References [64, 65] extend the results to the anisotropic case and [38] writes the integrals of [64–66] in terms of trigonometric instead of elliptic functions. We list here the expression of [38]:

$$H^{(j)} = \left(\prod_{1 \le i < k \le j} h_{ik} \right)^2 \tag{10.192}$$

with

$$h_{ik} = \cot\alpha \frac{\sin\frac{1}{2}(\omega_i - \omega_k)}{\sinh\frac{1}{2}(\gamma_i - \gamma_k)} = \frac{1}{\cot\alpha} \frac{\sinh\frac{1}{2}(\gamma_i - \gamma_k)}{\sin\frac{1}{2}(\omega_i + \omega_k)}. \tag{10.193}$$

For $j = 1, 2$ the integrals for $\hat{\chi}^{(j)}(T)$ may be explicitly evaluated the isotropic case $E^v = E^h = E$ we have, with t given by (10.53)

$$\hat{\chi}^{(1)}_{\text{iso}}(T) = \frac{t^{1/4}}{(1 - t^{1/4})^2} \tag{10.194}$$

and with t given by (10.50)

$$\hat{\chi}^{(2)}_{\rm iso}(T) = \frac{(1+t)E(\sqrt{t}) - (1-t)K(\sqrt{t})}{3\pi(1-t^{1/2})(1-t)} \tag{10.195}$$

where $K(\sqrt{t})$ and $E(\sqrt{t})$ are the complete elliptic integrals of the first and second kind (10.58) and (10.59). More generally for the anisotropic case we have [38, (3.21)]

$$\hat{\chi}^{(1)}(T) = \frac{t^{1/4}}{(1+t^{1/4})^2}\left[2\csc 2\alpha + \sqrt{(t^{1/4}-t^{-1/4})^2 + 4\csc^2 2\alpha}\right]\hat{\chi}^{(1)}_{\rm iso}(T) \tag{10.196}$$

and [38, (3.23)]

$$\hat{\chi}^{(2)}(T) = \frac{t^{1/4}}{1+t^{1/4}}\sqrt{(t^{1/4}-t^{-1/4})^2 + 4\csc^2 2\alpha}\;\hat{\chi}^{(2)}_{\rm iso}(T) \tag{10.197}$$

If we use the scaling form of the two-point function in (10.183) we see that as $T \to T_c\pm$

$$k_B T\chi(T)_\pm \sim M_\pm^2 \xi_{h\pm}\xi_{v\pm} 2\pi \int_0^\infty dr\; r\{G_\pm(r) - \begin{bmatrix}0\\1\end{bmatrix}\} \tag{10.198}$$

where by $\begin{bmatrix}0\\1\end{bmatrix}$ we mean 0 (1) for $T > T_c$ ($T < T_c$). Then if we further use the fact that the corrections to the $T = T_c$ correlation function are of order $O(R^{-9/4})$ (and not $O(R^{-5/4})$) we find that [21, 22]

$$k_B T\chi(T) \sim C_{0\pm}|1 - T_c/T|^{-7/4} + C_{1\pm}|1 - T_c/T|^{-3/4} + D + o(1) \tag{10.199}$$

where

$$\frac{C_{1+}}{C_{0+}} = -\frac{C_{1-}}{C_{0-}} = -R_0 \tag{10.200}$$

with

$$R_0 = \frac{E_h^2 z_{vc}^2(1 + 6z_{hc}^2 + z_{hc}^4) + E_v^2 z_{hc}^2(1 + 6z_{vc}^2 + z_{vc}^4) - 8E_vE_h z_{vc}z_{hc}(z_{vc} + z_{hc})^2}{8z_{vc}z_{hc}(z_{vc} + z_{hc})k_BT_c[E_h(1 - z_{hc}) + E_v(1 - z^{vc})]}, \tag{10.201}$$

and

$$C_{0\pm} = 2^{-1/2}\coth 2K_c^v \coth 2K_c^h [K_c^v \coth 2K_c^v + K_c^h \coth 2K_c^h]^{-7/4} I_\pm \tag{10.202}$$

where

$$I_\pm = \pi 2^{-1/2}\int_0^\infty dr\; r\{G_\pm(r) - \begin{bmatrix}0\\1\end{bmatrix}\} \tag{10.203}$$

and the integrals have been numerically evaluated to 52 digits in [38]:

$$I_+ = 1.000815260440212647119476363047210236937534925597789 \tag{10.204}$$

$$I_- = \frac{1.000960328725262189480934955172097320572505951770117}{12\pi} \tag{10.205}$$

To compute the constant D in (10.199) short distance contributions not included in the scaling functions $G_\pm(r)$ must be taken into account. This was done numerically in [67] where it was found that

$$D = -0.05365771128w^{-1} + 0.006362291w - 0.00000132w^3 \tag{10.206}$$

with $w = \tanh 2K_c^v \tanh 2K_c^h$ and all coefficients are accurate to nine significant figures.

Finally we note that all correlations $\langle\sigma_{0,0}\sigma_{M,N}\rangle$ have logarithmic singularities at $T = T_c$ which must be present in the susceptibility and that there will in addition be an infinite number of "corrections to scaling". These two effects have been studied in detail in [38] where it is shown that, for the isotropic lattice in terms of the temperature variable $\tau = (s^{-1} - s)/2$,

$$k_B T\chi_{\pm}(t) = I_{\pm}(2K_c\sqrt{2})^{-7/4}|\tau|^{-7/4}F_{\pm}(\tau) + B \qquad (10.207)$$

where

$$F_{\pm}(\tau) = 1 + \frac{\tau}{2} + \frac{5\tau^2}{8} + \frac{3\tau^3}{16} - \frac{23\tau^4}{384} - \frac{35\tau^5}{768} + \sum_{j=6} f_{\pm}^{(j)}\tau^j \qquad (10.208)$$

where $f_+^{(j)} \neq f_-^{(j)}$ and terms as far as $f_+^{(14)}$ and $f_-^{(15)}$ have been computed and

$$B = \sum_{q=0}^{\infty}\sum_{p=0}^{[\sqrt{q}]} b^{p,q}\tau^q(\ln|\tau|)^p \qquad (10.209)$$

where the terms in (10.209) have been numerically evaluated up to order $\ln^3 \tau$.

But what is most interesting about the susceptibility $\chi(T)$ and makes it much more complicated than either the free energy or the spontaneous magnetization is that in addition to the singularity at T_c there are many other singularities in $\hat{\chi}^{(j)}(T)$ in the complex T plane. These new singularities were first discovered in the isotropic case $E^v = E^h$ by Nickel [36] in 1999 and found in the anisotropic case by Orrick, Nickel, Guttmann and Perk [38] in 2001. These singularities occur because the integrals $\hat{\chi}^{(j)}(T)$ of (10.186) will be singular at the symmetry points of the integrand and where the denominator factor $1 - \prod_{n=1}^{j} x_n$ vanishes. The symmetry condition requires all ω_n to be equal and given by $\omega_n = \omega = 2\pi m'/j$ with $m' = 1, 2, \cdots, j$. The vanishing of the denominator factor requires the x_n, now all equal, to be given by $x_n = x = \exp(2\pi i m/j)$ with $m = 1, 2, \cdots j$. From the explicit formula for x_n (10.187) we find

$$\cot^2\alpha(\xi - \cos(2\pi m'/j)) = \cos(2\pi m/j) \qquad (10.210)$$

or substituting (10.189) for α and (10.190) for ξ we find that there are singularities in $\hat{\chi}^{(j)}$ at (complex) temperatures where

$$\begin{aligned}&\cosh 2E^v/k_BT \cosh 2E^h/k_BT \\ &- \sinh 2E^h/k_BT \cos(2\pi m'/j) - \sinh 2E^v/k_BT \cos(2\pi m/j) = 0\end{aligned} \qquad (10.211)$$

which remarkably is exactly the condition (10.9) for the location of the zeros of the partition function. If we call ϵ the deviation from the singular temperatures $T_{m,m'}^{(j)}$ determined by (10.211) then for $T > T_c$ the singularity in $\hat{\chi}^{(2j+1)}(T)$ is shown in [36] to be

$$\epsilon^{2j(j+1)-1}\ln\epsilon \qquad (10.212)$$

and for $T < T_c$ the singularity in $\hat{\chi}^{(2j)}(T)$ shown in [37] to be

$$\epsilon^{2j^2-3/2}. \tag{10.213}$$

These singularities become dense as $j \to \infty$ and consequently if there are no cancellations there will be a natural boundary in $\chi(T)$ at the location of the zeros of the partition function. For arbitrary values of the parameter λ in the form factor expansions such as (10.76) and (10.78) where each n particle contribution is weighted by λ^n such cancellations are generically impossible. But for the Ising model the value of λ is not arbitrary, and it is surely possible that for this value cancellations can occur. The study of this problem is most important because the existence of a natural boundary would be a new phenomenon not previously seen.

We conclude this section by noting that in [38] it is proven that $\hat{\chi}^{(j)}$ must satisfy a linear differential equation in the temperature. Such a function is called D-finite. The theorem in [38] is an existence proof, and in that paper no examples beyond the elementary cases of $\chi^{(1)}(T)$ and $\chi^{(2)}(T)$ were known. However, in 2004 it was found [41] that in terms of the variable

$$w = \frac{1}{2(s+s^{-1})} \tag{10.214}$$

$\hat{\chi}^{(3)}(w)$ satisfies the seventh-order equation

$$\sum_{n=0}^{7} P_n(w)\frac{d^n}{dw^n}\hat{\chi}^{(3)}(w) = 0 \tag{10.215}$$

where $P_0(w), \cdots, P_7(w)$ are polynomials in w of degrees 36, 41, 42, 43, 44, 45, 46 and 47 respectively. All the singularities in (10.215) are regular and thus the differential equation is Fuchsian. For $\chi^{(4)}(T)$ a Fuchsian equation of tenth order has been also found [43].

The equation (10.215) was discovered by generating the first 490 terms in the power series expansion of $\hat{\chi}^{(3)}(w)$ in terms of w. The equation is determined by the first 359 of these coefficients and thus there are 131 verifications that (10.215) is correct. The challenge now is to find an analytic way to produce differential equations for all $\hat{\chi}^{(j)}(w)$ and to see if it is possible to find some equation (differential, difference or functional) that can characterize the full susceptibility in the way in which the Painlevé VI equation characterizes the diagonal two-point function.

10.1.10 The diagonal susceptibility

The expression (10.186) for $\hat{\chi}^{(j)}(T)$ is cumbersome and to gain further insight it is useful to consider the simpler problem where the magnetic field interacts with the spins of only one diagonal of the lattice [45]. The susceptibility of a spin on the diagonal in response to this diagonal magnetic field is expressed in terms of the diagonal spin correlations in analogy with (10.183) as

$$k_B T \chi_d(T) = \sum_{N=-\infty}^{\infty} \{C(N,N) - M_-^2(T)\}. \tag{10.216}$$

Thus if we use the form factor expansions (10.76)–(10.79) we find in analogy to (10.184)–(10.186) that for $T > T_c$

$$k_BT\chi_{d+}(T) = (1-t)^{1/4}\sum_{n=0}^{\infty}\hat{\chi}_d^{(2n+1)}(t) \tag{10.217}$$

where

$$\hat{\chi}_d^{(2n+1)}(t) = \frac{t^{n(n+1)}}{n!(n+1)!\pi^{2n+1}}\int_0^1\prod_{k=1}^{2n+1}dx_k\frac{1+t^{n+1/2}\prod_{k=1}^{2n+1}x_k}{1-t^{n+1/2}\prod_{k=1}^{2n+1}x_k}$$
$$\prod_{j=1}^{n+1}[x_{2j-1}(1-tx_{2j-1})(1-x_{2j-1}^{-1})]^{-1/2}\prod_{j=1}^{n}[x_{2j}(1-tx_{2j})(1-x_{2j})]^{1/2}$$
$$\prod_{1\le j\le n+1}\prod_{1\le k\le n}\left(\frac{1}{1-tx_{2j-1}x_{2k}}\right)^2\prod_{1\le j<k\le n+1}(x_{2j-1}-x_{2k-1})^2\prod_{1\le j<k\le n}(x_{2j}-x_{2k})^2 \tag{10.218}$$

and for $T < T_c$

$$k_BT\chi_{d-}(T) = (1-t)^{1/4}\sum_{n=0}^{\infty}\hat{\chi}_d^{(2n)}(t) \tag{10.219}$$

where

$$\hat{\chi}_d^{(2n)}(t) = \frac{t^{n^2}}{(n!)^2\pi^{2n}}\int_0^1\prod_{k=1}^{2n}dx_k\frac{1+t^n\prod_{k=1}^{2n}x_k}{1-t^n\prod_{k=1}^{2n}x_k}$$
$$\prod_{j=1}^{n}\left(\frac{x_{2j-1}(1-tx_{2j})(1-x_{2j})}{x_{2j}(1-tx_{2j-1})(1-x_{2j-1})}\right)^{1/2}$$
$$\prod_{1\le j\le n}\prod_{1\le k\le n}\left(\frac{1}{1-tx_{2k-1}x_{2j}}\right)^2\prod_{1\le j<k\le n}(x_{2j-1}-x_{2k-1})^2(x_{2j}-x_{2k})^2 \tag{10.220}$$

The expressions (10.218) and (10.220) are indeed simpler than (10.186).

All the $\hat{\chi}_d^{(n)}(t)$ have singularities at $t = 1$ which come from the vanishing of the denominators in (10.218) and (10.220). For $\hat{\chi}_d^{(1)}(t)$ and $\hat{\chi}_d^{(2)}(t)$ these singularities are seen in the explicit evaluations done by contour integration [45]

$$\hat{\chi}_d^{(1)}(t) = \frac{1}{1-t^{1/2}} \tag{10.221}$$

and

$$\hat{\chi}_d^{(2)}(t) = \frac{t}{4(1-t)} \tag{10.222}$$

which are to be compared with the corresponding expressions for the susceptibility of the bulk (10.194) and (10.195). In particular we note that the bulk susceptibility $\hat{\chi}^{(2)}(T)$ has a logarithmic singularity at $T = T_c$ while $\hat{\chi}_d^{(2)}(t)$ does not.

For $n \geq 3$ all $\hat{\chi}_d^{(n)}(t)$ have logarithmic singularities at $t = 1$ which are inherited from the logarithmic singularities of the form factors. However, there are additional singularities in $\hat{\chi}_d^{(2n)}(t)$ coming from the vanishing of the denominator

$$1 - t^n x_1 x_2 \cdots x_{2n} \tag{10.223}$$

and in $\hat{\chi}_d^{(2n+1)}(t)$ coming from the vanishing of the denominator

$$1 - t^{n+1/2} x_1 x_2 \cdots x_{2n+1} \tag{10.224}$$

which are not present in $f_N^{(n)}(t)$. For $\tilde{\chi}_d^{(2n)}$ the vanishing of (10.223) occurs for $t^n = 1$ at the endpoints $x_1 = x_2 = \cdots = x_{2n} = 1$ and, for $\hat{\chi}_d^{(2n+1)}$, the vanishing of (10.224) occurs for $t^{n+1/2} = 1$ at the endpoints $x_1 = x_2 = \cdots = x_{2n+1} = 1$.

When $t \to 1$ this additional singularity is the simple pole $(1-t)^{-1}$ for both $\hat{\chi}_d^{(2n)}(t)$ and $\hat{\chi}_d^{(2n+1)}(t)$. When t approaches the roots of unity $t^n = 1$ (with $t \neq 1$)

$$t_{l,n} = e^{2\pi i l/n} \qquad \text{with} \qquad l = 1, 2, \cdots, n-1 \tag{10.225}$$

then, calling $\epsilon = t - t_{l,n}$, $\hat{\chi}_d^{(2n)}(t)$ has a singularity of the form

$$\epsilon^{2n^2-1} \ln \epsilon. \tag{10.226}$$

and $\hat{\chi}_d^{(2n+1)}(t)$ has a singularity of the form

$$\epsilon^{2n(n+1)-1/2}. \tag{10.227}$$

These singularities are to be compared with the corresponding singularities (10.212) and (10.213) of the bulk susceptibility.

There are no other values of t in the complex plane for which the functions $\hat{\chi}_d^{(n)}(t)$ are singular. This is in distinct contrast with the $\hat{\chi}^{(n)}(s)$ of the bulk which have singularities at many other places on the complex s plane [41].

Linear differential equations have been obtained in [45], where it is shown that the differential operators $\mathcal{L}^{(n)}$, which for $n = 3, 4$ annihilate $\hat{\chi}_d^{(n)}(t)$, have direct sum decompositions

$$\mathcal{L}^{(3)} = L_1^{(3)} \oplus L_2^{(3)} \oplus L_3^{(3)} \tag{10.228}$$

$$\mathcal{L}^{(4)} = L_1^{(4)} \oplus L_3^{(4)} \oplus L_4^{(4)} \tag{10.229}$$

where $L_m^{(n)}$ is an operator of order m. The solutions of $L_1^{(3)}$ and $L_1^{(4)}$ are precisely the functions $\hat{\chi}_d^{(1)}(t)$ and $\hat{\chi}_d^{(2)}(t)$ of (10.221) and (10.222). In [45] the solution of $L_2^{(3)}$ is, using the notation of (10.58) and (10.59) found to be

$$\frac{1}{t^{1/2}-1} K(t^{1/2}) + \frac{1}{(t^{1/2}-1)^2} E(t^{1/2}) \tag{10.230}$$

and $L^{(4)}$ is

$$-K^2(t^{1/2}) + \frac{t+1}{(t-1)^2}E^2(t^{1/2}) + \frac{2t}{t-1}K(t^{1/2})E(t^{1/2}) \tag{10.231}$$

and in [68] the solutions of $L_3^{(3)}$ are given in terms of the variable

$$Q = \frac{27}{4}\frac{(1+x)^2x^2}{(x^2+x+1)^3} \tag{10.232}$$

with $x = t^{1/2}$ as the two Meijer's G functions [55, chapter 5]

$$\begin{aligned}&\rho(x)G([-1/2, 1/3, 2/3], [], [[0,0],[0]]; Q)\\ &\rho(x)G([[-1/2, 1/3, 2/3], []], [[0,0,0],[]]; -Q)\end{aligned} \tag{10.233}$$

where

$$\rho(x) = \frac{(1+2x)(x+2)}{(1-x)(1+x+x^2)} \tag{10.234}$$

and the generalized hypergeometric function [55, chapter 4]

$$\rho(x)_3F_2([1/2, 2/3, 3/2], [1,1]; Q). \tag{10.235}$$

All of these solutions may be reduced further to sums of products of hypergeometric functions with arguments which are rational functions of x.

10.2 Boundary properties of the homogeneous lattice at H = 0

In this section we study the Ising model with a magnetic field on a free edge

$$\mathcal{E}_b = -\sum_{j=1}^{L_v}\sum_{k=1}^{L_h}\{E^h\sigma_{j,k}\sigma_{j,k+1} + E^v\sigma_{j.k}\sigma_{j+1,k}\} - \sum_{k=1}^{L_h} H_b\sigma_{1,k} \tag{10.236}$$

where periodic boundary conditions are applied only in the horizontal direction, and the magnetic field is applied only to the boundary row called $j = 1$.

As L_v and $L_h \to \infty$ the partition function for (10.236) behaves as

$$-k_BT\ln Z(H_b) \sim L_vL_hF + 2L_hF_{b0} + L_hF_b(H_b) \tag{10.237}$$

and therefore in addition to a bulk free energy F (which is the same as that studied in the previous section) there are contributions F_{b0}, from the two free boundaries, that are independent of the boundary magnetic field H_b, and a contribution $F_b(H_b)$ (with $F_b(0) = 0$) from the row $j = 1$ on which the magnetic field H_b acts.

10.2.1 Boundary free energy at $H_b = 0$

The boundary free energy at $H_b = 0$ is given by [15]:

$$F_b(0)/k_BT = \frac{1}{2}\ln\cosh E^v/k_BT - \frac{1}{4\pi}\int_{-\pi}^{\pi} d\theta \ln\frac{1}{2}\left(1 + \left[\frac{(1-\alpha_1e^{i\theta})(1-e^{-i\theta}/\alpha_2)}{(1-\alpha_1e^{-i\theta})(1-e^{i\theta}/\alpha_2)}\right]^{1/2}\right) \tag{10.238}$$

From this we compute the boundary contribution to the entropy S_b and specific heat c_b

$$S_b/k_B = -\frac{\partial F_{b0}}{\partial T} \quad c_b/k_B = -T\frac{\partial^2 F_{b0}}{\partial T^2}. \tag{10.239}$$

The behavior of these quantities as $T \to T_c$ is

$$S_b/k_B = -\frac{s_1}{2\pi}\ln|1 - T/T_c| + O(1) \tag{10.240}$$

$$\lim_{\delta T \to 0}[S_b(T_c + \delta T) - S_b(T_c - \delta T)] = -\frac{1}{2}s_1 \tag{10.241}$$

and

$$c_b = -\frac{1}{2\pi}\left[\frac{s_1}{T/T_c - 1} + s_2 \ln|1 - T/T_c|\right] + O(1) \tag{10.242}$$

where

$$s_1 = (1 - z_{vc})^{-1}[(1 - z_{hc})K_c^h + (1 - z_{vc})K_c^v] \tag{10.243}$$

$$s_2 = \frac{z_{vc} + z_{hc}}{(1 - z_{hc})^2}\left[3K_c^{h2}(1 - z_{hc}) + 2K_c^v K_c^h(1 - z_{vc}) + K_c^{v2}\frac{(1 - z_{vc})^2}{1 - z_{hc}}\right]. \tag{10.244}$$

We see from (10.242) that the boundary contribution to the specific heat diverges at a simple pole as $T \to T_c$ and thus the boundary specific heat exponent is $\alpha_b = 1$. This singularity is not integrable. We also note that c_b is not always positive. These two properties are impossible for a bulk specific heat. In particular the infinite nonintegrable divergence of c_b as $T \to T_c$ is an indication that at $T = T_c$ the influence of the boundary extends deep into the bulk.

We also note that the discontinuity in the entropy (10.241) can be interpreted as a latent heat associated with the boundary. Such a phenomenon does not occur in the bulk.

10.2.2 Boundary magnetization $M_1(H_b)$

The additional free energy due to the boundary magnetic field H_b is [15]

$$F_b(H_b)/k_B T = -\ln\cosh H_b/k_B T$$
$$-\frac{1}{4\pi}\int_{-\pi}^{\pi} d\theta \ln\left[1 - \frac{z^2 z_h z_v^{-1}|1 + e^{i\theta}|^2}{z^v(1 + z_h^2 + 2z_h\cos\theta) - (1 - z_h^2)\alpha(\theta)}\right] \tag{10.245}$$

with

$$z = \tanh H_b/k_B T \tag{10.246}$$

and

$$\alpha(\theta)^{\pm 1} = \frac{1}{z^v(1 - z_h^2)}\{(1 + z_v^2)(1 + z_h^2) - z_h(1 - z_v^2)(e^{i\theta} + e^{-i\theta})$$
$$\pm(1 - z_v^2)[(1 - \alpha_1 e^{i\theta})(1 - \alpha_1 e^{-i\theta})(1 - \alpha_2^{-1}e^{i\theta})(1 - \alpha_2^{-1}e^{-i\theta})]^{1/2}\}. \tag{10.247}$$

We obtain the boundary magnetization from the boundary free energy as

$$M_1(H_b) = \langle \sigma_{1,0} \rangle_{H_b} = k_B T \frac{\partial F_b(H_b)}{\partial T} \tag{10.248}$$

which from (10.245) may be written in several equivalent forms

$$\begin{aligned} M_1(H_b) &= z + \frac{1-z^2}{2\pi} z z_h \int_{-\pi}^{\pi} \frac{|1+e^{i\theta}|^2}{z^2 z_h |1+e^{i\theta}|^2 - z_v^2(1+z_h^2+2z_h\cos\theta) + z_v(1-z_h^2)\alpha(\theta)} \\ &= z + \frac{1-z^2}{2\pi} z \int_{-\pi}^{\pi} d\theta \frac{1+z_h - z_v(1-z_h)\alpha(\theta)}{z_v(1-z_h)(1-z^2)\alpha(\theta) - (1+z_h)(z_v^2 - z^2)}. \end{aligned} \tag{10.249}$$

The integrand in (10.249) contains, in addition to the square root branch points present in $\alpha(\theta)$, simple poles at those values of $e^{i\theta}$ where the denominator vanishes. From the last line in (10.249) we see that this happens when

$$\alpha(\theta) = \frac{(1+z_h)(z_v^2 - z^2)}{z_v(1-z_h)(1-z^2)} \tag{10.250}$$

and using the definition of $\alpha(\theta)$ (10.247) we see that the poles are at $e^{i\theta} = r, r^{-1}$ where using the definitions of $\alpha_1,\ \alpha_2$ of (10.26) and defining

$$\alpha_3 = \frac{2z}{1+z_v} \tag{10.251}$$

we have for $|r| < 1$

$$\begin{aligned} r = {} & [(1-\alpha_1\alpha_2)^2 - 2(1+\alpha_1\alpha_2)\alpha_3^2 - \alpha_3^4]^{-1} \\ & \times\{(1-\alpha_1\alpha_2)^2 - 2(\alpha_1+\alpha_2)\alpha_3^2 - \alpha_3^4 \\ & -2\sqrt{\alpha_3^2[\alpha_3^2 - (1-\alpha_1\alpha_2)][(1+\alpha_1)(1+\alpha_2)\alpha_3^2 - (1-\alpha_1\alpha_2)^2]}\}. \end{aligned} \tag{10.252}$$

The qualitative motion of r is of interest. In the $e^{i\theta}$ plane $\alpha(\theta)$ has four branch points at $\alpha_1, 1/\alpha_1, \alpha_2$ and $1/\alpha_2$.We define the cut plane for $e^{i\theta}$ by joining these branch points pair-wise along the real axis; thus the unit circle does not intersect the branch cuts unless $|\alpha_2| = 1$. In this cut plane $|\alpha(\theta)| > 1$. Therefore from (10.250) there is a pair of poles at r and $1/r$ in the cut plane if and only if

$$\left| \frac{(1+z_h)(z_v^2 - z^2)}{z_v(1-z)h)(1-z^2)} \right| \geq 1. \tag{10.253}$$

Since $|z| \leq 1$. (10.253) holds if and only if either (for all T)

$$z^2 \geq z_v \frac{1-\alpha_1}{1+\alpha_1} \tag{10.254}$$

in which case r is real and

$$0 \leq r \leq \alpha_1 \tag{10.255}$$

or if $T \leq T_c$ and

$$z^2 \leq z_h \frac{1-\alpha_2}{1+\alpha_2} \tag{10.256}$$

in which case

$$0 \leq \alpha_2 \leq r \leq 1. \tag{10.257}$$

The boundary spontaneous magnetization is defined as

$$M_1(0^+) = \lim_{H_b \to 0+} M_1(H_b) \tag{10.258}$$

and from (10.249) we see that as $H_b \to 0$ the only nonvanishing contribution comes from the pole which pinches the real unit circle for $T < T_c$ and thus we find

$$M_1(0^+) = \left[\frac{\cosh 2E^v/k_BT - \coth 2E^h/k_BT}{\cosh 2E^v/k_BT - 1}\right]^{1/2}. \tag{10.259}$$

Using the T_c condition in the form (10.31) we see that $M_1(0^+)$ vanishes at T_c, and expanding for $T \sim T_c$ we see that the $M_1(0+)$ vanishes as a square root as $T \to T_c$ and thus the boundary critical exponent is $\beta_b = 1/2$.

For $T \sim T_c^-$ and $H \sim 0$ we have

$$M_1(H_b) \sim \frac{(1-\alpha_2)^{1/2}}{z_v^{1/2}} \text{sgn } z - \frac{2z}{\pi z_v} \ln(1-\alpha_2+z^2) \tag{10.260}$$

and for $T \sim T_c^+$ and $H \sim 0$

$$M_1(H_b) \sim -\frac{2z}{\pi z_v} \ln(1-\alpha_2+z^2). \tag{10.261}$$

As $z \to 0^+$ we see that (10.260) agrees with (10.259) and explicitly exhibits the square root singularity. At $T = T_c$ it follows from both (10.260) and (10.261) that

$$M_1(H_b) \sim -\frac{4z}{\pi z_v} \ln|z| \tag{10.262}$$

and at $H_b = 0$ the boundary susceptibility at $H_b = 0$ is

$$k_BT \frac{\partial M_1(H_b)}{\partial H_b}\bigg|_{H_b=0} = -\frac{2}{\pi} \coth E^v/k_BT_c \ln|1-\alpha_2| \tag{10.263}$$

both above and below the critical temperature. Therefore the boundary critical exponents are $\delta_b = 1$ and $\gamma_b = 0$.

10.2.3 Boundary spin correlations

The correlation of two spins on the boundary row was computed in [15] as

$$
\begin{aligned}
&\langle\sigma_{1,0}\sigma_{1,N}\rangle = M_1(H_b)^2 \\
&-(1-z^2)^2\Big\{\left[\frac{z}{2\pi}\int_{-\pi}^{\pi} d\theta \frac{e^{iN\theta}|1+e^{i\theta}|^2 z_h}{z_v^2|1+z_h e^{i\theta}|^2 - z^2 z_h|1+e^{i\theta}|^2 - z_v(1-z_h^2)\alpha(\theta)}\right]^2 \\
&-\left[\frac{1}{2\pi}\int_{-\pi}^{\pi} d\theta \frac{e^{iN\theta}(e^{i\theta}-e^{-i\theta})}{z_v^2|1+z_h e^{i\theta}|^2 - z^2 z_h|1+e^{i\theta}|^2 - z_v(1-z_h^2)\alpha(\theta)}\right] \\
&\times \int_{-\pi}^{\pi} d\theta e^{iN\theta}\frac{e^{i\theta}+1}{e^{i\theta}-1}\left[1+\frac{z_h z^2|1+e^{i\theta}|^2}{z_v^2|1+z_h e^{i\theta}|^2 - z^2 z_h|1+e^{i\theta}|^2 - z_v(1-z_h^2)\alpha(\theta)}\right]\Big\}.
\end{aligned}
\tag{10.264}
$$

This result shares with the boundary free energy the property that they can be computed in the presence of a boundary magnetic field explicitly in terms of (a product of) a finite number of integrals with algebraic integrands. It is therefore straightforward to study the behavior of $\langle\sigma_{1,0}\sigma_{1,N}\rangle$ for large N for fixed T and fixed H_b. It is also straightforward to compute scaling functions but now instead of the Painlevé functions which satisfy nonlinear equations the scaling limit involves nothing more complicated than the well-known Bessel and incomplete gamma functions. In this sense the boundary spin correlations are much simpler than correlations in the bulk.

The large N behavior is easily obtained from the integrals in (10.264). The only complicating feature is that in some cases the behavior is dominated by the branch cut at $\alpha_2^{\pm}$ and in other cases by the pole at r.

For fixed $T > T_c$ and $N \to \infty$ we have

$$\langle\sigma_{1,0}\sigma_{1,N}\rangle \sim M_1^2(H_b) + A_{b+}\alpha_2^N/N^{3/2}. \tag{10.265}$$

For fixed $T < T_c$ and $N \to \infty$ there are two cases. If $z^2 < z_v(1-\alpha_2)/(1+\alpha_2)$ we have

$$\langle\sigma_{1,0}\sigma_{1,N}\rangle \sim M_1^2(H_b) + A_{b-1}\alpha^N r^N/N^{3/2} \tag{10.266}$$

while for $z^2 > z_v(1-\alpha_2)/(1+\alpha_2)$ we find

$$\langle\sigma_{1,0}\sigma_{1,N}\rangle \sim M_1^2(H_b) + A_{b-2}\alpha^{2N}/N^5 \tag{10.267}$$

where the amplitudes A_{b+}, A_{b-1}, A_{b-2} are rather tedious functions of E^h/k_BT, E^v/k_BT and H_b/k_BT.

For $T = T_c$ if $H_b = 0$ we have

$$\langle\sigma_{1,0}\sigma_{1,N}\rangle \sim \frac{1}{\pi z_{vc} N} \tag{10.268}$$

and if $H_b \neq 0$

$$\langle\sigma_{1,0}\sigma_{1,N}\rangle \sim M_1^2(H_b) + \frac{4z^2 z_{vc}^2}{\pi^2(z_{vc}^2 - z^2)^2(1-r)^2(1-r^{-1})^2 N^4}. \tag{10.269}$$

These large N expansions are not uniform, and to connect the various regions together appropriate scaling variables and functions must be used. To connect the expansion of $T > T_c$ and $T < T_c$ with the expansions for $T - T_c$ the scaling variable $(1-\alpha_2)N$ is used, and the scaling functions are expressed in terms of Bessel functions. To connect the $T = T_c$ expansions for $H_b = 0$ with $H_b \neq 0$ the scaling variable Nz^2 is used, and the scaling functions are given in terms of functions such as the cosine integral

$$Ci(z) = -\int_z^\infty dt \frac{\cos t}{t}. \tag{10.270}$$

10.2.4 Analytic continuation and hysteresis

The boundary magnetization $M_1(H_b)$ has one further property that deserves to be discussed. Namely, even though for $T < T_c$ the magnetization is discontinuous in the sense that

$$\lim_{H_b \to 0^-} M_1(H_b) = -\lim_{H_b \to 0^+} M_1(H_b), \tag{10.271}$$

the boundary magnetization $M_1(H_b)$ defined for $H_b > 0$ is analytic at $H_b = 0$ and can be analytically continued into the region $H_b < 0$ to a function $M_1^c(H_b)$ which is not the same as $M_1(H_b)$.

This analytic continuation may be made by writing $M_1(H_b)$ in the form

$$M_1(H_b) = z + \frac{1-z^2}{2\pi}\int_{-\pi}^{\pi} d\theta \left(\frac{1+e^{i\theta}}{s(e^{i\theta})} + \frac{1+e^{-i\theta}}{s(e^{-i\theta})}\right) \tag{10.272}$$

where

$$\begin{aligned} s(e^{i\theta}) &= 2z(1+e^{i\theta}) \\ &- (1+z_v)\{[(1-\alpha_1 e^{i\theta})(1-\alpha_2 e^{i\theta})]^{1/2} - e^{i\theta}[(1-\alpha_1 e^{-i\theta})(1-\alpha_2 e^{-i\theta})]^{1/2}\}. \end{aligned} \tag{10.273}$$

The functions $s(e^{\pm i\theta})$ have at most one zero in the cut $e^{i\theta}$ plane. For H_b small and positive $s(e^{i\theta})$ has a zero inside the unit circle at r where $|r| < 1$ and $s(e^{-i\theta})$ has a zero outside the unit circle at $1/r$. After analytic continuation to small negative values of H_b we have $r > 1$. For $H_b < 0$ the difference between $M_1^c(H_b)$ and $M_1(H_b)$ is due to the residues at r and $1/r$ and we find from (10.273) that

$$M_1^c(H_b) - M_1(H_b) = \frac{2z[(1+z_h)^2(z_v^2-z^2)^2 - z_v^2(1-z_h)^2(1-z^2)^2]}{(r^{-1}-r)(1-z^2)z_h(z_v^2-z^2)^2}. \tag{10.274}$$

When H_b is small and negative the right-hand side of (10.274) is positive and decreases as H_b becomes more negative. The right-hand side of (10.273) vanishes when H_b reaches a critical value H_b^c defined by

$$\tanh^2 H_b^c/k_BT = z_v \frac{1-\alpha_2}{1+\alpha_2}. \tag{10.275}$$

We plot $M_1(H_b)$ and the analytic continuation $M_1^c(H_b)$ together in Fig. 10.4. It is natural to interpret this figure as a hysteresis loop. The loop shrinks to a point at

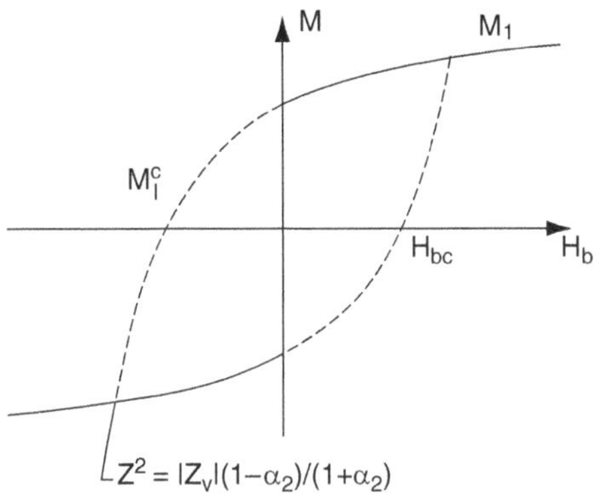

Fig. 10.4 Hysteresis loop for the magnetization in the first row for $E^v = E^h$ at $T/T_c = 0.9$. The solid curve is M_1; the dotted curve shows the analytic continuation.

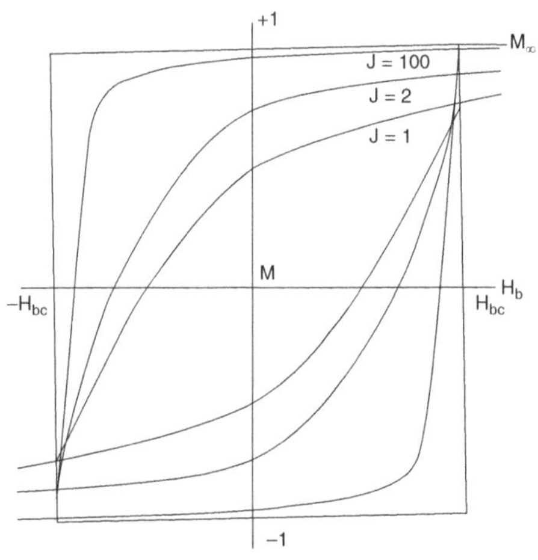

Fig. 10.5 A schematic plot of $M_j(H_b)$ and its analytic continuation for $E^v = E^h$ versus H_b for various j.

$H_b = M_1 = 0$ as $T \to T_c$ and as $T \to 0$ the loop becomes a rectangle that continues as far into the unstable regime as $|H_b| = E_v$.

Further insight into this hysteresis phenomenon can be obtained by computing the magnetization $M_j(H_b)$ on the row j in from the boundary. The results of this computation are plotted in Fig. 10.5 where we give the hysteresis loops for $M_j(H_b)$ as a function of H_b and in Fig. 10.6 where we plot $M_j(H_b)$ as a function of j for various H_b. The critical value H_b^c at which $M_j^c(H_b^c) = M_j(H_b^c)$ is independent of j, and in Fig. 10.6 we see that as $H_b \to H_b^c$ that the region of overturned spins penetrates into the bulk. When H_b reaches H_b^c no further continuation is possible because the bulk magnetization flips from positive to negative. This phenomenon of the boundary field affecting the bulk spins in the interior is sometimes referred to as "wetting."

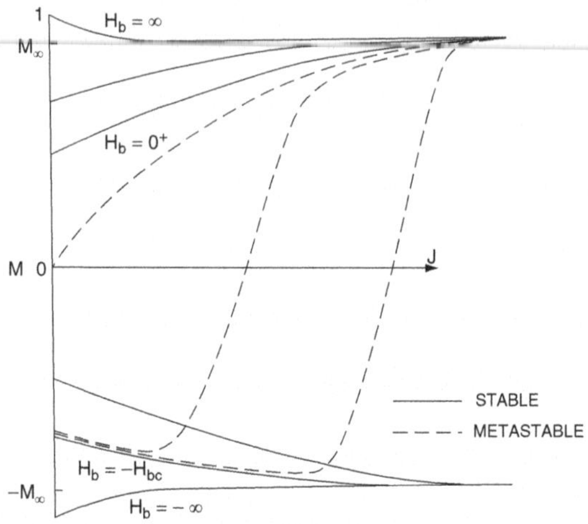

Fig. 10.6 A schematic plot of $M_j(H_b)$ and its analytic continuation for $E^v = E^h$ versus j for various H_b. Metastable values between $H_b = 0$ and $-H_B^c$ which are reached from positive values of H_b are shown in dotted lines.

10.3 The layered random lattice

Thus far we have considered only homogeneous models where the interaction energies E^v and E^h are constant throughout the entire lattice. But real systems will contain impurities and will not have this homogeneous property. Therefore it is very desirable to be able to study what new effects can be present if the assumption of homogeneous interactions is lifted.

The effect of inhomogeneities can be studied in the Ising model by letting the interaction constants depend on the position in the lattice. The case which is best studied is the case of the layered Ising model where E^v and E^h are the same within a row where the values in different rows are allowed to be different. For this layered model the interaction energy is

$$\mathcal{E} = -\sum_{j=1}^{L_v}\sum_{k=1}^{L_h}\{E^h(j)\sigma_{j,k}\sigma_{j,k+1} + E^v(j)\sigma_{j,k}\sigma_{j+1,k} + H\sigma_{j,k}\}. \tag{10.276}$$

The case where the energies $E^v(j)$ and $E^h(j)$ depend on j but still have a periodicity $E^v(j+J) = E^v(j)$ and $E^v(j+J) = E^v(j)$ has been studied in [69]. It is found that the specific heat still has a logarithmic singularity at a well defined T_c which is determined from

$$\prod_{j=1}^{J} e^{4E^h(j)/k_BT_c}\tanh^2 E^v(j)/k_bT_c = 1 \tag{10.277}$$

which is invariant under permutations of the $E^v(j)$ and $E^h(j)$. The amplitude of the logarithmic singularity depends in detail on the values of $E^v(j)$ and $E^h(j)$ and is not invariant under permutations.

A more interesting case, however, is when the $E^v(j)$ and $E^h(j)$ are quasiperiodic functions of j. A particularly nice example is the Fibonacci lattice [70] which is defined as follows. Consider a set of sequences S_n of the letters A and B defined recursively by

$$S_{n+1} = S_n S_{n-1} \text{ with } S_0 = 1,\ S_1 = A. \tag{10.278}$$

For example $S_2 = AB$, $S_3 = ABA$ and $S_4 = ABAAB$. The sequence S_n contains F_{n-1} $A's$ and F_{n-2} $B's$ where F_n are the Fibonacci numbers defined by

$$F_{n+1} = F_n + F_{n-1}, \quad F_0 = F_1 = 1. \tag{10.279}$$

Now consider two values of (positive) energies $E^v = E^v_A$ and E^v_B, place them in the lattice according to the sequence S_n and then repeat the sequence to build up the entire lattice. Each sequence S_n thus defines a set of interactions with periodic F_{n+1} from which T_c is computed from (10.277) as

$$e^{-2F_n E^h/k_B T_c} = \tanh^{F_{n-1}} E^v_A/k_B T_c \tanh^{F_{n-2}} E^h_B/k_B T_c \tag{10.280}$$

and for each of these lattices the specific heat has a logarithmic divergence. In the limit where the sequence length $n \to \infty$ the critical temperature is found from (10.280) as

$$e^{-2E^h/k_B T_c} = \tanh^{\alpha} E^v/k_B T_c \tanh^{\alpha^2} E^v_B/k_B T_c \tag{10.281}$$

where

$$\alpha = (\sqrt{5} - 1)/2 \tag{10.282}$$

and the specific heat still has a logarithmic divergence with the amplitude

$$A = \frac{1}{4\pi(k_b T_c)^2}(z_{hc}^{-1} - z_{hc})\left|\frac{x^2 \ln x^2}{1 - x^2}\right| \{2E^h + E^v_A \alpha(z_{vAc}^{-1} - z_{vAc}) + E^v_B \alpha^2(z_{vBc} - z_{vBc})\}^2 \tag{10.283}$$

where $x = z_{vBc}/z_{vAc}$. Thus the mere destruction of periodicity is not sufficient to destroy the logarithmic divergence in the specific heat.

The most interesting case of all is when these energies $E^h(j)$ and $E^v(j)$ are allowed to be chosen in a random fashion with probability distributions $P_v(E^v)$ and $P_h(E^h)$. This is the case of frozen (or quenched) random impurities which should be most relevant for real ferromagnets on impure lattices.

These frozen random layered Ising models have been extensively investigated [16–18], and many interesting properties have been found.

There is now a new length scale in the problem determined by the width of the probability distributions. As an example we consider the case where E^h is fixed and the distribution function for E^v is $P(E^v)$. Then there is still a T_c which is determined from (10.277) in the random limit as

$$0 = \int_0^\infty dE^h P(E^v) \ln(e^{4E^h/k_B T_c} \tanh^2 E^v/k_B T_c). \tag{10.284}$$

Near this T_c we define a temperature variable δ which is useful for sufficiently narrow $P(E^v)$

$$\delta = \frac{2\int_0^\infty dE^v P(E^v) \ln(e^{4E^h/k_B T} \tanh^2 E^v/k_B T)}{\int_0^\infty dE^v P(E^v)[\ln(e^{4E^h/k_B T} \tanh^2 E^v/k_B T)]^2} \tag{10.285}$$

and we find that for narrow $P(e^v)$ that when δ is of order one that the leading contribution to the specific heat which depends on δ is

$$\int_0^\infty d\phi[\frac{\partial^2}{\partial\delta^2} \ln K_\delta(\phi) - (1+\phi)^{-1}]. \tag{10.286}$$

This function is analytic except at $\delta = 0$ where there is an essential singularity because, while all derivatives exist at $\delta = 0$, the Taylor series fails to converge.

Further insight into the behavior of this random system is provided by a general theorem proven specifically by Griffiths [19] for a lattice where all interactions (not just in layers) are random with the specific probability distribution

$$P(E) = p\delta(E - E_0) + (1-p)\delta(E). \tag{10.287}$$

Griffiths proved, that for all temperatures below the temperature where the lattice would be critical for the pure case $p = 0$, the magnetization $M(H)$ will not be analytic at $H = 0$ even if $T > T_c$.

There is nothing in this theorem which is limited to the specific distribution (10.287) or which cannot be extended to the layered case. The essence of the physics is that if the probability distribution $P(E)$ is nonzero only for $E^L < E < E^U$ then, if

$$T^L < T < T^U \tag{10.288}$$

where T^L (T^U) is the critical temperature the lattice would have if all interactions E take on the single value E^U (E^L), there will be large regions of the random lattice which are locally above (below) the actual global value of T_c.

It is in the temperature range (10.288) that the computation of the specific heat (10.286) is valid and Griffiths theorem shows that $M(H)$ is not analytic at $H = 0$. These two results strongly suggest that the zeros of the finite size partition function are no longer pinching the real temperature axis at a single point T_c but are instead pinching the entire line segment (10.288).

What is not shown by these computations is 1) the connection of T_c defined by (10.284) with the temperature below which there is spontaneous magnetization and 2) the nature of the nonanalyticity in $M(H)$.

The first question has been studied [71] for the layered random lattice by means of renormalization group techniques where he finds that the temperature for the onset of spontaneous magnetization is indeed the T_c computed from (10.284), but that the exponent β for the random layered lattice is now $(3-\sqrt{5})/2$.

The question of the nature of the nonanalyticity has only been studied for the average magnetization on the boundary [17,18] where it is found that when the temperature variable δ is of order one that there are terms in the average boundary magnetization $\langle M_1(H_b)\rangle$ which depend on $H^{2|\delta|}$ which are not analytic (unless $2|\delta|$ is an integer) and which even fail to be differentiable for $|\delta| < 1/2$. Thus there is an entire region around T_c where the boundary magnetic susceptibility is infinite. Furthermore in the temperature region (10.288) the average over $P(E^v)$ of the boundary spin correlations falls off at large separations as a power of the separation of the spins which depends on δ. We thus conclude that in the temperature region (10.288) the standard phenomenological description of critical phenomena does not apply.

10.4 The Ising model for $H \neq 0$

The desire to extend the exact computation of the two-dimensional Ising model from $H = 0$ to $H \neq 0$ has been one of the outstanding problems in statistical mechanics ever since Onsager's computation of 1944.

10.4.1 The circle theorem

The first result obtained for $H \neq 0$ is the famous circle theorem of Lee and Yang [46] which states that in terms of the variable $z = e^{-2H/k_BT}$ the zeroes of the partition function of an Ising model with all interaction energies E positive (i.e. ferromagnetic) and with boundary conditions which are invariant under the reflection $H \to -H$ for real temperatures on a finite size lattice in D dimensions lie on the unit circle $|z| = 1$. In other words, in the magnetic field plane the partition function is periodic with a period $i\pi k_BT$ and the zeros of the partition function all lie exactly on the imaginary H axis. When $T > T_c$ there is a strip surrounding the real H axis which is zero-free but for $T < T_c$ the zeros pinch the real H axis at $H = 0$ and the density of these zeros of the spontaneous magnetization.

10.4.2 The imaginary magnetic field $H/k_BT = i\pi/2$

Lee and Yang also found [46] that for

$$H/k_BT = i\pi/2 \tag{10.289}$$

the free energy of the Ising model can also be exactly solved for $E^v = E^h$. This solution was extended to the anisotropic case in 1966 where [47] it was found that free energy is

$$\begin{aligned}-F(ik_BT\pi/2)/k_BT &= \ln(2\cosh E^v/k_BT\cosh E^h/k_BT) + \frac{1}{16\pi^2}\int_{-\pi}^{\pi}d\phi_1\int_{-\pi}^{\pi}d\phi_2 \\ &\times \ln[4(z_v^2+z_h^2)(1+z_v^2z_h^2) - 4z_v^2(z_h^2-1)^2\cos^2\phi_1 - 4z_h^2(z_v^2-1)^2\cos^2\phi_2]\end{aligned} \tag{10.290}$$

with z_j defined by (10.125), and that the magnetization is

$$M(ik_BT\pi/2) = \left[\frac{(z_v^{-1}+z_v)^2(z_h^{-1}+z_h)^2}{4(z_v^{-2}+z_v^2+z_h^{-2}+z_h^2)}\right]^{1/8}. \tag{10.291}$$

The free energy (10.290) is analytic for all real $T > 0$. The magnetization (10.291) is the density of zeros at $H/k_BT = i\pi/2$ $(y = -1)$ and for positive T is a monotonic function with the limiting values

$$M(ik_BT\pi/2) \to 1 \text{ as } T \to \infty \tag{10.292}$$

$$\to \infty \text{ as } T \to 0. \tag{10.293}$$

The results for the free energy extend to the finite size partition function with toroidal boundary conditions as the sum of four terms. For the isotropic lattice we find in analogy with the $H = 0$ result that the partition function normalized to unity at $u = x^2 = e^{-4K} = 0$ is in the form (10.2) with Z_j replaced by

$$Z_j(u, z=-1) = (1-u)^{L_vL_h/2}\prod_{\theta_1}\prod_{\theta_2}\{1+u^2+u[6-4\cos^2\theta_1-4\cos^2\theta_2]\}. \tag{10.294}$$

For Brascamp–Kunz boundary conditions we have [72] a result analogous to (10.19) for the partition function normalized to unity at $u = 0$ when both L_v and L_h are even:

$$Z_{BK}(u, z=-1) = (1-u)^{L_vL_h/2}\prod_{j=1}^{L_h/2}\prod_{k=1}^{L_v/2}\{1+u^2 + u[6-4\cos^2((2j-1)\pi/L_h)-4\cos^2((2k-1)\pi/(2(L_v+1)))]\} \tag{10.295}$$

If $L_h/4$ is an integer this further reduces to a perfect square

$$Z_{BK}(u, z=-1) = (1-u)^{L_vL_h/2}\prod_{j=1}^{L_h/4}\prod_{k=1}^{L_v/2}\{1+u^2 + u[6-4\cos^2((2j-1)\pi/L_h)-4\cos^2((2k-1)\pi/(2(L_v+1)))]\}^2. \tag{10.296}$$

We note that, because of the factor $(1-u)^{L_vL_h/2}$, the partition functions for both toroidal and Brascamp–Kunz boundary conditions have half of the zeros at the point $u = 1$. We also note that for the Brascamp–Kunz boundary conditions all the remaining zeros lie either in the circle

$$|u| = 1 \text{ where } -1 \le 3-2\cos^2\theta_1-2\cos^2\theta_2 \le 1 \tag{10.297}$$

or on the segment of the negative real axis

$$-3-2\sqrt{2} \le u \le -3+2\sqrt{2} \text{ where } 1 \le 3-2\cos^2\theta_1-2\cos^2\theta_2. \tag{10.298}$$

A plot of the zeros of a 12×12 lattice with Brascamp-Kunz boundary conditions is given in Fig. 10.7. Partition function zeros for finite size lattices at $H/k_BT = i\pi/2$ with other boundary conditions have been studied in [54].

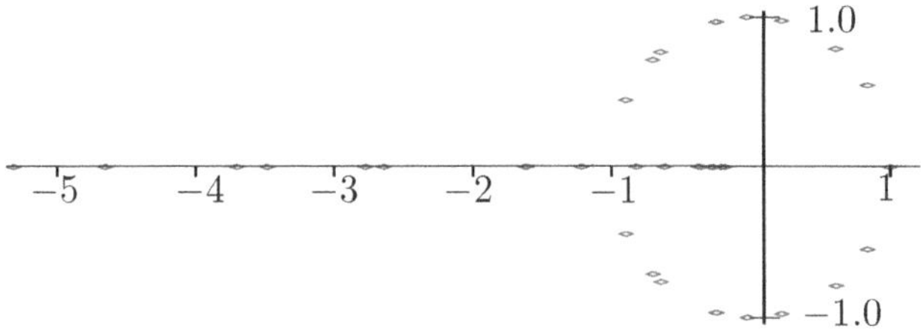

Fig. 10.7 Zeros in the complex $u = x^2$ plane of the Ising partition function at $H/k_BT = i\pi/2$ for Brascamp-Kunz boundary conditions on a 12×12 isotropic lattice. The zero at $x = 1$ has multiplicity 72 and all other zeros have multiplicity 2.

10.4.3 Expansions for small H

For $H \neq 0$ there are no further results known, and to obtain further information we turn to expansions of quantities such as the free energy and correlation functions as power series in H.

Free energy. The free energy $F(T;H)$ is expanded as

$$-[F(T;H) - F(T;0)]/k_BT = \sum_{n=0}^{\infty}\left(\frac{H}{k_BT}\right)^{n+1} \sum_{\mathbf{R}_1,\cdots,\mathbf{R}_n} \langle \sigma_{0,0}\sigma_{\mathbf{R}_1}\cdots\sigma_{\mathbf{R}_n}\rangle^c \quad (10.299)$$

where $\langle \sigma_{0,0}\sigma_{\mathbf{R}_1}\cdots\sigma_{\mathbf{R}_n}\rangle^c$ denotes the connected part of the correlation function which vanishes when the separation of any two spins goes to infinity. The first two terms in this expansion are the magnetization and the susceptibility which have been previously discussed.

For $T > T_c$ all terms in (10.299) with n even vanish. Furthermore the circle theorem of Lee and Yang [46] guarantees that $F(T,H)$ is analytic at $H = 0$ and thus the series (10.299) will converge.

For $T < T_c$ it was shown by Isakov [49] in 1984 that in dimension two and greater that the derivatives $\partial^k F(T,H)/\partial H^k$ increase with k sufficiently rapidly that the series in (10.299) will not converge. Therefore for $T < T_c$ the free energy and magnetization are not analytic at $H = 0$ and unlike the boundary magnetization as a function of the boundary magnetic field the magnetization $M(T;H)$ cannot be analytically continued from $H > 0$ into a metastable (hysteresis) phase for $H < 0$.

Two-point function. The two-point function in a magnetic field $G(\mathbf{R}_1 - \mathbf{R}_2; H)$ is expanded in terms of the n point correlations at $H = 0$ as

$$G(\mathbf{R}_1 - \mathbf{R}_2; H) - M^2(T,H) = \sum_{n=0}^{\infty}(H/k_BT)^n \sum_{\mathbf{R}_3\cdots\mathbf{R}_{n+2}} \langle \sigma_{\mathbf{R}_1}\sigma_{\mathbf{R}_2}\sigma_{\mathbf{R}_3}\cdots\sigma_{\mathbf{R}_{n+2}}\rangle^c \quad (10.300)$$

and we may use the extensive information available [28–30] about n point functions at $H = 0$ to study the region $T \sim T_c$.

In the scaling limit

$$T \to T_c \;\text{ and }\; \mathbf{R}_j - \mathbf{R}_k \to \infty, \quad (10.301)$$

assuming with no loss of generality that $E^v = E^h$,

$$(\mathbf{R}_j - \mathbf{R}_k)|T - T_c| = \mathbf{r}_j - \mathbf{r}_k \quad \text{fixed} \tag{10.302}$$

it was shown in [29, 30] that at $H = 0$ the correlation functions have the form

$$\langle \sigma_{\mathbf{R}_1} \cdots \sigma_{\mathbf{R}_n} \rangle \sim M_{\pm}^n g(\mathbf{r}_1, \cdots, \mathbf{r}_n). \tag{10.303}$$

Therefore by using the scaling form (10.303) in (10.300) and recalling that $M_{\pm}(T) \sim |T - T_c|^{1/8}$ as $T \to T_c$ we see that if we define a scaled magnetic field as

$$h = H/|T - T_c|^{15/8} \tag{10.304}$$

then

$$\lim_{\text{scaling}} M_{\mp}^{-2} G(\mathbf{R}_1, \cdots \mathbf{R}_2; H)^c = g^c(r; h) \tag{10.305}$$

is a function of r and h alone.

The most interesting case is $T < T_c$ where the connected two-point function $g^c(r; h)$ for $h \sim 0$ and large r was found in [48] to be

$$g^c(r; h) = \sum_j a_j(h) K_0[(2 + \kappa_j(h))r] \sim \pi^{1/2} r^{-1/2} e^{-2r} \sum_j a_j(h) e^{r\kappa_j(h)} \tag{10.306}$$

with

$$\kappa_j(h) = h^{2/3} \lambda_j \quad \text{and} \quad a_j(h) = ha \tag{10.307}$$

where a is a constant and λ_j are the solutions of

$$J_{1/3}(\lambda^{3/2}/3) + J_{-1/3}(\lambda^{2/3}/3) = 0 \tag{10.308}$$

where $J_n(z)$ is the Bessel function of order n. There are an infinite number of solutions to (10.308). As $h \to 0$ the spacing between these zeros vanishes and the large r behavior (10.306) reduces to the behavior $r^{-2} e^{-2r}$ of the two-point function at $H = 0$ for $T < T_c$. If we consider the two-dimensional Fourier transform of $g^c(r, h)$

$$\tilde{g}^c(k, h) = \int d^2\mathbf{r} e^{i\mathbf{k}\cdot\mathbf{r}} = \int_0^{2\pi} d\theta \int_0^{\infty} dr\, r e^{kr\cos\theta} g^c(r; h) \tag{10.309}$$

we see from (10.306) that the singularities in $g^c(k; h)$ nearest the real k axis are poles. In the language of field theory the positions of these poles may be interpreted as particle masses.

10.4.4 $T = T_c$ with $H > 0$

We have now exhausted the known results for the Ising model (10.1) on a lattice with $H \neq 0$. However, if we are willing to consider only the scaling limit where the lattice has disappeared and continuum methods are used then in a most remarkable paper Zamolodchikov [50] discovered in 1988 that the locations m_i of the poles in the k plane of the Fourier transform (10.309) of the two-point function are given by the

Table 10.5 The location of the poles in the Fourier transform of the two-point function of the scaled Ising model for $T = T_c$ and $H > 0$.

m_1	$= m$	$=m$
m_2	$= 2m \cos \pi/5$	$= m1.61803 \cdots$
m_3	$= 2m \cos \pi/30$	$= m1.98904 \cdots$
m_4	$= 4m \cos \pi/5 \cos 7\pi/30$	$= m2.40487 \cdots$
m_5	$= 4m \cos \pi/5 \cos 2\pi/15$	$= m2.95629 \cdots$
m_6	$= 4m \cos \pi/5 \cos \pi/30$	$= m3.21834 \cdots$
m_7	$= 8m \cos^2 \pi/5 \cos 7\pi/30$	$= m3.89116 \cdots$
m_8	$= 8m \cos^2 \pi/5 \cos 2\pi/15$	$= m4.78338 \cdots$

eight components S_k of the Perron–Frobenius vector of the Cartan matrix of the Lie algebra E_8 as $m_j/m_k = S_i/S_k$. We list these poles in Table 10.5.

In the language of field theory this spectrum of poles has three particles below the first two-particle threshold and five particles above threshold. Nevertheless the five particles above the two-body threshold are stable and do not decay. However, if (in the field theory limit) the system with $H > 0$ is perturbed from $T = T_c$ it is found [73] that most of the decays compatible with energy conservation do occur. The exceptional cases that are still forbidden are $m_7 \to m_1 m_1$ and the four decays $m_8 \to m_1 m_1,\ m_1 m_2,\ m_2 m_3,\ m_3 m_4$.

10.4.5 Extended analyticity

We conclude by noting that if it is assumed [51] that the free energy in the scaling limit, which depends on the single variable h and not on the two separate variables H and T, can be analytically continued through the position of the Lee–Yang zeros then a scenario is obtained which joins together all of the regions previously discussed. To fully evaluate this conjecture it is necessary to determine whether or not the singularities found in $\chi^{(j)}$ discovered in [36–39] actually lead to a natural boundary.

References

[1] W. Lenz, Phys. ZS. 21 (1920) 613.
[2] E. Ising, Beitrag zur theorie des ferromagnetismus, Z. Physik 31 (1925) 253–258.
[3] R. Peierls, On Ising's model of ferromagnetism, Proc. Cambridge Phil. Soc. 32 (1936) 477–481,
[4] H.A. Kramers and G.H. Wannier, Statistics of the two-dimensional ferromagnet. Part I, Phys. Rev. 60 (1941) 252–262.
[5] H.A. Kramers and G.H. Wannier, Statistics of the two-dimensional ferromagnet. Part II, Phys. Rev. 60 (1941) 263–276.
[6] L. Onsager. Crystal statistics I. A two dimensional model with an order disorder transition, Phys. Rev. 65 (1944) 117–149.
[7] L. Onsager, discussion, Nuovo Cimento 6 Suppl. (1949) 261.
[8] B. Kaufman, Crystal statistics II. Partition function evaluated by spinor analysis, Phys. Rev. 76 (1949) 1232–1243.
[9] B. Kaufman and L. Onsager, Crystal statistics III short range order in a binary Ising lattice, Phys. Rev. 76 (1949) 1244–1252.
[10] C.N. Yang, The spontaneous magnetization of the two dimensional Ising model, Phys. Rev. 85 (1952) 808–816.
[11] P.W. Kastelyn, Dimer statistics and phase transitions, J. Math. Phys. 4 (1963) 287–293.
[12] E.W. Montroll, R.B. Potts and J.C. Ward, Correlations and spontaneous magnetization of the two dimensional Ising model, J. Math. Phys. 4 (1963) 308–322
[13] T.T. Wu, Theory of Toeplitz determinants and the spin correlations of the two dimensional Ising model I, Phys. Rev. 149 (1966) 380–401.
[14] H. Cheng and T.T. Wu, Theory of Toeplitz determinants and the spin correlations of the two dimensional Ising model III, Phys. Rev. 164 (1967) 719–735.
[15] B.M. McCoy and T.T. Wu, Theory of Toeplitz determinants and the spin correlations of the two dimensional Ising model IV, Phys. Rev. 162 (1967) 436–475.
[16] B.M. McCoy and T.T. Wu, Theory of a two dimensional Ising model with random impurities, Phys. Rev. 176 (1968) 631–643.
[17] B.M. McCoy, Incompleteness of the critical exponent description for ferromagnetic systems containing random impurities, Phys. Rev. Letts, 23 (1969) 383–386.
[18] B.M. McCoy, Theory of a two dimensional Ising model with random impurities III. Boundary effects, Phys. Rev. 188 (1969) 1014–1031.
[19] R.B. Griffiths, Nonanalytic behavior above the critical points in a random Ising ferromagnet, Phys. Rev. Letts. 23 (1969) 17–19.
[20] E. Barouch, B.M. McCoy and T.T. Wu, Zero-field susceptibility of the two dimensional Ising model near T_c, Phys. Rev. Letts. 31 (1973) 1409–1411.

[21] C.A. Tracy and B.M. McCoy, Neutron scattering and the correlations functions of the Ising model near T_c, Phys. Rev. Letters. 31 (1973) 1500–1504.
[22] T.T. Wu, B.M. McCoy, C.A. Tracy and E. Barouch, Spin-spin correlation functions for the two dimensional Ising model: exact theory in the scaling region, Phys. Rev. B13 (1976) 315–374.
[23] M. Sato, T. Miwa and M. Jimbo, Holonomic quantum fields I, Pub. Res. Math. Sci. 14 (1978) 223–267.
[24] M. Sato, T. Miwa and M. Jimbo, Holonomic quantum fields II, Pub. Res. Math. Sci. 15 (1979) 201–278.
[25] M. Sato, T. Miwa and M. Jimbo, Holonomic quantum fields III, Pub. Res. Math. Sci. 15 (1979) 577–629.
[26] M. Sato, T. Miwa and M. Jimbo, Holonomic quantum fields IV, Pub. Res. Math. Sci. 15 (1978) 871–972.
[27] M. Sato, T. Miwa and M. Jimbo, Holonomic quantum fields V, Pub. Res. Math. Sci. 16 (1980) 531–584.
[28] B.M.McCoy, C.A. Tracy and T.T. Wu, Two-dimensional Ising model as an exactly solved relativistic quantum field theory: Explicit formulas for the $N-$ point functions, Phys. Rev. Letts. 38 (1977) 793–796.
[29] B.M. McCoy and T.T. Wu, Two-dimensional Ising field theory for $T < T_c$: string structure of the three point function, Phys. Rev. D 18 (1978) 1243–1252.
[30] B.M. McCoy and T.T. Wu, Two-dimensional Ising field theory for $T < T_c$: Greens function strings in strings in n point functions, Phys. Rev. D 18 (1978) 1253–1258.
[31] B.M. McCoy and T.T. Wu, Nonlinear partial difference equations for the two-dimensional Ising model, Phys. Rev. Letts. 45 (1980) 675–678.
[32] J.H.H. Perk, Quadratic identities for the Ising model, Phys. Letts.A 79 (1980) 3–5.
[33] B.M. McCoy and T.T. Wu, Non-linear partial difference equations for the two-spin correlation function of the two dimensional Ising model, Nucl. Phys. B 180[FS2] (1981) 89–115.
[34] B.M. McCoy, J.H.H. Perk and T.T. Wu, Ising field theory: quadratic difference equations for the n-point Green's functions on the lattice, Phys. Rev. Letts. 46 (1981) 757–760.
[35] M. Jimbo and T. Miwa, Studies on holonomic quantum fields XVII, Proc. Jpn. Acad. 56A (1980) 405; 57A (1981) 347.
[36] B.G. Nickel, On the singularity structure of the susceptibility of the 2D Ising model, J. Phys. A32 (1999) 3889–3906.
[37] B.G. Nickel, Addendum to "On the singularity structure of the susceptibility of the 2D Ising model," J. Phys. A 33 (2000) 1693–1711.
[38] W.P. Orrick, B.G. Nickel, A.J. Guttmann, J.H.H. Perk, The susceptibility of the square lattice Ising model: new developments, J. Stat. Phys. 102 (2001) 795–841.
[39] W.P. Orrick, B.G. Nickel, A.J. Guttmann, J.H.H. Perk, Critical behavior of the two-dimensional Ising susceptibility, Phys. Rev. Letts. 86 (2001) 4120–4123.
[40] H. Au-Yang and J.H.H. Perk, Correlation functions and susceptibility in the Z-invariant Ising model, in *MathPhys Odyssey 2001: Integrable models and beyond* M. Kashiwara and T. Miwa, eds (Birkhäuser, Boston, 2002) 23–48

[41] N. Zenine, S. Boukraa, S. Hassani and J.M. Maillard, The Fuchsian differential equation of the square lattice Ising $\chi^{(3)}$ susceptibility, J. Phys. A 37 (2004) 9651–9668.
[42] N. Zenine, S. Boukraa, S. Hassani and J.M. Maillard, Square lattice Ising model susceptibility: Series expansion method and differential equation for $\chi^{(3)}$, J. Phys. A 38 (2005) 1875–1899.
[43] N. Zenine, S. Boukraa, S. Hassani and J.M. Maillard, Ising model susceptibility: The Fuchsian equation for $\chi^{(4)}$ and its factorization properties, J.Phys. A 38 (2005) 4149–4173.
[44] S. Boukraa, S. Hassani, J-M. Maillard, B.M. McCoy, W.P.Orrick and N. Zenine, Holonomy of the Ising model form factors, J. Phys. A40 (2007) 75–112.
[45] S. Boukraa, S. Hassani, J-M. Maillard, B.M. McCoy, J-A Weil and N. Zenine, The diagonal Ising susceptibility, J. Phys. A 40 (2007) 8219–8236.
[46] T.D. Lee and C.N. Yang, Statistical theory of equations of state and phase transitions II. Lattice gas and Ising models, Phys. Rev. 87 (1952) 410–419.
[47] B.M. McCoy and T.T. Wu, Theory of Toeplitz determinants and the spin correlations of the two-dimensional Ising model. II, Phys. Rev. 155 (1967) 438–452.
[48] B.M. McCoy and T.T. Wu, The two dimensional Ising model in a magnetic field: breakup of the cut in the two point function, Phys. Rev. D18 (1974) 1259–1267.
[49] S.N. Isakov, Nonanalytic features of the first order phase transition in the Ising model, Comm. Math. Phys. 95 (1984) 427–443.
[50] A.B. Zamolodchikov, Integrals of the motion and the S-matrix of the (scaled) $T = T_c$ Ising model with a magnetic field, Int. J. Mod. Phys. A4 (1989) 4235–4248.
[51] P. Fonseca and A. Zamolodchikov, Ising field theory in a magnetic field: analytic properties of the free energy, J. Stat. Phys. 110 (2002) 527–590.
[52] B.M. McCoy and T.T. Wu, *The two dimensional Ising model* (Harvard University Press 1973).
[53] H.J. Brascamp and H. Kunz, Zeros of the partition function for the Ising model in the complex temperature plane, J. Math. Phys. 15 (1974) 65–66.
[54] V. Matveev and R. Shrock, Complex temperature properties of the two-dimensional Ising model for nonzero magnetic field, Phys. Rev. E 53 (1996) 254–266.
[55] A. Erdelyi, W. Magnus, F. Oberhettinger and T.G. Tricomi, Higher Transcendental Functions, Vol 1 (McGraw-Hill, NY 1953) .
[56] I. Lyberg and B.M. McCoy, Form factor expansion of the row and diagonal correlation functions of the two dimensional Ising model, J. Phys. A 40 (2007) 3329–3346.
[57] M.Jimbo and T. Miwa, Monodromy preserving deformation of ordinary differential equations with rational coefficients II. Physica 2D (1981) 407–448.
[58] K. Okamoto, Studies of the Painlevé equations I. Sixth equation PVI, Annali di Mathematici Pura et Appl. 146 (1987) 337–381.
[59] B.M. McCoy and J.H.H. Perk, The relation of conformal field theory and deformation theory for the Ising model, Nucl. Phys. B285[FS19] (1987) 279–294.
[60] A.E. Ferdinand and M.E. Fisher, Bounded and inhomogeneous Ising models II. Specific heat anomaly of a finite lattice, Phys. Rev. 185 (1969) 832–846.

[61] B.M. McCoy, C.A. Tracy and T.T. Wu, Painlevé equations of the third kind, J.Math. Phys. 18 (1977) 1058–1092.
[62] C.A. Tracy, Asymptotics of τ function arising in the two-dimensional Ising model, Comm. Math. Phys. 142 (1991) 297–311.
[63] J.Palmer and C.A. Tracy, Two-dimensional Ising correlations: convergence of the scaling limit Adv. Appl. Math. 2 (1981) 329–388.
[64] K. Yamada, On the spin-spin correlation function in the Ising square lattice and the zero field susceptibility, Prog. Theo. Phys. 71 (1984) 1416–1418.
[65] K. Yamada, Pair correlation function in the Ising square lattice: determinental form, Prog. Theo.Phys. 72 (1984) 922–930.
[66] K. Yamada, Pair correlation function in the Ising square lattice: generalized Wronskian form, Prog. Theo.Phys. 74 (1986) 602–612.
[67] X-P. Kong, H. Au-Yang and J.H.H. Perk, New results for the susceptibility of the two-dimensional Ising model at criticality, Phys.Letts. A 116 (1986) 54–57.
[68] A. Bostan, S. Boukraa, S. Hassani, J.-M. Maillard, J-A. Weil and N. Zenine, Global nilpotent differential operators and the square Ising model, arXive:0812.4931.
[69] H. Au-Yang and B.M. McCoy, Theory of layered Ising models I.Thermodynamics, Phys. Rev. B10 (1974) 886–891.
[70] C.A. Tracy, Universality class of a Fibonacci Ising model, J. Stat. Phys. 51 (1988) 481–490.
[71] D. Fisher, Random transverse field Ising spin chains. Phys. Rev. Letts. 69 (1992) 534–537.
[72] I. Lyberg, The Ising lattice with Brascamp-Kunz boundary conditions and an external magnetic field, arXive:0805.2497
[73] G. Mussardo, Off critical statistical models: factorized scattering theories and the bootstrap program, Physics Reports 218 (1992) 215–382.

11
The Pfaffian solution of the Ising model

The calculation of the free energy of the two-dimensional Ising model in zero magnetic field is one of the most important computations of theoretical physics of the 20th century. It is one of those few problems that is so important that every physicist should have seen the solution during their graduate education. The present chapter is devoted to the explicit presentation of this calculation.

Onsager first computed the free energy of the Ising model at $H = 0$ in 1944. Since then there have been at least four other methods of solution given. We give the chronology of these methods of computation in Table 11.1.

Table 11.1 History of the methods of the computation of the two-dimensional Ising model free energy at $H = 0$.

1944	Onsager [1]	Onsager's algebra
1949,1964	Kaufman, Onsager [2, 3]	Fermionization
	Schultz, Mattis, Lieb [4]	
1952-1963	Kac, Ward [5]	Combinatorial
	Potts, Ward [6]	
	Hurst, Green [7]	
	Kasteleyn [8–10]	
1978	Baxter, Enting [11]	The 399th solution
1982	Baxter [12]	Functional TQ equation

The 1944 paper of Onsager [1] is one of the most inventive computations in 20th century physics. In it Onsager invents the concept of an infinite dimensional loop algebra and uses it to compute the eigenvalues of the transfer matrix of the Ising model. This algebraic method remained almost totally un-understood until 1985 when it was used to analyze the chiral Potts model. We will touch on this method in chapter 15 but there are still aspects of this profound method which remain to be explored.

A partial reason why this 1944 method has not been extensively developed is that in 1949 Kaufman [2] reduced the computation of the partition function to a problem involving free fermions. In 1964 the method was further simplified by Schulz, Mattis and Lieb [4] by the use of the Jordan–Wigner and Bogoliubov transformations. This method is more specialized than Onsager's original method but in compensation for this restriction in generality the same method computes correlation functions as well as partition functions.

The last two methods shown in Table 11.1 rely on the use of what is called the star–triangle equation. The original star–triangle equation was found by Onsager [1], [13] in the Ising model, and in the hands of Baxter has been developed into a method of far-reaching power and generality. These methods will be described in chapters 13 and 14. However, the method lacks the simplicity of the fermionization and combinatorial methods and does not lead by itself to computations of the correlation functions.

In this chapter we will present the combinatorial solution of the Ising model. This method is, if you will, a geometrization of the fermionic method of Kaufman. Kaufman's method relies heavily on group theory and operator methods whereas the combinatorial method uses only a few formulas from linear algebra. Furthermore Kaufman's method uses the technique of transfer matrices which treat the vertical and horizontal interactions in very different ways whereas the combinatorial method explicitly maintains the symmetry between the vertical and horizontal interactions at all states of the computation. Furthermore the combinatorial method is instantly generalizable to inhomogeneous lattices.

We will follow the presentation of the *The Two Dimensional Ising model* [14, chapters 4 and 5]. In section 11.1 we introduce and solve the dimer problem first solved by Kasteleyn in 1960 [8]. In section 11.2 we reduce the Ising model to a dimer problem and solve for the partition function for several different boundary conditions. In section 11.3 we extend the method to compute correlation functions in terms of determinants. The final results and their analysis have already been given in chapter 10.

11.1 Dimers

A dimer is a figure drawn on a lattice which occupies two sites and the bond connecting them. A dimer configuration is said to be allowed if a site is occupied by no more than one dimer.

An allowed close packed dimer configuration is an allowed collection of dimers which occupies all the sites of the underlying lattice

We will often be interested in splitting the full set of dimers into several classes. For example of a square lattice we may specialize the bonds into vertical and horizontal. Let the number of dimer configurations of N types of bonds with n_j bonds of class j be denoted by $g(n_1, \cdots, n_N)$. Then the generating function for the number of configurations is

$$Z(z_1, \cdots, z_N) = \sum_{n_1, \cdots, n_N} g(n_1, \cdots, n_N) \prod_{j=1}^{N} z_j^{n_j} \tag{11.1}$$

where the sum is over all allowed close packed dimer configurations. In this section we will compute an explicit formula for this generating function.

The Pfaffian of a set of numbers a_{jk} with $1 \leq j, k \leq 2N$ satisfying the antisymmetry condition

$$a_{jk} = -a_{kj}, \quad \text{and } a_{kk} = 0 \tag{11.2}$$

is defined as

$$\text{Pf}A = \sum_p{}' \delta_p a_{p_1 p_2} a_{p_3 p_4} \cdots a_{p_{2N-1} p_{2N}} \tag{11.3}$$

where $p_1, \cdots, p_{2N}$ is a permutation of the numbers $1, 2, \cdots, 2N$, the sum $\sum_p'$ is over all permutations which satisfy the restrictions

$$p_{2m-1} < p_{2m} \quad 1 \leq m \leq N \tag{11.4}$$

$$p_{2m-1} < p_{2m+1} \quad 1 \leq m \leq N-1 \tag{11.5}$$

and δ_p, the parity of the permutation, is $+1$ if the permutation p is made up of an even number of transpositions and -1 if the permutation is made up of an odd number of transpositions. We note that, because of (11.2), PfA may be written in the alternative form

$$\text{Pf}A = \frac{1}{N!2^N} \sum_p \delta_p a_{p_1 p_2} a_{p_3 p_4} \cdots a_{p_{2N-1} p_{2N}} \tag{11.6}$$

where the sum is over all permutations. For $2N = 4$ we explicitly have

$$\text{Pf}A = a_{12}a_{34} - a_{13}a_{24} + a_{14}a_{23}. \tag{11.7}$$

For an arbitrary value of N the number of terms in the Pfaffian is

$$(2N-1)(2N-3)(2N-5)\cdots 5 \cdot 3 \cdot 1 = (2N-1)!!. \tag{11.8}$$

The usefulness of the Pfaffian stems from the formula

$$[\text{Pf}A]^2 = \det A \tag{11.9}$$

(where det A is the determinant of the antisymmetric matrix A). Therefore Pfaffians may be studied by the methods of linear algebra.

We will demonstrate in 11.1.1 that the generating function for close packed dimers on a planar lattice with free boundary conditions can be written as the Pfaffian of a suitably chosen matrix A and in 11.1.2 the result is extended to cylindrical lattices. In 11.1.3 we extend our considerations to planar lattices of genus g and on these lattices the generating function can be written as the sum of the Pfaffians of 2^{2g} different matrices. Proofs for arbitrary planar lattices are given and for concreteness the results are specifically illustrated for the square lattice

In 11.1.4 we show that for square lattices with free, cylindrical and periodic boundary conditions the resulting Pfaffians may be explicitly evaluated in terms of finite products using (11.9) and in 11.1.5 the thermodynamic limit is explicitly computed. We conclude in 11.1.6 with a survey of explicit results for other cases such as the triangular lattice and the square lattice with Moebius strip and Klein bottle boundary conditions.

11.1.1 Dimers on lattices with free boundary conditions

We begin the discussion of reducing the generating function of closest packed dimers to the evaluation of a Pfaffian by considering planar lattices with free boundary conditions. It will be most important that this derivation will be valid for any planar lattice and not just for a square or triangular lattice because more general lattices will be needed for the application to the Ising model in section 11.2. Thus, while for concreteness we will use the square lattice to illustrate the method all theorems will be stated in a form which applies to the arbitrary planar lattice. We trust that the reader will see the generalization to arbitrary lattices without the need of the introduction of a general (and cumbersome) notation.

We have previously labeled the sites of a square lattice by giving the number of the row j and column k where $1 \leq j \leq L_v$ and $1 \leq k \leq L_h$. However they may also be labeled by a single index using the map

$$(j,k) \leftrightarrow p = k + (j-1)L_h. \tag{11.10}$$

We shall refer to the arrangement of dimers which occupies the pairs of sites p_1 and p_2, p_3 and $p_4, \cdots, p_{L_vL_h-1}$ and $p_{L_vL_h}$ as

$$C = |p_1, p_2|p_3, p_4| \cdots |p_{L_vL_h-1}, p_{L_vL_h}|. \tag{11.11}$$

For example the dimer configuration shown in Fig. 11.1 is specified by

$$C_0 = |1,2|3,4|5,6| \cdots |L_vL_h - 1, L_vL_h| \tag{11.12}$$

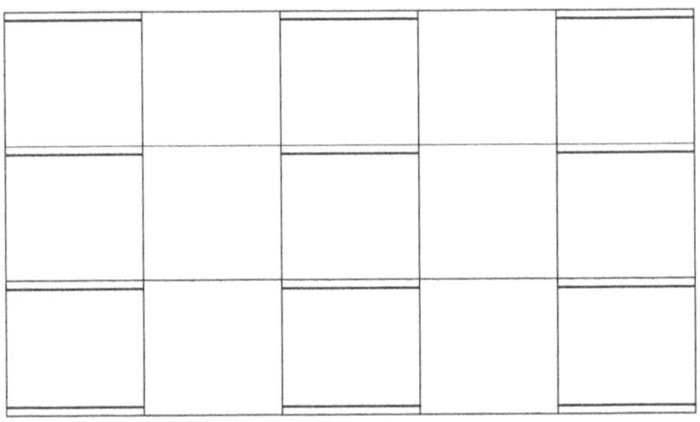

Fig. 11.1 The configuration C_0.

Suppose we let the nonzero elements of a'_{p_1,p_2} be

$$a'_{p_1,p_2} = z_i \tag{11.13}$$

where $p_1 < p_2$ and p_1 and p_2 are connected by a bond of class i. As an example for a square lattice, if the classes of bonds are vertical and horizontal z_v refers to the vertical and z_h refers to the horizontal bonds. It is then obvious that we can write the generating function for closest packed dimers with free boundary conditions $Z^F_{L_v,L_h}$ as

$$Z^F_{L_v,L_h} = \sum_p{}' a'_{p_1p_2} a'_{p_3p_4} \cdots a'_{p_{L_vL_h-1}p_{L_vL_h}} \tag{11.14}$$

where the summation is over all permutations satisfying (11.4) and (11.5). This expression is called a Haffnian but, because there is no connection with a determinant, there is no efficient way to study this expression when L_v and L_h become large.

The expression (11.14) would be a Pfaffian (11.3) if in each term in (11.14) there were an extra factor of δ_p. Then we could use (11.9) to aid in evaluating the generating function for large L_v, L_h. Therefore we will investigate the possibility that instead of (11.13) we can choose the nonzero elements of $a_{p_1 p_2}$ to satisfy

$$a_{p_1 p_2} = s(p_1, p_2) z_i \tag{11.15}$$

where $|s(p_1, p_2)| = 1$ and

$$s(p_1, p_2) = -s(p_2, p_1) \tag{11.16}$$

such that

$$Z^F_{L_v, L_h} = \mathrm{Pf} A = \sum_p{}' \delta_P a_{p_1 p_2} a_{p_3 p_4} \cdots a_{p_{L_v L_h - 1} p_{L_v L_h}}. \tag{11.17}$$

For (11.17) to hold we need to show that if $p^{(1)}$ and $p^{(2)}$ are any two permutations satisfying (11.4) and (11.5) that

$$\delta_{p^{(1)}} s(p_1^{(1)}, p_2^{(1)}) \cdots s(p^{(1)}_{L_v L_h - 1}, p^{(1)}_{L_v L_h}) = \delta_{p^{(2)}} s(p_1^{(2)}, p_2^{(2)}) \cdots s(p^{(2)}_{L_v L_h - 1}, p^{(2)}_{L_v L_h}). \tag{11.18}$$

However, the restrictions (11.4) and (11.5) are awkward, and therefore it is useful to note that if we let $\bar{p}$ be any one of the permutations which, by violating (11.4) and (11.5) may be obtained from a given permutation which does satisfy (11.4) and (11.5) then because of (11.16) we see that (11.18) will hold if we can find one such $\bar{p}^{(1)}$ and $\bar{p}^{(2)}$ such that

$$\delta_{\bar{p}^{(1)}} s(\bar{p}_1^{(1)}, \bar{p}_2^{(1)}) \cdots s(\bar{p}^{(1)}_{L_v L_h - 1}, p^{(1)}_{L_v L_h}) = \delta_{\bar{p}^{(2)}} s(\bar{p}_1^{(2)}, \bar{p}_2^{(2)}) \cdots s(\bar{p}^{(2)}_{L_v L_h - 1}, \bar{p}^{(2)}_{L_v L_h}). \tag{11.19}$$

With these definitions we proceed to show that for any permutation p there is (at least) one related permutation $\bar{p}$ for which δ_p may be computed from geometric considerations.

Consider any two arrangements of dimers specified by permutations $p^{(1)}$ and $p^{(2)}$. Draw the dimers of $p^{(1)}$ on the lattice as dotted lines and the dimers of $p^{(2)}$ as solid lines. The resulting set of figures is referred to as a transition graph. An example is given in Fig. 11.2.

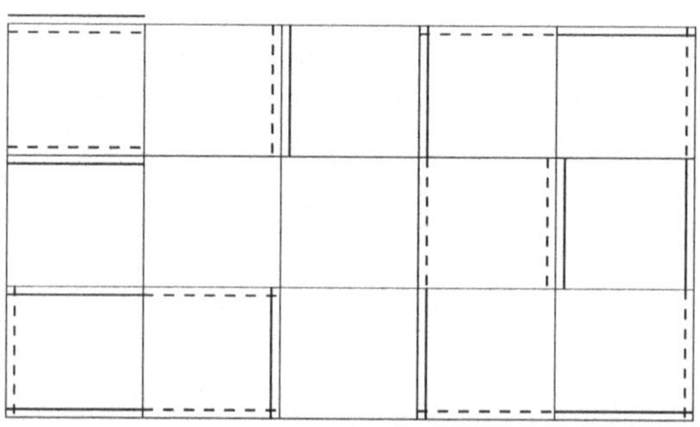

Fig. 11.2 An example of a transition graph.

Every lattice point is the endpoint of one and only one line of each type of dimer and therefore the figures in the transition graph are of two possible types: 1) Two sites connected by a dotted and by a solid line (these figures are called double bonds) and 2) closed polygons with an even number of bonds in which the dotted and solid lines alternate. We call such closed polygons transition cycles because if the bonds of $p^{(1)}$ are permuted clockwise or counterclockwise one step around this cycle they go over to bonds of $p^{(2)}$.

First consider a case in which two permutations differ from each other by only one transition cycle. The preceding discussion showed that (11.18) holds if (11.19) holds where we replace $p^{(1)}$ and $p^{(2)}$ which do obey (11.4) and (11.5) by equivalent permutations $\bar{p}^{(1)}$ and $\bar{p}^{(2)}$ which do not obey (11.4) and (11.5). In particular (11.19) will guarantee (11.18) if we replace $p^{(1)}$ and $p^{(2)}$ by those $\bar{p}^{(1)}$ and $\bar{p}^{(1)}$ that arrange the sites in a clockwise order as we go around the graph. For example consider the transition cycle in Fig. 11.3.

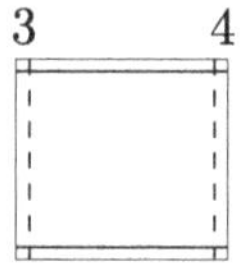

Fig. 11.3 A transition cycle of four vertices.

The permutation $p^{(1)}$ obeying (11.4) and (11.5) which specifies the solid dimers is $|1,2|3,4|$ and the permutation specifying the dashed dimers is $|1,3|2,4|$. However, a permutation $\bar{p}^{(1)}$ which arranges the sites of the configuration $p^{(1)}$ in clockwise order is $|2,1|3,4|$. Similarly a permutation $\bar{p}^{(2)}$ which arranges the sites of $p^{(2)}$ in clockwise order is $|1,3|4,2|$. Thus the shift from $\bar{p}^{(1)}$ to $\bar{p}^{(2)}$ is a cyclic permutation of one step of four objects. More generally the same argument shows that for any transition cycle there are permutations $\bar{p}^{(1)}$ and $\bar{p}^{(2)}$ such that the shift from one to the other is a cyclic permutation of one step of an even number of objects. Hence this shift involves an odd number of transpositions and accordingly if there is only one transition cycle

$$\delta_{\bar{p}^{(1)}} = -\delta_{\bar{p}^{(2)}}. \tag{11.20}$$

In general, if there are t transition cycles, we apply this argument one cycle at a time and find that

$$\delta_{\bar{p}^{(1)}} = (-1)^t \delta_{\bar{p}^{(2)}}. \tag{11.21}$$

Therefore the requirement that the terms associated with the two permutations $p^{(1)}$ and $p^{(2)}$ (or equivalently $\bar{p}^{(1)}$ and $\bar{p}^{(2)}$) have the same sign will be satisfied if for each transition cycle

$$s(\bar{p}_1^{(1)}, \bar{p}_2^{(1)}) s(\bar{p}_3^{(1)}, \bar{p}_4^{(1)}) \cdots s(\bar{p}_{2N-1}^{(1)}, \bar{p}_{2N}^{(1)}) = -s(\bar{p}_2^{(1)}, \bar{p}_3^{(1)}) s(\bar{p}_4^{(1)}, \bar{p}_5^{(1)}) \cdots s(\bar{p}_{2N}^{(1)}, \bar{p}_1^{(1)}). \tag{11.22}$$

Thus for each transition cycle

$$\prod_{k=1}^{2N} s(\bar{p}_k^{(1)}, \bar{p}_{k+1}^{(1)}) = -1 \tag{11.23}$$

or, in other words the product of the factors $s(p_1, p_1)$ in any transition cycle must be -1.

It is possible to satisfy (11.23) with complex numbers [15] but in order to obtain the most geometric picture possible we restrict our attention $s(p_1, p_2) = \pm 1$. Then we can represent $s(p_1, p_2)$ by drawing an arrow on the lattice such that if $s(p_1, p_2) = +1$ an arrow points from site p_1 to site p_2 and $s(p_1, p_2) = -1$ an arrow points from site p_2 to site p_1. This geometric picture is consistent with the antisymmetry of $s(p_1, p_2)$. A lattice on which these arrows are drawn will be called an oriented lattice.

We define the orientation parity of a transition cycle to be $+1$ (-1) if, as we traverse this cycle in either direction the number of arrows pointing in the direction of motion is even (odd). The previous discussion therefore proves

Theorem A: If the orientation parity of every transition cycle is odd, all terms in the Pfaffian will have the same sign.

It is not possible to draw arrows on a general planar lattice so that the orientation parity of every polygon with an even number of sides is odd. However, not all polygons with an even number of sides are transition cycles.

To obtain a characterization of transition cycles on a lattice with free boundary conditions it is useful to introduce the concept of inside and outside. For the square lattice drawn in its "natural" configuration Fig 11.4(a) this concept is obvious. However this "natural" definition of inside and outside is not topologically invariant as is demonstrated by the topologically equivalent form of the square lattice shown in Fig. 11.4(b). Our informal definition can be made mathematically unambiguous [16] but for our purposes this degree of rigor is not necessary.

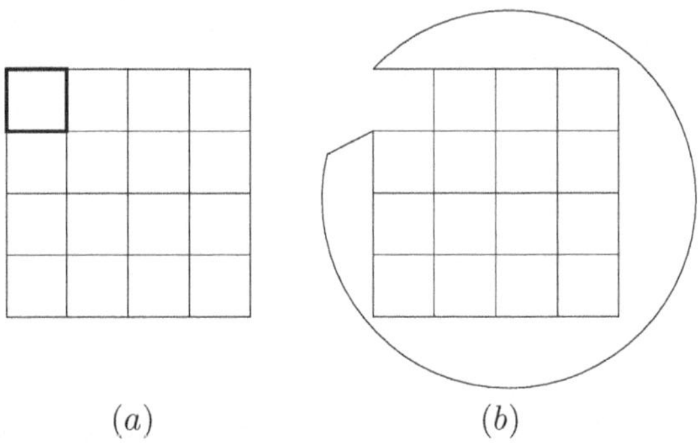

Fig. 11.4 (a) The square lattice in its "natural" configuration. The outlined square has no points or bonds of the lattice in its interior. (b) A lattice which is topologically equivalent to the lattice in (a) where the same outlined square now contains many sites and bonds.

Using this concept of inside and outside we may now easily characterize transition cycles. The only figures that occur in transition graphs are double bonds and transition cycles. Both of these figures have an even number of sites and they completely cover the lattice as illustrated in Fig. 11.2. Therefore we have:

Theorem B: The number of points contained within any transition cycle on a planar lattice is even.

To use the property of transition cycles given in theorem B to choose a set of arrows on a planar lattice to satisfy theorem A we make the following definitions:

An **elementary polygon** is a polygon drawn on the lattice (in its natural configuration) which has no points in its interior.

An elementary polygon is said to be **clockwise odd (even)** if the number of arrows pointing in the clockwise direction is odd (even). This even/odd property will be referred to as the **orientation parity**, and a lattice with arrows drawn on the bonds will be called an **oriented lattice**.

Note that elementary polygons may have either an even or an odd number of sides and that elementary polygons with an odd number of sides which are clockwise odd (even) are counterclockwise even (odd).

With these definitions we now state:

Theorem C: On any planar lattice (in its "natural" configuration) we always choose a set of arrows such that every elementary polygon is clockwise odd.

We will not prove this theorem in general but will rather verify it for the cases of interest. In particular in Fig. 11.5 we exhibit an oriented lattice for the square lattice. It is elementary to verify that the arrows on this lattice satisfy the condition of theorem C. However, we note that this orientation is not unique and that many different orientations exist.

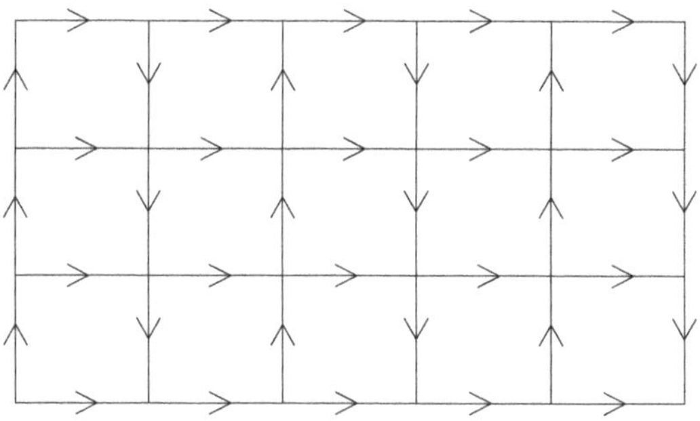

Fig. 11.5 An oriented square lattice.

It is now easy to see that the property of transition cycles of theorem B guarantees that the oriented lattices of Fig. 11.5 satisfy the conditions of theorem A. More generally the proof that theorem B guarantees, for all planar lattices with free boundary conditions, that the orientations specified by theorem C satisfy the conditions of theorem A follows from:

Theorem D: Once arrows have been specified such that every elementary polygon is clockwise odd, then for any polygon the number of clockwise bonds is odd if the number of enclosed lattice points is even and is even if the number of enclosed lattice points is odd.

We prove theorem D by first remarking that any polygon on the lattice is made up of elementary polygons and that theorem C guarantees that theorem D holds on these elementary polygons. Therefore theorem D will follow by induction if we assume it to be true on all polygons made up of n elementary polygons and prove that if Γ_n is one of those polygons then the theorem also holds on the polygon Γ_{n+1} obtained by enlarging Γ_n to include any adjacent elementary polygon Γ_1.

Suppose that Γ_n surrounds p lattice points, contains a clockwise arrows, and has c arrows in common with Γ_1 and that the polygon Γ_1 contains a' clockwise arrows, which from theorem C must be odd. The number of clockwise arrows in Γ_{n+1} is the number of clockwise arrows in Γ_n plus the number of clockwise arrows in Γ_1 minus the number of clockwise arrows lost from Γ_n, and from Γ_1 by omitting common arrows. Now if an arrow on a common bond is clockwise for Γ_n it will be counter clockwise for Γ_1 because Γ_1 is outside of Γ_n. Therefore, the number of clockwise arrows in Γ_{n+1} is $a + a' - c$. Furthermore, the number of enclosed lattice points in Γ_{n+1} is $p + c - 1$, since when we omit c common bonds we must gain $c - 1$ new points in the interior. Now by assumption a is even (odd) if p is odd (even) and also a' is odd. Therefore $a + a' - c$ must be even (odd) if $p + c - 1$ is odd (even). Hence theorem D follows by induction.

The proof of this theorem is illustrated in Fig. 11.6 for the square lattice, but the proof given above is valid for the general case as well.

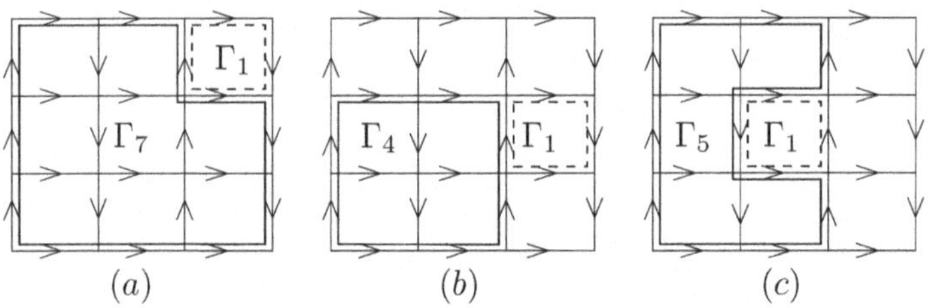

Fig. 11.6 Special cases of the proof of theorem D where on a square lattice an elementary polygon Γ_1 is added to a polygon Γ_n to produce the polygon Γ_{n+1}.

By combining theorems A–D we have now proven that, for any planar lattice with free boundary conditions, the factors $s(p_1, p_2)$ in (11.15) may be chosen such that all terms in (11.17) have the same sign and thus that

$$Z^F_{L_v,L_h} = \pm \mathrm{Pf} A^F. \tag{11.24}$$

The sign $\pm$ must be chosen by the requirement that $Z^F_{L_v,L_h}$ be positive when the weights z_i are positive. For the square (and triangular) lattice this can be done by noting that the configuration C_0 of (11.12) shown in Fig 11.1 has a positive sign. Thus we find for the square lattice with free boundary conditions that

$$Z^F_{L_v,L_h} = \mathrm{Pf} A^F \tag{11.25}$$

where the matrix A is determined from (11.15) and the orientation lattices of Fig. 11.5.

It remains to explicitly write the matrix A^F for the square lattice where we separate the bonds into the classes vertical and horizontal for which we use the weight variables z_v and z_h respectively. To write the matrix we return to the convention of specifying row and column indices by the pair (j,k) Thus from (11.15) and the orientation lattice of Fig. 11.5 we have

$$\begin{aligned} a(j,k;j,k+1) &= -a(j,k+1;j,k) = z_h \\ &\quad \text{for } 1 \leq j \leq L_v,\ 1 \leq k \leq L_h - 1 \qquad (11.26) \\ a(j,k;j+1,k) &= -a(j+1,k;j,k) = (-1)^{k+1} z_v \\ &\quad \text{for } 1 \leq j \leq L_v - 1,\ 1 \leq k \leq L_h. \qquad (11.27) \end{aligned}$$

11.1.2 Dimers on a cylinder

The considerations of the previous subsection can readily be extended from free to cylindrical boundary conditions. For concreteness we consider the case where locally the lattice is square or triangular and impose cylindrical boundary conditions in the horizontal direction. There are then two cases to consider depending on whether L_h is even or odd.

L_h odd

As in the previous section we may consider reducing the generating function of close packed dimers to a Pfaffian by considering transition cycles between two dimer configurations specified by any two permutations $p^{(1)}$ and $p^{(2)}$. However when L_h is odd there can be no transition cycle which completely loops the lattice and therefore the considerations of the previous section guarantee that for the square lattice the generating function for closest packed dimer configurations on the cylindrical lattice is

$$Z^c_{L_v,L_h} = \text{Pf} A^c \qquad (11.28)$$

where the elements of the matrices A^c are given by (11.26) and (11.27) and by the additional nonzero elements

$$a(j,L_h;j,1) = -a(j,1;j,L_h) = z_h \quad \text{for } 1 \leq j \leq L_v \qquad (11.29)$$

L_h even

Now there will be two distinct classes of transition cycles: 1) those which do not loop the cylinder and 2) those which loop the cylinder precisely once. One particularly simple class two transition cycle has no vertical bonds and is shown in Fig. 11.7. There are, in addition, $L_v - 1$ transition cycles which differ from this one only by a vertical translation. These L_v transition cycles will be called elementary transition cycles of class two.

The class one transition cycles will be properly included in the Pfaffian if the matrix A is given as in the previous case by (11.26), (11.27) and (11.29). However, this specification of arrows is not unique and it is easily verified that because all class

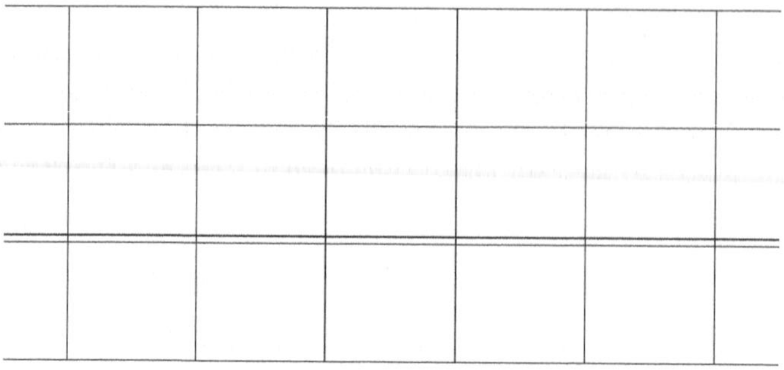

Fig. 11.7 An elementary class two transition cycle on a cylinder.

one transition cycles must have an even number of bonds which connect column L_h with column 1 that the matrix given by (11.26), (11.27) and

$$a(j, L_h; j, 1) = -a(j, 1; j, L_h) = -z_h \quad \text{for } 1 \leq j \leq L_v \tag{11.30}$$

works just as well. However, class two transition cycles have only one bond between column L_h and column 1 and thus the orientation parity of an elementary class 2 transition cycle is negative only with the choice (11.30). Furthermore it is straightforward to demonstrate that if class one transition cycles and elementary class two transition cycles are counted properly that all class two transition cycles are counted properly. Therefore we have shown that (11.28) holds with A^c given by (11.26), (11.27) and (11.30).

11.1.3 Dimers on lattices of genus $g \geq 1$

The final type of boundary conditions we need to investigate are toroidal boundary conditions where the lattice is periodic in both the vertical and horizontal directions. This case proves to be at the same time both harder and easier than free or cyclic boundary conditions. It is easier because the periodic boundary conditions allow a simple evaluation of Pfaffians and determinants by means of Fourier transforms. It is harder because for toroidal boundary conditions the generating function is no longer one Pfaffian but is instead the sum of four Pfaffians.

We will present the four Pfaffian formula in this subsection and explicitly evaluate the related determinants in 11.1.4. However, our methods are more general than merely dealing with a torus which has genus 1 but will also be applicable to boundary conditions which can be embedded on a surface of genus g in which case the resulting sum of Pfaffians will have 2^{2g} terms. This generalization was noted in the original articles [9, 10] which derived the results for the toroidal case but until rather recently [16, 17] they have not been greatly studied because 1) no one has discovered how to explicitly evaluate the Pfaffians and 2) they are of marginal interest in statistical mechanics. However, the identical problem has more recently arisen in contexts other than statistical mechanics where surfaces of higher genus are of physical interest and thus we will state the general results for future reference.

For lattices with toroidal boundary conditions we now may have, in addition to transition cycles which do not loop the lattice, transition cycles which loop the torus v

times in the vertical direction and h times in the horizontal direction where is it easily seen that h and v must be relatively prime (if they are not both zero).

It is a simple matter to extend the considerations of 11.1.1 and 11.1.2 to ensure that all dimer configurations connected by transitions cycles with $v = h = 0$ are counted with the same sign. Indeed, there are many ways to do this and in particular we will need the following four different matrices $A_{\pm\pm'}$ where the elements $a(j, k; j', k')_{\pm,\pm'}$ are given by (11.26) and (11.27) for all bonds which do not connect row L_v with row 1 or column L_h with column 1. For the rest of the elements we have the following

$$\begin{aligned} a_{\pm\pm'}(j, L_h; j, 1) &= -a_{\pm\pm'}(j, 1; j, L_h) = \pm z_h \quad 1 \leq j \leq L_v \\ a_{\pm\pm'}(L_v, k; 1, k) &= -a_{\pm\pm'}(1, k; L_v, k) = \pm'(-1)^{k+1} z_v \quad 1 \leq k \leq L_h. \end{aligned} \tag{11.31}$$

Unlike the case of cylindrical boundary conditions the Pfaffians of none of the matrices $A_{\pm\pm'}$ will count all configurations with the same sign. These signs may be efficiently determined by comparing any particular dimer configuration with the configuration C_0 (11.12) of Fig. 11.1. Transition cycles where one of the dimer configurations is C_0 are called C_0 transition cycles and to correctly compute the generating function it is sufficient to include all C_0 transition cycles with the same sign. It is shown in [9], [10], and [14] that for v and h relatively prime that the Pfaffians of the four matrices $A_{\pm\pm'}$ include C_0 transition cycles with the signs shown in Table 11.2.

Table 11.2 Signs with which C_0 transition cycles are included in the Pfaffian of $A_{\pm\pm'}$.

(h, v)	A_{++}	A_{+-}	A_{-+}	A_{--}
$(0, 0)$	+	+	+	+
(odd,even)	−	−	+	+
(even,odd)	−	+	−	+
(odd,odd)	−	+	+	−

Thus although none of the four matrices includes all C_0 transition cycles with the same sign there is a linear combination of the four Pfaffians which will correctly give the generating function

$$Z^P_{L_v,L_h} = \frac{1}{2}\{-\mathrm{Pf}A_{++} + \mathrm{Pf}A_{+-} + \mathrm{Pf}A_{-+} + \mathrm{Pf}A_{--}\}. \tag{11.32}$$

The result (11.32) can be extended to genus $g > 1$ surfaces there are 2^{2g} matrices analogous to (11.31) with one matrix for each of the 2^{2g} ways that $\pm$ signs can be assigned to the $2g$ loops of the surface and there is a linear combination of Pfaffians which correctly gives the generating function [9, 10]. Explicit examples for $g = 2$ given in [17].

11.1.4 Explicit evaluation of the Pfaffians

Strictly speaking, if all we are interested in is the Ising model we do not need to pursue the dimer problem further, However, because dimers are interesting as a statistical problem in their own right and because the techniques used will be needed for the Ising model we will complete the study of the dimer problem by explicitly evaluating the

Pfaffians by means of the determinental formula (11.9). We proceed in the opposite order from that used in the derivation of the generating functions in terms of the Pfaffians and consider toroidal (periodic) boundary conditions first, and cyclic and free boundary conditions later.

Toroidal (periodic) boundary conditions

The four matrices $A_{\pm,\pm'}$ (with L_h even) are written in a compact form by defining four $N \times N$ matrices, I_N the identity matrix, F_N (with N even) and $J_N^{\pm}$ where

$$F_N = \begin{bmatrix} 1 & 0 & 0 & 0 \cdots & 0 \\ 0 & -1 & 0 & 0 \cdots & 0 \\ 0 & 0 & 1 & 0 \cdots & 0 \\ 0 & 0 & 0 & -1 \cdots & 0 \\ \vdots & \vdots & \vdots & \vdots \quad \vdots & \vdots \\ 0 & 0 & 0 & 0 \cdots & -1 \end{bmatrix}, \quad J_N^{\pm} = \begin{bmatrix} 0 & 1 & 0 \cdots 0 & 0 & \mp 1 \\ -1 & 0 & 1 \cdots 0 & 0 & 0 \\ 0 & -1 & 0 \cdots 0 & 0 & 0 \\ \vdots & \vdots & \vdots \quad \vdots \vdots & \vdots & \vdots \\ \pm 1 & 0 & 0 \cdots 0 & -1 & 0 \end{bmatrix} \tag{11.33}$$

Then the matrices $A_{\pm\pm'}$ are written in a direct product notation as

$$A_{\pm\pm'} = z_h I_{L_v} \otimes J_{L_h}^{\pm} + z_v J_{L_v}^{\pm'} \otimes F_{L_h} \tag{11.34}$$

where the labeling of the basis is such that

$$a_{\pm\pm'}(j,k;j',k') = z_h [I_{L_v}]_{j,j'} [J_{L_h}^{\pm}]_{k,k'} + z_v [J_{L_v}^{\pm'}]_{j,j'} [F_{L_h}]_{k,k'}. \tag{11.35}$$

To proceed further it is convenient to define for even N one more $N \times N$ matrix:

$$T_N = \begin{bmatrix} i & 0 & 0 & 0 \cdots 0 \\ 0 & 1 & 0 & 0 \cdots 0 \\ 0 & 0 & i & 0 \cdots 0 \\ 0 & 0 & 0 & 1 \cdots 0 \\ \vdots & \vdots & \vdots & \vdots \quad \vdots \\ 0 & 0 & 0 & 0 \cdots 1 \end{bmatrix} \tag{11.36}$$

Then, multiplying $A_{\pm\pm'}$ on the right by $I_{L_v} \otimes T_{L_h}$ and on the left by $I_{L_v} \otimes (-iT_{L_h})$, and calling the resulting matrix $\bar{A}_{\pm\pm'}$ we use

$$\det[I_{L_v} \otimes T_{L_h}] = (i)^{L_v L_h/2}, \quad \det[I_{L_v} \otimes (-iT_{L_h})] = (-i)^{L_v L_h/2} \tag{11.37}$$

to obtain

$$\det A_{\pm\pm'} = \det \bar{A}_{\pm\pm'} \tag{11.38}$$

where

$$\bar{A}_{\pm\pm'} = z_h I_{L_v} \otimes J_{L_h}^{\pm} + i z_v J_{L_v}^{\pm'} \otimes I_{L_h}. \tag{11.39}$$

Let $\lambda_k^{(N;\pm)}$ with $k = 1, \cdots, N$ be the N eigenvalues of $J_N^{\pm}$. Then the $L_v L_h$ eigenvalues of $\bar{A}_1$ are $z_h \lambda_k^{(L_h;\pm)} + i z_v \lambda_j^{(L_v;\pm)}$ where $1 \leq j \leq L_v$ and $1 \leq k \leq L_h$. Then

$$\det A_{\pm\pm'} = \prod_{j=1}^{L_v}\prod_{k=1}^{L_h}[z_h\lambda_k^{(L_h;\pm)} + iz_v\lambda_j^{(L_v;\pm')}]. \tag{11.40}$$

It remains to calculate the eigenvalues of $J_N^{\pm}$. We note that $J_N^{\pm}$ has two important properties. First, the matrix elements are of the form

$$a_{j,k} = a_{j-k}. \tag{11.41}$$

Matrices of this form are called Toeplitz. Second, we note for J_N^+ that if the first row is transposed to the bottom of the matrix and the first column is transposed to the extreme right of the matrix the matrix is transformed into itself. Toeplitz matrices of this form are called cyclic. Correspondingly J_N^- is called near cyclic.

The eigenvalues of (near) cyclic matrices are easily evaluated. For the specific case at hand eigenvalue equation $J_N^{\pm}v = \lambda v$ is explicitly written in component form as

$$v_2 \mp v_N = \lambda v_1 \tag{11.42}$$

$$-v_{k-1} + v_{k+1} = \lambda v_k \quad \text{for } 2 \le k \le N-1 \tag{11.43}$$

$$\pm v_1 - v_{N-1} = \lambda v_N. \tag{11.44}$$

We seek a solution of the form

$$v_k = \alpha^k \tag{11.45}$$

and then find that the N equations in (11.42)–(11.44) reduce to the three equations

$$\alpha \mp \alpha^{N-1} = \lambda \tag{11.46}$$

$$-\alpha^{-1} + \alpha = \lambda \tag{11.47}$$

$$\pm\alpha^{-N+1} - \alpha^{-1} = \lambda \tag{11.48}$$

and these three equations are identical if

$$\alpha^N = \pm 1. \tag{11.49}$$

Therefore the N eigenvectors with $1 \le n \le N$ are

$$v_k^{(n;+)} = e^{2\pi ikn/N} \tag{11.50}$$

$$v_k^{(n;-)} = e^{\pi ik(2n-1)/N} \tag{11.51}$$

and the corresponding eigenvalues as

$$\lambda_n^{(N;+)} = e^{2\pi in/N} - e^{-2\pi in/N} = 2i\sin(2\pi n/N) \tag{11.52}$$

$$\lambda_n^{(N;-)} = e^{\pi i(2n-1)/N} - e^{-\pi i(2n-1)/N} = 2i\sin(\pi(2n-1)/N). \tag{11.53}$$

Using (11.53) in (11.40) we obtain

$$\det A_{++} = \prod_{j=1}^{L_v}\prod_{k=1}^{L_h}\left[2iz_h\sin\frac{2\pi k}{L_h} - 2z_v\sin\frac{2\pi j}{L_v}\right] \tag{11.54}$$

$$\det A_{+-} = \prod_{j=1}^{L_v}\prod_{k=1}^{L_h}\left[2iz_h \sin\frac{2\pi k}{L_h} - 2z_v \sin\frac{\pi(2j-1)}{L_v}\right] \tag{11.55}$$

$$\det A_{-+} = \prod_{j=1}^{L_v}\prod_{k=1}^{L_h}\left[2iz_h \sin\frac{\pi(2k-1)}{L_h} - 2z_v \sin\frac{2\pi j}{L_v}\right] \tag{11.56}$$

$$\det A_{--} = \prod_{j=1}^{L_v}\prod_{k=1}^{L_h}\left[2iz_h \sin\frac{\pi(2k-1)}{L_h} - 2z_v \sin\frac{\pi(2j-1)}{L_v}\right]. \tag{11.57}$$

In A_{++} the term in the product with $j = L_v$ and $k = L_h$ vanishes for all z_v and z_h and hence

$$\mathrm{Pf}A_{++'} = (\det A_{++'})^{1/2} = 0. \tag{11.58}$$

Thus using

$$\mathrm{Pf}A = \pm(\det A)^{1/2} \tag{11.59}$$

and (11.55)–(11.57) in (11.32), choosing the signs in (11.59) to make the three nonzero Pfaffians positive for $z_v, z_h > 0$ and using the fact that L_h is even by definition we obtain the final result

$$\begin{aligned} Z^P_{L_v,L_h} = \frac{1}{2}\{ &\prod_{j=1}^{L_v}\prod_{k=1}^{L_h/2}\left[4z_h^2 \sin^2\frac{2\pi k}{L_h} + 4z_v^2 \sin^2\frac{\pi(2j-1)}{L_v}\right]^{1/2} \\ &+\prod_{j=1}^{L_v}\prod_{k=1}^{L_h/2}\left[4z_h^2 \sin^2\frac{\pi(2k-1)}{L_h} + 4z_v^2 \sin^2\frac{2\pi j}{L_v}\right]^{1/2} \\ &+\prod_{j=1}^{L_v}\prod_{k=1}^{L_h/2}\left[4z_h^2 \sin^2\frac{\pi(2k-1)}{L_h} + 4z_v^2 \sin^2\frac{\pi(2j-1)}{L_v}\right]^{1/2} \}. \end{aligned} \tag{11.60}$$

The square roots in this expression are only apparent because, in each of the three terms, the factors in the bracket in the double products are either perfect squares or they occur in pairs.

Free and cylindrical boundary conditions

To explicitly evaluate the Pfaffians of A^F and A^c for the generating functions of free (11.25) and cylindrical (11.28) boundary conditions we define one further $N \times N$ matrix

$$J'_N = \begin{bmatrix} 0 & 1 & 0\cdots & 0 & 0 \\ -1 & 0 & 1\cdots & 0 & 0 \\ 0 & -1 & 0\cdots & 0 & 0 \\ \vdots & \vdots & \vdots \quad \vdots & \vdots & \vdots \\ 0 & 0 & 0\cdots & -1 & 0 \end{bmatrix} \tag{11.61}$$

and write

$$A^c = z_h I_{L_v} \otimes J^-_{L_h} + z_v J'_{L_v} \otimes F_{L_h}, \qquad A^F = z_h I_{L_v} \otimes J'_{L_h} + z_v J'_{L_v} \otimes F_{L_h}. \tag{11.62}$$

As for the case of the previous subsection we compute the determinants of these matrices by first multiplying on the right by $I_{L_v} \otimes T_{L_h}$ and on the left by $I_{L_v} \otimes (-iT_{L_h})$. Thus denoting the eigenvalues of J'_N by $\lambda_k^{(N;')}$ we have

$$\det A^c = \prod_{j=1}^{L_v} \prod_{k=1}^{L_h} (z_h \lambda_k^{(L_h;-)} + i z_v \lambda_j^{(L_v;')}), \quad \det A^F = \prod_{j=1}^{L_v} \prod_{k=1}^{L_h} (z_h \lambda_k^{(L_h;')} + i z_v \lambda_j^{(L_v;')}). \tag{11.63}$$

The eigenvalues of J'_N are obtained by writing the eigenvalue equation in component form

$$\begin{aligned} v_2 &= \lambda v_1 \\ -v_{k-1} + v_{k+1} &= \lambda v_k \quad \text{for } 2 \leq k \leq N-1 \\ -v_{N-1} &= \lambda v_N \end{aligned} \tag{11.64}$$

which is equivalent to

$$-v_{k-1} + v_{k+1} = \lambda v_k \quad \text{for } 1 \leq k \leq N \tag{11.65}$$

with the boundary conditions

$$v_0 = v_{N+1} = 0. \tag{11.66}$$

The most general solution to the second-order difference equation (11.65) is

$$v_k = A\alpha_+^k + B\alpha_-^k \tag{11.67}$$

where α_+ and α_- satisfy

$$\alpha - \alpha^{-1} = \lambda. \tag{11.68}$$

Therefore

$$\alpha_\pm = \frac{1}{2}[\lambda \pm (\lambda^2 + 4)^{1/2}] \tag{11.69}$$

and in particular

$$\alpha_+ \alpha_- = -1. \tag{11.70}$$

To satisfy the boundary conditions (11.66) we need

$$v_0 = A + B = 0 \tag{11.71}$$

$$v_{N+1} = A\alpha_+^{N+1} + B\alpha_-^{N+1} = 0. \tag{11.72}$$

From (11.70), (11.71) and (11.72) we have

$$\alpha_+^{2(N+1)} = (-1)^{N+1} \tag{11.73}$$

and thus we find for $1 \leq n \leq 2(N+1)$

$$\alpha_\pm = \begin{cases} \mp \exp \frac{\mp \pi i (2n-1)}{2(N+1)} & N \text{ even} \\ \mp \exp \frac{\mp \pi i n}{N+1} & N \text{ odd.} \end{cases} \tag{11.74}$$

so the eigenvalues $\lambda_n^{N;'}$ are contained in the set for $1 \leq n \leq 2(N+1)$

$$2i\sin\frac{\pi(2n-1)}{2(N+1)}, \text{ for } N \text{ even}, \quad 2i\sin\frac{\pi n}{N+1}, \text{ for } N \text{ odd}. \tag{11.75}$$

From (11.69) we see that each $\lambda \neq 2i$ has two distinct eigenvectors. Each eigenvalue is counted twice in the set (11.75) and therefore we extract a nondegenerate set by setting for $1 \leq n' \leq N+1$

$$n = \begin{cases} n' + \frac{N+2}{2}, & N \text{ even} \\ n' + \frac{N+1}{2}, & N \text{ odd}. \end{cases} \tag{11.76}$$

However, if $n' = N+1$ then $\lambda = -2i$ and this value of λ has only the trivial eigenvector $v_k = 0$. Therefore we find the desired result

$$\lambda_n^{N;'} = 2i\cos\frac{\pi n}{N+1}, \quad 1 \leq n \leq N. \tag{11.77}$$

We now use (11.53) and (11.77) in (11.63) and choose the sign of the square root to make the Pfaffians in (11.25) and (11.28) positive to obtain the desired results for L_v and L_h even

$$Z^c_{L_v,L_h} = [\det \mathrm{A}^c]^{1/2} = \prod_{k=1}^{L_h/2}\prod_{j=1}^{L_v/2}\left[4z_h^2\sin^2\frac{\pi(2k-1)}{L_h} + z_v^2\cos^2\frac{\pi j}{L_v+1}\right] \tag{11.78}$$

$$Z^F_{L_v,L_h} = [\det \mathrm{A}^c]^{1/2} = \prod_{k=1}^{L_h/2}\prod_{j=1}^{L_v/2}\left[4z_h^2\cos^2\frac{\pi k}{L_h+1} + z_v^2\cos^2\frac{\pi j}{L_v+1}\right] \tag{11.79}$$

and for L_v odd and L_h even

$$Z^c_{L_v,L_h} = \prod_{k=1}^{L_h/2} 2z_h\sin\frac{\pi(2k-1)}{L_h}\prod_{j=1}^{(L_v-1)/2}\left[4z_h^2\sin^2\frac{\pi(2k-1)}{L_h} + z_v^2\cos^2\frac{\pi j}{L_v+1}\right] \tag{11.80}$$

$$Z^F_{L_v,L_h} = \prod_{k=1}^{L_h/2} 2z_h\cos\frac{\pi k}{L_h+1}\prod_{j=1}^{(L_v-1)/2}\left[4z_h^2\cos^2\frac{\pi k}{L_h+1} + z_v^2\cos^2\frac{\pi j}{L_v+1}\right]. \tag{11.81}$$

11.1.5 Thermodynamic limit

It remains to compute the thermodynamic limit $L_v, L_h \to \infty$ of the dimer generating functions computed above. The products may all be written as the exponential of the sum of the logarithms and to leading order the sums may be replaced by integrals in the thermodynamic limit. The result is the same for all boundary conditions and thus we find that

$$
\begin{aligned}
f_d^{SQ} &= \lim_{\substack{L_v \to \infty \\ L_h \to \infty}} (L_v L_h)^{-1} \ln Z^F_{L_v, L_h} \\
&= \lim_{\substack{L_v \to \infty \\ L_h \to \infty}} (L_v L_h)^{-1} \ln Z^c_{L_v, L_h} = \lim_{\substack{L_v \to \infty \\ L_h \to \infty}} (L_v L_h)^{-1} \ln Z^P_{L_v, L_h} \\
&= (2\pi)^{-2} \int_0^{\pi} d\theta_1 \int_0^{\pi} d\theta_2 \ln[4(z_v^2 \sin^2 \theta_1 + z_h^2 \sin^2 \theta_2]. \qquad (11.82)
\end{aligned}
$$

However, if we expand these generating functions beyond leading order they are no longer equal. In particular, the generating function for toroidal boundary conditions, which on the lattice consists of three separate terms, has been more accurately expanded [18] as

$$
Z^P_{L_v, L_h} = \frac{1}{2} \left[\left(\frac{\theta_2(0|\tau)}{\eta(\tau)} \right)^2 + \left(\frac{\theta_3(0|\tau)}{\eta(\tau)} \right)^2 + \left(\frac{\theta_4(0|\tau)}{\eta(\tau)} \right)^2 \right] \exp(f_d^{SQ} L_v L_h) \qquad (11.83)
$$

where $\eta(\tau)$ and the theta functions of zero argument $\theta_i(0|\tau)$ are defined by (10.37)–(10.40), the modulus is

$$
\tau = i L_v z_h / L_h z_v \qquad (11.84)
$$

and multiplicative corrections are of order $(\ln L_v L_h)^3 / L_v L_h$.

11.1.6 Other lattices and boundary conditions

We have illustrated the general theory of close packed dimer statistics with the explicit computation for the square lattice. However, many other problems have been solved, and we conclude this section by surveying the results.

Triangular lattice

The Pfaffian for the triangular lattice with free boundary conditions is obtained from the orientation lattice of Fig. 11.8.

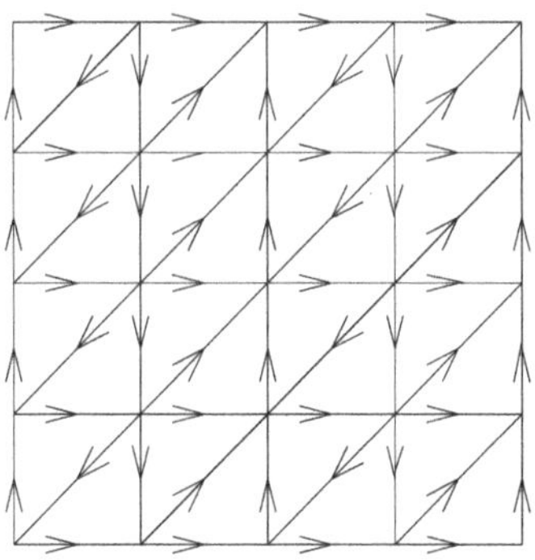

Fig. 11.8 The orientation lattice for the triangular lattice with free boundary conditions.

Cylindrical and toroidal boundary conditions are treated exactly as for the square lattice. The generating function for the toroidal lattice is computed to be [19]

$$
\begin{aligned}
&Z^{PT}_{L_v,L_h} = \\
&\frac{1}{2}\{-\prod_{j=1}^{L_v}\prod_{k=1}^{L_h/2}\left[4z_h^2\sin^2\frac{2\pi k}{L_h}+4z_v^2\sin^2\frac{2\pi j}{L_v}+4z_d^2\cos^2\left(\frac{2\pi k}{L_h}+\frac{2\pi j}{L_v}\right)\right]^{1/2} \\
&+\prod_{j=1}^{L_v}\prod_{k=1}^{L_h/2}\left[4z_h^2\sin^2\frac{2\pi k}{L_h}+4z_v^2\sin^2\frac{\pi(2j-1)}{L_v}+4z_d^2\cos^2\left(\frac{2\pi k}{L_h}+\frac{\pi(2j-1)}{L_v}\right)\right]^{1/2} \\
&+\prod_{j=1}^{L_v}\prod_{k=1}^{L_h/2}\left[4z_h^2\sin^2\frac{\pi(2k-1)}{L_h}+4z_v^2\sin^2\frac{2\pi j}{L_v}+4z_d^2\cos^2\left(\frac{\pi(2k-1)}{L_h}+\frac{2\pi j}{L_v}\right)\right]^{1/2} \\
&+\prod_{j=1}^{L_v}\prod_{k=1}^{L_h/2}\left[4z_h^2\sin^2\frac{\pi(2k-1)}{L_h}+4z_v^2\sin^2\frac{\pi(2j-1)}{L_v}\right. \\
&\qquad\left.+4z_d^2\cos^2\left(\frac{\pi(2k-1)}{L_h}+\frac{\pi(2j-1)}{L_v}\right)\right]^{1/2}\}. \qquad (11.85)
\end{aligned}
$$

This differs from the corresponding result for the square lattice (11.60) in that now all four Pfaffians are nonzero.

In the thermodynamic limit the leading term in the generating function is the same for all boundary conditions and thus for the triangular lattice

$$
\begin{aligned}
f_d^T &= \lim_{\substack{L_v\to\infty \\ L_h\to\infty}} (L_vL_h)^{-1}\ln Z^F_{L_v,L_h} \\
&= \lim_{\substack{L_v\to\infty \\ L_h\to\infty}} (L_vL_h)^{-1}\ln Z^c_{L_v,L_h} = \lim_{\substack{L_v\to\infty \\ L_h\to\infty}} (L_vL_h)^{-1}\ln Z^P_{L_v,L_h} \\
&= (2\pi)^{-2}\int_0^\pi d\theta_1\int_0^\pi d\theta_2 \ln[4(z_v^2\sin^2\theta_1+z_h^2\sin^2\theta_2+z_d^2\cos(\theta_1+\theta_2)]. \quad (11.86)
\end{aligned}
$$

A more accurate expansion for $L_v, L_h \to \infty$ for toroidal boundary conditions with $z_v, z_h, z_d > 0$ is

$$
Z^T_{L_v,L_h} = e^{f_b^T L_v L_h}(1+O(e^{\text{const}L_vL_h})) \qquad (11.87)
$$

which does not have the theta function terms of the result (11.83) for the square lattice with $z_d = 0$.

Moebius strips and Klein bottles

The discussion of boundary conditions of the previous sections can be extended to lattices on nonorientable surfaces [16].

The Moebius strip has boundary conditions twisted as shown on the left in Fig. 11.9 where the horizontal bonds at the end of the lattice are identified as shown. A Klein bottle is obtained by imposing periodic boundary conditions on the free boundary of the Moebius strip as shown on the right in Fig. 11.9.

The generating function for dimer configurations on both these lattices is given as a single Pfaffian [16, 20]. The Pfaffians may be explicitly evaluated in (60) of [20] as

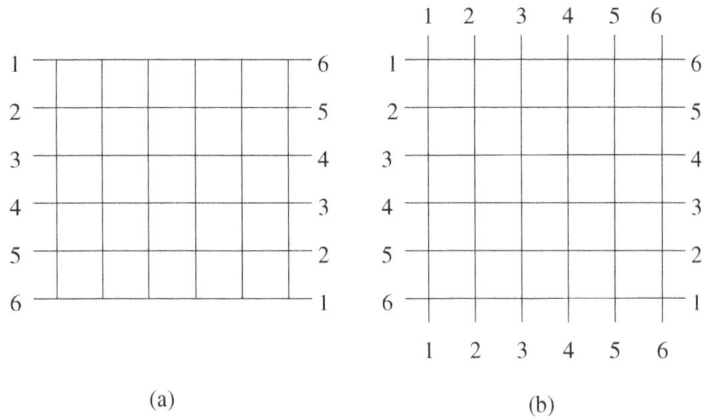

Fig. 11.9 Moebius strip (a) and Klein bottle (b) boundary conditions on a square lattice.

$$Z_{L_v,L_h} = z_h^{L_v L_h/2}\mathrm{Re}\left[(1-i)\prod_{j=1}^{[(L_v+1)/2]}\prod_{k=1}^{L_h}\left(2i(-1)^{L_v/2+j+1}\sin\frac{\pi(4k-1)}{2L_h}+X_j\right)\right] \tag{11.88}$$

where $[(L_v+1)/2]$ is the integer part and

$$X_j = \begin{cases} 2(z_v/z_h)\cos\frac{\pi j}{L_v+1} & \text{for the Moebius strip} \\ 2(z_v/z_h)\sin\frac{\pi(2j-1)}{L_v} & \text{for the Klein bottle.} \end{cases} \tag{11.89}$$

11.2 The Ising partition function

We now turn our attention to the topic of principal interest, the computation of the partition function for the Ising model at $H = 0$ defined by the interaction energy

$$\mathcal{E} = -E^h\sum_{j=1}^{L_v}\sum_{k=1}^{L_h}\sigma_{j,k}\sigma_{j,k+1} - E^v\sum_{j=1}^{L_v}\sum_{k=1}^{L_h}\sigma_{j,k}\sigma_{j+1,k} \tag{11.90}$$

where $\sigma_{j,k} = \pm 1$. We will solve this problem by reducing it to the computation of the generating function of dimers on a related "counting" lattice. We treat the case of toroidal and cylindrical boundary conditions separately and for cylindrical boundary conditions we will be able to let an external magnetic field interact with the boundary row of spins.

11.2.1 Toroidal (periodic) boundary conditions

We impose periodic boundary conditions by setting $\sigma_{j,L_h+1} \equiv \sigma_{j,k}$ and $\sigma_{L_v+1,k} \equiv \sigma_{1,k}$. and write the partition function as

$$\begin{aligned} Z^{IP}_{L_v,L_h} &= \sum_{\sigma=\pm 1} e^{-\beta\mathcal{E}} = \sum_{\sigma=\pm 1}\exp\left[\beta E^h\sum_{j=1}^{L_v}\sum_{k=1}^{L_h}\sigma_{j,k}\sigma_{j,k+1} + \beta E^v\sum_{j=1}^{L_v}\sum_{k=1}^{L_h}\sigma_{j,k}\sigma_{j+1,k}\right] \\ &= \sum_{\sigma=\pm 1}\left[\prod_{j=1}^{L_v}\prod_{k=1}^{L_h} e^{\beta E^h\sigma_{j,k}\sigma_{j,k+1}}\right]\left[\prod_{j=1}^{L_v}\prod_{k=1}^{L_h} e^{\beta E^v\sigma_{j,k}\sigma_{j+1,k}}\right]. \end{aligned} \tag{11.91}$$

Now, since σ may only take on the values ± 1, we have

$$e^{\beta E \sigma\sigma'} = \begin{cases} e^{\beta E} & \text{if } \sigma\sigma' = 1 \\ e^{-\beta E} & \text{if } \sigma\sigma' = -1 \end{cases} = \cosh \beta E + \sigma\sigma' \sinh \beta E. \tag{11.92}$$

Therefore we may write the partition function as

$$Z^{IP}_{L_v,L_h} = (\cosh \beta E^h \cosh \beta E^v)^{L_v L_h} \sum_{\sigma=\pm 1} \prod_{j=1}^{L_v} \prod_{k=1}^{L_h} (1 + z_h \sigma_{j,k}\sigma_{j,k+1})(1 + z_v \sigma_{j,k}\sigma_{j+1,k}) \tag{11.93}$$

where

$$z_v = \tanh \beta E^v \quad \text{and} \quad z_h = \tanh \beta E^h. \tag{11.94}$$

We now expand the products over j and k. Any term with a factor $\sigma_{j,k}$ or $\sigma^3_{j,k}$ vanishes when the sum over $\sigma_{j.k} = \pm 1$ is taken. For all other terms we use $\sigma^2_{j,k} = \sigma^4_{j,k} = 1$ and

$$\sum_{\text{all } \sigma=\pm 1} 1 = 2^{L_v L_h} \tag{11.95}$$

and thus we find

$$Z^{IP}_{L_v,L_h} = (2 \cosh \beta E^h \cosh \beta E^v)^{L_v L_h} \sum_{p,q} N_{p,q} z_h^p z_v^q \tag{11.96}$$

where $N_{p,q}$ is the number of configurations with the following properties: i) each bond between nearest neighbors can be used at most once; ii) an even number of bonds terminate at each vertex; iii) each figure has p horizontal and q vertical bonds. An example of such a figure on a portion of a square lattice is given in Fig. 11.10.

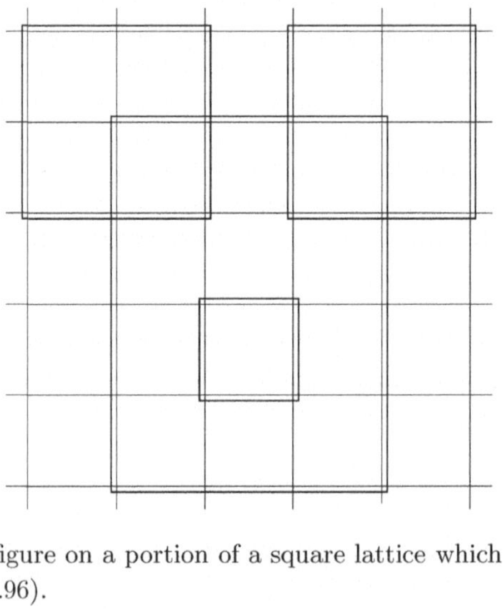

Fig. 11.10 A polygon figure on a portion of a square lattice which contributes to the Ising partition function of (11.96).

We evaluate this generating function by mapping it to a problem of close packed dimers. This is done by replacing each site of the square lattice where four bonds intersect by the cluster of six sites shown in Fig. 11.11. We call this dimer lattice the "counting lattice".

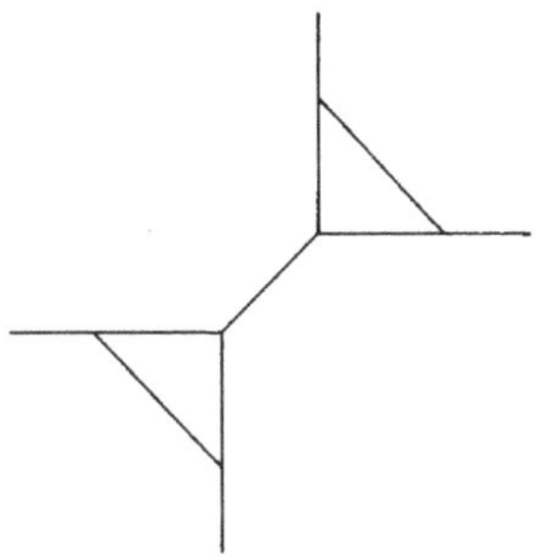

Fig. 11.11 The six-site cluster used to convert the Ising problem into a dimer problem.

There is a one to one correspondence between closed polygon configurations on the Ising lattice and close packed dimers on the counting lattice as is verified by the map of Fig. 11.12.

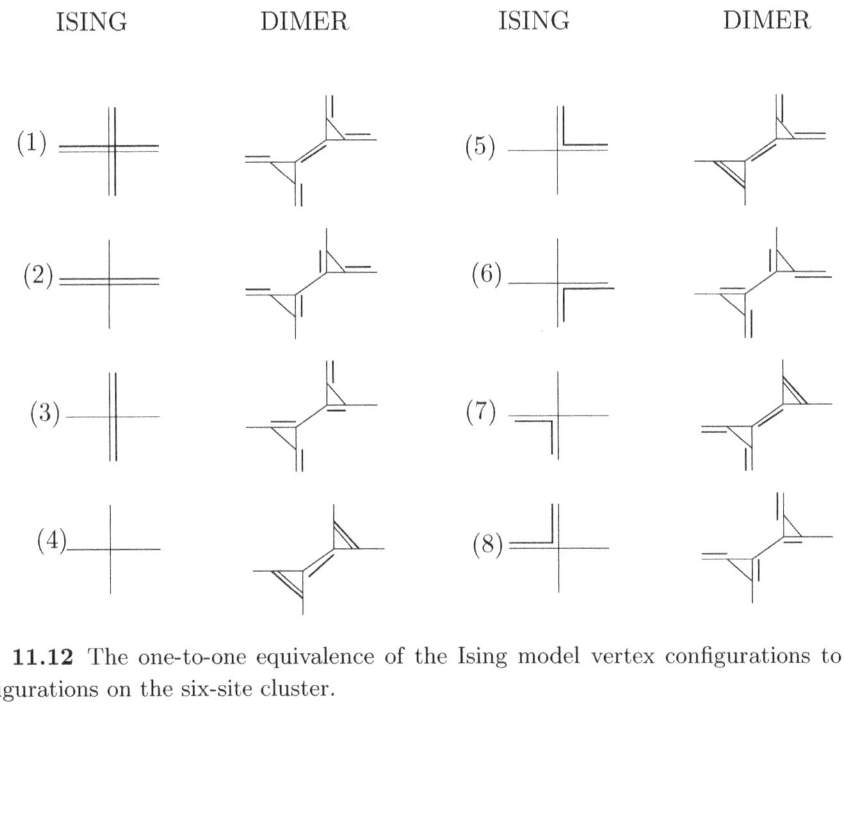

Fig. 11.12 The one-to-one equivalence of the Ising model vertex configurations to dimer configurations on the six-site cluster.

Therefore if the weights on the six internal sites of the dimer counting lattice are set equal to one and the weights of the original horizontal and vertical bonds are called z_h and z_v then the generating function for close packed dimers with these weights on the counting lattice is equal to the generation function for polygons on the Ising lattice. Hence we have reduced the computation of the partition function of the Ising model to a problem in dimer statistics.

This dimer problem is solved by the methods introduced in the previous section. Because we are studying periodic boundary conditions the generating function is given by the linear combination of four Pfaffians

$$Z^{IP}_{L_v,L_h} = \frac{1}{2}\{-\mathrm{Pf}\bar{A}^{IP}_{++} + \mathrm{Pf}\bar{A}^{IP}_{+-} + \mathrm{Pf}\bar{A}^{IP}_{-+} + \mathrm{Pf}\bar{A}^{IP}_{--}\} \qquad (11.97)$$

where the matrices $\bar{A}^{IP}_{\pm\pm'}$ are $6L_vL_h \times 6L_vL_h$ matrices specified in the interior by the oriented lattice of Fig. 11.13.

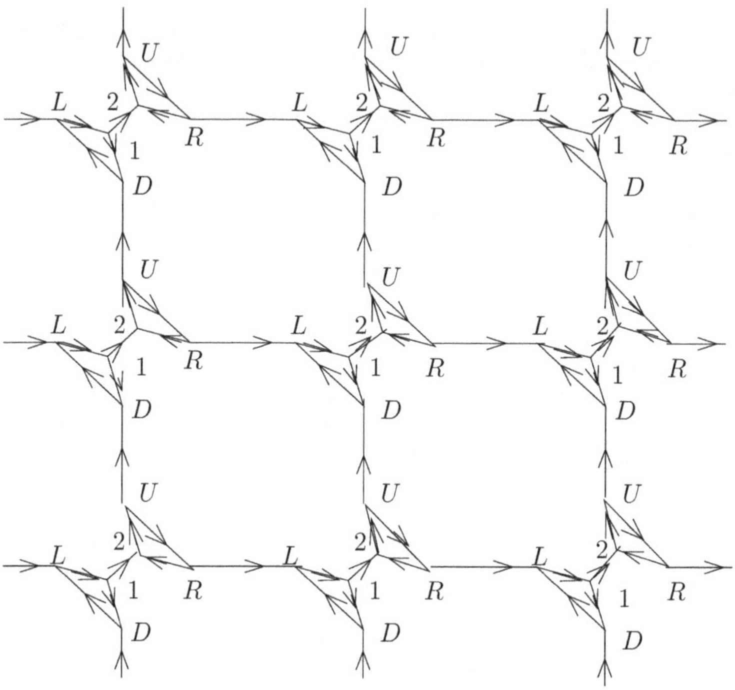

Fig. 11.13 A portion of the dimer counting lattice with six-site clusters. The bonds internal to the six-site clusters have weight 1, the vertical bonds between U and D sites are z_v and the horizontal bonds between R and L sites are z_h.

We label the elements of the matrix by the site (j,k) of the original Ising lattice and the six indices $R, L, U, D, 1, 2$ for the six sites of the cluster as indicated in Fig. 11.13. Thus we have

$$\bar{A}^{IP}(j,k;j,k)_{\pm\pm'} = \begin{array}{c} R \\ L \\ U \\ D \\ 1 \\ 2 \end{array} \begin{bmatrix} 0 & 0 & -1 & 0 & 0 & 1 \\ 0 & 0 & 0 & -1 & 1 & 0 \\ 1 & 0 & 0 & 0 & 0 & -1 \\ 0 & 1 & 0 & 0 & -1 & 0 \\ 0 & -1 & 0 & 1 & 0 & 1 \\ -1 & 0 & 1 & 0 & -1 & 0 \end{bmatrix} \quad \text{for} \quad \begin{array}{l} 1 \le j \le L_v \\ 1 \le k \le L_h \end{array} \tag{11.98}$$

$$\begin{aligned} \bar{A}^{IP}(j,k;j,k+1)_{\pm\pm'} &= -(\bar{A}^{IP})^T(j,k+1;j,k)_{\pm\pm'} \\ &= \begin{array}{c} R \\ L \\ U \\ D \\ 1 \\ 2 \end{array} \begin{bmatrix} 0 & z_h & 0 & 0 & 0 & 0 \\ 0 & 0 & 0 & 0 & 0 & 0 \\ 0 & 0 & 0 & 0 & 0 & 0 \\ 0 & 0 & 0 & 0 & 0 & 0 \\ 0 & 0 & 0 & 0 & 0 & 0 \\ 0 & 0 & 0 & 0 & 0 & 0 \end{bmatrix} \quad \text{for} \quad \begin{array}{l} 1 \le j \le L_v \\ 1 \le k \le L_h - 1 \end{array} \end{aligned} \tag{11.99}$$

$$\begin{aligned} \bar{A}^{IP}(j,k;j+1,k)_{\pm\pm'} &= -(\bar{A}^{IP})^T(j+1,k;j,k)_{\pm\pm'} \\ &= \begin{array}{c} R \\ L \\ U \\ D \\ 1 \\ 2 \end{array} \begin{bmatrix} 0 & 0 & 0 & 0 & 0 & 0 \\ 0 & 0 & 0 & 0 & 0 & 0 \\ 0 & 0 & 0 & z_v & 0 & 0 \\ 0 & 0 & 0 & 0 & 0 & 0 \\ 0 & 0 & 0 & 0 & 0 & 0 \\ 0 & 0 & 0 & 0 & 0 & 0 \end{bmatrix} \quad \text{for} \quad \begin{array}{l} 1 \le j \le L_v - 1 \\ 1 \le k \le L_h \end{array} \end{aligned} \tag{11.100}$$

$$\begin{aligned} \bar{A}^{IP}(j,L_h;j,1)_{\pm\pm'} &= -(\bar{A}^{IP})^T(j,1;j,L_h)_{\pm\pm'} = \pm\bar{A}^{IP}(j,1;j,2)_{\pm\pm'},\ 1 \le j \le L_v \\ \bar{A}^{IP}(L_v,k;1,k)_{\pm\pm'} &= -(\bar{A}^{IP})^T(1,k;L_v,k)_{\pm\pm'} = \pm'\bar{A}^{IP}(1,k;2,k)_{\pm\pm'},\ 1 \le k \le L_h. \end{aligned} \tag{11.101}$$

We will compute the Pfaffians of $\bar{A}^{IP}_{\pm\pm'}$ by using

$$\mathrm{Pf}\bar{A}^{IP}_{\pm\pm'} = \pm(\det\bar{A}^{IP}_{\pm\pm'})^{1/2}. \tag{11.102}$$

The dimensions of the matrices $\bar{A}^{IP}_{\pm\pm'}$ are $6L_vL_h \times 6L_vL_h$. However, it is trivially possible to find $4L_vL_h \times 4L_vL_h$ matrices $A^{I}_{\pm\pm'}$ such that

$$\mathrm{Pf}\bar{A}^{IP}_{\pm\pm'} = \pm(\det A^{IP}_{\pm\pm'})^{1/2}. \tag{11.103}$$

This is done by noticing that if in every 6×6 submatrix $\bar{A}^{IP}(j,k;j'k')_{\pm\pm'}$ we subtract column 1 from column R, add column 1 to column U, add column 2 to column L and subtract column 2 from column D we have

$$\det\bar{A}^{IP}_{\pm\pm'} = \det A^{IP}_{\pm\pm'} \tag{11.104}$$

where

$$A^{IP}(j,k;j,k)_{\pm\pm'} = \begin{matrix} R \\ L \\ U \\ D \end{matrix}\begin{bmatrix} 0 & 1 & -1 & -1 \\ -1 & 0 & 1 & -1 \\ 1 & -1 & 0 & 1 \\ 1 & 1 & -1 & 0 \end{bmatrix} \tag{11.105}$$

and all other elements $A^{IP}(j,k;j',k')_{\pm\pm'}$ are identical to $\bar{A}^{IP}(j,k;j',k')_{\pm\pm'}$ with the rows and columns labeled 1 and 2 removed. The matrix $A^{IP}(j,k;j,k)_{\pm\pm'}$ can be given an interpretation in terms of a nonplanar graph as indicated in Fig 11.14.

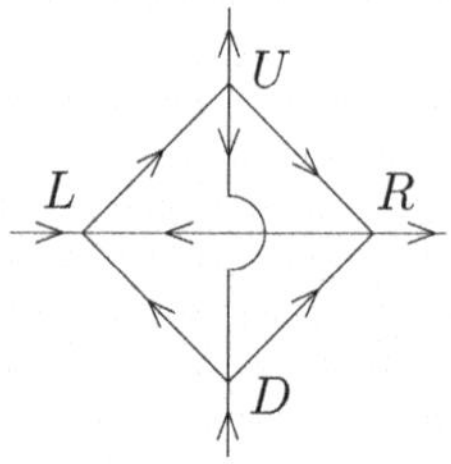

Fig. 11.14 A representation of the matrix $A^{IP}(j,k;j,k)_{\pm\pm'}$ as an oriented nonplanar graph.

We now turn to the evaluation of $\det A^{IP}_{\pm\pm'}$. We do this by first defining

$$H_N^{\pm} = \begin{bmatrix} 0 & 1 & 0 \cdots 0 \\ 0 & 0 & 1 \cdots 0 \\ \vdots & \vdots & \vdots \quad \vdots \\ 0 & 0 & 0 \cdots 1 \\ \pm 1 & 0 & 0 \cdots 0 \end{bmatrix} \tag{11.106}$$

and then writing $A^{IP}_{\pm\pm'}$ in a direct product notation as

$$\begin{aligned} A^{IP}_{\pm\pm'} &= \begin{bmatrix} 0 & 1 & -1 & -1 \\ -1 & 0 & 1 & -1 \\ 1 & -1 & 0 & 1 \\ 1 & 1 & -1 & 0 \end{bmatrix} \otimes I_{L_v} \otimes I_{L_h} \\ &+ \begin{bmatrix} 0 & z_h & 0 & 0 \\ 0 & 0 & 0 & 0 \\ 0 & 0 & 0 & 0 \\ 0 & 0 & 0 & 0 \end{bmatrix} \otimes I_{L_v} \otimes H^{\pm}_{L_h} + \begin{bmatrix} 0 & 0 & 0 & 0 \\ -z_h & 0 & 0 & 0 \\ 0 & 0 & 0 & 0 \\ 0 & 0 & 0 & 0 \end{bmatrix} \otimes I_{L_v} \otimes H^{\pm T}_{L_h} \\ &+ \begin{bmatrix} 0 & 0 & 0 & 0 \\ 0 & 0 & 0 & 0 \\ 0 & 0 & 0 & z_v \\ 0 & 0 & 0 & 0 \end{bmatrix} \otimes H^{\pm'}_{L_v} \otimes I_{L_h} + \begin{bmatrix} 0 & 0 & 0 & 0 \\ 0 & 0 & 0 & 0 \\ 0 & 0 & 0 & 0 \\ 0 & 0 & -z_v & 0 \end{bmatrix} \otimes H^{\pm' T}_{L_v} \otimes I_{L_h}. \end{aligned} \tag{11.107}$$

The matrices H^+ (H^-) are (near) cyclic matrices and therefore their eigenvalues are computed by the methods of section 11.1 to be $e^{2\pi i n/N}$ and $e^{\pi i(2n+1)/N}$ respectively for $n = 1, \cdots, N$. Furthermore the matrices $H_N^{\pm}$ are unitary

$$H^{\pm}H^{\pm T} = 1 \tag{11.108}$$

and therefore the diagonalizing matrix U has the property that if

$$UH_N^{+}U^{-1} = \begin{bmatrix} e^{2\pi i/N} & 0 & \cdots & 0 \\ 0 & e^{4\pi i/N} & \cdots & 0 \\ \vdots & \vdots & & \vdots \\ 0 & 0 & \cdots & 1 \end{bmatrix} \tag{11.109}$$

then

$$UH_N^{+T}U^{-1} = \begin{bmatrix} e^{-2\pi i/N} & 0 & \cdots & 0 \\ 0 & e^{-4\pi i/N} & \cdots & 0 \\ \vdots & \vdots & & \vdots \\ 0 & 0 & \cdots & 1 \end{bmatrix} \tag{11.110}$$

and similarly if

$$UH_N^{-}U^{-1} = \begin{bmatrix} e^{\pi i/N} & 0 & \cdots & 0 \\ 0 & e^{3\pi i/N} & \cdots & 0 \\ \vdots & \vdots & \vdots & \vdots \\ 0 & 0 & \cdots & -1 \end{bmatrix} \tag{11.111}$$

then

$$UH_N^{-T}U^{-1} = \begin{bmatrix} e^{-\pi i/N} & 0 & \cdots & 0 \\ 0 & e^{-3\pi i/N} & \cdots & 0 \\ \vdots & \vdots & \vdots & \vdots \\ 0 & 0 & \cdots & -1 \end{bmatrix}. \tag{11.112}$$

Therefore we evaluate $\det A^{IP}_{\pm\pm'}$ by first transforming $H^{\pm}_{L_v}$ and $H^{\pm}_{L_h}$ into diagonal form by a similarity transformation to obtain

$$\det A^{IP}_{\pm\pm'} = \prod_{\theta_h^{\pm}} \prod_{\theta_v^{\pm'}} \det A^{IP}(\theta_h, \theta_v) \tag{11.113}$$

where the products over $\theta_k^{+} = 2\pi n/L_k$ and $\theta_k^{-} = \pi(2n-1)/L_k$ for $k = v, h$ and

$$A^{IP}(\theta_h, \theta_v) = \begin{bmatrix} 0 & 1 + z_h e^{i\theta_h} & -1 & -1 \\ -1 - z_h e^{-i\theta_h} & 0 & 1 & -1 \\ 1 & -1 & 0 & 1 + z_v e^{i\theta_v} \\ 1 & 1 & -1 - z_v e^{-i\theta_v} & 0 \end{bmatrix}. \tag{11.114}$$

It is now straightforward to compute the determinant of the 4×4 matrix $A^{IP}(\theta_h, \theta_v)$ and thus we find

$$\det A^{I}_{\pm\pm'} = \prod_{\theta_h^{\pm}} \prod_{\theta_v^{\pm'}} [(1+z_h^2)(1+z_v^2) - 2z_h(1-z_v^2)\cos\theta_h - 2z_v(1-z_h^2)\cos\theta_v]. \tag{11.115}$$

Using this in (11.97) thus completes the exact evaluation of the partition function of the two-dimensional Ising model at $H = 0$ first obtained by Kaufman [2] which was presented in chapter 10. For real z_v and z_h the factors in the product (11.115) are nonnegative and bounded below by

$$(|z_v z_h| + |z_v| + |z_h| - 1)^2. \tag{11.116}$$

When the temperature is chosen such that

$$z_v z_h + z_v + z_h - 1 = 0 \tag{11.117}$$

there is a vanishing factor in the product over θ_h^+, θ_v^+. For the value of the temperature T determined by (11.117) we have

$$\det A_{++}^{IP} = 0 \tag{11.118}$$

and the expression for the partition function reduced to the sum of three terms just as it did for dimers on the square lattice.

The thermodynamic limit can now be explicitly taken and thus we find the famous result of Onsager [1] presented in chapter 10 that the free energy is

$$\begin{aligned}
-F_I/k_BT &= \lim_{\substack{L_v\to\infty \\ L_h\to\infty}} (L_v L_h)^{-1} \ln Z_{L_v,L_h}^{IP} \\
&= \ln(2\cosh\beta E^h \cosh\beta E^v) + \frac{1}{2}(2\pi)^{-2}\int_0^{2\pi} d\theta_h \int_0^{2\pi} d\theta_v \ln[(1+z_h^2)(1+z_v^2) \\
&\qquad - 2z_h(1-z_v^2)\cos\theta_h - 2z_v(1-z_h^2)\cos\theta_v] \\
&= \ln 2 + \frac{1}{2}(2\pi)^{-2}\int_0^{2\pi} d\theta_h \int_0^{2\pi} d\theta_v \ln[\cosh 2\beta E^h \cosh 2\beta E^v \\
&\qquad - \sinh 2\beta E^h \cos\theta_h - \sinh 2\beta E^v \cos\theta_v].
\end{aligned} \tag{11.119}$$

This integral fails to be analytic as a function of T at the point determined by (11.117) which is the condition for the critical temperature T_c discussed in chapter 10.

11.2.2 Cylindrical boundary conditions

When, instead of using toroidal boundary conditions, we impose cyclic boundary conditions only in the horizontal direction we can let the boundary row interact with an external magnetic field and the problem is still solvable by dimer methods. Thus we consider

$$\mathcal{E} = -E^h \sum_{j=1}^{L_v}\sum_{k=1}^{L_h} \sigma_{j,k}\sigma_{j,k+1} - E^v \sum_{j=1}^{L_v-1}\sum_{k=1}^{L_h} \sigma_{j,k}\sigma_{j+1,k} - H_b \sum_{k=1}^{L_h} \sigma_{1,k} \tag{11.120}$$

The methods used above for the toroidal lattice reduce this problem to a dimer problem on the counting lattice of Fig. 11.15 where the magnetic field is included by introducing an extra row of spins at $j = 0$ with horizontal weights of unity and vertical weights of

$$z = \tanh\beta H_b. \tag{11.121}$$

Thus we find that the partition function is given as a single Pfaffian

$$Z^{Ic}_{l_v,L_h} = (2\cosh\beta E^h)^{L_vL_h}(\cosh\beta E^v)^{L_h(L_v-1)}\mathrm{Pf}A^{Ic} \tag{11.122}$$

where the matrix is determined from Fig. 11.15 with the horizontal arrows connecting columns L_h and 1 in the direction opposite to all other horizontal arrows. The evaluation of this Pfaffian was carried out in [21]. The results were presented in chapter 10.

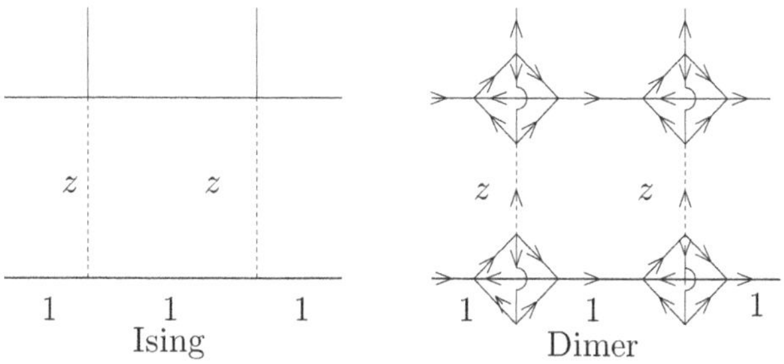

Fig. 11.15 The counting lattice for the Ising model on a cylinder with a magnetic field on the boundary row $j = 1$ and the corresponding oriented dimer lattice with four-site clusters. The magnetic field is included by introducing an extra row of spins at $j = 0$ with horizontal weights of unity and vertical weights of $z = \tanh\beta H_b$.

11.3 Correlation functions

The great virtue which the combinatorial method as presented above has over the original method of Onsager [1] and over the methods based on the star–triangle equation [11, 12] is that, in addition to computing the partition function, we may also obtain expressions for correlation functions of any number of spins in terms of determinants [22]. We illustrate the method in detail for the correlation of two spins in the same row. We also consider the correlation on the diagonal $\langle\sigma_{N',N'}\sigma_{N,N}\rangle$ and near the boundary of a cylinder.

11.3.1 The correlation $\langle\sigma_{M,N'}\sigma_{M,N}\rangle$

The two-point function $\langle\sigma_{M',N'}\sigma_{M,N}\rangle$ is defined by

$$\langle\sigma_{M',N'}\sigma_{M,N}\rangle = \frac{1}{Z^{Ic}_{L_v,L_h}}\sum_{\sigma=\pm1}\sigma_{M',N'}\sigma_{M,N}e^{-\beta\mathcal{E}} \tag{11.123}$$

where for purposes of illustration we have (temporarily) chosen cylindrical boundary conditions. The exponentials in the numerator of (11.123) may be expanded using (11.92) and thus we find

$$\langle\sigma_{M',N'}\sigma_{M,N}\rangle = (\cosh\beta E^h)^{L_hL_v}(\cosh\beta E^v)^{L_h(L_v-1)}Z^{Ic-1}_{L_v,L_h}$$
$$\sum_{\sigma=\pm1}\sigma_{M',N'}\sigma_{M,N}\prod_{j=1}^{L_v}\prod_{k=1}^{L_h}(1+z_h\sigma_{j,k}\sigma_{j,k+1})\prod_{j=1}^{L_v-1}\prod_{k=1}^{L_h}(1+z_v\sigma_{j,k}\sigma_{j+1,k}) \tag{11.124}$$

The key step is now to show that the numerator in (11.124) can be rewritten as a partition function on a new modified lattice which can also be expressed as a Pfaffian. Connect the spins at M, N' and M, N by *any* path on the lattice we wish. (For example we may consider the straight line path which uses the sites M, k with $k = N'+1, \cdots, N-1$). On each of the sites j, k on the path between $\sigma_{M,N'}$ and $\sigma_{M,N}$ insert the factor of $1 = \sigma_{j,k}^2$ and associate the factors into nearest neighbor pairs as (for the example of the straight line path)

$$\sigma_{M,N'}\sigma_{M,N} = (\sigma_{M,N'}\sigma_{M,N'+1})(\sigma_{M,N'+1}\sigma_{M,N'+2})\cdots(\sigma_{M,N-1}\sigma_{M,N}). \tag{11.125}$$

Then we may multiply (11.125) by the factors $1 + z_v\sigma_{j,k}\sigma_{j+1,k}$ and $1 + z_h\sigma_{j,k}\sigma_{j,k+1}$ in (11.123) which involving the same nearest neighbor pairs using the identity

$$\sigma\sigma'(1 + z\sigma\sigma') = z(1 + z^{-1}\sigma\sigma'). \tag{11.126}$$

Thus for the straight line path we may write the correlation function $\langle\sigma_{M,N'}\sigma_{M,N}\rangle$ as

$$\begin{aligned}\langle\sigma_{M,N'}\sigma_{M,N}\rangle &= (\cosh\beta E^h)^{L_h L_v}(\cosh\beta E^v)^{L_h(L_v-1)} z_h^{N-N'}\\ \frac{1}{Z^{Ic}_{L_v,L_h}} &\times \sum_{\sigma=\pm1}\left[\prod_{k=0}^{N-N'-1}(1 + z_h^{-1}\sigma_{M,k}\sigma_{m,k+1})\right]\\ &\times \left[\prod_{j=1}^{L_v}\prod_{k=1}^{L'_h}(1 + z_h\sigma_{j,k}\sigma_{j,k+1})\right]\left[\prod_{j=1}^{L_v-1}\prod_{k=1}^{L_h}(1 + z_v\sigma_{j,k}\sigma_{j+1,k})\right]\end{aligned} \tag{11.127}$$

where $\prod_{k=1}^{L'_h}$ means that the terms with $j = M$ and $N' \leq k \leq N-1$ which are on the path are omitted.

The expression in the numerator of (11.127) is of the form of a partition function for an Ising counting lattice with the bonds z_k on the path connecting the sites M, N' with M, N replaced by z_h^{-1} as illustrated in Fig. 11.16.

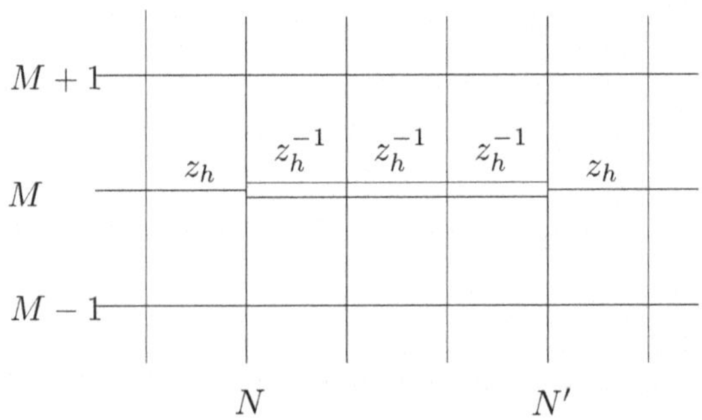

Fig. 11.16 The Ising counting lattice with the bonds z_h on the straight line path connecting sites (M, N) and (M, N') replaced by z_h^{-1}.

It follows that the numerator can be written as the Pfaffian of a matrix $A^{c'}$. Thus we have

$$\langle \sigma_{M,N'} \sigma_{M,N} \rangle = z_h^{N-N'} \mathrm{Pf} A^{c'} / \mathrm{Pf} A^{c} \tag{11.128}$$

where the difference $\delta = A^{c'} - A^{c}$ is given by

$$\delta(0,k;0,k+1) = -\delta^T(0,k+1;0,k) = \begin{array}{c} R \\ L \\ U \\ D \end{array} \begin{bmatrix} 0 & z_h^{-1} - z_h & 0 & 0 \\ 0 & 0 & 0 & 0 \\ 0 & 0 & 0 & 0 \\ 0 & 0 & 0 & 0 \end{bmatrix} \tag{11.129}$$

if $N' \leq k \leq N-1$ and zero otherwise. Thus, if we define y as the $2(N-N') \times 2(N-N')$ submatrix of δ in the subspace where δ does not vanish identically and if we define Q as the $2(N-N') \times 2(N-N')$ submatrix of A^{Ic-1} in this same subspace, we find

$$\begin{aligned} \langle \sigma_{M,N'} \sigma_{M,N} \rangle^2 &= z_h^{2(N-N')} \det(A^{Ic} + \delta)/\det A^{Ic} = z_h^{2(N-N')} \det(1 + A^{Ic-1}\delta) \\ &= z_h^{2(N-N')} \det y \, \det(y^{-1} + Q). \end{aligned} \tag{11.130}$$

Explicitly we see from (11.129)

$$y = \begin{array}{ll} MN' & R \\ MN'+1 & R \\ \vdots & \vdots \\ MN-1 & R \\ MN'+1 & L \\ MN'+2 & L \\ \vdots & \vdots \\ MN & L \end{array} \begin{bmatrix} 0 & 0 \cdots & 0 & y_0 & 0 \ldots & 0 \\ 0 & 0 \cdots & 0 & 0 & y_0 \cdots & 0 \\ \vdots & \vdots \;\; \vdots & \vdots & \vdots & \vdots \;\; \vdots & \vdots \\ 0 & 0 \cdots & 0 & 0 & 0 \cdots & y_0 \\ -y_0 & 0 \ldots & 0 & 0 & 0 \cdots & 0 \\ 0 & -y_0 \cdots & 0 & 0 & 0 \cdots & 0 \\ \vdots & \vdots \;\; \vdots & \vdots & \vdots & \vdots \;\; \vdots & \vdots \\ 0 & 0 \cdots & -y_0 & 0 & 0 \cdots & 0 \end{bmatrix} \tag{11.131}$$

with $y_0 = z_h^{-1} - z_h$. Therefore

$$\det y = (z_h^{-1} - z_h)^{2(N-N')} \tag{11.132}$$

and y^{-1} is trivially computed. Therefore we obtain

$$\langle \sigma_{M,N'} \sigma_{M,N} \rangle^2 = (1 - z_h^2)^{2(N-N')}$$
$$\times \left| \begin{array}{ccc} 0 & \cdots & A^{-1}(M,N';M;N-1)_{RR} \\ A^{-1}(M,N'+1;M,N')_{RR} & \cdots & A^{-1}(M,N'+1,M,N-1)_{RR} \\ \vdots & \vdots & \vdots \\ A^{-1}(M,N-1;M,N')_{RR} & \cdots & 0 \\ A^{-1}(M,N'+1;M,N')_{LR} + (z_h^{-1} - z_h)^{-1} & \cdots & A^{-1}(M,N'+1;M,N-1)_{LR} \\ A^{-1}(M,N'+2;M,N')_{LR} & \cdots & A^{-1}(M,N'+2;M,N-1)_{LR} \\ \vdots & \vdots & \vdots \\ A^{-1}(M,N;M,N')_{LR} & \cdots & A^{-1}(M,N;M,N-1)_{LR} + (z_h^{-1} - z_h)^{-1} \end{array} \right.$$

$$
\left.\begin{array}{ccc}
A^{-1}(M,N';M,N'+1)_{RL} & \cdots & A^{-1}(M,N';M,N)_{RL} \\
-(z_h^{-1}-z_h)^{-1} & & \\
A^{-1}(M,N'+1;M,N'+1)_{RL} & \cdots & A^{-1}(M,N'+1;M,N)_{RL} \\
\vdots & \vdots & \vdots \\
A^{-1}(M,N-1;M,N'+1)_{RL} & \cdots & A^{-1}(M,N-1;M,N)_{RL} \\
 & & -(z_h^{-1}-z_h)^{-1} \\
0 & \cdots & A^{-1}(M,N'+1;M,N)_{LL} \\
A^{-1}(M,N'+2;M,N+1)_{LL} & \cdots & A^{-1}(M,N+2;M,N)_{LL} \\
\vdots & \vdots & \vdots \\
A^{-1}(M,N;M,N'+1)_{LL} & \cdots & 0
\end{array}\right| \qquad (11.133)
$$

where $A^{-1} \equiv A^{Ic-1}$.

It remains to compute the inverse matrix elements needed in (11.133). For arbitrary M, N and N' this is done in [21] and the results have been used to compute the boundary correlations presented in chapter 10. We are most interested here in the case (say $M = [L_v/2]$) where in the thermodynamic limit the row M is infinitely far from the boundaries of the cylinder. In this limit the matrix elements of A^{Ic-1} needed for (11.133) are identical to the corresponding matrix elements of A_{++}^{IP-1}. This matrix is cyclic in both the vertical and horizontal directions and it is easily verified that the inverse matrix elements (in the thermodynamic limit) are

$$
A^{IP-1}(j',k,;j,k) = \frac{1}{2\pi}\int_0^{2\pi} d\theta_h \frac{1}{2\pi}\int_0^{2\pi} d\theta_v e^{i\theta_h(k'-k)+i\theta_v(j'-j)} A^{IP-1}(\theta_h.\theta_v) \qquad (11.134)
$$

where the 4×4 matrix $A^{IP}(\theta_h,\theta_v)$ is given by (11.114). The inverse of $A^{IP}(\theta_h,\theta_v)$ is readily computed as

$$
A^{IP-1}(\theta_h,\theta_v) = \frac{1}{\Delta(\theta_h,\theta_v)} \times
$$
$$
\begin{array}{c} R \\ L \\ U \\ D \end{array}
\begin{bmatrix}
2iz_v\sin\theta_v & b+b^*-abb^* & 2-ab^* & 2-ab \\
-b*-b+a^*b^*b & -2iz_v\sin\theta_v & -2+a^*b^* & 2-a^*b \\
-2+a^*b & 2-ab & -2iz_h\sin\theta_h & a+a^*-aa^*b \\
-2+a^*b^* & -2+ab^* & -a^*-a+a^*ab^* & 2iz_h\sin\theta_h
\end{bmatrix} \qquad (11.135)
$$

where

$$
\begin{aligned}
& a = 1+z_h e^{i\theta_h} \quad b = 1+z_v e^{i\theta_v} \\
& b+b^*-abb^* = 1-z_v^2-z_h(1+z_v^2+2z_v\cos\theta_h)
\end{aligned} \qquad (11.136)
$$

and

$$
\Delta(\theta_h,\theta_v) = (1+z_h^2)(1+z_v^2) - 2z_h(1-z_v^2)\cos\theta_h - 2z_v(1-z_h^2)\cos\theta_v. \qquad (11.137)
$$

Thus we see that, in (11.133),

$$A^{-1}(M,N';M,N)_{RR} = A^{-1}(M,N';M,N)_{LL} = 0 \tag{11.138}$$

so that $\langle\sigma_{M,N'}\sigma_{M,N}\rangle$ reduces to an $(N-N')\times(N-N')$ determinant. Thus, using the RL element of (11.135), carrying out the integral over θ_v, simplifying the result and noting that because of translational invariance the correlation depends only on $N-N'$ we obtain the final result (10.45)–(10.47) that in the thermodynamic limit in the interior of the lattice

$$\langle\sigma_{0,0}\sigma_{0,N}\rangle = \begin{vmatrix} a_0 & a_{-1} & \cdots & a_{-N+1} \\ a_1 & a_0 & \cdots & a_{-N+2} \\ \vdots & \vdots & & \vdots \\ a_{N-1} & a_{N-2} & \cdots & a_0 \end{vmatrix} \tag{11.139}$$

where

$$a_n = \frac{1}{2\pi}\int_0^{2\pi} d\theta e^{-in\theta} C(e^{i\theta}) \tag{11.140}$$

where

$$C(e^{i\theta}) = \left[\frac{(1-\alpha_1 e^{i\theta})(1-\alpha_2 e^{-i\theta})}{(1-\alpha_1 e^{-i\theta})(1-\alpha_2 e^{i\theta})}\right]^{1/2} \tag{11.141}$$

with

$$\begin{aligned} \alpha_1 &= z_h \frac{1-z_v}{1+z_v} = e^{-2E^v\beta}\tanh E^h\beta \\ \alpha_2 &= z_h^{-1}\frac{1-z_v}{1+z_v} = e^{-2E^v\beta}\coth E^h\beta \end{aligned} \tag{11.142}$$

and the square roots are defined to be positive at $\theta = \pi$.

11.3.2 The diagonal correlation $\langle\sigma_{0,0}\sigma_{N,N}\rangle$

The result (11.139) which expresses $\langle\sigma_{0,0}\sigma_{0,N}\rangle$ as an $N\times N$ determinant is by no means unique since even the size of the determinant depends on the choice of the path joining the two spins. This arbitrariness in the choice of path may often be exploited to find useful forms of determinental representations which are difficult to see by merely looking at the determinants themselves.

As an example of this phenomenon consider the correlation on the diagonal of the square lattice with periodic boundary conditions $\langle\sigma_{0,0}\sigma_{N,N}\rangle$. If the path connecting the two spins is chosen to be the stairstep path which goes from $(0,0)$ to (N,N) in $2N$ steps then the construction of the previous subsection gives a determinant of size $4N\times 4N$. However, a representation in terms of a determinant of size $N\times N$ may be obtained if we first consider a triangular lattice with diagonal bonds of strength E_d between (j,k) and $(j+1,k+1)$. This problem may be reduced to a dimer problem on a lattice obtained from the triangular Ising lattice by replacing each Ising site by a cluster of six dimer sites [14]. On this triangular lattice we then consider a path of length N which goes from $(0,0)$ to (N,N) along the diagonal bonds. Using this representation the correlation is computed as a $2N\times 2N$ determinant. The result for the square lattice is then regained by letting $E^d\to 0$. In this limit the $2N\times 2N$

determinant for $\langle\sigma_{0,0}\sigma_{N,N}\rangle^2$ reduces to the square of a $N \times N$ determinant and the simple result (10.47) is obtained that the diagonal correlation $\langle\sigma_{0,0}\sigma_{N,N}\rangle$ is given by the Toeplitz determinant (11.139) with

$$\alpha_1 = 0, \quad \alpha_2 = (\sinh 2E^h\beta \sinh 2E^v\beta)^{-1}. \tag{11.143}$$

Full details are given in [14, pp186–199]

11.3.3 Correlations near the boundary

As the final example of the arbitrariness in the choice of path in the computation of correlation functions consider $\langle\sigma_{M,N'}\sigma_{M,N}\rangle$ on row M which is a fixed distance from the boundary of the cylindrical lattice which has a magnetic field interacting with the spins in row 1. Again, if a straight line path is drawn from (M, N') to (M, N) the correlation is computed as a $2(N - N') \times 2(N - N')$ determinant. However, on the expanded lattice of Fig. 11.15 where an extra row (called zero) of spins which interact with themselves with infinite strength has been added to incorporate the magnetic field we may draw an alternative path consisting of a vertical line from (M, N') to $(0, N')$, a horizontal line from $(0, N')$ to $(0, N)$ and a vertical line from $(0, N)$ to (M, N). On the segment of this path from $(0, N')$ to $(0, N)$ the replacement of the weight $x \to x^{-1}$ results in no change at all because $x = 1$. Therefore the size of the determinant instead of being $2(2M + N - N') \times 2(2M + N - N')$ is only $4M \times 4M$ which has the extremely useful property that it is independent of the separation of the spins $N - N'$. This is the reason why correlation functions at a finite distance from the boundary of the cylindrical lattice may be studied in an elementary manner. This representation was first used in [21] to obtain the results presented in chapter 10.

References

[1] L. Onsager, Crystal statistics I. A two dimensional model with an order-disorder transition, Phys. Rev. 65 (1944) 117–149.
[2] B. Kaufman, Crystal statistics II. Partition function evaluated by spinor analysis, Phys. Rev. 76 (1949) 1232–1243.
[3] B. Kaufman and L. Onsager, Crystal statistics III. Short-range order in a binary Ising lattice, Phys. Rev. 76 (1949) 1244–1252.
[4] T.D. Schultz, D.C. Mattis and E.H. Lieb, Two-dimensional Ising model as a soluble problem in many fermions, Rev. Mod. Phys. 36 (1964) 856–871.
[5] M. Kac and J.C. Ward, A combinatorial solution of the two-dimensional Ising model, Phys. Rev. 88 (1952) 1332–1337.
[6] R.B. Potts and J.C. Ward, The combinatorial method and the two dimensional Ising model, Prog. Theo. Phys. (Kyoto) 13 (1955) 38–46.
[7] C.A. Hurst and H.S. Green, New solution of the Ising problem for a rectangular lattice, J.Chem.Phys. 33 (1960) 1059.
[8] P.W. Kasteleyn, The statistics of dimers on a lattice 1. The number of dimer arrangements on a quadratic lattice, Physica 27 (1961) 1209–1225.
[9] P.W. Kasteleyn, Dimer statistics and phase transitions, J.Math. Phys. 4 (1963) 287–293.
[10] P.W. Kasteleyn in *Graph theory and theoretical physics*, ed. F. Harary,(Academic Press, New York 1967) 43.
[11] R.J. Baxter and I.G.Enting, 399th solution of the Ising model, J. Phys. A11 (1978) 2463–2473.
[12] R.J. Baxter, *Exactly solved models in Statistical Mechanics*, Academic Press, London (1982).
[13] G.H. Wannier, The statistical problem in cooperative phenomena, Rev. Mod. Phys.17 (1945) 50–60.
[14] B.M. McCoy and T.T. Wu, *The two dimensional Ising model* (Harvard University Press 1973).
[15] M.E. Fisher, On the dimer solution of planar Ising models, J.Math. Phys. 7 (1966) 1776–1781.
[16] G. Tesler, Matchings in graphs on non-orientable surfaces, J. Comb. Theo. B 78 (2002) 198–231.
[17] R. Costa-Santos and B.M. McCoy, Dimers and the critical Ising model on lattices of genus > 1, Nucl. Phys. B623 [FS](2002) 439–473.
[18] A.E. Ferdinand, Statistical mechanics of dimers on a quadratic lattice, J.Math. Phys. 8 (1967) 2332–2339.
[19] P. Fendley, R. Moessner and S.L. Sondhi, Classical dimers on the triangular lattice, Phys. Rev. B66 (2002) 214513-(1–14).

[20] W.T. Lu and F.Y. Wu, Close-packed dimers on nonorientable surfaces, Phys. Lett. A 293 (2002) 235–246.
[21] B.M. McCoy and T.T. Wu, Theory of Toeplitz determinants and the spin correlations of the two-dimensional Ising model. IV, Phys. Rev. 162 (1967) 436–475.
[22] E.W. Montroll, R.B. Potts and J.C. Ward, Correlations and spontaneous magnetization of the two-dimensional Ising model, J.Math. Phys. 4 (1963) 308–322.

12

Ising model spontaneous magnetization and form factors

In the preceding chapter we demonstrated that every correlation function $\langle\sigma_{0,0}\sigma_{M,N}\rangle$ of the two-dimensional Ising model in zero external field can be expressed a determinant (in an infinite number of ways). The simplest such results are for the correlation $\langle\sigma_{0,0}\sigma_{N,N}\rangle$ on the diagonal and $\langle\sigma_{0,0}\sigma_{0,N}\rangle$ in a row which were both shown to be given as an $N \times N$ Toeplitz determinant

$$D_N = \begin{vmatrix} c_0 & c_{-1} & c_{-2} & \cdots & c_{-N+1} \\ c_1 & c_0 & c_{-1} & \cdots & c_{-N+2} \\ c_2 & c_1 & c_0 & \cdots & c_{-N+3} \\ \vdots & \vdots & \vdots & \vdots & \vdots \\ c_{N-1} & c_{N-2} & c_{N-3} & \cdots & c_0 \end{vmatrix} \tag{12.1}$$

with

$$c_n = \frac{1}{2\pi}\int_0^{2\pi} d\theta e^{-in\theta} C(e^{i\theta}) \tag{12.2}$$

where

$$C(e^{i\theta}) = \left[\frac{(1-\alpha_1 e^{i\theta})(1-\alpha_2 e^{-i\theta})}{(1-\alpha_1 e^{-i\theta})(1-\alpha_2 e^{i\theta})}\right]^{1/2} \tag{12.3}$$

and the square root is defined to be positive at $\theta = \pi$. For the diagonal correlation

$$\alpha_1 = 0, \quad \alpha_2 = (\sinh 2E^v\beta \sinh 2E^h\beta)^{-1} \tag{12.4}$$

and for the row correlation

$$\alpha_1 = e^{-2E^v\beta}\tanh E^h\beta, \quad \alpha_2 = e^{-2E^v\beta}\coth E^h\beta. \tag{12.5}$$

These determinants give a very efficient method of computing the correlations when the separation N is small. Examples of these results were given in chapter 10. However, the direct expansion of these determinants is not an efficient way to obtain the behavior of the correlations when the separation N is large.

In chapter 10 we gave alternative expressions for the correlations in terms of the form factor expansion which are useful in the large N regime. In this chapter we will prove these form factor expansions for $\langle\sigma_{0,0}\sigma_{N,N}\rangle$ and $\langle\sigma_{0,0}\sigma_{0,N}\rangle$ by the use of

Wiener–Hopf sum equations. The solution of Wiener–Hopf sum equations is presented in section 12.1.

The leading term in the form factor expansion for $T < T_c$ is the spontaneous magnetization. This is computed in section 12.2 by proving first a much more general theorem due to Szegö. In section 12.3 we compute the form factor expansions both above and below T_c. In section 12.4 we compute the leading behavior as $N \to \infty$ for the three cases of $T > T_c$, $T = T_c$ and $T < T_c$. We conclude in section 12.5 by presenting the method whereby the diagonal form factor $f_N^{(n)}(t)$ which is given as an n dimensional integral can be reduced to the sum of products of the hypergeometric functions $F(-1/2, 1/2; 1; t)$ and $F(1/2, 1/2; 1; t)$ where $t = \alpha_2^2$ for $T < T_c$ and α_2^{-2} for $T > T_c$ which were given in chapter 10.

12.1 Wiener–Hopf sum equations

A Wiener–Hopf sum equation is the set of simultaneous linear equations

$$\sum_{m=0}^{\infty} c_{n-m} x_m = y_n \quad 0 \leq n. \tag{12.6}$$

These sum equations are distinguished from the most general set of linear equations in two respects: i) The numbers c_{m-n} depend only on the difference $m - n$ and not on m and n separately and ii) the upper limit of of summation runs to infinity. Because of condition ii) we need some conditions on c_n, x_n and y_n to make the problem well defined and for our applications it is sufficient to impose the conditions

$$\sum_{n=-\infty}^{\infty} |y_n| < \infty$$
$$\sum_{n=-\infty}^{\infty} |c_n| < \infty$$
$$\sum_{n=-\infty}^{\infty} |x_n| < \infty. \tag{12.7}$$

We will solve these equations under the further condition that $\ln C(e^{i\theta})$, with $C(e^{-i\theta})$ defined by (12.11), is continuous, periodic and zero free for $0 \leq \theta \leq 2\pi$. We follow the procedure in [1, chapter 9].

12.1.1 Fourier transforms

We begin by considering an equation simpler than (12.6) where the lower limit of summation extends to minus infinity

$$\sum_{m=-\infty}^{\infty} c_{n-m} x_m = y_n \quad -\infty < n < \infty. \tag{12.8}$$

To solve (12.8) we define the three Fourier transforms on the unit circle $|\xi| = 1$

$$X(\xi) = \sum_{n=-\infty}^{\infty} x_n \xi^n \tag{12.9}$$

$$Y(\xi) = \sum_{n=-\infty}^{\infty} y_n \xi^n \tag{12.10}$$

$$C(\xi) = \sum_{n=-\infty}^{\infty} c_n \xi^n \tag{12.11}$$

which, because of (12.7), exist and are continuous on $|\xi| = 1$.

Multiply (12.8) by ξ^n and sum n from $-\infty$ to ∞. Using (12.7) we may interchange the order of the double sums to find

$$Y(\xi) = \sum_{n=-\infty}^{\infty} \xi^n \sum_{m=-\infty}^{\infty} c_{n-m} x_m = \sum_{m=-\infty}^{\infty} x_m \xi^m \sum_{n=-\infty}^{\infty} c_{n-m} \xi^{n-m} = C(\xi) X(\xi). \tag{12.12}$$

Then if we make the further restriction that $C(\xi) \neq 0$ for $|\xi| = 1$ we may solve (12.12) as

$$X(\xi) = Y(\xi)/C(\xi). \tag{12.13}$$

and thus we may invert (12.9) to find the solution

$$x_n = \frac{1}{2\pi i} \oint_{|\xi|=1} d\xi \xi^{-n-1} \frac{Y(\xi)}{C(\xi)}. \tag{12.14}$$

We now consider applying this same procedure of Fourier transform to the original Wiener–Hopf sum equation (12.6). Now x_n, y_n and the equations (12.6) exist only for $n \geq 0$. In order to execute a Fourier transform we thus need to make the further definitions

$$x_n = 0 \quad \text{for} \quad n \leq -1 \tag{12.15}$$

$$y_n = 0 \quad \text{for} \quad n \leq -1 \tag{12.16}$$

$$v_n = \sum_{m=0}^{\infty} c_{n-m} x_m \quad \text{for} \quad n \leq -1 \tag{12.17}$$

$$= \quad 0 \quad \text{for} \quad n \geq 0 \tag{12.18}$$

and thus we may write (12.6) as

$$\sum_{m-\infty}^{\infty} c_{n-m} x_m = y_n + v_n \quad \text{for} \quad -\infty < n < \infty. \tag{12.19}$$

We may now define Fourier transforms on $|\xi| = 1$ by (12.11) and

$$X(\xi) = \sum_{n=0}^{\infty} x_n \xi^n \tag{12.20}$$

$$Y(\xi) = \sum_{n=0}^{\infty} y_n \xi^n \tag{12.21}$$

$$V(\xi) = \sum_{n=-\infty}^{-1} v_n \xi^n \tag{12.22}$$

and multiply (12.19) by ξ^n and sum n from $-\infty$ to ∞ to obtain

$$C(\xi)X(\xi) = Y(\xi) + V(\xi). \tag{12.23}$$

Equation (12.23) differs from (12.12) in that it contains two unknown functions $X(\xi)$ and $V(\xi)$ instead of only one unknown. Thus further information is needed to solve the problem.

12.1.2 Splitting and factorization

Consider any function $F(\xi)$ which has a convergent Laurent expansion on the unit circle $|\xi| = 1$

$$F(\xi) = \sum_{n=-\infty}^{\infty} f_n \xi^n \tag{12.24}$$

with

$$\sum_{n=-\infty}^{\infty} |f_n| < \infty. \tag{12.25}$$

Then we may define what we call a + function $[F(\xi)]_+$ which is continuous on $|\xi| = 1$, and analytic for $|\xi| < 1$ by

$$[F(\xi)]_+ = \sum_{n=0}^{\infty} f_n \xi^n \tag{12.26}$$

and we can define a $-$ function $[F(\xi)]_-$ continuous on $|\xi| = 1$, analytic for $|\xi| > 1$ and vanishing as $\xi \to \infty$ as

$$[F(\xi)]_- = \sum_{n=-\infty}^{-1} f_n \xi^n. \tag{12.27}$$

Clearly for $|\xi| = 1$

$$F(\xi) = [F(\xi)]_+ + [F(\xi)]_-. \tag{12.28}$$

Furthermore we have from Cauchy's theorem for $|\xi| > 1$

$$[F(\xi)]_- = -\frac{1}{2\pi i} \oint_{|\xi'|=1} d\xi' \frac{[F(\xi')]_-}{\xi' - \xi} \tag{12.29}$$

and

$$0 = -\frac{1}{2\pi i} \oint_{|\xi'|=1} d\xi' \frac{[F(\xi')]_+}{\xi' - \xi} \tag{12.30}$$

where the integration contour is counterclockwise on $|\xi'| = 1$ and thus using (12.28) we find that, for $|\xi| > 1$,

$$[F(\xi)]_- = -\frac{1}{2\pi i}\oint_{|\xi'|=1} d\xi' \frac{F(\xi')}{\xi'-\xi}. \tag{12.31}$$

Similarly, for $|\xi| < 1$

$$[F(\xi)]_+ = \frac{1}{2\pi i}\oint_{|\xi'|=1} d\xi' \frac{F(\xi')}{\xi'-\xi}. \tag{12.32}$$

The decomposition (12.28) is called a(n) (additive) splitting of the function $F(\xi)$.

We now consider making a factorization of the function $C(\xi)$ on $|\xi| = 1$ as

$$C(\xi) = P^{-1}(\xi)Q^{-1}(\xi^{-1}) \tag{12.33}$$

where both $P(\xi)$ and $Q(\xi)$ are both + functions (continuous for $|\xi| = 1$ and analytic for $|\xi| < 1$) which are nonzero for $|\xi| < 1$. When $C(\xi)$ has the property that $\ln C(\xi)$ is continuous and periodic on $|\xi| = 1$ we may obtain the factorization (12.33) by first taking the logarithm

$$\ln C(\xi) = -\ln P(\xi) - \ln Q(\xi^{-1}) \tag{12.34}$$

where $\ln P(\xi)$ is a + function and $\ln Q(\xi^{-1})$ is a − function. Thus by making an additive splitting of $\ln C(\xi)$

$$\ln C(\xi) = [\ln C(\xi)]_+ + [\ln C(\xi)]_-. \tag{12.35}$$

Thus by comparing (12.34) and (12.35) we find that

$$P(\xi) = \exp(-[\ln C(\xi)]_+) \tag{12.36}$$

$$Q(\xi^{-1}) = \exp(-[\ln C(\xi)]_-). \tag{12.37}$$

which satisfies the requirement that $P(\xi)$ and $Q(\xi)$ are + functions which are nonzero for $|\xi| < 1$. It follows from (12.26) and (12.37) that

$$[Q(\xi^{-1})]_+ = Q(0) = 1. \tag{12.38}$$

12.1.3 Solution

We may now use the factorization (12.33) in (12.23) to write

$$P^{-1}(\xi)Q^{-1}(\xi^{-1})X(\xi) = Y(\xi) + V(\xi) \tag{12.39}$$

and multiply by $Q(\xi^{-1})$ to find

$$P^{-1}(\xi)X(\xi) = Q(\xi^{-1})Y(\xi) + Q(\xi^{-1})V(\xi). \tag{12.40}$$

We now recognize that $P^{-1}(\xi)X(\xi)$ is a + function (because both $P^{-1}(\xi)$ and $X(\xi)$ are + functions) and that $Q(\xi^{-1})V(\xi)$ is a − function (because $Q(\xi)$ is a + function and $V(\xi)$ is a − function). However, $Q(\xi^{-1})Y(\xi)$ contains both + and − parts. Thus we make the additive splitting of $Q(\xi^{-1})Y(\xi)$

$$Q(\xi^{-1})Y(\xi) = [Q(\xi^{-1})Y(\xi)]_+ + [Q(\xi^{-1})Y(\xi)]_- \tag{12.41}$$

and write (12.40) for $|\xi| = 1$

$$P^{-1}(\xi)X(\xi) - [Q(\xi^{-1})Y(\xi)]_+ = [Q(\xi^{-1})Y(\xi)]_- + Q(\xi^{-1})V(\xi). \tag{12.42}$$

The left-hand side defines a function analytic for $|\xi| < 1$ and continuous on $|\xi| = 1$ and the right-hand side defines a function which is analytic for $|\xi| > 1$ and is continuous for $|\xi| = 1$. Taken together they define a function $E(\xi)$ analytic for all ξ except possibly for $|\xi| = 1$ and continuous everywhere. But these properties are sufficient to prove [1, pp.211-213] that $E(\xi)$ is an entire function which vanishes at $|\xi| = \infty$ and thus, by Liouville's theorem, must be zero everywhere. Therefore both the right-hand side and the left-hand side of (12.42) vanish separately and thus we have

$$X(\xi) = P(\xi)[Q(\xi^{-1})Y(\xi)]_+ \tag{12.43}$$

and

$$V(\xi) = -Q^{-1}(\xi^{-1})[Q(\xi^{-1})Y(\xi)]_-. \tag{12.44}$$

Thus inverting the Fourier transform we have the desired solution

$$x_n = \frac{1}{2\pi i}\oint_{|\xi|=1} d\xi \frac{1}{\xi^{n+1}} P(\xi)[Q(\xi^{-1})Y(\xi)]_+. \tag{12.45}$$

12.2 Spontaneous magnetization and Szegö's theorem

The spontaneous magnetization $M(T)_-$ is obtained from the two-point function as

$$\lim_{M^2+N^2\to\infty} \langle\sigma_{0,0}\sigma_{M,N}\rangle = M(T)_-^2. \tag{12.46}$$

In particular because this limit should be independent of how M and N go to infinity it should be sufficient to compute either

$$\lim_{N\to\infty} \langle\sigma_{0,0}\sigma_{N,N}\rangle \tag{12.47}$$

or

$$\lim_{N\to\infty} \langle\sigma_{0,0}\sigma_{0,N}\rangle. \tag{12.48}$$

Both of these correlations are given in (12.1) as $N \times N$ Toeplitz determinants. We thus need to compute the value of this determinant when its size becomes infinite.

The mathematics needed to compute this limit was specifically invented to study this Ising model problem. However, once the technique was discovered it was found to lead to results of much greater generality. We will therefore compute the spontaneous magnetization of the Ising model by a very general theorem found by Szegö [2].

Szegö's theorem. If the generating function $C(\xi)$ and $\ln C(\xi)$ are continuous on the unit circle $|\xi| = 1$ then the behavior for large N of the Toeplitz determinant

$$D_N = \begin{vmatrix} c_0 & c_{-1} & c_{-2} & \cdots & c_{-N+1} \\ c_1 & c_0 & c_{-1} & \cdots & c_{-N+2} \\ c_2 & c_1 & c_0 & \cdots & c_{-N+3} \\ \vdots & \vdots & \vdots & \vdots & \vdots \\ c_{N-1} & c_{N-2} & c_{N-3} & \cdots & c_0 \end{vmatrix} \tag{12.49}$$

with

$$c_n = \frac{1}{2\pi}\int_0^{2\pi} d\theta e^{-in\theta} C(e^{i\theta}) \tag{12.50}$$

is given by

$$\lim_{N\to\infty} D_N/\mu^N = \exp\left(\sum_{n=1}^{\infty} n g_{-n} g_n\right) \tag{12.51}$$

where

$$\mu = \exp\left[\frac{1}{2\pi}\int_0^{2\pi} d\theta \ln C(e^{i\theta})\right] \tag{12.52}$$

and

$$g_n = \frac{1}{2\pi}\int_0^{2\pi} d\theta e^{-in\theta} \ln C(e^{i\theta}) \tag{12.53}$$

wherever the sum in (12.51) converges.

12.2.1 Proof of Szegö's theorem

The proof of Szegö's theorem has a long history which is given in [3]. The result actually proven by Szegö [2] required that the generating function $C(e^{i\theta})$ be real and positive which is not satisfied in the Ising case. We follow the presentation in [1, chapter 10] and separate the derivation of the result (12.51) into two parts by first computing the ratio D_N/D_{N+1} as $N \to \infty$ to find μ of (12.52) and then computing the limit (12.51).

Behavior of D_N/D_{N+1} for $N \to \infty$
We define μ_N as the ratio

$$\mu_N = D_{N+1}/D_N \tag{12.54}$$

and define the quantities $x_n^{(N)}$ as the solution of the set of $N+1$ linear equations

$$\sum_{m=0}^{N} c_{n-m} x_m^{(N)} = \delta_{n,0} \quad \text{for} \quad 0 \le n \le N. \tag{12.55}$$

These equations have a unique solution for $x_n^{(N)}$ if

$$D_{N+1} \neq 0 \tag{12.56}$$

and thus by Cramer's rule

$$\mu_N = D_{N+1}/D_N = [x_0^{(N)}]^{-1}. \tag{12.57}$$

Therefore, assuming that the limit

$$x_0^{(\infty)} = \lim_{N\to\infty} x_0^{(N)} \tag{12.58}$$

exists, this limit may be computed by solving the Wiener–Hopf sum equation for $n \ge 0$

$$\sum_{m=0}^{\infty} c_{n-m} x_m^{(\infty)} = \delta_{n,0}. \tag{12.59}$$

The solution of (12.59) is given by (12.45) with

$$Y(\xi) = \sum_{n=0}^{\infty} \xi^n \delta_{n,0} = 1 \tag{12.60}$$

and thus we obtain

$$x_n^{(\infty)} = \frac{1}{2\pi i} \oint_{|\xi|=1} d\xi \frac{1}{\xi^{n+1}} P(\xi)[Q(\xi^{-1})]_+. \tag{12.61}$$

Thus, using (12.38) for $[Q(\xi^{-1})]_+$, we obtain

$$x_n^{(\infty)} = \frac{1}{2\pi i} \oint_{|\xi|=1} d\xi \frac{1}{\xi^{n+1}} P(\xi). \tag{12.62}$$

Therefore if we set $n = 0$ and use the fact that $P(\xi)$ is analytic for $|\xi| \leq 1$ we find that

$$x_0^{(\infty)} = P(0) \tag{12.63}$$

from which, using the computation (12.36) for $P(\xi)$ in terms of $C(\xi)$ we find that when $\ln C(\xi)$ is continuous on $|\xi| = 1$ that

$$\begin{aligned}\mu^{-1} &= \lim_{N\to\infty} D_N/D_{N+1} = x_0^{(\infty)} = P(0) = \exp\left(-\frac{1}{2\pi i}\oint_{|\xi|=1} \frac{d\xi}{\xi} \ln C(\xi)\right) \\ &= \exp\left(-\frac{1}{2\pi}\int_0^{2\pi} d\theta \ln C(e^{i\theta})\right)\end{aligned} \tag{12.64}$$

which is (12.52).

If we assume that there exists an N_0 such that $D_N \neq 0$ for $N \geq N_0$ then from (12.57) it follows that

$$\lim_{N\to\infty} D_N/\mu^N = \lim_{N\to\infty} D_{N_0}/\mu^{N_0+1} \prod_{n=N_0}^{N-1} \mu_n/\mu. \tag{12.65}$$

Thus the limit in (12.51) will exist and be nonzero if

$$\sum_{N=N_0}^{\infty} |1 - \mu/\mu_N| < \infty. \tag{12.66}$$

Computation of $\lim_{N\to\infty} D_N/\mu^N$

We will prove (or at least discover) (12.51) by studying the dependence of the limiting value of D_N/μ^N as $N \to \infty$ on the generating function $C(e^{i\theta})$. For this purpose, we compare D_N defined by (12.49) with the determinant

$$\bar{D}_N = \begin{vmatrix} \bar{c}_0 & \bar{c}_{-1} & \bar{c}_{-2} & \cdots & \bar{c}_{-N+1} \\ \bar{c}_1 & \bar{c}_0 & \bar{c}_{-1} & \cdots & \bar{c}_{-N+2} \\ \bar{c}_2 & \bar{c}_1 & c_0 & \cdots & \bar{c}_{-N+3} \\ \vdots & \vdots & \vdots & \vdots & \vdots \\ \bar{c}_{N-1} & \bar{c}_{N-2} & \bar{c}_{N-3} & \cdots & \bar{c}_0 \end{vmatrix} \tag{12.67}$$

with

$$\bar{c}_n = \frac{1}{2\pi}\int_0^{2\pi} d\theta e^{-in\theta}\bar{C}(e^{i\theta}) \tag{12.68}$$

and

$$\bar{C}(e^{i\theta}) = C(e^{i\theta})(1 - \alpha e^{-i\theta}) \tag{12.69}$$

with

$$|\alpha| < 1. \tag{12.70}$$

From the condition (12.70) it follows that

$$\exp\left[\frac{1}{2\pi}\int_0^{2\pi} d\theta \ln \bar{C}(e^{i\theta})\right] = \exp\left[\frac{1}{2\pi}\int_0^{2\pi} d\theta \ln C(e^{i\theta})\right] = \mu. \tag{12.71}$$

It follows from (12.50), (12.68) and (12.69) that

$$\bar{c}_n = c_n - \alpha c_{n+1} \tag{12.72}$$

and thus from (12.67) that

$$\bar{D}_N = \begin{vmatrix} c_0 - \alpha c_1 & c_{-1} - \alpha c_0 & \cdots & c_{-N+1} - \alpha c_{-N+2} \\ c_1 - \alpha c_2 & c_0 - \alpha c_1 & \cdots & c_{-N+2} - \alpha c_{-N+3} \\ c_2 - \alpha c_3 & c_1 - \alpha c_2 & \cdots & c_{-N+3} - \alpha c_{-N+4} \\ \vdots & \vdots & \vdots & \vdots \\ c_{N-1} - \alpha c_N & c_{N-2} - \alpha c_{N-1} & \cdots & c_0 - \alpha c_1 \end{vmatrix} \tag{12.73}$$

and this may be written as an $(N+1) \times (N+1)$ determinant

$$\bar{D}_N = \begin{vmatrix} 1 & \alpha & \alpha^2 & \alpha^3 & \cdots & \alpha^N \\ c_1 & c_0 & c_{-1} & c_{-2} & \cdots & c_{-N+1} \\ c_2 & c_1 & c_0 & c_{-1} & \cdots & c_{-N+2} \\ c_3 & c_2 & c_1 & c_0 & \cdots & c_{-N+3} \\ \vdots & \vdots & \vdots & \vdots & \vdots & \vdots \\ c_N & c_{N-1} & c_{N-2} & c_{N-3} & \cdots & c_0 \end{vmatrix} \tag{12.74}$$

which may be verified by subtracting α times column N from column $N+1$, then α times column $N-1$ from column N and continuing until we subtract α times column one from column 2.

We now define the $N+1$ quantities $\bar{x}_j^{(N)}$ with $0 \leq j \leq N$ as the solutions of the $N+1$ linear equations

$$\sum_{m=0}^{N} \alpha^m \bar{x}_m^{(N)} = 1 \tag{12.75}$$

and

$$\sum_{m=0}^{N} c_{n-m}\bar{x}_m^{(N)} = 0, \quad \text{for } 1 \leq n \leq N. \tag{12.76}$$

Then again by Cramer's rule

$$\bar{x}_0^{(N)} = D_N/\bar{D}_N. \tag{12.77}$$

We now let $N \to \infty$ and assume as before that

$$\lim_{N\to\infty} x_0^{(N)} = x_0^{(\infty)} \tag{12.78}$$

where from (12.75) and (12.76) the $\bar{x}_j^{(\infty)}$ for $0 \le j$ satisfy

$$\sum_{m=0}^{\infty} \alpha^m \bar{x}_m^{(\infty)} = 1 \tag{12.79}$$

and

$$\sum_{m=0}^{\infty} c_{n-m}\bar{x}_m^{(\infty)} = 0, \quad \text{for } 1 \le n. \tag{12.80}$$

Equations (12.79) and (12.80) taken together are not a set of Wiener–Hopf sum equations. However, if we define y_0 by

$$\sum_{m=0}^{\infty} c_{-m}\bar{x}_m^{(\infty)} = y_0 \tag{12.81}$$

then (12.80) and (12.81) taken together do form a set of Wiener–Hopf sum equations and thus if we define

$$\bar{X}^{(\infty)}(\xi) = \sum_{n=0}^{\infty} \bar{x}_n^{(\infty)}\xi^n \tag{12.82}$$

we find from (12.43) that

$$\bar{X}^{(\infty)}(\xi) = P(\xi)[Q(\xi^{-1})Y(\xi)]_+ = y_0 Q(0)P(\xi). \tag{12.83}$$

In order to determine y_0 we note that from (12.79)

$$\bar{X}^{(\infty)}(\alpha) = 1 \tag{12.84}$$

and thus using (12.83) we find

$$y_0 Q(0)P(\alpha) = 1. \tag{12.85}$$

Therefore from (12.83)

$$\bar{X}^{(\infty)}(\xi) = \frac{P(\xi)}{P(\alpha)} \tag{12.86}$$

and thus, from (12.77), we obtain the result

$$\lim_{N\to\infty} \frac{D_N}{\bar{D}_N} = \bar{x}_0^{(\infty)} = \bar{X}^{(\infty)}(0) = \frac{P(0)}{P(\alpha)}. \tag{12.87}$$

Finally we use the expression (12.36) for $P(\xi)$ in the form

$$P(\xi) = P(0) \exp\left(-\sum_{n=1}^{\infty} g_n \xi^n\right), \tag{12.88}$$

where g_n is defined by (12.53), to find that (12.87) reduces to

$$\lim_{N\to\infty} \frac{D_N}{\bar{D}_N} = \exp \sum_{n=1}^{\infty} g_n \alpha^n. \tag{12.89}$$

We now use the definition of g_n of (12.53) and make the analogous definition

$$\bar{g}_n = \frac{1}{2\pi} \int_0^{2\pi} d\theta e^{-in\theta} \ln \bar{C}(e^{i\theta}) \tag{12.90}$$

to find

$$\bar{g}_n = g_n \quad \text{for} \quad n \geq 0 \tag{12.91}$$

$$= g_n + \alpha^{-n}/n \quad \text{for} \quad n < 0 \tag{12.92}$$

and thus we may rewrite (12.89) as

$$\lim_{N\to\infty} \frac{D_N}{\bar{D}_N} = \exp \sum_{n=1}^{\infty} n(g_{-n} g_n - \bar{g}_{-n} \bar{g}_n). \tag{12.93}$$

We next consider the same problem but instead of (12.69) we have

$$\bar{C}(e^{i\theta}) = C(e^{i\theta})(1 - \bar{\alpha} e^{i\theta}) \tag{12.94}$$

with $|\bar{\alpha}| < 1$ and for this we can apply (12.93) to the complex conjugate functions to find that (12.93) applies in this case as well.

We now repeat the process a finite number of times to find that for

$$\bar{C}(e^{i\theta}) = C(e^{i\theta}) \prod_{n=1}^{n_1} (1 - \alpha^{(n)} e^{-i\theta}) \prod_{n=1}^{n_2} (1 - \bar{\alpha}^{(n)} e^{i\theta}) \tag{12.95}$$

with $|\alpha^{(n)}| < 1$ and $|\bar{\alpha}^{(n)}| < 1$ that

$$\lim_{N\to\infty} \frac{D_N}{\bar{D}_N^{(n_1,n_2)}} = \exp \sum_{k=1}^{\infty} k(g_{-k} g_k - \bar{g}_{-k}^{(n_1,n_2)} g_k^{(n_1,n_2)}). \tag{12.96}$$

To obtain Szegö's theorem we now need to consider the special limiting case $n_1, n_2 \to \infty$ such that $\bar{C}(e^{i\theta}) = 1$. Then $\bar{g}_n = 0$ and we obtain from (12.96) the desired result (12.51). To justify this we must interchange the limit $N \to \infty$ with the limits $n_1, n_2 \to \infty$. The validity of this interchange will depend on the smoothness of the generating function $C(e^{i\theta})$ on $0 \leq \theta \leq 2\pi$. A proof of this interchange under conditions which are sufficient for the generating function (12.3) which is analytic on $0 \leq \theta \leq 2\pi$ is given in [1]. The most general restrictions known are fully presented in [3]. Thus Szegö's theorem is proven.

12.2.2 The spontaneous magnetization

It remains to apply Szegö's theorem to the diagonal and row correlations given by (12.1) with $C(e^{i\theta})$ given by (12.3). This function has the property that

$$C(e^{-i\theta}) = 1/C(e^{i\theta}) \tag{12.97}$$

and hence we find from (12.52) that

$$\mu = 1. \tag{12.98}$$

Furthermore because $0 \leq \alpha_1 \leq \alpha_2 < 1$ we use the formula

$$\ln(1 - \alpha_j e^{i\theta}) = -\sum_{n=1}^{\infty} n^{-1}(\alpha_j e^{i\theta})^n \tag{12.99}$$

in the definition of g_n (12.53) to find for $n > 0$

$$g_n = \frac{1}{2n}[\alpha_2^n - \alpha_1^n] \tag{12.100}$$

$$g_{-n} = -\frac{1}{2n}[\alpha_2^n - \alpha_1^n] \tag{12.101}$$

Thus we find

$$\begin{aligned}\sum_{n=1}^{\infty} n g_{-n} g_n &= -\frac{1}{4}\sum_{n=1}^{\infty} n^{-1}[\alpha_2^n - \alpha_1^n]^2 \\ = -\frac{1}{4}\sum_{n=1}^{\infty} n^{-1}[\alpha_2^{2n} &+ \alpha_1^{2n} - 2\alpha_1^n\alpha_2^n] = \frac{1}{4}\ln\left[\frac{(1-\alpha_2^2)(1-\alpha_1^2)}{(1-\alpha_1\alpha_2)^2}\right].\end{aligned} \tag{12.102}$$

Therefore we find from Szegö's theorem (12.51) that for the diagonal correlation where α_1 and α_2 are given by (12.4)

$$M^2 = \lim_{N\to\infty}\langle\sigma_{0,0}\sigma_{N,N}\rangle = [1 - (\sinh 2E^v\beta \sinh 2E^h\beta)^{-2}]^{1/4} \tag{12.103}$$

Similarly when we consider the row correlation function $\langle\sigma_{0,0}\sigma_{0,N}\rangle$ where α_1 and α_2 are given by (12.5) we also find that

$$\begin{aligned}\lim_{N\to\infty}\langle\sigma_{0,0}\sigma_{0,N}\rangle &= \left[\frac{(1-\alpha_2^2)(1-\alpha_1^2)}{(1-\alpha_1\alpha_2)^2}\right]^{1/4} \\ = [1 - (\sinh 2E^v\beta \sinh 2E^h\beta)^{-2}]^{1/4}.\end{aligned} \tag{12.104}$$

Thus, as expected, the limit $N \to \infty$ of both the diagonal and the row correlation are the same, and we conclude that the spontaneous magnetization of the Ising model is

$$M = [1 - (\sinh 2E^v\beta \sinh 2E^h\beta)^{-2}]^{1/8}. \tag{12.105}$$

This result was first announced by Onsager [4] in 1949 and proven by Yang [5] in 1952.

12.3 Form factor expansions of $C(N,N)$ and $C(0,N)$

The representation of the diagonal $C(N,N)$ and row $C(0,N)$ correlation functions as $N \times N$ determinants is very efficient for computation when N is small. However, this representation is not efficient when N is large, and this large N behavior is mandatory for the microscopic understanding of critical behavior presented in chapter 5. Therefore in this section we will extend the considerations which we used to derive Szegö's theorem to recast the row and diagonal correlations into the form factor forms of (10.76)–(10.79) from which the large N limit for both $T < T_c$ and $T > T_c$ can be easily determined. The calculations are different for $T < T_c$ and $T > T_c$ and will be treated separately. We follow the treatment of [6] which extends the Wiener–Hopf methods of [7].

12.3.1 Expansion for $T < T_c$

The row and diagonal correlations are given in (12.1) by a determinant D_N of the form (12.49) with $C(e^{i\theta})$ given by (12.3) and for $T < T_c$ we have seen that $\ln C(\xi)$ is continuous on $|\xi| = 1$. We begin the derivation of the form factor expansions of chapter 10 by writing

$$D_N = M^2 \prod_{n=N}^{\infty} \left(\frac{D_n}{D_{n+1}} \right) \tag{12.106}$$

where

$$M^2 = (1-t)^{1/4} \quad \text{with} \quad t = (\sinh 2E^h\beta \sinh 2E^v\beta)^{-2} \tag{12.107}$$

is the limiting value as $N \to \infty$ previously computed from Szegö's theorem. We will proceed in three steps. First we extend the calculation of the previous section to obtain an exact expression for the ratio D_N/D_{N+1} valid for all N. We will then show how the product in (12.106) can be put in the exponential form of (10.67)–(10.69). Finally we will expand the exponential to obtain the form factor expression (10.76).

The ratio D_N/D_{N+1}

The ratio D_N/D_{N+1} has already been seen in the previous section to be given by

$$x_0^{(N)} = D_N/D_{N+1} \tag{12.108}$$

where the $x_n^{(N)}$ are determined from the set of linear equations

$$\sum_{m=0}^{N} c_{n-m} x_m^{(N)} = \delta_{n,0} \quad \text{for} \quad 0 \leq n \leq N. \tag{12.109}$$

We will prove in this subsection that

$$D_N/D_{N+1} = x_0^{(N)} = \sum_{n=0}^{\infty} \phi_N^{(2n)} \tag{12.110}$$

where $\phi_N^{(0)} = 1$ and for $k \geq 1$ $\phi_N^{(2k)}$ is defined as

$$\phi_N^{(2k)} = (-1)^{k+1}\frac{1}{(2\pi)^{2k}}\oint\prod_{j=1}^{2k}dz_j\ z_j^{N+1}\prod_{j=1}^{k}\{Q(z_{2j-1})Q(z_{2j-1}^{-1})P(z_{2j})P(z_{2j}^{-1})\}$$
$$\times z_1^{-1}z_{2k}^{-1}\prod_{l=1}^{2k-1}\frac{1}{1-z_l z_{l+1}} \qquad (12.111)$$

where the contours are such that $|z_i| = 1-\epsilon$, and the functions $P(\xi)$ and $Q(\xi)$ are the functions in the Wiener–Hopf splitting (12.33) of $C(e^{i\theta})$ (12.3) which for our problem with $T < T_c$ are explicitly given by

$$Q(\xi) = \left(\frac{1-\alpha_1\xi}{1-\alpha_2\xi}\right)^{1/2} = 1/P(\xi). \qquad (12.112)$$

We follow the procedure of [7] and generalize the Wiener–Hopf procedure by defining

$$x_n^{(N)} = y_n = 0, \quad \text{for } n \leq -1 \text{ and } n \geq N+1 \qquad (12.113)$$

$$y_n = \sum_{m=0}^{N} c_{n-m}x_m^{(N)} \quad \text{for } 0 \leq n \leq N \qquad (12.114)$$

$$\begin{aligned} v_n^{(N)} &= \sum_{m=0}^{N} c_{-n-m}x_m^{(N)} \quad \text{for } n \geq 1 \\ &= 0 \quad \text{for } n \leq 0 \end{aligned} \qquad (12.115)$$

$$\begin{aligned} u_n^{(N)} &= \sum_{m=0}^{N} c_{N+n-m}x_m^{(N)} \quad \text{for } n \geq 1 \\ &= 0 \quad \text{for } n \leq 0. \end{aligned} \qquad (12.116)$$

We further define

$$X_N(\xi) = \sum_{n=0}^{N} x_n^{(N)}\xi^n \qquad (12.117)$$

$$Y(\xi) = \sum_{n=0}^{N} y_n\xi^n \qquad (12.118)$$

$$U_N(\xi) = \sum_{n=1}^{\infty} u_n^{(N)}\xi^n \qquad (12.119)$$

$$V_N(\xi) = \sum_{n=1}^{\infty} v_n^{(N)}\xi^n \qquad (12.120)$$

and we note from (12.106) and (12.117) that

$$D_N/D_{N+1} = x_0^{(N)} = X_N(0). \qquad (12.121)$$

We multiply (12.109) by ξ^n and sum on n from $-\infty$ to ∞ to obtain

$$C(\xi)X_N(\xi) = Y(\xi) + U_N(\xi)\xi^N + V_N(\xi^{-1}) \tag{12.122}$$

and then use the factorization (12.33) in (12.122) and decompose into + and − parts to find for $|\xi| = 1$ that

$$\begin{aligned} &P^{-1}(\xi)X_N(\xi) - [Q(\xi^{-1})Y(\xi)]_+ - [Q(\xi^{-1})U_N(\xi)\xi^N]_+ \\ &\quad = Q(\xi^{-1})V_N(\xi^{-1}) + [Q(\xi^{-1})Y(\xi)]_- + [Q(\xi^{-1})U_N(\xi)\xi^N]_-. \end{aligned} \tag{12.123}$$

Then we make the identical argument made in section 12.1 to show that each side of this equation vanishes separately and thus we have the two equations

$$X_N(\xi) = P(\xi)\{[Q(\xi^{-1})Y(\xi)]_+ + [Q(\xi^{-1})U_N(\xi)\xi^N]_+\} \tag{12.124}$$
$$V_N(\xi^{-1}) = -Q^{-1}(\xi^{-1})\{[Q(\xi^{-1})Y(\xi)]_- + [Q(\xi^{-1})U_N(\xi)\xi^N]_-\}. \tag{12.125}$$

Furthermore a second set of equations is found by considering $\xi^N X_N(\xi^{-1})$ in place of $X_N(\xi)$ and thus we also find

$$X_N(\xi^{-1})\xi^N = Q(\xi)\{[P(\xi^{-1})Y(\xi^{-1})\xi^N]_+ + [P(\xi^{-1})V_N(\xi)\xi^N]_+\} \tag{12.126}$$
$$U_N(\xi^{-1}) = -P^{-1}(\xi^{-1})\{[P(\xi^{-1})Y(\xi^{-1})\xi^N]_- + [P(\xi^{-1})V_N(\xi)\xi^N]_-\}. \tag{12.127}$$

For any function $F(\xi)$ we will need, in addition to the functions $[F(\xi)]_+$ and $[F(\xi)]_-$ defined in (12.26) and (12.27), the function

$$[F(\xi)]'_+ = \sum_{n=1}^{\infty} f_n \xi^n. \tag{12.128}$$

From (12.24) and (12.27) we have

$$[F(\xi^{-1})]_- = [F(\xi)]'_+ \tag{12.129}$$

and $[F(\xi)]'_+$ has the integral representation

$$[F(\xi)]'_+ = [F(\xi)]_+ - \frac{1}{2\pi i}\oint d\xi' \frac{F(\xi')}{\xi'} = \frac{1}{2\pi i}\xi \oint_{|\xi'|=1} d\xi' \frac{F(\xi')}{\xi'(\xi' - \xi)} \tag{12.130}$$

where the contour of integration is indented outward at $\xi' = \xi$.

Thus noting that $[Q(\xi^{-1})]_+ = 1$ and $Y(\xi) = 1$, and using (12.112) and (12.129) we rewrite equations (12.124),(12.125) and (12.127) as

$$X_N(\xi) = P(\xi)\{1 + [Q(\xi^{-1})U_N(\xi)\xi^N]_+\} \tag{12.131}$$
$$V_N(\xi^{-1}) = -P(\xi^{-1})\{[Q(\xi^{-1})]_- + [Q(\xi^{-1})U_N(\xi)\xi^N]_-\} \tag{12.132}$$

$$U_N(\xi) = -Q(\xi)\{[P(\xi)\xi^{-N}]'_+ + [P(\xi)V_N(\xi^{-1})\xi^{-N}]'_+\}. \qquad (12.133)$$

When $N \to \infty$ we see from the definition (12.116) that $U_N(\xi)$ vanishes. Therefore we solve the equations (12.131)–(12.133) iteratively by defining $V_N^{(1)}(\xi)$ as the first approximation to $V_N(\xi)$ obtained by replacing $U_N(\xi)$ by zero in (12.132). Thus

$$V_N^{(1)}(\xi^{-1}) = -P(\xi^{-1})[Q(\xi^{-1})]_-. \qquad (12.134)$$

Furthermore because $Q(\xi^{-1})$ is analytic for $|\xi| > 1$ and because $Q(0) = 1$ we have

$$[Q(\xi^{-1})]_- = Q(\xi^{-1}) - Q(0) = Q(\xi^{-1}) - 1 \qquad (12.135)$$

therefore it follows from (12.112) and (12.135) that (12.134) becomes

$$V_N^{(1)}(\xi^{-1}) = -P(\xi^{-1})[Q(\xi^{-1})]_- = P(\xi^{-1}) - 1. \qquad (12.136)$$

We define $U_N^{(1)}(\xi)$ by replacing $V_N(\xi^{-1})$ in (12.133) by $V_N^{(1)}(\xi^{-1})$ as given by equation (12.136). Thus we find

$$U_N^{(1)}(\xi) = -Q(\xi)[P(\xi^{-1})P(\xi)\xi^{-N}]'_+. \qquad (12.137)$$

It thus follows from equation (12.131) that the first approximation $X_N^{(1)}(\xi)$ to $X_N(\xi)$ is given by

$$\begin{aligned} X_N^{(1)}(\xi) &= P(\xi)\{1 - [Q(\xi^{-1})Q(\xi)[P(\xi^{-1})P(\xi)\xi^{-N}]'_+\xi^N]_+\} \\ &= P(\xi)\{1 - \frac{1}{2\pi i}\oint d\xi' \frac{\xi'^N}{\xi' - \xi} Q(\xi'^{-1})Q(\xi')[P(\xi'^{-1})P(\xi')\xi'^{-N}]'_+\}. \end{aligned} \qquad (12.138)$$

Letting $\xi = 0$ in equation (12.138), and using $P(0) = 1$, and writing $X^{(1)}(0) = 1 + \phi_N^{(2)}$, we obtain

$$\begin{aligned} \phi_N^{(2)} &= -\frac{1}{2\pi i}\oint d\xi\ Q(\xi^{-1})Q(\xi)[P(\xi^{-1})P(\xi)\xi^{-N}]'_+\xi^{N-1} \\ &= -\frac{1}{2\pi i}\oint d\xi_1\ Q(\xi_1^{-1})Q(\xi_1)\frac{1}{2\pi i}\xi_1^N \oint d\xi_2 \frac{1}{\xi_2}\frac{1}{\xi_2 - \xi_1}P(\xi_2^{-1})P(\xi_2)\xi_2^{-N} \end{aligned} \qquad (12.139)$$

where ξ_2 is indented outward at $\xi_2 = \xi_1$. Thus, if we set

$$\xi_{2k+1} = z_{2k+1} \qquad \xi_{2k} = z_{2k}^{-1} \qquad (12.140)$$

we obtain $\phi_N^{(2)}$ of (12.111).

We now continue by calculating $V_N^{(2)}(\xi^{-1})$, the second approximation to $V_N(\xi^{-1})$, by using (12.137) in (12.132) to find

$$\begin{aligned}V_N^{(2)}(\xi^{-1}) &= -P(\xi^{-1})\{[Q(\xi^{-1})]_- + [Q(\xi^{-1})U_N^{(1)}(\xi)\xi^N]\}\\ &= -P(\xi^{-1})[Q(\xi^{-1})]_- + P(\xi^{-1})[Q(\xi^{-1})Q(\xi)\xi^N[P(\xi^{-1})P(\xi)\xi^{-N}]'_+]_- . \end{aligned} \tag{12.141}$$

Next, we calculate $U_N^{(2)}(\xi)$ by using (12.141) in (12.133) with (12.135) to obtain

$$\begin{aligned}U_N^{(2)}(\xi) &= -(P(\xi))^{-1}\{[P(\xi)\xi^{-N}]'_+ + [P(\xi)V_N^{(2)}(\xi^{-1})\xi^{-N}]'_+\}\\ &= -Q(\xi)[P(\xi)P(\xi^{-1})\xi^{-N}]'_+\\ &\quad -Q(\xi)[P(\xi)P(\xi^{-1})\xi^{-N}[Q(\xi)Q(\xi^{-1})\xi^N[P(\xi)P(\xi^{-1})\xi^{-N}]'_+]_-]'_+ . \end{aligned} \tag{12.142}$$

We will now calculate $X_N^{(2)}(\xi)$ from (12.131) as

$$\begin{aligned}X_N^{(2)}(\xi) &= P(\xi)\{1 + [Q(\xi^{-1})U_N^{(2)}(\xi)\xi^N]_+\}\\ &= P(\xi) - P(\xi)[Q(\xi^{-1})Q(\xi)\xi^N[P(\xi)P(\xi^{-1})\xi^{-N}]'_+]_+\\ &\quad -P(\xi)[Q(\xi^{-1})Q(\xi)\xi^N[P(\xi)P(\xi^{-1})\xi^{-N}[Q(\xi)Q(\xi^{-1})\xi^N[P(\xi)P(\xi^{-1})\xi^{-N}]'_+]_-]'_+]_+ . \end{aligned} \tag{12.143}$$

Letting $\xi = 0$ in equation (12.143), we obtain $X_N^{(2)}(0) = 1 + \phi_N^{(2)} + \phi_N^{(4)}$ where

$$\begin{aligned}\phi_N^{(4)} &= -\frac{1}{2\pi i}\oint d\xi\ Q(\xi^{-1})Q(\xi)\\ &\quad \times [P(\xi^{-1})P(\xi)\xi^{-N}[Q(\xi^{-1})Q(\xi)\xi^N[P(\xi^{-1})P(\xi)\xi^{-N}]'_+]_-]'_+\xi^{N-1}\\ &= -\frac{1}{(2\pi i)^4}\oint d\xi_1\ \xi_1^N Q(\xi_1^{-1})Q(\xi_1)\oint d\xi_2\ \frac{1}{\xi_2-\xi_1}\xi_2^{-N-1}P(\xi_2^{-1})P(\xi_2)\\ &\quad \oint d\xi_3\ \frac{1}{\xi_3-\xi_2}\xi_3^{N+1}Q(\xi_3^{-1})Q(\xi_3)\oint d\xi_4\ \frac{1}{\xi_4-\xi_3}\xi_4^{-N-1}P(\xi_4^{-1})P(\xi_4) \end{aligned} \tag{12.144}$$

Using the change of variables (12.140) we obtain $\phi_N^{(4)}$ of (12.111).

In general, we iteratively define (from equation 12.132)

$$V_N^{(n+1)}(\xi^{-1}) = -P(\xi^{-1})\{[Q(\xi^{-1})]_- + [Q(\xi^{-1})U_N^{(n)}(\xi)\xi^N]_-\}. \tag{12.145}$$

It then follows from equation (12.133) that

$$\begin{aligned}&U_N^{(n)}(\xi) - U_N^{(n-1)}(\xi)\\ &= -Q(\xi^{-1})[P(\xi)P(\xi^{-1})\xi^{-N}[Q(\xi)Q(\xi^{-1})\xi^N[P(\xi)P(\xi^{-1})\xi^{-N}[Q(\xi)Q(\xi^{-1})\xi^N\ldots]_-]'_+]_-]'_+ \end{aligned} \tag{12.146}$$

where there are $2n-1$ brackets. It now follows from equation (12.131) that $\phi_N^{(2k)}$ is

$$\phi_N^{(2k)} = -\frac{1}{2\pi i}\oint d\xi\ \xi^{N-1}Q(\xi)Q(\xi^{-1})[P(\xi)P(\xi^{-1})\xi^{-N}$$
$$[Q(\xi)Q(\xi^{-1})\xi^N[P(\xi)P(\xi^{-1})\xi^{-N}[Q(\xi)Q(\xi^{-1})\xi^N ...]_-]'_+]_-]'_+ \qquad (12.147)$$

where there are $2k - 1$ brackets. By use of (12.140), we obtain equation (12.111) and thus from (12.106) and (12.108) we have shown that

$$D_N = (1-t)^{1/4} \prod_{m=N}^{\infty} \sum_{k=0}^{\infty} \phi_m^{(2k)}. \qquad (12.148)$$

Exponentiation

Our next step is to show that (12.148) can be put in the form of (10.67)–(10.69).

$$D_N = (1-t)^{1/4} e^{F_N} \qquad (12.149)$$

where

$$F_N = \sum_{n=1}^{\infty} F_N^{(2n)} \qquad (12.150)$$

with

$$F_N^{(2n)} = \frac{(-1)^{n+1}}{n} \frac{1}{(2\pi)^{2n}} \oint \prod_{j=1}^{2n} \frac{dz_j\ z_j^N}{1 - z_j z_{j+1}} \prod_{j=1}^{n} P(z_{2j})P(z_{2j}^{-1})Q(z_{2j-1})Q(z_{2j-1}^{-1}). \qquad (12.151)$$

and $z_{2n+1} \equiv z_1$. The path of integration is along the unit circle $|z_j| = 1 - \epsilon$.

We begin this demonstration by defining a function

$$\tilde{F}_N^{(2n)}$$
$$= \frac{(-1)^{n+1}}{n} \frac{1}{(2\pi)^{2n}} \oint \prod_{j=1}^{2n} \frac{dz_j\ z_j^N}{1 - z_j z_{j+1}} \prod_{j=1}^{n} \left(Q(z_{2j-1})Q(z_{2j-1}^{-1})P(z_{2j})P(z_{2j}^{-1})\right) (1 - \prod_{j=1}^{2n} z_j)$$
$$= F_N^{(2n)} - F_{N+1}^{(2n)} \qquad (12.152)$$

which has the property that

$$F_N^{(2n)} = \sum_{k=N}^{\infty} \tilde{F}_k^{(2n)}. \qquad (12.153)$$

We define the functions

$$\phi_N(\lambda) = \sum_{n=0}^{\infty} \phi_N^{(2n)} \lambda^n \qquad (12.154)$$

and

$$\tilde{F}_N(\lambda) = \sum_{n=1}^{\infty} \tilde{F}_N^{(2n)} \lambda^n \tag{12.155}$$

where $\phi_N(0) = 1$ and $\tilde{F}_N(0) = 0$. We will show that

$$\phi_N(\lambda) = e^{\tilde{F}_N(\lambda)}, \tag{12.156}$$

from which if we set $\lambda = 1$ and use (12.110), we obtain

$$D_N/D_{N+1} = \exp \sum_{k=1}^{\infty} \tilde{F}_N^{(2k)}. \tag{12.157}$$

Thus it follows from equations (12.106) and (12.153) that

$$\begin{aligned} D_N &= (1-t)^{1/4} \prod_{n=N}^{\infty} D_n/D_{n+1} = (1-t)^{1/4} \exp \sum_{k=N}^{\infty} \sum_{m=1}^{\infty} \tilde{F}_k^{(2m)} \\ &= (1-t)^{1/4} \exp \sum_{m=1}^{\infty} \sum_{k=N}^{\infty} \tilde{F}_k^{(2m)} = (1-t)^{1/4} \exp \sum_{n=1}^{\infty} F_N^{(2n)}. \end{aligned} \tag{12.158}$$

This proves equations (12.149)–(12.151).

It remains to show that equation (12.156) holds. Since $\phi(0) = 1$ and $F(0) = 0$, equation (12.156) is equivalent to the equation

$$\frac{d\phi(\lambda)}{d\lambda} = e^{\tilde{F}(\lambda)} \frac{d\tilde{F}(\lambda)}{d\lambda} = \phi(\lambda) \frac{d\tilde{F}(\lambda)}{d\lambda}. \tag{12.159}$$

Thus it follows from equations (12.154), (12.155) and (12.159), by equating like powers of λ, that equation (12.156) is equivalent to the following equation:

$$n\phi_N^{(2n)} = \sum_{l=1}^{n} l \tilde{F}_N^{(2l)} \phi_N^{(2n-2l)} \tag{12.160}$$

The left-hand side of (12.160) is

$$\begin{aligned} n\phi_N^{(2n)} &= n(-1)^{n+1} \frac{1}{(2\pi)^{2n}} \oint \prod_{j=1}^{2n} dz_j \; z_j^{N+1} \\ &\prod_{j=1}^{n} P(z_{2j}) P(z_{2j}^{-1}) Q(z_{2j-1}) Q(z_{2j-1}^{-1}) \prod_{j=1}^{2n-1} \frac{1}{1 - z_j z_{j+1}} \frac{1}{z_{2n} z_1} \end{aligned} \tag{12.161}$$

and the right-hand side is

$$\sum_{l=1}^{n} l\tilde{F}_N^{(2l)}\phi_N^{(2n-2l)} =$$

$$(-1)^n \frac{1}{(2\pi)^{2n}} \oint \prod_{j=1}^{2n} \frac{dz_j\, z_j^N}{1-z_jz_{j+1}} \prod_{j=1}^{n} P(z_{2j})P(z_{2j}^{-1})Q(z_{2j-1})Q(z_{2j-1}^{-1})$$
$$\left(\sum_{l=1}^{n-1} \frac{1}{1-z_1z_{2l}}(1-\prod_{k=1}^{2l} z_k)(1-z_{2l}z_{2l+1})(1-z_{2n}z_1)\prod_{m=2l+2}^{2n-1} z_m - (1-\prod_{p=1}^{2n} z_p)\right) \tag{12.162}$$

where the product $\prod_{m=2l+2}^{2n-1} z_m$ is defined as 1 when $l = n-1$. The product

$$\prod_{j=1}^{2n} \frac{z_j^N}{1-z_jz_{j+1}} \prod_{j=1}^{n} P(z_{2j})P(z_{2j}^{-1})Q(z_{2j-1})Q(z_{2j-1}^{-1}) \tag{12.163}$$

is symmetric both in even and in odd variables separately. Hence $1-\prod_{k=1}^{2n} z_k$ can be rewritten (under the integration sign) as

$$1-\prod_{k=1}^{2n} z_k \equiv (1-z_1z_{2n})(1+\sum_{q=1}^{n-1}\prod_{r=2}^{2q+1} z_r). \tag{12.164}$$

Next, note that the summand

$$\frac{1}{1-z_1z_{2l}}(1-z_{2l}z_{2l+1})(1-z_{2n}z_1)\prod_{m=2l+2}^{2n-1} z_m \tag{12.165}$$

does not involve any of the variables $\{z_i\}_{i=1}^{2l-1}$. Hence the product $1-\prod_{k=1}^{2l} z_k$ becomes

$$1-\prod_{k=1}^{2l} z_k \equiv (1-z_1z_{2l})(1+\sum_{q=1}^{l-1}\prod_{r=2}^{2q+1} z_r). \tag{12.166}$$

Then the relevant factor of the integrand of the right-hand side of (12.162) becomes

$$(1-z_{2n}z_1)\left(\sum_{l=1}^{n-1}(1-z_{2l}z_{2l+1})(1+\sum_{q=1}^{l-1}\prod_{r=2}^{2q+1} z_r)\prod_{m=2l+2}^{2n-1} z_m - (1+\sum_{q=1}^{n-1}\prod_{r=2}^{2q+1} z_r)\right)$$
$$= (1-z_{2n}z_1)\left(\sum_{l=1}^{n-1}(1+\sum_{q=1}^{l-1}\prod_{r=2}^{2q+1} z_r)(\prod_{m=2l+2}^{2n-1} z_m - \prod_{m=2l}^{2n-1} z_m) - (1+\sum_{q=1}^{n-1}\prod_{r=2}^{2q+1} z_r)\right). \tag{12.167}$$

After expansion of the first summand this becomes

$$(1-z_{2n}z_1)\left(\sum_{l=1}^{n-1}\prod_{m=2l+2}^{2n-1} z_m - \sum_{l=1}^{n-1}\prod_{r=2}^{2n-1} z_r - (1+\sum_{q=1}^{n-1}\prod_{r=2}^{2q+1} z_r)\right) \tag{12.168}$$

under integration which, after summation, reduces to

$$-n(1-z_{2n}z_1)\prod_{r=2}^{2n-1} z_r. \tag{12.169}$$

Thus the equality (12.160) holds, and hence we have proven the desired result (12.149)-(12.151).

The form factor expansion

It remains to show for $T < T_c$ that the exponential form of the correlation (12.149)-(12.151) can be rewritten in the form factor representation (10.76):

$$D_N = (1-t)^{1/4}\{1+\sum_{n=1}^{\infty} f_N^{(2n)}\} \tag{12.170}$$

with

$$f_N^{(2n)} = \frac{1}{(n!)^2}\frac{1}{(2\pi i)^{2n}}\oint \prod_{j=1}^{2n} dz_j z_j^N \prod_{j=1}^{n} Q(z_{2j-1})Q(z_{2j-1}^{-1})P(z_{2j})P(z_{2j}^{-1})$$
$$\prod_{1\le j\le n}\prod_{1\le k\le n}\left(\frac{1}{1-z_{2j-1}z_{2k}}\right)^2 \prod_{1\le j<k\le n}(z_{2j-1}-z_{2k-1})^2(z_{2j}-z_{2k})^2. \tag{12.171}$$

To do this we denote by π a partition of n into a set of $\nu(\pi)$ pairs $\{(n_i, m_i)\}_{i=1}^{\nu}$ such that $n_i \neq n_k$ for $i \neq k$ and

$$\sum_{i=1}^{\nu(\pi)} n_i m_i = n \tag{12.172}$$

and let $\mathcal{P}(n)$ represent the set of all such partitions. As examples we have:

$$\begin{array}{ll} \mathcal{P}(1) = \{(1,1)\} & \nu(\pi)=1 \\ \mathcal{P}(2) = \{(1,2)\},\ \{(2,1)\} & \nu(\pi)=1 \\ \mathcal{P}(3) = \{(1,3)\},\ \{(3,1)\} & \nu(\pi)=1 \\ \qquad\quad\ \{(1,1),(2,1)\} & \nu(\pi)=2 \\ \mathcal{P}(4) = \{(1,4)\},\ \{(4,1)\},\ \{(2,2)\} & \nu(\pi)=1 \\ \qquad\quad\ \{(1,2),(2,1)\},\{(3,1),(1,1)\} & \nu(\pi)=2 \end{array} \tag{12.173}$$

Using this decomposition the exponential in (12.149) may be expanded into the form (12.170) and we find

$$f_N^{(2n)} = \sum_{\pi\in\mathcal{P}(n)}\prod_{i=1}^{\nu(\pi)}\frac{1}{m_i!}\left(F_N^{(2n_i)}\right)^{m_i}. \tag{12.174}$$

Then using (12.151) in (12.174) we find

$$f_N^{(2n)} = \frac{(-1)^n}{(2\pi)^{2n}} \oint \prod_{j=1}^{2n} dz_j z_j^N \prod_{j=1}^{n} Q(z_{2j-1})Q(z_{2j-1}^{-1})P(z_{2j})P(z_{2j}^{-1}) \frac{1}{1 - z_{2j-1}z_{2j}}$$

$$\sum_{\pi \in P(n)} \prod_{k=1}^{\nu(\pi)} \frac{(-1)^{m_k}}{m_k! n_k^{m_k}} \prod_{p=1}^{m_k} \prod_{q=1}^{n_k} \frac{1}{1 - z_{\sum_{r=1}^{k-1} 2m_r n_r + 2(p-1)n_k + 2q} z_{\sum_{r=1}^{k-1} 2m_r n_r + 2(p-1)n_k + 2q \oplus_k 1}} \tag{12.175}$$

where we have split the $2n$ factors of $1/(1 - z_j z_k)$ into two sets of n factors each using (12.172) and in the last line the notation $2q \oplus_k 1$ is defined as

$$2q \oplus_k + 1 = \begin{matrix} 2q + 1 & \text{if } q < n_k \\ 1 & \text{if } q = n_k. \end{matrix} \tag{12.176}$$

As an illustration, for $n = 3$,

$$f_N^{(6)} = \frac{1}{(2\pi)^6} \oint \prod_{j+1}^{6} dz_j z_j^N \prod_{j=1}^{3} (Q(z_{2j-1})Q(z_{2j-1}^{-1})P(z_{2j})P(z_{2j}^{-1}))$$

$$\frac{1}{1 - z_1 z_2} \frac{1}{1 - z_3 z_4} \frac{1}{1 - z_5 z_6}$$

$$\times \left(\frac{1}{3!} \frac{1}{1 - z_2 z_1} \frac{1}{1 - z_4 z_3} \frac{1}{1 - z_6 z_5} + \frac{1}{3} \frac{1}{1 - z_2 z_3} \frac{1}{1 - z_4 z_5} \frac{1}{1 - z_6 z_1} \right.$$

$$\left. - \frac{1}{2} \frac{1}{1 - z_2 z_1} \frac{1}{1 - z_4 z_5} \frac{1}{1 - z_6 z_3} \right) \tag{12.177}$$

where the first term is from $\{(1,3)\}$, the second from $\{(3,1)\}$ and the third from $\{(1,1),(2,1)\}$.

We next wish to show that (12.175) can be written as

$$f_N^{(2n)} = \frac{1}{n!(2\pi)^{2n}} \oint \prod_{j+1}^{2n} dz_j z_j^N \prod_{j=1}^{n} Q(z_{2j-1})Q(z_{2j-1}^{-1})P(z_{2j})P(z_{2j}^{-1}) \frac{1}{1 - z_{2j-1}z_{2j}}$$

$$\times \sum_{\sigma \in S_n} \delta_\sigma \prod_{k=1}^{n} \frac{1}{1 - z_{2k} z_{\sigma(2k+1)}} \tag{12.178}$$

where S_n is the group of permutations σ of the n elements $\{2k-1\}_{k=1}^n$ and δ_σ is the parity of the permutation.

The permutations in S_n can be grouped into equivalence classes such that each member of the class has the same value of δ_σ. For example, when $n = 3$ the six permutations in S_3 are

$$S_3 = \{(1)(3)(5),\ (13)(5),\ (15)(3),\ (35)(1),\ (135),\ (153)\} \tag{12.179}$$

where the cycle (abc) denotes the cyclic permutation $a \to b \to c \to a$. These six permutations can be grouped into the set of three equivalence classes

$$E_3 = \{[(1)(3)(5)],\ [(13)(5)],\ [(135)]\} \tag{12.180}$$

and for $n = 4$ the 24 permutations of S_4 are grouped into the five equivalence classes

$$E_4 = \{[(1)(3)(5)(7)],\ [(13)(5)(7)],\ [(13)(57)],\ [(135)(7)],\ [(1357)]\} \tag{12.181}$$

Each member of the equivalence class E_n is characterized by the set (n_i, m_i) of the number m_i of cycles of length n_i. Thus the three permutations in E_3 are characterized by $[(1,3)], [(2,1),(1,1)], [(3,1)]$ and the 5 permutations in E_4 are $[(1,4)], [(2,1),(1,2)], [(2,2)], [(3,1),(1,1)], [(4,1)]$. Comparing with (12.173) we see that for $n = 1,2,3,4$ there is a bijection between $\mathcal{P}(n)$ and E_n. This bijection between $\mathcal{P}(n)$ and E_n can be shown to valid for all n. Thus the sum in (12.175) over the set $\mathcal{P}(n)$ can be reinterpreted as a sum over the equivalence classes of permutations E_n.

Every permutation in an equivalence class will give equal contributions to the integral in (12.178). Denote by $||[\sigma]||$ the number of permutations in each equivalence class $[\sigma]$. The number of permutations in any cycle of n_i elements is $(n_i - 1)!$. Therefore the number of permutations for m_i cycles each of length n_i is $[(n_i - 1)!]^{m_i}$. Furthermore consider freely choosing from n elements that are divided into a total of $\sum_{i=1}^{\nu} m_i$ cycles with m_i cycles of length n_i without distinguishing between loops with the same number of elements. There are

$$\frac{n!}{\prod_{i=1}^{\nu}[m_i!(n_i!)^{m_i}]} \tag{12.182}$$

ways of doing this and thus the total number of permutations in an equivalence class is

$$||[\sigma]|| = \frac{n!\prod_{i=1}^{\nu}[(n_i-1)!]_i^m}{\prod_{i=1}^{\nu}[m_i!(n_i!)^{m_i}]} = \frac{n!}{\prod_{i=1}^{\nu} n_i^{m_i} m_i!}. \tag{12.183}$$

Finally we note that, for any permutation σ in the equivalence class $[\sigma]$,

$$\delta_\sigma = (-1)^n \prod_{k=1}^{\nu(\pi)} (-1)^{m_k}. \tag{12.184}$$

Thus from (12.175) we have proven that

$$f_N^{(2n)} = \frac{1}{n!(2\pi)^{2n}} \oint \prod_{j=1}^{2n} dz_j z_j^N \prod_{j=1}^{n} Q(z_{2j-1})Q(z_{2j-1}^{-1})P(z_{2j})P(z_{2j}^{-1})\frac{1}{1 - z_{2j-1}z_{2j}}$$
$$\sum_{\sigma\in E_n} \delta_\sigma ||[\sigma]|| \prod_{k=1}^{n} \frac{1}{1 - z_{2k}z_{\sigma(2k+1)}} \tag{12.185}$$

and thus if we extend the sum in (12.185) from all equivalence classes in E_n to all permutations in S_n we obtain the desired result (12.178).

We then symmetrize the right-hand side of (12.178) over all $n!$ permutations of the n odd variables z_{2j-1} to find

$$f_N^{(2n)} = \frac{1}{(n!)^2(2\pi)^{2n}} \oint \prod_m dz_m z_m^N \prod_{j=1}^{n} \{Q(z_{2j-1})Q(z_{2j-1}^{-1})P(z_{2j})P(z_{2j}^{-1})$$
$$\times \left(\sum_{\sigma \in S_n} \delta_\sigma \prod_{k=1}^{n} \frac{1}{1 - z_{2j} z_{\sigma(2k-1)}} \right)^2 \}. \tag{12.186}$$

Finally we note that the expression in the integrand of (12.186)

$$\sum_{\sigma \in S_n} \delta_\sigma \prod_{k=1}^{n} \frac{1}{1 - z_{2j} z_{\sigma(2k-1)}} \tag{12.187}$$

will vanish if $z_{2j} = z_{2k}$ or if $z_{2j-1} = z_{2k-1}$ for all j, k. Thus

$$\sum_{\sigma \in S_n} \delta_\sigma \prod_{k=1}^{n} \frac{1}{1 - z_{2j} z_{\sigma(2k-1)}}$$
$$= A_n \prod_{1 \le j \le n} \prod_{1 \le k \le n} \left(\frac{1}{1 - z_{2j-1} z_{2k}} \right) \prod_{1 \le j < k \le n} (z_{2j-1} - z_{2k-1})(z_{2j} - z_{2k}). \tag{12.188}$$

By letting $z_{2n} = z_{2n-1} = 0$ we find that

$$A_n = A_{n-1}. \tag{12.189}$$

It is obvious that $A_1 = 1$ and therefore $A_n = 1$ for all n. Thus using (12.188) in (12.186) we see that we have proven the desired result (12.171).

12.3.2 Expansion for $T > T_c$

When $T > T_c$ the row and diagonal correlations are still given by the Toeplitz determinant (12.1) with the generating function $C(e^{i\theta})$ given by (12.3). However, in contrast to $T < T_c$ we see from (12.4) and (12.5) that $\alpha_2 > 1$ and therefore $\ln C(\xi)$ is not continuous on the unit circle $|\xi| = 1$.

We define a new function $\hat{C}(\xi)$ for which $\ln \hat{C}(\xi)$ is continuous by

$$\hat{C}(\xi) = \xi C(\xi) = \left[\frac{(1 - \alpha_1 \xi)(1 - \alpha_2^{-1} \xi)}{(1 - \alpha_1 \xi^{-1})(1 - \alpha_2^{-1} \xi^{-1})} \right]^{1/2} \tag{12.190}$$

which has the Wiener–Hopf factorization

$$\hat{C}(\xi) = \hat{P}(\xi)^{-1} \hat{Q}(\xi^{-1})^{-1} \tag{12.191}$$

with

$$\hat{Q}(\xi) = [(1 - \alpha_1 \xi)(1 - \alpha_2^{-1} \xi)]^{1/2} = 1/\hat{P}(\xi) \tag{12.192}$$

where $\hat{P}(\xi)$ and $\hat{Q}(\xi)$ are analytic and nonzero for $|\xi| < 1$.

From $\hat{C}(\xi)$ we define a new $N+1 \times N+1$ Toeplitz determinant $\hat{D}_{N+1}$

$$\hat{D}_{N+1} = \begin{vmatrix} b_0 & b_{-1} & \cdots & b_{-N} \\ b_1 & b_0 & \cdots & b_{-N+1} \\ \vdots & \vdots & & \vdots \\ b_N & b_{N-1} & \cdots & b_0 \end{vmatrix} \tag{12.193}$$

where

$$b_n = \frac{1}{2\pi} \int_0^{2\pi} d\theta e^{-in\theta} \hat{C}(e^{i\theta}) = a_{n-1}. \tag{12.194}$$

We see from (12.193) that if we remove the first row and the last column from the $N+1$-dimensional determinant (12.193) we obtain the N-dimensional determinant (12.49). Therefore we may write

$$D_N = \frac{D_N}{\hat{D}_{N+1}} \hat{D}_{N+1} \tag{12.195}$$

and because the ratio $\frac{D_N}{\hat{D}_{N+1}}$ is $(-1)^N$ times the cofactor of the element in the first row and last column of $\hat{D}_{N+1}$ we may write

$$D_N = (-1)^N x_N^{(N)} \hat{D}_{N+1} \tag{12.196}$$

where $x_n^{(N)}$ with $0 \le n \le N$ satisfies the linear equations

$$\sum_{m=0}^{N} b_{n-m} x_m^{(N)} = \delta_{n,0}. \tag{12.197}$$

The logarithm of the generating function $\hat{C}(\xi)$ for $\hat{D}_N$ is continuous and periodic on the unit circle $|\xi| = 1$ and thus may be expanded in exactly the same fashion as we expanded the determinant D_N in the previous section. We obtain the limit $N \to \infty$ by use of Szegö's theorem to find

$$\lim_{N\to\infty} (-1)^N \hat{D}_N = \left[(1-\alpha_1^2)(1-\alpha_2^{-2})(1-\alpha_1/\alpha_2)^2\right]^{1/4} = M_*^2 \tag{12.198}$$

which defines M_*^2 and we note that, for the diagonal correlation,

$$M_*^2 = (1-t)^{1/4} \tag{12.199}$$

where for $T > T_c$ we are defining

$$t = (\sinh 2E^v\beta \sinh 2E^h\beta)^2. \tag{12.200}$$

Thus we immediately find from (12.149) that

$$D_N = -x_N^{(N)} M_*^2 e^{\hat{F}_{N+1}} \tag{12.201}$$

where

$$\hat{F}_N = \sum_{n=1}^{\infty} \hat{F}_N^{(2n)} \tag{12.202}$$

with

$$\hat{F}_N^{(2n)} = \frac{(-1)^{n+1}}{n} \frac{1}{(2\pi)^{2n}} \oint \prod_{j=1}^{2n} \frac{dz_j \, z_j^N}{1 - z_j z_{j+1}} \prod_{j=1}^{n} \hat{P}(z_{2j}) \hat{P}(z_{2j}^{-1}) \hat{Q}(z_{2j-1}) \hat{Q}(z_{2j-1}^{-1}) \tag{12.203}$$

The path of integration is along the unit circle $|z_j| = 1 - \epsilon$.

Computation of the ratio $D_N/\hat{D}_{N+1}$

The ratio $x_N^{(N)}$ is computed from the linear equations (12.197) and because this equation is identical in form to (12.110) we may use the definitions of Fourier transform (12.113)–(12.120) to again derive by means of the Wiener–Hopf procedure equations (12.124)–(12.127). The desired element $x_N^{(N)}$ is obtained as

$$x_N^{(N)} = \lim_{\xi \to 0} X(\xi^{-1}) \xi^N \tag{12.204}$$

and this will be obtained by using (12.125) with $\xi \to \xi^{-1}$, (12.126), (12.127) and the factorization (12.191), (12.192). Thus we consider the three equations

$$\begin{aligned} V_N(\xi) &= -\hat{P}(\xi)\{[\hat{Q}(\xi)]'_+ + [\hat{Q}(\xi) U_N(\xi^{-1}) \xi^{-N}]'_+\} \\ &= \hat{P}(\xi) - 1 - \hat{P}(\xi)[\hat{Q}(\xi) U_N(\xi^{-1}) \xi^{-N}]'_+ \end{aligned} \tag{12.205}$$

$$X_N(\xi^{-1}) \xi^N = \hat{Q}(\xi)\{[\hat{P}(\xi^{-1}) \xi^N]_+ + [\hat{P}(\xi^{-1}) V_N(\xi) \xi^N]_+\} \tag{12.206}$$

$$U_N(\xi^{-1}) = -\hat{Q}(\xi^{-1})\{[\hat{P}(\xi^{-1}) \xi^N]_- + [\hat{P}(\xi^{-1}) V_N(\xi) \xi^N]_-\}. \tag{12.207}$$

We obtain a first approximation for $X_N^{(N)}$ by replacing $U_N(\xi)$ by zero in (12.205) to write

$$V_N^{(1)}(\xi) = \hat{P}(\xi) - 1. \tag{12.208}$$

We then use $V_N^{(1)}(\xi)$ in (12.206) to obtain

$$X_N^{(1)}(\xi^{-1}) \xi^N = \hat{Q}(\xi)[\hat{P}(\xi^{-1}) \hat{P}(\xi) \xi^N]_+. \tag{12.209}$$

Thus letting $\xi \to 0$ in (12.209) we find from (12.204) the first approximation to $x_N^{(N)}$ which we denote as $\hat{G}_N^{(1)}$

$$\hat{G}_N^{(1)} = \frac{1}{2\pi i} \oint d\xi \hat{P}(\xi^{-1}) \hat{P}(\xi) \xi^{N-1}. \tag{12.210}$$

We continue to compute the second approximation $U_N^{(2)}(\xi^{-1})$ by using $V_N^{(1)}(\xi)$ as given by (12.208) in (12.207) to obtain

$$U_N^{(2)}(\xi^{-1}) = -\hat{Q}(\xi^{-1})[\hat{P}(\xi^{-1})\hat{P}(\xi)\xi^N]_-. \tag{12.211}$$

Then we use (12.211) in (12.205) to obtain

$$V_N^{(2)}(\xi) = \hat{P}(\xi) - 1 + \hat{P}(\xi)[\hat{Q}(\xi)\hat{Q}(\xi^{-1})\xi^{-N}[\hat{P}(\xi^{-1})\hat{P}(\xi)\xi^N]_-]'_+ \tag{12.212}$$

and using (12.212) in (12.206) we find

$$\begin{aligned} X_N^{(2)}(\xi^{-1})\xi^N &= \hat{Q}(\xi)\{[\hat{P}(\xi^{-1})\hat{P}(\xi)\xi^N]_+ \\ &+[\hat{P}(\xi^{-1})\hat{P}(\xi)\xi^N[\hat{Q}(\xi^{-1})\hat{Q}(\xi)\xi^{-N}[\hat{P}(\xi^{-1})\hat{P}(\xi)\xi^N]_-]'_+]_+\}. \end{aligned} \tag{12.213}$$

Thus letting $\xi \to 0$ in (12.213) and using (12.140) we find from (12.204)

$$x_N^{(N)(2)} = \hat{G}_N^{(1)} + \hat{G}_N^{(3)} \tag{12.214}$$

where

$$\begin{aligned} \hat{G}_N^{(3)} &= \frac{1}{(2\pi i)^3}\oint dz_1 z_1^N \hat{P}(z_1)\hat{P}(z_1^{-1})\oint dz_2 \frac{z_2^{N+1}}{1-z_1z_2}\hat{Q}(z_2)\hat{Q}(z_2^{-1}) \\ &\times \oint dz_3 \frac{z_3^N}{1-z_2z_3}\hat{P}(z_3)\hat{P}(z_3^{-1}). \end{aligned} \tag{12.215}$$

Continuing in this fashion we find that the nth approximation to $x_N^{(N)}$ is given as

$$x_N^{(n)}(N) = \sum_{j=0}^{n-1} \hat{G}_N^{(2j+1)} \tag{12.216}$$

with

$$\begin{aligned} \hat{G}_N^{(2n+1)} &= \frac{1}{(2\pi i)^{2n+1}}\oint dz_1 z_1^N \hat{P}(z_1)\hat{P}(z_1^{-1})\oint dz_2 \frac{z_2^{N+1}}{1-z_1z_2}\hat{Q}(z_2)\hat{Q}(z_2^{-1}) \\ \cdots &\oint dz_{2n}\frac{z_{2n}^{N+1}}{1-z_{2n-1}z_{2n}}\hat{Q}(z_{2n})\hat{Q}(z_{2n}^{-1})\oint dz_{2n+1}\frac{z_{2n+1}^N}{1-z_{2n}z_{2n+1}}\hat{P}(z_{2n+1})\hat{P}(z_{2n+1}^{-1}). \end{aligned} \tag{12.217}$$

Thus from (12.201) we see that we have proven that, for $T > T_c$,

$$D_N = M_*^2 G_N e^{\hat{F}_{N+1}} \tag{12.218}$$

where

$$G_N = -\sum_{j=0}^{\infty} \hat{G}_N^{(2j+1)} \tag{12.219}$$

with $\hat{G}_N^{(2j+1)}$ given by (12.217), $\hat{F}_N$ by (12.202) and (12.203) and M_*^2 by (12.198).

The form factor expansion

It remains to show that for $T > T_c$ the determinant D_N given by (12.217)–(12.219) can be put into the form factor representation (10.78)

$$D_N = M_*^2 \sum_{n=0}^{\infty} f_N^{(2n+1)} \tag{12.220}$$

with

$$f_N^{(2n+1)} = \frac{1}{n!}\frac{1}{(n+1)!}\frac{1}{(2\pi i)^{2n+1}} \oint \prod_{j=1}^{2n+1} (dz_j z_j^N) \prod_{j=1}^{n+1} z_{2j-1}^{-1} \hat{P}(z_{2j-1}) \hat{P}(z_{2j-1}^{-1})$$
$$\prod_{j=1}^{n} z_{2j} \hat{Q}(z_{2j}) \hat{Q}(z_{2j}^{-1}) \prod_{1 \le j \le n+1} \prod_{1 \le k \le n} \left(\frac{1}{1 - z_{2j-1} z_{2k}} \right)^2$$
$$\prod_{1 \le j < k \le n+1} (z_{2j-1} - z_{2k-1})^2 \prod_{1 \le j < k \le n} (z_{2j} - z_{2k})^2. \tag{12.221}$$

To do this we first use (12.217)-(12.219) and the expansion of $e^{\hat{F}_{N+1}}$ which is obtained from 12.3.2 to write

$$f_N^{(2n+1)} = \sum_{k=0}^{n} G_N^{(2k+1)} \hat{f}_{N+1}^{(2n-2k)} \tag{12.222}$$

where $\hat{f}_N^{(2n)}$ is given by (12.178) with P and Q replaced by $\hat{P}$ and $\hat{Q}$. Hence it follows that

$$f_N^{(2n+1)} = -\frac{i}{(2\pi)^{2n+1}} \prod_{i=1}^{2n+1} \oint_{|z_i|=1-\epsilon} dz_i z_i^{N+1} \prod_{l=1}^{n+1} \hat{P}(z_{2l-1}) \hat{P}(z_{2l-1}^{-1}) \prod_{m=1}^{n} \hat{Q}(z_{2m}) \hat{Q}(z_{2m}^{-1})$$
$$\times \frac{1}{z_{2n+1}} \prod_{p=1}^{n} \frac{1}{1 - z_{2p-1} z_{2p}} \sum_{k=0}^{n} \frac{(-1)^k}{(n-k)! z_{2n-2k+1}}$$
$$\times \sum_{\sigma \in S_{n-k}} \text{sign}(\sigma) \prod_{q=1}^{n-k} \frac{1}{1 - z_{2q-1} z_{\sigma(2q)}} \prod_{s-n-k+1}^{n} \frac{1}{1 - z_{2s} z_{2s+1}} \tag{12.223}$$

As an example

$$f_n^{(5)} = -\frac{i}{(2\pi)^5} \prod_{i=1}^{5} \oint_{|z_i|=1-\epsilon} dz_i z_i^{N+1} \prod_{l=1}^{3} \hat{P}(z_{2l-1}) \hat{P}(z_{2l-1}^{-1}) \prod_{m=1}^{2} \hat{Q}(z_{2m}) \hat{Q}(z_{2m}^{-1})$$
$$\times \frac{1}{z_5(1 - z_1 z_2)(1 - z_3 z_4)} \left(\frac{1}{2 z_5} \left(\frac{1}{(1 - z_1 z_2)(1 - z_2 z_4)} - \frac{1}{(1 - z_1 z_4)(1 - z_2 z_3)} \right) \right.$$
$$\left. - \frac{1}{z_3(1 - z_1 z_2)(1 - z_4 z_5)} + \frac{1}{z_1(1 - z_2 z_3)(1 - z_4 z_5)} \right). \tag{12.224}$$

Let $(i_1^{(k)}, \ldots, i_n^{(k)}) = (1, \ldots, n-k, n-k+2, \ldots, n+1)$. It follows by symmetry that (12.223) can be rewritten as

$$f_N^{(2n+1)} = -\frac{i}{(2\pi)^{2n+1}} \prod_{i=1}^{2n+1} \oint_{|z_i|=1-\epsilon} dz_i z_i^{N+1} \prod_{l=1}^{n+1} \hat{P}(z_{2l-1})\hat{P}(z_{2l-1}^{-1}) \prod_{m=1}^{n} \hat{Q}(z_{2m})\hat{Q}(z_{2m}^{-1})$$
$$\frac{1}{n+1}\sum_{r=0}^{n} \frac{(-1)^r}{z_{2n-2r+1}} \prod_{p=1}^{n} \frac{1}{1 - z_{2i^{(r)}{}_p-1} z^{2i_p^{(r)}}}$$
$$\frac{1}{n!}\sum_{k=0}^{n} \frac{1}{z_{2n-2k+1}} \sum_{\sigma\in s_n} \text{sign}(\sigma) \prod_{q=1}^{n} \frac{1}{1 - z_{2i_q^{(k)}-1} z_{\sigma(2i_q^{(k)})}}. \qquad (12.225)$$

In particular

$$f_N^{(5)} = -\frac{i}{(2\pi)^5} \prod_{i=1}^{5} \oint_{|z_i|=1-\epsilon} dz_i \; z_i^{N+1} \prod_{l=1}^{3} \hat{P}(z_{2l-1})\hat{P}(z_{2l-1}^{-1}) \prod_{m=1}^{2} \hat{Q}(z_{2m})\hat{Q}(z_{2m}^{-1})$$
$$\frac{1}{3}\left(\frac{1}{z_5}\frac{1}{1-z_1z_2}\frac{1}{1-z_3z_4} - \frac{1}{z_3}\frac{1}{1-z_1z_2}\frac{1}{1-z_4z_5} + \frac{1}{z_1}\frac{1}{1-z_2z_3}\frac{1}{1-z_4z_5}\right)$$
$$\left\{\frac{1}{2}\frac{1}{z_5}\left(\frac{1}{1-z_1z_2}\frac{1}{1-z_3z_4} - \frac{1}{1-z_1z_4}\frac{1}{1-z_2z_3}\right)\right.$$
$$-\frac{1}{2}\frac{1}{z_3}\left(\frac{1}{1-z_1z_2}\frac{1}{1-z_4z_5} - \frac{1}{1-z_1z_4}\frac{1}{1-z_2z_5}\right)$$
$$\left.+\frac{1}{2}\frac{1}{z_1}\left(\frac{1}{1-z_2z_3}\frac{1}{1-z_4z_5} - \frac{1}{1-z_3z_4}\frac{1}{1-z_2z_5}\right)\right\}. \qquad (12.226)$$

Since all permutations of the even elements are present in the sum $\sum_{k=0}^{n}$ symmetry allows the permutation of all even elements in the sum $\sum_{r=0}^{n}$. But the sum $\sum_{k=0}^{n}\sum_{\sigma S_n}$ may be rewritten as the sum $\sum_{\sigma\in S_{n-1}}$ of permutations of the odd elements. Therefore we find

$$f_N^{(2n+1)} =$$
$$-\frac{i}{(2\pi)^{2n+1}} \lim_{\epsilon\to 0} \prod_{i=1}^{2n+1} \oint_{|z_i|=1-\epsilon} dz_i \; z_i^{N+1} \prod_{l=1}^{n+1} \hat{P}(z_{\ l})\hat{P}(z_l^{-1}) \prod_{m=1}^{n} \hat{Q}(z_m)\hat{Q}(z_m^{-1})$$
$$\frac{1}{n!(n+1)!}\left(\sum_{\sigma\in S_{n+1}} \text{sign}(\sigma)\frac{1}{z_{\sigma(2n+1)}} \prod_{q=1}^{n} \frac{1}{1 - z_{\sigma(2q-1)}z_{2q}}\right)^2. \qquad (12.227)$$

An argument similar to the one given in 12.3.1 shows that

$$\sum_{\sigma\in S_{n+1}} \text{sign}(\sigma)\frac{1}{z_{\sigma(2n+1)}} \prod_{q=1}^{n} \frac{1}{1 - z_{\sigma(2q-1)}z_{2q}} = \prod_{j=1}^{n+1} \frac{1}{z_{z_j-1}}$$
$$\prod_{k=1}^{n} \frac{1}{1 - z_{2j-1}z_{2k}} \prod_{1\le l<m\le n+1} (z_{2l-1} - z_{2m-1}) \prod_{1\le p<q\le n} (z_{2p} - z_{2q}). \quad (12.228)$$

Thus $f_N^{(2n+1)}$ is given by (12.221) as desired.

12.4 Asymptotic expansions of $C(N,N)$ and $C(0,N)$ for $N \to \infty$

For $T < T_c$ and $T > T_c$ the large N behavior of the correlation functions may now be very efficiently obtained from the form factor expansions. However, for $T = T_c$ each form factor $(1-t)^{1/4} f_N^{(n)}$ vanishes and a separate computation, starting from the determinental representation, must be carried out. Thus we treat the three cases of $T < T_c$, $T > T_c$ and $T = T_c$ separately.

12.4.1 Large N for $T < T_c$

To obtain the large N expansion of $C(N,N)$ and $C(0,N)$ for $T < T_c$ we use the expression (12.170) and (12.171) of the form factors $(1-t)^{1/4} f_N^{(2n)}$ and deform the contours of integration from the unit circles $|z_j| = 1$ to the branch cuts which run from $z_j = \alpha_1$ to $z_j = \alpha_2$. Thus setting $z_j = \alpha_2 x_j$ we find

$$(1-t)^{1/4} f_N^{(2n)} = (1-t)^{1/4} \frac{\alpha_2^{2n(N+n)}}{(n!)^2 \pi^{2n}} \int_{\alpha_1/\alpha_2}^{1} \prod_{k=1}^{2n} dx_k x_k^N$$
$$\prod_{j=1}^{n} \left(\frac{(1-\alpha_1\alpha_2 x_{2j-1})(1-\alpha_1\alpha_2^{-1} x_{2j-1}^{-1})(1-\alpha_2^2 x_{2j})(x_{2j}^{-1}-1)}{(1-\alpha_2^2 x_{2j-1})(x_{2j-1}^{-1}-1)(1-\alpha_1\alpha_{2j} x_{2j})(1-\alpha_1\alpha_2^{-1} x_{2j}^{-1})} \right)^{1/2}$$
$$\prod_{1\le j\le n} \prod_{1\le k\le n} \left(\frac{1}{1-\alpha_2^2 x_{2k-1} x_{2j}} \right)^2 \prod_{1\le j<k\le n} (x_{2j-1}-x_{2k-1})^2 (x_{2j}-x_{2k})^2. \tag{12.229}$$

For large N the leading contribution to the integrals comes from the region $x_j \sim 1$. Thus setting

$$x_j = 1 - \zeta_j/N \tag{12.230}$$

we obtain the result that to leading order as $N \to \infty$

$$(1-t)^{1/4} f_N^{(2n)}$$
$$\sim (1-t)^{1/4} \frac{\alpha_2^{2n(N+n)}}{(n!)^2 \pi^{2n} (1-\alpha_2^2)^{2n^2} N^{2n^2}} \int_0^\infty \prod_{j=1}^{n} d\zeta_{2j} e^{-\zeta_{2j}} \zeta_{2j}^{1/2} \prod_{1\le j<k\le n} (\zeta_{2j}-\zeta_{2k})^2$$
$$\times \int_0^\infty \prod_{j=1}^{n} d\zeta_{2j-1} e^{-\zeta_{2j-1}} \zeta_{2j-1}^{-1/2} \prod_{1\le j<k\le n} (\zeta_{2j-1}-\zeta_{2k-1})^2 \tag{12.231}$$

Thus we find that as $N \to \infty$

$$C(N,N) \sim (1-t)^{1/4} \left[1 + \frac{t^{N+1}}{2\pi N^2 (1-t)^2} + \cdots \right] \tag{12.232}$$

and

$$C(0,N) \sim (1-t)^{1/4} \left[1 + \frac{\alpha_2^{2(N+1)}}{2\pi N^2 (1-\alpha_2^2)^2} + \cdots \right] \tag{12.233}$$

12.4.2 Large N for $T > T_c$

To obtain the large N expansion of $C(N,N)$ and $C(0,N)$ for $T > T_c$ we use the expressions (12.220) and (12.221) for the form factors $M_*^2 f_N^{(2n+1)}$ and deform the contours of integration from the unit circles $|z_j| = 1$ to the branch cuts which run from $z_j = \alpha_1$ to $z_j = \alpha_2^{-1}$. Thus setting $z_j = \alpha_2^{-1} x_j$ we find

$$
\begin{aligned}
M_*^2 f_N^{(2n+1)} &= M_*^2 \frac{\alpha_2^{-(2n+1)N-2n(n+1)}}{n!(n+1)!\pi^{2n+1}} \int_{\alpha_1\alpha_2}^{1} \prod_{k=1}^{2n+1} dx_k x_k^N \\
&\prod_{j=1}^{n+1} x_{2j-1}^{-1}[(1-\alpha_1\alpha_2^{-1}x_{2j-1})(1-\alpha_2^{-2}x_{2j-1})(1-\alpha_1\alpha_2 x_{2j-1}^{-1})(x_{2j-1}^{-1}-1)]^{-1/2} \\
&\prod_{j=1}^{n} x_{2j}[(1-\alpha_1\alpha_2^{-1}x_{2j})(1-\alpha_2^{-2}x_{2j})(1-\alpha_1\alpha_2 x_{2j}^{-1})(x_{2j}^{-1}-1)]^{1/2} \\
&\prod_{1\le j\le n+1}\prod_{1\le k\le n}\left(\frac{1}{1-\alpha_2^{-2}x_{2j-1}x_{2k}}\right)^2 \prod_{1\le j<k\le n+1}(x_{2j-1}-x_{2k-1})^2 \prod_{1\le j<k\le n}(x_{2j}-x_{2k})^2.
\end{aligned}
\tag{12.234}
$$

As was the case for $T < T_c$ the leading contribution for large N comes from the region $x_j \sim 1$ and thus using (12.230) we find the leading term as $N \to \infty$

$$
\begin{aligned}
&M_*^2 f_N^{(2n+1)} \\
&\sim \left[\frac{(1-\alpha_1^2)(1-\alpha_2^{-2})}{(1-\alpha_1\alpha_2)^2}\right]^{1/4} \frac{\alpha_2^{-(2n+1)N-2n(n+1)}}{n!(n+1)!\pi^{2n+1}(1-\alpha_2^{-2})^{2n(n+1)+1/2}N^{2n(n+1)+1/2}} \\
&\int_0^\infty \prod_{j=1}^{n+1} d\zeta_{2j-1} e^{-\zeta_{2j-1}} \zeta_{2j-1}^{-1/2} \prod_{1\le j<k\le n+1} (\zeta_{2j-1}-\zeta_{2k-1})^2 \\
&\int_0^\infty \prod_{j=1}^{n} d\zeta_{2j} e^{-\zeta_{2j}} \zeta_{2j}^{1/2} \prod_{1\le j<k\le n} (\zeta_{2j}-\zeta_{2k})^2.
\end{aligned}
\tag{12.235}
$$

Thus we find that as $N \to \infty$

$$
C(N,N) \sim (1-t)^{1/4} \frac{t^{N/2}}{\pi^{1/2}(1-t)^{1/2}N^{1/2}}
\tag{12.236}
$$

and

$$
C(0,N) \sim \left[\frac{(1-\alpha_1^2)(1-\alpha_2^{-2})}{(1-\alpha_1\alpha_2)^2}\right]^{1/4} \frac{\alpha_2^{-N}}{\pi^{1/2}(1-\alpha_2^{-2})^{1/2}N^{1/2}}.
\tag{12.237}
$$

12.4.3 Large N for $T = T_c$

When $T = T_c$ the form factor expansion is not an efficient way to compute the large N behavior of the correlation functions because, for each finite value of n, the form factor $(1-t)^{1/4} f_N^{(n)}$ vanishes because of the factor of $(1-t)^{1/4}$. We must therefore return to the original determinental form of the correlation (12.1)–(12.4) and specialize directly to the case $T = T_c$. However, unlike the case for $T \neq T_c$ the analysis for the row and diagonal correlations are not the same and thus we treat the two cases separately.

The diagonal correlation $C(N, N)$

For the diagonal correlation at $T = T_c$ see from (12.4) that $\alpha_1 = 0$ and $\alpha_2 = 1$ and therefore the generating function (12.3) specializes to

$$C_d(e^{i\theta})) = ie^{-i\theta/2} \tag{12.238}$$

where we have recalled that the square root in $C(e^{i\theta})$ is defined to be positive at $\theta = \pi$. Thus from (12.2)

$$c_n = \frac{i}{2\pi} \int_0^{2\pi} d\theta e^{-i\theta(n+1/2)} = \frac{1}{\pi(n+1/2)} \tag{12.239}$$

and from (12.1) the correlation written in the form

$$C(N, N) = \left(\frac{2}{\pi}\right)^N \det_N M \tag{12.240}$$

where

$$\det_N M = \begin{vmatrix} \frac{1}{\mu_0+\nu_0} & \frac{1}{\mu_0+\nu_1} & \cdots & \frac{1}{\mu_0+\nu_{N-1}} \\ \frac{1}{\mu_1+\nu_0} & \frac{1}{\mu_1+\nu_1} & \cdots & \frac{1}{\mu_1+\nu_{N-1}} \\ \vdots & \vdots & \vdots & \vdots \\ \frac{1}{\mu_{N-1}+\nu_0} & \frac{1}{\mu_{N-1}+\nu_1} & \cdots & \frac{1}{\mu_{N-1}+\nu_{N-1}} \end{vmatrix} \tag{12.241}$$

with

$$\mu_j = 2j + 1, \quad \nu_k = -2k. \tag{12.242}$$

The determinant (12.241) is known as a Cauchy determinant and we will evaluate it for arbitrary μ_j and ν_j.

When $\mu_j = \mu_k$ for any set $j \neq k$ then (12.241) has two rows equal and thus vanishes. Furthermore when $\nu_j = \nu_k$ for any set $j \neq k$ then (12.241) has two columns equal and thus vanishes. Therefore when expanded the determinant must have the factor

$$\prod_{0 \leq j < k \leq N-1} (\mu_j - \mu_k)(\nu_j - \nu_k). \tag{12.243}$$

Furthermore when expanded the common denominator of (12.241) is

$$\prod_{j=0}^{N-1} \prod_{k=0}^{N-1} (\mu_j + \nu_k). \tag{12.244}$$

Therefore by comparing the powers of μ_k and ν_k we conclude that

$$\det_N M = A_N \frac{\prod_{0 \leq j < k \leq N-1} (\mu_j - \mu_k)(\nu_j - \nu_k)}{\prod_{j=0}^{N-1} \prod_{k=0}^{N-1} (\mu_j + \nu_k)} \tag{12.245}$$

where A_N is independent of μ_k and ν_k but may depend on N.

To determine A_N we note that, if we multiply the last row of $\det_N M$ by μ_{N-1} and allow first μ_{N-1} and then ν_{N-1} to approach ∞,

$$\mu_{N-1}\det_N M \to \det_{N-1}M. \tag{12.246}$$

On the other hand, under this same limit process

$$\mu_{N-1}A_N\frac{\prod_{0\le j<k\le N-1}(\mu_j-\mu_k)(\nu_j-\nu_k)}{\prod_{j=0}^{N-1}\prod_{k=0}^{N-1}(\mu_j+\nu_k)} \to A_N\frac{\prod_{0\le j<k\le N-2}(\mu_j-\mu_k)(\nu_j-\nu_k)}{\prod_{j=0}^{N-2}\prod_{k=0}^{N-2}(\mu_j+\nu_k)}. \tag{12.247}$$

Thus we conclude that $A_N = A_{N-1}$ and since it is obvious that $A_0 = 1$ we have $A_N = 1$ and hence we have the general result

$$\det_N M = \frac{\prod_{0\le j<k\le N-1}(\mu_j-\mu_k)(\nu_j-\nu_k)}{\prod_{j=0}^{N-1}\prod_{k=0}^{N-1}(\mu_j+\nu_k)}. \tag{12.248}$$

When we specialize this to the case (12.242) we find that, when $T = T_c$,

$$C(N,N) = \left(\frac{2}{\pi}\right)^N \frac{\prod_{0\le j<k\le N-1}[2(j-k)][2(j-k)]}{\prod_{j=0}^{N-1}\prod_{k=0}^{N-1}[2(j-k)+1]}. \tag{12.249}$$

Finally noting that

$$\prod_{j=0}^{N-1}\prod_{j=0}^{N-1}[2(j-k)+1] = \prod_{0\le j<k\le N-1}[1-4(j-k)^2] \tag{12.250}$$

we obtain the closed form expression

$$\begin{aligned} C(N,N) &= \left(\frac{2}{\pi}\right)^N \prod_{0\le j<k\le N-1}\left[1-\frac{1}{4(j-k)^2}\right]^{-1} \\ &= \left(\frac{2}{\pi}\right)^N \prod_{m=1}^{N-1}\left[1-\frac{1}{4m^2}\right]^{m-N}. \end{aligned} \tag{12.251}$$

When N is small we use this directly and find, for example

$$C(1,1) = \frac{2}{\pi} \sim 0.636,619,722,\cdots \tag{12.252}$$

$$C(2,2) = \frac{16}{3\pi^2} \sim 0.540,379,646,\cdots \tag{12.253}$$

$$C(3,3) = \frac{2048}{135\pi^3} \sim 0.489,267,722,\cdots \tag{12.254}$$

We note in particular that $C(N,N)$ is a rational multiple of the single transcendental number π^{-N}.

When N is large we use the identity

$$\frac{\sin \pi\delta}{\pi\delta} = \prod_{m=1}^{\infty}\left(1 - \frac{\delta^2}{m^2}\right) \tag{12.255}$$

which, when $\delta = 1/2$, specializes to

$$\frac{2}{\pi} = \prod_{m=1}^{\infty}\left(1 - \frac{1}{4m^2}\right). \tag{12.256}$$

Thus using (12.256) in the result (12.251) we find

$$C(N,N) = \prod_{m=1}^{N-1}\left(1 - \frac{1}{4m^2}\right)^m \prod_{m=N}^{\infty}\left(1 - \frac{1}{4m^2}\right)^N. \tag{12.257}$$

We now obtain the behavior for large N by writing

$$\ln C(N,N) = \sum_{m=1}^{N-1} m \ln(1 - \frac{1}{4m^2}) + N \sum_{m=N}^{\infty} \ln(1 - \frac{1}{4m^2}). \tag{12.258}$$

The second sum is a convergent infinite series and thus its leading behavior as $N \to \infty$ can be obtained by replacing the sum by an integral. Thus

$$\lim_{N\to\infty} N \sum_{m=N}^{\infty} \ln(1 - \frac{1}{4m^2}) = \lim_{N\to\infty} N \int_N^{\infty} dm \left(-\frac{1}{4m^2}\right) = 1/4. \tag{12.259}$$

The first sum diverges logarithmically as $N \to \infty$ and to extract this divergence we write

$$\sum_{m=1}^{N-1} m \ln(1 - \frac{1}{4m^2}) = -\sum_{m=1}^{N-1}\frac{1}{4m} + \sum_{m=1}^{N-1} m\left[\ln(1 - \frac{1}{4m^2}) + \frac{1}{4m^2}\right]. \tag{12.260}$$

Recalling the definition of Euler's constant γ

$$\gamma = \lim_{n\to\infty}\left[\sum_{m=1}^{N}\frac{1}{m} - \ln N\right] = 0.577,255,665\cdots \tag{12.261}$$

we thus find, for large N,

$$\sum_{m=1}^{N-1} m \ln(1 - \frac{1}{4m^2}) = -\frac{1}{4}\ln N + \gamma + \sum_{m=1}^{\infty} m\left[\ln(1 - \frac{1}{4m^2}) + \frac{1}{4m^2}\right]. \tag{12.262}$$

Therefore as $N \to \infty$

$$C(N,N) \sim \frac{A}{N^{1/4}} \tag{12.263}$$

where

$$\ln A = \bar{A} - \gamma/4 - 1/4 \tag{12.264}$$

where

$$\bar{A} = \sum_{m=1}^{\infty} m \left[\ln(1 - \frac{1}{4m^2}) + \frac{1}{4m^2} \right] \tag{12.265}$$

is another transcendental constant. Numerically [1, 4.31 of page 264]

$$A = 0.645002448 \cdots \tag{12.266}$$

and thus, even for $N = 1$, the large N result agrees with the exact answer (12.252) to an accuracy of better than 2 percent.

The row correlation function $C(0, N)$

The row correlation function at $T = T_c$ is obtained from the determinant (12.1) by setting $\alpha_2 = 1$ in (12.3) to find

$$C_r(e^{i\theta}) = ie^{-i\theta/2} \left(\frac{1 - \alpha_1 e^{i\theta}}{1 - \alpha_1 e^{-i\theta}} \right)^{1/2} \tag{12.267}$$

with

$$\alpha_1 = \tanh^2 E^h / k_B T_c. \tag{12.268}$$

This generating function shares with the diagonal correlation function the property that it is discontinuous on $0 \le \theta \le 2\pi$. However, the ratio of the generating function (12.267) to the generating function (12.238) is analytic on the unit circle and thus we may calculate the ratio

$$\lim_{N\to\infty} \frac{C(0,N)}{C(N,N)} \tag{12.269}$$

by using the formula for the ratio of Toeplitz determinants (12.96) where we found in the derivation of Szegö's theorem that

$$\lim_{N\to\infty} \frac{C(0,N)}{C(N,N)} = \exp \sum_{k=1}^{\infty} k(g^r_{-k} g^r_k - g^d_{-k} g^d_k) \tag{12.270}$$

where

$$g^d_n = \frac{1}{2\pi} \int_0^{2\pi} d\theta e^{-in\theta} \ln C_d(e^{i\theta}) \tag{12.271}$$

and

$$g^r_n = \frac{1}{2\pi} \int_0^{2\pi} d\theta e^{-in\theta} \ln C_r(e^{i\theta}). \tag{12.272}$$

Using the explicit forms (12.238) and (12.267) of $C_d(e^{i\theta})$ and $C_r(e^{i\theta})$ in (12.271) and (12.272) we find

$$g^r_n = \frac{1}{2n} - \frac{\alpha_1^{|n|}}{2n} \quad \text{for} \quad n \neq 0 \tag{12.273}$$

$$0 \quad \text{for} \quad n = 0 \tag{12.274}$$

and

$$g_n^d = \frac{1}{2n} \quad \text{for} \quad n \neq 0 \tag{12.275}$$

$$0 \quad \text{for} \quad n = 0. \tag{12.276}$$

Therefore from (12.270) we find

$$\lim_{N\to\infty} \frac{C(0,N)}{C(N,N)} = \left(\frac{1+\alpha_1}{1-\alpha_1}\right)^{1/4} \tag{12.277}$$

and thus for large N the leading term in the row correlation at $T = T_c$ is

$$C(0,N) \sim \left(\frac{1+\alpha_1}{1-\alpha_1}\right)^{1/4} \frac{A}{N^{1/4}}. \tag{12.278}$$

12.5 Evaluation of diagonal form factor integrals

The diagonal form factors are explicitly obtained by specializing the results (12.229) for $T < T_c$ and (12.234) for $T > T_c$ to $\alpha_1 = 0$. Thus we find for $T < T_c$ that

$$f_N^{(2n)}(t) = \frac{t^{n(N+n)}}{(n!)^2\pi^{2n}} \int_0^1 \prod_{k=1}^{2n} dx_k x_k^N \prod_{j=1}^{n} \left(\frac{(1-tx_{2j})(x_{2j}^{-1}-1)}{(1-tx_{2j-1})(x_{2j-1}^{-1}-1)}\right)^{1/2}$$
$$\prod_{1\le j\le n} \prod_{1\le k\le n} \left(\frac{1}{1-tx_{2k-1}x_{2j}}\right)^2 \prod_{1\le j<k\le n} (x_{2j-1}-x_{2k-1})^2 (x_{2j}-x_{2k})^2 \tag{12.279}$$

with $t = \alpha_2^2$ and for $T > T_c$

$$f_N^{(2n+1)}(t) =$$
$$\frac{t^{(n+1/2)N+n(n+1)}}{n!(n+1)!\pi^{2n+1}} \int_0^1 \prod_{k=1}^{2n+1} dx_k x_k^N \prod_{j=1}^{n+1} x_{2j-1}^{-1}[(1-tx_{2j-1})(x_{2j-1}^{-1}-1)]^{-1/2}$$
$$\prod_{j=1}^{n} x_{2j}[(1-tx_{2j})(x_{2j}^{-1}-1)]^{1/2} \prod_{1\le j\le n+1} \prod_{1\le k\le n} \left(\frac{1}{1-tx_{2j-1}x_{2k}}\right)^2$$
$$\prod_{1\le j<k\le n+1} (x_{2j-1}-x_{2k-1})^2 \prod_{1\le j<k\le n} (x_{2j}-x_{2k})^2 \tag{12.280}$$

with $t = \alpha_2^{-2}$. In particular

$$f_N^{(1)}(t) = t^{N/2} \frac{\Gamma(N+1/2)}{\pi^{1/2} N!} F(\frac{1}{2}, N+\frac{1}{2}; N+1; t) \tag{12.281}$$

where $F(a,b;c;t)$ is the hypergeometric function.

For all $n \geq 2$ it has recently been discovered [8], by use of differential algebra as implemented on the computer, that the n-fold integrals of $f_N^{(n)}(t)$ can all be reduced to sums of products of the hypergeometric functions $F(-1/2, 1/2; 1; t)$ and $F(1/2, 1/2; 1; t)$ with coefficients that are polynomials in t. The results of these evaluations were presented in chapter 10 and here we will present the method by which the results have been obtained. Ultimately the only tool required is numerical linear algebra with integer coefficients as implemented in the DE tools and gfun packages of MAPLE and thus the results may be considered as rigorous and exact even though no "analytic" derivation has yet been published.

12.5.1 Differential equations

The first step in the reduction procedure is to obtain a linear Fuchsian differential operator $F_n(N;t)$ in the variable t which annihilates the form factor $f_N^{(n)}(t)$

$$F_n(N;t) f_N^{(n)}(t) = 0. \tag{12.282}$$

Such an operator is guaranteed to exist [9] because the integrands in the integrals are algebraic functions. The problem is to actually obtain the operator.

The annihilating operator $F_n(N;t)$ is obtained for each n and N by first expanding the integral representation of $f_N^{(n)}(t)$ as a power series in t. This is straightforward in principle but it must be realized that in practice many hundreds (if not thousands) of terms are needed to actually obtain the differential operator and the generation of the series is usually the limiting factor in the analysis. Once the (long) series is obtained we seek an operator $F_n(N;t)$ of the form

$$F_n(N;t) = \sum_{j=0}^{m} P_j^k(t) D_t^j \tag{12.283}$$

where $P_j^k(t)$ is a polynomial of maximum degree k, and D_t^j denotes the jth derivative with respect to t. This requires the solution of a linear algebra problem in a (usually) very large number of variables and is implemented in MAPLE by the use of the command seriestodiffeq in the package gfun. Once this operator is found for a series of length L_0 then if it continues to annihilate the series for (many but finite) values of length $L > L_0$ the operator (surely) annihilates the form factor $f_N^{(n)}(t)$.

12.5.2 Factorization and direct sums

The next step is to search for factorizations of $F_n(N;t)$. One such factorization is implemented on MAPLE by the command DFactor in the package DEtools. For all values of N and n which have been studied there is a factorization of the form:

for n even

$$F_2(N;t) = L_3(N;t) \cdot L_1(N;t) \tag{12.284}$$

$$F_4(N;t) = L_5(N;t) \cdot L_3(N;t) \cdot L_1(N;t) \tag{12.285}$$

$$\cdots$$

for n odd

$$F_1(N;t) = L_2(N;t) \tag{12.286}$$
$$F_3(N;t) = L_4(N;t) \cdot L_2(N;t) \tag{12.287}$$
$$F_5(N;t) = L_6(N;t) \cdot L_4(N;t) \cdot L_2(N;t) \tag{12.288}$$
$$\cdots$$

where the operator $L_n(N;t)$ is of order n. We thus see that the operator $F_n(N;t)$ right divides the operator $F_{n+2}(N,t)$. This iterative structure is called a Russian doll factorization.

The operators $L_n(N;t)$ have only been computed by MAPLE for a finite number of values of N but from these finite number of values a result for general N may be inferred. In particular, for $n = 1, 2, 3, 4$, we have

$$L_1(N;t) = D_t \tag{12.289}$$
$$L_2(N;t) = D_t^2 + \frac{2t-1}{t(t-1)}D_t + \frac{1}{4t(t-1)} - \frac{N^2}{4t^2} \tag{12.290}$$
$$L_3(N;t) = D_t^3 + 4\frac{2t-1}{t(t-1)}D_t^2 + \frac{14t^2-15t+2}{t^2(t-1)^2}D_t + \frac{8t^2-15t+5}{2t^2(t-1)^3} - N^2\left(\frac{D_t}{t^2} + \frac{1}{t^3}\right) \tag{12.291}$$
$$L_4(N;t) = L_{4,0}(t) - N^2 L_{4,2} + N^4\frac{9}{16t^4} \tag{12.292}$$

with

$$L_{4,0}(t) = D_t^4 + 10\frac{22-1}{t(t-1)}D_t^3 + \frac{241t^2-241t+46}{2t^2(t-a)^2}D_t^2 + \frac{(2t-1)(122t^2-122t+9)}{t^3(t-1)^3}D_t + \frac{81(5t-1)(5t-4)}{16t^3(t-1)^3} \tag{12.293}$$
$$L_{4,2}(t) = \frac{5}{t^2}D_t^2 + \frac{32t-23}{2t^3(t-1)}D_t + \frac{9(17t-8)}{8t^4(t-1)} \tag{12.294}$$

The operators $L_n(N;t)$ for $5 \leq n \leq 10$ are given in [8].

The factorizations (12.286) and (12.289) are not unique. For example it is found by use of MAPLE that the operators $F_3(N;t)$ and $F_2(N;t)$ have a second factorization

$$P_3(N;t)F_2(N;t) = L_1(N;t)M_3(N;t) \tag{12.295}$$
$$P_4(N;t)F_3(N;t) = L_2(N;t)M_4(N;t) \tag{12.296}$$

where $P_n(N;t)$ is a polynomial in t and $M_n(N;t)$ is an operator of degree n. For example

$$\begin{aligned} M_3(0;t) &= t^2(t^2-1)D_t^3 + 2(2t^2+2t-1)tD_t^2 \\ &+ \frac{2t^2+3t-3}{t-1}tD_t - \frac{1}{2} \end{aligned} \tag{12.297}$$

$$\begin{aligned} M_3(1;t) &= t(t^2-1)(2t-1)(t-2)D_t^3 + 3(t^4-2t^3-2t^2+3t-3)D_t^2 \\ &+ \frac{2t^5-5t^4-10t^3+20-11t+2}{t(t-1)}D_t - \frac{3}{2} \end{aligned} \tag{12.298}$$

and

$$\begin{aligned} M_4(0;t) &= t^2(t-1)^2(t^2-t+1)D_t^4 + 2(2t-1)(2t^2-2t+3)tD_t^3 \\ &+ \frac{1}{2}(29t^4-58t^3+102t^2-73t+14)D_t^2 \\ &+ \frac{(2t-1)(5t^4-10t^3+27t^2-22t+2)}{2t(t-1)}D_t \\ &+ \frac{t^4-2t^3+42t^2-41+4}{16t(t-1)} \end{aligned} \tag{12.299}$$

with

$$P_3(0;t) = t^2(t^2-1) \tag{12.300}$$

$$P_3(1;t) = t(t^2-1)(2t^2-5t+2) \tag{12.301}$$

$$P_4(0;t) = t^2(t-1)^2(t^2-t+1). \tag{12.302}$$

The operator $F_2(N;t)$ which annihilates $f_N^{(2)}(t)$ is fourth order; from (12.284) one solution to the $F_2(N;t)$ equation is the solution of the first order equation for $L_1(t)$; from (12.295) we see that any solution of the third order equation for $M_3(N;t)$ is also a solution for $F_2(N;t)$. Thus taken together the four solutions of $L_1(t)$ and $M_3(N;t)$ exhaust all solutions of $F_2(N;t)$ and thus we may write that the solution space of $F_2(N;t)$ is the direct sum of the solution spaces of $L_1(t)$ and $M_3(N;t)$ which we write as

$$F_2(N;t) = L_1(t) \oplus M_3(N;t). \tag{12.303}$$

Similarly by comparing the Russian doll factorization (12.287) of the sixth order operator $F_3(N;t)$ with the factorization (12.296) in terms of the fourth order operator $M_4(N;t)$ we see that

$$F_3(N;t) = L_2(N;t) \oplus M_4(N;t). \tag{12.304}$$

For $F_n(N;t)$ with $n \geq 5$ direct sum decompositions may be determined by use of the Maple command DFactorLCLM, and it has been discovered that, for all values of n and N investigated, all the operators F_n decompose into the direct sum of operators as

$$F_{2n-1}(N;t) = \sum_{l=0}^{n} \oplus M_{2l}(N;t) \tag{12.305}$$

$$F_{2n}(N;t) = \sum_{l=0}^{n} \oplus M_{2l+1}(N;t) \tag{12.306}$$

where the fact that the $M_l(N;t)$ in (12.305) and (12.306) are independent of n is guaranteed by the Russian doll decomposition of the operators $F_n(N;t)$.

There is a curious property of the operators $M_n(N;t)$ in the direct sum decompositions which should be explicitly noted. The differential operators $L_n(N;t)$ only have three singular points $(0, 1, \infty)$. However, in the operators $M_3(0;t)$, $M_3(1;t)$ and $M_4(0;t)$, displayed in (12.297)–(12.299), the coefficients of the highest derivatives vanish at several other values of t besides $(0, 1, \infty)$. Consequently these other vanishing points must NOT lead to genuine singularities in the solution. These "pesudosingular points" are referred to as "apparent singularities".

12.5.3 Homomorphisms of operators

To proceed further we need the notion of homomorphism of operators which generalizes the notion of similarity of matrices.

Two operators O_1 and O_2 are said to be homomorphic if there exist operators A and B such that

$$AO_1 = O_2B. \tag{12.307}$$

If there is such a homomorphism the operator B is obtained in MAPLE by use of the command Homomorphism(O_1, O_2) in the package DEtools. From (12.307) we see that if a function $f(t)$ is a solution of

$$O_1(t)f(t) = 0 \tag{12.308}$$

then the function

$$g(t) = B(t)f(t) \tag{12.309}$$

is a solution of

$$O_2(t)g(t) = 0. \tag{12.310}$$

For example, we note that

$$\text{Homomorphism}(F_2(N;t), L_1(t)) = M_3(N;t) \tag{12.311}$$

and

$$\text{Homomorphism}(F_3(N;t), L_2(N;t)) = M_4(N;t) \tag{12.312}$$

from which (12.295) and (12.296) follow by use of (12.307).

12.5.4 Symmetric powers

The last step needed in the reduction is to learn how to recognize when a solution of a differential equation can be written as the product of solutions of differential equations of lower order. We illustrate this reduction by considering the linear second order equation

$$\{\frac{d^2}{dt^2} + a(t)\frac{d}{dt} + b(t)\}y_a = 0 \tag{12.313}$$

where y_a with $a = 1, 2$ are two linearly independent solutions and determining the coefficients $c(t)$, $d(t)$ and $e(t)$ in the third order differential equation

$$\{\frac{d^3}{dt^3} + c(t)\frac{d^2}{dt^2} + d(t)\frac{d}{dt} + e(t)\}y_a y_b = 0. \tag{12.314}$$

To determine these coefficients we note that

$$\begin{aligned}
\frac{d}{dt}y_a y_b &= y_a y_b' + y_b y_a' \\
\frac{d^2}{dt^2}y_a y_b &= y_a y_b'' + y_b y_a'' + 2y_a' y_b' \\
\frac{d^3}{dt^3}y_a y_b &= y_a y_b''' + y_b y_a''' + 3(y_a' y_b'' + y_b y_a'')
\end{aligned} \tag{12.315}$$

and thus (12.315) is rewritten as

$$\begin{aligned}
y_a y_b''' + y_b y_a''' &+ 3(y_a' y_b'' + y_b y_a'') + c(t)(y_a y_b'' + y_b y_a'') \\
&+ 2c(t)y_a' y_b' + d(t)(y_a y_b' + y_b y_a') + e(t)y_a y_b = 0.
\end{aligned} \tag{12.316}$$

Now eliminate y''' and y'' from (12.316) using (12.313) in the form

$$\begin{aligned}
y_a'' &= -a(t)y_a' - b(t)y_a \\
y_a''' &= -a(t)y_a'' - [a'(t) + b(t)]y_a' - b'(t)y_a \\
&= [a^2(t) - a'(t) - b(t)]y_a' + [a(t)b(t) - b'(t)]y_a,
\end{aligned} \tag{12.317}$$

where prime indicates differentiation with respect to t, to obtain

$$\begin{aligned}
&[2c(t) - 6a(t)]y_a' y_b' + [d(t) - a(t)c(t) - a'(t) + a^2(t) - 4b(t)](y_a y_b' + y_b y_a') \\
&+[e(t) - 2b(t)(c(t) - 2a(t)) - 2b'(t)]y_a y_b = 0
\end{aligned} \tag{12.318}$$

Thus, by noting that $y_a' y_b'$, $(y_a y_b' + y_b y_a')$ and $y_a y_b$ are linearly independent, we obtain from (12.318) the three equations

$$\begin{aligned}
&c(t) - 3a(t) = 0 \\
&d(t) - a(t)c(t) - a'(t) + a^2(t) - 4b(t) = 0 \\
&e(t) - 2b(t)(c(t) - a(t)) - 2b'(t) = 0.
\end{aligned} \tag{12.319}$$

These are three linear equations for the three unknowns $c(t)$, $d(t)$ and $e(t)$, and are readily solved to give

$$\begin{aligned} c(t) &= 3a(t) \\ d(t) &= 2a(t)^2 + 4b(t) + a'(t) \\ e(t) &= 4a(t)b(t) + 2b'(t). \end{aligned} \tag{12.320}$$

Therefore we have shown that the three linearly independent solutions of the third order equation

$$g''' + 3a(t)g'' + [2a(t)^2 + 4b(t) + a'(t)]g' + (4a(t)b(t) + 2b'(t)g = 0 \tag{12.321}$$

are

$$g = y_a y_b \tag{12.322}$$

where y_a are two linearly independent solutions of the second order equation (12.314). We have used nothing more than linear algebra in this derivation.

The identical procedure will obtain the equation of order $n+1$ which annihilates the $n+1$ functions $y_{a_1} \cdots y_{a_n}$ where y_a satisfies second order equation (12.313). This linear procedure is implemented in MAPLE by the command symmetricpower in the package DEtools.

12.5.5 Results

Thus far we have proceeded in a constructive fashion; i.e. the differential operators $F_n(N;t)$ are determined from the series expansion of the form factor $f_N^{(n)}(t)$; the operators $F_n(N;t)$ are obtained from the MAPLE command DFactor and operators $M_n(N;t)$ in the direct sum decomposition are determined from the MAPLE command DFactorLCLM in the package DEtools. However, to decompose $M_n(N;t)$ we do not proceed in a "constructive" fashion but will make a "guess" and prove it to be true. The "guess" is that $M_n(N;t)$ is homomorphic with the $n-1$ symmetric product of the operator $L_2(t)$.

Testing this "guess" on the operators $M_3(0;t), M_3(1;t)$ and $M_4(0;t)$ we find that it is indeed correct and that calling

$$S_n(t) = \text{Symmetric n}^{\text{th}}\text{power}(L_2(t) \tag{12.323}$$

we find

$$\text{Homomorphisms}(S_2(t), M_3(0,t)) = t(t-1)D_t + t \tag{12.324}$$

$$\text{Homomorphisms}(S_2(t), M_3(1;t)) = t(t-1)D_t + \frac{2}{3}t - \frac{1}{3} \tag{12.325}$$

$$\text{Homomorphisms}(S_3(t), M_4(0;t)) = t(t-1)D_t + t - \frac{1}{2}. \tag{12.326}$$

However, once we have discovered that the $M_n(N;t)$ in the direct sum decomposition are homomorphic to the $(n-1)^{th}$ symmetric powers of $L_2(N;t)$ it immediately follows that the $F_n(N;t)$ themselves are homomorphic to the same symmetric powers with the same homomorphism and thus we may dispense with the need for an explicit computation of the $M_n(N;t)$ operators all together which obviates the need for dealing with the apparent singularities which appeared in the explicit constructions of $M_n(N;t)$. This is most helpful because the computing of the direct sum decomposition using DFactorLCLM takes much more time and memory than does the computations of homomorphisms. For completeness we list several of the homomorphisms of $F_n(N;t)$ with $S_n(t)$:

$$\text{Homomorphisms}(S_2(t), F_2(0,t)) = t(t-1)D_t + t \tag{12.327}$$

$$\text{Homomorphisms}(S_2(t), F_2(1;t)) = t(t-1)D_t + \frac{2}{3}t - \frac{1}{3} \tag{12.328}$$

$$\begin{aligned}\text{Homomorphisms}(S_2(t), F_2(2;t)) &= t(t-1)^2(t+1)D_t^2 \\ &+ (t-1)(3t^2 - \frac{11}{4} - 1) + t^2 - \frac{11}{4}t + 1\end{aligned} \tag{12.329}$$

$$\text{Homomorphisms}(S_3(t), F_3(0;t)) = t(t-1)D_t + t - \frac{1}{2} \tag{12.330}$$

$$\begin{aligned}\text{Homomorphisms}(S_4(t), F_4(0;t)) &= t^2(t-1)^2 D_t^2 + t(t-1)(5t-1)D_t \\ &+ 4t^2 - 3t\end{aligned} \tag{12.331}$$

In order to obtain the results for $f_N^{(n)}(t)$ given in chapter 10 it now remains to apply the homomorphism operator to the powers of the solution $F(1/2, 1/2; 1; t)$ which is the solution of the operator $L_2(0;t)$ which is regular at the origin and to determine the correct linear combination of the terms in the direct sum by matching a finite number of terms in the expansion at $t = 0$. (Note, that for the functions $f_{2N+1}^{(2n+1)}(t)$, to use MAPLE to find homomorphisms the variable $x = t^{1/2}$ must be used because MAPLE will only recognize homomorphisms with integer powers of the independent variable.) This will give a solution in terms of the basis

$$F(1/2, 1/2; 1; t) \text{ and } \frac{d}{dt}F(1/2, 1/2; 1; t). \tag{12.332}$$

Equivalently we may use as a basis the complete elliptic integrals (normalized to unity at $t = 0$)

$$\tilde{K}(t^{1/2}) = F(1/2, 1/2; 1; t) \text{ and } \tilde{E}(t^{1/2}) = F(-1/2, 1, 2; 1, t) \tag{12.333}$$

by using the relation

$$2\frac{d\tilde{K}}{dt} = (1-t)^{-1}t^{-1}(\tilde{E} - (1-t)\tilde{K}]. \tag{12.334}$$

In this basis we find from (12.327)–(12.331) that

$$f_0^{(2)}(t) = \frac{1}{2}\tilde{K}[\tilde{K} - \tilde{E}] \tag{12.335}$$

$$f_1^{(2)}(t) = \frac{1}{2}[1 - 3\tilde{K}\tilde{E} - (t-2)\tilde{K}^2] \tag{12.336}$$

$$f_2^{(2)}(t) = 1 - \frac{1}{6t}\{(6t^2 - 11t + 2)\tilde{K}^2 + (15t - 4)\tilde{K}\tilde{E} + 2(t+1)\tilde{E}^2\} \quad (12.337)$$

$$f_0^{(3)}(t) = \frac{1}{6}[\tilde{K} - (t-2)\tilde{K}^3 - 3\tilde{K}^2\tilde{E}] \quad (12.338)$$

$$f_0^{(4)}(t) = \frac{1}{24}\{4(\tilde{K} - \tilde{E})\tilde{K} - (2t-3)\tilde{K}^4 - 6\tilde{K}^3\tilde{E} + 3\tilde{K}^2\tilde{E}^2\} \quad (12.339)$$

12.5.6 Discussion

The reduction of the form factor integrals given above is both surprising and incomplete.

The reduction must be considered surprising because the integrals for $f_N^{(2)}(t)$ have been in the literature [1, 7] since 1966 whereas the reduction was only discovered [8] in 2007.

The reduction is incomplete for at least two reasons.

First of all, there is no reason that $\tilde{K}(t^{1/2})$ and $\tilde{E}(t^{1/2})$ or $F(1/2, 1/2; 1; t)$ and $F'(1/2, 1/2; 1; t)$ must be used as a basis to express the results. This is because there are homomorphisms with symmetric powers of *any* $L_2(m; t)$ and not just $L_2(0; t)$. The results given here, in chapter 10 and in [8] in the basis of $\tilde{K}(t^{1/2})$ and $\tilde{E}(t^{1/2})$ do not have a recognizable form which can be generalized to arbitrary N. It would be most desirable if the results could be rewritten in terms of solutions of $L_2(m; t)$ which would allow the identification of a general form.

Secondly the derivation given here does not reveal the structure of the integrals which is responsible for the reduction. It is most unsatisfactory that an analytic derivation of the operators $L_n(N; t)$ and an analytic derivation of the direct sum structure has not been found. Furthermore, is almost certain that related reductions for the row correlations must exist but so far they have not been found either.

References

[1] B.M. McCoy and T.T. Wu, *The two dimensional Ising model* (Harvard University Press 1973).

[2] G. Szegö, *Commun. Seminaire. Math. Univ. Lund.* suppl. dédié a Marcel Riesz. (1952) 228.

[3] A. Bottcher and P. Silberman, *Analysis of Toeplitz Operators*, (Springer 2 ed. 2005).

[4] L. Onsager, discussion, Nuovo Cimento 6, suppl. (1949) 261.

[5] C.N. Yang, The spontaneous magnetization of a two-dimensional Ising model, Phys. Rev. 85 (1952) 808–816.

[6] I. Lyberg and B.M. McCoy, Form factor expansion of the row and diagonal correlation functions of the two-dimensional Ising model, J. Phys. A 40 (2007) 3329–3346.

[7] T.T. Wu, Theory of Toeplitz determinants and spin correlations in the two-dimensional Ising model I, Phys. Rev. 149 (1066) 380–401.

[8] S. Boukraa, S. Hassani, J.-M. Maillard, B.M. McCoy, W.P. Orrick and N. Zenine, Holonomy of the Ising model form factors, J. Phys. A40 (2007) 75–111.

[9] R.P. Stanley, Differentiably finite power series, European J. Combin. 1 (1980) 175–188.

13
The star–triangle (Yang–Baxter) equation

The Pfaffian solution of the Ising model is very elementary and is sufficiently powerful that we could use it to calculate the free energy, order parameter and correlation functions of the model. However, that method gave no insight into why the Ising model could be solved and no indication that there might be other models for which exact computations could be carried out. There are, however, many other models for which exact calculations have been carried out and the tool for obtaining these models is the notion of a local relation between Boltzmann weights known as the star–triangle (Yang–Baxter) relation which originated in the study of electrical networks in 1899. In this chapter we will explain in detail the applications of the star–triangle equation in statistical mechanics. However, before giving precise definitions and deriving explicit results, it is helpful to give a historical survey of the subject. An overview of the chronology of the development of the star–triangle equation is given in Table 13.1.

13.1 Historical overview

The star–triangle equation first appeared as an equivalence of electrical circuits in [1] in 1899. However, it did not appear in statistical mechanics until a parenthetical remark was made in the 1944 paper of Onsager [2] in which the Ising model free energy was first computed. This parenthetical remark was explicitly elaborated the next year in the review paper of Wannier [3].

The star–triangle equation next appears not in the context of two-dimensional statistical mechanical models but in a paper by McGuire [4] on the scattering of bosons in one dimension which interact with the two-body delta function potential

$$V(x) = c\delta(x). \tag{13.1}$$

In this paper McGuire shows that the N-body scattering matrix factorizes into products of two-body amplitudes because the two-body scattering matrix satisfies a set of overdetermined equations. Several years later Yang [5] generalized this to multispecies scattering.

To connect this idea of factorized scattering matrices with the star–triangle equation of Onsager it is necessary to change the point of view which was used in the Pfaffian solution of the Ising model which keeps the symmetry between the vertical

Table 13.1 Historical overview of the development of the star–triangle equation from electrical networks to models with genus greater than one.

date	author(s)	contribution
1899	Kennelly [1]	Star–triangle for electrical networks
1944	Onsager [2, 3]	Star–triangle for Ising
1964	McGuire [4]	Factorized scattering of bosons with δ function interactions
1968	Yang [5]	Factorized scattering of several species with δ function interactions
1968	McCoy, Wu [6]	$[T(v), H] = 0$ for six-vertex
1970	Sutherland [7]	$[T(v), H] = 0$ for eight-vertex
1971	Baxter [8–11]	Star–triangle for eight-vertex
1973	Baxter [12–14]	Star–triangle for face models
1980	Baxter [15, 16]	The hard hexagon model
1980	Zamolodchikov [17, 18]	Tetrahedral equation
1982	Fateev, Zamalodchikov [19]	Self-dual Z_N spin model
1983	Baxter [20]	Solution of tetrahedral equations
1984	Andrews, Baxter, Forrester [21]	The RSOS models
1985	von Gehlen, Rittenberg [22]	Hamiltonian for superintegrable chiral Potts spin chain
1987	Au-Yang, McCoy, Perk, Tang, Yan [23]	Genus > 1 for the three-state chiral Potts model
1987	McCoy, Perk, Tang, Sah [24]	Fermat curve for the four-state self dual chiral Potts model
1988	Au-Yang, McCoy, Perk, Tang [25]	Product form for five-state self-dual chiral Potts model
1988	Baxter, Perk, Au-Yang [26] Au-Yang, Perk [27]	Star–triangle for N-state chiral Potts
1991	Bazhanov, Kashaev Mangazeev, Stroganov [28]	$sl(n)$ chiral Potts model
1993	Bazhanov, Baxter [29]	Connection of $sl(n)$ chiral Potts with the tetrahedral equation

and horizontal interaction energies E^v and E^h. Instead we will build up the two-dimensional lattice by defining a matrix T which adds Boltzmann weights to the lattice one row at a time. This matrix is called the transfer matrix. Its detailed definition and construction is given in section 13.2.

The connection between the star–triangle equation of the Boltzmann weights found by Onsager with the factorized scattering found by McGuire and Yang is seen by considering the dependence of the transfer matrix on the anisotropy of the interactions E^v and E^h. We parametrize this anisotropy by some variable u (called the spectral variable) and consider the dependence of the transfer matrix $T(u)$ on this variable. We will then look for models in which there is some variable in the Boltzmann weights analogous to the anisotropy in the Ising model for which the transfer matrix satisfies the very improbable equation of commuting transfer matrices

$$[T(u), T(u')] = 0. \tag{13.2}$$

The emergence of the significance of (13.2) begins with the paper of McCoy and Wu [6] in 1967 where it was shown that the transfer matrix of the then newly discovered solvable six-vertex model [30–36] (defined in section 13.3) and the XXZ anisotropic Heisenberg spin chain with periodic boundary conditions

$$H_{XXZ} = -\sum_{j=1}^{N}\{\sigma_j^x\sigma_{j+1}^x + \sigma_j^y\sigma_{j+1}^y + \Delta\sigma_j^z\sigma_{j+1}^z\} \tag{13.3}$$

whose solvability [37–41] goes back to the work of Bethe on the one-dimensional Heisenberg chain [42] with $\Delta = \pm 1$, satisfy the commutation relation

$$[T(u), H] = 0 \tag{13.4}$$

where u is a parameter in the Boltzmann weights. This was extended to the commutation of the symmetric eight-vertex model (defined also in section 13.3) with the XYZ spin chain

$$H_{XYZ} = -\sum_{j=1}^{N}\{J^x\sigma_j^x\sigma_{j+1}^x + J^y\sigma_j^y\sigma_{j+1}^y + J^z\sigma_j^z\sigma_{j+1}^z\} \tag{13.5}$$

by Sutherland [7] in 1970.

The significance of the commutation relation (13.4) may be seen by making a comparison with the classical mechanics of a system of N particles with momenta p_k and coordinates q_k specified by a Hamiltonian $H(\{p\}, \{q\})$. For such a system a set of operators J_k are called constants of the motion if they satisfy

$$\{J_k, H\}_{q,p} = 0 \tag{13.6}$$

where $\{A, B\}_{p,q}$ is called the Poisson bracket [43, (8-42) on page 252] defined by

$$\{A, B\}_{q,p} = \sum_k \left(\frac{\partial A}{\partial q_k}\frac{\partial B}{\partial p_k} - \frac{\partial A}{\partial p_k}\frac{\partial B}{\partial q_k}\right). \tag{13.7}$$

Quantum mechanics is obtained from classical mechanics by the replacement of the classical Poisson bracket of X and Y by the commutator. Thus if we consider the transfer matrix as a generating function in an appropriate variable

$$T(u) = \sum_k J_k u^k \tag{13.8}$$

we recognize and interpret (13.4) as saying that the Hamiltonian has as many constants of the motion as the Hamiltonian has degrees of freedom.

In classical mechanics a system of N particles is said to be integrable if there are N constants of the motion J_k (13.6) which satisfy for all $1 \leq j, k \leq N$

$$\{J_j, J_k\}_{p,q} = 0 \tag{13.9}$$

where by definition

$$J_1 = H. \tag{13.10}$$

When the classical constants of the motion J_k satisfy (13.9) they are said to be *in involution* or *compatible*, and the classical system is said to be *integrable*. This criterion originates in the work of Liouville in 1836 and for these systems the equations of motion may be explicitly solved in terms of action angle variables.

When the transfer matrix is interpreted as the generating function of constants of the motion (13.8) the commutation relation (13.2) is recognized as the quantum mechanical analogue of the classical integrability condition (13.9) and analogue of (13.10) is the statement that there is some value of the spectral variable u where the transfer matrix $T(u)$ reduces to the Hamiltonian.

The relevance of the star–triangle equation to the six- and eight-vertex model was recognized by Baxter in 1971 in truly monumental papers [8, 10] where it is shown that the commutation relation (13.2) of transfer matrices holds if the Boltzmann weights satisfy a set of overdetermined equations. Baxter also found [9, 11] that for the parametrization of the eight-vertex model given in [8, 10] that

$$T(u) \to T(0)\{1 + uH_{XYZ} + O(u^2)\} \tag{13.11}$$

which provides the analogue of (13.10).

Baxter recognized that the set of overdetermined equations is a vertex generalization of the spin system star–triangle equation of Onsager and for this reason the conditions of the local Boltzmann weights which allow (13.2) to hold are referred to as the star–triangle equations. These equations are also referred to in the literature as Yang–Baxter equations. These considerations were extended the following year by Baxter [12–14] to face models which subsequently led to the discovery of the solvability of the hard hexagon [15, 16] and RSOS [21] models.

The star–triangle equations were generalized from two to three dimensions by Zamolodchikov [17, 18] in 1980 where a set of equations called "tetragonal" equations were derived and solutions conjectured. A set of solutions was proven to satisfy these tetragonal equations by Baxter [20] in 1983.

In all of the preceding papers the spectral variable lies either on a genus one elliptic curve (for the eight-vertex and Ising model) or its degeneration to a genus zero trigonometric curve (for the 6 vertex model) and for many years it was folk wisdom that these were the only spectral curves possible. However, in 1987 this was proven false by Au-Yang, McCoy, Perk, Tang and Yan [23] who showed that the three state chiral Potts model has a spectral variable v which lies on a curve of genus 10 and that for the special "self dual" case this curve degenerates to the genus one (elliptic) Fermat curve

$$x^3 + y^3 = z^3. \tag{13.12}$$

This was very soon extended [24] to the self-dual four-state chiral Potts model where the spectral curve is the Fermat curve

$$x^4 + y^4 = z^4 \tag{13.13}$$

and to the self-dual five state model [25] where in addition to the fifth order Fermat curve a product form for the Boltzmann weights was obtained. The Boltzmann weights for the general case of the N-state chiral Potts model which satisfy the star–triangle equation was presented in 1988 by Baxter, Perk and Au-Yang [26] with a detailed proof published in 1989 by Au-Yang and Perk [27]. The further important generalization to the $sl(n)$ chiral Potts model was made by Bazhanov, Kashaev, Mangazeev and Stroganov [28] in 1991 and this model was shown to satisfy tetragonal equations for an $n-1$ layer chiral Potts model in three dimensions by Bazhanov and Baxter [29] in 1993.

In this chapter we will explain the derivation and solutions of star–triangle equations in detail. In section 13.2 we define three different types of models: spin (Ising, self-dual Z_N and chiral Potts models), vertex (six and eight vertex models) and face (RSOS and hard hexagons) and construct their transfer matrices in terms of the Boltzmann weights. In section 13.3 we introduce the important notion of a family of transfer matrices depending on a parameter u (which depends on the various interaction constants of the model) such that the commutation relation (13.2) holds.

In the subsequent three sections we show that the star–triangle equation implements the condition (13.2) of commuting transfer matrices and we find solutions to these equations. In section 13.4 we consider vertex models and obtain the six-vertex, eight-vertex and free fermion solutions of the star–triangle equations. In section 13.5 we consider spin models and obtain the solution of the chiral Potts model (and the special case of the self dual Z_N model). In section 13.6 we consider face models and present the RSOS and hard hexagon models. In section 13.7 we derive for the six- and eight-vertex models and the chiral Potts model the Hamiltonians H of one dimensional quantum spin chains which satisfy the commutation relation (13.4) with the transfer matrix $T(u)$ of the two-dimensional statistical model. The solutions of the three-dimensional tetragonal equations and the $sl(n)$ chiral Potts model are beyond the scope of this book.

13.2 Transfer matrices

The use of transfer matrices to solve two-dimensional models in statistical mechanics originates in the work of Kramers and Wannier [44, 45] and is the starting point of Onsager's [2] 1944 computation of the free energy of the Ising model. We will study three different types of models for which the star–triangle equation will be applied: 1) vertex models, 2) spin models and 3) face models.

Vertex models

Vertex models have state variables which lie on the links of the lattice and interaction energies, $\mathcal{E}(j', j; \mu, \nu; u)$, which depend on a parameter u, and are associated with the vertices where the variables j', j, μ, ν intersect as in Fig. 13.1. The associated local Boltzmann weight is written in the matrix form

$$W(j', j|u)|_{\mu,\nu} = \exp(-\mathcal{E}(j', j; \mu, \nu; u)/k_B T) \tag{13.14}$$

as illustrated in Fig. 13.1.

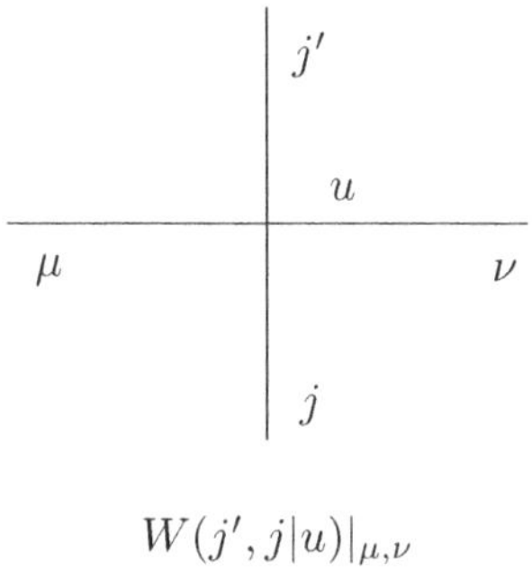

Fig. 13.1 Local Boltzmann weights of vertex models.

Spin models

Spin models, like the Ising model, have state variables which lie on the vertices of the lattice, and the interaction energies $\mathcal{E}^h(j, j'|u)$ and $\mathcal{E}^h(j, j'|u)$ are associated with the links (vertical and horizontal) which join nearest neighbors. The local Boltzmann weights

$$\begin{aligned} W^h(j, j')|u) &= \exp(-\mathcal{E}^h(j, j'|u)/k_B T) \\ W^v(j, j')|u) &= \exp(-\mathcal{E}^v(j, j'|u)/k_B T) \end{aligned} \tag{13.15}$$

of two neighboring spins are parameterized as shown in Fig. 13.2 where we have adopted the convention that for $W^h(j, j'|u)$ the site for the state variable j lies to the left of the site for j', and for $W^v(j, j'|u)$ the site for the state variable j lies below of the site for j'. We will at times indicate this convention by drawing an arrow on the bond which points from j to j'.

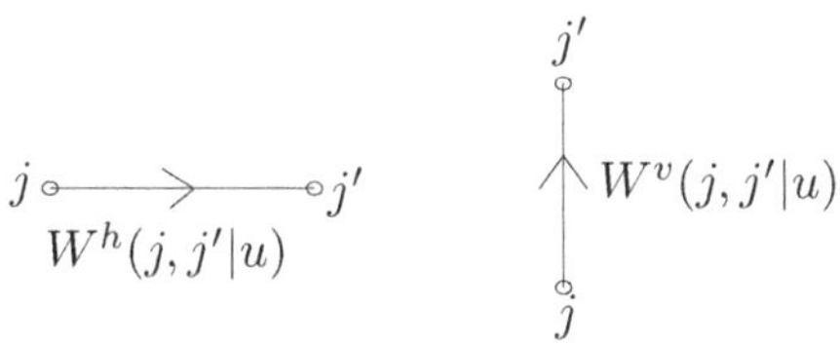

Fig. 13.2 Local Boltzmann weights of spin models. The arrows point from j to j'.

Face models

Face models have state variable which lie on the vertices of the lattice and the interaction energies $\mathcal{E}(a_1, a_2; b_1, b_2|u)$ are associated with all four spins around an elementary face as shown in Fig. 13.3. The associated local Boltzmann weights in Fig.13.3

are

$$W(a_1, a_2; b_1, b_2|u) = \exp(-\mathcal{E}(a_1, a_2; b_1, b_2|u)/k_B T). \tag{13.16}$$

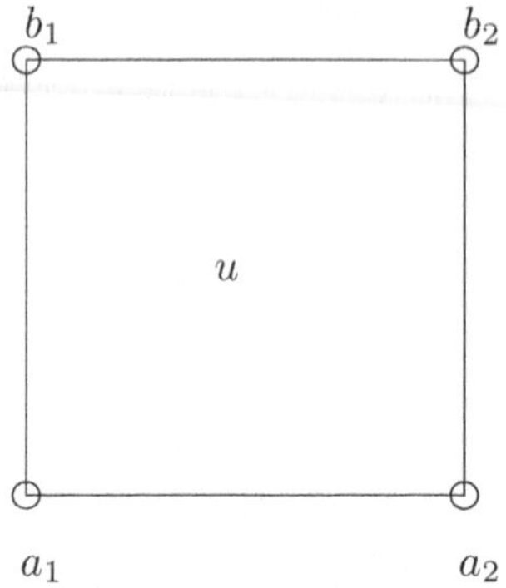

Fig. 13.3 Local Boltzmann weights for face models.

In each of the three cases the local Boltzmann weights are allowed to depend on an arbitrary parameter u.

The partition function is the sum over all the state variables of the product of all the local Boltzmann weights. Such a sum is obviously invariant under the rotation of the lattice by 90 degrees and in our treatment of the Ising model we used a method of solution which preserved this invariance. However, for the more general models of this chapter we will not use such a symmetric method of solution and instead we consider building up the full two-dimensional lattice of N_v rows and N_h columns by adding successive rows. This is the method used by Onsager [2] in his original computation of the free energy of the Ising model.

Thus, if we call $T_{\{j_2\},\{j_1\}}(u)$ the matrix of Boltzmann weights between the states $\{j_1\}$ in row 1 and the states $\{j_2\}$ in row 2, and if we impose periodic boundary conditions in the vertical direction, we may express the partition function as

$$\begin{aligned} Z &= \sum_{\{j_1\},\{j_2\},\cdots,\{j_{N_v}\}} T_{\{j_1\},\{j_2\}}(u) T_{\{j_2\},\{j_3\}}(u) \cdots T_{\{j_{N_v}\},\{j_1\}}(u) \\ &= \mathrm{Tr}\, T(u)^{N_v}. \end{aligned} \tag{13.17}$$

The matrix $T(u)$ is called the transfer matrix. If the number of states per site or bond is n, the dimension of the transfer matrix is n^{N_h}.

When the transfer matrix can be diagonalized then, calling $t_k(u)$ the eigenvalues of $T(u)$ we have

$$Z = \sum_k t_k(u)^{N_v}. \tag{13.18}$$

We are interested in the thermodynamic limit where $N_v \to \infty$. In this limit the sum in (13.18) is dominated by the largest eigenvalue and thus, calling this maximum eigenvalue $t_0(u)$, we have

$$\lim_{N_v \to \infty} \frac{1}{N_v} \ln Z = \ln t_0(u). \tag{13.19}$$

In the thermodynamic limit the number of columns N_h as well as the number of rows N_v goes to infinity. Thus the free energy per site F is given by

$$-F/k_B T = \lim_{N_v, N_h \to \infty} \frac{1}{N_v N_h} \ln Z = \lim_{N_h \to \infty} \frac{1}{N_h} \ln t_0(u). \tag{13.20}$$

Therefore the study of the thermodynamics is reduced to the computation of the largest eigenvalue of a very large matrix.

13.2.1 Explicit forms of the transfer matrix

The precise form of the transfer matrix is different for the three classes of models under consideration.

Vertex models

For vertex models with periodic boundary conditions in the horizontal direction, the row to row transfer matrix is

$$T|_{\{j'\},\{j\}} = \mathrm{Tr} W(j'_1, j_1) W(j'_2, j_2) \cdots W(j'_N, j_N) \tag{13.21}$$

where the trace is over the state variables on the horizontal links of the lattice. This is illustrated in Fig. 13.4

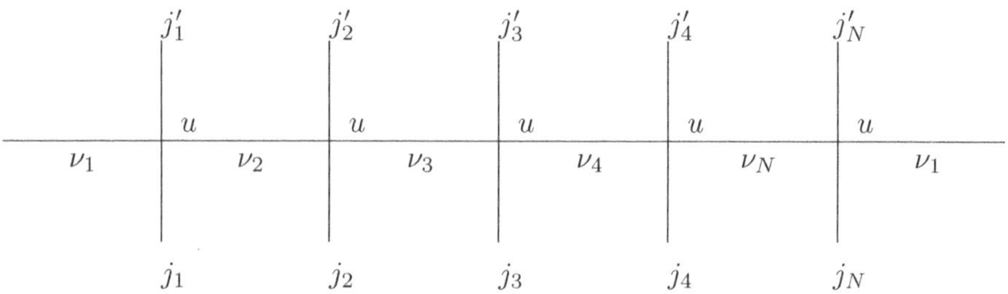

$$T(u)|_{\{j'\},\{j\}} = \mathrm{Tr} W(j'_1, j_1) W(j'_2, j_2) \cdots W(j'_N, j_N)$$

Fig. 13.4 Construction of the transfer matrix of vertex models from the local Boltzmann weights.

Spin models

For spin models we will find it convenient to consider the diagonal to diagonal transfer matrix with periodic boundary conditions in the horizontal direction

$$T(u)|_{\{j'\},\{j\}} = W^v(j_1, j'_1) W^h(j_1, j'_2) W^v(j_2, j'_2) \cdots W^v(j_N, j'_N) W^h(j_N, j'_1) \tag{13.22}$$

which is illustrated in Fig. 13.5.

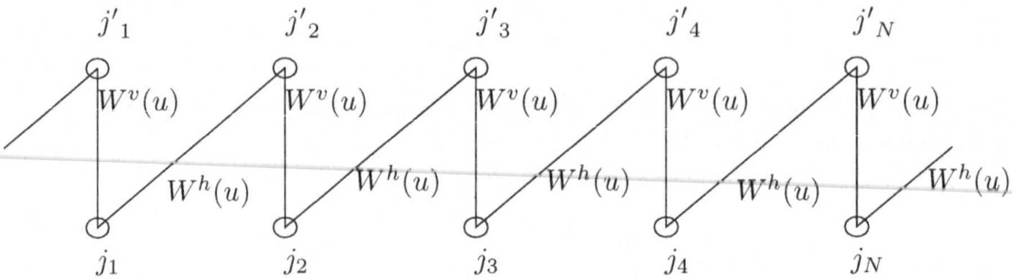

$$T(u)|_{\{j'\},\{j\}} = W^v(j_1, j'_1)W^h(j_1, j'_2)W^v(j_2, j'_2)W^h(j_2, j'_3) \cdots W^v(j_N, j'_N)W^h(j_N, j'_1)$$

Fig. 13.5 Construction of the "diagonal" transfer matrix of spin models from the local Boltzmann weights.

Face models

For face models with periodic boundary conditions in the horizontal direction the transfer matrix is

$$T|_{\{b_i\},\{a_i\}} = W(a_1, a_2; b_1, b_2)W(a_2, a_3; b_2, b_3) \cdots W(a_N, a_1; b_N, b_1) \tag{13.23}$$

which is illustrated in Fig. 13.6.

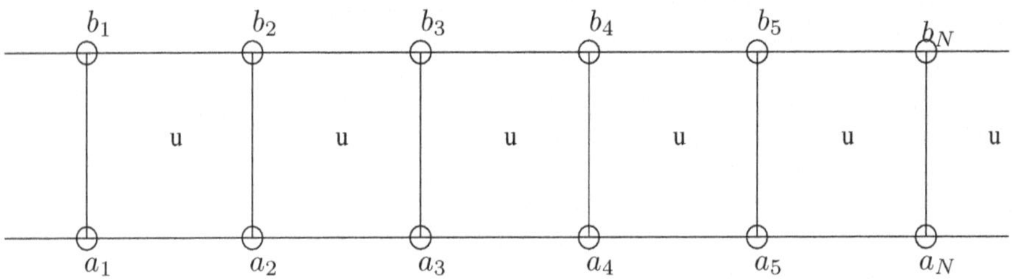

$$T|_{\{b_i\},\{a_i\}} = W(a_1, a_2; b_1, b_2)W(a_2, a_3; b_2, b_3) \cdots W(a_N, a_1; b_N, b_1)$$

Fig. 13.6 Construction of the transfer matrix face models from the local Boltzmann weights.

13.2.2 The physical regime

For a statistical system to be physically realizable the total configurational interaction energy of the system, which is the sum of all the local interaction energies, must be real. This will be guaranteed if the local interaction energies themselves are real, in which case the local Boltzmann weights will be real and positive. This is the situation considered for continuum models in chapter 3 and in the chapters in virial expansions.

However, for vertex and face models the concept of the physical region can be slightly extended when there is a global restriction on the configuration of the variables. This can at times lead to situations where for all configurations of the variables the interaction energy is real and the Boltzmann weight of each configuration is real and positive even though there may be some local Boltzmann weights which are negative.

These sets of local Boltzmann weights will also be called "physical" even though some of them are negative. However, unlike the case where the local Boltzmann weights are all positive, this more extended notion will in general depend on the boundary conditions as well as on the local Boltzmann weights. We will see that such an extended notion of "physical regime" occurs on the eight-vertex model.

In this book we will refer to a "physical regime" if the total interaction energy is real. However, this convention is not universally used in the literature and some authors require all the local Boltzmann weights to be nonnegative in order for the term "physical" to be applied.

When all the local Boltzmann weights are positive all elements of the transfer matrix are real and nonnegative. Moreover, this nonnegativity property can at times occur even if some local Boltzmann weights are negative. When all of the matrix elements of the transfer matrix are real and positive the Perron–Frobenius theorem guarantees that the largest eigenvalue is positive and unique.

If, in addition to having all nonnegative matrix elements the transfer matrix is symmetric

$$T^T(u) = T(u), \tag{13.24}$$

then $T(u)$ is also Hermitian and thus is diagonalizable with real eigenvalues. However, in general, transfer matrices are not symmetric and thus are not Hermitian in the physical regime with all matrix elements nonnegative. However, if the commutation relation holds of

$$[T^T(u), T(u)] = 0 \tag{13.25}$$

and thus because of the reality of the matrix elements

$$[T^\dagger(u), T(u)] = 0 \tag{13.26}$$

then the transfer matrix is normal . This is a sufficient condition to guarantee that the transfer matrix is diagonalizable but in this case, even though the Perron–Frobenius theorem guarantees that the largest eigenvalue must be real there are in general many eigenvalues that are complex.

13.3 Integrability

From the transfer matrix we define the important property we call **integrability**.

Definition

A model is called integrable if it allows a one-parameter family of commuting transfer matrices

$$[T(u), T(u')] = 0. \tag{13.27}$$

This definition is analogous to Liouville's definition in classical mechanics that a system with $2N$ degrees of freedom specified by a Hamiltonian H is called integrable if there are N operators J_k (with $J_1 = H$) such that

$$\{J_k, H\} = 0 \text{ and } \{J_k, J_j\} = 0 \tag{13.28}$$

for all k and j where $\{A, B\}$ is the Poisson bracket. The transfer matrix $T(u)$ can be thought of as the generating function of the constants of the motion J_k, with u as the generating variable.

The condition of commuting transfer matrices (13.27) was first introduced by Baxter [8, 10] in his solution to the eight-vertex model. It is a very overdetermined set of equations and it is not at all obvious if there are any solutions and how such solutions may be obtained. It is also not obvious why the condition (13.27) will allow the exact evaluation of the eigenvalues and eigenvectors of the transfer matrix. We study the problem of existence of solutions of (13.27) in this chapter. The computation of eigenvalues will be done in the next chapter.

The commutation relation (13.27) is a global relation which involves all the sites in one row. However, in the three classes of models being considered here the transfer matrix has been constructed out of local interactions at sites and vertices and thus it is natural to look for a local condition involving only a small number of Boltzmann weights which will guarantee that (13.27) holds. Such a local condition on Boltzmann weights was found by Baxter in [10] for the eight-vertex model. This condition was seen to be a vertex analogue of a property found by Onsager for the Boltzmann weight of the Ising model and named by him the star–triangle equation [2, 3]. These local conditions are alternatively referred to as Yang–Baxter equations. The form of the local relation is different for the three different classes of models, and each will be treated in a separate section below.

It must be stressed that the star–triangle equations presented below are only proven to be sufficient conditions for the commutation relation (13.27) to hold. It is unknown if the star–triangle equations are necessary for commutativity of the transfer matrices.

13.4 Star–triangle equation for vertex models

It is far more transparent to discuss star–triangle equations and commutation of transfer matrices graphically instead of using equations wherever possible. For vertex models we use the convention that state variables on external lines are fixed while state variables in internal lines are summed over. With these conventions the vertex model star–triangle equation is given by Fig. 13.7 where the "intertwining" matrix $R(u'')$ has an inverse as shown in Fig. 13.8.

We prove that the star–triangle equation of Fig. 13.7 and the inversion relation of Fig. 13.8 are sufficient to guarantee the commutation of the transfer matrices in Fig. 13.9 and Fig. 13.10. It is hoped that the figures provide a self-explanatory proof.

For vertex models with n states per bond we see from Fig. 13.1 that there are n^4 Boltzmann weights. In Fig. 13.7 we see that because there are six external legs in the star–triangle equation there will be n^6 equations to be satisfied. The star–triangle equation is clearly a very overdetermined system. Nevertheless it has been found that for each n there are solutions. We will here confine our attention to the case $n = 2$ and present the important solutions of the six- and eight-vertex model and the free fermion model.

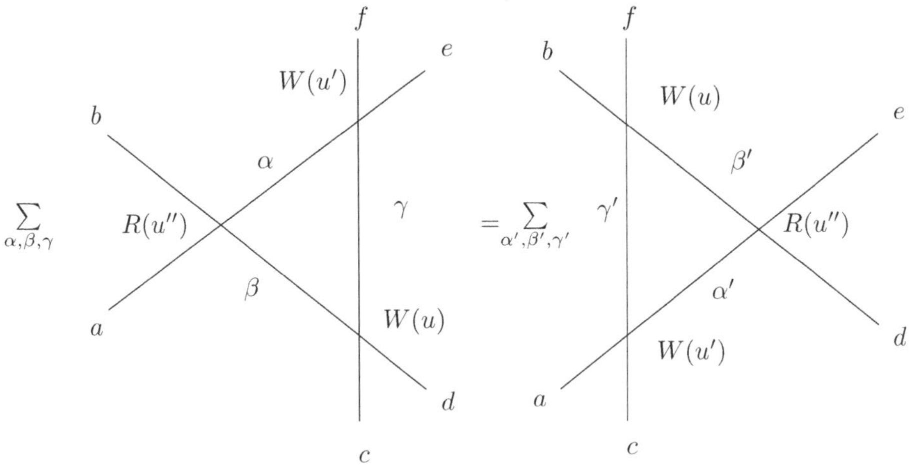

Fig. 13.7 The star–triangle equation for vertex models.

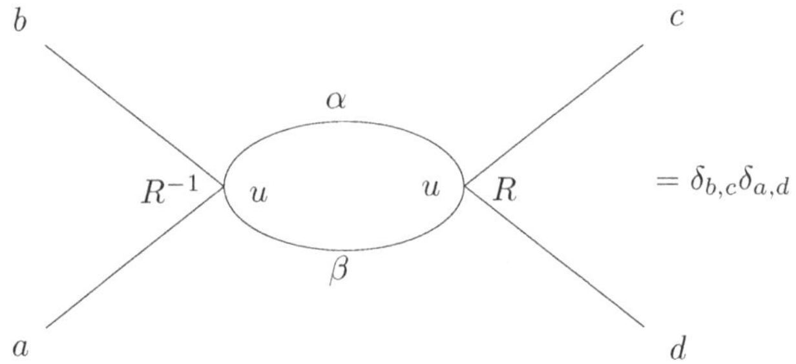

Fig. 13.8 The inversion relation for vertex models.

13.4.1 Boltzmann weights for two-state vertex models

For vertex models with two states per bond we may denote the states either by means of arrows or by variables ± 1 as shown in Fig. 13.11.

In general, a two-state model will have 16 nonvanishing Boltzmann weights at each site. We will here, however, restrict our attention to models where the Boltzmann weights are nonzero only when an even number (0, 2, 4) of arrows pointing towards (and away from) each vertex. The eight allowed Boltzmann weights are shown in Fig. 13.12.

It is easily seen that because of the periodic boundary conditions the restriction to this eight-vertex model requires that in the transfer matrix the weights w_5 and w_6, and the weights w_7 and w_8 can only occur in the combination w_5w_6 and w_7w_8. Therefore we may set them equal without loss of generality and thus we will denote the eight Boltzmann weights by a, $\bar{a}$, b, $\bar{b}$, c, d as shown in Fig. 13.12.

The partition function and the transfer matrix of the eight-vertex model have many symmetries. For the partition function the symmetries are explicitly derived in [46] and for the transfer matrix in [47]. We will restrict our attention here to the symmetric (or zero field case) where $a = \bar{a}$ and $b = \bar{b}$.

The partition function symmetries are of two types. The first type is those symmetries which follows directly by considering configurations of arrows. For example, because only the combinations $\omega_5\omega_6$ and $\omega_7\omega_8$ can occur, the partition function, (and the transfer matrix)

$$Z(a,b,c,d) = Z(a,b,\pm c,\pm d). \tag{13.29}$$

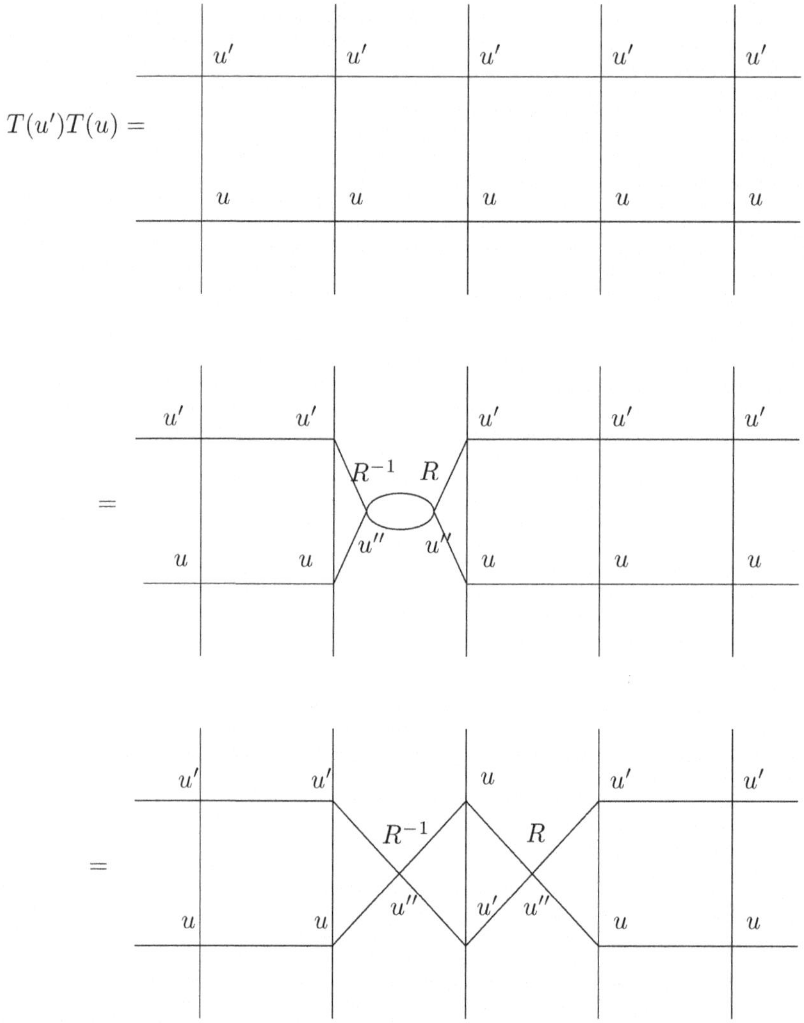

Fig. 13.9 Commutation of transfer matrices for vertex models.

If we reverse the direction of all horizontal arrows we find

$$Z(a,b,c,d) = Z(b,a,d,c), \tag{13.30}$$

and if we rotate the lattice through 90 degrees we find

Fig. 13.10 Commutation of transfer matrices for vertex models continued.

$$Z(a, b, c, d) = Z(b, a, c, d). \tag{13.31}$$

In addition if both N_h and N_v are even we can divide the lattice into two sublattices A and B. Then by reversing all arrows on horizontal (vertical) which have an A site on the left (top) edges we find

$$Z(a, b, c, d) = Z(c, d, a, b). \tag{13.32}$$

Therefore, because Z contains only even powers of c and d it is also an even function of a and b. Thus when N_h and N_v are both even and we have periodic boundary

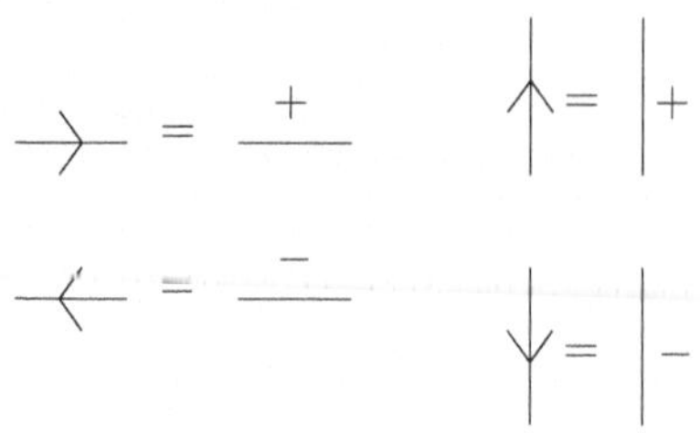

Fig. 13.11 The relation between the arrow and ± labeling for two-state vertex models.

conditions

$$Z(a, b, c, d) = Z(\pm a, \pm b, \pm c, \pm d). \tag{13.33}$$

This is an example of the previously discussed phenomenon that local Boltzmann weights can be negative, and the model is still physical.

There is a second symmetry which is more subtle. Namely if we define

$$\begin{aligned} a' &= (a+b+c+d)/2, \quad b' = (a+b-c-d)/2 \\ c' &= (a-b+c-d)/2, \quad d' = (a-b-c+d)/2 \end{aligned} \tag{13.34}$$

then it is shown in [46] and [48] that

$$Z(a, b, c, d) = Z(a', b', c', d'). \tag{13.35}$$

If we define as in [10]

$$w_1 = \frac{c+d}{2}, \quad w_2 = \frac{c-d}{2}, \quad w_3 = \frac{a-b}{2}, \quad w_4 = \frac{a+b}{2} \tag{13.36}$$

these symmetries are summarized (with an abuse of notation) as

$$Z(w_1, w_2, w_3, w_4) = Z(\pm w_i, \pm w_j, \pm w_k, \pm w_l) \tag{13.37}$$

where $i, j, k, 1$ are any of the 4! permutations of $1, 2, 3, 4$.

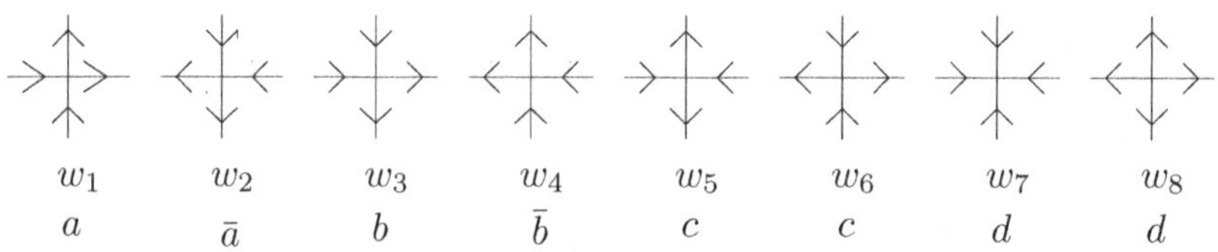

Fig. 13.12 The eight allowed vertices of the eight vertex model. The weights w_5 and w_6 (and w_7 and w_8) may be set equal with no loss of generality.

The transfer matrix has related symmetry properties. In particular in analogy with (13.29)–(13.31)

$$T(a, b, c, d) = T(b, a, c, d) \tag{13.38}$$

$$T^T(a, b, c, d) = T(b, a, c, d) \tag{13.39}$$

$$T(a,b,c,d) = T(a,b,\pm c,\pm d). \tag{13.40}$$

It follows from (13.39) when a, b, c, d are real that

$$[T^{\dagger}(a,b,c,d), T(a,b,c,d)] = 0 \tag{13.41}$$

and thus that $T(a,b,c,d)$ is diagonalizable.

We will search for solutions of the star–triangle equation of Fig. 13.7 where R has the same form as W, as indicated in Fig. 13.13.

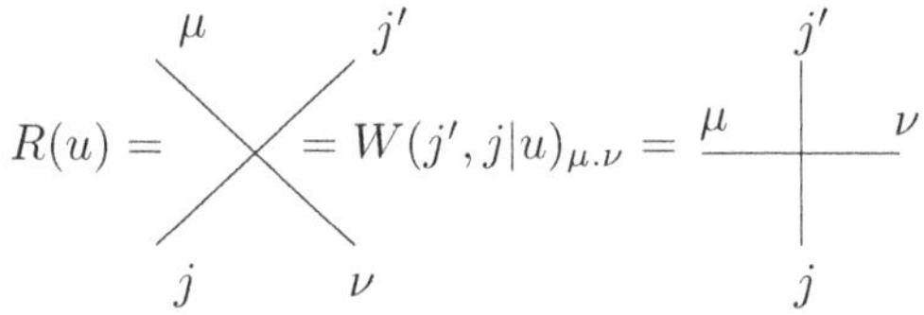

Fig. 13.13 The relation of $R(u)$ to $W(u)$ used in the solution of the six-vertex star–triangle equation Fig. 13.7.

The star–triangle equation of Fig. 13.7 is then explicitly written as

$$\sum_{\alpha,\beta,\gamma} w''(\alpha,a)_{b,\beta} w'(f,\gamma)_{\alpha,e} w(\gamma,c)_{\beta,d} = \sum_{\alpha',\beta',\gamma'} w(f,\gamma')_{b,\beta'} w'(\gamma',c)_{a,\alpha'} w''(e,\alpha')_{\beta',d}. \tag{13.42}$$

It is easily checked that the-eight vertex restriction of Fig. 13.12 makes both sides of the star–triangle equation (13.42) vanish if the external indices a, b, c, d, e, f have an odd number of minus signs. Thus there are only 32 nonvanishing equations of (13.42) to be considered. Of these the following eight are satisfied identically:

a	b	c	d	e	f
+	+	+	+	+	+
−	+	+	+	−	+
+	−	+	−	+	+
−	−	+	−	−	+
−	−	−	−	−	−
+	−	−	−	+	−
−	+	−	+	−	−
+	+	−	+	+	−

The remaining 24 equations in (13.42) are equal in pairs and thus we may restrict our attention to the six equations given by:

a	b	c	d	e	f	α	β	γ	α'	β'	γ'
$+$	$-$	$+$	$+$	$+$	$-$	$\pm$	$\mp$	$\mp$	$\pm$	$\pm$	$\pm$
$+$	$-$	$+$	$+$	$-$	$+$	$\pm$	$\mp$	$\mp$	$\mp$	$\pm$	$\pm$
$-$	$+$	$+$	$+$	$+$	$-$	$\pm$	$\mp$	$\mp$	$\pm$	$\pm$	$\mp$
$+$	$+$	$+$	$-$	$+$	$-$	$\pm$	$\pm$	$\mp$	$\pm$	$\mp$	$\pm$
$+$	$+$	$+$	$-$	$-$	$+$	$\pm$	$\pm$	$\mp$	$\mp$	$\mp$	$\mp$
$+$	$-$	$+$	$-$	$-$	$-$	$\mp$	$\pm$	$\mp$	$\mp$	$\mp$	$\mp$

and six more obtained by the interchange $+ \leftrightarrow -$.

Thus the 12 independent star–triangle equations are explicitly given as

$$\begin{aligned}
cb'\bar{b}'' + ac'c'' &= ca'a'' + \bar{a}d'd'' \\
a\bar{b}'c'' + cc'\bar{b}'' &= \bar{b}a'c'' + dd'b'' \\
ac'b'' + cb'c'' &= bc'a'' + d\bar{b}'d'' \\
db'a'' + \bar{b}c'd'' &= da'\bar{b}'' + bd'c'' \\
\bar{b}\bar{b}'d'' + dc'a'' &= aa'd'' + cd'\bar{a}'' \\
\bar{b}d'\bar{b}'' + d\bar{a}'c'' &= \bar{a}d'\bar{a}'' + ca'd''
\end{aligned} \tag{13.43}$$

and

$$\begin{aligned}
c\bar{b}'b'' + \bar{a}c'c'' &= c\bar{a}'\bar{a}'' + ad'd'' \\
\bar{a}b'c'' + cc'b'' &= b\bar{a}'c'' + dd'\bar{b}'' \\
\bar{a}c'\bar{b}'' + c\bar{b}'c'' &= \bar{b}c'\bar{a}'' + db'd'' \\
d\bar{b}'\bar{a}'' + bc'd'' &= d\bar{a}'b'' + \bar{b}d'c'' \\
bb'd'' + dc'\bar{a}'' &= \bar{a}\bar{a}'d'' + cd'a'' \\
bd'b'' + da'c'' &= ad'a'' + c\bar{a}'d''
\end{aligned} \tag{13.44}$$

where the last six equations are obtained from the first 6 by the interchanges of the barred and unbarred a's and b's.

We will consider separately four special cases where these equations are satisfied:

1) The symmetric six-vertex model where

$$d = 0, \quad a = \bar{a}, \quad b = \bar{b} \tag{13.45}$$

2) The asymmetric six-vertex model where

$$d = 0 \tag{13.46}$$

3) The symmetric eight-vertex model where

$$a = \bar{a}, \quad b = \bar{b} \tag{13.47}$$

4) The free fermion model where

$$a\bar{a} + b\bar{b} - c^2 - d^2 = 0 \tag{13.48}$$

The symmetric six-vertex model

When the Boltzmann weights are restricted to the symmetric six-vertex model (13.45) the 12 star–triangle equations reduce to the three equations (9.6.12) of [48]:

$$\begin{aligned} cb'b'' + ac'c'' &= ca'a'' \\ cc'b'' + ab'c'' &= ba'c'' \\ cb'c'' + ac'b'' &= bc'a'' \end{aligned} \tag{13.49}$$

We consider a'', b'', c'' as the unknowns and thus write (13.49) as the homogeneous system

$$\begin{aligned} -ca'a'' + cb'b'' + ac'c'' &= 0 \\ cc'b'' + (ab' - ba')c'' &= 0 \\ -bc'a'' + ac'b'' + cb'c'' &= 0 \end{aligned} \tag{13.50}$$

and for this system to have a nonvanishing solution the determinant of the coefficients must vanish. This determinant is directly computed to be

$$-ca'cc'cb' - bc'cb'(ab' - ba') + bc'cc'ac' + ca'ac'(ab' - ba') = 0. \tag{13.51}$$

This has a common factor of cc' which can be cancelled and thus we find

$$a'b'[a^2 + b^2 - c^2] - ab[a'^2 + b'^2 - c'^2] = 0 \tag{13.52}$$

which may be rearranged to have all primed variables on one side and all unprimed variables on the other side as

$$\frac{a^2 + b^2 - c^2}{ab} = \frac{a'^2 + b'^2 - c'^2}{a'b'}. \tag{13.53}$$

We have thus demonstrated that when the three Boltzmann weights satisfy the constraint which we parameterize in terms of Δ as

$$\frac{a^2 + b^2 - c^2}{2ab} = \Delta \tag{13.54}$$

that we do in fact have a one-parameter family of Boltzmann weights which satisfy the star–triangle equations and thus the symmetric six-vertex model has a one-parameter family of commuting transfer matrices.

The Boltzmann weights a, b and c may be parametrized in a useful fashion by setting

$$a = \rho \sin(u + \lambda), \quad b = \rho \sin(u - \lambda), \quad c = \rho \sin 2\lambda \tag{13.55}$$

from which we find that

$$\frac{a^2 + b^2 - c^2}{2ab} = \cos 2\lambda = \Delta \tag{13.56}$$

is independent of u as desired.

We note from (13.56) that the invariant Δ is unchanged under operations

$$a \leftrightarrow b, \quad c \to c \tag{13.57}$$

$$a \to -a, \quad b \to -b, \quad c \to c \tag{13.58}$$

$$a \to a, \quad b \to b, \quad c \to -c \tag{13.59}$$

where (13.58) and (13.59) are equivalent under $\rho \to -\rho$.

In the physical regime of the statistical model where the Boltzmann weights a, b and c are all positive there are three distinct regions of Δ shown in Table 13.2 with the appropriate regions for λ, u and ρ.

Table 13.2 The three cases of Δ, the corresponding restrictions of λ, u and ρ, and the four inequalities of a, b, and c for which the Boltzmann weights of the six-vertex model, as parametrized by (13.55), are real and positive.

Δ	λ	u	ρ
	$a, b, c < (a+b+c)/2$		
$-1 < \Delta < 1$	$0 < \lambda < \pi/2$	$\lambda < u < \pi - \lambda$	$0 < \rho$
	$c > a + b$		
$\Delta < -1$	$\lambda = \pi/2 + i\bar{\lambda}$	$u = \pi/2 + i\bar{u}$	$\rho = i\bar{\rho}$
	$\bar{\lambda} > 0$	$-\bar{\lambda} < \bar{u} < \bar{\lambda}$	$0 < \bar{\rho}$
	$a > b + c$		
$\Delta > 1$ (1)	$\lambda = i\bar{\lambda}$	$u = i\bar{u}$	$\rho = -i\bar{\rho}$
	$\bar{\lambda} > 0$	$\bar{\lambda} < \bar{u}$	$0 < \bar{\rho}$
	$b > a + c$		
$\Delta > 1$ (2)	$\lambda = i\bar{\lambda}$	$u = \pi - i\bar{u}$	$\rho = -i\bar{\rho}$
	$\bar{\lambda} > 0$	$\bar{\lambda} < \bar{u}$	$0 < \bar{\rho}$

In the region $-1 < \Delta < 1$ the form of the parametrization (13.55) is manifestly positive when

$$\lambda < u < \pi - \lambda \tag{13.60}$$

and the weights a, b and c satisfy

$$a, b, c < (a+b+c)/2. \tag{13.61}$$

In the region $\Delta < -1$ we set

$$\begin{aligned} &\lambda = \pi/2 + i\bar{\lambda} \quad \text{with} \quad \bar{\lambda} > 0 \\ &\rho = i\bar{\rho} \quad \text{with} \quad 0 < \bar{\rho} \\ &u = \pi/2 + i\bar{u} \quad \text{with} \quad \bar{\lambda} < \bar{u} \end{aligned} \tag{13.62}$$

and write the parametrization (13.55) as

$$a = \bar{\rho}\sinh(\bar{u} + \bar{\lambda}), \quad b = -\bar{\rho}\sinh(\bar{u} - \bar{\lambda}), \quad c = \bar{\rho}\sinh 2\bar{\lambda}. \tag{13.63}$$

These weights are manifestly positive for $-\bar{\lambda} < \bar{u} < \bar{\lambda}$ and they satisfy

$$c > a + b. \tag{13.64}$$

In the region $\Delta > 1$ we identify two subcases depending on whether a or b is greater.

When a is greater we set

$$\begin{aligned} \lambda &= i\bar{\lambda} \quad \text{with} \quad \bar{\lambda} > 0 \\ \rho &= -i\bar{\rho} \quad \text{with} \quad \bar{\rho} > 0 \\ u &= i\bar{u} \quad \text{with} \quad \bar{\lambda} < \bar{u} \end{aligned} \tag{13.65}$$

and write the parametrization (13.55) as

$$a = \bar{\rho} \sinh(\bar{u} + \bar{\lambda}), \quad b = \bar{\rho} \sinh(\bar{u} - \bar{\lambda}), \quad c = \bar{\rho} \sinh 2\bar{\lambda}. \tag{13.66}$$

These weights are manifestly positive for $\bar{\lambda} < \bar{u}$ and they satisfy

$$a > b + c. \tag{13.67}$$

The opposite case where

$$b > a + c \tag{13.68}$$

is obtained from the interchange $a \leftrightarrow b$ effected by $u \rightarrow \pi - u$ given in Table 13.3. Thus setting

$$\begin{aligned} \lambda &= i\bar{\lambda} \quad \text{with} \quad \bar{\lambda} > 0 \\ \rho &= -i\bar{\rho} \quad \text{with} \quad \rho > 0 \\ u &= \pi - i\bar{u} \quad \text{with} \quad \bar{u} > \bar{\lambda} \end{aligned} \tag{13.69}$$

the parametrization (13.55) becomes

$$a = \bar{\rho} \sinh(\bar{u} - \bar{\lambda}), \quad b = \bar{\rho} \sinh(\bar{u} + \bar{\lambda}), \quad c = \bar{\rho} \sinh 2\bar{\lambda}. \tag{13.70}$$

These weights are manifestly positive for $\bar{\lambda} < \bar{u}$.

The symmetry operations (13.57) and (13.58) are obtained from the parametrization (13.55) as shown in Table (13.3).

Table 13.3 The operations on the variables λ and u which correspond to the symmetries (13.57) and (13.58).

$u \rightarrow \pi - u$	$a \rightarrow b$	$b \rightarrow a$	$c \rightarrow c$
$u \rightarrow \pi + u$	$a \rightarrow -a$	$b \rightarrow -b$	$c \rightarrow c$

The transpose of the transfer matrix $T(u)$ is obtained (13.39) by the interchange $a \leftrightarrow b$ which is the symmetry operation (13.57). Thus in the three regions of Δ we see from several forms (13.55), (13.63), (13.66) and (13.70) that

$$
\begin{aligned}
&\text{for } -1<\Delta<1, \quad T^T(u)=T(\pi-u)\\
&\text{for } \Delta<-1, \quad T^T(\bar{u})=T(-u)\\
&\text{for } \Delta>1, \quad T^T(\bar{u})=T(i\pi-\bar{u}).
\end{aligned}
\tag{13.71}
$$

Thus it follows from the fundamental commutation relation (13.2) that $T(u)$ satisfies

$$[T(u),T^T(u)]=0. \tag{13.72}$$

Therefore when the Boltzmann weights are real the transfer matrix $T(u)$ is normal and hence may be diagonalized.

It remains to use the parametrization (13.55) of the Boltzmann weights in the three equations of (13.49) to determine u'' in terms of u and u'. All three equations of (13.49) must lead to the same relation and for example by substituting (13.55) in the first equation in (13.49) we find

$$\sin(u'-\lambda)\sin(u''-\lambda)+\sin 2\lambda\sin(u+\lambda)-\sin(u'+\lambda)\sin(u''+\lambda)=0 \tag{13.73}$$

which may be rewritten in the form

$$(e^{2i\lambda}-e^{-2i\lambda})\{e^{i(u+\lambda)}-e^{-i(u+\lambda)}-e^{i(u''+u')}+e^{-i(u''+u')}\}=0 \tag{13.74}$$

from which we find

$$u''=u-u'+\lambda. \tag{13.75}$$

Thus u'' depends on the difference of u and u'. This difference property is a special feature and does not need to hold in order that the star–triangle equations be valid.

The asymmetric six-vertex model

When the Boltzmann weights are restricted to the asymmetric six-vertex model (13.46) the 12 star–triangle equations reduce the six equations

$$
\begin{aligned}
cb'\bar{b}''+ac'c'' &= ca'a''\\
a\bar{b}'c''+cc'\bar{b}'' &= \bar{b}a'c''\\
ac'b''+cb'c'' &= bc'a''\\
c\bar{b}'b''+\bar{a}c'c'' &= c\bar{a}'\bar{a}''\\
\bar{a}b'c''+cc'b'' &= b\bar{a}'c''\\
\bar{a}c'\bar{b}''+c\bar{b}'c'' &= \bar{b}c'\bar{a}''.
\end{aligned}
\tag{13.76}
$$

If here we make the replacements

$$a\to hva, \quad \bar{a}\to h^{-1}v^{-1}a, \quad a'\to hva', \quad \bar{a}'\to h^{-1}v^{-1}a' \tag{13.77}$$

$$b\to hv^{-1}b, \quad \bar{b}\to h^{-1}vb, \quad b'\to hv^{-1}b', \quad \bar{b}'\to h^{-1}vb' \tag{13.78}$$

$$a''\to a'', \quad \bar{a}''\to a, \quad b''\to v^{-2}b'', \quad \bar{b}''\to v^2b'' \tag{13.79}$$

we see that the six equations in (13.76) reduce to the three equations (13.49) of the symmetric six-vertex model. Therefore from the solution (13.55) it follows that

$$a = hv\rho\sin(u+\lambda) \quad \bar{a} = h^{-1}v^{-1}\rho\sin(u-\lambda)$$
$$b = hv^{-1}\rho\sin(u-\lambda) \quad \bar{b} = h^{-1}v\sin(u-\lambda).$$
$$c = \rho\sin 2\lambda \tag{13.80}$$

The new variables h and v can be interpreted as the contributions to the Boltzmann weights from fields interacting with the horizontal and vertical arrows (considered as dipoles). The simple dependence of the Boltzmann weights (13.80) on v and h follows from the conservation property of arrows at each vertex.

The eight-vertex model

When the Boltzmann weights are restricted to the eight-vertex model (13.47) the 12 star–triangle equations reduce to the six equations [48, (10.4.1)]

$$\begin{aligned}
ca'a'' - cb'b'' - ac'c'' + ad'd'' &= 0\\
(dd' - cc')b'' + (ba' - ab')c'' &= 0\\
bc'a'' - ac'b'' - cb'c'' + db'd'' &= 0\\
-db'a'' + da'b'' + bd'c'' - bc'd'' &= 0\\
(cd' - dc')a'' + (aa' - bb')d'' &= 0\\
ad'a'' - bd'b'' - da'c'' + ca'd'' &= 0
\end{aligned} \tag{13.81}$$

which we consider as a system of homogeneous equations with a'', b'', c'', d'' as unknowns.

We first set the determinant of equations 1, 3, 4, 6 in (13.81) equal to zero, and note that it has the magic property that it can be written in the factorized form

$$(cda'b' - abc'd')[(a^2 - b^2)(c'^2 - d'^2) + (a'^2 - b'^2)(c^2 - d^2)] = 0 \tag{13.82}$$

which vanishes if either of the two factors vanishes. When the first factor vanishes we thus have the restriction

$$\frac{cd}{ab} = \frac{c'd'}{a'b'}. \tag{13.83}$$

To proceed further we solve equations 1, 3, 4, 6 of (13.81) as

$$\begin{aligned}
a'' &= a'(cc' - dd')(b'^2c^2 - c'^2a^2)/c'\\
b'' &= b'(dc' - cd')(a'^2c^2 - d'^2a^2)/d'\\
c'' &= c'(aa' - bb')(a'^2c^2 - d'^2a^2)/a'\\
d'' &= d'(ba' - ab')(b'^2c^2 - c'^2a^2)/b'
\end{aligned} \tag{13.84}$$

put this into equations 2 and 5 of (13.81) and use (13.83). In both cases we discover that we obtain the same equations when condition (13.83) is used and thus we obtain a second restriction which may be written in either of the two equivalent forms

$$\frac{a^2 + b^2 - c^2 - d^2}{ab} = \frac{a'^2 + b'^2 - c'^2 - d'^2}{a'b'} \tag{13.85}$$

or

$$\frac{a^2+b^2-c^2-d^2}{ab+cd} = \frac{a'^2+b'^2-c'^2-d'^2}{a'b'+c'd'}. \tag{13.86}$$

Thus when the four Boltzmann weights a, b, c, d are constrained by the two restrictions

$$\frac{cd}{ab} = \gamma \tag{13.87}$$

$$\frac{a^2+b^2-c^2-d^2}{2(ab+cd)} = \Delta \tag{13.88}$$

with γ and Δ constant we have a one-parameter family of Boltzmann weights which satisfy the star–triangle equations and thus the symmetric eight-vertex model has a one-parameter family of commuting transfer matrices. We note that if $d = 0$ then $\gamma = 0$ and the restriction (13.88) reduces to the restriction (13.56) of the six-vertex model.

We now wish to parameterize the Boltzmann weights explicitly in terms of a spectral variable u in such a way that the ratios (13.87) and (13.88) are constant. Unlike the six-vertex model a parameterization in terms of hyperbolic or trigonometric functions is not possible. Instead elliptic functions must be used. These functions are generalizations of trigonometric functions and we give proofs in the appendix of some of the properties of these functions which we will use in this and the following chapter. In particular we use the Jacobi theta functions

$$H(u;k) = 2\sum_{n=1}^{\infty}(-1)^{n-1}q^{(n-\frac{1}{2})^2}\sin[(2n-1)\pi u/(2K)] \tag{13.89}$$

$$\Theta(u;k) = 1 + 2\sum_{n=1}^{\infty}(-1)^n q^{n^2}\cos(nu\pi/K) \tag{13.90}$$

where

$$q = e^{-\pi K'/K} \tag{13.91}$$

is called the nome, K is the complete elliptic integral of the first kind

$$K(k) = \int_0^{\pi/2} \frac{d\theta}{(1-k^2\sin^2\theta)^{1/2}} \tag{13.92}$$

and

$$K' = K(k') \tag{13.93}$$

where

$$k'^2 = 1 - k^2. \tag{13.94}$$

The parameters k (which is called the modulus and k' (which is called the complementary modulus) are defined as

$$k^{1/2} = H(K)/\Theta(K), \quad k'^{1/2} = \Theta(0)/\Theta(K). \tag{13.95}$$

The modulus k is related to the nome q by means of the identity proven in the appendix

$$k = 4q^{1/2} \prod_{n=1}^{\infty} \left(\frac{1+q^{2n}}{1+q^{2n-1}} \right)^4 . \tag{13.96}$$

The dependence on k will be suppressed in the notation unless needed.

The theta functions (13.89) and (13.90) satisfy

$$\Theta(u) = \Theta(-u), \quad H(u) = -H(u) \tag{13.97}$$

and the quasi periodicity relations which are proven in the appendix

$$H(u+2K) = -H(v) \tag{13.98}$$

$$H(u+2iK') = -q^{-1}e^{-\pi iu/K}H(u) \tag{13.99}$$

and

$$\Theta(u+2K) = \Theta(u) \tag{13.100}$$

$$\Theta(u+2iK') = -q^{-1}e^{-i\pi u/K}\Theta(u). \tag{13.101}$$

From (13.90) we see that $\Theta(u)$ and $H(u)$ are related by

$$\Theta(u \pm iK') = \pm iq^{-1/4}e^{\mp\frac{\pi iu}{2K}}H(u) \tag{13.102}$$

$$H(u \pm iK') = \pm iq^{-1/4}e^{\frac{\mp\pi iu}{2K}}\Theta(u). \tag{13.103}$$

Furthermore we find that in the rectangle $0 \leq \mathrm{Re}u < 2K$ and $0 \leq \mathrm{Im}u < 2K'$ that the only zeros of $H(u)$ and $\Theta(u)$ are at

$$H(0) = \Theta(iK') = 0. \tag{13.104}$$

When $q \to 0$ we see from (13.89)–(13.96) that

$$k \to 4q^{1/2} \tag{13.105}$$

$$K \to \pi/2, \quad \text{and} \quad K' \to \infty \tag{13.106}$$

$$H(u) \sim 2q^{1/4}\sin u \quad \text{and} \quad \Theta(u) \to 1 \tag{13.107}$$

and hence the quasiperiodic theta functions reduce to the singly periodic trigonometric functions.

By taking quotients of products of these quasiperiodic theta functions we may construct doubly periodic functions which satisfy

$$\begin{aligned} f(u+2K) &= \pm f(u) \\ f(u+2iK') &= \pm f(u). \end{aligned} \tag{13.108}$$

These functions will in general be meromorphic (i.e. their only singularities are poles). There are three such functions which are particularly useful,

$$\mathrm{sn}(u,k) = k^{-1/2}H(u)/\Theta(u) = \frac{H(u)\Theta(K)}{\Theta(u)H(K)} \tag{13.109}$$

$$\mathrm{cn}(u,k) = (k'/k)^{1/2} H(u+K)/\Theta(u) = \frac{H(u+K)\Theta(0)}{\Theta(u)H(K)} \tag{13.110}$$

$$\mathrm{dn}(u,k) = k'^{1/2}\Theta(u+K)/\Theta(u) = \frac{\Theta(u+K)\Theta(0)}{\Theta(u)\Theta(K)} \tag{13.111}$$

(We will often suppress the modulus k in the notation.) These functions obviously have the properties that

$$\mathrm{sn}K = \mathrm{cn}0 = \mathrm{dn}0 = 1 \tag{13.112}$$

$$\mathrm{sn}0 = \mathrm{cn}K = 0 \tag{13.113}$$

and we prove in the appendix that they satisfy the identities

$$\mathrm{sn}^2 u + \mathrm{cn}^2 u = 1 \tag{13.114}$$

$$k^2\mathrm{sn}^2 u + \mathrm{dn}^2 u = 1. \tag{13.115}$$

Furthermore by using (13.102) and (13.103) in the definitions (13.109)–(13.111) we find

$$\mathrm{sn}(u+iK') = \frac{1}{k\mathrm{sn}u} \tag{13.116}$$

$$\mathrm{cn}(u+iK') = -\frac{i\mathrm{dn}u}{k\mathrm{sn}u} \tag{13.117}$$

$$\mathrm{dn}(u+iK') = -\frac{i\mathrm{cn}u}{\mathrm{sn}u}. \tag{13.118}$$

In the limit $k \to 0$ we find from (13.105)-(13.107) that

$$\mathrm{sn}(u,k) \to \sin u \tag{13.119}$$

$$\mathrm{cn}(u,k) \to \cos u \tag{13.120}$$

$$\mathrm{dn}(u,k) \to 1. \tag{13.121}$$

We now recall a most useful theorem from complex variable theory.

Liouville's theorem

Every bounded entire function must be a constant.

In particular we will use the following:

Corollary to Liouville's theorem

A doubly periodic function which has no poles must be a constant.

We may now parameterize the Boltzmann weights of the eight-vertex model in terms of these theta functions

$$
\begin{aligned}
a &= \rho\Theta(2\eta)\Theta(u-\eta)H(u+\eta)\\
b &= \rho\Theta(2\eta)H(u-\eta)\Theta(u+\eta)\\
c &= \pm\rho H(2\eta)\Theta(u-\eta)\Theta(u+\eta)\\
d &= \pm\rho H(2\eta)H(u-\eta)H(u+\eta)
\end{aligned}
\tag{13.122}
$$

where, because the transfer matrix depends only on c^2 and d^2, the $\pm$ signs in c and d may be arbitrarily chosen and the normalizing factor ρ may depend on the spectral variable u.

If we choose

$$
\rho = \frac{\rho'}{k^{1/2}\Theta(2\eta)\Theta(u-\eta)\Theta(u+\eta)}
\tag{13.123}
$$

we may write (13.122) as

$$
\begin{aligned}
a &= \rho'\mathrm{sn}(u+\eta)\\
b &= \rho'\mathrm{sn}(u-\eta)\\
c &= \pm\rho'\mathrm{sn}(2\eta)\\
d &= \pm\rho' k\mathrm{sn}(2\eta)\mathrm{sn}(u-\eta)\mathrm{sn}(u+\eta).
\end{aligned}
\tag{13.124}
$$

From (13.122) we see that the ratio (13.87) is

$$
\gamma = cd/ab = \pm\frac{H^2(2\eta)}{\Theta^2(2\eta)} = \pm k\mathrm{sn}^2(2\eta)
\tag{13.125}
$$

which is independent of u as required.

To demonstrate that the ratio Δ (13.88) is independent of u we note that from (13.125) it is sufficient to demonstrate that the ratio

$$
\begin{aligned}
\frac{a^2+b^2-c^2-d^2}{ab} &= \{\Theta^2(2\eta)[\Theta^2(u-\eta)H^2(u+\eta)+H^2(u-\eta)\Theta^2(u+\eta)]\\
&\quad -H^2(2\eta)[\Theta^2(u-\eta)\Theta^2(u+\eta)+H^2(u-\eta)H^2(u+\eta)]\}\\
&\quad \times\{\Theta^2(2\eta)\Theta(u-\eta)H(u+\eta)\Theta(u+\eta)H(u-\eta)\}^{-1}
\end{aligned}
\tag{13.126}
$$

is independent of u. We first note that, by use of the quasi periodicity properties (13.98)–(13.101), this ratio is indeed a doubly periodic function. Furthermore, by use of (13.104) it is straightforward to see that the numerator vanishes at the points $u=\pm\eta,\ \pm\eta+iK'$ which are precisely the points where the denominator $2ab$ vanishes. Therefore the ratio (13.126) has no poles and therefore by the corollary to Liouville's theorem it must be a constant, independent of u. We note further that by use of the identities on elliptic functions given in the appendix that (13.126) may be written as

$$
\frac{a^2+b^2-c^2-d^2}{ab} = 2\mathrm{cn}(2\eta)\mathrm{dn}(2\eta)
\tag{13.127}
$$

and thus (13.88) becomes

$$\Delta = \frac{\text{cn}(2\eta)\text{dn}(2\eta)}{1 \pm k\text{sn}^2(2\eta)}. \tag{13.128}$$

In the limit when $q \to 0$ we find from (13.124) that

$$\begin{aligned} a &\to \rho' \sin(u + \eta_0) \\ b &\to \rho' \sin(u - \eta_0) \\ c &\to \rho' \sin 2\eta_0 \\ d &\to 0 \end{aligned} \tag{13.129}$$

where $\eta_0 = \lim_{q \to 0} \eta$. Thus (13.129) reduces to the six-vertex Boltzmann weights (13.55) with the identification

$$\eta_0 = \lim_{q \to 0} \eta = \lambda. \tag{13.130}$$

The invariants Δ (13.87) and γ (13.191) are invariant under the three independent symmetry operations

$$a \leftrightarrow b, \quad c \to c, \quad d \to d \tag{13.131}$$

$$a \to -a, \quad b \to -b, \quad c \to c, \quad d \to d \tag{13.132}$$

$$a \leftrightarrow b, \quad c \leftrightarrow d \tag{13.133}$$

which correspond to the transfer matrix symmetries (13.38)–(13.40). The operations (13.131)–(13.133) are obtained from the parametrization (13.122) as shown in Table 13.4.

Table 13.4 The operations on the variable u which correspond to the symmetries (13.131)–(13.132).

$u \to 2K - u$	$a \leftrightarrow b$		$c \to c$	$d \to d$
$u \to 2K + u$	$a \to -a$	$b \to -b$	$c \to c$	$d \to d$
$u \to 2K + iK' - u \quad \rho \to -\rho q^{1/2} e^{-\pi i u/K}$	$a \to a$	$b \to b$	$c \leftrightarrow d$	

Because of the large number of symmetries of the partition function (13.37) it is impractical to write out all regions of the parameter space where the partition function represents a “physical” system. We restrict ourselves in Table 13.5 to those regions which correspond to the regions of the six-vertex model shown in Table 13.2

It remains to substitute the parametrization (13.122) into (13.81) to express u'' in terms of u and u'. To make this demonstration we write (13.81) in the symmetrical form

$$\begin{aligned} c(a'a'' - b'b'') &= a(c'c'' - d'd'') \\ b''(dd' - cc') &= c''(ab' - ba') \\ c'(ba'' - ab'') &= b'(cc'' - dd'') \\ d(a'b'' - b'a'') &= b(c'd'' - d'c'') \\ a''(cd' - dc') &= d''(bb' - aa') \\ d'(aa'' - bb'') &= a'(dc'' - cd''). \end{aligned} \tag{13.134}$$

Table 13.5 The regions of the parameter space η and u for $0 \leq q \leq 1$ for the five cases for the physical region where a, b, c, d are positive which correspond to the regions of the six-vertex model given in Table 13.2. We indicate in the first the signs of c and d which are used in (13.122.

$c\,,\,d$	Δ	η	u	ρ
$+,+$			$a, b, c, d < (a+b+c+d)/2$	
	$-1 < \Delta < 1$	$0 < \eta < K$	$\eta < u < 2K - \eta$	$0 < \rho$
$+,-$			$c > a+b+d$	
	$\Delta < -1$ (1)	$\eta = K + i\bar{\eta}$	$u = K + i\bar{u}$	$\rho = i\bar{\rho}$
		$0 < \bar{\eta} < K'$	$-\bar{\eta} < \bar{u} < \bar{\eta}$	$0 < \bar{\rho}$
$-,+$			$d > a+b+c$	
	$\Delta < -1$ (2)	$\eta = K + i\bar{\eta}$	$u = K + iK' - i\bar{u}$	$\rho = iq^{1/2}e^{\pi\bar{u}/K}\bar{\rho}$
		$0 < \bar{\eta} < K'$	$-\bar{\eta} < \bar{u} < \bar{\eta}$	$0 < \bar{\rho}$
$+,-$			$a > b+c+d$	
	$\Delta > 1$ (1)	$\eta = i\bar{\eta}$	$u = i\bar{u}$	$\rho = -i\bar{\rho}$
		$0 < \bar{\eta} < K'$	$\bar{\eta} < \bar{u} < K'$	$0 < \bar{\rho}$
$+,-$			$b > a+c+d$	
	$\Delta > 1$ (2)	$\eta = i\bar{\eta}$	$u = 2K - i\bar{u}$	$\rho = -i\bar{\rho}$
		$0 < \bar{\eta} < K'$	$\bar{\eta} < \bar{u} < K'$	$0 < \bar{\rho}$

To demonstrate the first equation in (13.134) we use the Boltzmann weights (13.122) to write

$$\begin{aligned} a'a'' - b'b'' &= \rho^2\Theta^2(2\eta)\{\Theta(u'-\eta)H(u'+\eta)\Theta(u''-\eta)H(u''+\eta) \\ &\quad -H(u-\eta)\Theta(u'+\eta)H(u''-\eta)\Theta(u''+\eta)\} \\ &= \rho^2\Theta(0)\Theta^2(2\eta)H(2\eta)\Theta(u'-u'')H(u'+u'') \end{aligned} \tag{13.135}$$

where in the last line we have used (13.397) with $v \to u'+\eta$, $u \to u'-\eta$, $a \to u'-u''$ and we write

$$\begin{aligned} c'c'' - d'd'' &= \rho^2H^2(2\eta)\{\Theta(u'-\eta)\Theta(u'+\eta)\Theta(u''-\eta)\Theta(u''+\eta) \\ &\quad -H(u'-\eta)H(u'+\eta)H(u''-\eta)H(u''+\eta)\} \\ &= \rho^2\Theta(0)H^2(2\eta)\Theta(2\eta)\Theta(u'-u'')\Theta(u''+u') \end{aligned} \tag{13.136}$$

where in the last line we have used (13.396) with $u \to u'+\eta$, $v \to u'-\eta$, $a = u'-u''$. Thus using (13.135) and (13.136) in the first equation of (13.134) we find

$$\Theta(u+\eta)H(u'+u'') = H(u+\eta)\Theta(u'+u'') \tag{13.137}$$

which holds if

$$u'' = u - u' + \eta. \tag{13.138}$$

To demonstrate the remainder of the equations in (13.134) we need

$$\begin{aligned}ab' - ba' &= \rho^2\Theta^2(2\eta)\{\Theta(u-\eta)H(u+\eta)H(u'-\eta)\Theta(u'+\eta)\\ &-H(u-\eta)\Theta(u+\eta)\Theta(u'-\eta)H(u'+\eta)\}\\ &= \rho^2\Theta(0)\Theta^2(2\eta)H(2\eta)\Theta(u+u')H(u'-u)\end{aligned} \tag{13.139}$$

where in the last line we used (13.397) with $u \to u-\eta, \quad v \to u+\eta, \quad a \to u+u'$ and finally

$$\begin{aligned}cd' - dc' &= \rho^2 H^2(2\eta)\{\Theta(u-\eta)\Theta(u+\eta)H(u'-\eta)H(u'+\eta)\\ &-H(u-\eta)H(u+\eta)\Theta(u'-\eta)\Theta(u'+\eta)\}\\ &= \rho^2\Theta(0)H^2(2\eta)\Theta(2\eta)H(u'-u)H(u+u')\end{aligned} \tag{13.140}$$

where in the last line we have used (13.397) with $u \to u-\eta, \ v \to u'-\eta, \ a \to -2\eta$. Thus we find that the fourth equation of (13.134) reduces to (13.137) and that the second, third, fifth and sixth equations of (13.134) reduce to

$$H(u''-\eta)\Theta(u-u') = \Theta(u''-\eta)H(u-u') \tag{13.141}$$

which is also satisfied if u'' satisfies (13.138). Thus we have a verification that the Boltzmann weights (13.122) satisfy the star–triangle equation (13.81).

The free fermion model

The final two state case where the star–triangle equation has been demonstrated to hold [48, 46] is the free fermion model where

$$a\bar{a} + b\bar{b} - c^2 - d^2 = 0. \tag{13.142}$$

This model includes the Ising model on a triangular lattice as a special case and is solved [46] by the dimer and Pfaffian methods of chapters 11 and 12.

13.4.2 Vertex–spin correspondence

The eight-vertex model which has the state variables on the bonds and the interaction energies on vertices as given in Fig. 13.12 has a two-to-one correspondence with a spin model with state variables $\sigma = \pm 1$ which lie on the faces of the lattice. To obtain this correspondence we use the two-to-one mapping between arrow configurations on bonds and spins on faces given in Fig. 13.14 where the spins on opposite sides of an up (down) pointing arrow are the same (opposite) and the spins on opposite sides of a right (left) pointing arrow are the same (opposite). If we specify the states on the eight-vertex lattice by a variable $\mu = \pm 1$ instead of an arrow as shown in Fig. 13.11 the relation between the spins σ and σ' on faces separated by the bond which carries the variable μ is

$$\mu = \sigma\sigma'. \tag{13.143}$$

Using the arrow–vertex correspondence of Fig. 13.14 we obtain from Fig. 13.12 the spin–arrow correspondence of vertex weights shown in Fig. 13.15.

We may reinterpret Fig. 13.15 as a spin model with interaction energies J_1 between horizontal spins, J_2 between vertical spins, J between spins on the NW to SE diagonal,

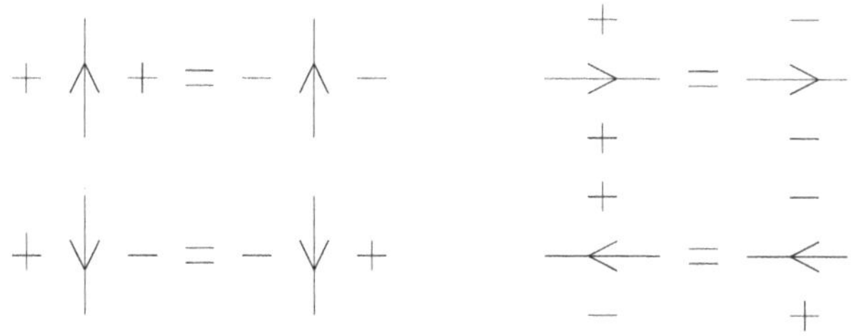

Fig. 13.14 The correspondence of spins σ, σ' and arrow configurations μ for eight-vertex models. For each spin configurations shown there is a second configuration obtained by sending σ and σ' into their negatives.

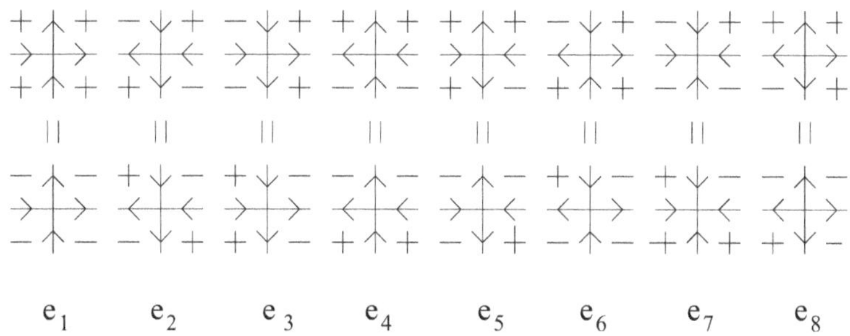

Fig. 13.15 The correspondence of the arrow configurations with spin configurations of the vertex energies of Fig. 13.12 where e_i indicates the interaction energy as opposed to $w_i = e^{-e_i/k_BT}$ which is the Boltzmann weight.

J' between spins on the NE to SW diagonal and the interaction J_4 for the product of all four spins. This interaction energy is written

$$\mathcal{E} = -J_0 - \sum_{j,k}\{J_1\sigma_{j,k}\sigma_{j,k+1} + J_2\sigma_{j,k}\sigma_{j+1,k} + J\sigma_{j,k+1}\sigma_{j+1,k} + J'\sigma_{j,k}\sigma_{j+1,k+1} + J_4\sigma_{j,k}\sigma_{j,k+1}\sigma_{j+1,k+1}\sigma_{j+1,k}\}. \tag{13.144}$$

The spin interaction energy (13.144) is shown graphically in Fig. 16 where we share the vertical and horizontal bond strengths equally between the two adjacent faces.

We thus find the following relation between vertex and spin energies:

$$\begin{aligned}
e_1 &= -J_0 - J_1 - J_2 - J - J' - J_4\\
e_2 &= -J_0 + J_1 + J_2 - J - J' - J_4\\
e_3 &= -J_0 + J_1 - J_2 + J + J' - J_4\\
e_4 &= -J_0 - J_1 + J_2 + J + J' - J_4\\
e_5 = e_6 &= -J_0 + J - J' + J_4\\
e_7 = e_8 &= -J_0 - J + J' + J_4
\end{aligned} \tag{13.145}$$

For the symmetric eight-vertex model $e_1 = e_2$ and $e_3 = e_4$. Therefore from (13.145) we find $J_1 = J_2 = 0$ and hence

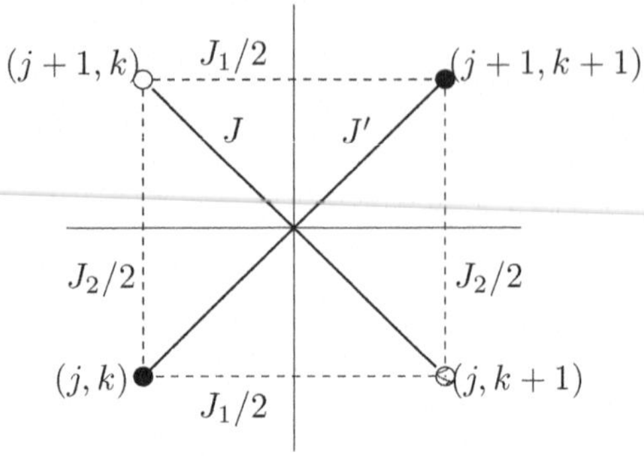

Fig. 13.16 The spin interaction energy (13.144).

$$\begin{aligned} e_1 = e_2 &= -J_0 - J - J' - J_4 \\ e_3 = e_4 &= -J_0 + J + J' - J_4 \\ e_5 = e_6 &= -J_0 + J - J' + J_4 \\ e_7 = e_8 &= -J_0 - J + J' + J_4 \end{aligned} \tag{13.146}$$

and thus the Boltzmann weights are

$$\begin{aligned} a &= \rho e^{K+K'+K_4} \\ b &= \rho e^{-K-K'+K_4} \\ c &= \rho e^{-K+K'-K_4} \\ d &= \rho e^{K-K'-K_4} \end{aligned} \tag{13.147}$$

where $K = J/k_BT$, $K' = J'/k_BT$, $K_4 = J_4/k_BT$. The invariants γ (13.87) and Δ (13.88) of the eight-vertex model are

$$\gamma = \frac{ab}{cd} = e^{-4K_4} \tag{13.148}$$

$$\begin{aligned} \Delta &= \frac{a^2 + b^2 - c^2 - d^2}{2(ab + cd)} \\ &= \frac{e^{2K_4}\cosh 2(K + K') - e^{-2K_4}\cosh 2(K - K')}{2\cosh 2K_4}. \end{aligned} \tag{13.149}$$

The decoupling point

Of particular interest is the special case $K_4 = 0$ where we see from (13.145) and Fig. 13.15 that the model reduces to two noninteracting Ising models as shown in Fig. 13.17. This is referred to as the **decoupling point**, and the invariants (13.148), (13.149) reduce to

$$\gamma = 1 \quad \text{and} \quad \Delta = \sinh 2K \sinh 2K' \tag{13.150}$$

where Δ is recognized as the modulus of the elliptic functions which appears in the diagonal Ising model correlations presented in chapters 11 and 12.

Furthermore we may use the relation (13.143) to express at the decoupling point the correlations of arrows (the state variables μ) in terms of the correlation functions of spins in the Ising model. If we call G_R the correlation of two vertical arrows $\mu_{0,0}$ and $\mu_{0,R}$ separated by a distance R in the same row and C_R the correlation of two vertical arrows $\mu_{0,0}$ and $\mu_{R,0}$ separated by a distance R in the same column we find from (13.143) and Fig. 13.17 that

$$\begin{aligned} G_{2R} &= \langle \mu_{0,0}\mu_{2R,0}\rangle = \langle \sigma_{0,0}\sigma'_{0,1}\sigma_{2R,0}\sigma'_{2R,1}\rangle \\ &= \langle \sigma_{0,0}\sigma_{2R,0}\rangle\langle \sigma'_{0,1}\sigma'_{2R,1}\rangle = C(R,R)^2 \end{aligned} \tag{13.151}$$

$$\begin{aligned} G_{2R+1} &= \langle \mu_{0,0}\mu_{2R+1,0}\rangle = \langle \sigma_{0,0}\sigma'_{0,1}\sigma'_{2R+1,0}\sigma_{2R+1,1}\rangle \\ &= \langle \sigma_{0,0}\sigma_{2R+1,0}\rangle\langle \sigma'_{0,1}\sigma'_{2R+1,1}\rangle = C(R,R+1)C(R+1,R) \end{aligned} \tag{13.152}$$

and similarly

$$\begin{aligned} C_{2R} &= \langle \mu_{0,0}\mu_{0,2R}\rangle = \langle \sigma_{0,0}\sigma'_{0,1}\sigma_{0,2R}\sigma'_{0,2R+1}\rangle \\ &= \langle \sigma_{0,0}\sigma_{0,2R}\rangle\langle \sigma'_{0,1}\sigma'_{0,2R+1}\rangle = C(R,R)^2 \end{aligned} \tag{13.153}$$

$$\begin{aligned} C_{2R+1} &= \langle \mu_{0,0}\mu_{0,2R+1}\rangle = \langle \sigma_{0,0}\sigma'_{0,1}\sigma'_{0,2R+1}\sigma_{0,2R+2}\rangle \\ &= \langle \sigma_{0,0}\sigma_{0,2R+2}\rangle\langle \sigma'_{0,1}\sigma'_{0,2R+1}\rangle = C(R+1,R+1)C(R,R) \end{aligned} \tag{13.154}$$

where we have denoted the spins on the two distinct noninteracting lattices by σ and σ', and $C(M,N)$ is the Ising correlation of spins at $(0,0)$ and at (M,N) which has been computed in detail in chapters 11 and 12.

13.4.3 Inhomogeneous lattices

Thus far we have considered lattices that are translationally invariant. However, we may extend our considerations to an inhomogeneous lattice with the transfer matrix

$$T(u,\{v_k\})_{\{j'\},\{j\}} = \mathrm{Tr}W(j'_1,j_1|u-v_1)W(j'_2,j_2|u-v_2)\cdots W(j'_1,j_1|u-v_N) \tag{13.155}$$

which reduces to (13.21) in the homogeneous limit

$$v_k = 0 \quad \text{for} \quad 1 \le k \le N. \tag{13.156}$$

Then, because the star–triangle equation is a local relation which involves only one column at a time an argument identical to that which proved (13.27) proves that

$$[T(u,\{v_k\}),T(u',\{v_k\})] = 0. \tag{13.157}$$

We thus may think of the lattice as having a set of N_h rapidities v_j in each of the N_h columns and rapidities u_k in each of the N_v rows and the Boltzmann weight at the site in the j column and k row depends on the rapidity $u_k - v_j$. This parametrization of the Boltzmann weights by two independent rapidities is a general property which extends to all models satisfying a star–triangle equation.

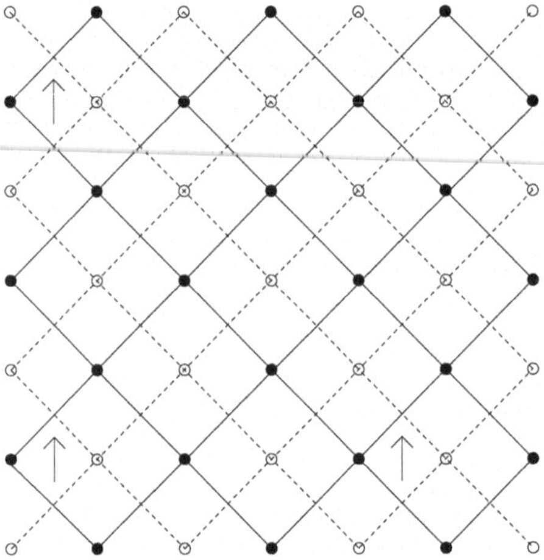

Fig. 13.17 The two interpenetrating Ising sublattices are represented by the filled and open circles. The vertical arrows correspond to the arrow of the eight-vertex model. We have illustrated the case where the vertical arrows are separated by an even number of sites in either the vertical or horizontal directions. In the decoupling limit both correlations for arrows separated by an even number of sites are equal to the square of the corresponding diagonal Ising correlation.

13.5 Star–triangle equation for spin models

For spin models the star–triangle equation is actually a pair of equations as shown in Fig. 13.18 where R is a scalar (normalizing) factor. When $W^v(u)$ and $W^h(v)$ are the weights of the Ising model this is the original star–triangle equation stated by Onsager [2] and proven in [3]. These two equations can be combined into the compound equation of Fig. 13.19.

13.5.1 Chiral Potts model

The most important solution of the star–triangle equation for spin models Fig. 13.18 is the N-state chiral Potts model whose integrability was first discovered in 1987 [23]. This model generalizes the Ising model from variables $\sigma_{j,k} = \pm 1$ to variables

$$\sigma_{j,k} = \omega^n \quad \text{with} \;\; n = 0, 1, \cdots N-1 \tag{13.158}$$

where

$$\omega = e^{2\pi i/N} \tag{13.159}$$

which satisfy

$$\sigma_{j,k}^N = 1 \tag{13.160}$$

and generalizes the interaction energy of the Ising model to

$$\mathcal{E}_{cp} = -\sum_{j,k}\sum_{n=1}^{N-1}\{E_n^h(\sigma_{j,k}\sigma_{j,k+1}^*)^n + E_n^v(\sigma_{j,k}\sigma_{j+1,k}^*)^n\} \tag{13.161}$$

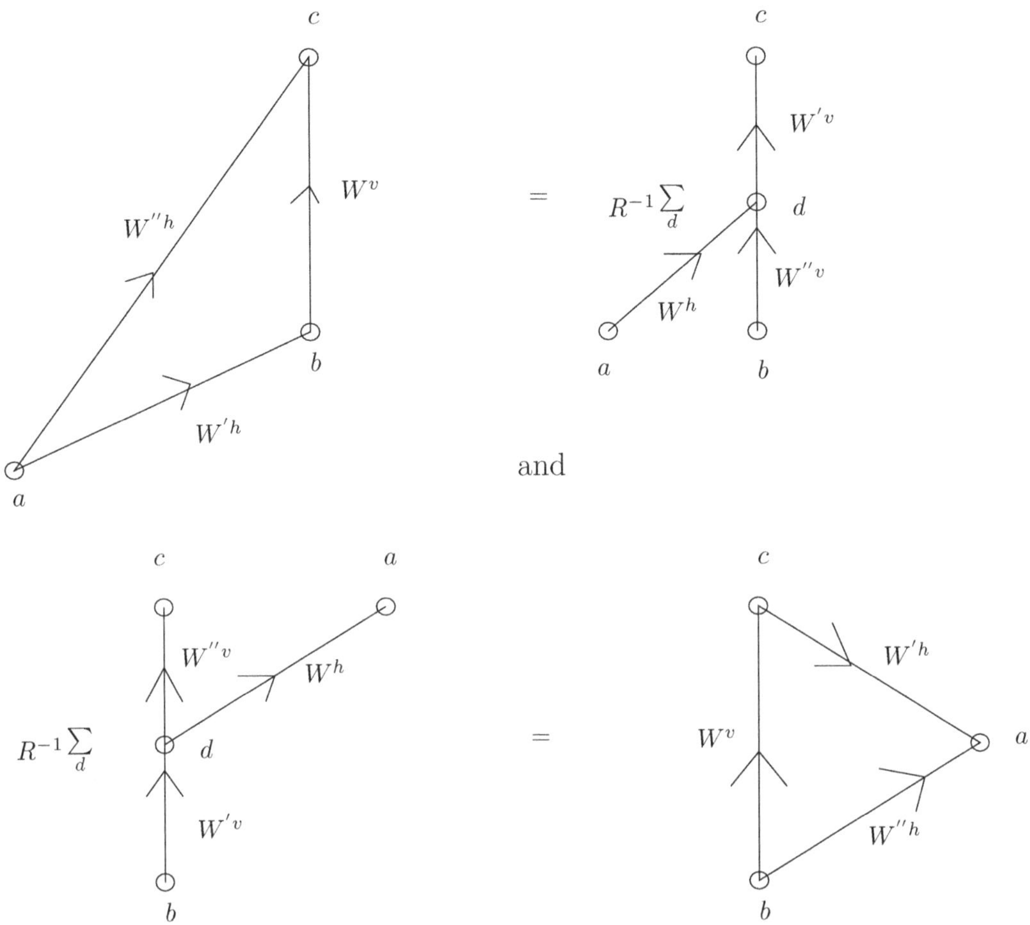

Fig. 13.18 The star–triangle equation for spin models. The convention that for $W^v(j, j')$ the state variable j' lies above the variable j and that for $W^h(j, j')$ the state variable j' lies to the right of j is graphically indicated by an arrow pointing from j to j'.

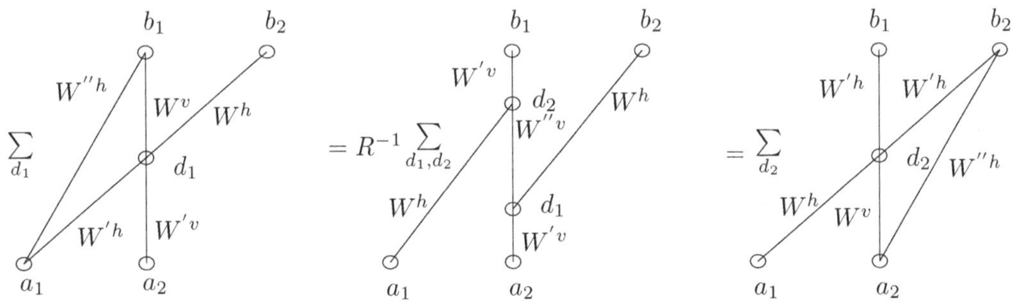

Fig. 13.19 The compound star–triangle equation for spin models.

from which we find Boltzmann weights

$$W^{v,h}(n) = \exp \beta \sum_{j=1}^{N-1} E_j^{v,h} \omega^{jn}. \tag{13.162}$$

There is a most important difference between the Boltzmann weights of the chiral Potts model and the Boltzmann weights of the eight-vertex model; namely that while the Boltzmann weights of the eight-vertex model are parametrized in terms of elliptic functions this is not the case for the chiral Potts model. This has the important consequence that while the Boltzmann weights for the inhomogeneous eight-vertex model depend only on the difference of the horizontal and vertical rapidities the Boltzmann weights for the chiral Potts model will depend on these two variables separately. We will indicate this by writing the Boltzmann weights as $W_{pq}^{v,h}(n)$.

The proof that the commutation relation (13.27) follows from the equation of Fig. 19 is given in Figs. 13.20 and 13.21.

The Boltzmann weights $W_{pq}^{v,h}(n)$ of the chiral Potts model specialize the general Boltzmann weights of spin models $W(j,j')$ by depending only on the difference $n = j - j'$ and by having the periodicity constraint

$$W_{pq}^{v,h}(j+N) = W_{pq}^{v,h}(j). \tag{13.163}$$

For spin models with these two constraints the pair of star–triangle equations in Fig. 13.18 are identical. This can be seen by 1) interchanging the primed and double primed variables and 2) noting for the two constraints that in the term with the sum over d the three arrows may be reversed. These two operations send the second equation into the first and from Fig. 13.18 we see that this equation is explicitly written as

$$\sum_{d=1}^{N} W_{qr}^{v}(b-d) W_{pr}^{h}(a-d) W_{pq}^{v}(d-c) = R_{pqr} W_{pq}^{h}(a-b) W_{pr}^{v}(b-c) W_{qr}^{h}(a-c) \tag{13.164}$$

where the unprimed weight is parametrized by the pair pr, the primed by pq and the double primed by qr. This equation is graphically expressed in Fig. 13.22.

We will show that the following Boltzmann weights of the N-state chiral Potts model first given for arbitrary N by Baxter, Perk and Au-Yang [26] satisfy the star–triangle equation (13.164)

$$\frac{W_{pq}^{h}(n)}{W_{pq}^{h}(0)} = \prod_{j=1}^{n} \left(\frac{d_p b_q - a_p c_q \omega^j}{b_p d_q - c_p a_q \omega^j} \right) \tag{13.165}$$

and

$$\frac{W_{pq}^{v}(n)}{W_{pq}^{v}(0)} = \prod_{j=1}^{n} \left(\frac{\omega a_p d_q - d_p a_q \omega^j}{c_p b_q - b_p c_q \omega^j} \right) \tag{13.166}$$

where a_p, b_p, c_p, d_p and a_q, b_q, c_q, d_q lie on the generalized elliptic curve

$$a_p^N + \lambda b_p^N = \lambda' d_p^N, \quad \lambda a_p^N + b_p^N = \lambda' c_p^N \tag{13.167}$$

with

$$\lambda' = (1 - \lambda^2)^{1/2}. \tag{13.168}$$

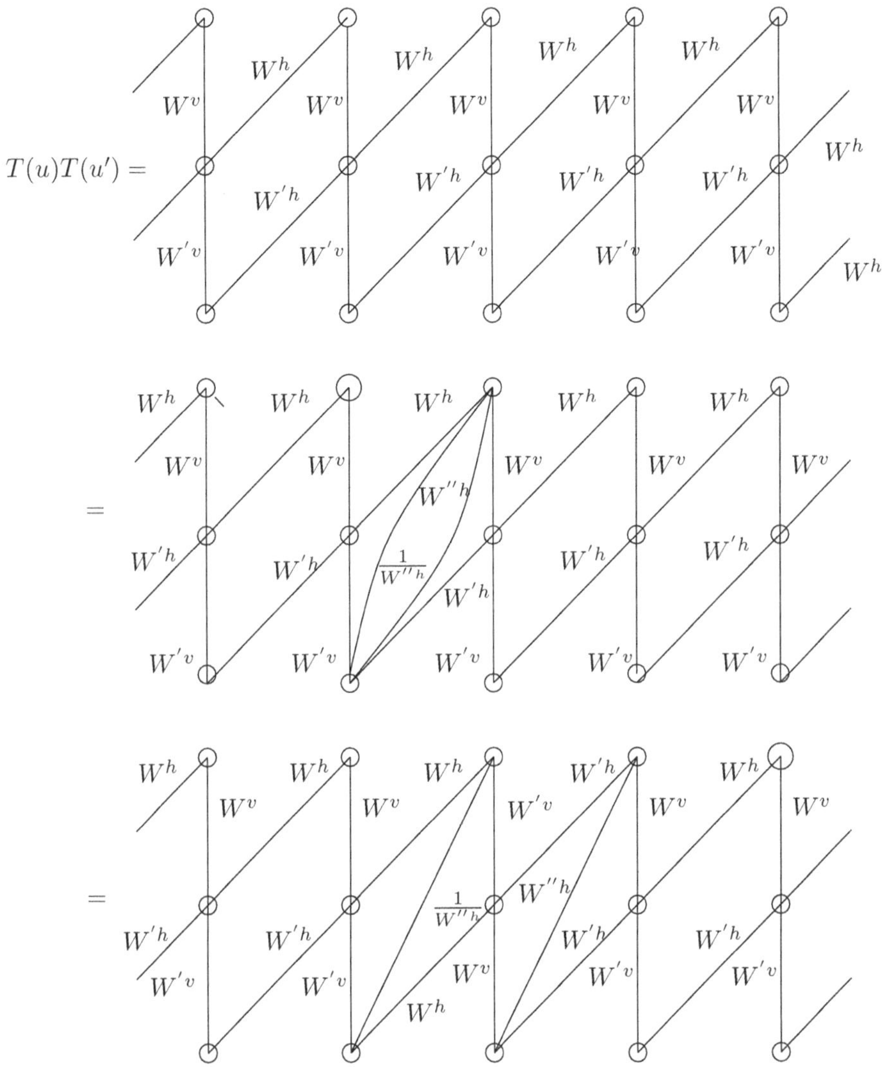

Fig. 13.20 Commutation of transfer matrices for spin models.

For $\lambda \neq 0, 1$ and ∞ the curve (13.167) has genus $N^3 - 2N^2 + 1 = (N-1)(N^2-N-1)$. The symbols p and q stand for points on the curve (13.167) and play the role of the spectral variable u in the vertex models. When $N = 2$ the curve (13.167) reduces to the elliptic curve of genus one with modulus λ and the model reduces to the Ising model. We refer to (13.167) as a spectral curve which generalizes the concept of the spectral variable u of the six- and eight-vertex model.

The relation (13.167) follows from the imposition of periodicity condition (13.163). To obtain this we follow [27] and impose the periodic constraint (13.163) on the weights (13.165) and (13.166) to find

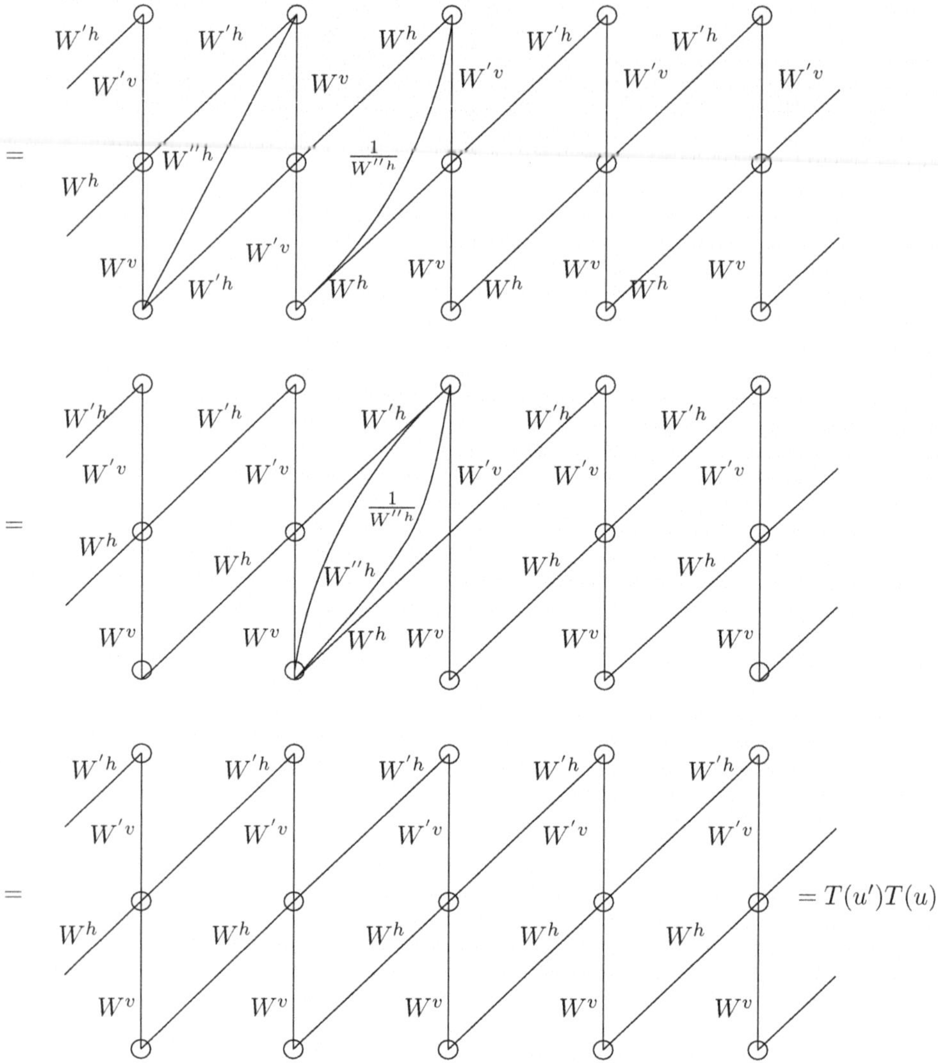

Fig. 13.21 Commutation of transfer matrices for spin models continued.

$$\frac{(d_p b_q)^N - (a_p c_q)^N}{(b_p d_q)^N - (c_p a_q)^N} = 1 \tag{13.169}$$

$$\frac{(a_p d_q)^N - (d_p a_q)^N}{(c_p b_q)^N - (b_p c_q)^N} = 1 \tag{13.170}$$

which we write in the form

$$(d_p b_q)^N + (c_p a_q)^N = (d_q b_p)^N + (c_q a_p)^N \tag{13.171}$$

$$(d_p a_q)^N + (c_p b_q)^N = (d_q a_p)^N + (c_q b_p)^N. \tag{13.172}$$

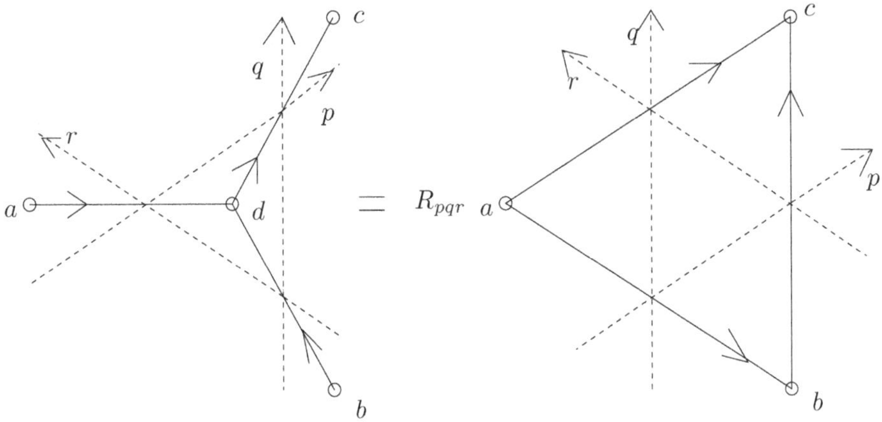

Fig. 13.22 The star–triangle equation (13.164) for the chiral Potts model where the rapidities p, q and r are indicated by dotted lines.

If we add these two equations we find

$$(d_p^N + c_p^N)(b_q^N + a_q^N) = (d_q^N + c_q^N)(b_p^N + a_p^N) \tag{13.173}$$

and if we subtract them we find

$$(d_p^N - c_p^N)(b_q^N - a_q^N) = (d_q^N - c_q^N)(b_q^N - a_q^N) \tag{13.174}$$

and thus

$$\frac{d_p^N \pm c_p^N}{b_p^N \pm a_p^N} = \frac{d_q^N \pm c_q^N}{b_q^N \pm a_q^N} = \lambda_\pm \tag{13.175}$$

is the same for both the p and the q variables. We now rewrite (13.175) as

$$\begin{aligned} d_p^N + c_p^N &= \lambda_+(b_p^N + a_p^N) \\ d_p^N - c_p^N &= \lambda_-(b_p^N - a_p^N) \end{aligned} \tag{13.176}$$

and by adding and subtracting the equations in (13.176) we obtain

$$\begin{aligned} d_p^N &= \frac{1}{2}(\lambda_+ + \lambda_-)b_p^N + \frac{1}{2}(\lambda_+ - \lambda_-)a_p^N \\ c_p^N &= \frac{1}{2}(\lambda_+ - \lambda_-)b_p^N + \frac{1}{2}(\lambda_+ + \lambda_-)a_p^N \end{aligned} \tag{13.177}$$

which reduces to (13.167) if we define

$$\lambda' = \frac{2}{\lambda_+ - \lambda_-}, \quad \lambda = \frac{\lambda_+ + \lambda_-}{\lambda_+ - \lambda_-}. \tag{13.178}$$

Finally we note that we may fix the normalization of a_p, b_p relative to c_p, d_p by choosing

$$\lambda_+\lambda_- = -1 \tag{13.179}$$

and with this choice we see that λ and λ' defined by (13.178) satisfy (13.168).

It is often convenient to express (13.165)–(13.167) in terms of the inhomogeneous variables (for both p and q)

$$x_p = \frac{a_p}{d_p}, \quad y_p = \frac{b_p}{c_p}, \quad \mu_p = \frac{d_p}{c_p} \tag{13.180}$$

so that

$$\frac{W^h_{pq}(n)}{W^h_{pq}(0)} = \left(\frac{\mu_p}{\mu_q}\right)^n \prod_{j=1}^{n} \left(\frac{y_q - x_p\omega^j}{y_p - x_q\omega^j}\right). \tag{13.181}$$

and

$$\frac{W^v_{pq}(n)}{W^v_{pq}(0)} = (\mu_p\mu_q)^n \prod_{j=1}^{n} \left(\frac{\omega x_p - x_q\omega^j}{y_q - y_p\omega^j}\right). \tag{13.182}$$

Using the periodicity condition (13.163) we find from (13.181) that

$$\left(\frac{\mu_p}{\mu_q}\right)^N = \frac{y_p^N - x_q^N}{y_q^N - x_p^N} \tag{13.183}$$

and from (13.182) that

$$(\mu_p\mu_q)^N = \frac{y_q^N - y_p^N}{x_p^N - x_q^N}. \tag{13.184}$$

Furthermore by using x_p and y_p of (13.180) we may write (13.167) as

$$\lambda' - x_p^N = \lambda\left(\frac{b_p}{d_p}\right)^N, \quad \lambda' - y_p^N = \lambda\left(\frac{a_p}{c_p}\right)^N. \tag{13.185}$$

Then by multiplying the two equations in (13.185) together we find

$$(x_p^N - \lambda')(y_p^N - \lambda') = \lambda^2 x_p^N y_p^N \tag{13.186}$$

which, by use of the definition of λ' (13.168), reduces to

$$x_p^N + y_p^N = \lambda'(1 + x_p^N y_p^N). \tag{13.187}$$

When $\lambda \neq 0, 1, \infty$ the genus of the curve (13.187) is $(N-1)^2$.
In addition, from (13.167) we find

$$x_p^N + \lambda y_p^N \mu_p^{-N} = \lambda' \text{ and } \lambda\mu_p^N x_p^N + y_p^N = \lambda'. \tag{13.188}$$

Therefore

$$\mu_p^N = \frac{\lambda y_p^N}{\lambda' - x_p^N} \text{ and } \mu_p^N = \frac{\lambda' - y_p^N}{\lambda x_p^N} \tag{13.189}$$

and by use of (13.187) we obtain

$$\mu_p^N = \frac{\lambda}{1 - \lambda' x_p^N} \text{ and } \mu_p^N = \frac{1 - \lambda' y_p^N}{\lambda}. \tag{13.190}$$

When $\lambda \to 1$ the left-hand sides of the equations in (13.167) become equal. From (13.168) we see that $\lambda' \to 0$ and thus from (13.190) we see that $\mu_p^N \to 1$. In this limit the genus of both the curves (13.167) and (13.187) will change in a discontinuous fashion. However, it is most important to observe that the resulting curve is not unique and depends on the manner in which the limit is approached.

Consider first letting $\lambda' \to 0$ while keeping $x_p,\ y_p \neq 0$. Then from (13.187) we find

$$x_p^N + y_p^N = 0, \quad \mu_p^N = 1. \tag{13.191}$$

Thus the curve (13.187) of genus $(N-1)^2$ has degenerated into a collection of disjoint curves of genus zero (rational curves). From the set (13.191) we are able to choose for both p and q

$$y_{p,q} = \omega^{1/2} x_{p,q}, \quad \mu_{p,q} = 1 \tag{13.192}$$

and using (13.192) we find that the formulas for the Boltzmann weights (13.181) and (13.182) reduce to

$$\frac{W_{pq}^h(n)}{W_{pq}^h(0)} \to \prod_{j=1}^{n} \left(\frac{\omega^{1/2} x_q - x_p \omega^j}{\omega^{1/2} x_p - x_q \omega^j} \right) \tag{13.193}$$

and

$$\frac{W_{pq}^v(n)}{W_{pq}^v(0)} \to \prod_{j=1}^{n} \left(\frac{\omega x_p - x_q \omega^j}{\omega^{1/2} x_q - \omega^{1/2} x_p \omega^j} \right) \tag{13.194}$$

which depend only on x_q/x_p. Then, writing

$$x_q/x_p = e^{2i\theta} \tag{13.195}$$

and denoting the limiting results on the right-hand side of (13.193) and (13.194) as $W_{FZ}^{h,v}(n;\theta)$ we obtain

$$\frac{W_{pq}^h(n)}{W_{pq}^h(0)} \to \frac{W_{FZ}^h(n;\theta)}{W_{FZ}^h(0;\theta)} == \prod_{j=1}^{n} \left(\frac{\sin[\frac{\pi}{N}(j-1/2)-\theta]}{\sin[\frac{\pi}{N}(j-1/2)+\theta]} \right). \tag{13.196}$$

and

$$\frac{W_{pq}^v(n)}{W_{pq}^v(0)} \to \frac{W_{FZ}^v(n;\theta)}{W_{FZ}^v(0;\theta)} == \prod_{j=1}^{n} \left(\frac{\sin[\frac{\pi}{N}(j-1)+\theta]}{\sin[\frac{\pi}{N}j-\theta]} \right). \tag{13.197}$$

The weights $W_{FZ}^{v,h}(n;\theta)$ are the weights of the self-dual Z_N model obtained by Fateev and Zamolodchikov [19] in 1982. When θ is real the Boltzmann weights (13.196), (13.197) are real, and when

$$0 \leq \theta \leq \frac{\pi}{2N} \tag{13.198}$$

the weights are positive. If $\theta = \frac{\pi}{4N}$ then

$$\frac{W_{FZ}^v(n;\pi/4N)}{W_{FZ}^v(0;\pi/4N)} = \frac{W_{FZ}^h(n;\pi/4N)}{W_{FZ}^h(0;\pi/4N)} \tag{13.199}$$

and the model is spatially isotropic.

There is, however, a second way to obtain the limit $\lambda \to 1$ if in (13.187) we let x_p^N and y_p^N both be proportional to λ'.

$$x_p^N = \lambda' \bar{x}_p^N, \quad y_p^N = \lambda' \bar{y}_p^N \tag{13.200}$$

Then (13.187) degenerates to

$$\bar{x}_p^N + \bar{y}_p^N = 1, \quad \mu_p^N = 1 \tag{13.201}$$

which is the famous Fermat curve. The genus of (13.201) is $(N-1)(N-2)/2$ and the Boltzmann weights (13.181) and (13.182) become

$$\frac{W_{pq}^h(n)}{W_{pq}^h(0)} = \prod_{j=1}^{n} \left(\frac{\bar{y}_q - \bar{x}_p \omega^j}{\bar{y}_p - \bar{x}_q \omega^j} \right) \tag{13.202}$$

and

$$\frac{W_{pq}^v(n)}{W_{pq}^v(0)} = \prod_{j=1}^{n} \left(\frac{\omega \bar{x}_p - \bar{x}_q \omega^j}{\bar{y}_q - \bar{y}_p \omega^j} \right). \tag{13.203}$$

Unlike the weights (13.196) and (13.197) of the Fateev–Zamolodchikov model there is no region where the weights (13.202) and (13.203) are real.

Alternatively in (13.167) we can rescale c_p^N and d_p^N by defining

$$\lambda' d_p^N = \bar{d}_p^N, \quad \lambda' c_p^N = \bar{c}_p^N \tag{13.204}$$

to find that in the limit $\lambda \to 1$ and $\lambda' \to 0$ that $\bar{c}_p^N = \bar{d}_p^N$ and (13.167) reduces to the Fermat curve in terms of homogeneous coordinates

$$a_p^N + b_p^N = \bar{c}_p^N. \tag{13.205}$$

13.5.2 Proof of the star–triangle equation

We must now demonstrate that the Boltzmann weights (13.165) and (13.166) actually do satisfy the star–triangle equation (13.164). Instead of deriving the Boltzmann weights for the chiral Potts model as we did for the six- and eight-vertex model we will here instead present the verification first given by Au-Yang and Perk in [27].

The Boltzmann weights in the star–triangle equation (13.164) depend only of differences and are periodic. Thus it suffices to prove that (13.164) holds for $b = 0$. Then if we let $c \to -c$, $d \to -d$, multiply (13.164) by ω^{nc} and sum $1 \le c \le N$ we find

$$\begin{aligned} &\sum_{d=1}^{N} W_{qr}^v(d) W_{pr}^h(a+d) \omega^{nd} \sum_{c=1}^{N} \omega^{n(c-d)} W_{pq}^v(-d+c) \\ &= R_{pqr} W_{pq}^h(a) \sum_{c=1}^{N} W_{pr}^v(c) W_{qr}^h(a+c) \omega^{nc}. \end{aligned} \tag{13.206}$$

Thus if we define the Fourier transform of $W_{pq}^v(n)$ as

$$\hat{W}^v_{pq}(m) = \sum_{n=1}^{N} \omega^{mn} W^v_{pq}(n) \tag{13.207}$$

and define

$$V_{m,n} = V_{m,n}(p,q,r) \equiv \sum_{d=1}^{N} \omega^{nd} W^v_{qr}(d) W^h_{pr}(m+d) \tag{13.208}$$

$$\bar{V}_{n,m} = \bar{V}_{n,m}(p,q,r) \equiv \sum_{c=1}^{N} \omega^{nc} W^v_{pr}(c) W^h_{qr}(m+c) \tag{13.209}$$

we may write (13.206) as

$$V_{m,n} \hat{W}^v_{pq}(n) = R_{pqr} \bar{V}_{n,m} W^h_{pq}(m) \tag{13.210}$$

and we note from (13.208) and (13.209) the symmetry relation

$$V_{m,n}(p,q,r) = \bar{V}_{n,m}(q,p,r). \tag{13.211}$$

To prove that the star–triangle equation (13.164) holds, it is sufficient to prove that (13.210) holds. We will prove that $\bar{V}_{n,m}$ and $V_{m,n}\hat{W}^v_{p,q}(n)/W^h_{p,q}(m)$ satisfy the same first order homogeneous difference equation. Therefore $\bar{V}_{n,m}$ and $V_{m,n}\hat{W}^v_{pq}(n)/W^h_{pq}(m)$ will have to be proportional to each other and thus if we call the constant of proportionality R_{pqr} we will have established that (13.210) holds.

From (13.166) with $(p,q) \to (q,r)$ we find

$$(c_q b_r - b_q c_r \omega^k) W^v_{qr}(k) = (\omega a_q d_r - d_q a_r \omega^k) W^v_{qr}(k-1). \tag{13.212}$$

Multiply (13.212) by $\omega^{nk} W^h_{pr}(m+k)$ and sum $1 \leq k \leq N$ and use the definition (13.208) of $V_{mj,n}$ to find

$$c_q b_r V_{m,n} - b_q c_r V_{m,n+1} = a_q d_r \omega^{n+1} V_{m+1,n} - d_q a_r \omega^{n+1} V_{m+1,n+1}. \tag{13.213}$$

Similarly from (13.165) with $(p,q) \to (p,r)$ we find

$$(b_p d_r - c_p a_r \omega^{m+k+1}) W^h_{pr}(m+k+1) = (d_r b_r - a_p c_r \omega^{m+k+1}) W^h_{pr}(m+k) \tag{13.214}$$

and multiplying by $\omega^{nk} W^v_{qr}(k)$ and summing $1 \leq k \leq N$ we find

$$d_p b_r V_{m,n} - b_p d_r V_{m+1,n} = a_p c_r \omega^{m+1} V_{m,n+1} - c_p a_r \omega^{m+1} V_{m+1,n+1}. \tag{13.215}$$

The pair of equations (13.213) and (13.215) are a set of first order homogeneous difference relations in the two variables m and n with $1 \leq m, n \leq N$ which are periodic in both m and n. It is thus expected that the solution of this pair of equations will be unique up to a multiplicative constant. This is demonstrated by defining

$$Y_{m+k,k} = \frac{1}{N}\sum_{n=1}^{N} V_{m,n}\omega^{-nk} \tag{13.216}$$

and then we see from (13.213) and (13.215) that $Y_{j,k}$ satisfies

$$\frac{Y_{j,k-1}}{Y_{j,k}} = \frac{W_{qr}^{v}(k-1)}{W_{qr}^{v}(k)} \quad \text{and} \quad \frac{Y_{j+1,k}}{Y_{j,k}} = \frac{W_{pr}^{h}(j+1)}{W_{pr}^{h}(j)} \tag{13.217}$$

which obviously has the unique solution

$$Y_{j,k} = \text{const } W_{p,r}^{h}(j)W_{q,r}^{v}(k). \tag{13.218}$$

We now find an equivalent pair of equations by eliminating $V_{m,n}$ between (13.213) and (13.215) to find

$$\begin{aligned} &c_r(d_pb_q - a_pc_q\omega^{m+1})V_{m,n+1} - d_r(b_pc_q - d_pa_q\omega^{n+1})V_{m+1,n} \\ &+a_r(c_pc_q\omega^{m+1} - d_pd_q\omega^{n+1})V_{m+1,n+1} = 0 \end{aligned} \tag{13.219}$$

by eliminating $V_{m+1,n+1}$ to obtain

$$\begin{aligned} &d_r(b_pd_q - c_pa_q\omega^{m+1})\omega^{n}V_{m+1,n} - c_r(c_pb_q - a_pd_q\omega^{n+1})\omega^{m}V_{m,n+1} \\ &+b_r(c_pc_q\omega^{m} - d_pc_q\omega^{n})V_{m,n} = 0. \end{aligned} \tag{13.220}$$

Furthermore we may let $p \leftrightarrow q$ and $m \leftrightarrow n$ in the pair (13.219) and (13.220) and use the relation (13.211) to obtain the pair

$$\begin{aligned} &d_r(c_pb_q - a_pd_q\omega^{m+1})\bar{V}_{m,n+1} - c_r(b_pd_q - c_pa_q\omega^{n+1})\bar{V}_{m+1,n} \\ &+a_r(d_pd_q\omega^{m+1} - c_pc_q\omega^{n+1})\bar{V}_{m+1,n+1} = 0 \end{aligned} \tag{13.221}$$

and

$$\begin{aligned} &c_r(b_pc_q - d_pa_q\omega^{m+1})\omega^{n}\bar{V}_{m+1,n} - d_r(d_pb_q - a_pc_q\omega^{n+1})\omega^{m}\bar{V}_{m,n+1} \\ &+b_r(d_pd_q\omega^{m} - c_pc_q\omega^{n})\bar{V}_{m,n} = 0. \end{aligned} \tag{13.222}$$

To complete the proof we need to compute the Fourier transform $\hat{W}_{pq}^{v}(m)$. This is done by first writing (13.166) as

$$(c_pb_q - b_pc_q\omega^{n})W_{pq}^{v}(n) = (\omega a_pd_q - d_pa_q\omega^{n})W_{pq}^{v}(n-1) \tag{13.223}$$

and then multiplying by ω^{mn}, summing $1 \le n \le N$ and using the definition (13.207) of the Fourier transform to write

$$c_pb_q\hat{W}_{pq}^{v}(m) - b_pc_q\hat{W}_{pq}^{v}(m+1) = \omega^{m+1}a_pd_q\hat{W}_{pq}^{v}(m) - \omega^{m+1}d_pa_q\hat{W}_{pq}^{v}(m+1) \tag{13.224}$$

from which we obtain

$$\frac{\hat{W}_{pq}^{v}(m)}{\hat{W}_{pq}^{v}(0)} = \prod_{j=1}^{m}\frac{c_pb_q - a_pd_q\omega^{j}}{b_pc_q - d_pa_q\omega^{j}}. \tag{13.225}$$

We now note that the coefficient of c_r in (13.219) is a factor in the numerator in (13.165) and the coefficient of d_r in (13.219) is a factor in the denominator in (13.225), the coefficient of d_r in (13.220) is the denominator in (13.165) and the coefficient of c_r in (13.220) is the numerator in (13.225). Therefore we see from (13.219) and (13.220) that $V_{n,m}\hat{W}^v_{p,q}(m)/W^h_{p,q}(n)$ satisfies the same equations (13.221) and (13.222) that are satisfied by $\bar{V}_{m,n}$ Thus we have shown that $\bar{V}_{m,n}$ and $V_{n,m}\hat{W}^v_{pq}(m)/W^h_{pq}(n)$ satisfy the same set of first order difference equations. The solution of this set of equations is unique up to a multiplicative constant and thus $\bar{V}_{m,n}$ and $V_{n,m}\hat{W}^v_{pq}(m)/W^h_{pq}(n)$ must be proportional to each other. Thus we have demonstrated that (13.210) holds and hence we have proven that the star–triangle equation (13.164) is satisfied by the Boltzmann weights (13.165) and (13.166) of the chiral Potts model.

13.5.3 Determination of R_{pqr}

It remains to compute the proportionality constant R_{pqr} in the star–triangle equation (13.164). This is easily done by the method of [49]. First set $a-d=k,\ a-c=j,\ a-b=l$ to rewrite (13.164) as

$$\sum_{k=1}^{N} W^v_{pq}(j-k)W^h_{pr}(k)W^v_{qr}(k-l) = R_{pqr}W^h_{pq}(j)W^v_{qr}(j-l)W^h_{pq}(l) \tag{13.226}$$

which by defining the $N \times N$ matrices

$$\mathbf{W}^v(pq)_{j,k} = W^v_{pq}(j-k) \tag{13.227}$$

$$\mathbf{W}^h(pq)_{j,k} = \delta_{j,k}W^h_{pq}(k) \tag{13.228}$$

is rewritten in a matrix form as

$$\mathbf{W}^v(p,q)\mathbf{W}^h(p,r)\mathbf{W}^v(q,r) = R_{pqr}\mathbf{W}^h(q,r)\mathbf{W}^v(p,r)\mathbf{W}^h(p,q). \tag{13.229}$$

Thus by taking the determinant of both sides of (13.229) we find

$$R^N_{pqr} = \frac{\det\mathbf{W}^v(p,q)\det\mathbf{W}^h(p,r)\det\mathbf{W}^v(q,r)}{\det\mathbf{W}^h(q,r)\det\mathbf{W}^v(p,r)\det\mathbf{W}^h(p,q)} \tag{13.230}$$

and thus we have the factorized form

$$R_{pqr} = \frac{f_{p,q}f_{q,r}}{f_{p,r}} \tag{13.231}$$

where

$$f_{p,q} = \left[\frac{\det\mathbf{W}^v(p,q)}{\det\mathbf{W}^h(p,q)}\right]^{1/N}. \tag{13.232}$$

The matrix $\mathbf{W}^h(p,q)$ is diagonal so from (13.228) we find

$$\det\mathbf{W}^h(p,q) = \prod_{j=1}^{N} W^h_{pq}(j). \tag{13.233}$$

The matrix $\mathbf{W}^v(p,q)$ is seen from (13.227) to be cyclic. Therefore the eigenvalues of $\mathbf{W}^v(p,q)$ are

$$\lambda_j = \sum_{k=1}^{N} \omega^{kj} w^v_{p,q}(k) = \hat{W}^v_{pq}(j) \tag{13.234}$$

and thus

$$\det \mathbf{W}^v(p,q) = \prod_{j=1}^{N} \hat{W}^v_{p,q}(j). \tag{13.235}$$

Therefore using (13.233) and (13.235) in (13.232) we find

$$f_{p,q} = \left[\frac{\prod_{j=1}^{N} \hat{W}^v_{pq}(j)}{\prod_{j=1}^{N} W^h_{pq}(j)} \right]^{1/N}. \tag{13.236}$$

Finally we note that we may take the trace of (13.229) to find an alternative formula for R_{pqr}

$$R_{pqr} = \frac{\mathrm{Tr}\{\mathbf{W}^v(p,q)\mathbf{W}^h(p,r)\mathbf{W}^v(q,r)\}}{\mathrm{Tr}\{\mathbf{W}^h(q,r)\mathbf{W}^v(p,r)\mathbf{W}^h(p,q)\}} \tag{13.237}$$

which demonstrates that even though the formula (13.231) for R_{pqr} involves an Nth root through the formula (13.236) for f_{pq} that R_{pqr} is a single-valued function on the curve (13.167).

13.6 Star–triangle equation for face models

The star–triangle equation for face models is given in Fig. 13.23 and the inversion relation in Fig. 13.24. The proof that the commutation relation (13.27) follows from Fig. 13.23 and 13.24 is given in Fig. 13.25 and Fig. 13.26.

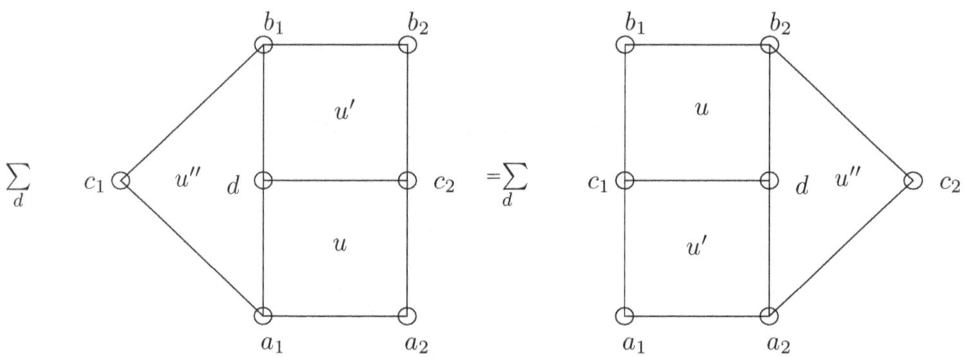

Fig. 13.23 Star–triangle equation for face models.

13.6.1 SOS and RSOS models

Two of the most important examples of integrable face models are the solid on solid (SOS) and restricted solid on solid (RSOS) models first introduced by Baxter [12–14] in 1973 in connection with the solution of the eight-vertex model and studied in detail in 1984 by Andrews, Baxter and Forrester [21].

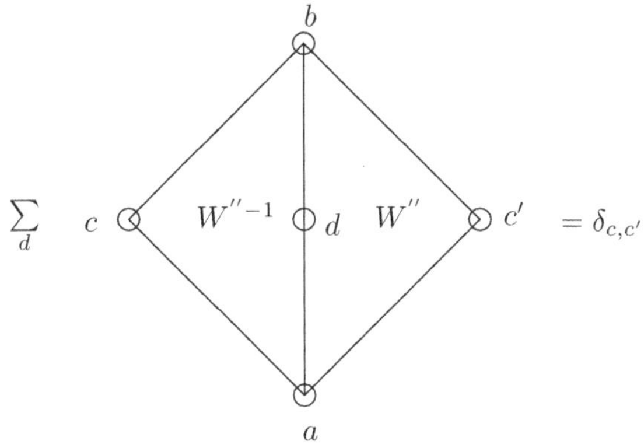

Fig. 13.24 Inversion for face models.

The SOS model

The (unrestricted) solid-on-solid model has a "height" variable l_i at each site of a square lattice which can take on all integer values

$$-\infty < l_i < \infty. \tag{13.238}$$

The Boltzmann weights have the property that if l_i and l_j are nearest neighbors (either vertical or horizontal) that

$$l_i - l_j = \pm 1. \tag{13.239}$$

There are thus six possible types of Boltzmann weights

$$W(l, l+1|l-1, l) = W(l, l-1|l+1, l) = \alpha_l \tag{13.240}$$
$$W(l+1, l|l, l-1) = W(l-1, l|l, l+1) = \beta_l \tag{13.241}$$
$$W(l+1, l|l, l+1) = \gamma_l \tag{13.242}$$
$$W(l-1, l|l, l-1) = \delta_l. \tag{13.243}$$

where each weight will in general depend on the height l. These weights are shown in Fig. 13.27. They have the symmetry property that

$$W(l, m'|l', m) = W(l, l'|m', m) = W(m, m'|l', l). \tag{13.244}$$

In order for the model to be integrable it must satisfy the star–triangle equation of Fig. 13.23 which we write explicitly as

$$\sum_d W''(c_1, a_1|b_1, d)W(a_1, a_2|d, c_2)W'(d, c_2|b_1, b_2)$$
$$= \sum_d W(c_1, d|b_1, b_2)W'(a_1, a_2|c_1, d)W''(d, a_2|b_2, c_2). \tag{13.245}$$

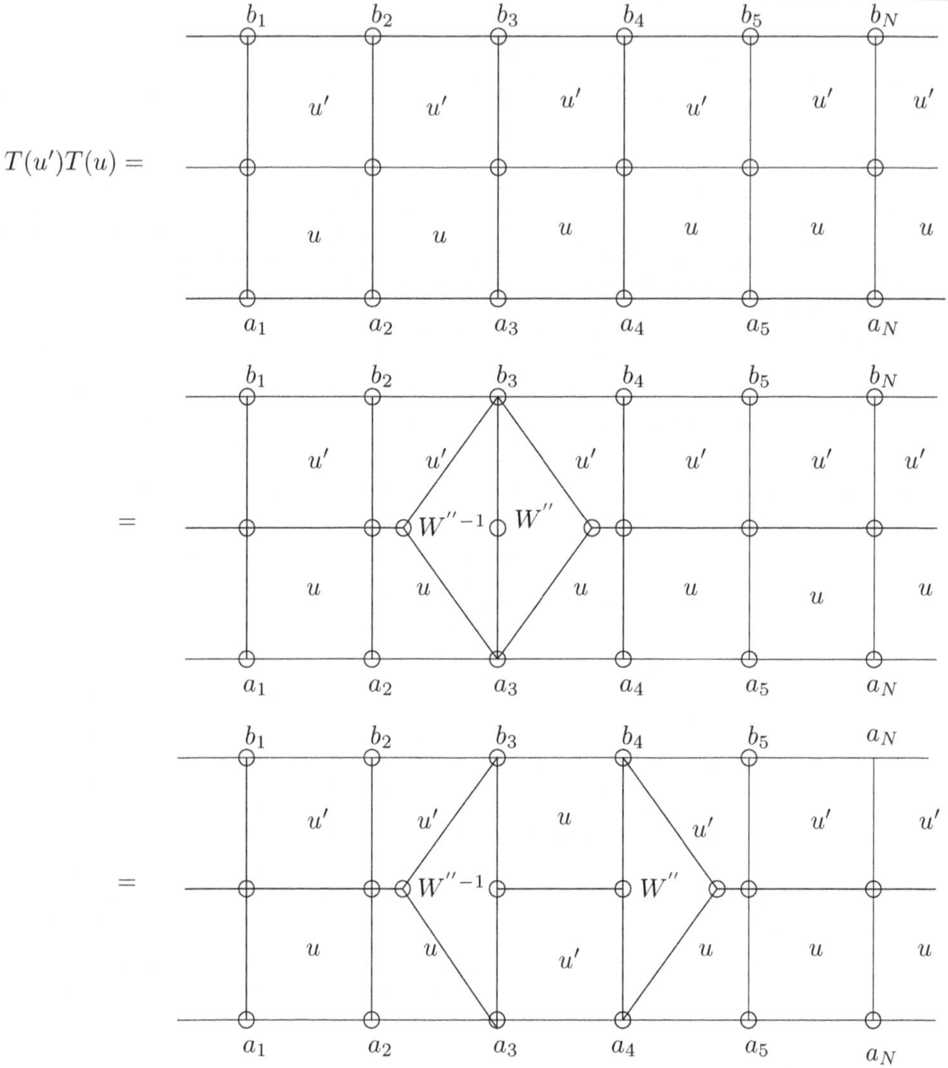

Fig. 13.25 Commutation of transfer matrices for face models.

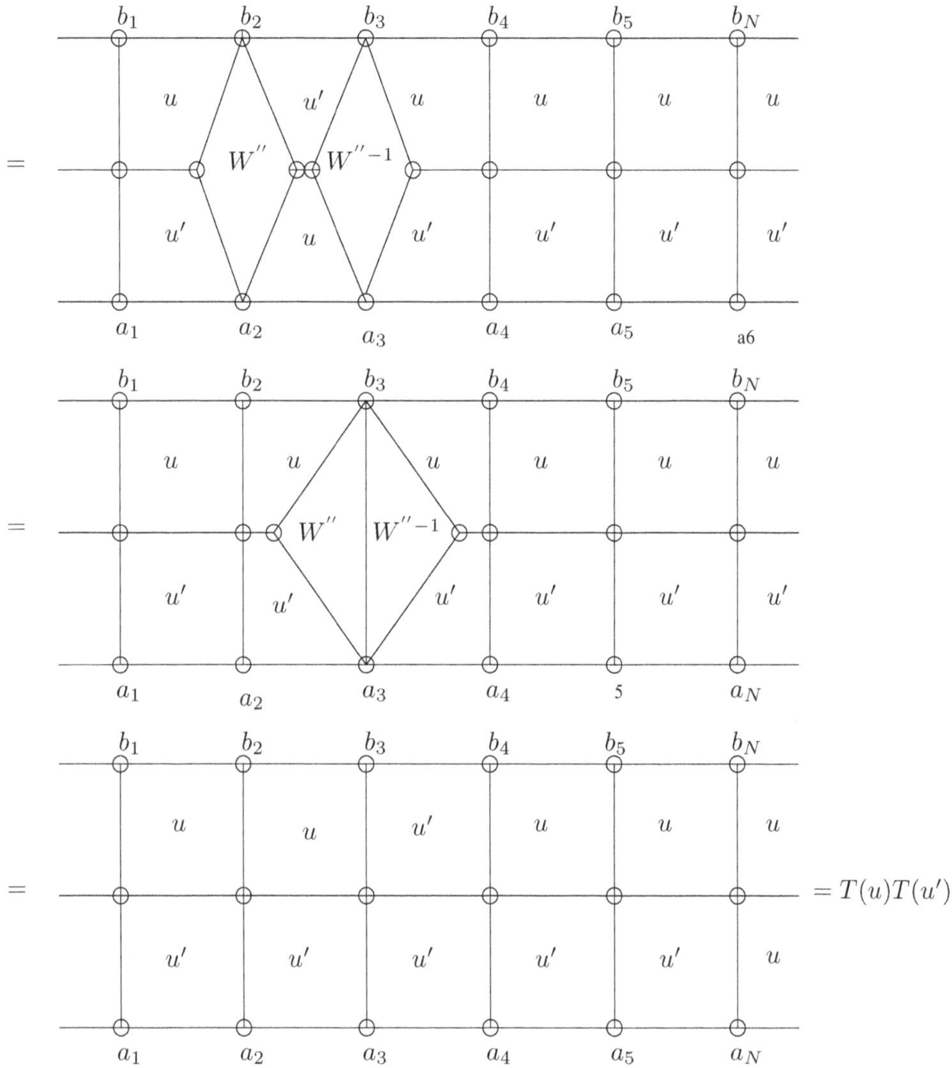

Fig. 13.26 Commutation of transfer matrices for face models.

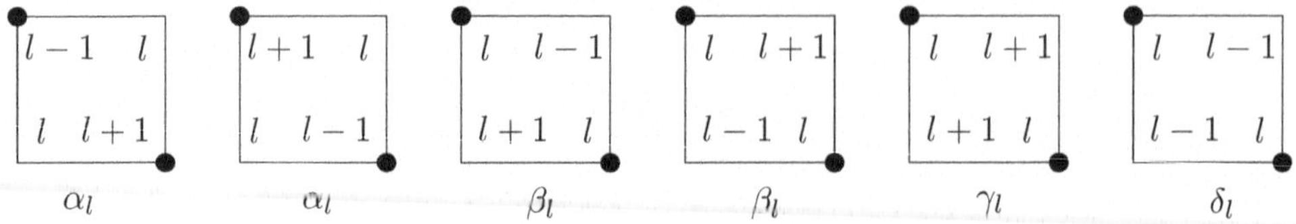

Fig. 13.27 The Boltzmann weights of the solid-on-solid models.

Because of the restriction (13.239) there are 20 choices for the indices in (13.245). These equations occur in pairs and thus there are 10 distinct equations to be satisfied. Seven of these are (1.4.7 of [21]):

$$\alpha_l\beta'_l\beta''_l + \delta_{l+1}\gamma'_l\gamma''_l = \gamma'_l\delta_{l+1}\delta''_{l+1} + \alpha_{l+1}\beta'_{l+1}\beta''_{l+1} \tag{13.246}$$
$$\gamma_{l-1}\beta'_l\beta''_l + \alpha_l\gamma'_l\gamma''_l = \gamma_l\alpha_l\alpha''_l \tag{13.247}$$
$$\alpha_l\beta_l\delta''_l + \delta'_{l+1}\beta''_l = \beta_l\alpha_;\delta''_{l+1} \tag{13.248}$$
$$\alpha_l\delta_l\delta''_l + \delta_{l+1}\beta'_l\beta''_l = \delta_l\alpha'_l\alpha''_l \tag{13.249}$$
$$\gamma_{l-1}\delta'_l\beta''_l + \alpha_l\beta'_l\gamma''_l = \beta_l\alpha'_l\gamma''_{l-1} \tag{13.250}$$
$$\alpha_l\alpha'_{l+1}\alpha''_{l+1} = \alpha_{l+1}\alpha'_l\alpha''_l \tag{13.251}$$
$$\beta'_{l+1}\alpha'_{l+1}\beta''_l = \beta_l\alpha'_l\beta''_{l+1} \tag{13.252}$$

and three more equations are obtained by interchanging primed and double primed symbols. The solution to these equations is given in 1.2.12 of [21]:

$$\alpha_l = \rho h(v+\eta) \tag{13.253}$$
$$\beta_l = \rho h(\eta - v)[h(w_{l-1})h(w_{l+1})]^{1/2}/h(w_l) \tag{13.254}$$
$$\gamma_l = \rho h(2\eta)h(w_l + \eta - v)/h(w_l) \tag{13.255}$$
$$\delta_l = \rho h(2\eta)h(w_l - \eta + v)/h(w_l) \tag{13.256}$$

where

$$w_l = w_0 + 2l\eta \tag{13.257}$$

and

$$h(v) = H(v)\Theta(v) \tag{13.258}$$

which from the product formulas for theta functions (13.381) and (13.382) of the appendix may be written as

$$h(v) = 2q^{1/4}\sin\frac{\pi v}{2K}\prod_{n=1}^{\infty}(1-q^n e^{i\pi v/K})(1-q^n e^{-i\pi v/K})(1-q^{2n})^2. \tag{13.259}$$

The solution (13.253)–(13.256) of (13.246)–(13.252) is obtained ab initio in [21] and may be verified by direct substitution by using the properties of theta functions in the appendix. Thus (13.240)–(13.243) with (13.253)–(13.256) is a solution of the face version of the star–triangle equations.

The RSOS model

The restricted solid-on-solid model is a restriction of the SOS model to

$$\eta = K/r \tag{13.260}$$

$$w_0 = 0 \tag{13.261}$$

and each height l_i lies in the interval

$$l_i = 1, 2, \cdots, r-1 \tag{13.262}$$

where $r \geq 3$ is an integer.

For the restricted model each $a_1, \cdots, c_2$ must lie in the interval (13.262). This means that l takes the values $1, \cdots, r-2$ in the first equation in (13.252). The values are $2, \cdots, r-2$ in the next four equations and $2, \cdots, r-3$ in the last two. Furthermore, the summation variable d must also lie in the interval (13.262) which means that in the first equation the terms

$$\alpha_1 \beta_1' \beta_1'', \quad \alpha_{r-1} \beta_{r-1}' \beta_{r-1}'' \tag{13.263}$$

should be deleted. However, from (13.261) it is clear that

$$h(w_0) = h(w_r) = 0 \tag{13.264}$$

and thus from (13.254)

$$\beta_1' = \beta_1'' = \beta_{r-1}' = \beta_{r-1}'' = 0 \tag{13.265}$$

and thus the unwanted terms (13.263) vanish for the restricted model and thus the RSOS model also satisfies the star–triangle equation (13.245).

13.6.2 The hard hexagon model

In this subsection we demonstrate that the hard hexagon model discussed in chapter 7 in terms of low density virial expansion satisfies a star–triangle equation.

The hard hexagon model may be represented as in Fig. 13.28 by placing particles on a square lattice with the restriction that there be no nearest neighbor pairs and that there be no pairs connected by a diagonal which goes from the NW to the SE corner of the square. The grand partition function for a lattice of N sites is defined as

$$Q^{gr}(z) = \sum_{n=0}^{[N/3]} z^n g_n \tag{13.266}$$

where g_n is the number of allowed configurations containing n particles.

To represent the hard hexagon model as a face model we define a variable σ_j at site j to be zero if the site is unoccupied and unity if the site is occupied and we share the fugacity z equally among all four faces which contain the given site. Then we may represent the hard hexagon model as a face model whose nonvanishing Boltzmann weights are

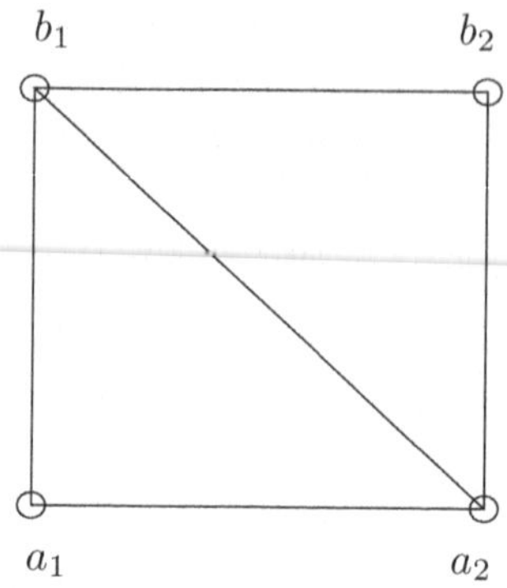

Fig. 13.28 The Boltzmann weights $W_{HH}(a_1, a_2; b_1, b_2)$ for the hard hexagon model represented on a square lattice with a diagonal interaction. No pair of particles is allowed to occupy sites connected by a solid line.

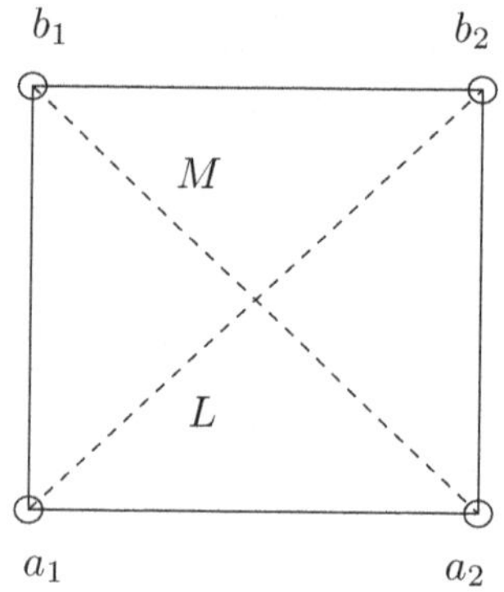

Fig. 13.29 The Boltzmann weights $W_{HSQ}(a_1, a_2; b_1, b_2)$ for hard squares with diagonal interactions. No pair of particles is allowed to occupy sites connected by a solid line.

$$\begin{aligned}
&W_{HH}(0,0;0,0) = 1\\
&W_{HH}(1,0;0,0) = W_{HH}(0,0;0,1) = W_{HH}(0,1;0,0) = W_{HH}(0,0;1,0) = z^{1/4}\\
&W_{HH}(1,0;0,1) = z^{1/2}.
\end{aligned} \tag{13.267}$$

The Boltzmann weights (13.267) contain only the single parameter z and thus by itself cannot be in a one-parameter family of commuting transfer matrices. Therefore in order for hard hexagons to be treated by the method of star–triangle equations it is necessary to embed the hard hexagon model into a larger family of models which specializes to hard hexagons at some particular point. Baxter [15, 16] found in 1980 that such a model is hard squares with diagonal interactions shown in Fig. 13.29.

In this model there is still nearest neighbor exclusion but now, in contrast to the hard hexagon model, there are nonzero Boltzmann weights for pairs on particles on both the NW-SE and on the NE-SW diagonals of the faces. The Boltzmann weights for this model may be written as

$$\begin{aligned}
&W(0,0;0,0) = 1\\
&W(1,0;0,0) = W(0,0;0,1) = W(0,1;0,0) = W(0,0;1,0) = z^{1/4}\\
&W(1,0;0,1) = z^{1/2}e^{L}\\
&W(0,1;1,0) = z^{1/2}e^{M}
\end{aligned} \tag{13.268}$$

which reduces to (13.267) when $L \to 0$ and $M \to -\infty$.

It is important to note, however, that these local face weights are not unique in the sense that there is a parameter which may be included in the weights which will cancel out in the partition function and that the "sharing" of the factor z between the lattice sites may be done in several different ways. We will follow [15] and exploit this lack of uniqueness to write the face weights for the hard square model with diagonal interactions in what appears to be an asymmetric form

$$\begin{aligned}
&W_{HSQ}(0,0;0,0) = m\\
&W_{HSQ}(1,0;0,0) = W_{HSQ}(0,0;0,1) = -mz^{1/4}t^{-1}\\
&W_{HSQ}(0,1;0,0) = W_{HSQ}(0,0;1,0) = mz^{1/4}t\\
&W_{HSQ}(1,0;0,1) = mz^{1/2}t^{-2}e^{L}\\
&W_{HSQ}(0,1;1,0) = mz^{1/2}t^{2}e^{M}
\end{aligned} \tag{13.269}$$

where the factors of t and the minus sign will cancel in the partition function and the factor of m merely multiplies the partition function by m^N where N is the number of sites. Solving (13.269) for z and t we find

$$z^{1/2} = -W_{HSQ}(1,0;0,0)W_{HSQ}(0,1;0,0)/W^2_{HSQ}(0,0;0,0) \tag{13.270}$$

$$t^2 = -W_{HSQ}(0,1;0,0)/W_{HSQ}(1,0;0,0). \tag{13.271}$$

It is shown in [15] and [16] that the face weights (13.269) satisfy the star–triangle equation if z, L and M are related

$$z(e^{L+M} - e^{L} - e^{M}) = (1-e^{-L})(1-e^{-M}). \tag{13.272}$$

The face weights (13.269) with the restriction (13.272) are parametrized in [21] in terms of elliptic functions as

$$\begin{aligned}
&W_{HSQ}(0,0;0,0) = h(5\eta - v)/h(4\eta)\\
&W_{HSQ}(1,0;0,0) = W_{HSQ}(0,0;0,1) = h(\eta - v)/[h(2\eta)h(4\eta)]^{1/2}\\
&W_{HSQ}(0,1;0,0) = W_{HSQ}(0,0;1,0) = h(\eta + v)/h(2\eta)\\
&W_{HSQ}(1,0;0,1) = h(3\eta + v)/h(4\eta)\\
&W_{HSQ}(0,1;1,0) = h(3\eta - v)/h(2\eta)
\end{aligned} \tag{13.273}$$

where

$$\eta = K/5. \tag{13.274}$$

In the hard hexagon limit $L \to 0$, $M \to -\infty$ the face weights (13.269) reduce to the nonvanishing face weights of hard hexagons as

$$\begin{aligned}
&\tilde{W}_{HH}(0,0;0,0) = m\\
&\tilde{W}_{HH}(1,0;0,0) = \tilde{W}_{HH}(0,0;0,1) = -mz^{1/4}t^{-1}\\
&\tilde{W}_{HH}(0,1;0,0) = \tilde{W}_{HH}(0,0;1,0) = mz^{1/4}t\\
&\tilde{W}_{HH}(1,0;0,1) = mz^{1/2}t^{-2}.
\end{aligned} \tag{13.275}$$

In terms of the parametrization (13.273) we see that the weight $W_{HSQ}(0,1;1,0)$ vanishes when

$$v = 3\eta \tag{13.276}$$

and the remaining face weights reduce to the hard hexagon weights $\tilde{W}_{HH}(a_1, a_2; b_1, b_2)$ as

$$\begin{aligned}
&\tilde{W}_{HH}(0,0;0,0) = h(2\eta)/h(4\eta)\\
&\tilde{W}_{HH}(1,0;0,0) = \tilde{W}_{HH}(0,0;01) = -[h(2\eta)/h(4\eta)]^{1/2}\\
&\tilde{W}_{HH}(0,1;0,0) = \tilde{W}_{HH}(0,0;1,0) = h(4\eta)/h(2\eta)\\
&\tilde{W}_{HH}(1,0;0,1) = h(6\eta)/h(4\eta) = 1
\end{aligned} \tag{13.277}$$

where in the last line we have used the identity $h(2K - v) = h(v)$ and from (13.270) and (13.271) we find

$$z^{1/2} = [h(4\eta)/h(2\eta)]^{5/2} \tag{13.278}$$

$$t^2 = [h(4\eta)/h(2\eta)]^{3/2}. \tag{13.279}$$

By use of the definition (13.259) of $h(v)$ the fugacity (13.278) is explicitly written as

$$z = \left[\frac{\sin(2\pi/5)}{\sin(\pi/5)} \prod_{n=1}^{\infty} \frac{(1 - q^n e^{i\pi 4/5})(1 - q^n e^{-i\pi 4/5})}{(1 - q^n e^{i\pi 2/5})(1 - q^n e^{-i\pi 2/5})}\right]^5. \tag{13.280}$$

When $q = 0$ (13.280) reduces to

$$z_c = \left(\frac{\sin(2\pi/5)}{\sin(\pi/5)}\right)^5 = [(1+\sqrt{5})/2]^5 = (11 + 5\sqrt{5})/2 = 11.09017\cdots \tag{13.281}$$

This value was conjectured (but not published) by Gaunt [48, page 405] as the value of the fugacity at which freezing begins for the hard hexagon model.

The fugacity z in the grand partition function (13.266) takes on all values in $0 \leq z < \infty$. This is achieved in the expression (13.280) by allowing the modulus q to lie in the range $-1 \leq q \leq 1$. To see this it is useful to make a transformation of the theta functions. The procedure for $z_c < z$ and $0 \leq z < z_c$ is slightly different and the two cases will be treated separately.

The nome q lies in the interval $0 \leq q \leq 1$ when the modulus k lies in the interval $0 \leq k \leq 1$ and in this case we define a new modulus $\tilde{k}$ by

$$e^{-\frac{\pi \tilde{K}'}{\tilde{K}}} = \tilde{q} = q^{1/2} = e^{-\frac{\pi K'}{2K}} \tag{13.282}$$

where

$$\tilde{K} = K(\tilde{k}) \quad \text{and} \quad \tilde{K}' = K'(\tilde{k}) \tag{13.283}$$

and note from either (13.89) or (13.381) that (13.278) is written as

$$z = \left[\frac{H(4\eta\tilde{K}/K;\tilde{k})}{H(2\eta\tilde{K}/K;\tilde{k})}\right]^5 = \left[\frac{H(4\tilde{K}/5;\tilde{k})}{H(2\tilde{K}/5;\tilde{k})}\right]^5. \tag{13.284}$$

We then use the complementary modulus transformation (13.411) to write

$$H(v;\tilde{k}) = -i(\tilde{K}/\tilde{K}')^{1/2} e^{-\frac{\pi v^2}{4\tilde{K}\tilde{K}'}} H(iv;\tilde{k}'). \tag{13.285}$$

Then using (13.285) in (13.284) we obtain

$$z = e^{-\frac{6\pi K}{5K'}} \left[\frac{H(4i\tilde{K}/5;\tilde{k}')}{H(2i\tilde{K}/5;\tilde{k}')}\right]^5 \tag{13.286}$$

which, if we use the product representation (13.381) for $H(v;\tilde{q}_T)$ reduces to

$$z = \frac{1}{x[g(x)]^5} \tag{13.287}$$

where

$$x = e^{-\frac{4\pi K}{5K'}} \tag{13.288}$$

and

$$g(x) = \prod_{n=1}^{\infty} \frac{(1-x^{5n-4})(1-x^{5n-1})}{(1-x^{5n-3})(1-x^{5n-2})}. \tag{13.289}$$

From (13.287) it is obvious when $k \to 1$ that $q \to 1$, $x \to 0$ and $z \to \infty$. Furthermore by construction when $k = 0$ we have $q = 0$, $x = 1$ and $z = z_c$. Thus the region $z_c \leq x < \infty$ is obtained for $0 \leq q \leq 1$ for which the modulus k is in the interval $0 \leq k \leq 1$.

In order to obtain the low density regime of fugacity where $0 \leq z \leq z_c$ we need the nome q to lie in the interval $-1 \leq q \leq 0$. This is achieved by defining a new modulus $\hat{k}$ such that

$$q = e^{-\pi K'/K} = -e^{-\pi\hat{K}'/\hat{K}} \equiv -\hat{q} \tag{13.290}$$

where

$$\hat{K} = K(\hat{k}) \text{ and } \hat{K}' = K'(\hat{k}) \tag{13.291}$$

and the last line defines the nome $\hat{q}$, which in the region

$$0 \leq \hat{k} \leq 1 \tag{13.292}$$

is real and positive. From the properties of complete elliptic integrals it can be shown that the new modulus $\hat{k}$ is related to the original modulus by

$$k = i\hat{k}/\hat{k}' \tag{13.293}$$

and that

$$K(k) = \hat{k}'\hat{K} \tag{13.294}$$
$$K'(k) = \hat{k}'(\hat{K}' - i\hat{K}). \tag{13.295}$$

Thus from the representations (13.381), (13.382) and (13.385) we obtain

$$H(v;k)/q^{1/4} = H(v\hat{K}/K;\hat{k})/\hat{q}^{1/4} \tag{13.296}$$
$$\Theta(v;k) = \Theta_1(v\hat{K}/K;\hat{k}) \tag{13.297}$$

and hence we may write (13.278) as

$$z = \left[\frac{H(4\hat{K}/5;\hat{k})\Theta_1(4\hat{K}/5;\hat{k})}{H(2\hat{K}/5;\hat{k})\Theta_1(2\hat{K}/5;\hat{k})}\right]^5. \tag{13.298}$$

We now use the complementary modulus transformations (13.411) and (13.412) of the appendix

$$H(v;\hat{k}) = -i[\hat{K}/\hat{K}']^{1/2}e^{-\pi v^2/(4\hat{K}\hat{K}')}H(iv;\hat{k}') \tag{13.299}$$
$$\Theta_1(v;\hat{k}) = [\hat{K}/\hat{K}']^{1/2}e^{-\pi v^2/(4\hat{K}\hat{K}')}\Theta_1(iv;\hat{k}') \tag{13.300}$$

in (13.298) to find

$$z = e^{-\frac{6\pi\hat{K}}{5\hat{K}'}}\left[\frac{H(4i\hat{K}/5;\hat{k}')\Theta_1(4i\hat{K}/5;\hat{k}')}{H(2i\hat{K}/5;\hat{k}')\Theta_1(2i\hat{K}/5;\hat{k}')}\right]^5. \tag{13.301}$$

Thus, using the product representation of the theta functions (13.381) and (13.382) and defining

$$\hat{x} = -e^{-\frac{\pi\hat{K}}{5\hat{K}'}} \tag{13.302}$$

we find for $-1 \leq q \leq 0$ that

$$z = -\hat{x}[g(\hat{x})]^5 \tag{13.303}$$

which is obviously positive and vanishes for $x = 0$. By construction $z = z_c$ when $\hat{x} = -1$.

We remark that if both of the diagonal interactions L and M vanish that the model of hard squares with diagonal interactions reduces to the model of hard squares introduced in chapter 9. However, when $L = M = 0$ the restriction (13.272) requires that $z = 0$ and thus, in contrast with the hard hexagon model, the hard square model is not a special case of the model of hard squares with diagonal interactions.

The model of hard squares with diagonal interactions is, in fact, closely related to the RSOS model introduced in the previous subsection with four states. This is seen

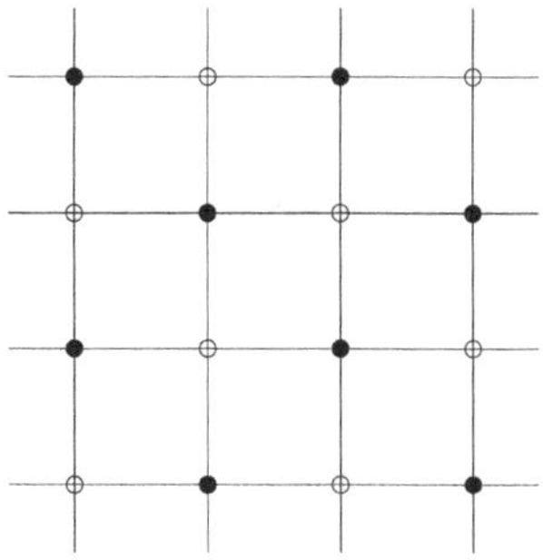

Fig. 13.30 The decomposition of the square lattice into two sublattices denoted by the filled and open circles.

if we consider decomposing the square lattice into two sublattices as indicated in Fig. 13.30 by the filled and open circles. Then if on the open circles we define a height variable l which takes on odd values by

$$l = 3 - 2\sigma \tag{13.304}$$

and on the closed circles a height variable which takes on even values

$$l = 2 + 2\sigma \tag{13.305}$$

where we recall that $\sigma = 0, 1$ then we have the correspondence between the allowed vertices of the model of hard squares with diagonal interactions and the configurations of the four-state RSOS model shown in Fig. 13.31. If we further note that we can just as well associate odd (even) values of the height variables l with the filled (open) circles we see that for every configuration of the model of hard squares with diagonal interactions there are associated two configurations of the RSOS model.

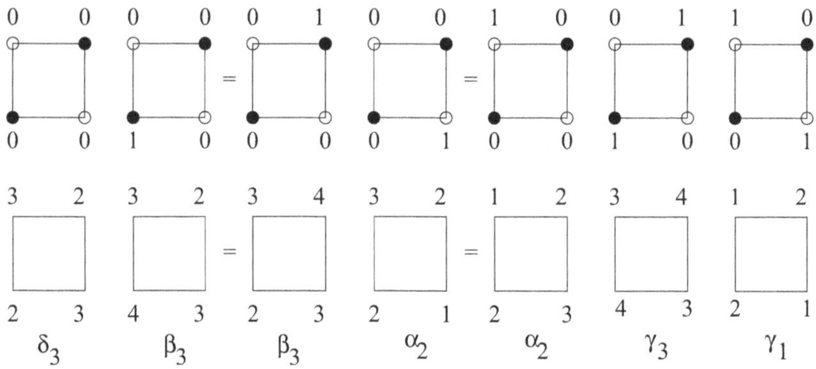

Fig. 13.31 The correspondence between the weights of the model of hard squares with diagonal interactions in the upper row with the four-state RSOS model in the lower row.

We finally note that the RSOS weights defined by (13.246)–(13.262) agree with the weights (13.273) of the model of hard squares with diagonal interactions if we use the identity $h(2K - v) = h(v)$ and set $\rho = 1/h(2\eta)$ in (13.253)–(13.256). Thus we have shown that the partition function Z_{4RSOS} for the 4 state RSOS model is related to the partition function Z_{HSQ} for hard squares with diagonal interactions by

$$Z_{4RSOS} = 2Z_{HSQ}. \tag{13.306}$$

13.7 Hamiltonian limits

The six- and eight-vertex model have the property presented in the introduction to this chapter that there exists a Hamiltonian H of a one-dimensional quantum spin chain which commutes with the transfer matrix of the two-dimensional classical statistical mechanical model

$$[T(u), H] = 0 \tag{13.307}$$

where for the eight-vertex model the Hamiltonian for the XYZ spin chain is

$$H_{XYZ} = -\sum_{j=1}^{N}\{J^x\sigma_j^x\sigma_{j+1}^x + J^y\sigma_j^y\sigma_{j+1}^y + J^z\sigma_j^z\sigma_{j+1}^z\} \tag{13.308}$$

where σ_j^i is the direct product notation

$$\sigma_j^i = I_2 \otimes \cdots \otimes \sigma^i \otimes \cdots I_2 \tag{13.309}$$

where I_2 is the 2×2 identity matrix and σ^i are the three Pauli spin matrices located at site j of the chain.

The Hamiltonian for the symmetric six-vertex model is the Hamiltonian of the XXZ (or Heisenberg–Ising) model

$$H_{XXZ} = -\sum_{j=1}^{N}\{\sigma_j^x\sigma_{j+1}^x + \sigma_j^y\sigma_{j+1}^y + J^z\sigma_j^z\sigma_{j+1}^z\} \tag{13.310}$$

which generalizes to the asymmetric six-vertex model as

$$H_{XXZ}^{as} = -\sum_{j=1}^{N}\{\sigma_j^x\sigma_{j+1}^x+\sigma_j^y\sigma_{j+1}^y+J^z\sigma_j^z\sigma_{j+1}^z+H\sigma_j^z+J_d i(\sigma_j^x\sigma_{j+1}^y-\sigma_j^y\sigma_{j+1}^x)\} \tag{13.311}$$

where the last term in (13.311) is called a Dzyaloshinski term. The commutation relation (13.307) for the symmetric and asymmetric six-vertex model was first derived in [6] and for the eight-vertex model in [7].

We note that the correlations of spins in the quantum spin chain depend only on the eigenvectors of the Hamiltonian and because of the commutation relation (13.307) these eigenvectors are independent of the spectral variable u and are identical with the eigenvectors of the transfer matrix. Therefore when we are able to identify the ground state eigenvector of the spin chain with the eigenvector of the transfer matrix which has the largest eigenvalue we are thus able to identify the correlation functions of the statistical model which depend only on the eigenvectors with the correlation functions of the spin chain.

The transfer matrix for the chiral Potts model T_{pq} depends on both the indices p and q of the Boltzmann weights, and the associated Hamiltonian is obtained from the commutation relation

$$[T_{pq}, T_{pq'}] = 0 \tag{13.312}$$

by expanding about the point $q' = p$. The resulting spin Hamiltonian $H_{cp}(p)$ which depends on p satisfies

$$[T_{pq}, H_{cp}(p)] = 0 \tag{13.313}$$

with

$$H_{cp}(p) = -\sum_{j=1}^{\mathcal{N}} \sum_{n=1}^{N-1} \{\bar{\alpha}_n (X_j)^n + \alpha_n (Z_j Z_{j+1}^{\dagger})^n\} \tag{13.314}$$

where to avoid confusion the length of the chain is $\mathcal{N}$, the matrices X_j and Z_j are in the direct product notation (13.309) with I_N the $N \times N$ identity matrix and the $N \times N$ matrices Z and X generalize the Pauli matrices σ^z and σ^x as

$$Z_{l,m} = \delta_{l,m} \omega^{l-1}, \quad X_{l,m} = \delta_{l,m+1} \pmod N. \tag{13.315}$$

In (13.314) we have

$$\alpha_n = \frac{e^{i(2n-N)\phi/N}}{\sin \pi n/N} \tag{13.316}$$

$$\bar{\alpha}_n = \lambda \frac{e^{i(2n-N)\bar{\phi}/N}}{\sin \pi n/N} \tag{13.317}$$

where the angles ϕ and $\bar{\phi}$ are expressed in terms of the variables of the chiral Potts curve (13.167)

$$e^{2i\phi/N} = \omega^{1/2} \frac{a_p c_p}{b_p d_p} \tag{13.318}$$

$$e^{2i\bar{\phi}/N} = \omega^{1/2} \frac{a_p d_p}{b_p c_p} \tag{13.319}$$

and from (13.318) and (13.319) and the chiral Potts curve (13.167) it follows that ϕ, $\bar{\phi}$ and k are related by

$$\cos \phi = \lambda \cos \bar{\phi}. \tag{13.320}$$

Using the identities

$$(X_j^n)^{\dagger} = X_j^{N-n} \text{ and } (Z_j^n)^{\dagger} = Z_j^{N-n} \tag{13.321}$$

which follow from (13.315) we see that the Hamiltonian $H_{cp}(p)$ of (13.314) is Hermitian when α_n and $\bar{\alpha}_n$ are given by (13.316) and (13.317) with ϕ and $\bar{\phi}$ real (but not necessarily constrained by (13.320)).

The Hamiltonian (13.314) with $N = 3$ but without the restriction (13.320) which is required for integrability was introduced in 1983 by Howes, Kadanoff and den Nijs [50] as a quantum model to study commensurate–incommensurate transitions. The Hamiltonian (13.314) in the special case $\phi = \bar{\phi} = \pi/2$ for arbitrary N was derived in 1985 by von Gehlen and Rittenberg [23] as a quantum spin chain which satisfies the algebra used by Onsager [2] in 1944 to solve the Ising model.

13.7.1 Spin chains for the eight- and six-vertex models

To derive the XYZ spin chain Hamiltonian (13.308) which commutes (13.307) with the transfer matrix of the eight-vertex model we follow [9, 11] and note that when $u = \eta$ the Boltzmann weights (13.122) specialize to

$$b_0 = d_0 = 0, \qquad a_0 = c_0 = \rho\Theta(0)\Theta(2\eta)H(2\eta). \tag{13.322}$$

Therefore in the notation of Fig. 13.1

$$W(j', j|\eta)_{\mu,\nu} = c_0\delta_{\mu,j'}\delta_{j,\nu} \tag{13.323}$$

and thus the transfer matrix (13.21) reduces to the operator

$$T(\eta)|_{\{j'\},\{j\}} = c_0^N\delta_{j'_2,j_1}\delta_{j'_3,j_2}\cdots\delta_{j'_1,j_N} \tag{13.324}$$

which shifts the configuration $(j_1, j_2, \cdots, j_N)$ one step to the right to $(j_2, j_3, \cdots, j_1)$.

We now expand the Boltzmann weights $W(j', j)_{\mu,\nu}$ about the point $u = \eta$. Thus we set

$$a = c_0 + \delta a, \quad b = \delta b, \quad c = c_0 + \delta c, \quad d = \delta d \tag{13.325}$$

and

$$T(u) = T(\eta) + \delta T \tag{13.326}$$

with

$$\delta T_{\{j'\},\{j\}} = c_0^{N-1}\sum_{k=1}^{N}\cdots\delta_{j'_{k-1},j_{k-2}}\delta W(j'_k, j_k)|_{j_{k-1},j'_{k+1}}\delta_{j'_{k+1},j_k}\cdots \tag{13.327}$$

If we now multiply (13.326) on the left by the operator

$$T^{-1}(\eta)|_{\{j'\},\{j\}} = c_0^{-N}\delta_{j'_1,j_2}\delta_{j'_2,j_3}\cdots\delta_{j'_N,j_1}, \tag{13.328}$$

which shifts the configuration $(j_1, j_2, \cdots, j_N)$ one step to the left to $(j_N, j_1, \cdots, j_{N-1})$, we obtain to first order as $u \to \eta$

$$T^{-1}(\eta)T(u) \to I + \sum_{k=1}^{N} H_k \tag{13.329}$$

where H_k is the 4×4 matrix at the site k

$$H_{\{j'_k,j'_{k+1}\},\{j_k,j_{k+1}\}} = c_0^{-1}\delta W(j'_k, j_{k+1})|_{j_k,j'_{k+1}}. \tag{13.330}$$

It is straightforward to use the conventions of Fig. 13.1 and Fig. 13.12 to verify that, for the eight-vertex model with $\bar{a} = a$ and $\bar{b} = b$,

$$W(j'_k, j_{k+1})|_{j_k,j'_{k+1}} = \frac{1}{2}\{(a+c)I + (b+d)\sigma_k^x\sigma_{k+1}^x + (b-d)\sigma_k^y\sigma_{k+1}^y + (a-c)\sigma_k^z\sigma_{k+1}^z\} \tag{13.331}$$

where the indices are such that

$$\sigma_k^i = \sigma^i|_{j'_k, j_k}, \qquad \sigma_{k+1}^i = \sigma^i|_{j'_{k+1}, j_{k+1}} \tag{13.332}$$

and thus we have

$$H_k = (2c_0)^{-1}\{(\delta a + \delta c)I + (\delta b + \delta d)\sigma_k^x\sigma_{k+1}^x + (\delta b - \delta d)\sigma_k^y\sigma_{k+1}^y + (\delta a - \delta c)\sigma_k^z\sigma_{k+1}^z\}. \tag{13.333}$$

We may write (13.333) in terms of the invariants (13.87) and (13.88) by noting that, because the invariants are independent of the expansion variable u and are indeterminate for $0/0$, when the values in (13.322) are substituted into the definitions (13.87) and (13.88), we have

$$\Delta = \frac{\delta a - \delta c}{\delta b + \delta d} = \frac{\mathrm{cn}2\eta\mathrm{dn}2\eta}{1 + k\mathrm{sn}^2 2\eta} \tag{13.334}$$

$$\Gamma = \frac{1-\gamma}{1+\gamma} = \frac{\delta b - \delta d}{\delta b + \delta d} = \frac{1 - k\mathrm{sn}^2 2\eta}{1 + k\mathrm{sn}^2 2\eta}. \tag{13.335}$$

Thus we find from (13.333) that

$$H_k = \frac{\delta b + \delta d}{2c_0}\{\frac{\delta a + \delta c}{\delta b + \delta d} + \sigma_k^x\sigma_{k+1}^x + \Gamma\sigma_k^y\sigma_{k+1}^y + \Delta\sigma_k^z\sigma_{k+1}^z\}. \tag{13.336}$$

Furthermore, writing the variations δ as derivatives we have

$$\frac{\delta b + \delta d}{2c_0} \to \frac{1}{2}\frac{\partial}{\partial u}\left(\frac{b}{c} + \frac{d}{c}\right)_{u=\eta} = \frac{1}{2\mathrm{sn}(2\eta)}[1 + k\mathrm{sn}^2(2\eta)] \tag{13.337}$$

and

$$\frac{\delta a + \delta c}{2c_0} = \frac{1}{2}\left(\frac{H'(2\eta)}{H(2\eta)} + \frac{\Theta'(2\eta)}{\Theta(2\eta)}\right) \tag{13.338}$$

where to obtain (13.337) we have used the derivative

$$\frac{d}{du}\mathrm{sn}u = \mathrm{cn}u\mathrm{dn}u. \tag{13.339}$$

Thus we find

$$\begin{aligned} H_{XYZ} &= -\frac{1}{2}\sum_{j=1}^{N}\{J^x\sigma_j^x\sigma_{j+1}^x + J^y\sigma_j^y\sigma_{j+1}^y + J^z\sigma_j^z\sigma_{j+1}^z\} \\ &= -\mathrm{sn}(2\eta)\left[\frac{\partial}{\partial u}\ln T(u) - N\frac{1}{2}\left(\frac{H'(2\eta)}{H(2\eta)} + \frac{\Theta'(2\eta)}{\Theta(2\eta)}\right)\right]_{u=\eta} \end{aligned} \tag{13.340}$$

with

$$J^x = 1 + k\mathrm{sn}(2\eta) \tag{13.341}$$

$$J^y = 1 - k\mathrm{sn}(2\eta) \tag{13.342}$$

$$J^z = \text{cn}(2\eta)\text{dn}(2\eta). \tag{13.343}$$

We note that

$$\text{cn}K = 0, \quad \text{sn}K = 1 \tag{13.344}$$

and thus when

$$2\eta = K \tag{13.345}$$

then (13.334) and (13.335) reduce to

$$\Delta = 0, \quad \Gamma = \frac{1-k}{1+k} \tag{13.346}$$

and thus (13.308) reduces to the Hamiltonian of the XY model

$$H_{XY} = -\frac{1}{2}\sum_{j=1}^{N}\{(1+k)\sigma_j^x\sigma_{j+1}^x + (1-k)\sigma_j^y\sigma_{j+1}^y\}. \tag{13.347}$$

The six-vertex specialization

When $d = 0$ the eight-vertex model reduces to the symmetric six-vertex model. In this limit the invariant $\Gamma = 1$ and the Hamiltonian of the XYZ model (13.308) reduces to the Hamiltonian of the XXZ model (13.310) with

$$J^z = \Delta. \tag{13.348}$$

The commutation of the asymmetric six-vertex model with the asymmetric XXZ model follows if we note that the two operators

$$-\sum_{j=1}^{N}\sigma_j^z, \quad \sum_{j=1}^{N}\{\sigma_j^x\sigma_{j+1}^y - \sigma_j^y\sigma_{j+1}^x\} \tag{13.349}$$

separately commute both with H_{XXZ} and with the transfer matrix of the six-vertex model. These commutations follow from the conservation separately of the number of spin up arrows and the number of right pointing arrows in the statistical mechanical model.

The decoupling specialization

In 13.4.2 we found that at the decoupling point $K_4 = 0$ discussed where the eight-vertex model reduces to two noninteracting Ising models with vertical and horizontal coupling constants $K = E/k_BT$ and $K' = E'/k_BT$, the invariants γ and Δ are given by (13.150) and (13.150). Thus from (13.340) we find that at the decoupling point the eight-vertex model commutes with the Hamiltonian

$$H = -\sum_{j=1}^{N}\{\sigma_j^x\sigma_{j+1}^x + \sinh 2K \sinh 2K'\sigma_j^z\sigma_{j+1}^z\} \tag{13.350}$$

which is in the form of the XY model Hamiltonian (13.347) previously found. Thus the ground state of the XY model corresponds to the eigenvector of the decoupled eight-vertex model with the maximum eigenvalue of the transfer matrix and thus the factorization of the arrow correlations in a row of the decoupled eight-vertex model implies a corresponding factorization of the spin correlations of the XY model into products of Ising correlation functions. In particular we find from (13.153) and (13.154) that for XY model (13.347) with $\sinh 2K \sinh 2K' = k < 1$ the correlations in the ground state are given in terms of Ising correlations at a temperature T by

$$\langle \sigma_0^x \sigma_{2R}^x \rangle = C_-(R,R)^2 \tag{13.351}$$

$$\langle \sigma_0^x \sigma_{2R+1}^x \rangle = C_-(R,R)C_-(R+1,R+1) \tag{13.352}$$

where $C_-(N,N)$ is the diagonal Ising correlation for $T < T_c$ and by symmetry

$$\langle \sigma_0^y \sigma_{2R}^y \rangle = C_+(R,R)^2 \tag{13.353}$$

$$\langle \sigma_0^y \sigma_{2R+1}^y \rangle = C_+(R,R)C_+(R+1,R+1) \tag{13.354}$$

where $C_+(N,N)$ is the diagonal Ising correlation function for $T > T_c$. The value of $k = 1$ where the Ising model has a critical point is often referred to as a quantum critical point of the quantum spin chain.

13.7.2 Spin chain for the chiral Potts model

To derive the chiral Potts spin chain (13.314) it is most convenient to write the Boltzmann weights of the chiral Potts model in a matrix form in terms of the matrices X_k and Z_k (13.315). It is easily seen for any function $A(n)$ that satisfies $A(n+N) = A(n)$ that

$$\sum_{n=0}^{N=1} A(n)[X^n]_{j_k,j'_k} = A(j_k - j'_k) \tag{13.355}$$

$$\sum_{n=0}^{N-1} A(n)[(Z_k Z_{k+1}^\dagger)^n]_{j,j'} = \sum_{n=0}^{N-1} \omega^{n(j_k - j'_{k+1})} A(n). = \hat{A}(j_k - j'_{k+1}) \tag{13.356}$$

where the last line in (13.356) defines the Fourier transform of $A(n)$ which has the inverse relation

$$A(m) = \frac{1}{N} \sum_{n=0}^{N-1} \omega^{-mn} \hat{A}(n). \tag{13.357}$$

Thus with $A(n) = W_{pq}^v(n)$ in (13.355) and $\hat{A}(n) = W_{pq}^h(n)$ in (13.356) and by computing the inverse Fourier transform in a manner identical with that used to compute the Fourier transform of $W_{pq}^v(n)$ we find

$$W_{pq}^v(j_k, j'_k) = \sum_{n=1}^{N-1} W_{pq}^v(n)[X_k^n]_{j,j'} \tag{13.358}$$

$$W^h_{pq}(j_k, j'_{k+1}) = \sum_{n=0}^{N-1} \hat{W}^h_{pq}(n)[(Z_k Z^\dagger_{k+1})^n]_{j,j'} \tag{13.359}$$

where

$$\frac{W^v_{pq}(n)}{W^v_{pq}(0)} = \prod_{j=1}^{n} \frac{\omega a_p d_q - d_p a_q \omega^j}{c_p b_q - b_p c_q \omega^j} \tag{13.360}$$

and

$$\frac{\hat{W}^h_{pq}(n)}{\hat{W}^h_{pq}(0)} = \prod_{j=1}^{n} \frac{\omega a_p c_q - c_p a_q \omega^j}{d_p b_q - b_p d_q \omega^j} \tag{13.361}$$

with

$$\hat{W}^h_{pq}(0) = \frac{1}{N} \sum_{n=0}^{N-1} W^h_{pq}(n). \tag{13.362}$$

To obtain the spin chain Hamiltonian we first consider the point $p = q$ where the Boltzmann weights (13.165) and (13.166) of the chiral Potts model reduce to

$$\frac{W^h_{p,p}(n)}{W^h_{p,p}(0)} = \prod_{j=1}^{n} \left(\frac{d_p b_p - a_p c_p \omega^j}{b_p d_p - c_p a_p \omega^j} \right) = 1 \tag{13.363}$$

and

$$\frac{W^v_{p,p}(n)}{W^v_{p,p}(0)} = \prod_{j=1}^{n} \left(\frac{\omega a_p d_p - d_p a_p \omega^j}{c_p b_p - b_p c_p \omega^j} \right) = \delta_{n,0} \tag{13.364}$$

and the transfer matrix (13.22) reduces to the identity matrix

$$T_{p,p} = I. \tag{13.365}$$

We now expand the variables $(a_q,\ b_q,\ c_q,\ d_q)$ about the point $q = p$ as

$$\begin{aligned} a_q &= a_p + a'\epsilon \\ b_q &= b_p + b'\epsilon \\ c_q &= c_p + c'\epsilon \\ d_q &= d_p + d'\epsilon \end{aligned} \tag{13.366}$$

where a', b', c' and d' are constrained by the chiral Potts curve (13.167). For $1 \leq n \leq N-1$ we have the expansions

$$\frac{W^v_{pq}(n)}{W^v_{pq}(0)} \sim \frac{\omega^n}{1-\omega^n} \left(\frac{a_p d_p}{b_p c_p} \right)^n (d'/d_p - a'/a_p)\epsilon \tag{13.367}$$

$$\frac{\hat{W}^h_{pq}(n)}{\hat{W}^h_{pq}(0)} \sim \frac{\omega^n}{1-\omega^n}\left(\frac{a_p c_p}{b_p d_p}\right)^n (c'/c_p - a'/a_p)\epsilon \tag{13.368}$$

which, using the definitions (13.318) and (13.319) of the angles ϕ and $\bar{\phi}$ and setting

$$\epsilon = \frac{2k\epsilon'}{d'/d_p - a'/a_p}\left(\frac{a_p d_p}{b_p c_p}\right)^{-N/2}, \tag{13.369}$$

are written as

$$\frac{W^v_{pq}(n)}{W^v_{pq}(0)} \sim -\frac{e^{i\bar{\phi}(2n-N)/N}}{\sin \pi n/N} k\epsilon' \tag{13.370}$$

$$\frac{\hat{W}^h_{pq}(n)}{\hat{W}^h_{pq}(0)} \sim -\frac{e^{i\phi(2n-N)/N}}{\sin \pi n/N}\frac{(c'/c_p - a'/a_p)c_p^N}{(d'/d_p - a'/a_p)d_p^N} k\epsilon'. \tag{13.371}$$

From the chiral Potts curve (13.167) we find

$$\lambda' a_p^N = d_p - \lambda c_p^N \tag{13.372}$$

and

$$\lambda' a_p^N a'/a_p = d_p^N d'/d_p - \lambda c_p^N c'/c_p \tag{13.373}$$

and from (13.372) and (13.373) it follows that

$$\frac{(c'/c_p - a'/a_p)c_p^N}{(d'/d_p - a'/a_p)d_p^N}\lambda = 1. \tag{13.374}$$

Therefore (13.371) reduces to

$$\frac{\hat{W}^h_{pq}(n)}{\hat{W}^h_{pq}(0)} \sim -\frac{e^{i\phi(2n-N)/N}}{\sin \pi n/N}\epsilon'. \tag{13.375}$$

Thus using the normalization $W^{v,h}_{p,q}(0) = 1$ and noting that, from (13.362), for $p \sim q$

$$\hat{W}^h_{pq}(0) \sim 1 + \delta\hat{W}^h_p \epsilon', \tag{13.376}$$

we expand the transfer matrix T_{pq} defined from (13.22) by use of (13.358), (13.359), (13.370) and (13.375) to obtain for $q \to p$

$$T_{pq} \sim I(1 + \mathcal{N}\delta\hat{W}^h_p(0)) + H_{cp}\epsilon' \tag{13.377}$$

where H_{cp} is given by (13.314) and thus the commutation relation (13.313) holds.

We also note that the shift operator may be obtained from the transfer matrix as

$$\lim_{q\to p} T_{p,Rq} = e^{-iP} \tag{13.378}$$

where R is the automorphism

$$R(a_q, b_q, c_q, d_q) = (b_q, \omega a_q, d_q, c_q) \tag{13.379}$$

and P is the momentum operator which has the eigenvalues $P = 2\pi k/\mathcal{N}$. Furthermore, because the Boltzmann weights depend on the difference of the spin values, the spin translation operator

$$e^{2\pi i Q/N} = \prod_{k=1}^{\mathcal{N}} X_k \tag{13.380}$$

commutes with $T_{p,q}$ and that the eigenvalues of Q are $0, 1, \cdots N-1$.

13.8 Appendix: Properties of theta functions

In this appendix we derive the various properties of the quasiperiodic theta functions and the doubly periodic functions $\mathrm{sn}u, \mathrm{cn}u$ and $\mathrm{dn}u$ used in the text.

Product forms

The theta functions $H(u)$ and $\Theta(u)$ are defined by the infinite series (13.89) and (13.90). We note that by means of an identity called the triple product identity they have equivalent representations as infinite products (which are often given as the definition)

$$H(u) = 2q^{1/4} \sin \pi u/(2K) \prod_{n=1}^{\infty} (1 - q^{2n} e^{i\pi u/K})(1 - q^{2n} e^{-i\pi u/K})(1 - q^{2n}) \tag{13.381}$$

$$\Theta(u) = \prod_{n=1}^{\infty} (1 - q^{2n-1} e^{i\pi u/K})(1 - q^{2n-1} e^{-i\pi u/K}))(1 - q^{2n}). \tag{13.382}$$

with

$$q = e^{-\pi K'/K}. \tag{13.383}$$

We also define

$$\begin{aligned} H_1(u) &= H(u+K) \\ &= 2q^{1/4} \cos \pi u/(2K) \prod_{n=1}^{\infty} (1 + q^{2n} e^{i\pi u/K})(1 + q^{2n} e^{-i\pi u/K})(1 - q^{2n}) \end{aligned} \tag{13.384}$$

$$\begin{aligned} \Theta_1(u) &= \Theta(u+K) \\ &= \prod_{n=1}^{\infty} (1 + q^{2n-1} e^{i\pi u/K})(1 + q^{2n-1} e^{-i\pi u/K}))(1 - q^{2n}). \end{aligned} \tag{13.385}$$

Periodicity

The periodicity properties (13.97) follow immediately from the definitions (13.89) and (13.90) by using $\cos(-u) = \cos u$ and $\sin(-u) = -\sin u$ and the periodicity properties (13.98) and (13.100) follow from the definitions (13.89) and (13.90) by use of the properties $\cos n(u+2\pi) = \cos nu$ and $\sin(2n-1)(u+2\pi) = -\sin(2n-1)u$ which hold for integer n.

Quasiperiodicity

To prove the quasiperiodicity property (13.101) we write the definition (13.90) as

$$\Theta(u) = \sum_{n=-\infty}^{\infty} (-1)^n q^{n^2} e^{i\pi nu/K} \tag{13.386}$$

and thus using the definition (13.91) of q we find

$$\begin{aligned}\Theta(u+2iK') &= \sum_{n=-\infty}^{\infty} (-1)^n q^{n^2} e^{-2n\pi K'/K} e^{inu\pi/K}\\ &= \sum_{n=-\infty}^{\infty} (-1)^n q^{n^2+2n} e^{inu\pi/K} = q^{-1} e^{-i\pi u/K} \sum_{n=-\infty}^{\infty} (-1)^n q^{(n+1)^2} e^{i(n+1)u\pi/K}\\ &= -q^{-1} e^{-i\pi u/K}\Theta(u)\end{aligned} \tag{13.387}$$

as desired.

Relation of $H(u)$ and $\Theta(u)$

To prove the relation (13.102) between $\Theta(u)$ and $H(u)$ we write

$$\begin{aligned}\Theta(u+iK') &= \sum_{n=-\infty}^{\infty} (-1)^n q^{n^2} q^n e^{inu\pi/K}\\ &= q^{-1/4} e^{-\pi iu/(2K)} \sum_{n=-\infty}^{\infty} (-1)^n q^{(n+\frac{1}{2})^2} e^{i(2n+1)u\pi/(2K)}\end{aligned} \tag{13.388}$$

and thus if we combine together the terms with n and $-n-1$ we find

$$\Theta(u+iK') = iq^{-1/4} e^{-\pi iu/(2K)} H(u) \tag{13.389}$$

as desired. This also follows directly from the product representation (13.381) and (13.382). Similarly

$$H(u+iK') = iq^{-1/4} e^{-\pi iu/(2K)}\Theta(u). \tag{13.390}$$

The remaining relations (13.99) and (13.103) now follow as consequences of (13.387) and (13.389).

Zeros

From (13.381) and (13.382) and from the periodicity and quasiperiodicity properties we see that the only zeros of $H(u)$ and $\Theta(u)$ are given by (13.104). In particular

$$H(2nK + i2mK') = 0 \quad \text{for all integer m and n} \tag{13.391}$$

$$\Theta(2nK + (2m+1)iK') = 0 \quad \text{for all integer m and n.} \tag{13.392}$$

Modulus

From (13.381) and (13.382) we find

$$H(K) = 2q^{1/4} \prod_{n=1}^{\infty} (1 + q^{2n})^2 (1 - q^{2n}) \tag{13.393}$$

$$\Theta(K) = \prod_{n=1}^{\infty} (1 + q^{2n-1})^2 (1 - q^{2n}) \tag{13.394}$$

and thus from the definition (13.95) of the modulus k we find

$$k = \frac{H^2(K)}{\Theta^2(K)} = 4q^{1/2} \prod_{n=1}^{\infty} \left(\frac{1 + q^{2n}}{1 + q^{2n-1}} \right)^4 \tag{13.395}$$

which proves (13.96).

Identities

We next prove three identities from which all other identities used in the text will be derived:

$$\begin{aligned} &\Theta(u)\Theta(v)\Theta(a-u)\Theta(a-v) - H(u)H(v)H(a-u)H(a-v) \\ &= \Theta(0)\Theta(a)\Theta(u-v)\Theta(a-u-v) \end{aligned} \tag{13.396}$$

$$\begin{aligned} &H(v)H(a-v)\Theta(u)\Theta(a-u) - \Theta(v)\Theta(a-v)H(u)H(a-u) \\ &= \Theta(0)\Theta(a)H(v-u)H(a-u-v) \end{aligned} \tag{13.397}$$

and

$$\begin{aligned} &H(u)\Theta(u)H(v+K)\Theta(v+K) - H(v)\Theta(v)H(u+K)\Theta(u+K) \\ &= H(u-v)\Theta(u+v)H(K)\Theta(K). \end{aligned} \tag{13.398}$$

The equations (13.396), (13.397) and (13.398) obviously hold at $u = 0$. Furthermore, for each equation the periodicity under $u \to u + 2K$ and the quasiperiodicity under $u \to u + 2iK'$ of the right- and left-hand sides are the same and it is easily checked that, for each equation, the right- and left-hand sides vanish at the same values of u. Therefore it follows from Liouville's theorem that the two equations hold identically as stated.

To prove (13.114) set $a = 0,\ v = K$ in (13.397) to find

$$\Theta^2(K)H^2(u) - H^2(K)\Theta^2(u) = -\Theta^2(0)H^2(u+K) \tag{13.399}$$

and divide by $\Theta^2(u)H^2(K)$ to obtain

$$\frac{\Theta^2(K)H^2(u)}{H^2(K)\Theta^2(u)} + \frac{\Theta^2(0)H^2(u+K)}{H^2(K)\Theta^2(u)} = 1. \tag{13.400}$$

Using the definitions (13.109) and (13.110) of $\mathrm{sn}u$ and $\mathrm{cn}u$ in (13.400) we find

$$\mathrm{sn}^2u + \mathrm{cn}^2u = 1 \tag{13.401}$$

which is the identity (13.114).

To prove (13.115) set $a = 0,\ v = K$ in (13.396) to find

$$\Theta^2(K)\Theta^2(u) - H^2(K)H^2(u) = \Theta^2(0)\Theta^2(u+K) \tag{13.402}$$

and divide by $\Theta^2(u)\Theta^2(K)$ to obtain

$$\frac{H^2(K)H^2(u)}{\Theta^2(K)\Theta^2(u)} + \frac{\Theta^2(0)\Theta^2(u+K)}{\Theta^2(K)\Theta^2(u)} = 1. \tag{13.403}$$

Using the definitions (13.109) and (13.111) of $\mathrm{sn}u$ and $\mathrm{dn}u$ and definition (13.95) of k in (13.403) we find

$$k^2\mathrm{sn}^2u + \mathrm{dn}^2u = 1 \tag{13.404}$$

which is the identity (13.115).

If we further specialize (13.402) by setting $u = K$ and use the periodicity condition (13.100) we find

$$\Theta^4(K) - H^4(K) = \Theta^4(0) \tag{13.405}$$

from which when we divide by $\Theta^4(K)$ and use the definitions (13.95) for k and k', we obtain

$$k^2 + k'^2 = 1 \tag{13.406}$$

which is the relation (13.94).

We conclude by proving an addition theorem for $\mathrm{sn}u$:

$$\mathrm{sn}(u-v) = \frac{\mathrm{sn}u\mathrm{cn}v\mathrm{dn}v - \mathrm{sn}v\mathrm{cn}u\mathrm{dn}u}{1 - k^2\mathrm{sn}^2u\mathrm{sn}^2v} \tag{13.407}$$

This is proven by first setting $a = 0$ in (13.396) to find

$$\Theta^2(u)\Theta^2(v) - H^2(u)H^2(v) = \Theta^2(0)\Theta(u+v) \tag{13.408}$$

and then dividing each side of (13.398) by the corresponding side of (13.408) to obtain

$$\frac{H(u-v)H(K)\Theta(K)}{\Theta(u-v)\Theta^2(0)} = \frac{H(u)\Theta(u)H(v+K)\Theta(v+K) - H(v)\Theta(v)H(u+K)\Theta(u+K)}{\Theta^2(u)\Theta^2(v) - H^2(u)H^2(v)} \tag{13.409}$$

Then if we multiply (13.409) by $\Theta^2(0)/H^2(K)$, write

$$\begin{aligned} &\frac{H(u)}{\Theta(u)}\frac{H(v+K)\Theta(v+K)}{\Theta^2(v)}\frac{\Theta^2(0)}{H^2(K)} \\ &= \left(\frac{H(u)\Theta(K)}{\Theta(u)H(K)}\right)\left(\frac{H(v+K)\Theta(0)}{\Theta(u)H(K)}\right)\left(\frac{\Theta(v+K)\Theta(0)}{\Theta(u)\Theta(K)}\right) \end{aligned} \tag{13.410}$$

and use the definitions (13.109)–(13.111) of sn$u$, cn$u$ and dn$u$ and definition (13.95) of k, we obtain the desired result (13.407).

Complementary modulus transformation

The complementary modulus transformation for $H(v;k)$ is

$$H(v;k) = -i(K/K')^{1/2}e^{-\frac{\pi v^2}{4KK'}}H(iv;k'). \tag{13.411}$$

From these it follows by use of (13.385) and (13.389) that

$$\Theta_1(v;k) = (K/K')^{1/2}e^{-\frac{\pi v^2}{4KK'}}.\Theta_1(iv;ik'). \tag{13.412}$$

To prove (13.411) it suffices by Liouville's theorem to show that both sides have the same quasiperiodicity properties, have the same zeros and agree at one point. The zeros of the left-hand side of (13.411) are at $2mK + 2niK'$ and the zeros of the right-hand side are $i(2mK' + 2niK)$ and hence both sides have the same zeros. Furthermore we see from (13.98) and (13.99) that

$$H(i(v+2K);k') = -e^{\pi K/K'}e^{\pi v/K'}H(iv;k') \tag{13.413}$$
$$H(i(v+2iK');k') = -iH(iv;k') \tag{13.414}$$

and thus it follows that the right-hand side of (13.411) has the same quasiperiodicity properties (13.98) and (13.99) as does the right-hand side. Thus by Liouville's theorem the two sides of (13.411) must be proportional. This proportionality constant is not needed for the purposes of the text because only ratios of the theta functions appear and thus the proof of this constant is omitted. A proof can be found, for example, in [48].

References

[1] A.E. Kennelly, The equivalence of triangles and three-pointed stars in conducting networks, Electrical World and Engineer, 34 (1899) 413–414.
[2] L. Onsager, Crystal statistics I. A two dimensional model with an order disorder transition, Phys. Rev. 65 (1944) 117–149.
[3] G.H. Wannier, The statistical problem in cooperative phenomena, Rev. Mod. Phys. 17 (1945) 50–60.
[4] J.B. McGuire, Study of exactly soluble one dimensional N–body problem, J. Math. Phys. 5 (1964) 622–636.
[5] C.N. Yang, S matrix for the one dimensional N-body problem with repulsive or attractive δ-function interaction, Phys. Rev. 168 (1968) 1920–1923.
[6] B.M. McCoy and T.T. Wu, Hydrogen bonded crystals and the anisotropic Heisenberg chain, Il Nuovo Cimento, 56 (1968) 311–315.
[7] B. Sutherland, Two-dimensional hydrogen bonded crystals without the ice rule, J. Math. Phys. 11 (1970) 3183–3186.
[8] R.J. Baxter, Eight-vertex model in lattice statistics, Phys. Rev. Lett. 26 (1971) 832–833.
[9] R.J. Baxter, One-dimensional anisotropic Heisenberg chain, Phys. Rev. Lett. 26 (1971) 834.
[10] R.J. Baxter, Partition function of the eight-vertex lattice model, Ann. Phys. 70 (1972) 193–228.
[11] R.J. Baxter, One-dimensional anisotropic Heisenberg chain, Ann. Phys, 70 (1972) 323–337.
[12] R.J. Baxter, Eight-vertex model in lattice statistics and one-dimensional anisotropic Heisenberg chain I. Some fundamental eigenvectors, Ann. Phys. 76 (1973) 1–24.
[13] R.J. Baxter, Eight-vertex model in lattice statistics and one dimensional anisotropic Heisenberg chain II. Equivalence to a generalized ice-type model, Ann. Phys. 76 (1973) 25–47
[14] R.J. Baxter, Eight-vertex model in lattice statistics and one dimensional anisotropic Heisenberg chain III. Eigenvectors of the transfer matrix and Hamiltonian, Ann. Phys. 76 (1973) 48–71.
[15] R.J. Baxter, Hard Hexagons: Exact solution, J. Phys. A 13 (1980) L61–L70.
[16] R.J. Baxter, Rogers–Ramanujan identities in the hard hexagon model, J. Stat. Phys. 26 (1981) 427–452.
[17] A.B. Zamolodchikov, Soviet Physics JETP 52 (1980) 325–336.
[18] A.B. Zamolodchikov, Tetrahedron equations and the relativistic S-matrix of straight-strings in 2+1 dimensions, Comm. Math. Phys. 79 (1981) 489–505.
[19] V.A. Fateev and A.B. Zamolodchikov, Self dual solutions of star–triangle relations in Z_N models, Phys. Letts. 92A (1982) 37–39.

[20] R.J. Baxter, On Zamolodchikov's solution of the tetrahedron equations, Comm. Math. Phys. 88 (1983) 185–205.

[21] G.E. Andrews, R.J. Baxter and P.J. Forrester, Eight-vertex SOS model and generalized Rogers-Ramanujan-type identities, J. Stat. Phys. 35 (1984) 193–266

[22] G. von Gehlen and V. Rittenberg, Z_n-symmetric quantum chains with an infinite set of conserved charges and Z_n zero modes, Nucl. Phys. B257 [FS14] (1985) 351–370.

[23] H. Au-Yang, B.M. McCoy, J.H.H. Perk, S. Tang and M.L. Yan, Commuting transfer matrices in the chiral Potts models: Solutions of star–triangle equations for genus > 1, Phys. Letts A123 (1987) 219–223.

[24] B.M. McCoy, J.H.H. Perk, S. Tang and C.H. Sah, Commuting transfer matrices for the four-state self-dual chiral Potts model with a genus-three uniformizing Fermat curve, Phys. Lett. A. 125 (1987) 9–14.

[25] H. Au-Yang, B.M. McCoy, J.H.H. Perk and S. Tang, Solvable models in statistical mechanics and Riemann surfaces of genus greater than one, in *Algebraic Analysis*, vol. 1 M. Kashiwara and T. Kawai, eds (Academic Press, San Diego, 1988) 29–40.

[26] R.J. Baxter, J.H.H. Perk and H. Au-Yang, New solutions of the star–triangle relations for the chiral Potts model, Phys. Lett. A 128 (1988) 138–142.

[27] H. Au-Yang and J.H.H. Perk, Onsager's star–triangle equation: master key to integrability, Advanced Studies in Pure Mathematics vol. 19 (1989) 57–94.

[28] V.V. Bazhanov, R.M. Kashaev, V.V. Mangazeev and Yu. G. Stroganov, $(Z_N\times)^{n-1}$ generalization of the chiral Potts model, Comm. Math. Phys. 138 (1991) 393–408.

[29] V.V. Bazhanov and R.J. Baxter, Star–triangle equation for a three dimensional model, J. Stat. Phys. 71 (1993) 839–864.

[30] E.H. Lieb, Exact solution of the problem of the entropy of two–dimensional ice, Phys. Rev. Lett. 18 (1967) 692–694.

[31] E.H. Lieb, Exact solution of the F model of an antiferroelectric, Phys. Rev. Lett. 18, (1967) 1046–1048.

[32] E.H. Lieb, Exact solution of the two-dimensional Slater KDP model of a ferroelectric, Phys. Rev. Lett. 19 (1967) 108–110.

[33] E.H. Lieb, Residual entropy of square ice, Phys. Rev. 162 (1967) 162–172.

[34] B. Sutherland, Exact solution of a two-dimensional model for hydrogen bonded crystals, Phys. Rev. Lett. 19 (1967) 103–104.

[35] C.P. Yang, Exact solution of two dimensional ferroelectrics in an arbitrary external electric field, Phys. Rev. Lett. 19 (1967) 586–588.

[36] B. Sutherland, C.N. Yang and C.P. Yang, Exact solution of two dimensional ferroelectrics in an arbitrary external electric field, Phys. Rev. Lett. 19 (1967) 588–591.

[37] R. Orbach, Linear antiferromagnetic chain with anisotropic coupling, Phys. Rev. 112 (1958) 309–316.

[38] L.R. Walker, Antiferromagnetic linear chain, Phys. Rev. 116 (1959) 1089–1090.

[39] J. des Cloizeaux and M. Gaudin, Anisotropic linear magnetic chain, J. Math. Phys. 7 (1966) 1384–1400.

[40] C.N. Yang and C.P. Yang, One-dimensional chain of anisotropic spin–spin interactions I. Proof of Bethe's hypothesis for ground state in a finite system, Phys.

Rev. 150 (1966) 321–327.

[41] C.N. Yang and C.P. Yang, One-dimensional chain of anisotropic spin-spin interactions II. Properties of the ground-state energy per lattice site for an infinite system, Phys. Rev. 150 (1966) 327–339.

[42] H.A. Bethe, Zur Theorie der Metalle, Z. Physik 71 (1931) 205–226.

[43] H. Goldstein, *Classical Mechanics* (Addison Wesley, Reading Massachusetts 1959).

[44] H.A. Kramers and G.H. Wannier, Statistics of the two-dimensional ferromagnet, Part I, Phys. Rev. 60 (1941) 252–262.

[45] H.A. Kramers and G.H. Wannier, Statistics of the two-dimensional ferromagnet, Part II, Phys. Rev. 60 (1941) 263–276.

[46] C. Fan and F.Y. Wu, General lattice model of phase transitions, Phys. Rev. B2 (1970) 723–733.

[47] J.D. Johnson, S. Krinsky and B.M. McCoy, Vertical-arrow correlation length in the eight-vertex model and the low lying excitations of the X-Y-Z Hamiltonian, Phys. Rev. A8 (1973) 2526–2547.

[48] R.J. Baxter, *Exactly solved models in statistical mechanics* (Academic Press, 1982)

[49] V.B. Matveev and A.O. Smirnov, Some comments on the solvable chiral Potts model, Lett. Math. Phys. 19 (1990) 179–185.

[50] S. Howes, L.P. Kadanoff and M. den Nijs, Quantum model for commensurate-incommensurate transitions, Nucl. Phys. B215[FS7] (1983) 169–208.

14

The eight-vertex and XYZ model

The method of commuting transfer matrices and star–triangle equations introduced in the previous chapter gave a prescription for obtaining models which have very special properties. However, the statement that a set of Boltzmann weights satisfies the star- triangle equation does not provide a method of actually computing the partition function, free energy, order parameter or correlation functions.

The free energy of all the statistical models and the ground state energy and the low lying excitations of all the spin chains introduced in chapter 13 have been computed. Indeed the ability to have actually carried out an exact computation of the free energy is often taken as the definition of being an "exactly solved model" as contrasted with calling a model "integrable" if it satisfies the star–triangle equation. Furthermore the order parameters for all the models of chapter 13 have been computed (although computations of the order parameters always have occurred some years after the free energy computations). On the other hand the computation of correlation functions of the models of chapter 13 is still under development.

Each model of chapter 13 has its own separate special features and the solution of each one is quite distinct. Therefore in the space of one chapter it is quite impractical to present all the exact computations that have been done for all the models of chapter 13. Consequently we will focus on the computation of the eigenvalues of the transfer matrix of the eight-vertex model.

In section 14.1 we give a historical overview of the development of methods to compute these eigenvalues and eigenvectors. These methods fall into two broad categories: the Bethe ansatz which originates with the solutions on the isotropic Heisenberg anti-ferromagnet by Bethe in 1931, and functional equations which originate with Baxter's solution of the eight-vertex model in 1971.

We will concentrate on the method of functional equations, and, in section 14.2, will present in detail the derivation of the functional equation for the eigenvalues of the eight-vertex model as it has evolved since 1972. This functional equation is derived from a matrix equation which relates the transfer matrix $T(v)$ to an auxiliary matrix $Q(v)$. In this derivation the continuous parameter η of chapter 13 is specialized to the discrete values

$$2L\eta = 2m_1K + im_2K' \tag{14.1}$$

called (elliptic) roots of unity. For generic values of η for which (14.1) does not hold numerical studies on small systems indicates that the transfer matrix is nondegenerate. However, when (14.1) does hold, the transfer matrix has an extensive degeneracy. When N, the number of sites of the transfer matrix, is even and when m_2 is even and

m_1 is odd or when m_2 is odd and m_1 is unrestricted it is found numerically that there are doublets of multiplets of size 2^{2n-1} where the eigenvalues in the two multiplets have opposite signs. However, when m_1 and m_2 are both even there are multiplets of size 2^{2n}. When N is odd, all states are at least doubly degenerate but there are cases such as $m_2 = 0$ and m_1 even where (as $N \to \infty$) there are multiplets with a larger degeneracy.

These degeneracies must have an explanation in terms of a symmetry algebra, and we will find an explanation for some of these degeneracies from the properties of the functional equations. However, a complete explanation will require an understanding of the symmetry algebra of the model. In the six-vertex limit the symmetry algebra is the loop algebra of sl_2 and has been extensively studied, but the corresponding algebra for the 8 vertex model is unknown at the time of writing.

In section 14.3 we use this functional equation to explicitly compute the free energy of the eight-vertex model. We conclude in section 14.4 with a summary of results for eigenvalues, order parameters and correlation functions for the six- and eight-vertex models.

14.1 Historical overview

The Heisenberg model was first introduced by Heisenberg [1] in 1928 as a model for magnetism in three dimensions. In 1931 Hans Bethe [2] found the exact eigenvectors of the isotropic Heisenberg antiferromagnetic spin chain, long before the methods presented in chapter 13 of star–triangle equations and commuting transfer matrices were known. Bethe's method of solution makes an ansatz for the eigenvectors of the spin chain of finite length which gives all the eigenvectors, and from these eigenvectors the eigenvalues of the spin chain are computed. Sommerfeld and Bethe [3] and Hulthén [4] used this to compute the ground state energy of the Heisenberg antiferromagnet in the thermodynamic limit.

Subsequent to the work of Bethe the ansatz for the eigenvectors was extended to the XXZ spin chain and the corresponding eigenvalues were computed in a series of papers [5–10] given in Table 14.1. This ansatz for the eigenvectors is called Bethe's ansatz and states that the eigenvectors of the XXZ model in the sector that has n spins with $\sigma^z = -1$ at the site x_j are given as a linear superposition of plane waves

$$\psi(x_1, x_2, \cdots, x_n) = \sum_P A_P e^{i(x_1 k_{P1} + x_2 k_{P2} + \cdots x_n k_{Pn})} \tag{14.2}$$

where the sum is over all $n!$ permutations P of $1, \cdots, n$. Many properties of the spin chain at $T = 0$ were computed in [5–10] by use of the ansatz (14.2). Extensions to $T > 0$ were made in [19–21].

The integrability of the related six-vertex model was discovered in 1967 by Lieb [11–14] for several cases of the symmetric six-vertex model and extended soon after to the fully anisotropic case by Sutherland, Yang and Yang [15–17]. These computations also make the Bethe ansatz (14.2) for the eigenvectors of the transfer matrix of the six-vertex model.

The commutation of the six-vertex model transfer matrix and the Hamiltonian of the XXZ spin chain derived in the previous chapter was found in 1968 [18]. The

Table 14.1 Historical overview of major developments in the computation of the eigenvectors and eigenvalues of the XXZ model and the six-vertex model transfer matrix.

Date	Author(s)	Property
1928	Heisenberg [1]	Introduction of the Heisenberg model
1931	Bethe [2]	Eigenfunctions and eigenvalues of the isotropic Heisenberg antiferromagnet finite spin chain
1933	Sommerfeld, Bethe [3]	Ground state energy of the
1938	Hulthén [4]	antiferromagnetic Heisenberg chain
1958	Orbach [5]	Eigenfunctions and eigenvalues
1959	Walker [6]	of the XXZ chain
1966	des Cloizeaux, Gaudin [7]	
1966	Yang, Yang [8–10]	
1967	Lieb [11–14]	Eigenvectors and eigenvalues of the symmetric six-vertex model
1967	Sutherland, Yang, Yang [15–17]	Eigenvectors and eigenvalues of the asymmetric six-vertex model
1968	McCoy, Wu [18]	Commutation of six-vertex transfer matrix and XXZ Hamiltonian
1971	Takahashi [19]	Antiferromagnetic Heisenberg chain for $T > 0$
1971	Gaudin [20]	XXZ for $\|\Delta\| \geq 1$ for $T > 0$
1972	Takahashi, Suzuki [21]	XXZ for $\|\Delta\| < 1$ for $T > 0$
1979	Takhtajan, Faddeev [22]	Algebraic Bethe ansatz for nondegenerate six-vertex eigenvectors
2001	Deguchi, Fabricius, McCoy [23]	sl_2 loop algebra symmetry at roots of unity
2001	Fabricius, McCoy [24]	Algebraic Bethe ansatz for degenerate six-vertex eigenvectors
2001	Stroganov, Razumov [25, 26]	$\Delta = -1/2, \quad N$ odd
2002	de Gier, Batchelor Nienhuis, Mitra [27]	
2002	Baxter [28]	Completeness of Bethe's ansatz for generic Δ

connection with commuting transfer matrices which was the starting point in chapter 13 was in fact developed only after the ground state energy of the spin chain and the free energy of the statistical mechanical model had been computed.

The Bethe's ansatz solution of the six-vertex model is valid for all values of the anisotropy parameter Δ of the XXZ spin chain and for generic values the only degeneracy of the eigenvalues is the degeneracy under $S^z \to -S^z$ for $H = 0$ where

$$S^z = \frac{1}{2}\sum_{j=1}^{N} \sigma_j^z. \tag{14.3}$$

However, at the special points

$$\Delta = \cos \pi m_1/2L \tag{14.4}$$

the transfer matrix of the six-vertex model has many degenerate multiplets. The origin of these multiplets is an sl_2 loop algebra whose generators commute with the transfer matrix and spin chain Hamiltonian which was first discovered [23] in 2001. Further very special properties which only happen for N odd were found for $\Delta = -1/2$ in [25–27] and are still under active investigation.

The method of solution based on the wave function (14.2) given in coordinate space is often referred to as the "coordinate Bethe ansatz". A related powerful operator algebra method of solution given in 1979 by Takhtajan and Faddeev [22] for nondegenerate eigenvectors is referred to as the "algebraic Bethe ansatz". The algebraic Bethe ansatz for degenerate eigenvectors of the XXZ chain was first presented by Fabricius and McCoy [24] in 2001.

The solution of the eight-vertex model has a completely different development. It begins with the demonstration of the commutation relation of the eight-vertex transfer matrix with the Hamiltonian of the XYZ model found by Sutherland [29] in 1970 and was followed in 1971 by the pioneering and fundamental papers of Baxter [30, 32] who invented the principle of the commuting transfer matrix and star–triangle equation presented in the chapter 13. However, unlike the six-vertex model which was solved by the using the coordinate Bethe's ansatz 14.2 for the eigenvectors for all values of the anisotropy parameter Δ, Baxter's solution [30–33] for the eight-vertex free energy and the XYZ ground state energy makes and uses the root of unity restriction (14.1) with L, m_1 and m_2 relatively prime integers. Furthermore, the computation of [30, 32] is not for the eigenvectors of the transfer matrix but instead is for the eigenvalues (not the eigenvectors) of the transfer matrix which are computed by means of discovering an auxiliary matrix $Q(u)$ which has the properties that

$$T(v)Q(v) = [h(v+\eta)]^N Q(v-2\eta) + [h(v-\eta)]^N Q(v+2\eta) \tag{14.5}$$

with

$$[T(v), Q(v')] = 0 \tag{14.6}$$

$$[Q(v), Q(v')] = 0 \tag{14.7}$$

where $h(v)$ is a suitable quasiperiodic function. The equation (14.5) with the commutation relations (14.6) and (14.7) is referred to as the matrix TQ equation. This TQ equation will be derived in detail in section 14.2

The nondegenerate eigenvectors of the transfer matrix at roots of unity (14.1) were computed in the following year by Baxter in a separate computation [34–36, 28] which extends the method of the coordinate Bethe's ansatz. In addition in [34, section 6] Baxter derives a second matrix $Q_{73}(v)$ which is valid for all η that satisfies a TQ equation (14.5)–(14.7) but which is not the same the Q_{72} derived in [32]. The algebraic Bethe's ansatz solution for the nondegenerate eigenvectors at roots of unity was presented by Takhtajan and Faddeev [22] in 1979 and for the degenerate eigenvectors by Fabricius and McCoy [42, 47] in 2007.

Table 14.2 Historical overview of developments of the matrix TQ equation, eigenvectors and eigenvalues for the 8 vertex model and the XYZ spin chain.

Date	Author(s)	Property
1970	Sutherland [29]	Eight-vertex commutation with H_{XYZ}
1971	Baxter [30]	Announcement of eight-vertex free energy
1971	Baxter [31]	Announcement of XYZ ground state energy
1972	Baxter [32]	Matrix TQ equation at roots of unity eight-vertex free energy
1972	Baxter [33]	XYZ ground state energy
1972	Takahashi, Suzuki [21]	XYZ chain at $T > 0$
1973	Baxter [34–36, 28]	Eight-vertex eigenvectors at roots of unity; matrix TQ equation for generic η
1973	Johnson, Krinsky, McCoy [37]	Spectrum of the XYZ chain
1979	Takhtajan, Faddeev [22]	Algebraic Bethe's ansatz for nondegenerate 8-vertex eigenvectors at roots of unity
1989	Baxter [38]	$\eta = 2K/3$ with N odd
2003	Fabricius, McCoy [39]	Matrix TQ equation for m_1 odd, $m_2 = 0$
2005	Bazhanov, Mangazeev [40]	$\eta = 2K/3$ with N odd
2005	Fabricius, McCoy [41]	Matrix TQ equation for $m_2 = 0$, N odd
2006	Fabricius, McCoy [42]	Algebraic Bethe's ansatz for degenerate 8-vertex eigenvectors at roots of unity
2006	Bazhanov, Mangazeev [43]	Painlevé VI for $\eta = 2K/3$ with N odd
2007	Fabricius [44] Roan [45]	Matrix TQ equation for m_1 even, $m_2 = 0$, N even
2007–2009	Fabricius, McCoy [46, 47]	Matrix TQ equation for all m_1 and m_2

The principles of the computation of the TQ equation were established in the 1972 paper of Baxter [32] and we will refer to all matrices constructed from these principles as $Q_{72}(v)$. However, in the decades subsequent to the publication of this original paper it has been found that there are several details not covered in the original paper. The first of these details is the discovery by Baxter [34–36] in 1973 that the parametrization used in the 1972 paper [32] is not well adapted to the case $m_2 \neq 0$ and it was realized in 2003 [39] that the construction of [32] does not cover the case m_1 even and $m_2 = 0$. A construction which does cover this case was found [44, 45] in 2007. The general case for $m_2 \neq 0$ was treated in [46, 47] and this is the method we will follow in the presentation in section 14.2. All of these several constructions of Q lead to the same result for the free energy originally computed in [10].

These many years of developments are summarized in Table 14.2.

14.2 The matrix TQ equation for the eight-vertex model

The transfer matrix $T(v)$ of the eight-vertex model has two discrete symmetries expressed by the commutation relations

$$[T(v), S] = 0 \qquad [T(v), R] = 0 \tag{14.8}$$

where

$$S = \sigma_1^z \otimes \sigma_2^z \otimes \cdots \otimes \sigma_N^z \qquad R = \sigma_1^x \otimes \sigma_2^x \otimes \cdots \otimes \sigma_N^x. \tag{14.9}$$

These commutation relations follow directly from the definition of the Boltzmann weights in terms of arrows given in Fig. 13.12 with $a = \bar{a}$ and $b = \bar{b}$. The first relation in (14.8) follows from the fact that the number of arrows which point toward (and away) from each vertex is even, and the second relation follows from the invariance of the Boltzmann weights under inversion of all arrows.

We note that

$$RS = (-1)^N SR \tag{14.10}$$

and from (14.8) it follows that

$$[T(v), RS] = 0. \tag{14.11}$$

Furthermore, it follows from (14.10) that when N is even the two discrete symmetries of (14.8) are compatible, and the transfer matrix may be diagonalized in a basis where both S and R are diagonal. However, for N odd the operators S and R do not commute. Therefore even though we may diagonalize the transfer matrix in a basis where one of the discrete symmetry operators is diagonal the commutation of T with the other independent symmetry operator mandates that all eigenvalues of T must be doubly degenerate.

If the transfer matrix $T(v)$ were nondegenerate then the matrix $Q(v)$ of (14.5)–(14.7) would also satisfy (14.8) and (14.11). However, when (14.1) holds, the matrix $T(v)$ has degenerate eigenvalues and the matrix $Q(v)$ no longer needs to satisfy these commutation relations. For this reason the matrix Q which satisfies (14.5)–(14.7) is not uniquely defined when (14.1) holds.

The $Q(v)$ matrix can be chosen to have the same discrete symmetries (14.8) and (14.11) as the transfer matrix. This choice is made in the 1973 paper of Baxter [34]. We call this matrix $Q_{73}(v)$. However, the matrices $Q_{72}(v)$ constructed from the principles of [32] do not commute with all three of the operators R, S and RS and we will find that there are different commutation relations depending on the parity of m_1 and m_2. For the three cases where m_1 and m_2 are not both even the construction of [32] will directly apply and we will denote these cases by writing $Q^{(1)}_{72oe}(v), Q^{(1)}_{72oo}(v), Q^{(1)}_{72eo}(v)$. For the case where m_1 and m_2 are both even the construction of [44] must be used and this construction contains a parameter t which will be seen to take on the two possible values $t = n\eta$ and $(n + 1/2)\eta$ where n is an integer. We denote the Q matrix in this case as $Q^{(2)}_{72ee}(v;t)$.

These cases are distinguished by different commutation relations with the operators S, R and RS as follows.

Case 1 m_1 odd, m_2 even, N unrestricted

$$[Q^{(1)}_{72oe}(v), S] = 0, \;\; [Q^{(1)}_{72oe}(v), R] \neq 0, \;\; [Q^{(1)}_{72oe}(v), RS] \neq 0 \tag{14.12}$$

Case 2 m_1 odd, m_2 odd, N unrestricted

$$[Q^{(1)}_{72oo}(v), S] \neq 0,\ [Q^{(1)}_{72oo}(v), R] \neq 0,\ [Q^{(1)}_{72oo}(v), RS] = 0 \tag{14.13}$$

Case 3 m_1 even, m_2 odd, N unrestricted

$$[Q^{(1)}_{72eo}(v), S] \neq 0,\ [Q^{(1)}_{72eo}(v), R] = 0,\ [Q^{(1)}_{72eo}(v), RS] \neq 0 \tag{14.14}$$

Case 4 m_1 even, m_2 even, N even

We find that there are matrices $Q^{(2)}_{72ee}(v;t)$ for $t = n\eta$ and $t = (n+1/2)\eta$ with n an integer and thus there are several subcases to be distinguished.

Case 4A $t = n\eta$

$$[Q^{(2)}_{72ee}(v;0), S] = 0,\ [Q^{(2)}_{72ee}(v;0), R] \neq 0,\ [Q^{(2)}_{72ee}(v;0), RS] \neq 0 \tag{14.15}$$

Case 4B $t = (n+1/2)\eta$, $m_1 \equiv 0$ (mod4), and $m_2 \equiv 0$ (mod4)

$$\begin{aligned}&[Q^{(2)}_{72ee}(v;(n+1/2)\eta), S] = 0,\ [Q^{(2)}_{72ee}(v;(n+1/2)\eta), R] \neq 0,\\ &[Q^{(2)}_{72ee}(v;(n+1/2)\eta), RS] \neq 0\end{aligned} \tag{14.16}$$

Case 4C $t = (n+1/2)\eta$, $m_1 \equiv 2 \pmod 4$, $m_2 \equiv 2 \pmod 4$

$$\begin{aligned}&[Q^{(2)}_{72ee}(v;(n+1/2)\eta), S] \neq 0,\ [Q^{(2)}_{72ee}(v;(n+1/2)\eta), R] = 0,\\ &[Q^{(2)}_{72ee}(v;(n+1/2)\eta/2), RS] \neq 0\end{aligned} \tag{14.17}$$

Case 4D $t = (n+1/2)\eta$, $m_1 \equiv 0 \pmod 4$, $m_2 \equiv 2 \pmod 4$

$$\begin{aligned}&[Q^{(2)}_{72ee}(v;(n+1/2)\eta), S] \neq 0,\ [Q^{(2)}_{72ee}(v;(n+1/2)\eta), R] \neq 0,\\ &[Q^{(2)}_{72ee}(v;(n+1/2)\eta), RS] = 0\end{aligned} \tag{14.18}$$

No matrix $Q^{(2)}_{72ee}(v,(n+1/2)\eta)$ exists for $m_1 \equiv 2 \pmod 4$, $m_2 \equiv 0 \pmod 4$.

There remains one case for which no matrix has yet been found by use of the methods of [32] or [44] which satisfies (14.5)–(14.7): this is m_1 even, m_2 even and N odd. It has been seen in [41] from numerical computations that a TQ equation for eigenvalues holds and the eigenvalues of $Q(v)$ have unique properties not seen in cases 1–4. Similar unique properties also exist for the six-vertex limit [25–27]. A full understanding of this case is still lacking.

14.2.1 Modified theta functions

The parametrization of [32] is sufficient for the case $m_2 = 0$. However, to deal with the general case of (14.1) with $m_2 \neq 0$ Baxter in [34] introduces the "modified" theta functions:

$$H_m(v) = \exp\left(\frac{i\pi m_2}{8KL\eta}(v-K)^2\right) H(v) \qquad \Theta_m(v) = \exp\left(\frac{i\pi m_2}{8KL\eta}(v-K)^2\right)\Theta(v) \tag{14.19}$$

In terms of $\Theta_m(v)$ and $H_m(v)$ the Boltzmann weights are parametrized as

$$
\begin{aligned}
a &= \quad \Theta_m(-2\eta)\Theta_m(\eta - v)H_m(\eta + v) \\
b &= -\Theta_m(-2\eta)H_m(\eta - v)\Theta_m(\eta + v) \\
c &= -H_m(-2\eta)\Theta_m(\eta - v)\Theta_m(\eta + v) \\
d &= \quad H_m(-2\eta)H_m(\eta - v)H_m(\eta + v).
\end{aligned} \tag{14.20}
$$

When $m_2 = 0$ the parametrization (14.20) reduces to the parametrization of [32].

The phase factor in (14.19) is chosen so that the modified theta functions have the periodicity

$$H_m(v + 4L\eta) = H_m(v) \qquad \Theta_m(v + 4L\eta) = \Theta_m(v). \tag{14.21}$$

In appendix A we demonstrate that these modified theta functions are in fact quasiperiodic functions but that their fundamental parallelogram is no longer spanned by $2K$ and $2iK'$ (the quasiperiods of $H(v)$ and $\Theta(v)$). Instead we find the quasiperiodicity properties

$$H_m(v + \omega_1) = (-1)^{r_1}(-1)^{r_1 r_2}H_m(v) \tag{14.22}$$

$$\Theta_m(v + \omega_1) = (-1)^{r_1 r_2}\Theta_m(v) \tag{14.23}$$

$$
\begin{aligned}
H_m(v + \omega_2) &= (-1)^{b}(-1)^{ab}q'^{-1}e^{-2\pi i(v-K)/\omega_1}H_m(v) \\
&= (-1)^{a+b}(-1)^{ab}q'^{-1-r_2}e^{-2\pi iv/\omega_1}H_m(v)
\end{aligned} \tag{14.24}
$$

$$
\begin{aligned}
\Theta_m(v + \omega_2) &= (-1)^{ab}q'^{-1}e^{-2\pi i(v-K/\omega_1)}\Theta_m(v) \\
&= (-1)^{a}(-1)^{ab}q'^{-1-r_2}e^{-2\pi iv/\omega_1}\Theta_m(v)
\end{aligned} \tag{14.25}
$$

where the quasiperiods are given by

$$\omega_1 = 2(r_1K + ir_2K') \qquad \omega_2 = 2(bK + iaK'). \tag{14.26}$$

Here, with r_0 defined as the greatest common factor in $2m_1$ and m_2, the quantities r_1 and r_2 are given by

$$2m_1 = r_0 r_1 \qquad m_2 = r_0 r_2, \tag{14.27}$$

where the integers a and b are determined as the solutions to

$$ar_1 - br_2 = 1. \tag{14.28}$$

From (14.28) the area of the fundamental period parallelogram

$$0,\ \omega_1,\ \omega_1 + \omega_2,\ \omega_2 \tag{14.29}$$

is $4KK'$. We thus see that the modified theta functions are in fact theta functions of nome

$$q' = e^{i\pi\omega_2/\omega_1} \tag{14.30}$$

which are modular transforms of the original theta functions $\Theta(v)$ and $H(v)$. We also note that

$$2L\eta = r_0\omega_1/2. \tag{14.31}$$

14.2.2 Formal construction of the matrices $Q_{72}(v)$

The construction of [32] of a matrix Q which satisfies (14.5)–(14.7) under the condition (14.1) consists of three steps:

1. Construction of matrices $Q_R(v)$ and $Q_L(v)$

The first step begins with an assumption that there exists a matrix $Q_R(v)$ of the form

$$Q_R(v)|_{\alpha,\beta} = \text{Tr}S_R(\alpha_1,\beta_1)S_R(\alpha_2,\beta_2)\cdots S_R(\alpha_N,\beta_N) \tag{14.32}$$

with $S_R(\alpha,\beta)$ an $L \times L$ matrix with elements $s_{m,n}(\alpha,\beta)$ which satisfies

$$T(v)Q_R(v) = [h(v+\eta)]^N Q_R(v-2\eta) + [h(v-\eta)]^N Q_R(v+2\eta) \tag{14.33}$$

where

$$h(v) = \Theta_m(0)\Theta_m(-v)H_m(v). \tag{14.34}$$

This matrix $Q_R(v)$ cannot be unique because if (14.33) is multiplied on the left by any matrix A which commutes with $T(v)$ then $AQ_R(v)$ also satisfies (14.33). Furthermore we note that the matrix $e^{av}Q_R(v)$ will satisfy (14.33) with $h(v\pm\eta)^N$ replaced by $e^{\pm 2a\eta}h(v\pm\eta)^N$.

Similarly we construct a matrix $Q_L(v)$

$$Q_L(v)|_{\alpha,\beta} = \text{Tr}S_L(\alpha_1,\beta_1)S_L(\alpha_2,\beta_2)\cdots S_L(\alpha_N,\beta_N) \tag{14.35}$$

which satisfies

$$Q_L(v)T(v) = [h(v+\eta)]^N Q_L(v-2\eta) + [h(v-\eta)]^N Q_L(v+2\eta). \tag{14.36}$$

This matrix is also non-unique because (14.36) may be multiplied on the right by any matrix which commutes with $T(v)$.

The matrices $Q_R(v)$ and $Q_L(v)$ are independently defined and can be independently constructed by analogous procedures. However, it is also instructive to note that the matrix $Q_L(v)$ can also be obtained by taking the transpose of (14.33), using the symmetry properties of the transfer matrix

$$T^T(v) = (-1)^N T(-v) \tag{14.37}$$

and the property

$$h(v) = -h(-v) \tag{14.38}$$

to find

$$Q_L(v) = Q_R^T(-v). \tag{14.39}$$

The matrices $Q_R(v)$ and $Q_L(v)$ will not in general satisfy either (14.6) or (14.7)

2. The interchange relation
To satisfy commutators (14.6) and (14.7) we impose the interchange relation

$$Q_L(v_1)AQ_R(v_2) = Q_L(v_2)AQ_R(v_1) \tag{14.40}$$

where the matrix A is independent of v_1 and v_2, satisfies $A^2 = 1$ and commutes with the transfer matrix $T(v)$. We will consider the four choices:

$$A = I, S, R, RS \tag{14.41}$$

These choices may be thought of as representing the arbitrariness in the construction of $Q_R(v)$ and/or $Q_L(v)$. We will see below that, for cases 1–3 where m_1 and m_2 are not both even, (14.40) holds for only two of the four choices of A whereas for case 4 where m_1 and m_2 are both even (14.40) holds for all four choices (14.41).

3. The nonsingularity condition
The final requirement is that the matrices $Q_R(v)$ and $Q_L(v)$ possess at least one value $v = v_0$ such that $Q_R(v_0)^{-1}$ and $Q_L(v_0)^{-1}$ exist. Under this nonsingularity assumption we use (14.40) with $v_1 = v_0$ and $v_2 = v$ to construct the matrices

$$Q_{72}(v) = Q_R(v)Q_R^{-1}(v_0) = AQ_L^{-1}(v_0)Q_L(v)A \tag{14.42}$$

and

$$AQ_{72}(v)A = AQ_R(v)Q_R^{-1}(v_0)A = Q_L^{-1}(v_0)Q_L(v). \tag{14.43}$$

From (14.33), (14.36) and (14.40) the matrices both satisfy the conditions (14.5) and (14.6). Furthermore

$$\begin{aligned} Q_{72}(v)Q_{72}(v') &= AQ_L^{-1}(v_0)Q_L(v)AQ_R(v')Q_R^{-1}(v_0) \\ &= AQ_L^{-1}(v_0)Q_L(v')AQ_R(v)Q_R^{-1}(v_0) = Q_{72}(v')Q_{72}(v) \end{aligned} \tag{14.44}$$

and thus (14.7) is also satisfied.

We will see below in cases 1–3 that $Q_R(v)$ is generically nonsingular but for case 4 where m_1 and m_2 are both even and (14.40) holds for all four choices (14.41) $Q_R(v)$ is singular for all v. In cases 1–3 where the interchange relation (14.40) holds for two and only two matrices A_1 and A_2 we may write

$$\begin{aligned} A_1A_2Q_{72}(v) &= A_1A_2[A_2Q_L^{-1}(v_0)Q_L(v)A_2] \\ &= [A_1Q_L^{-1}(v_0)Q_L(v)A_1]A_1A_2 = Q_{72}(v)A_1A_2 \end{aligned} \tag{14.45}$$

and thus $Q_{72}(v)$ commutes with the discrete symmetry operator A_1A_2.

14.2.3 Explicit construction of $Q_R(v)$ and $Q_L(v)$

To construct the matrix $Q_R(v)$ which satisfies (14.33) by the method of [32] we write

$$T(v)Q_R(v) = \mathrm{Tr}U(\alpha_1, \beta_1)\cdots U(\alpha_N, \beta_N) \tag{14.46}$$

with

$$U(+, \beta) = \begin{pmatrix} aS_R(+,\beta) & dS_R(-,\beta) \\ cS_R(-,\beta) & bS_R(+,\beta) \end{pmatrix} \tag{14.47}$$

$$U(-,\beta) = \begin{pmatrix} bS_R(-,\beta) & cS_R(+,\beta) \\ dS_R(+,\beta) & aS_R(-,\beta) \end{pmatrix} \tag{14.48}$$

and seek a similarity transformation

$$\begin{pmatrix} A(\alpha,\beta) & 0 \\ C(\alpha,\beta) & B(\alpha,\beta) \end{pmatrix} = \begin{pmatrix} I & -P \\ 0 & I \end{pmatrix} U(\alpha,\beta) \begin{pmatrix} I & P \\ 0 & I \end{pmatrix} \tag{14.49}$$

with

$$P_{m,n} = p_n \delta_{m,n} \tag{14.50}$$

so that

$$T(v)Q_R(v) = \mathrm{Tr}A(\alpha_1,\beta_1)\cdots A(\alpha_N,\beta_N) + \mathrm{Tr}B(\alpha_1,\beta_1)\cdots B(\alpha_N,\beta_N). \tag{14.51}$$

In order for (14.49) to hold, the matrix elements $S_R(\alpha,\beta)_{m,n}$ of S_R and the quantities p_n must satisfy

$$(ap_n - bp_m)S_R(+,\beta)_{m,n} + (d - cp_m p_n)S_R(-,\beta)_{m,n} = 0 \tag{14.52}$$
$$(c - dp_m p_n)S_R(+\beta)_{m,n} + (bp_n - ap_m)S_R(-,\beta)_{m,n} = 0. \tag{14.53}$$

This set of homogeneous linear equations will have a nontrivial solution provided

$$(a^2 + b^2 - c^2 - d^2)p_m p_n = ab(p_m^2 + p_n^2) - cd(1 + p_m^2 p_n^2). \tag{14.54}$$

This can only happen for certain values of m and n. For all other values we have

$$S_R(\alpha,\beta)_{m,n} = 0. \tag{14.55}$$

Using the parameterizations (14.20) we have

$$\frac{a^2 + b^2 - c^2 - d^2}{ab} = 2\mathrm{cn}2\eta\mathrm{dn}2\eta, \quad cd/ab = k\mathrm{sn}^2 2\eta \tag{14.56}$$

where $\mathrm{sn}(v), \mathrm{cn}(v), \mathrm{dn}(v)$ are the conventional doubly periodic functions with periods $2K$ and $2iK'$, and $\mathrm{sn}(v)$ is given in terms of the modified theta functions as

$$k^{1/2}\mathrm{sn}(v) = H_m(v)/\Theta_m(v). \tag{14.57}$$

Thus (14.54) becomes

$$2\mathrm{cn}(2\eta)\mathrm{dn}(2\eta)p_m p_n = p_m^2 + p_n^2 - k\mathrm{sn}^2(2\eta)(1 + p_m^2 p_n^2). \tag{14.58}$$

We solve for p_n in terms of p_m to obtain

$$p_n = \frac{1}{1 - k^2\mathrm{sn}^2 2\eta p_m^2}\{\mathrm{cn}2\eta\mathrm{dn}2\eta p_m$$
$$\pm[p_m^2\mathrm{cn}^2 2\eta\mathrm{dn}^2 2\eta - (p_m^2 - k\mathrm{n}^2 2\eta)(1 - p_m^2 k\mathrm{sn}^2 2\eta)]^{1/2}\} \tag{14.59}$$

which by use of the identities proven in the last chapter

$$\mathrm{cn}^2 u = 1 - \mathrm{sn}^2 u, \quad \mathrm{dn}^2 u = 1 - k^2\mathrm{sn}^2 u \tag{14.60}$$

may be rewritten as

$$p_n = \frac{1}{1 - kp_m^2\mathrm{sn}^2 2\eta}\{\mathrm{cn}2\eta\mathrm{dn}2\eta p_m \pm \mathrm{sn}2\eta[(k - p_m^2)(1 - kp_m^2)]^{1/2}\}. \tag{14.61}$$

Now set

$$p_m = \pm k^{1/2}\mathrm{sn}u. \tag{14.62}$$

Then (14.61) becomes

$$p_n = \frac{\pm k^{1/2}}{1 + k^2\mathrm{sn}^2 2\eta\mathrm{sn}^2 u}\{\mathrm{cn}2\eta\mathrm{dn}2\eta\mathrm{sn}u \pm \mathrm{sn}2\eta[(1 - \mathrm{sn}^2 u)(1 - k^2\mathrm{sn}^2 u)]^{1/2}\} \tag{14.63}$$

which by use, once again, of the identities (14.60) reduces to

$$p_n = \frac{\pm k^{1/2}}{1 - k^2\mathrm{sn}^2 2\eta\mathrm{sn}^2 u}\{\mathrm{cn}2\eta\mathrm{dn}2\eta\mathrm{sn}u \pm \mathrm{sn}2\eta\mathrm{cn}u\mathrm{dn}u\} \tag{14.64}$$

which we are able to recognize as the addition formula (13.407) for $\mathrm{sn}u$. Thus we obtain

$$p_n = \pm k^{1/2}\mathrm{sn}(u \pm 2\eta) \tag{14.65}$$

and hence we can choose an ordering of $p_1, \cdots, p_L$ such that, for any v_r, p_m may be written as

$$p_m = \pm k^{1/2}\mathrm{sn}(v_r + 2m\eta) = \pm\frac{H_m(v_r + 2m\eta)}{\Theta_m(v_r + 2m\eta)}. \tag{14.66}$$

(We note that (14.66) can equivalently be written as

$$p_m = \pm\frac{\Theta_m(v_r + 2m\eta)}{H_m(v_r + 2m\eta)} \tag{14.67}$$

which corresponds to multiplying $Q_R(v)$ in the left by the operator R.) Using (14.66) with the plus sign and making use of the identity (14.515) we find

$$ap_{k+1} - bp_k = \Theta_m(0)\Theta_m(-2\eta)H_m(2\eta)$$
$$\times\frac{\Theta_m(v_r - v + (2k+1)\eta)H_m(v_r + v + (2k+1)\eta)}{\Theta_m(v_r + 2k\eta)\Theta_m(v_r + 2(k+1)\eta)} \tag{14.68}$$
$$ap_k - bp_{k+1} = \Theta_m(0)\Theta_m(-2\eta)H_m(2\eta)$$
$$\times\frac{\Theta_m(v_r + v + (2k+1)\eta)H_m(v_r - v + (2k+1)\eta)}{\Theta_m(K + 2k\eta)\Theta_m(K + 2(k+1)\eta)} \tag{14.69}$$

$$d - cp_{k+1}p_k = \Theta_m(0)H_m(-2\eta)\Theta_m(2\eta) \times \frac{H_m(v_r + v + (2k+1)\eta)H_m(v_r - v + (2k+1)\eta)}{\Theta_m(K + 2k\eta)\Theta_m(K + 2(k+1)\eta)}. \tag{14.70}$$

The expressions (14.68)–(14.70) may now be used in (14.52) and (14.53) to compute the two ratios $S_R(+,\beta)_{k,k+1}/S_R(-,\beta)_{k,k+1}$ and $S_R(+,\beta)_{k+1,k}/S_R(-,\beta)_{k+1,k}$ which thus determines $S_R(\pm,\beta)_{k,k+1}$ and $S_R(\pm,\beta)_{k+1,k}$ up to arbitrary multiplicative functions $f(v)_{k,k+1}$ and $f(v)_{k+1,k}$. We will consider here two choices which will lead to matrices $Q(v)$ which satisfy (14.5)–(14.7) which we call $S_R^{(1)}(v)$ and $S_R^{(2)}(v)$

$$\begin{aligned}
S_R^{(1)}(+,\beta)_{k,k+1} &= H_m(v_r - v + (2k+1)\eta)\tau_{\beta,-k}\\
S_R^{(1)}(-,\beta)_{k,k+1} &= \Theta_m(v_r - v + (2k+1)\eta)\tau_{\beta,-k}\\
S_R^{(1)}(+,\beta)_{k+1,k} &= H_m(v_r + v + (2k+1)\eta)\tau_{\beta,k}\\
S_R^{(1)}(-,\beta)_{k+1,k} &= \Theta_m(v_r + v + (2k+1)\eta)\tau_{\beta,k}
\end{aligned} \tag{14.71}$$

and

$$\begin{aligned}
S_R^{(2)}(+,\beta)_{k,k+1} &= -H_m(v - v_r - (2k+1)\eta)\tau_{\beta,-k}\\
S_R^{(2)}(-,\beta)_{k,k+1} &= \Theta_m(v - v_r - (2k+1)\eta)\tau_{\beta,-k}\\
S_R^{(2)}(+,\beta)_{k+1,k} &= H_m(v_r + v + (2k+1)\eta)\tau_{\beta,k}\\
S_R^{(2)}(-,\beta)_{k+1,k} &= \Theta_m(v_r + v + (2k+1)\eta)\tau_{\beta,k}
\end{aligned} \tag{14.72}$$

where $1 \leq k \leq L-1$ and $\tau_{\beta,\pm k}$ are arbitrary constants (not functions). We note that if $H_m(v)$ is an odd function of v, which is the case for $m_2 = 0$, these two choices are identical. However, in the general case $m_2 \neq 0$ these two choices are not the same.

We now may compute $A(\alpha,\beta)_{m,n}$ and $B(\alpha,\beta)_{m,n}$ from (14.49) as

$$\begin{aligned}
A(+,\beta)_{m,n} &= aS_R(+,\beta)_{m,n} - cp_m S_R(-,\beta)_{m,n}\\
A(-,\beta)_{m,n} &= bS_R(-,\beta)_{m,n} - dp_m S_R(+,\beta)_{m,n}\\
B(+,\beta)_{m,n} &= bS_R(+,\beta)_{m,n} + cS_R(-,\beta)_{m,n}p_n\\
B(-,\beta)_{m,n} &= aS_R(-,\beta)_{m,n} + dS_R(+,\beta)_{m,n}p_n
\end{aligned} \tag{14.73}$$

which is valid for both $S_R^{(1)}$ and $S_R^{(2)}$. Thus, using (14.71) and the identities (14.514), (14.515) and (14.506) we find for all v_r

$$\begin{aligned}
A^{(1)}(\alpha,\beta)_{k,k+1}(v) &= h(v-\eta)\left[\frac{\Theta_m(v_r + 2(k+1)\eta)}{\Theta_m(v_r + 2k\eta)}\right] S_R^{(1)}(\alpha,\beta)_{k,k+1}(v + 2\eta)\\
A^{(1)}(\alpha,\beta)_{k+1,k}(v) &= h(v-\eta)\left[\frac{\Theta_m(v_r + 2k\eta)}{\Theta_m(v_r + 2(k+1)\eta)}\right] S_R^{(1)}(\alpha,\beta)_{k+1,k}(v + 2\eta)\\
B^{(1)}(\alpha,\beta)_{k,k+1}(v) &= h(v+\eta)\left[\frac{\Theta_m(v_r + 2k\eta)}{\Theta_m(v_r + 2(k+1)\eta)}\right] S_R^{(1)}(\alpha,\beta)_{k,k+1}(v - 2\eta)\\
B^{(1)}(\alpha,\beta)_{k+1,k}(v) &= h(v+\eta)\left[\frac{\Theta_m(v_r + 2(k+1)\eta)}{\Theta_m(v_r + 2k\eta)}\right] S_R^{(1)}(\alpha,\beta)_{k+1,k}(v - 2\eta)
\end{aligned} \tag{14.74}$$

and using (14.72) and the identity which follows from (14.506)

$$\frac{\Theta_m(-v_r-2(k+1)\eta)}{\Theta_m(-v_r-2k\eta)}=e^{\pi i m_2/L}\frac{\Theta_m(v_r+2(k+1)\eta)}{\Theta_m(v_r+2k\eta)} \tag{14.75}$$

we have

$$\begin{aligned}
A^{(2)}(\alpha,\beta)_{k,k+1}(v)&=h(v-\eta)e^{\pi i m_2/L}\left[\frac{\Theta_m(v_r+2(k+1)\eta)}{\Theta_m(-v_r-2k\eta)}\right]S_R^{(2)}(v+2\eta)_{k,k+1}\\
A^{(2)}(\alpha,\beta)_{k+1,k}(v)&=h(v-\eta)\left[\frac{\Theta_m(v_r+2k\eta)}{\Theta_m(v_r+2(k+1)\eta)}\right]S_R^{(2}(v+2\eta)_{k+1,k}\\
B^{(2)}(\alpha,\beta)_{k,k+1}(v)&=h(v+\eta)e^{-\pi i m_2/L}\left[\frac{\Theta_m(v_r+2k\eta)}{\Theta_m(v_r+2(k+1)\eta)}\right]S_R^{(2)}(\alpha,\beta)_{k,k+1}(v-2\eta)\\
B^{(2)}(\alpha,\beta)_{k+1,k}(v)&=h(v+\eta)\left[\frac{\Theta_m(v_r+2(k+1)\eta)}{\Theta_m(v_r+2k\eta)}\right]S_R^{(2)}(\alpha,\beta)_{k+1,k}(v-2\eta).
\end{aligned} \tag{14.76}$$

From (14.74) we see that there is a diagonal matrix

$$M^{(1)}_{k,k'}=\delta_{k.k'}\Theta_m(K+(2k+1)\eta) \tag{14.77}$$

such that

$$\begin{aligned}
M^{(1)}A^{(1)}(\alpha,\beta)(v)M^{(1)-1}&=h(v-\eta)S_R^{(1)}(\alpha,\beta)(v+2\eta)\\
M^{(1)-1}B^{(1)}(\alpha,\beta)(v)M^{(1)}&=h(v+\eta)S_R^{(1)}(\alpha,\beta)(v-2\eta)
\end{aligned} \tag{14.78}$$

and thus from (14.32) we see that (14.33) is satisfied.

Similarly from (14.76) we see that there is a diagonal matrix

$$M^{(2)}_{k,k'}=\delta_{k,k'}e^{i\pi m_2/2L}\Theta_m(K+(2k+1)\eta) \tag{14.79}$$

such that

$$\begin{aligned}
M^{(2)}A^{(1)}(\alpha,\beta)(v)M^{(2)-1}&=h(v-\eta)e^{i\pi m_2/2L}S_R^{(2)}(\alpha,\beta)(v+2\eta)\\
M^{(2)-1}B^{(2)}(\alpha,\beta)(v)M^{(2)}&=h(v+\eta)e^{-i\pi m_2/2L}S_R^{(2)}(\alpha,\beta)(v-2\eta)
\end{aligned} \tag{14.80}$$

and thus from (14.32) we see that

$$T(v)Q_R^{(2)}(v)=[h(v+\eta)]^N\omega^NQ_R^{(2)}(v-2\eta)+[h(v-\eta)]^N\omega^{-N}Q_R^{(2)}(v+2\eta) \tag{14.81}$$

where

$$\omega=e^{i\pi m_2/2L}. \tag{14.82}$$

We will also refer to (14.81) as a TQ equation and note that it can be reduced to (14.33) by defining $\tilde{Q}^{(2)}(v)=e^{i\pi m_2 Nv/4L\eta}Q_R^{(2)}(v)$. In analogy with (14.36) we consider

$$Q_L^{(2)}(v)T(v)=[h(v+\eta)]^N\omega^NQ_L^{(2)}(v-2\eta)+[h(v-\eta)]^N\omega^{-N}Q_L^{(2)}(v+2\eta) \tag{14.83}$$

and using the properties

$$T^T(v) = e^{\pi i m_2(v-K)/L\eta} T(2K - v) \tag{14.84}$$

and

$$h(v) = e^{i\pi m_2(v-K)/L\eta} h(2K - v) \tag{14.85}$$

we find

$$Q_L^{(2)}(v) = Q_R^{(2)T}(2K - v). \tag{14.86}$$

If the matrices $Q_R^{(j)}(v)$ obtained from $S_R^{(j)}(v)$ were nonsingular we could proceed immediately to the study of the interchange relation (14.40). However, the matrices $S_R^{(1)}(v)$ and $S_R^{(2)}(v)$ defined by (14.71) and (14.72) do not have sufficient nonvanishing elements to produce nonsingular matrices $Q_R^{(1)}(v)$ and $Q_R^{(2)}(v)$. The additional elements needed will turn out to be very different in the two cases and will thus be treated separately.

The matrices $Q_R^{(1)}$ and $Q_L^{(1)}$

For the matrix $Q_R^{(1)}(v)$ we follow [32] and require

$$S_R^{(1)}(\alpha,\beta)_{1,1} \neq 0, \quad S_R^{(1)}(\alpha,\beta)_{L,L} \neq 0. \tag{14.87}$$

Thus we require that (14.58) holds for $m = n = 1$ and $m = n = L$

$$2\text{sn}^2(v_r + 2\eta) - \text{sn}^2 2\eta[1 + k^2\text{sn}^4(v_r + 2\eta)] - 2\text{sn}^2(v_r + 2\eta)\text{cn}2\eta\text{dn}2\eta = 0 \tag{14.88}$$

$$2\text{sn}^2(v_r + 2L\eta) - \text{sn}^2 2\eta[1 + k^2\text{sn}^4(v_r + 2L\eta)] - 2\text{sn}^2(v_r + 2L\eta)\text{cn}2\eta\text{dn}2\eta = 0. \tag{14.89}$$

By use of the relations

$$\text{sn}(v + 2K) = \text{sn}v, \quad \text{sn}(v + iK') = (k\text{sn}v)^{-1} \tag{14.90}$$

we see that if we use the root of unity condition (14.1) then (14.89) reduces to

$$\text{sn}^2 v - \text{sn}^2 2\eta[1 + k^2\text{sn}^4 v] - 2\text{sn}^2 v\text{cn}2\eta\text{dn}2\eta = 0. \tag{14.91}$$

Thus if we set

$$v_r = K - \eta \tag{14.92}$$

and use

$$\text{sn}(K - \eta) = \text{sn}(K + \eta) \tag{14.93}$$

we see that (14.88) and (14.91) are identical. Furthermore we also may use (14.93) to write both (14.88) and (14.91) with v_r given by (14.92) as

$$\text{sn}(K-\eta)\text{sn}(K+\eta) - \text{sn}^2 2\eta[1+k^2\text{sn}^2(K+\eta)\text{sn}^2(K-\eta)] - 2\text{sn}(K-\eta)\text{sn}(K+\eta)\text{cn}2\eta\text{dn}2\eta = 0. \tag{14.94}$$

But if we multiply by k then (14.94) is exactly of the form (14.58) which we have already shown is satisfied by (14.66). Therefore v_r is indeed given by (14.92).

We thus compute $S_R^{(1)}(\alpha,\beta)_{1,1}$ from (14.52) with (14.505) and (14.515) using

$$\begin{aligned}(a-b)p_1 &= \Theta_m(0)\Theta_m(-2\eta)H(2\eta)\frac{\Theta_m(K-v)H_m(K+v)}{\Theta_m^2(K+\eta)}\\ d-cp_1^2 &= \Theta_m(0)\Theta_m(2\eta)H_m(-2\eta)\frac{H_m(K-v)H_m(K+v)}{\Theta_m^2(K+\eta)}\end{aligned} \tag{14.95}$$

and $S_R^{(1)}(\alpha,\beta)_{L,L}$ from (14.52) with (14.21), (14.505) and (14.515) using

$$\begin{aligned}(a-b)p_L &= \Theta_m(0)\Theta_m(-2\eta)H_m(2\eta)\frac{\Theta_m(K+2L-v)H_m(K+2L+v)}{\Theta_m^2(K+(2L-1)\eta)}\\ d-cp_L^2 &= \Theta_m(0)\Theta_m(2\eta)H_m(-2\eta)\frac{H_m(K+2L\eta-v)H(K+2L\eta_v)}{\Theta_m^2(K+(2L-1)\eta)}\end{aligned} \tag{14.96}$$

and the remaining elements of $S_R^{(1)}(\alpha,\beta)$ from (14.71) with (14.92) to obtain the desired result for $1\le k\le L-1$

$$\begin{aligned}
S_R^{(1)}(+,\beta)_{k,k+1}(v) &= H_m(K-v+2k\eta)\tau_{\beta,-k} = H_m(K+v-2k\eta)\tau_{\beta,-k}\\
S_R^{(1)}(+,\beta)_{k+1,k}(v) &= H_m(K+v+2k\eta)\tau_{\beta,\ k}\\
S_R^{(1)}(+,\beta)_{1,1}(v) &= H_m(K-v)\tau_{\beta,0} = H_m(K+v)\tau_{\beta,0}\\
S_R^{(1)}(+,\beta)_{L,L}(v) &= H_m(K-v+2L\eta)\tau_{\beta,L} = H_m(K+v-2L\eta)\tau_{\beta,L}\\
S_R^{(1)}(-,\beta)_{k,k+1}(v) &= \Theta_m(K-v+2k\eta)\tau_{\beta,-k} = \Theta_m(K+v-2k\eta)\tau_{\beta,-k}\\
S_R^{(1)}(-,\beta)_{k+1,k}(v) &= \Theta_m(K+v+2k\eta)\tau_{\beta,\ k}\\
S_R^{(1)}(-,\beta)_{1,1}(v) &= \Theta_m(K-v)\tau_{\beta,0} = \Theta_m(K+v)\tau_{\beta,0}\\
S_R^{(1)}(-,\beta)_{L,L}(v) &= \Theta_m(K-v+2L\eta)\tau_{\beta,L} = \Theta_m(K+v-2L\eta)\tau_{\beta,L}
\end{aligned} \tag{14.97}$$

where in the cases where two expressions are given, the identity (14.505) has been used.

We now use (14.73) to compute

$$\begin{aligned}
A^{(1)}(\alpha,\beta)_{1,1}(v) &= h(v-\eta)S_R^{(1)}(\alpha,\beta)_{1,1}(v+2\eta)\\
A^{(1)}(\alpha,\beta)_{L,L}(v) &= h(v-\eta)S_R^{(1)}(\alpha,\beta)_{L,L}(v+2\eta)\\
B^{(1)}(\alpha,\beta)_{1,1}(v) &= h(v+\eta)S_R^{(1)}(\alpha,\beta)_{1,1}(v-2\eta)\\
B^{(1)}(\alpha,\beta)_{L,L}(v) &= h(v+\eta)S_R^{(1)}(\alpha,\beta)_{L,L}(v-2\eta).
\end{aligned} \tag{14.98}$$

Therefore the diagonal similarity transformation (14.78) demonstrates that $Q_R^{(1)}(v)$ defined from (14.97) satisfies (14.33).

The matrix $Q_L^{(1)}(v)$ is constructed by using (14.97) in (14.39) to find for $1\le k\le L-1$

$$\begin{aligned}
S_L^{(1)}(\alpha,+)_{k,k+1}(v) &= H_m(K+v+2k\eta)\tau'_{\alpha,-k}\\
S_L^{(1)}(\alpha,+)_{k+1,k}(v) &= H_m(K-v+2k\eta)\tau'_{\alpha,\ k} = H_m(K+v-2k\eta)\tau'_{\alpha,\ k}\\
S_L^{(1)}(\alpha,+)_{1,1}(v) &= H_m(K+v)\tau'_{\alpha,0}\\
S_L^{(1)}(\alpha,+)_{L,L}(v) &= H_m(K+v+2L\eta)\tau'_{\alpha,L}\\
S_L^{(1)}(\alpha,-)_{k,k+1}(v) &= \Theta_m(K+v+2k\eta)\tau'_{\alpha,-k}\\
S_L^{(1)}(\alpha,-)_{k+1,k}(v) &= \Theta_m(K-v+2k\eta)\tau'_{\alpha,\ k} = \Theta_m(K+v-2k\eta)\tau'_{\alpha,\ k}\\
S_L^{(1)}(\alpha,-)_{1,1}(v) &= \Theta_m(K+v)\tau'_{\alpha,0}\\
S_L^{(1)}(\alpha,-)_{L,L}(v) &= \Theta_m(K+v+2L\eta)\tau'_{\alpha,L}
\end{aligned} \tag{14.99}$$

The matrices $Q_R^{(2)}(v;t)$ and $Q_L^{(2)}(v;t)$ for N even

To complete the construction of $Q_R^{(2)}(v;t)$ we follow the construction found in [44] which sets

$$S_R(\alpha,\beta)_{1,L}\neq 0, \quad S_R(\alpha,\beta)_{L,1}\neq 0. \tag{14.100}$$

This choice will always give a vanishing matrix $Q_R^{(2)}(v;t)$ when used in (14.32) when N is odd. Consequently whenever we consider $Q_R^{(2)}(v;t)$ we will always assume that N is even.

To obtain this case we need to have (14.58) hold $m=1, n=L$ and $m=L, n=1$ which because of the symmetry in (14.58) in m and n gives the single equation

$$\begin{aligned}
&\mathrm{sn}^2(v_r+2\eta)+\mathrm{sn}^2(v_r+2L\eta)-\mathrm{sn}^2 2\eta(1+k^2\mathrm{sn}^2(v_r+2\eta)\mathrm{sn}^2(v_r+2L\eta))\\
&-2\mathrm{sn}(v_r+2\eta)\mathrm{sn}(v_r+2L\eta)\mathrm{cn}2\eta\mathrm{dn}2\eta=0.
\end{aligned} \tag{14.101}$$

This equation will hold if p_n is given by (14.66) with $p_1=p_{L+1}$ and thus

$$\mathrm{sn}(v_r+2\eta)=\mathrm{sn}(v_r+2(L+1)\eta) \tag{14.102}$$

which, using the periodicity properties $\mathrm{sn}(v+2K)=-\mathrm{sn}v$ and $\mathrm{sn}(v+2iK')=\mathrm{sn}v$, is satisfied for all v if

$$2L\eta=4\tilde{m}_1K+2i\tilde{m}_2K' \tag{14.103}$$

which is the root of unity condition (14.1) with $m_1=2\tilde{m}_1,\ m_2=2\tilde{m}_2$. In other words we are restricted to m_1 and m_2 even in the root of unity condition (14.1).

We set

$$v_r=t-\eta \tag{14.104}$$

in (14.66) so that

$$p_n=k^{1/2}\mathrm{sn}[t+(2n-1)\eta]=H_m(t+(2n-1)\eta)/\Theta_m(t+(2n-1)\eta). \tag{14.105}$$

Then noting the relation for m_1 and m_2 even (14.512)

$$H_m(v+2L\eta)=H_m(v) \tag{14.106}$$

we compute $S_R^{(2)}(\alpha,\beta)(v)_{1,L}$ and $S_R^{(2)}(\alpha,\beta)(v)_{L,1}$ from (14.52) using

$$ap_1-bp_L=\Theta_m(0)\Theta_m(-2\eta)H_m(2\eta)\frac{\Theta_m(t-v)H_m(t+v)}{\Theta_m(t+\eta)\Theta_m(t-\eta)} \tag{14.107}$$

$$ap_L - bp_1 = \Theta_m(0)\Theta_m(-2\eta)H(2\eta)\frac{\Theta_m(t+v)H_m(t-v)}{\Theta_m(t+\eta)\Theta_m(t-\eta)} \tag{14.108}$$

$$d - cp_1p_L = \Theta_m(0)\Theta_m(2\eta)H_m(-2\eta)\frac{H_m(t-v)H_m(t+v)}{\Theta_m(t+\eta)\theta_m(t-\eta)} \tag{14.109}$$

and the remaining elements from (14.72) to find

$$\begin{aligned}
S_R^{(2)}(+,\beta)_{k,k+1}(v) &= -H_m(v-t-2k\eta)\tau_{\beta,-k}\\
S_R^{(2)}(+,\beta)_{k+1,k}(v) &= \quad H_m(v+t+2k\eta)\tau_{\beta,\ k}\\
S_R^{(2)}(-,\beta)_{k,k+1}(v) &= \quad \Theta_m(v-t-2k\eta)\tau_{\beta,-k}\\
S_R^{(2)}(-,\beta)_{k+1,k}(v) &= \quad \Theta_m(v+t+2k\eta)\tau_{\beta,\ k}\\
S_R^{(2)}(+,\beta)_{1,L}(v) &= \quad H_m(v+t)\tau_{\beta,\ L}\\
S_R^{(2)}(+,\beta)_{L,1}(v) &= -H_m(v-t)\tau_{\beta,-L}\\
S_R^{(2)}(-,\beta)_{1,L}(v) &= \quad \Theta_m(v+t)\tau_{\beta,L}\\
S_R^{(2)}(-,\beta)_{L,1}(v) &= \quad \Theta_m(v-t)\tau_{\beta,-L}.
\end{aligned} \tag{14.110}$$

We now compute from (14.73)

$$\begin{aligned}
A^{(2)}(\alpha,\beta)_{1,L}(v) &= h(v-\eta)\frac{\Theta_m(t-\eta)}{\Theta_m(t+\eta)}S_R^{(2)}(\alpha,\beta)_{1,L}(v+2\eta)\\
A^{(2)}(\alpha,\beta)_{L,1}(v) &= h(v-\eta)e^{i\pi m_2/L}\frac{\Theta_m(t+\eta)}{\Theta_m(t-\eta)}S_R^{(2)}(\alpha,\beta)_{L,1}(v+2\eta)\\
B^{(2)}(\alpha,\beta)_{1,L}(v) &= h(v+\eta)\frac{\Theta_m(t+\eta)}{\Theta_m(t-\eta)}S_R^{(2)}(\alpha,\beta)_{1,L}(v-2\eta)\\
B^{(2)}(\alpha,\beta)_{L,1}(v) &= h(v+2\eta)e^{-i\pi m_2/L}\frac{\Theta_m(t-\eta)}{\Theta_m(t+\eta)}S_R^{(2)}(\alpha,\beta)_{L,1}(v-2\eta).
\end{aligned} \tag{14.111}$$

Thus using the similarity transformation (14.80) we find that (14.81) is satisfied as desired.

The companion matrix $S_L^{(2)}(\alpha,\beta)(v)$ is found from (14.110) by use of (14.86) and (14.106) as

$$\begin{aligned}
S_L^{(2)}(\alpha,+)_{k,k+1}(v) &= \quad H_m(v+t+2k\eta)\tau'_{\alpha,-k}\\
S_L^{(2)}(\alpha,+)_{k+1,k}(v) &= -H_m(v-t-2k\eta)\tau'_{\alpha,\ k}\\
S_L^{(2)}(\alpha,-)_{k,k+1}(v) &= \quad \Theta_m(v+t+2k\eta)\tau'_{\alpha,-k}\\
S_L^{(2)}(\alpha,-)_{k+1,k}(v) &= \quad \Theta_m(v-t-2k\eta)\tau'_{\alpha,\ k}\\
S_L^{(2)}(\alpha,+)_{1,L}(v) &= -H_m(v-t)\tau'_{\alpha,L}\\
S_L^{(2)}(\alpha,+)_{L,1}(v) &= \quad H_m(v+t)\tau'_{\alpha,-L}\\
S_L^{(2)}(\alpha,-)_{1,L}(v) &= \quad \Theta_m(v-t)\tau'_{\alpha,L}\\
S_L^{(2)}(\alpha,-)_{L,1}(v) &= \quad \Theta_m(v+t)\tau'_{\alpha,-L}.
\end{aligned} \tag{14.112}$$

14.2.4 The interchange relation

We next examine the interchange relation (14.40) for the four operators A of (14.41) for the pairs $Q_R^{(1)}(v)$, $Q_L^{(1)}(v)$ and $Q_R^{(2)}(v;t)$, $Q_L^{(2)}(v;t)$. There are thus eight separate cases to consider. Furthermore, we will see that for the pair $Q_R^{(2)}(v;t)$, $Q_L^{(2)}(v;t)$ there are two values of t for which the interchange relation holds and the properties of the two cases are different. All twelve computations are carried out by similar methods and are treated in detail in [46]. Here we will present the computation in detail for the cases of the pairs $Q_R^{(1)}(v)$, $Q_L^{(1)}(v)$ and $Q_R^{(2)}(v;t)$, $Q_L^{(2)}(v;t)$ for $A = I$ in order to present the method, but the other cases will be treated in a summary fashion.

Interchange for $Q_R^{(1)}(v)$ and $Q_L^{(1)}(v)$

We consider first the interchange relation (14.40) for $Q_R^{(1)}(v)$, $Q_L^{(1)}(v)$ with $A = I$ and write

$$Q_L^{(1)}(v')Q_R^{(1)}(v)|_{\alpha,\beta} = \mathrm{Tr}W^{(1)}(\alpha_1,\beta_1|v',v)\cdots W^{(1)}(\alpha_N,\beta_N|v',v) \qquad (14.113)$$

where $W^{(1)}(\alpha,\beta|v',v)$ are $L^2 \times L^2$ matrices with elements

$$W^{(1)}(\alpha,\beta|v',v)_{k,k';l,l'} = \sum_{\gamma=\pm} S_L^{(1)}(\alpha,\gamma|v')_{k,l}S_R^{(1)}(\gamma,\beta|v)_{k',l'}. \qquad (14.114)$$

Thus the interchange relation (14.40) with $A = I$ will follow if we can show that there exists an $L^2 \times L^2$ diagonal matrix Y with elements

$$y^{(1)}_{k,k';l,l'} = y^{(1)}_{k,k'}\delta_{k,l}\delta_{k',k'} \qquad (14.115)$$

such that

$$W^{(1)}(\alpha,\beta|v',v) = Y^{(1)}W^{(1)}(\alpha,\beta|v,v')Y^{(1)-1}. \qquad (14.116)$$

To examine the possibility of the existence of such a diagonal similarity transformation we need to explicitly compute $W(\alpha,\beta|v',v)$ from (14.114). To do this we use the identity

$$\Theta_m(z')\Theta_m(z) + H_m(z')H_m(z) = f_+(z+z')g_+(z'-z) \qquad (14.117)$$

with

$$f_+(z) = N(z)g_-(z) \qquad (14.118)$$

$$N(z) = -\frac{2q^{1/4}}{H(K)\Theta(K)}\exp\left(\frac{i\pi m_2}{8KL\eta}(K'^2 + 2iKK' - 2Kz)\right) \qquad (14.119)$$

$$g_-(z) = H_m((iK'+z)/2)H_m((iK'-z)/2) \qquad (14.120)$$

$$g_+(z) = H_m((iK'+z)/2+K)H_m((iK'-z)/2+K) \qquad (14.121)$$

where we note the following properties

$$g_{\pm}(-z) = g_{\pm}(z) \tag{14.122}$$
$$g_{+}(z+4L\eta) = (-1)^{m_1 m_2} g_{+}(z) \tag{14.123}$$
$$g_{-}(z+4L\eta) = (-1)^{m_1 m_2}(-1)^{m_2} g_{-}(z) \tag{14.124}$$

and for m_1 and m_2 both even

$$g_{+}(z+2L\eta) = (-1)^{m_1 m_2/4} g_{+}(z) \tag{14.125}$$
$$g_{-}(v+2L\eta) = (-1)^{m_1 m_2/4}(-1)^{m_2/2} g_{-}(z). \tag{14.126}$$

The properties in (14.122) are obvious from the definitions (14.121) and (14.121). The relations (14.123) and (14.123) follow from (14.506), (14.507), (14.508) and (14.510).

Using (14.117) we find explicitly by using the expressions furthest to the right in (14.97) and (14.99)

$$\begin{aligned}
&W^{(1)}(\alpha,\beta|v',v)_{k,k';l,l'} \\
&= \delta_{k+1,l}\delta_{k'+1,l'}\tau_{\alpha,-k}\tau_{\beta,-k'}f_{-}(v'+v+2K+2(k-k')\eta)g_{+}(v'-v+2(k+k')\eta) \\
&+\delta_{k+1,l}\delta_{k',l'+1}\tau'_{\alpha,-k}\tau_{\beta,l'}f_{-}(v'+v+2K+2(k+l')\eta)g_{+}(v'-v+2(k-l')\eta) \\
&+\delta_{k+1,l}\delta_{k',1'}\delta_{l',1}\tau'_{\alpha,-k}\tau_{\beta,0}f_{-}(v'+v+2K+2k\eta)g_{+}(v'-v+2k\eta) \\
&+\delta_{k+1,l}\delta_{k',L}\delta_{l',L}\tau'_{\alpha,-k}\tau_{\beta,L}f_{-}(v'+v+2K+2(k+L)\eta)g_{+}(v'-v+2(k-L)\eta) \\
&+\delta_{k,l+1}\delta_{k'+1,l'}\tau'_{\alpha,l}\tau_{\beta,-k'}f_{-}(v'+v+2K-2(l+k')\eta)g_{+}(v'-v-2(l-k')\eta) \\
&+\delta_{k,l+1}\delta_{k',l'+1}\tau'_{\alpha,l}\tau_{\beta,l'}f_{-}(v'+v+2K-2(l-l')\eta)g_{+}(v'-v-2(l+l')\eta) \\
&+\delta_{k,l+1}\delta_{k',1'}\delta_{l',1}\tau'_{\alpha,l}\tau_{\beta,0}f_{-}(v'+v+2K-2l\eta)g_{+}(v'-v-2l\eta) \\
&+\delta_{k,l+1}\delta_{k',L}\delta_{l',L}\tau'_{\alpha,l}\tau_{\beta,L}f_{-}(v'+v+2K-2(l-L)\eta)g_{+}(v'-v-2(l+L)\eta) \\
&+\delta_{k,1}\delta_{l,1}\delta_{k'+1,l'}\tau'_{\alpha,0}\tau_{\beta,-k'}f_{-}(v'+v+2K-2k'\eta)g_{+}(v'-v+2k'\eta) \\
&+\delta_{k,1}\delta_{l,1}\delta_{k',l'+1}\tau'_{\alpha,0}\tau_{\beta,l'}f_{-}(v'+v+2K+2l'\eta)g_{+}(v'-v-2l'\eta) \\
&+\delta_{k,L}\delta_{l,L}\delta_{k'+1,l'}\tau'_{\alpha,L}\tau_{\beta,-k'}f_{-}(v'+v+2K-2(L+k')\eta)g_{+}(v'-v-2(L-k')\eta) \\
&+\delta_{k,L}\delta_{l,L}\delta_{k',l'+1}\tau'_{\alpha,L}\tau_{\beta,l'}f_{-}(v'+v+2K-2(L-l')\eta)g_{+}(v'-v-2(L+l')\eta) \\
&+\delta_{k,1}\delta_{l,1}\delta_{k',1'}\delta_{l',1}\tau'_{\alpha,0}\tau_{\beta,0}f_{-}(v'+v+2K)g_{+}(v'-v) \\
&+\delta_{k,1}\delta_{l,1}\delta_{k',L}\delta_{l',L}\tau'_{\alpha,0}\tau_{\beta,L}f_{-}(v'+v+2K+2L\eta)g_{+}(v'-v-2L\eta) \\
&+\delta_{k,L}\delta_{l,L}\delta_{k',1'}\delta_{l',1}\tau'_{\alpha,L}\tau_{\beta,0}f_{-}(v'+v+2K-2L\eta)g_{+}(v'-v-2L\eta) \\
&+\delta_{k,L}\delta_{l,L}\delta_{k',L}\delta_{l',L}\tau'_{\alpha,L}\tau_{\beta,L}f_{-}(v'+v+2K)g_{+}(v'-v-4L\eta).
\end{aligned} \tag{14.127}$$

A necessary condition for the existence of a diagonal similarity transformation is that the diagonal elements $W^{(1)}(\alpha,\beta|v',v)_{k,k';k,k'}$ and $W^{(1)}(\alpha,\beta|v,v')_{k.k':k,k'}$ be equal. From the last four terms in (14.127) we find that these diagonal elements are

$$W^{(1)}(\alpha,\beta|v',v)_{11,11} = f_{-}(v'+v+2K)g_{+}(v'-v)\tau'_{\alpha,0}\tau_{\beta,0} \tag{14.128}$$

$$\begin{aligned}
&W^{(1)}(\alpha,\beta|v',v)_{1,L;1,L} \\
&\quad = f_{-}(v'+v+2K+2L\eta)g_{+}(v'-v-2L\eta)\tau'_{\alpha,0}\tau_{\beta,L}
\end{aligned} \tag{14.129}$$

$$W^{(1)}(\alpha,\beta|v',v)_{L,1;L,1}$$
$$= f_-(v'+v+2K-2L\eta)g_+(v'-v-2L\eta)\tau'_{\alpha,L}\tau_{\beta,0} \tag{14.130}$$
$$W(\alpha,\beta|v',v)_{L,L;L,L} = f_-(v'+v+2K)g_+(v'-v-4L\eta)\tau'_{\alpha,L}\tau_{\beta,L}. \tag{14.131}$$

It follows from (14.128) and (14.131) by use of (14.123) that

$$W^{(1)}(\alpha,\beta|v',v)_{1,1;1,1} = W^{(1)}(\alpha,\beta|v,v')_{1.1;1,1} \tag{14.132}$$
$$W^{(1)}(\alpha,\beta|v',v)_{L,L;L,L} = W^{(1)}(\alpha,\beta|v,v')_{L,L;L,L}. \tag{14.133}$$

To examine the elements $W^{(1)}(\alpha,\beta|v',v)_{1,L;1,L}$ and $W^{(1)}(\alpha,\beta|v',v)_{L,1;L,1}$ we use the identity (14.123) in (14.129) and (14.130) to find

$$W^{(1)}(\alpha,\beta|v',v)_{1,L;1,L} = (-1)^{m_1m_2}W^{(1)}(\alpha,\beta|v,v')_{1,L;1,L} \tag{14.134}$$
$$W^{(1)}(\alpha,\beta|v',v)_{L,1;L,1} = (-1)^{m_1m_2}W^{(1)}(\alpha,\beta|v,v')_{L,1;L,1}. \tag{14.135}$$

We thus conclude that the interchange relation (14.40) is not satisfied if m_1 and m_2 are both odd.

To proceed further we note that the arguments of $f_-(z)$ in (14.127) are all symmetric in v and v'. Thus by writing (14.116) in component form

$$W^{(1)}(\alpha,\beta|v',v)_{k,k';l,l'} = y^{(1)}_{k,k'}W^{(1)}(\alpha,\beta|v',v)_{k,k';l,l'}/y^{(1)}_{l,l'} \tag{14.136}$$

we find from the terms in (14.127) with $l=k+1,\ l'=k'+1$ and with $k=l+1,\ k'=l'+1$ that $y^{(1)}_{k,k'}$ must satisfy the two equations

$$g_+(v'-v'+2(k+k')\eta) = \frac{y^{(1)}_{k,k'}}{y^{(1)}_{k+1,k'+1}}g_+(v-v'+2(k+k')\eta) \tag{14.137}$$

$$g_+(v'-v-2(l+l')\eta) = \frac{y^{(1)}_{l+1,l'+1}}{y^{(1)}_{l,l'}}g(v-v'-2(l+l')\eta) \tag{14.138}$$

which, using (14.123) reduces to the single relation

$$y^{(1)}_{k+1,k'+1} = y^{(1)}_{k,k'}\frac{g_+(v'-v-2(k+k')\eta)}{g_+(v'-v+2(k+k')\eta)}. \tag{14.139}$$

Similarly using the elements of (14.127) with $l=k+1,\ k'=l'+1$ and with $k=l+1,\ l'=k'+1$ we find by use of (14.123) the single equation

$$g_+(v'-v+2(k-l')\eta) = \frac{y^{(1)}_{k,l'+1}}{y^{(1)}_{k+1,l'}}g_+(v-v'+2(k-l')\eta) \tag{14.140}$$

from which by sending $l'\to k'-1$ we find the second relation

$$y^{(1)}_{k+1,k'-1} = y^{(1)}_{k,k'}\frac{g_+(v'-v-2(k-k'+1)\eta)}{g_+(v'-v+2(k-k'+1)\eta)}. \tag{14.141}$$

From (14.139) and (14.141) it follows that

$$y^{(1)}_{k,k'} = t_{k+k'}t_{k-k'+1} \tag{14.142}$$

where

$$t_{k+2} = t_k\frac{g_+(v'-v-2k\eta)}{g_+(v'-v+2k\eta)} \tag{14.143}$$

from which we see that t_{2m} and t_{2m+1} are independent.

It remains to consider the elements of (14.127) in the four cases where either $k=l=1$, $k'=l'=1$, $k=l=L$ or $k'=l'=L$ which lead to the four recursion relations

$$y^{(1)}_{1,k'+1} = y^{(1)}_{1,k'}\frac{g_+(v'-v-2k'\eta)}{g_+(v'-v+2k'\eta)} \tag{14.144}$$

$$y^{(1)}_{k+1,1} = y^{(1)}_{k,1}\frac{g_+(v'-v-2k\eta)}{g_+(v'-v+2k\eta)} \tag{14.145}$$

$$y^{(1)}_{L,k'+1} = y^{(1)}_{L,k'}\frac{g_+(v'-v-2(k'-L)\eta)}{g_+(v'-v+2(k'-L)\eta)} \tag{14.146}$$

$$y^{(1)}_{k+1,L} = y^{(1)}_{L,k}\frac{g_+(v'-v-2(k-L)\eta)}{g_+(v'-v+2(k-L)\eta)} \tag{14.147}$$

where to obtain (14.146) and (14.147) we have used (14.123) which is only valid in the cases m_1m_2 is even. We must show that these four recursion relations are consistent with (14.139) and (14.141).

Consider first (14.144) which is rewritten in terms of $t_{k'}$ by use of (14.142) as

$$t_{2+k'}t_{1-k'} = t_{2+k'}t_{1+k'}\frac{g_+(v'-v-2k'\eta)}{g_+(v'-v+2k'\eta)}. \tag{14.148}$$

However, from (14.143) it follows that

$$t_{1+k'} = t_{1-k'} \tag{14.149}$$

$$t_{2+k'} = t_{2-k'}\frac{g_+(v'-v-2k'\eta)}{g_+(v'-v+2k'\eta)} \tag{14.150}$$

and thus (14.148) follows from (14.139) and (14.141). The remaining relations (14.145)-(14.147) follow from (14.139) and (14.141) in a similar fashion. Thus we have established that when m_1m_2 is even the similarity transformation (14.116) exists and therefore the interchange relation (14.40) with $A=I$ is established.

The three other interchange relations with $A=S,R$ and RS are all studied in a similar fashion with the following replacements for (14.117)–(14.125)

The case $A = S$

The case $A = S$ is studied by sending $z \to z'$ in (14.121) and using (14.506) to find

$$H_m(z')H_m(z) - \Theta_m(z')\Theta_m(z) = f_+(z'+z)g_-(z'-z) \tag{14.151}$$

with

$$f_+(z) = N(z)g_+(z) \tag{14.152}$$

where $N(z)$ is given by (14.119) and $g_+(z)$ by (14.121) and the proof just given for $A = I$ holds with the interchange of $g_+(z)$ with $g_-(z)$. In particular the the property (14.124) is used in place of (14.123) and with these relations the proof given above shows that the interchange relation (14.40) holds for m_1 and m_2 both odd or m_2 even.

The case $A = R$

In this case we use the identity

$$\Theta_m(z')H_m(z) + H_m(z')\Theta_m(z) = N^R(z'+z)g_-^R(z'+z)g_+^R(z'-z) \tag{14.153}$$

with

$$N^R(z) = 2\frac{e^{-i\pi m_2(z'+z)/(4L\eta)}}{H_m(K)\Theta_m(K)} \tag{14.154}$$

$$g_-^R(z) = H_m(z/2)\Theta_m(-z/2) \tag{14.155}$$

$$g_+^R(z) = H_m(K+z/2)\Theta_m(K+z/2) \tag{14.156}$$

where we note the following properties

$$g_-^R(-z) = -g_-^R(z) \tag{14.157}$$

$$g_+^R(-z) = g_+^R(z) \tag{14.158}$$

$$g_+^R(z+4L\eta) = (-1)^{m_1 m_2}(-1)^{m_1}g_+^R(z) \tag{14.159}$$

$$g_-^R(z+4L\eta) = (-1)^{m_1 m_2}(-1)^{m_1+m_2}g_+^R(z) \tag{14.160}$$

and for m_1 and m_2 both even

$$g_+^R(z+2L\eta) = (-1)^{m_1 m_2/4}(-1)^{m_1/2}g_+^R(z) \tag{14.161}$$

$$g_-^R(z+2L\eta) = (-1)^{m_1 m_2/4}(-1)^{(m_1+m_2)/2}g_-^R(z). \tag{14.162}$$

Using (14.157) and (14.158) and (14.159) the proof given above shows that the interchange relation (14.40) holds for m_1 and m_2 both odd or m_1 even.

The case $A = RS$

For the final case of $A = RS$ we send $z \to -z$ in (14.153) and use (14.506) to find

$$\Theta_m(z')H_m(z) - H_m(z')\Theta_m(z) = -N^R(z'+z)g_+^R(z'+z)g_-^R(z'-z). \tag{14.163}$$

With (14.163) and the relations (14.157) and (14.160) the proof given above shows that the interchange relation (14.40) holds only for N even and both m_1 and m_2 even.

Summary

The results obtained above for the validity of the interchange relation (14.40) for $A = I, S, R$ and RS are summarized in the table 14.3 where Y (N) indicates that the relation holds (fails).

Table 14.3 Summary of the values of the matrix A for which the interchange relation (14.40) holds for the matrices $Q_L^{(1)}(v)$ and $Q_R^{(1)}(v)$.

m_1	m_2	I	S	R	RS
o	e	Y	Y	N	N
o	o	N	Y	Y	N
e	o	Y	N	Y	N
e	e	Y	Y	Y	Y

Interchange for $Q_R^{(2)}(v;t)$ and $Q_L^{(2)}(v;t)$

The matrices $Q_R^{(2)}(v;t)$ and $Q_L^{(2)}(v;t)$, in contrast to the matrices $Q_R^{(1)}(v)$ and $Q_L^{(1)}(v)$ depend on a free parameter t. However, we will find that the interchange relations (14.40) will be satisfied only in the two cases $t = n\eta$ and $t = (n+1/2)\eta$ where n is integer. As in the previous subsection we treat the case $A = I$ in detail and only sketch the modifications needed in the other three cases.

We now write

$$Q_L^{(2)}(v';t)Q_R^{(2)}(v;t)|_{\alpha,\beta} = \mathrm{Tr}W^{(2)}(\alpha_1,\beta_1|v',v;t)\cdots W^{(2)}(\alpha_N,\beta_N|v',v;t) \tag{14.164}$$

where $W^{(2)}(\alpha,\beta|v',v;t)$ are $L^2 \times L^2$ matrices with elements

$$W^{(2)}(\alpha,\beta|v',v;t)_{k,k';l,l'} = \sum_{\gamma=\pm} S_L^{(2)}(\alpha,\gamma|v';t)_{k,l}S_R^{(2)}(\gamma,\beta|v;t)_{k',l'} \tag{14.165}$$

and use (14.117) with (14.110) and (14.112) to find in analogue to (14.127)

$$\begin{aligned}
W^{(2)}_{k,k';k+1,k'+1}(\alpha,\beta|v',v;t) &= f_-(v'+v+2(k-k')\eta)g_+(v'-v+2t+2(k+k')\eta)\\
W^{(2)}_{k+1,k'+1;k,k'}(\alpha,\beta|v',v;t) &= f_-(v'+v-2(k-k')\eta)g_+(v'-v+2t-2(k-k')\eta)\\
W^{(2)}_{k,k'+1;k+1,k'}(\alpha,\beta|v',v;t) &= -f_+(v+v'+2t+2(k+k')\eta)g_-(v'-v+2(k-k')\eta)\\
W^{(2)}_{k+1,k';k,k'+1}(\alpha,\beta|v',v;t) &= -f_+(v'+v-2t-2(k+k')\eta)g_-(v'-v-2(k-k')\eta)
\end{aligned} \tag{14.166}$$

where $f_+(z)$ is defined by (14.152) and we have the identification $k = L+1 \equiv 1$.

As in the case of $W^{(1)}$ we seek a diagonal similarity transformation of the form (14.115) and (14.116) and find from the four equations in (14.166) that $y^{(2)}_{k,k'}$ must satisfy

$$\begin{aligned}
&g_+(v'-v+2t+2(k+k')\eta)\\
&= \frac{y^{(2)}_{k,k'}}{y^{(2)}_{k+1,k'+1}} g_+(2t+v-v'+2(k+k')\eta)
\end{aligned} \tag{14.167}$$

$$g_-(v'-v+2(k-k')\eta)=\frac{y^{(2)}_{k,k'+1}}{y^{(2)}_{k+1,k'}}g_-(v-v'+2(k-k')\eta) \tag{14.168}$$

or equivalently using the relation $g_\pm(-z)=g_\pm(z)$

$$y^{(2)}_{k+1,k'+1}=y^{(2)}_{k,k'}\frac{g_+(v'-v-2(k+k')\eta-2t)}{g_+(v'-v+2(k+k')\eta+2t)} \tag{14.169}$$

$$y^{(2)}_{k+1,k'-1}=y^{(2)}_{k,k'}\frac{g_-(v'-v-2(k-k'+1)\eta)}{g_-(v'-v+2(k-k'+1)\eta)}. \tag{14.170}$$

For the recursions (14.169) and (14.170) to be consistent with the requirement that (14.166) hold for $k=L+1=1$ we need the constraints

$$y_{k+L,k'+L}=y_{k,k'} \tag{14.171}$$

$$y_{k-L,k'+L}=y_{k,k'}. \tag{14.172}$$

Thus by using (14.169) in (14.171) and (14.170) in (14.172) we find that we must satisfy

$$1=\prod_{j=1}^{L}\frac{g_+(v'-v-2(k+k')\eta-4(L-j)\eta-2t)}{g_+(v'-v+2(k+k')\eta+4(L-j)\eta+2t)} \tag{14.173}$$

$$1=\prod_{j=1}^{L}\frac{g_-(v'-v-2(k-k')\eta-4(L-j)\eta)}{g_-(v'-v+2(k+k')\eta+4(L-j)\eta)}. \tag{14.174}$$

These relations will hold if the factors $g_+(v'-v-2(k+k')\eta-4c_1\eta-2t)$ and $g_-(v'-v-2(k+k')\eta-4c_1\eta)$ in the numerator cancel the factors $g_+(v'-v+2(k+k')\eta+4c_2\eta+2t)$ and $g_-(v'-v+2(k+k')\eta+4c_2\eta)$ in the denominator. For this we need the periodicity properties of $g_\pm(v)$.

From the definitions (14.121) and (14.121) of $g_\pm(v)$ and the periodicity of $H_m(v)$ (14.22) it follows that

$$g_\pm(v+4(r_1K+r_2K')\eta)=g_\pm(v) \tag{14.175}$$

Furthermore by use of the definitions (14.1) and (14.27) in (14.125) and (14.126) we find for m_1 and m_2 even

$$g_+(v+r_0(r_1K+ir_2K'))=(-1)^{m_1m_2/4}g_+(v) \tag{14.176}$$

$$g_-(v+r_0(r_1K+ir_2K'))=(-1)^{m_1m_2/4}(-1)^{m_2/2}g_{-1}(v). \tag{14.177}$$

When $m_2\equiv 2$ (mod4) we see from (14.27) that $r_0\equiv 2$ (mod4) and thus it follows from (14.175)–(14.177) that we have the additional periodicity conditions

$$g_+(v+2(r_1K+ir_2K'))=(-1)^{m_1/2}g_+(v) \tag{14.178}$$

$$g_-(v+2(r_1K+ir_2K'))=-(-1)^{m_1/2}g_-(v). \qquad (14.179)$$

Consider first the consistency condition (14.173) which depends on t. The periodicity condition (14.175) will provide the needed cancellations if an integer I can be found such that

$$-2(k+k')\eta-4c_1\eta-2t=2(k+k')\eta+4c_2\eta+2t+4I(r_1K+ir_2K') \qquad (14.180)$$

which by multiplying by L and defining $\bar{t}$ by $t=\bar{t}\eta$ becomes

$$-(k+k')r_0-c_1-\bar{t}r_0=c-2r_0=2LI. \qquad (14.181)$$

The equation (14.181) can be satisfied in two different ways. First we note that because m_2 is even it follows from (14.27) that r_0 is even. Thus (14.181) can always be satisfied in integers for $\bar{t}=n$ because c_2 can be shifted into the interval $0\leq c_2<L$. Second we note that if $m_2\equiv 0$ (mod4) then for even m_1 we have $r_0\equiv 0$ (mod4) and thus (14.181) may be satisfied in integers for $\bar{t}=n+1/2$.

We next note that the relation (14.178) which is valid for $m_2\equiv 2$ (mod)4 and $r_0\equiv 2$ (mod4) with the additional restriction $m_1\equiv 0$ (mod4) specializes to

$$g_+(v+2(r_1K+ir_2K'))=g_+(v) \qquad (14.182)$$

and this will give cancellation in (14.173) if instead of (14.181) we have

$$-(k+k')r_0-c_1-\bar{t}r_0=c_2+LI. \qquad (14.183)$$

Using the fact that $r_0\equiv 2$ (mod4) we see that (14.183) is satisfied in integers for $\bar{t}=n+1/2$.

It remains to demonstrate that (14.174) holds. This is easily done by noting that (14.174) is identical to (14.173) with $g_+(v)$ replaced by $g_-(v)$ and t set equal to zero. But the proof of (14.173) with $t=0$ depended only on the identity (14.175) which holds both for $g_+(v)$ and $g_-(v)$ and thus the validity of (14.174) follows identically from the proof of (14.173) with $t=0$.

In summary we have proven that the interchange relation (14.40) with $A=I$ holds for the cases

$$t=n\eta \;\text{ for }\; m_1,m_2 \text{ even} \qquad (14.184)$$

$$t=(n+1/2)\eta \;\text{ for }\; m_2\equiv 0 \text{ (mod4)},\;\; m_1 \text{ even}$$
$$m_2\equiv 2 \text{ (mod4)},\;\; m_1\equiv 0 \text{ (mod4)}. \qquad (14.185)$$

The case $A=S$

The proof of the interchange relation for $A=S$ follows from the proof for $A=I$ with the replacements of $f_\pm(v)$ by $f_\mp(v)$ and $g_\pm(v)$ by $g_\mp(v)$ everywhere. Thus we find that for $A=S$ the relation (14.40) holds for

$$t=n\eta \;\text{ for }\; m_1,m_2 \text{ even} \qquad (14.186)$$

$$t=(n+1/2)\eta \;\text{ for }\; m_2\equiv 0 \text{ (mod4)},\;\; m_1 \text{ even}$$
$$m_2\equiv 2 \text{ (mod4)},\;\; m_1\equiv 2 \text{ (mod4)}. \qquad (14.187)$$

The case $A = RS$

To investigate (14.40) with $A = RS$ we explicitly write

$$W^{(2)RS}(\alpha,\beta|v',v)_{k,k';l,l'} = \sum_{\gamma=\pm} S_L^{(2)}(\alpha,\gamma|v')_{k,l}(\gamma)S_R^{(2)}(-\gamma,\beta|v)_{k',l'} \tag{14.188}$$

using (14.153) and (14.163) as

$$\begin{aligned}
W^{(2)R}_{k,k';k+1,k'+1}(\alpha,\beta|v',v;t) &= f_+^R(v'+v+2(k-k')\eta)g_-^R(v'-v+2t+2(k+k')\eta)\\
W^{(2)R}_{k+1,k'+1;k,k'}(\alpha,\beta|v',v;t) &= f_+^R(v'+v-2(k-k')\eta)g_-^R(v'-v+2t-2(k-k')\eta)\\
W^{(2)R}_{k,k'+1;k+1,k'}(\alpha,\beta|v',v;t) &= -f_-^R(v+v'+2t+2(k+k')\eta)g_+^R(v'-v+2(k-k')\eta)\\
W^{(2)R}_{k+1,k';k,k'+1}(\alpha,\beta|v',v;t) &= -f_-^R(v'+v-2t-2(k+k')\eta)g_+^R(v'-v-2(k-k')\eta)
\end{aligned} \tag{14.189}$$

where

$$f_\pm^R(v) = N^R(v)g_\mp^R(v). \tag{14.190}$$

We again follow the procedure of the previous subsection and look for a diagonal similarity transformation. However, because of the antisymmetry of $g_-^R(z)$ and recalling that L is odd the consistency conditions analogous to (14.173) and (14.181) are

$$-1 = \prod_{j=1}^{L} \frac{g_-^R(v'-v-2(k+k')\eta-4(L-j)\eta-2t)}{g_-^R(v'-v+2(k+k')\eta+4(L-j)\eta+2t)} \tag{14.191}$$

$$1 = \prod_{j=1}^{L} \frac{g_+^R(v'-v-2(k+k')\eta-4(L-j)\eta)}{g_+^R(v'-v+2(k+k')\eta+4(L-j)\eta)}. \tag{14.192}$$

These conditions will be satisfied by pairwise cancellation of terms in the numerator and denominator by establishing

$$\begin{aligned}
g_-^R(v'-v-2(k+k')\eta-4c_1\eta-2t) =&\\
-g_-^R(v'-v+2(k+k')\eta+4c_2\eta+2t)&
\end{aligned} \tag{14.193}$$

$$g_+^R(v'-v+2(k-k')\eta-4c_1\eta) = g_+^R(v'-v+2(k+k')\eta+c_2\eta). \tag{14.194}$$

To study (14.193) and (14.194) we use the periodicity properties of $g_\pm^R(v)$. From the definitions (14.156) and (14.156) and the periodicity properties (14.22) and (14.23) of $H_m(v)$ and $\Theta_m(v)$ it follows that

$$g_\pm^R(v+4(r_1K+ir_2K')) = (-1)^{r_1}g_\pm^R(v). \tag{14.195}$$

Consider first the condition (14.194). In the case that r_1 is even we have from (14.195)

$$g_+^R(v+4(r_1K+ir_2K')) = g_+^R(v) \tag{14.196}$$

and thus (14.194) will hold if

$$4(k-k'+1)\eta - 4c_1 = 4c_2 + 4I(r_1K + ir_2K') \tag{14.197}$$

which after multiplying by L and using the root of unity condition (14.1) becomes

$$(k-k'+1)r_0 - c_1r_0 = c_2r_0 + 2IL. \tag{14.198}$$

This equation can always be satisfied in integers because for m_1 and m_2 even r_0 is always even.

In the opposite case where r_1 is odd we have from (14.195)

$$g_+^R(v + 8(r_1K + ir_2K')) = g_+^R(v) \tag{14.199}$$

and thus (14.194) will be satisfied if

$$(k-k'+1)r_0 - c_1r_0 = c_2r_0 + 4LI. \tag{14.200}$$

This can only hold if $r_0 \equiv 0$ (mod4). However, it follows from the definition (14.27) of r_1 that when m_1 is even and r_1 is odd that $r_0/4$ must be an integer and thus the condition (14.194) is always satisfied.

To complete the derivation we must determine the conditions under which (14.193) is satisfied. For this to be the case we need the antiperiodicity properties of $g_-^R(v)$.

We first consider the use of the antiperiodicity property of (14.195) with r_1 odd where

$$g_-^R(v + 4(2I+1)(r_1K + ir_2K')) = -g_-^R(v) \tag{14.201}$$

and find that (14.193) will be satisfied if

$$-(k+k')r_0 - c_1r_0 - r_0\bar{t} = c_2r_0 + 2(2I+1)L. \tag{14.202}$$

However, we have just seen that when r_1 is odd and m_1 is even that $r_0/4$ and $m_2/4$ must be integers. Therefore because $2I+1$ and L are odd (14.202) cannot be satisfied in integers for any integer $\bar{t}$. However, if $\bar{t} = n + 1/2$ the condition (14.202) can be satisfied in integers if $r_0/4$ is an odd integer. Thus we conclude that the interchange relation (14.40) with $A = RS$ will hold for $t = (n+1/2)\eta$ when

$$m_1 \equiv 2 \text{ (mod4)}, \quad m_2 \equiv 2 \text{ (mod4)} \tag{14.203}$$

It remains to investigate the possibility of satisfying (14.193) by use of (14.162) which is an antiperiodicity condition in the three cases where $m_1/4$ and $m_2/4$ are not both integers. We have already considered the case $m_2 \equiv 0$ (mod4) and thus need only consider $m_2 \equiv 2$ (mod4). In this case $r_0 \equiv 2$ (mod4) and thus because m_1 is even r_1 must be odd. Thus we find from (14.195) and (14.162) that

$$g_-^R(v + 2(2I+1)(r_1K + ir_2K')) = -g_-^R(v) \tag{14.204}$$

and from this it follows that (14.193) holds for $t = (n+1/2)\eta)$.

The case $A = R$

The case $A = R$ is treated by similar methods and we find that the only case where the interchange relation holds is for $t = (n + 1/2)\eta$ with $m_1 \equiv 2$ (mod4) and $m_2 \equiv 0$ (mod4)

Summary

The results obtained above for the validity of the interchange relation (14.40) with $t = n\eta$ and $(n+1/2)\eta$ for $A = I, S, R$ and RS are summarized in Tables 14.4 and 14.5 where Y (N) indicates that the relation holds (fails).

Table 14.4 Summary of the values of the matrix A for which the interchange relation (14.40) with $t = n\eta$ holds where the notation 0(2) stands for $\equiv 0(2)$ (mod4).

m_1	m_2	I	S	R	RS
0	0	Y	Y	N	N
2	0	Y	Y	N	N
0	2	Y	Y	N	N
2	2	Y	Y	N	N

Table 14.5 Summary of the values of the matrix A for which the interchange relation (14.40) with $t = (n + 1/2)\eta$ holds where the notation 0(2) stands for $\equiv 0(2)$ (mod4).

m_1	m_2	I	S	R	RS
0	0	Y	Y	N	N
2	0	Y	Y	Y	Y
0	2	Y	N	N	Y
2	2	N	Y	N	Y

14.2.5 Nonsingularity and nondegeneracy

In order to complete the construction of $Q_{72}(v)$ by use of (14.42) we must show that there exists at least one value v_0 such that $Q_R(v_0)$ exists. This nonsingularity condition has been investigated numerically in [39] and [46] for small values of N for several cases of m_1, m_1 and L. The results of this study are as follows: $Q_R^{(1)}(v)$ is nonsingular for generic values of v except for the case m_1 and m_2 both even where it is singular for all values of v; $Q_R^{(2)}(v; n\eta)$ is generically nonsingular in all cases; $Q_R^{(2)}(v; (n + 1/2)\eta)$ is generically nonsingular except for $m_1 \equiv 2(\text{mod}4)$ and $m_2 \equiv 0(\text{mod}4)$. Thus we see that the only cases where the matrix $Q_R^{-1}(v)$ fails to exist are precisely those cases where $Q_R(v)$ commutes with all four choices of the matrix A.

We also observe numerically that in all cases $Q_{72}(v)$ is nondegenerate even though when the root of unity condition (14.1) holds the transfer matrix has many degenerate eigenvalues if the size of the system N is sufficiently large.

However, there is no analytic proof of these nonsingularity and nondegeneracy statements. More specifically the rank of $Q_R(v)$ in the cases where the inverse does not exist is not known except for a few numerical examples for small values of L and N.

14.2.6 Quasiperiodicity

We complete our discussion of the Q matrices by deriving their quasiperiodicity properties. The matrices $Q_{72}^{(1)}(v)$ and $Q_{72}^{(2)}(v)$ will be treated separately.

Quasiperiodicity of $Q_{72}^{(1)}(v)$

We begin the derivation of the quasiperiodicity properties of $Q_{72}^{(1)}(v)$ by using the quasiperiodic properties of the modified theta functions (14.22)–(14.25) in the definition (14.97) of $S_R^{(1)}(v)$ to find

$$S^{(1)}(\alpha),\beta)_{j,k}(v+\omega_1)=(-\alpha)^{r_1}(-1)^{r_1r_2}S_R^{(1)}(\alpha,\beta)_{j,k}(v) \tag{14.205}$$

and

$$\begin{aligned}
S_R^{(1)}(\alpha,\beta)_{k,k+1}(v+\omega_2)&=(-\alpha)^b(-1)^{ab}q'^{-1}e^{-2\pi iv/\omega_1}e^{4\pi ik\eta/\omega_1}S_R^{(1)}(\alpha,\beta)_{k,k+1}(v)\\
S_R^{(1)}(\alpha,\beta)_{k+1,k}(v+\omega_2)&=(-\alpha)^b(-1)^{ab}q'^{-1}e^{-2\pi iv/\omega_1}e^{-4\pi ik\eta/\omega_1}S_R^{(1)}(\alpha,\beta)_{k+1,k}(v)\\
S_R^{(1)}(\alpha,\beta)_{1,1}(v+\omega_2)&=(-\alpha)^b(-1)^{ab}q'^{-1}e^{-2\pi iv/\omega_1}S_R^{(1)}(\alpha,\beta)_{1,1}(v)\\
S_R^{(1)}(\alpha,\beta)_{L,L}(v+\omega_2)&=(-\alpha)^b(-1)^{ab}(-1)^{r_0}q'^{-1}e^{-2\pi iv/\omega_1}S_R^{(1)}(\alpha,\beta)_{L,L}(v).
\end{aligned} \tag{14.206}$$

The dependence on r_0 in the last line of (14.206) distinguishes the case with m_2 even from the cases with m_2 odd.

The case m_2 even and m_1 odd

When m_1 is odd and m_2 is even we see from (14.27) that r_0 is even r_1 is odd but r_2 is unrestricted. Therefore we find directly from the quasiperiodicity (14.205) and from (14.32) and (14.42) that

$$Q_{72oe}^{(1)}(v+\omega_1)=-S(-1)^{Nr_2}Q_{72oe}^{(1)}(v). \tag{14.207}$$

For quasiperiodicity under $v\rightarrow v+\omega_2$ we make a diagonal similarity transformation to write (14.206) as

$$S_R^{(1)}(\alpha,\beta)(v+\omega_2)=(-\alpha)^b(-1)^{ab}q'^{-1}e^{-2\pi iv/\omega_1}M^{(1)}S_R^{(1)}(\alpha,\beta)(v)M^{(1)-1} \tag{14.208}$$

with

$$M_{k,k'}^{(1)}=e^{-2\pi i\eta k(k-1)/\omega_1}\delta_{k,k'}. \tag{14.209}$$

Thus we find from (14.32) and (14.42) that

$$Q_{72oe}^{(1)}(v+\omega_2)=(-S)^b(-1)^{Nab}e^{-2\pi iNv/\omega_1}q'^{-N}Q_{72oe}^{(1)}(v). \tag{14.210}$$

It follows from (14.207), (14.210) and the fact that the eigenvectors of $Q_{72}^{(1)}(v)$ are independent of v that $Q_{72oe}^{(1)}(v)$ commutes with S (as was previously shown directly).

The restriction to m_2 odd

When m_2 is odd we see from (14.27) that r_0 and r_2 are odd and r_1 is even for both m_1 even and odd. Therefore because r_0 is odd we find instead of the quasiperiodicity condition (14.208) under $v \to v + \omega_2$ we have a quasiperiodicity under $v \to v + 2\omega_2$

$$S_R^{(1)}(\alpha,\beta)(v+2\omega_2) = q'^{-4}e^{-4\pi i v/\omega_1}M^{(1)2}S_R^{(1)}(\alpha,\beta)(v)M^{(1)-2} \tag{14.211}$$

and thus

$$Q_R^{(1)}(v+2\omega_2) = q'^{-4N}e^{-4\pi i N v/\omega_1}Q_R^{(1)}(v). \tag{14.212}$$

Therefore by use of (14.42) we find instead of (14.220) that

$$Q_{72xo}^{(1)}(v+2\omega_2) = q'^{-4N}e^{-4\pi i N v/\omega_1}Q_{72xo}^{(1)}(v) \tag{14.213}$$

where x is either e or o.

If the area of the fundamental parallelogram is to be $4KK'$ then the quasiperiodic property (14.213) mandates that instead of the parallelogram (14.29) we need to consider the parallelogram

$$0,\ \omega_1/2, \omega_1/2+2\omega_2,\ 2\omega_2. \tag{14.214}$$

To obtain the periodicity properties under $v \to \omega_1/2$ we write

$$\omega_1/2 = r_1 K + i r_2 K' \tag{14.215}$$

where $r_1/2$ is an integer because r_1 is even. It then follows from the definitions (14.19) and the properties (14.503), (14.507) and (14.508) that

$$H_m(v+\omega_1/2) = (-1)^{r_1/2}e^{\pi i r_1 r_2/4}\Theta_m(v) \tag{14.216}$$
$$\Theta_m(v+\omega_1/2) = e^{\pi i r_1 r_2/4}H_m(v). \tag{14.217}$$

We therefore obtain for m_2 odd and all m_1 that

$$S_R^{(1)}(v+\omega_1/2) = e^{\pi i r_1 r_2/4}RS^{r_1/2}S_R^{(1)}(v). \tag{14.218}$$

The case m_2 odd with m_1 even

When m_1 is further restricted to be even we find from (14.27) that $r_1/2$ is even and therefore (14.218) may be written as

$$S_R^{(1)}(v+\omega_1/2) = e^{\pi i r_1 r_2/4}RS_R^{(1)}(v). \tag{14.219}$$

Therefore we find from (14.32) and (14.42) that

$$Q_{72eo}^{(1)}(v+\omega_1/2) = e^{N\pi i r_1 r_2/4}RQ_{72eo}^{(1)}(v) \tag{14.220}$$

and from (14.220) and the fact that the eigenvectors of $Q_{72eo}^{(1)}(v)$ are independent of v it follows that

$$[Q_{72eo}^{(1)}(v), R] = 0 \tag{14.221}$$

which has been previously proven directly.

The case m_2 odd with m_1 odd

The quasiperiodicity relation (14.218) also holds but now for m_1 odd we have from (14.27) that $r_1/2$ is odd and thus instead of (14.220) we have

$$Q^{(1)}_{72oo}(v+\omega_1/2)=e^{N\pi i r_1 r_2/4}RSQ^{(1)}_{72oo}(v) \tag{14.222}$$

and therefore

$$[Q^{(1)}_{72oo}(v),RS]=0 \tag{14.223}$$

which has been previously proven directly.

Quasiperiodicity of $Q^{(2)}_{72}(v)$

We find from (14.110) and (14.22)-(14.25) that the quasiperiodicity properties of $S^{(2)}_R(v)$ are

$$S^{(2)}_R(\alpha,\beta)_{j,k}(v+\omega_1)=(-\alpha)^{r_1}(-1)^{r_1r_2}S^{(2)}_R(\alpha,\beta)_{j,k}(v) \tag{14.224}$$

$$S^{(2)}_R(\alpha,\beta)_{k,k+1}(v+\omega_2)=(-\alpha)^b(-1)^{ab}q'^{-1}e^{-2\pi i(v-2k\eta-t-K)/\omega_1}S^{(2)}_R(\alpha,\beta)_{k,k+1}(v) \tag{14.225}$$

$$S^{(2)}_R(\alpha,\beta)_{k+1,k}(v+\omega_2)=(-\alpha)^b(-1)^{ab}q'^{-1}e^{-2\pi i(v+2k\eta+t-K)/\omega_1}S^{(2)}_R(\alpha,\beta)_{k+1,k}(v). \tag{14.226}$$

From (14.224) we find for all t that $Q^{(2)}_R(v;t)$ has the periodicity property (recalling that N is even)

$$Q^{(2)}_R(v+\omega_1;t)=(-S)^{r_1}Q^{(2)}_R(v;t). \tag{14.227}$$

However,the quasiperiodicity of $Q^{(2)}_R(v;t)$ under $v\rightarrow v+\omega_2$ is different for the two cases $t=n\eta$ and $t=(n+1/2)\eta$ and will be treated separately.

Quasiperiodicity for $t=n\eta$

When $t=n\eta$ the matrix $S^{(2)}(\alpha,\beta)(v+\omega_2)$ may be written as

$$\begin{aligned}&S^{(2)}_R(\alpha,\beta)(v+\omega_2)\\&=(-1)^{nr_0/2}(-\alpha)^b(-1)^{ab}q'^{-1}e^{-2\pi i(v-K)/\omega_1}M^{(2;0)}S^{(2)}_R(\alpha,\beta)(v)M^{(2;0)-1}\end{aligned} \tag{14.228}$$

with $M^{(2;0)}$ given by

$$M^{(2;0)}_{k,k'}=\delta_{k,k'}e^{-\pi i r_0 k(k-1)/(2L)}(-1)^{nr_0k/2}e^{-\pi i n r_0 k/(2L)} \tag{14.229}$$

and thus we obtain from (14.32)

$$Q^{(2)}_R(v+\omega_2;n\eta)=(-S)^bq'^{-N}e^{-2\pi i N(v-K)/\omega_1}Q^{(2)}_R(v;t). \tag{14.230}$$

Thus for $t=n\eta$ we find from the definition (14.42) of $Q^{(2)}_{72ee}(v;n\eta)$ that

$$Q^{(2)}_{72ee}(v+\omega_1;n\eta)=(-S)^{r_1}Q^{(2)}_{72ee}(v;n\eta) \tag{14.231}$$

$$Q^{(2)}_{72ee}(v+\omega_2;n\eta)=(-S)^b q'^{-N}e^{-2\pi iN(v-K)/\omega_1}Q^{(2)}_{72ee}(v;n\eta). \tag{14.232}$$

It follows from (14.27) that b and r_1 cannot both be even and thus S appears in at least one of (14.231) or (14.232) and therefore, as previously found, it follows that $Q^{(2)}_{72ee}(v;n\eta)$ commutes with the operator S.

Quasiperiodicity for $t=(n+1/2)\eta$

When $t=(n+1/2)\eta$ we first consider the case $m_2\equiv 0$ (mod4). In this case $r_0\equiv 0$ (mod4) and we find from (14.225) and (14.226) that we may write

$$\begin{aligned}&S^{(2)}_R(\alpha,\beta)(v+\omega_2;(n+1/2)\eta)\\&=(-1)^{r_0/4}(-\alpha)^b(-1)^{ab}q'^{-1}e^{-2\pi i(v-K)/\omega_1}M^{(2;1)}S^{(2)}_R(\alpha,\beta)(v;(n+1/2)\eta)M^{(2;1)-1}\end{aligned} \tag{14.233}$$

where

$$M^{(2;1)}_{k,k'}=\delta_{k,k'}e^{-\pi i r_0 k(k-1)/(2L)}(-1)^{r_0k/4}e^{-\pi i(2n+1)r_0k/(4L)}. \tag{14.234}$$

Thus we find from (14.32) that

$$Q^{(2)}_R(v+\omega_2;(n+1/2)\eta)=(-S)^b q'^{-N}e^{-2\pi iN(v-K)/\omega_1}Q^{(2)}_R(v;(n+1/2)\eta). \tag{14.235}$$

This is identical to (14.230) for $Q^{(2)}(v;n\eta)$ and thus we conclude that for $r_0\equiv 0$ (mod4) that $Q^{(2)}_{72ee}(v;(n+1/2)\eta)$ has the same quasiperiodicity properties (14.231), (14.232) as $Q^{(2)}_{72ee}(v;n\eta)$.

We next consider $m_2\equiv 2$ (mod4) where $r_0\equiv 2$ (mod4) r_1 is even and r_2 is odd. In this case the similarity transformation in (14.233) will not exist. The reason for this is that in order for (14.233) to hold it was necessary that

$$e^{\pi i r_0/4}=\pm 1 \tag{14.236}$$

which is not the case when $r_0\equiv 2$ (mod4). In this case the analogous argument shows that $Q^{(2)}_{72ee}(v;(n+1/2)\eta)$ is quasiperiodic under $v\to v+2\omega_2$.

However, there is an additional quasiperiodicity under $v\to v+\omega_2+\omega_1/2$. To show this we use the relations which follow from (14.507)-(14.511) when r_1 is even and r_2 is odd

$$\begin{aligned}&H_m(v+\omega_1/2+\omega_2)\\&=-(-1)^{r_1/2+b}(-1)^{r_1r_2/4+ab}q'^{-1}e^{-2\pi i(v-K)/\omega_1}\Theta_m(v)\end{aligned} \tag{14.237}$$

$$\Theta_m(v+\omega_1/2+\omega_2)=-(-1)^{r_1r_2/4+ab}q'^{-1}e^{-2\pi i(v-K)/\omega_1}H_m(v) \tag{14.238}$$

and find from (14.110) that

$$\begin{aligned}&S^{(2)}_R(\alpha,\beta)_{k,k+1}(v+\omega_1/2+\omega_2)\\&=(-\alpha)^{r_1/2+b}f(v)e^{\pi i/2}e^{2\pi i(t+2k\eta)/\omega_1}S^{(2)}_R(-\alpha,\beta)_{k,k+1}(v)\end{aligned} \tag{14.239}$$

$$S_R^{(2)}(\alpha,\beta)_{k+1,k}(v+\omega_1/2+\omega_2)$$
$$=(-\alpha)^{r_1/2+b}f(v)e^{-\pi i/2}e^{-2\pi i(t+2k\eta)/\omega_1}S_R^{(2)}(-\alpha,\beta)_{k+1,k}(v) \quad (14.240)$$
$$S^{(2)}(\alpha,\beta)_{1,L}(v+\omega_1/2+\omega_2)$$
$$=(-\alpha)^{r_1/2+b}f(v)e^{-\pi i/2}e^{-2\pi i(t+2L\eta)/\omega_1}S_R^{(2)}(-\alpha,\beta)_{1,L}(v) \quad (14.241)$$
$$S_R^{(2)}(\alpha,\beta)_{L,1}(v+\omega_1/2+\omega_2)$$
$$=(-\alpha)^{r_1/2+b}f(v)e^{\pi i/2}e^{2\pi i(t+2L\eta)/\omega_1}S_R^{(2)}(-\alpha,\beta)_{L,1}(v) \quad (14.242)$$

where

$$f(v)=-i(-1)^{r_1r_2/4+ab}q'^{-1}e^{-2\pi i(v-K)/\omega_1}. \quad (14.243)$$

The expressions (14.239)–(14.242) can be written as

$$M^{(2;2)}S_R^{(2)}(\alpha,\beta)(v+\omega_1/2+\omega_2)M^{(2;2)-1}=\epsilon(-\alpha)^{r_1/2+b}f(v)S_R^{(2)}(-\alpha,\beta)(v) \quad (14.244)$$

with $\epsilon=\pm1$ and

$$M_{k,k'}^{(2;2)}=(i\epsilon)^{k-1}e^{2\pi i(k-1)(t+k\eta)/\omega_1}\delta_{k,k'} \quad (14.245)$$

where for consistency we need

$$(i\epsilon)^L e^{2\pi itL/\omega_1}e^{2\pi iL(L+1)\eta/\omega_1}=1. \quad (14.246)$$

Using (14.31) and the fact that $r_0/2$ and L are odd, (14.246) reduces to

$$(i\epsilon)^L e^{\pi itr_0/(2\eta)}=1 \quad (14.247)$$

which with an appropriate choice of $\epsilon=\pm1$ is satisfied for $t=(n+1/2)\eta$ as desired. Thus we obtain from (14.244) for $m_2\equiv 2$ (mod4) and N even

$$Q_{72ee}^{(2)}(v+\omega_1/2+\omega_2;(n+1/2)\eta)$$
$$=(-1)^{N/2}S^{r_1/2+b}(-1)^{Nr_1r_2/4}q'^{-N}e^{-2\pi iN(v-K)/\omega_1}RQ_{72ee}^{(2)}(v;(n+1/2)\eta). \quad (14.248)$$

Finally we recall from (14.28) that b must be odd because r_1 is even and r_2 is odd and that because $r_0\equiv 2$ (mod4) we have $r_1\equiv 2(0)$ (mod4) for $m_1\equiv 2(0)$ (mod4). Thus we find that for $m_1\equiv 2$ (mod4) and $m_2\equiv 2$ (mod4) that

$$Q_{72ee}^{(2)}(v+\omega_1/2+\omega_2;(n+1/2)\eta)=q'^{-N}e^{-2\pi iN(v-K)/\omega_1}RQ_{72ee}^{(2)}(v;(n+1/2)\eta) \quad (14.249)$$

from which it follows in agreement with (14.17) that $Q_{72ee}^{(2)}(v;(n+1/2)\eta)$ commutes with R. For $m_2\equiv 2$ (mod4) and $m_1\equiv 0$ (mod4)

$$Q_{72ee}^{(2)}(v+\omega_1/2+\omega_2;(n+1/2)\eta)=(-1)^{N/2}q'^{-N}e^{-2\pi iN(v-K)/\omega_1}RSQ_{72ee}^{(2)}(v;(n+1/2)\eta) \quad (14.250)$$

from which it follows in agreement with (14.18) that $Q_{72ee}^{(2)}(v;(n+1/2)\eta)$ commutes with RS.

14.3 Eigenvalues and free energy

The purpose of deriving the TQ equation (14.5) is to use it to compute the eigenvalues of the transfer matrix $T(v)$. In particular the free energy is derived from the maximum eigenvalue $t_{max}(v)$ of $T(v)$ as

$$F(v) = -k_B T \lim_{N\to\infty} N^{-1} \ln t_{max}(v). \tag{14.251}$$

The TQ equation (14.5) is a matrix equation. However, from the commutation relations (14.6) and (14.7) all the matrices in the equation may be simultaneously diagonalized and thus, calling $t(v)$ and $q(v)$ the eigenvalues of $T(v)$ and $Q(v)$ we obtain the scalar functional equation for eigenvalues

$$t(v)q(v) = [h(v+\eta)]^N q(v-2\eta) + [h(v-\eta)]^N q(v+2\eta). \tag{14.252}$$

14.3.1 The form of the eigenvalues

To make effective use of (14.252) we require qualitative information about the form of the eigenvalues which is obtained from the quasiperiodicity conditions derived in 14.2.6.

It follows from the fact that $T(v)$ and $T(v')$ commute that the eigenvectors of $T(v)$ are in fact independent of v. Therefore the eigenvalues are linear combinations independent of v of the v dependent matrix elements of $T(v)$ and because their matrix elements are all entire functions of v it follows that the eigenvalues $q(v)$ of $Q(v)$ are also entire functions of v. These entire functions must have the quasiperiodicity properties which follow from the matrix quasiperiodicity properties of 14.2.6. Every such entire function can be expressed in terms of its zeros which lie in the fundamental parallelogram where the number of zeros n is

$$n = \frac{1}{2\pi i}\int_C dv \frac{q'(v)}{q(v)} = \frac{1}{2\pi i}\int_c dv \frac{d}{dv}\ln q(v) \tag{14.253}$$

where the path C is the boundary of the fundamental parallelogram. This integral may be evaluated by use of the quasiperiodicity properties of 14.2.6 and for all cases we find

$$n = N. \tag{14.254}$$

To proceed further the specific form of the quasiperiodicity condition must be used.

The case m_1 odd and m_2 even

In this case the fundamental parallelogram is $(0, \omega_1, \omega_1+\omega_2, \omega_2)$ and thus the most general form of the eigenvalues is

$$q^{(1)}_{72oe}(v) = \mathcal{K}\exp(-i\nu\pi v/\omega_1)\prod_{j=1}^{N} H_m(v - v_j) \tag{14.255}$$

where $\mathcal{K}$ is independent of v and v_j in the parallelogram $(0, \omega_1, \omega_1+\omega_2, \omega_2)$. Using the form (14.255) we find from the quasiperiodicity condition (14.207)

$$1 = e^{\pi i(1+\nu_S+\nu+N)} \tag{14.256}$$

and thus

$$1 + \nu + N + \nu_S = \text{even integer} \tag{14.257}$$

where $(-1)^{\nu_S}$ is the eigenvalue of S.

From (14.210) we find

$$(-S)^b = e^{-\nu\pi i\omega_2/\omega_1}(-1)^{bN}\exp(2\pi i\sum_{j=1}^{N}(v_j + K)/\omega_1) \tag{14.258}$$

and thus

$$b(\nu_S + 1) - \nu\omega_2/\omega_1 + bN + 2\sum_{j=1}^{N}(v_j + K)/\omega_1 = \text{even integer.} \tag{14.259}$$

In particular, for the special case $m_2 = 0$ and m_1 odd where $\omega_1 = 2K$ and $\omega_2 = 2iK'$, (14.259) specializes to

$$\sum_{j=1}^{N}\text{Im}v_j/K' = \nu, \quad \text{and} \quad \sum_{j=1}^{N}\text{Re}(v_j + K)/K = \text{even integer.} \tag{14.260}$$

The case m_1 even and m_2 odd

In this case the fundamental parallelogram is (14.214) and for the general form we have

$$q^{(1)}_{72eo}(v) = \mathcal{K}e^{-\nu 2\pi iv/\omega_1}\prod_{j=1}^{N}H_m(v/2 - v_j/2)H_m(v/2 - v_j/2 + \omega_1/4)$$
$$H_m(v/2 - v_j/2 + \omega_1/2)H_m(v/2 - v_j/2 + 3\omega_1/4) \tag{14.261}$$

and we have the sum rules

$$e^{N\pi ir_1r_2/4}(-1)^{\nu_R} = (-1)^{\nu} \tag{14.262}$$

where $(-1)^{\nu_R}$ are the eigenvalues of R and

$$1 = \exp\left(-\nu 4\pi i\omega_2/\omega_1 + 4\pi i\sum_{j=1}^{N}(v_j + 2K)/\omega_1\right). \tag{14.263}$$

The case m_1 odd and m_2 odd

In this case the eigenvalues $q^{(1)}_{72oo}(v)$ are of the form (14.261) where the sum rule (14.262) is replaced by

$$e^{N\pi ir_1r_2/4}(-1)^{\nu_{RS}} = (-1)^{\nu} \tag{14.264}$$

where $(-1)^{\nu_{RS}}$ are the eigenvalues of RS.

The eigenvalues $q^{(2)}_{72ee}(v;n\eta)$ for m_1, m_2 and N even

Here the fundamental parallelogram is $(0,\omega_1,\omega_1+\omega_2,\omega_2)$ and for the general form we have

$$q^{(2)}_{72ee}(v;n\eta) = \mathcal{K}\exp(-i\nu\pi v/\omega_1)\prod_{j=1}^{N} H_m(v-v_j) \tag{14.265}$$

with the sum rules

$$e^{-i\pi\nu} = (-1)^{r_1(\nu_s-1)} \tag{14.266}$$

$$q'^{-\nu}\exp(2\pi i\sum_{j=1}^{N} v_j/\omega_1) = (-1)^{b(\nu_s-1)} \tag{14.267}$$

are satisfied. In particular if r_1 is even we see from (14.266) that $\nu = 0$.

The eigenvalues $q^{(2)}_{72ee}(v;(n+1/2)\eta)$ for $m_2 \equiv 0$ (mod4)

The eigenvalues in this case have the same form (14.265)–(14.267) as $q^{(2)}_{72ee}(v;n\eta)$.

The case $m_1 \equiv 2$ (mod4) and $m_2 \equiv 2$ (mod4)

In this case it follows from (14.249) that the eigenvalues $q^{(2)}_{72ee}(v;(n+1/2)\eta)$ are of the form

$$q^{(2)}_{72ee}(v;(n+1/2)\eta) = \mathcal{K}\prod_{j=1}^{N} H_m(\frac{v-v_j}{2})H_m(\frac{v-v_j+\omega_1/2+\omega_2}{2})$$
$$H_m(\frac{v-v_j+\omega_1+2\omega_2}{2})H_m(\frac{v-v_j+3\omega_1/2+3\omega_2}{2}) \tag{14.268}$$

with the sum rule

$$(-1)^{\nu_R} = q'^{-3N}e^{2\pi i\sum_{j=1}^{N}(v_j+K)/\omega_1} = q'^{-3N-Nr_2}e^{2\pi i\sum_{j=1}^{N} v_j/\omega_1}. \tag{14.269}$$

The case $m_1 \equiv 0$ (mod4) and $m_2 \equiv 2$ (mod4)

In this case it follows from (14.250) that the eigenvalues $q^{(2)}_{72ee}(v;(n+1/2)\eta)$ are of the form (14.268) with the sum rule

$$(-1)^{N/2}(-1)^{\nu_{RS}} = q'^{-3N}e^{2\pi i\sum_{j=1}^{N}(v_j+K)/\omega_1} = q'^{-3N-Nr_2}e^{2\pi i\sum_{j=1}^{N} v_j/\omega_1}. \tag{14.270}$$

14.3.2 Numerical study of the eigenvalues of $Q_{72}(v)$

For further progress we require more quantitative information about the locations of the roots v_j and for this purpose it is most useful to do a numerical study of the eigenvalues $q(u)$ using the explicit forms of the matrices $Q_{72}(u)$. We are particularly interested in determining the roots v_j for the largest (in magnitude) eigenvalue of the transfer matrix because this will give the free energy of the eight-vertex model in the thermodynamic limit.

Furthermore we found in section 13.7 that the eight-vertex transfer matrix commutes with the Hamiltonian of the XYZ model

$$H_{XYZ} = -\sum_{j=1}^{N}\{J^x\sigma_j^x\sigma_{j+1}^x + J^y\sigma_j^y\sigma_{j+1}^y + J^z\sigma_j^z\sigma_{j+1}^z\} \tag{14.271}$$

with

$$J^x = 1 + k\mathrm{sn}^2(2\eta), \quad J^y = 1 - k\mathrm{sn}^2(2\eta), \quad J^z = \mathrm{cn}(2\eta)\mathrm{dn}(2\eta) \tag{14.272}$$

which in the six-vertex limit $k \to 0$ reduces to the Hamiltonian of the XXZ spin chain

$$H_{XXZ} = -\sum_{j=1}^{N}\{\sigma_j^x\sigma_{j+1}^x + \sigma_j^y\sigma_{j+1}^y + \Delta\sigma_{j+1}^z\} \tag{14.273}$$

where in the case $m_2 = 0$

$$\Delta = \cos m_1\pi/L. \tag{14.274}$$

In section 13.7 we showed that the maximum eigenvalue of the transfer matrix corresponds to the ground state of H_{XYZ} of (14.271)

We will consider in detail the case $m_2 = 0$ and m_1 odd. The restriction $m_2 = 0$ means that η is real and this is referred to in the literature [32] as the "disordered region". In the six-vertex limit this corresponds to $-1 < \Delta < 1$ in the XXZ spin chain.

The root configurations are quite different for N even and N odd and will be treated separately.

N even

There are two distinct classes of roots when N is even which we call Bethe roots and complete L-strings.

Bethe roots

These roots occur in pairs

$$v_j^B, \quad \text{and} \quad v_j^B + iK' \tag{14.275}$$

and we denote the number of such pairs as n_B.

There are further restrictions which are satisfied by the v_j^B. One restriction is on individual values of j

$$\mathrm{Re}v_j^B = 0 \;\text{ or }\; \mathrm{Re}v_j^B = K. \tag{14.276}$$

Examples are given for $\eta = K/3$ with $q = 0.02$ in Table 14.6 where the root configuration for the three most negative eigenvalues of the XYZ chain are given for chains with $N = 6$ and 8 sites. We see here that the ground state has all roots satisfying $\mathrm{Re}v_j = K$. We also see illustrated the fact that the ground state has $S = +1$ for $N/2$ even and $S = -1$ for $N/2$ odd. Furthermore, there are exactly two states with $\mathrm{Re}v_j = K$. These two states have opposite values of S and become degenerate as $N \to \infty$. For comparison we give in Table 14.7 the root configurations of the same three states for $q = 0.002$.

Table 14.6 The roots v_j for the case $q = 0.02$ of the three most negative eigenvalues of H_{XYZ} for $\eta = K/3$ for chains with $N = 6$ and $N = 8$. This illustrates how the eigenvalue of S and the pattern of roots depends of whether $N/2$ is even or odd. The roots are chosen in the fundamental parallelogram $0 \leq \text{Re}v_j < 2K$ and $K' \leq \text{Im}v_j < K'$. The eigenvalues of H_{XYZ}, S, R and the momentum P of the state are given.

$N=8$					$N=6$				
roots	S	R	P	E_{xyz}	roots	S	R	P	E_{xy}
$K+0.8894iK'$	1	1	0	-11.3093	$K+0.6867iK'$	-1	1	0	-8.484
$K+0.6417iK'$					$K+0.3132iK'$				
$K+0.3582iK'$					$K+0.0iK'$				
$K+0.1105iK'$					$K-0.3132iK'$				
$K-0.1105iK'$					$K-0.6867iK'$				
$K-0.3582iK'$					$K-iK'$				
$K-0.6417iK'$									
$K-0.8894iK'$									
$K+0.7718iK'$	-1	1	0	-11.3087	$K+0.8513iK'$	1	1	0	-8.479
$K+0.5iK'$					$K+0.5iK'$				
$K+0.2281iK'$					$K+0.1486iK'$				
$K+0.0iK'$					$K-0.1486K'$				
$K-0.2281iK'$					$K-0.5iK'$				
$K-0.5iK'$					$K-0.8513iK'$				
$K-0.7718iK'$									
$K-iK'$									
$K+0.7159iK'$	-1	-1	0	-7.804	$K+0.8091iK'$	1	-1	0	-5.004
$+0.5iK'$					$+0.5iK'$				
$K+0.2840iK'$					$K+0.1908iK'$				
$K+0.0iK'$					$K-0.1908iK'$				
$K-0.2840iK'$					$-0.5iK'$				
$-0.5iK'$					$K-0.8091iK'$				
$K-0.7159iK'$									
$K-iK'$									

A second restriction groups together $n \leq L-1$ values of j such that

$$v_j^B = v_c^B + j2\eta + \epsilon_j \quad \text{for } j = 0, 1, \cdots n-1 \tag{14.277}$$

where ϵ_j vanishes as N, the size of the system, becomes large. We refer to such configurations as strings of length n (or merely as n-strings). Examples $\eta = K/3$ with $q = 0.002$ and $N = 8$ are given in Table 14.8 for $S = 1$. Here we give two of the several states which contain two 2-strings and the one unique state with four 2-strings. For $N \geq 10$ many states with four (or more) 2-strings occur. In Table 14.9 we present for $N = 8$ and $q = 0.002$ the five states with two 2-strings for $S = -1$ and $P = -\pi/4$ and their degenerate partners in the $q \rightarrow 0$ limit.

Table 14.7 The roots v_j for the case $q = 0.002$ of the three most negative eigenvalues of H_{XYZ} for $\eta = K/3$ for chains with $N = 8$. Comparison with the corresponding data for $q = 0.02$ in Table 14.6 shows how the second and third eigenvalues become degenerate in the limit $q \to 0$. The underlined roots remain finite as $q \to 0$. The roots are chosen in the fundamental parallelogram $0 \leq \mathrm{Re}v_j < 2K$ and $K' \leq \mathrm{Im}v_j < K'$. The eigenvalues of H_{XYZ}, S, R and the momentum P of the state are given.

roots	S	R	P	E_{xyz}
$K + 0.9171iK'$	1	1	0	-9.4031
$K + 0.6935iK'$				
$\underline{K + 0.3064iK'}$				
$\underline{K + 0.0828iK'}$				
$\underline{K - 0.0828iK'}$				
$\underline{K - 0.3064iK'}$				
$K - 0.6935iK'$				
$K - 0.9171iK'$				
$K + 0.8212iK'$	-1	1	0	-9.3580
$K + 0.5iK'$				
$\underline{K + 0.1787iK'}$				
$\underline{K + 0.0iK'}$				
$\underline{K - 0.1787iK'}$				
$K - 0.5iK'$				
$K - 0.8212iK'$				
$K - iK'$				
$K + 0.8025iK'$	-1	-1	0	-8.1004
$+0.5iK'$				
$\underline{K + 0.1974iK'}$				
$\underline{K + 0.0iK'}$				
$\underline{K - 0.1974iK'}$				
$-0.5iK'$				
$K - 0.8025iK'$				
$K - iK'$				

Complete L-strings

These roots are a set of L values of v_j which are exactly given as

$$v_j = v^c + j2\eta \quad \text{for} \quad j = 0, 1, \cdots, L-1. \tag{14.278}$$

Complete L-strings differ from Bethe roots (14.275) in that they do *not* occur in pairs. Furthermore the spacing of complete L-strings is exactly 2η and does *not* have the deviation ϵ_j of (14.277).

We denote the number of complete complete L-strings in a set v_j as n_L. It is observed that n_L is even for all values of L. Furthermore it is observed that for every complete L-string with the value v^c there is another configuration with $v^c + iK$. Examples of complete L-strings for $\eta = K/3$ and $q = 0.002$ are given in Table 14.10.

These two classes of roots exhaust all the N values of j and thus

$$n_B + Ln_L = N. \tag{14.279}$$

Table 14.8 The roots v_j for the case $q = 0.002$ for states containing 2-strings with $S = 1$ for $\eta = K/3$ and $N = 8$. The first and second states have two 2-strings and the third state has four 2-strings. The second and third eigenvalues become degenerate as $q \to 0$. The underlined roots remain finite as $q \to 0$. The roots are chosen in the fundamental parallelogram $0 \leq \mathrm{Re} v_j < 2K$ and $K' \leq \mathrm{Im} v_j < K'$. The eigenvalues of H_{XYZ}, S, R and the momentum P of the state are given.

roots	S	R	P	E_{xyz}
$K + 0.9071iK'$	1	1	0	-7.3123
$\underline{K + 0.0928iK'}$				
$\underline{K - 0.0928iK'}$				
$K - 0.9071iK'$				
$0.3868K + 0.5iK'$				
$1.6131K + 0.5iK'$				
$0.3863K - 0.5iK'$				
$1.6131K - 0.5iK'$				
$\underline{0.3333K + 0.0iK'}$	1	-1	π	-1.9642
$\underline{1.6666K + 0.0iK'}$				
$0.3333K - iK'$				
$1.6666K - iK'$				
$0.5iK'$				
$-0.5iK'$				
$K + 0.5iK'$				
$K - 0.5iK'$				
$\underline{0.3333K + 0.0iK'}$	1	1	π	-1.7652
$\underline{1.6666K + 0.0iK'}$				
$0.3333K - iK'$				
$1.6666K - iK'$				
$0.4133K + 0.5iK'$				
$1.5866K + 0.5iK'$				
$0.4133K - 0.5iK'$				
$1.5866K - 0.5iK'$				

From these results we may write the most general form of the eigenvalue as

$$q^{(1)}_{72oe}(v) = \mathcal{K}\exp(-i\nu\pi v/2K) \prod_{j=1}^{n_B} H(v - v_j^B) H(v - v_j^B - iK')$$

$$\times \prod_{j=1}^{n_L} H(v - v_j^c) H(v - v_j^c - 2\eta) \cdots H(v - v_j^c - 2(L-1)\eta). \quad (14.280)$$

Table 14.9 The roots v_j for the five states which have two 2-strings $S = -1$ and $P = -\pi/4$ for $\eta = K/3$ and $N = 8$ with $q = 0.002$ and their degenerate partners in the limit $q \to 0$. The underlined roots remain finite as $q \to 0$ while the remaining roots diverge to infinity. The roots are chosen in the fundamental parallelogram $0 \le \mathrm{Re}v_j < 2K$ and $K' \le \mathrm{Im}v_j < K'$. The eigenvalues of H_{XYZ} the state are given.

roots	R	E_{xyz}	roots	R	E_{xyz}
$K + 0.8213iK'$	1	-6.4536	$K + 0.8263iK'$	-1	-6.7054
$\underline{K + 0.0237iK'}$			$K + 0.3804iK'$		
$\underline{K - 0.1786iK'}$			$0.2697iK'$		
$K - 0.9762iK'$			$\underline{K + 0.0233iK'}$		
$0.2982K + 0.3275iK'$			$\underline{K - 0.1736iK'}$		
$1.7017K + 0.3275iK'$			$K - 0.6195iK'$		
$0.2982K - 0.6724iK'$			$-0.7302iK'$		
$1.7017K - 0.6724iK'$			$K - 0.9766iK'$		
$0.7363iK'$	1	-1.5932	$\underline{0.1309iK'}$	-1	-1.3905
$K + 0.6015iK'$			$\underline{K + 0.0332iK'}$		
$\underline{0.1294iK'}$			$-0.8690iK'$		
$\underline{K + 0.0323iK'}$			$K - 0.9667iK'$		
$-0.2636iK'$			$0.2768K + 0.6678iK'$		
$K - 0.3984iK'$			$1.7231K + 0.6678iK'$		
$-0.8705iK'$			$0.2768K - 0.3321iK'$		
$K - 0.9676iK'$			$1.7231K - 0.3321iK'$		
$0.9012iK'$	1	2.8194	$0.9055iK'$	-1	1.8648
$0.5083iK'$			$K + 0.5121iK'$		
$\underline{-0.1987iK'}$			$\underline{-0.0944iK'}$		
$-0.4916iK'$			$K - 0.4878iK'$		
$\underline{0.3333K + 0.0459iK'}$			$\underline{0.3333K + 0.0411iK'}$		
$\underline{1.6666K + 0.0459iK'}$			$\underline{1.6666K + 0.0411iK'}$		
$0.3333K - 0.9540iK'$			$0.3333K - 0.9588iK'$		
$1.6666K - 0.9540iK'$			$1.6666K - 0.9588iK'$		
$0.9898iK'$	1	8.1789	$0.9850iK'$	-1	8.2316
$\underline{0.0809iK'}$			$K + 0.2521iK'$		
$\underline{-0.0101iK'}$			$0.1954iK'$		
$-0.9190iK'$			$\underline{0.0673iK'}$		
$0.3160K + 0.2145iK'$			$\underline{-0.0145iK'}$		
$1.6839K + 0.2145iK'$			$K - 0.7478iK'$		
$0.3160K - 0.7854iK'$			$-0.8045iK'$		
$1.6839K - 0.7854iK'$			$-0.9326iK'$		

Table 14.10 The roots v_j for two examples of degenerate quartets which have complete L-strings for $\eta = K/3$ and $N = 8$ with $q = 0.002$. The eigenvalues on the right become degenerate with the eigenvalues on the left in the limit $q \to 0$. The underlined roots remain finite as $q \to 0$ while the remaining roots diverge to infinity. The roots are chosen in the fundamental parallelogram $0 \leq \mathrm{Re}v_j < 2K$ and $K' \leq \mathrm{Im}v_j < K'$. These states are not eigenstates of the operator R.

roots	S	P	E_{xyz}	roots	S	P	E_{xyz}
$0.5iK'$	1	0	-3.4654	$K+0.5iK'$	1	0	-4.5345
$-0.5iK'$				$K-0.5iK'$			
$\underline{K/3+0.1192iK'}$				$\underline{K/3+0.1095iK'}$			
$\underline{K+0.1192iK'}$				$\underline{K+0.1095iK'}$			
$\underline{4K/3+0.1192iK'}$				$\underline{4K/3+0.1095iK'}$			
$\underline{K/3-0.1192iK'}$				$\underline{K/3-0.1095iK'}$			
$\underline{K-0.1192iK'}$				$\underline{K-0.1095iK'}$			
$\underline{4K/3-0.1192iK'}$				$\underline{4K/3-0.1095iK'}$			
$0.5iK'$	1	0		$K+0.5iK'$	1	0	
$-0.5iK'$				$K-0.5iK'$			
$K/3+0.8807iK'$				$K/3+0.8904iK'$			
$K+0.8807iK'$				$K+0.8904iK'$			
$4K/3+0.8807iK'$				$4K/3+0.8904iK'$			
$K/3-0.8807iK'$				$K/3-0.8904iK'$			
$K-0.8807iK'$				$K-0.8904iK'$			
$4K/3-0.8807iK'$				$4K/3-0.8904iK'$			
$0.5iK'$	-1	π		$K+0.5iK'$	-1	π	
$-0.5iK'$				$K-0.5iK'$			
$K/3+0.8807iK'$				$K/3+0.8904iK'$			
$K+0.8807iK'$				$K+0.8904iK'$			
$4K/3+0.8807iK'$				$4K/3+0.8904iK'$			
$\underline{K/3+0.1192iK'}$				$\underline{K/3+0.1095iK'}$			
$\underline{K+0.1192iK'}$				$\underline{K+0.1095iK'}$			
$\underline{4K/3+0.1192iK'}$				$\underline{4K/3+0.1095iK'}$			
$0.5iK'$	-1	π		$K+0.5iK'$	-1	π	
$-0.5iK'$				$K-0.5iK'$			
$\underline{K/3-0.1192iK'}$				$\underline{K/3-0.1095iK'}$			
$\underline{K-0.1192iK'}$				$\underline{K-0.1095iK'}$			
$\underline{4K/3-0.1192iK'}$				$\underline{4K/3-0.1095iK'}$			
$K/3-0.8807iK'$				$K/3-0.8904iK'$			
$K-0.8807iK'$				$K-0.8904iK'$			
$4K/3-0.8807iK'$				$4K/3-0.8904iK'$			

Degeneracy of the eight-vertex eigenvalues of the transfer matrix for N even

It is easily seen from the definition (14.97) of the matrix $S_R^{(1)}(\alpha,\beta)(v)$ that for $m_2 = 0$ if we send $v \to v + iK'$ that $Q_{72oe}^{(1)}(v) \to RQ_{72oe}^{(1)}(v)$. Therefore the states which have only Bethe roots (and thus are invariant as $v \to v + iK'$) are eigenfunctions of R. However, the states that contain complete L-strings are not invariant as $v \to v + iK'$ and thus those eigenfunctions of $Q_{72oe}^{(1)}(v)$ are not eigenfunctions of R. However, the corresponding eigenfunctions of T do not depend on the complete L-strings. Thus the states with complete L-strings lead to eigenvalues of the transfer matrix $T(v)$ which are at least doubly degenerate. This is in agreement with the commutation relations (14.12).

However, there is more degeneracy than this. In Table 14.10 we see that for $n_L = 2$ there are four allowed values of the complete L-string centers

$$(v_1^c, v_2^c), \quad (v_1^c + iK', v_2^c + iK') \tag{14.281}$$

$$(v_1^c + iK', v_2^c), \quad (v_1^c, v_2^c + iK') \tag{14.282}$$

where the pairs (14.281) and (14.282) have opposite values of S and values of P which differ by π. The transfer matrix eigenvalues of these pairs have opposite sign and the values of the energy eigenvalues E_{xyz} are degenerate . More generally, studies for larger values N show that when there are n_L complete L-strings in the state that all 2^{n_L} sets of string centers $v_j^c + (0,1)iK'$ occur. This set is divided into two subsets each with 2^{n_L-1} members which have opposite values of S, momenta which differ by π, transfer matrix eigenvalues which differ by a minus sign if m_1 is odd and identical values of E_{xyz}. These observed degeneracies must be explained in terms of a symmetry algebra just as the double degeneracy is explained by the commutation relation (14.12). However, at the time of writing this symmetry algebra is unknown.

The six-vertex limit for N even

We complete the study of the eigenvalues of the transfer matrix and the matrix $Q_{72oe}(v)$ with $m_2 = 0$ by considering the limit $q \to 0$. In this limit the eight-vertex model reduces to the six-vertex model for which the operator

$$S^z = \frac{1}{2}\sum_{j=1}^{N} \sigma_j^z \tag{14.283}$$

commutes with the transfer matrix. Thus the six-vertex model has an additional symmetry which the eight-vertex model does not possess. This additional symmetry makes it possible to consider a basis for eigenvectors in which the value of S^z is the same for all components of the vector. This basis with a fixed number of down spins is universally used in the presentation of the Bethe's ansatz form of the eigenvectors of the six-vertex model.

However, the transfer matrix of the eight-vertex model does not commute with S^z but only commutes with the operator R. Therefore the only eigenvectors of the transfer matrix T which can possibly become an eigenvector of the six-vertex model in the basis with a fixed value of S^z in the limit $q \to 0$ will be those with $S^z = 0$. For all other eigenvectors of T there must be a degeneracy of the eigenvalues of the eight vertex model such that a linear combination of the limiting values of the eigenvectors will give the eigenvectors of the six-vertex model in the basis of fixed S^z.

This process of degeneration of eight-vertex eigenvalues in the six-vertex limit is illustrated in Tables 14.7–14.10 where we have indicated pairs of eigenvalues E_{xyz} which become degenerate as $q \to 0$. In this limit some of the roots go to infinity and the finite roots of the pairs become equal. Thus in the limit $q \to 0$ the matrix $Q^{(1)}_{72oe}(v)$ has many degenerate eigenvalues. This degeneracy of $Q^{(1)}_{72oe}(v)$ is accompanied by the fact that the matrix $Q^{(1)}_{Roe}(v)$ becomes singular in the limit $q \to 0$. It would be highly desirable to have an operator which classifies the states by how many roots go to infinity but such an operator is not known.

N odd and m_1 odd

The eigenvectors of $Q^{(1)}_{72oe}(v)$ for N odd have a completely different behavior from the case of N even. The most striking difference is that for odd N there are no eigenvectors which are eigenstates of the reflection operator R and that the roots v_j do *not* appear in the pairs (14.275 which we called Bethe roots. Instead all eigenvalues occur in degenerate pairs which have the property that if v_j is the set of roots in one member then the other member has the roots $v_j + iK.'$ There is a similarity to the N even case in that there are root configurations

$$\mathrm{Re} v_j = 0, K \tag{14.284}$$

and there are n strings

$$\mathrm{Re} v_j = v_c + 2j\eta + \epsilon_j \tag{14.285}$$

for $j = 0, 1, \cdots, n-1$ with $n \leq L-1$. However, complete L-strings do not occur. This behavior is illustrated in Table 14.11 for $N = 7$.

The six-vertex limit for N odd and m_1 odd

The six-vertex limit $q \to 0$ for N odd is similar to the N even case in that all roots which satisfy

$$K'/2 < \mathrm{Im} v_j < K'/2 \tag{14.286}$$

remain finite and the remaining roots go to infinity. However, unlike the N even case no eigenvalues of $Q^{(1)}_{72oe}(v)$ become degenerate as $q \to 0$. Instead in the six-vertex limit the degenerate pairs satisfy the condition that if there are n finite roots v_j for one member the other member of the pair has $N - n$ finite roots. This phenomenon is illustrated by the underlined roots in Table 14.11.

Comments on other cases

We have presented in detail the case of $m_2 = 0$ and m_1 odd. However, there are two other cases that are physically significant that need to be mentioned.

The first case is $2\eta = 2K + im_2K'/L = K + i\lambda$ with $\lambda > 0$. In the $q \to 0$ six-vertex limit of the spin chain Hamiltonian H_{XXZ} is given by (14.273) with

$$\Delta = -\cosh\lambda < 1. \tag{14.287}$$

In order for Δ to be finite λ must be remain finite in the limit $q \to 0$. Thus because $K' \to \infty$ we see that we must have $L \to \infty$. Therefore we conclude that there can be

Table 14.11 Examples of roots v_j for $\eta = K/3$ and $N = 7$ with $q = 0.02$. The energy eigenvalues on the right are degenerate with the energy eigenvalues on the left. The underlined roots remain finite as $q \to 0$. The roots are chosen in the fundamental parallelogram $0 \leq \mathrm{Re}v_j < 2K$ and $K' \leq \mathrm{Im}v_j < K'$. These states are not eigenstates of the operator R.

roots	S	P	E_{xyz}	roots	S	P	E_{xyz}
$K + 0.8732iK'$	1	0	-9.895	$K + 0.7361iK'$	-1	0	-9.895
$K + 0.5819iK'$				$\underline{K + 0.4180iK'}$			
$\underline{K + 0.2638iK'}$				$\underline{K + 0.1264iK'}$			
$\underline{K + 0.0iK'}$				$\underline{K - 0.1264iK'}$			
$\underline{K - 0.2638iK'}$				$\underline{K - 0.4180iK'}$			
$K - 0.5819iK'$				$K - 0.7361iK'$			
$K - 0.8732iK'$				$K - iK'$			
$K + 0.8443iK'$	1	0	-6.364	$K + 0.6452iK'$	-1	0	-6.364
$\underline{0.4925iK'}$				$0.5074iK'$			
$\underline{K + 0.3547iK'}$				$\underline{K + 0.1556iK'}$			
$\underline{K + 0.0iK'}$				$\underline{K - 0.1556iK'}$			
$\underline{K - 0.3547iK'}$				$-0.5074iK'$			
$\underline{K - 0.4925iK'}$				$K - 0.6452iK'$			
$K - 0.8443iK'$				$K - iK'$			
$K + 0.7639iK'$	1	$\frac{2\pi}{7}$	-5.540	$K + 0.9457iK'$	-1	$\frac{2\pi}{7}$	-5.540
$0.7009iK'$				$0.6931iK'$			
$\underline{K + 0.2805iK'}$				$\underline{K + 0.4955iK'}$			
$\underline{K - 0.0457iK'}$				$\underline{K + 0.1115iK'}$			
$\underline{-0.3068iK'}$				$\underline{4K/3 - 0.2360iK'}$			
$K + 0.5044iK'$				$\underline{-0.2990iK'}$			
$K + 0.8884iK'$				$K + 0.7194iK'$			
$K + 0.8329iK'$	1	0	-5.191	$\underline{K + 0.1670iK'}$	-1	0	-5.191
$\underline{K - 0.0iK'}$				$\underline{K - 0.1670iK'}$			
$K - 0.8320iK'$				$K - 1.0iK'$			
$0.335K + 0.6277iK'$				$\underline{0.335K + 0.3722iK'}$			
$1.664K + 0.6277iK'$				$\underline{1.664K + 0.3722iK'}$			
$0.335K - 0.6227iK'$				$\underline{0.335K - 0.3722iK'}$			
$1.664K - 0.6227iK'$				$\underline{1.664K - 0.3722iK'}$			

no multiplets in the XXZ chain for $\Delta < -1$ analogous to the complete L-strings of the case $|\Delta| < 1$.

The other case of particular interest is $m_2 = 0$ and m_1 even for N odd which is the one special case for which a matrix in the form of $Q_{72}(v)$ has not been found. However, on the assumption that there is such a matrix which satisfies the TQ equation the solutions of the TQ equation can be studied numerically for values of η to values of $\eta = m_1 K'/L$ for small L.

Such a study has been done in [41] for $\eta = 2K'/3$ and $2K'/5$ for $N = 9$. That study reveals the following fascinating properties. First of all, unlike the case of m_1 odd, where for odd N there were no Bethe roots and no complete L-strings, we find that there are many states with Bethe roots and an odd number of complete L-strings. However, there are a few states that have no Bethe roots and no complete L-strings and are degenerate in pairs just as for the m_1 odd case. These states always include the ground state and for $\eta = 2K'/3$ these two ground states are the only states with this property. This very striking property of systems with $m_2 = 0$, m_1 even, and odd N persists at $q \to 0$ where several further remarkable properties have been obtained [25–27] for the special case $\Delta = -1/2$. The ground state of the eight-vertex model with $\eta = 2K/3$ has been related to Painlevé VI by Bazhanov and Mangazeev [43] but full significance of the case m_1 even, $m_2 = 0$ with N odd is still under development.

14.3.3 Bethe's equation

We now may use the factored form of the eigenvalues (14.280) for N even, and (14.255) for N odd, in the TQ equation for eigenvalues (14.252) and then set $v = v_j$. Then because $q^{(1)}_{72oe}(v_j) = 0$ by definition the term in the TQ equation which involves $t(v)$ vanishes and we obtain an equation for the roots v_j which does not involve $t(v)$.

N odd m_1 odd

For N odd there are N such equations for each v_j and we find for $\eta = m_1K/L$ with m_1 odd

$$\left(\frac{h(v_j-\eta)}{h(v_j+\eta)}\right)^N = e^{2\pi i\nu m_1/L} \prod_{k=1,\neq j}^{N} \frac{H(v_j-v_k-2\eta)}{H(v_j-v_k+2\eta)} \tag{14.288}$$

and the eigenvalue $t(v)$ is

$$\begin{aligned} t(v) &= e^{i\pi\nu m_1/L}[h(v+\eta)]^N \prod_{j=1}^{N} \frac{H(v-v_j-2\eta)}{H(v-v_j)} \\ &\quad + e^{-i\pi\nu m_1/L}[h(v-\eta)]^N \prod_{j=1}^{N} \frac{H(v-v_j+2\eta)}{H(v-v_j)}. \end{aligned} \tag{14.289}$$

It follows from the parametrization of the Boltzmann weights (14.20) and from the property (14.503) that for $m_2 = 0$

$$T(v+iK') = [-q^{-1/2}e^{-\pi iv/K}]^N T(v). \tag{14.290}$$

Therefore if in (14.252) we set $v \to v + iK'$ and use

$$h(v+iK') = -q^{-1/2}e^{-\pi iv/K}h(v) \tag{14.291}$$

we see that both $q^{(1)}_{72oe}(v)$ and $e^{i\pi Nv/2K}q^{(1)}_{72oe}(v+iK')$ satisfy the TQ equation (14.252) with the same $t(v)$ and that if v_j satisfies the equation (14.288) then $v_j + iK'$ will satisfy (14.288) with $\nu \to \nu + N$. Therefore for odd N each eigenvalue $t(v)$ is doubly degenerate and the two eigenvalues $q^{(1)}_{72eo}(v)$ which correspond to this eigenvalue have opposite values of $(-1)^{\nu_S}$.

N even m_1 odd

For N even the situation is more complicated. This is because by the use of the periodicity $H(v+2K) = -H(v)$ we see that complete L-strings satisfy

$$\prod_{k=0}^{L-1} H(v - v_{j;k}) = -\prod_{k=0}^{L-1} H(v - v_{j;k} \pm 2\eta) \tag{14.292}$$

and therefore, recalling that n_L, the number of complete L-strings is even, we see that the factors in the TQ equation for eigenvalues which involve L strings cancels out from the equation (14.288). Furthermore the remaining roots occur in pairs (14.275). Thus the equation (14.288) for the N roots v_j^B may be reduced to an equation for the n_B roots v_j^B lying in

$$-K'/2 \leq \text{Im} v_j < K'/2 \tag{14.293}$$

by use of the identity which follows from (14.503)

$$\frac{H(v_j^B - v_k^B - 2\eta)H(v_j^B - (v_k^B \pm iK') - 2\eta)}{H(v_j^B - v_k^B + 2\eta)H(v_j^B - (v_k^B \pm iK') + 2\eta)} = e^{\mp 2\pi i \eta/K} \frac{h(v_j^B - v_k^B - 2\eta)}{h(v_j^B - v_k^B + 2\eta)}. \tag{14.294}$$

Therefore, calling n_+^B the number of v_j^B in

$$0 \leq \text{Im} v_j^B < K'/2 \tag{14.295}$$

and n_-^B the number of v_j^B in

$$-K'/2 \leq \text{Im} v_j^B < 0 \tag{14.296}$$

and using $\eta = m_1/L$ we find that the v_j^B satisfy

$$\left(\frac{h(v_j^B - \eta)}{h(v_j^B + \eta)}\right)^N = e^{2\pi i(\nu + n_+^B - n_-^B)m_1/L} \prod_{\substack{k=1 \\ k \neq j}}^{n_B} \frac{h(v_j^B - v_j^B - 2\eta)}{h(v_j^B - v_k^B + 2\eta)}. \tag{14.297}$$

We refer to this equation for v_j^B as a Bethe equation. The eigenvalue $t(v)$ is given as

$$t(v) = e^{i\pi(\nu + n_+^B - n_-^B)m_1/L}[h(v+\eta)]^N \prod_{j=1}^{n_B} \frac{h(v - v_j^B - 2\eta)}{h(v - v_j^B)}$$
$$+ e^{-i\pi(\nu + n_+^B - n_-^B)m_1/L}[h(v-\eta)]^N \prod_{j=1}^{n_B} \frac{h(v - v_j^B + 2\eta)}{h(v - v_j^B)}. \tag{14.298}$$

The sum rule (14.260) shows that if the center of an complete L-string is moved from v_j^c to $v_j^c + iK'$ the quantum number ν increases by L. From this we see that the Bethe equations (14.297) are the same for all the 2^{n_L} allowed positions of the string centers contained in the configuration v_j and thus the Bethe roots in v_j are independent of the L strings. Therefore the products in the expression for the eigenvalues (14.298) is the same for all the 2^{n_L} different configurations of the n_L complete L-strings. However, because m_1 is odd the phase factors in (14.298) change sign when $\nu \to \nu + L$ and thus these 2^{n_L} eigenvalues are split into two degenerate multiplets of size 2^{n_L-1} which have opposite signs. It follows from (14.257) and (14.260) that for L even that the quantum number ν_S is the same for both multiplets of eigenvalues but for L odd the two multiplets have opposite signs of the eigenvalue of S.

The functional equation for $Q_{72}(v)$

If our interest were restricted only to the eigenvalues of the transfer matrix $T(v)$ it would be sufficient to confine our attention to the computation of the Bethe roots v_j^B. However, if we wish to complete the study of the matrix $Q_{72}(v)$ we still require an equation which determines the values of the centers of the complete L-strings. Such an equation has been conjectured in [39] from the connection which the eight-vertex model has with the chiral Potts model

$$A' e^{-N\pi i v/2K} Q_{72}(v - iK') = \sum_{l=0}^{L-1} \frac{h^N(v-(2l+1)\eta) Q_{72}(v)}{Q_{72}(v-2l\eta) Q_{72}(v-2(l+1)\eta)} \tag{14.299}$$

where $\mathbf{A}'$ is a matrix which commutes with $\mathbf{Q}_{72}(v)$, is independent of v and depends on the normalization in the construction of $\mathbf{Q}_{72}(v)$.

It is to be noted that the left-hand side of (14.299) is manifestly an entire function and therefore the poles at the zeros of $Q_{72}(v)$ which appear in the individual terms on the right-hand side must cancel in the sum. This necessary cancellation leads to the Bethe equation (14.297) and thus (14.299) not only determines the centers of the complete L-strings but determines the Bethe roots as well. At the time of writing this more powerful equation remains to be proven.

14.3.4 Computation of the free energy

To complete the computation of the free energy of the eight-vertex model we need to use the Bethe equation to compute in the limit $N \to \infty$ the location of the roots v_k studied for small systems in s14.3.2. The cases of even and odd N need to be treated separately.

The case N is even

For even N it is seen in Tables 14.6 and 14.7 that all of the N roots v_j for the maximum eigenvalue of the transfer matrix (and the ground state of the XYZ chain given by (14.271)) have the property that

$$Re(v_j) = K \tag{14.300}$$

for all j and thus we set

$$v_j^B = K + ix_j \tag{14.301}$$

with

$$-K'/2 \leq x_j < K'/2. \tag{14.302}$$

For $N/2$ even we see that

$$\nu = \sum_{j=1}^{N} \mathrm{Im} v_j / K' = 0, \quad n_+^B = n_-^B = 0 \tag{14.303}$$

and for $N/2$ odd we have

$$\nu = -1, \quad n_+^B - n_-^B = 1 \tag{14.304}$$

and thus the Bethe equation (14.297) for v_j^B with N even reduces to

$$\left(\frac{h(K + ix_j - \eta)}{h(K + ix_j + \eta)} \right)^N = \prod_{\substack{k=1 \\ k \neq j}}^{N/2} \frac{h(ix_j - ix_k - 2\eta)}{h(ix_j - ix_k + 2\eta)}. \tag{14.305}$$

We wish to study the limiting behavior of x_j as $N \to \infty$. In this limit the roots x_j will become dense on the real line (14.302) We characterize this limiting solution for x_j by letting the number of roots in the interval $[x, x + dx]$ be given in terms of a density function $\rho(x)$ as $N\rho(x)dx$. Thus for any function $f(x)$ we have

$$\lim_{N \to \infty} \frac{1}{N} \sum_{k=1}^{N/2} f(v_k) = \int_{-K'/2}^{K'/2} f(x)\rho(x)dx \tag{14.306}$$

where setting $f(x) = 1$ we have the normalization condition

$$\int_{-K'/2}^{K'/2} \rho(x) = 1/2. \tag{14.307}$$

To compute the density function $\rho(x)$ we must convert the product in (14.305) into a sum by taking the logarithm. To do this we first use the properties $h(-v) = -h(v)$ and $h(v + 2K) = -h(v)$ and the fact that N is even to write (14.305) as

$$\left(\frac{h(\eta + K - ix_j)}{h(\eta + K + ix_j)} \right)^N = (-1)^{1+N/2} \prod_{k=1}^{N/2} \frac{h(2\eta - ix_j + ix_k)}{h(2\eta + ix_j - ix_k)}. \tag{14.308}$$

It is most important to recognize that the branch of the logarithm of $(-1)^{1+N/1}$ does not need to be zero for N odd or $i\pi$ for N even but may be

$$\ln(-1)^{1+N/2} = 2\pi i I_j \tag{14.309}$$

where I_j is a set of integers if $N/2$ is odd and integers plus $1/2$ if $N/2$ is even which must be appropriately chosen to correspond to the ground state configuration of roots. Thus we find for $1 \leq j \leq N/2$

$$N \ln \left(\frac{h(\eta + K - ix_j)}{h(\eta + K + ix_j)} \right) = 2\pi i I_j + \sum_{k=1}^{N/2} \ln \left(\frac{h(2\eta - ix_j + ix_k)}{h(2\eta + ix_j - ix_k)} \right) \tag{14.310}$$

where the logarithms are purely imaginary, are defined to be zero when $x_j = x_k = 0$ and lie in the interval $[-i\pi, i\pi]$.

To determine the proper values of I_j we note that it has been proven by Yang and Yang [9] that in the limit where $q \to 0$ the I_j for the ground state of the XXZ model are uniformly spaced with

$$I_j - I_{j-1} = 1. \tag{14.311}$$

These integers do not depend on q and thus (14.311) holds in general. The similar conclusion can be obtained by considering the case $\eta = K/2$ where the XYZ spin chain reduces to the XY model.

It remains to use x as the independent variable instead of the discrete integers j. This is done by considering the difference of two equations in (14.310) with j and $j-1$ and using the definition of the density function in the form

$$\frac{I_j - I_{j-1}}{N} = \rho(x)dx. \tag{14.312}$$

Thus from the discrete set of nonlinear equations (14.308) we obtain the linear integral equation

$$i\frac{d}{dx}\ln\left(\frac{h(\eta+K-ix)}{h(\eta+K+ix)}\right) = -2\pi\rho(x) + i\int_{-K'/2}^{K'/2} dx'\rho(x')\frac{d}{dx}\ln\left(\frac{h(2\eta-ix+ix')}{h(2\eta+ix-ix')}\right). \tag{14.313}$$

The functions which are the arguments of the logarithms are quasiperiodic in x with quasiperiod K'. Therefore the derivative of the logarithm is a periodic function of x. The range of integration is over a full period and therefore the equation can be solved by Fourier transforms.

We define the Fourier transform $\hat{f}_n$ of a function $f(x)$ which is periodic in the interval $-K'/2 \leq x \leq K'/2$ as

$$\hat{f}_n = \int_{-K'/2}^{K'/2} dx f(x) e^{2\pi i n x/K'} \tag{14.314}$$

with the inverse

$$f(x) = \frac{1}{K'}\sum_{n=-\infty}^{\infty} \hat{f}_n e^{-2\pi i n x/K'}. \tag{14.315}$$

Thus we multiply (14.313) by $e^{2\pi i n x/K'}$, integrate on x from $-K'/2$ to $K'/2$, use the property that

$$\int_{-K'/2}^{K'/2} dx e^{2\pi i n x/K'} \int_{-K'/2}^{K'/2} dx' f(x')g(x-x') =$$
$$\int_{-K'/2}^{K'/2} dx \int_{-K'/2}^{K'/2} dx' e^{2\pi i n x'/K'} f(x') e^{2\pi i n(x-x')/K'} g(x-x') = \hat{f}_n \hat{g}_n \tag{14.316}$$

and use the Fourier transforms (14.534) and (14.535) to find for $n = 0$

$$\frac{2\pi\eta}{K} = 2\pi\hat{\rho}_0 + \frac{2\pi(2\eta - K)}{K}\hat{\rho}_0 \tag{14.317}$$

and for $n \neq 0$

$$\frac{2\pi \sinh(2\pi n\eta/K')}{\sinh(2\pi nK/K')} = 2\pi\hat{\rho}_n + \frac{\sinh[2\pi n(2\eta - K)]}{\sinh(2n\pi K/K')}\hat{\rho}_n. \tag{14.318}$$

Equation (14.317) is easily solved to give

$$\hat{\rho}_0 = 1/2 \tag{14.319}$$

which is consistent with the normalization (14.307). By use of the identity

$$\sinh a + \sinh(2b - a) = 2\sinh b \cosh(b - a) \tag{14.320}$$

equation (14.318) is solved as

$$\hat{\rho}_n = \frac{1}{\cosh[2\pi n(K - \eta)/K']}. \tag{14.321}$$

To complete the computation of the free energy from (14.251) it remains to use the density of ground state roots $\rho(x)$ to determine the maximum eigenvalue $t_{max}(v)$ from (14.298). For $0 \leq \eta \leq v \leq 2K - \eta$ the first term in (14.298) is exponentially larger than the second and thus we obtain

$$\lim_{N\to\infty} \frac{1}{N} \ln t_{max}(v) = \ln h(v+\eta) + \int_{-K'/2}^{K'/2} dx\rho(x) \ln\left(\frac{h(ix + K - v + 2\eta)}{h(ix + K - v)}\right). \tag{14.322}$$

By use of (14.315) this is written in terms of $\hat{\rho}_n$ as

$$\lim_{N\to\infty} \frac{1}{N} \ln t_{max}(v) = \ln h(v + \eta)$$
$$+\frac{1}{K'} \sum_{n=-\infty}^{\infty} \hat{\rho}_n \int_{-K'/2}^{K'/2} dx e^{-2\pi inx/K'} \ln\left(\frac{h(ix + K - v + 2\eta)}{h(ix + K - v)}\right) \tag{14.323}$$

and thus recalling the normalization in definition of $h(v)$ (14.34) and using (14.321) and (14.536) we obtain the desired result for the free energy for $0 \leq \eta \leq K$

$$-F_8/k_B T = \lim_{N\to\infty} \frac{1}{N} \ln t_{max}(v) = \ln[\Theta(0)H(v + \eta)\Theta(v + \eta)]$$
$$+\frac{\pi\eta(v - \eta)}{KK'} + \sum_{n=1}^{\infty} \frac{\sinh[2n\pi(v - \eta)/K'] \sinh[2n\pi\eta/K']}{n \sinh[2n\pi K/K'] \cosh[2n\pi(K - \eta)/K']}. \tag{14.324}$$

We note that the expression (14.324) is not periodic in $0 \leq v \leq 2K$ because this expression is valid only in the restricted region $0 < \eta < v < 2K - \eta$ where the series converges. In the remainder of the region $2K - \eta \leq 2K$ a companion form is valid and taken together the two regions have the necessary periodicity.

The case $2\eta = 2K + i\lambda$

The free energy in the region $0 \leq \eta \leq K$ and the free energy in the region $2\eta = 2K + i\lambda$ are related by the symmetries of the partition function derived in 13.4.1 for N_v and N_h even

$$Z(w_1, w_2, w_3, w_4) = Z(\pm w_i, \pm w_j, \pm w_k, \pm w_l) \tag{14.325}$$

where i, j, k, l take on all permutations of 1, 2, 3, 4 and

$$w_1 = \frac{c+d}{2}, \quad w_2 = \frac{c-d}{2}, \quad w_3 = \frac{a-b}{2}, \quad w_4 = \frac{a+b}{2}. \tag{14.326}$$

The original computations of [32] for the free energy are explicitly carried out in the region $0 \leq \lambda \leq K'$ where $w_1 > w_2 > w_3 > |w_4|$ with the result

$$\lim_{N\to\infty} \frac{1}{N} \ln t_{max}(v) = \ln c + 2\sum_{n=1}^{\infty} \frac{\sinh^2[(K'-\lambda)n/2K][\cosh(n\lambda/2K) - \cosh(nV/2K)]}{n \sinh(nK'/K)\cosh(n\lambda/2K)} \tag{14.327}$$

where

$$2v = 2K + iV. \tag{14.328}$$

It is shown in [48] that the result (14.324) may be obtained from (14.327) by use of the substitutions

$$K'/K \to 4K/K' \tag{14.329}$$

$$\lambda \to 4(K-\eta)/K' \tag{14.330}$$

$$V \to 4(K-v)/K' \tag{14.331}$$

which are obtained from the symmetries (14.325).

The six-vertex limit

In the limit where $k \to 0$ ($q \to 0$) with η and v fixed, the sum in (14.324) reduces to an integral and we find the free energy of the six-vertex model for $0 \leq \eta \leq \pi/2$

$$-F_6/k_BT = \ln\sin(v+\eta) + \int_0^{\infty} dx \frac{\sinh(v-\eta)x \sinh\eta x}{x \sinh \pi x/2 \cosh(\pi/2-\eta)x}. \tag{14.332}$$

Similarly, in the region $2\eta = 2K + i\lambda$ we find from (14.327) that as $k \to 0$ the free energy reduces to

$$-F_6/k_BT = \ln c + \sum_{n=1}^{\infty} \frac{e^{-2n\lambda}[\sinh(n\lambda) - \cosh(nV)]}{n\cosh(n\lambda)}. \tag{14.333}$$

The XYZ spin chain

In 13.340 we found that the Hamiltonian of the XYZ model is given in terms of the transfer matrix of the eight-vertex model as

$$H_{XYZ} = -\frac{1}{2}\sum_{j=1}^{N}\{J^x\sigma_j^x\sigma_{j+1}^x + J^y\sigma_j^y\sigma_{j+1}^y + J^z\sigma_j^z\sigma_{j+1}^z\} \tag{14.334}$$

$$= -\mathrm{sn}(2\eta)\left[\frac{\partial}{\partial v}\ln T(v) - N\frac{1}{2}\left(\frac{H'(2\eta)}{H(2\eta)} + \frac{\Theta'(2\eta)}{\Theta(2\eta)}\right)\right] \tag{14.335}$$

with

$$J^x = 1 + k\mathrm{sn}^2(2\eta) \tag{14.336}$$
$$J^y = 1 - k\mathrm{sn}^2(2\eta) \tag{14.337}$$
$$J^z = \mathrm{cn}(2\eta)\mathrm{dn}(2\eta). \tag{14.338}$$

When $0 \leq \eta \leq K$ is real then $|J^z| \leq 1$ and we note the special cases

$$\eta = 0 \qquad J^x = J^y = J^z = 1 \tag{14.339}$$
$$\eta = K/2 \qquad J^x = 1 + k,\ J^y = 1 - k\ J^z = 0 \tag{14.340}$$
$$\eta = K \qquad J^x = J^y = -J^z = 1 \tag{14.341}$$

and for $k = 0$ (14.334) reduces to the XXZ model

$$H_{XXZ} = -\frac{1}{2}\sum_{j=1}^{N}\{\sigma_j^x\sigma_{j+1}^x + \sigma_j^y\sigma_{j+1}^y + J^z\sigma_j^z\sigma_{j+1}^z\} \tag{14.342}$$

with $|J^z| \leq 1$. There is no long range order in the XXZ chain for $|J^z| \leq 1$ and for all $0 \leq k \leq 1$ the region $0 \leq \eta \leq K$ is referred to as the disorder region.

When

$$2\eta = 2K + i\lambda \ \text{ with } \ 0 \leq \lambda \leq K' \tag{14.343}$$

we have

$$J^x = 1 + k\mathrm{sn}^2(i\lambda) \tag{14.344}$$
$$J^y = 1 - k\mathrm{sn}^2(i\lambda) \tag{14.345}$$
$$J^z = -\mathrm{cn}(i\lambda)\mathrm{dn}(i\lambda) \tag{14.346}$$

and for $k = 0$ this reduces to the XXZ spin chain (14.342) with $J^z < -1$. In this case there is long range antiferromagnetic order in the variable σ_j^z and for $0 \leq 0 \leq 1$ the entire region $0 < \lambda < K'$ is referred to as the ordered region.

Using (14.324) in (14.335) we find for the case $0 \leq \eta K$ that the ground state energy of the XYZ chain is

$$\lim_{N\to\infty} E_{XYX}/N = -\frac{1}{2}\mathrm{sn}2\eta\left(\frac{H'(2\eta)}{H(2\eta)} + \frac{\Theta'(2\eta)}{\Theta(2\eta)} + \frac{2\pi\eta}{KK'}\right.$$
$$\left. + \frac{4\pi}{K'}\sum_{n=1}^{\infty}\frac{\sinh[2n\pi\eta/K']}{n\sinh[2n\pi K/K']\cosh[2n\pi(K-\eta)/K']}\right). \tag{14.347}$$

This result was first found by Baxter [33]. In the limit $k \to 0$ this reduces to the ground state energy of the XXZ model

$$\lim_{N\to\infty} E_{XXZ}/N = -\frac{1}{2}\cos 2\eta - \sin 2\eta\int_0^{\infty} dx\frac{\sinh \eta x}{\sinh \pi x/2\cosh(\pi/2-\eta)x}. \tag{14.348}$$

In the ferromagnetic limit $\eta \to 0$, (14.348) reduces to

$$\lim_{N\to\infty} E_{XXZ}/N = -\frac{1}{2} \tag{14.349}$$

and in the antiferromagnetic limit $\eta \to K$ (14.348) reduces to

$$\lim_{N\to\infty} E_{XXZ}/N = \frac{1}{2} - 2\ln 2 \tag{14.350}$$

which is the result first obtained in [3,4]. An alternative form of (14.348) is given in [9]:

$$\lim_{N\to\infty} E_{XXZ}/N = -\frac{1}{2}\cos 2\eta - \frac{\sin^2 2\eta}{2}\int_0^\infty \frac{dx}{\cosh \pi x/2[\cosh((\pi - 2\eta)x + \cos 2\eta]}. \tag{14.351}$$

The reduction of (14.351) to (14.348) is given in [33].

Special cases

The starting point for these computations is the $Q(v)$ matrix for the case $\eta = m_1 K/L$ where m_1 is odd. However, in the final results for the free energy of the eight- and six-vertex models and the ground state energy for the XYZ and XXZ spin chain (14.324)-(14.348) this restriction on η no longer plays any role. Therefore by continuity in η these results (14.324)–(14.348) are valid for all real η in the range $0 \leq \eta \leq K$, even for the case where m_1 is even. What is special in (14.324)–(14.348) is that by a partial fraction decomposition the sums and integrals may all be reduced to simpler expressions.The simplest such reduction occurs for $\eta = 2K/3$ where [38,33] for all N

$$\frac{1}{N}\ln t_{max}(v) = \ln(a(v) + b(v)) = \ln h(v + \eta) = \ln[\Theta(0)H(v + \eta)\Theta(v + \eta)]. \tag{14.352}$$

Critical behavior as $q \to 0$

The most interesting property of the free energy is the approach as $q \to 0$ of the free energy F_8 (14.324) to the limiting value F_6 (14.332) of the six-vertex model at $q = 0$. This is because if q is interpreted as a "temperature-like" variable then $q = 0$ is analogous to $T - T_c$ and the approach to $q = 0$ gives the critical exponent α.

To extract this critical behavior from the result (14.324) we use the

Poisson summation formula

$$\sum_{n=-\infty}^{\infty} f(n\delta) = \delta^{-1} \sum_{n=-\infty}^{\infty} g(2\pi n/\delta) \tag{14.353}$$

where

$$g(k) = \int_{-\infty}^{\infty} dx e^{ikx} f(x). \tag{14.354}$$

We prove this formula by noting that $\sum_{n=-\infty}^{\infty} f(n\delta + t)$ is a periodic function of t with period δ. Therefore it can be expanded in a Fourier series and thus from (14.314) and the inverse (14.315) we have the identity

$$\sum_{n=-\infty}^{\infty} f(n\delta + t) = \frac{1}{\delta} \sum_{m=-\infty}^{\infty} e^{-2\pi itm/\delta} \int_0^{\delta} d\tau e^{2\pi im\tau/\delta} \sum_{n=-\infty}^{\infty} f(n\delta + \tau). \qquad (14.355)$$

Thus, by interchanging the integral with the second sum in (14.355) we have

$$\int_0^{\delta} \sum_{n=-\infty}^{\infty} e^{2\pi im\tau/\delta} f(n\delta + \tau) = \sum_{n=-\infty}^{\infty} \int_0^{\delta} e^{2\pi im\tau/\delta} f(n\delta + \tau)$$
$$= \sum_{n=-\infty}^{\infty} \int_{n\delta}^{(n+1)\delta} d\tau e^{2\pi im\tau/\delta} f(\tau) = \int_{-\infty}^{\infty} d\tau e^{2\pi im\tau/\delta} f(\tau). \qquad (14.356)$$

The desired result (14.353) follows by using (14.356) in (14.355) and then setting $t = 0$.

We use (14.353) to study (14.324) by first noting that (14.324) may be written as

$$-F_8/k_BT = \lim_{N\to\infty} \frac{1}{N} \ln t_{max}(v) = \ln[\Theta(0)H(v+\eta)\Theta(v+\eta)]$$
$$+\frac{1}{2} \sum_{n=-\infty}^{\infty} \frac{\sinh[2n\pi(v-\eta)/K'] \sinh[2n\pi\eta/K']}{n \sinh[2n\pi K/K'] \cosh[2n\pi(K-\eta)/K']} \qquad (14.357)$$

where the term $N = 0$ is interpreted as the limit $n \to 0$. Then by using (14.353) with

$$\delta = 2\pi/K' \qquad (14.358)$$

we find

$$-F_8/k_BT = \ln[\Theta(0)H(v+\eta)\Theta(v+\eta)] + \sum_{n=-\infty}^{\infty} g(K'n) \qquad (14.359)$$

where

$$g(k) = \frac{1}{2} \int_{-\infty}^{\infty} dx e^{ikx} \frac{\sinh(v-\eta)x \sinh \eta x}{x \sinh Kx \cosh(K-\eta)x}. \qquad (14.360)$$

The term in (14.359) with $n = 0$ is an integral which reduces to the six-vertex free energy (14.332) when $q = 0$. For the terms with $n \geq 1$ (≤ -1) we evaluate the integral in (14.360) by closing the contour of integration in the upper (lower) half planes and evaluating the residues at the poles at

$$x = \pm\pi i l_1/K \qquad (14.361)$$
$$x = \pm\pi i(l_2 - 1/2)/(K-\eta) \qquad (14.362)$$

where l_1 and l_2 are positive integers. When the poles (14.361) and (14.362) are all distinct we use this residue evaluation of (14.360) in (14.359) and evaluate the sums over n as a geometric series, to obtain the desired result

$$-F_8/k_BT = \ln[\Theta(0)H(v+\eta)\Theta(v+\eta)] + \frac{1}{2}\int_{-\infty}^{\infty} dx \frac{\sinh(v-\eta)x \sinh \eta x}{x \sinh Kx \cosh(K-\eta)x}$$
$$-2\sum_{l_1=1}^{\infty} \frac{q^{l_1}}{1-q^{l_1}} \frac{\sin l_1\pi(v-\eta)/K \sin l_1\pi\eta/K}{l_1 \cos l_1\pi\eta/K}$$
$$-2\sum_{l_2=1}^{\infty}(-1)^{l_2} \frac{q^{l_2/2(1-\eta/K)}}{1-q^{l_2/(1-\eta/K)}} \frac{\sin\frac{\pi(l_2-1/2)(v-\eta)}{K-\eta} \sin\frac{\pi(l_2-1/2)\eta}{K-\eta}}{(l_2-1/2)\sin\frac{K\pi(l_2-1/2)}{K-\eta}} \tag{14.363}$$

where $q = e^{-\pi K'/K}$ is the original nome used in the parametrization of the Boltzmann weights.

The first three terms in (14.363) for fixed v and η are analytic functions of q at $q \to 0$. The last term, however, has terms which at $q \to 0$ behave as

$$q^{\frac{l_2}{2(1-\eta/K)}} \tag{14.364}$$

which fails to be analytic values of l_2 which do not satisfy

$$\frac{l_2}{2(1-\eta/K)} = I \tag{14.365}$$

where I is an integer. Thus, when (14.365) does not hold, the leading singularity of the free energy (14.363) is

$$f_s = 4q^{\frac{1}{2(1-\eta/K)}} \sin\frac{\pi(v-\eta)}{2(K-\eta)} \sin\frac{\pi\eta}{2(K-\eta)} / \sin\frac{\pi K}{2(K-\eta)}. \tag{14.366}$$

When η/K is not rational the condition (14.365) is never satisfied for any l_2 and thus the leading singularity is indeed given by (14.366).

When the root of unity condition holds of

$$\eta/K = m_1/L \tag{14.367}$$

then (14.366) reduces to

$$f_s = 4q^{\frac{L}{2(L-m_1)}} \sin\frac{\pi(Lv-m_1)}{2(L-m_1)} \sin\frac{\pi m_1}{2(L-m_1)} / \sin\frac{\pi L}{2(L-m_1)}. \tag{14.368}$$

In particular consider the case $m_1 = L-1$ where (14.368) reduces to

$$f_s = 4q^{L/2} \sin\frac{\pi}{2}(Lv-L+1) \sin\frac{\pi}{2}(L-1)/ \sin\frac{\pi}{2}L \tag{14.369}$$

When L is even then $q^{L/2}$ is analytic in q at $q = 0$. However, the denominator in (14.369) $\sin\pi(L-1)/2$ vanishes for L even and thus the form (14.366) has broken down. This occurs because the poles (14.361) are no longer distinct from all the poles

(14.362), and the term $l_1 = L$ in the third term in (14.363) also diverges. The effect of this is that for L even and $m_1 = L - 1$ the leading singularity in f_s is of the form

$$f_s \sim q^{L/2} \ln q. \tag{14.370}$$

In particular, this is the case when $\eta/K = 1/2$ where the XYZ model reduces to the XY model.

Conversely, when L is odd and $m_1 = L - 1$ then the term in (14.363)

$$\sin \frac{\pi(l_2 - 1/2)\eta}{K - \eta} = \sin \frac{\pi}{2}(L-1)(2l_2 - 1) \tag{14.371}$$

vanishes for all l_2. Thus in this case the free energy is an analytic function of q. It is to be noted that this case of $\eta/K = (L-1)/L$ with L odd is one of those cases where the matrix $Q_R(v)$ of [32] is singular.

The case N is odd

For odd N all eigenvalues are (at least) doubly degenerate. For the ground state the roots v_j have the property that $\mathrm{Re} v_j = K$ for all j but unlike the case of N even instead of the roots being invariant under

$$v_j \to v_j + iK' \tag{14.372}$$

the two ground state root configurations transform into each other. With this modification the computation of the ground state free energy follows from (14.288) in a manner similar to the computation done for N even which starts with (14.297). The result [49] is that (14.324) holds for odd N as well as even N.

There is, however, a most interesting property for odd N which does not have a counterpart for N even. This is the case $\eta = 2K/3$ and N odd where we have the exact result for finite N [38]

$$t_{max}(v) = (a(v) + b(v))^N = (\Theta(0)H(v+\eta)\Theta(v+\eta))^N \tag{14.373}$$

which is to be compared with the corresponding result for even N (14.352) which only holds in the limit $N \to \infty$.

14.4 Excitations, order parameters and correlation functions of the eight- and six-vertex model

In section 14.3 we used the TQ equation to compute the free energy of the eight-vertex model and the ground state energy of the XYZ spin chain. However, the TQ equation is not restricted to the maximum eigenvalue of the transfer matrix, and the methods of section 14.3 may be extended to compute the low lying excitations of the eight-vertex transfer matrix and of the XYZ spin chain. This has been explicitly carried out in [37] as noted in Table 14.2. In particular, we note from (7.6) of [37] that for $0 \leq \eta \leq K$ the low-lying excitations of the XYZ model are obtained from root configurations where one of the Bethe roots has $\mathrm{Re}v_j = 0$ and the remaining $N/2 - 1$ roots have $\mathrm{Re}v_j = K$. These states have energy

$$\Delta E = \mathrm{sn}(2\eta; k)\frac{K_1}{K-\eta}(1 - k_1^2 \cos^2 p_1)^{1/2} \tag{14.374}$$

where the modulus k_1 is defined from

$$\frac{K'(k_1)}{K(k_1)} = \frac{K-\eta}{K'}. \tag{14.375}$$

For $0 \leq \eta \leq K/2$ it is shown in [37] that there are additional states which arise from the "string" type roots.

In the six-vertex limit where $k \to 0$, (14.375) reduces to $k_1 = 1$ for all $0 \leq \eta \leq K$ and thus (14.374) reduces to

$$\Delta E = \frac{\pi \sin 2\eta}{\pi - 2\eta}|\sin p_1|. \tag{14.376}$$

The goal in the investigation of all integrable models is to replicate for them all the computations that have been done for the Ising model. For example, the order parameter should be computed and the correlation functions should be characterized in as much detail as was done for the Ising model in chapters 10–12. However, for the eight-vertex model the study of the correlations is very far from complete and the computation of the order parameters requires more space than we have at our disposal. We will therefore restrict our attention to a summary presentation of the computations of order parameters and correlation functions is given in Table 14.12. However, as is almost always the case, the several papers in Table 14.12 do not all use the same notation. In particular we use the notations of the papers of 1971–1973 [30,32,34–36].

14.4.1 Eight-vertex polarization P_8 and XYZ order

We are interested in both the correlations of the spin variables σ_n^i in the XYZ chain and the correlations of the variables σ which specify the bond configurations of the eight-vertex model. We have seen in chapter 13 that these are closely related.

Order parameters were first studied in the language of the eight-vertex variables $\mu = \pm 1$ which lie on the bonds of the lattice in the regime $2\eta = 2K + i\lambda$ with $0 \leq \lambda \leq K'$ where the order parameter is called the polarization P_8 and is defined as

$$\langle \mu \rangle = P_8. \tag{14.377}$$

where the variable σ lies on a vertical bond of the lattice and the expectation is defined on a lattice with boundary conditions compatible with a staggered configuration in the ground state. This expectation was conjectured by Baxter and Kelland in 1974 [57] and proven by Jimbo, Miwa and Nakayashiki [64] in 1993 that

$$P_8 = \prod_{n=1}^{\infty} \left(\frac{1 + e^{-\pi n K'/K}}{1 - e^{-\pi n K'/K}} \frac{1 - e^{-\pi n \lambda/K}}{1 + e^{-\pi n \lambda/K}} \right)^2. \tag{14.378}$$

Table 14.12 Selected developements in the study of the order parameters and correlation functions of six- and eight-vertex models

Date	Author(s)	Development
1961	Lieb, Schultz, Mattis [50]	XY model correlations at $H = 0$
1968	McCoy [51]	Asymptotics of XY correlations
1971	Barouch, McCoy [52]	XY correlations for $H \neq 0$
1971	Suzuki [53, 54]	Ising–XY relation
1973	Baxter [55]	Polarization of six-vertex model
1973	Barber,Baxter [56]	Conjecture for eight-vertex "magnetic" order M_8
1974	Baxter, Kelland [57]	Conjecture for eight-vertex "electric" order P_8
1976	Baxter [58]	8 vertex corner transfer matrix
1977	Baxter [59]	Corner transfer matrix for Ising magnetization
1977	Takahashi [60]	$\langle\sigma_0^z\sigma_2^z\rangle$ for $\Delta = -1$ as $\zeta(3)$
1982	Baxter [61]	Proof of conjecture for M_8
1992	Smirnov [62]	Functional equations for form factors
1992	Jimbo, Miki Miwa,Nakayashiki [63]	Correlations for XXZ with $\Delta < -1$
1993	Jimbo, Miwa, Nakayashiki [64]	Derivation of P_8 and difference equations for eight-vertex correlations
1996	Jimbo,Miwa [65]	Conjecture for XXZ correlations for $-1 < \Delta < 1$
1998	Lashkevich, Pugai [66]	Free field construction for eight-vertex correlations
1999-2005	Kitanine, Maillet Slavnov, Terras [67–75]	Form factors and correlations for XXZ for all H, T, t and finite chains
2001-2002	Boos, Korepin Nishiyama, Shiroishi [76, 77]	Emptiness for $\Delta = -1$ as ζ(odd)
2003-2005	Boos, Nishiyama, Sakai,Sato Shiroishi, Takahashi [78–80]	$\langle\sigma_0^z\sigma_n^z\rangle$ for $\Delta = -1$ and $n = 3, 4, 5$ as $\zeta(3), \zeta(5), \zeta(7), \zeta(9)$
2005	Boos, Jimbo, Miwa Smirnov, Takeyama [81]	Conjecture for XYZ correlations

Because $P_8 > 0$ the regime $2\eta = 2K + i\lambda$ with $0 \leq \lambda \leq K'$ is called the ordered regime. In the six vertex limit $k \to 0$ this reduces to the six vertex polarization computed by Baxter [55] in 1973 by means of corner transfer matrices

$$P_6 = \prod_{n=1}^{\infty} \left(\frac{1 - e^{-2n\lambda}}{1 + e^{-2n\lambda}} \right)^2 . \tag{14.379}$$

In the regime $0 \leq \eta \leq K$ this expectation P_8 vanishes identically and this regime is called disordered.

For the XYZ chain there are three correlations of interest

$$G^i(n) = \langle \sigma_0^i \sigma_n^i \rangle \tag{14.380}$$

where $i = x, y, z$. These correlations have the property that when

$$|J^i| > \{|J^j|, |J^k|\} \tag{14.381}$$

for any i, j, k that

$$\lim_{n\to\infty} |G^i(n)| > 0 \tag{14.382}$$

$$\lim_{n\to\infty} G^{j,k}(n) = 0. \tag{14.383}$$

However, in the case

$$|J^i| = |J^j| > |J^k| \tag{14.384}$$

then

$$\lim_{n\to\infty} G^l(n) = 0 \quad \text{for all} \quad l = i, j, k. \tag{14.385}$$

If

$$\lim_{n\to\infty} G^i(n) > 0 \tag{14.386}$$

we say that the system has ferromagnetic long range order. If, however we have

$$\lim_{n\to\infty} (-1)^n G^i(n) > 0 \tag{14.387}$$

the system is said to have antiferromagnetic (or staggered) order.

In the ordered region where in (14.271) $-J^z > J^x, J^y$

$$\lim_{n\to\infty} (-1)^n G^z(n) = M_z^2 \tag{14.388}$$

where the spin expectation M_z is related to the eight-vertex expectation by

$$M_z = \langle \mu \rangle = P_8 = \prod_{n=1}^{\infty} \left(\frac{1 + e^{-\pi n K'/K}}{1 - e^{-\pi n K'/K}} \frac{1 - e^{-\pi n \lambda/K}}{1 + e^{-\pi n \lambda/K}} \right)^2 . \tag{14.389}$$

In the region where $J^x > J^y > |J^z|$ where $0 \leq \eta \leq K$ the transformation (14.329) and (14.330) from the ordered to the disordered regime sends $\sigma^z \to \sigma^x$ and thus using (14.330) and (14.330) in (14.378) we find

$$M_x = \prod_{n=1}^{\infty} \left(\frac{1 + e^{-4\pi n K/K'}}{1 - e^{-4\pi n K/K'}} \frac{1 - e^{-4\pi n (K-\eta)/K'}}{1 + e^{-4\pi n (K-\eta)/K'}} \right)^2 . \tag{14.390}$$

When $\eta = K/2$ then $J^z = 0$ and the Hamiltonian H_{XYZ} reduces to the Hamiltonian of the XY model and (14.390) reduces to

$$M_{XY} = \prod_{n=1}^{\infty} \left(\frac{1 - e^{-\pi 2(2n-1)K/K'}}{1 + e^{-\pi 2(2n-1)K/K'}} \right)^2 . \tag{14.391}$$

If we now use the Landen transformation [82, page 391]

$$K_l = \frac{1}{2}(1+k)K', \quad K_l' = (1+k)K \tag{14.392}$$

with

$$l = \frac{1-k}{1+k} \tag{14.393}$$

then (14.391) reduces to

$$M_{XY} = \prod_{n=1}^{\infty} \left(\frac{1 - e^{-(2n-1)\pi K'(l)/K(l)}}{1 + e^{-(2n-1)\pi K'(l)/K(l)}} \right)^2 . \tag{14.394}$$

Thus, by use of the identity [82, page 362]

$$l' = (1-l^2)^{1/2} = \prod_{n=1}^{\infty} \left(\frac{1 - e^{-(2n-1)\pi K'(l)/K(l)}}{1 + e^{-(2n-1)\pi K'(l)/K(l)}} \right)^4 \tag{14.395}$$

we find that

$$M_{XY} = \left(1 - \left[\frac{1-k}{1+k} \right]^2 \right)^{1/4} = \left[\frac{4k}{(1+k)^2} \right]^{1/4} \tag{14.396}$$

which agrees with the result for the XY model of [51] which is presented in 14.4.3.

14.4.2 Eight-vertex magnetization M_8

In section 13.4.2 we gave the two-to-one correspondence between the eight-vertex model with the variables $\mu = \pm 1$ on the bonds of the lattice and an equivalent model with variables $\sigma = \pm 1$ which lie on the faces of the lattice by means of the mapping

$$\mu = \sigma\sigma' \tag{14.397}$$

where σ and σ' are nearest neighbor sites. In this language the polarization P_8 is

$$P_8 = \langle \mu \rangle = \langle \sigma\sigma' \rangle. \tag{14.398}$$

It is also possible to consider

$$M_8 = \langle \sigma \rangle. \tag{14.399}$$

In the limit $K_4 = 0$ discussed in chapter 13 where the eight-vertex model reduces to two decoupled Ising models M_8 reduces to the spontaneous magnetization and thus in general we refer to M_8 as the magnetization of the eight-vertex model. In the regime $2\eta = 2K + i\lambda$ with $0 \leq \lambda \leq K'$ where the polarization P_8 has been seen to be nonzero the magnetization M_8 is also nonvanishing. It was conjectured in 1973 by Barber and Baxter [56] and proven by Baxter [61] in 1982 by corner transfer matrix methods that

$$M_8 = \prod_{n=1}^{\infty} \frac{1 - e^{-2(2n-1)\lambda}}{1 + e^{-2(2n-1)\lambda}}. \tag{14.400}$$

By use of the identity (14.395) we see that

$$M_8 = (1 - k^2)^{1/2} \tag{14.401}$$

where the modulus k is defined by

$$2\lambda = \pi \frac{K'(k)}{K(k)}. \tag{14.402}$$

14.4.3 Correlations for the XY model

When $J^z = 0$, the XYZ model is known as the XY model. The correlations were computed in 1961 by Lieb, Schultz and Mattis [50]. In particular, the correlations $\langle \sigma_j^x \sigma_{j+R}^x \rangle$ and $\langle \sigma_j^y \sigma_{j+R}^y \rangle$ for all temperatures T were computed as $R \times R$ Toeplitz determinants and, for $T = 0$, reduce to products of Ising model correlations for spins on the diagonal. The asymptotic behavior of these XY correlations was first given by McCoy [51] in 1968.

Unlike the case $J^z \neq 0$ the XY model may also be solved in the presence of a magnetic field and the correlations may be exactly computed [52]. If we write the Hamiltonian as

$$H_{xy} = -\sum_{k=1}^{N} \{ \frac{1+k}{2} \sigma_k^x \sigma_{k+1}^x + \frac{1-k}{2} \sigma_k^y \sigma_{k+1}^y + H \sigma_k^z \} \tag{14.403}$$

the ground state energy is

$$E_0 = -\frac{1}{2\pi} \int_0^{2\pi} d\theta \Lambda(\theta) \tag{14.404}$$

where

$$\Lambda(\theta) = [k^2 \sin^2 \theta + (H - \cos\theta)^2]^{1/2} > 0. \tag{14.405}$$

The correlations for all temperatures are given by

$$\langle \sigma_0^z \sigma_R^z \rangle = m_z^2 - G_R G_{-R} \tag{14.406}$$

$$\langle \sigma_0^x \sigma_R^x \rangle = \begin{vmatrix} G_{-1} & G_{-2} & \cdots & G_{-R} \\ G_0 & G_{-1} & \cdots & G_{-R+1} \\ \cdot & \cdot & \cdots & \cdot \\ G_{R-2} & G_{R-3} & \cdots & G_{-1} \end{vmatrix} \tag{14.407}$$

$$\langle\sigma_0^y\sigma_R^y\rangle = \begin{vmatrix} G_1 & G_0 & \cdots & G_{-R+2} \\ G_2 & G_1 & \cdots & G_{-R+3} \\ \cdot & \cdot & \cdots & \cdot \\ G_R & G_{R-1} & \cdots & G_1 \end{vmatrix} \tag{14.408}$$

where

$$G_n = \frac{1}{2\pi}\int_0^{2\pi} d\theta e^{-ni\theta}\left[\frac{\cos\theta - H - ik\sin\theta}{\cos\theta - H + ik\sin\theta)}\right]^{1/2} \tanh\frac{\beta}{2}\Lambda(\theta) \tag{14.409}$$

with $\beta = 1/k_B T$ and

$$m_z = G_0 = \frac{1}{2\pi}\int_0^{2\pi} d\theta\left[\frac{\cos\theta - H - ik\sin\theta)}{\cos\theta - H + ik\sin\theta)}\right]^{1/2} \tanh\frac{\beta}{2}\Lambda(\theta). \tag{14.410}$$

There are two important special cases which need to be separately discussed: $k = 1$ and $H = 0$

The transverse Ising chain $k = 1$

The special case $k = 1$ where the Hamiltonian (14.403) reduces to

$$H_{TI} = -\sum_{k=1}^{N}\{\sigma_k^x\sigma_{k+1}^x + H\sigma_k^z\} \tag{14.411}$$

is called the transverse Ising chain and in this case the ground state energy (14.404) becomes

$$E_{TI}(H) = -\frac{1}{2\pi}\int_{-\pi}^{\pi} d\theta[1 + H^2 - 2H\cos\theta]^{1/2} \tag{14.412}$$

which has the obvious symmetry property

$$E_{TI}(-H) = E_{TI}(H) \tag{14.413}$$

and what is called a duality relation

$$HE_{TI}(H^{-1}) = E_{TI}(H). \tag{14.414}$$

To express (14.412) as a hypergeometric function as defined by the integral representation (10.49) we write

$$\begin{aligned} E_{TI}(H) &= -(1+H)\frac{1}{2\pi}\int_{-\pi}^{\pi} d\theta[1 + \frac{2H}{(1+H)^2}(\cos\theta - 1)]^{1/2} \\ &= -(1+H)\frac{1}{\pi}\int_0^{\pi} d\theta[1 - \frac{4H}{(1+H)^2}\sin^2\frac{\theta}{2}]^{1/2} \end{aligned} \tag{14.415}$$

and then make the substitution

$$z = \sin^2\frac{\theta}{2}, \quad dz = \cos\frac{\theta}{2}\sin\frac{\theta}{2}d\theta = z^{1/2}(1-z)^{1/2}d\theta \tag{14.416}$$

to find

$$E_{TI}(H) = -(1+H)\frac{1}{\pi}\int_0^1 dz z^{-1/2}(1-z)^{-1/2}[1 - \frac{4H}{(1+H)^2}z]^{1/2}$$

$$= -(1+H)F(-\frac{1}{2},\frac{1}{2};1;\frac{4H}{(1+H)^2}) = -F(-\frac{1}{2},-\frac{1}{2};1;H^2) \tag{14.417}$$

where in the last line we have used the quadratic transformation of the hypergeometric function [83, page 64].

We also note for the transverse Ising model

$$\langle\sigma_0^x\sigma_1^z\rangle = G_{-1}(H) = m_z(H^{-1}) \tag{14.418}$$

which at $T = 0$ reduces to

$$\langle\sigma_0^x\sigma_1^x\rangle = m_z(H^{-1}) = \frac{1}{2\pi}\int_0^{2\pi} d\theta \left(\frac{1-He^{i\theta}}{1-He^{-i\theta}}\right)^{1/2} = F(-\frac{1}{2},\frac{1}{2};1;H^2) \tag{14.419}$$

where the square root is defined positive at $\theta = \pi$. Furthermore

$$m_z(H) = G_0(H) = \frac{H}{2}F(\frac{1}{2},\frac{1}{2};2;H^2). \tag{14.420}$$

From (14.419) it is straightforward to verify that

$$E_{TI}(H) = -\langle\sigma_0^x\sigma_1^x\rangle - Hm_z(H) \tag{14.421}$$

if the form (14.412) for $E_{TI}(H)$ is used. On the other hand if the hypergeometric form (14.417) is used the identity (14.421) takes the form

$$-(1+H)F(-\frac{1}{2},\frac{1}{2};1;\frac{4H}{(1+H)^2}) = -F(-\frac{1}{2},\frac{1}{2};1;H^2) - \frac{H^2}{2}F(\frac{1}{2},\frac{1}{2};2;H^2). \tag{14.422}$$

Furthermore, we see from (14.409) with $T = 0$ that if $k = 1$ then

$$G_n = \frac{1}{2\pi}\int_0^{2\pi} d\theta e^{-i(n+1)\theta}\left[\frac{1-He^{i\theta}}{1-He^{-i\theta}}\right]^{1/2} \tag{14.423}$$

which, by comparing with the expression for the diagonal correlation function of the Ising model $\langle\sigma_{0,0}\sigma_{N,N}\rangle$ given in (10.45)–(10.47) shows that with the identification

$$H = \alpha_2 = (\sinh 2K^v \sinh 2K^h)^{-1} \tag{14.424}$$

we have

$$\langle\sigma_0^x\sigma_N^x\rangle_{TI} = \langle\sigma_{0,0}\sigma_{N,N}\rangle_{Ising}. \tag{14.425}$$

The case $H = 0$

When $H = 0$ the XY model reduces to the special case $J_z = 0$ of the XYZ spin chain. This is the decoupling point discussed in 13.4.2 where the eight-vertex correlations reduce to products of diagonal Ising correlations. This reduction is seen from the determinental expressions (14.407) and (14.408) by noting that for $H = 0$

the integrands of (14.410) and of (14.409) with n even are odd under $\phi \to \phi + \pi$, and thus we obtain

$$G_{2n} = 0 \tag{14.426}$$
$$m_z = 0. \tag{14.427}$$

Therefore for $H = 0$ we have the results of [50]

$$\langle \sigma_0^z \sigma_R^z \rangle = -G_R G_{-R} \text{ for odd } R \tag{14.428}$$
$$0 \text{ for even } R \tag{14.429}$$

$$\langle \sigma_0^x \sigma_{2n}^x \rangle = D_n^2 \tag{14.430}$$
$$\langle \sigma_0^x \sigma_{2n-1}^x \rangle = D_n D_{n-1} \tag{14.431}$$

where

$$D_n = \begin{vmatrix} G_{-1} & G_{-3} & \cdots G_{-2n+1} \\ G_1 & G_{-1} & \cdots G_{-2n+3} \\ \cdot & \cdot & \cdots\cdot \\ G_{2n-3} & G_{2n-5} & \cdots G_{-1} \end{vmatrix} \tag{14.432}$$

and

$$\langle \sigma_0^y \sigma_{2n}^y \rangle = \tilde{D}_n^2 \tag{14.433}$$
$$\langle \sigma_0^y \sigma_{2n-1}^y \rangle = \tilde{D}_n \tilde{D}_{n-1} \tag{14.434}$$

where

$$\tilde{D}_n = \begin{vmatrix} G_1 & G_{-1} & \cdots G_{-2n+3} \\ G_3 & G_1 & \cdots G_{-2n+5} \\ \cdot & \cdot & \cdots\cdot \\ G_{2n-1} & G_{2n-3} & \cdots G_1 \end{vmatrix}. \tag{14.435}$$

Behavior for large R

For large R the behavior of the correlation $\langle \sigma_0^z \sigma_R^z \rangle$ is readily obtained by expanding G_R as given by (14.409). The correlations $\langle \sigma_0^x \sigma_R^x \rangle$ and $\langle \sigma_0^y \sigma_R^y \rangle$ of (14.407) and (14.408) are Toeplitz determinants and are expanded for large R using the methods developed for the Ising model in chapter 12.

For $T > 0$ all correlations decay exponentially to zero as $R \to \infty$.

For $T = 0$ the function G_n of (14.409) simplifies because the factor $\tanh \beta\Lambda(\theta)/2$ is replaced by unity and thus we may write G_n in the form

$$G_n = \frac{1}{2\pi} \int_0^{2\pi} d\theta e^{-ni\theta} \left[\frac{(1 - \lambda_+^{-1} e^{i\theta})(1 - \lambda_- e^{-i\theta})}{(1 - \lambda_+^{-1} e^{-i\theta})(1 - \lambda_- e^{i\theta})} \right]^{1/2} \tag{14.436}$$

where

$$\lambda_\pm = \frac{H \pm [H^2 - (1 - k^2)]^{1/2}}{1 - k} \tag{14.437}$$

with the square root defined positive for $H^2 > 1 - k^2$. Using (14.436) in (14.407) and (14.408) we obtain

$$\langle \sigma_0^x \sigma_R^x \rangle = \begin{vmatrix} a_0 & a_{-1} & \cdots & a_{-R+1} \\ a_1 & a_0 & \cdots & a_{-R+2} \\ . & . & \cdots & . \\ a_{R-1} & a_{R-2} & \cdots & a_0 \end{vmatrix} \tag{14.438}$$

with

$$a_n = \frac{1}{2\pi} \int_0^{2\pi} d\theta e^{-ni\theta} C_x(\theta) \tag{14.439}$$

where

$$C_x(\theta) = \left[\frac{(1 - \lambda_+^{-1} e^{i\theta})(1 - \lambda_-^{-1} e^{i\theta})}{(1 - \lambda_+^{-1} e^{-i\theta})(1 - \lambda_-^{-1} e^{-i\theta})} \right]^{1/2} \tag{14.440}$$

and

$$\langle \sigma_0^y \sigma_R^y \rangle = \begin{vmatrix} b_0 & b_{-1} & \cdots & b_{-R+1} \\ b_1 & b_0 & \cdots & b_{-R+2} \\ . & . & \cdots & . \\ b_{R-1} & b_{R-2} & \cdots & b_0 \end{vmatrix} \tag{14.441}$$

with

$$b_n = \frac{1}{2\pi} \int_0^{2\pi} d\theta e^{-ni\theta} C_y(\theta) \tag{14.442}$$

where

$$C_y(\theta) = \left[\frac{(1 - \lambda_+ e^{-i\theta})(1 - \lambda_- e^{-i\theta})}{(1 - \lambda_+ e^{i\theta})(1 - \lambda_- e^{i\theta})} \right]^{1/2}. \tag{14.443}$$

These correlations are related to the row-to-row correlations of the Ising model of chapter 12 by a result of Suzuki [53]

$$\langle \sigma_{0,0} \sigma_{0,R} \rangle_{\text{ising}} = \cosh^2 E_v^* \beta \langle \sigma_0^x \sigma_R^x \rangle - \sinh^2 E_v^* \beta \langle \sigma_0^y \sigma_R^y \rangle \tag{14.444}$$

where

$$\tanh 2E_h \beta = (1 - k^2)^{1/2}/h, \quad \cosh 2E_v^* \beta = 1/k \tag{14.445}$$

with $E_v^* \beta$ defined from

$$\tanh E_v^* \beta = e^{-2E_v \beta}. \tag{14.446}$$

The behavior of the correlations $\langle \sigma_0^x \sigma_R^x \rangle$ and $\langle \sigma_0^y \sigma_R^y \rangle$ is obtained by the methods presented in chapter 12 for the Ising correlation functions. These methods depend on the value of the index

$$I_j = [\ln C_j(2\pi) - \ln C_j(0)]/2\pi i. \tag{14.447}$$

We find that there are several cases to consider.

1) $H > 1$

In this case $\lambda_\pm$ are real with

$$\lambda_+ > 1 > \lambda_- \tag{14.448}$$

and the indices are

$$I_x = 1, \quad I_y = -1. \tag{14.449}$$

2) $0 \leq H < 1, k \neq 0$
In this case if $H^2 + k^2 > 1$ then λ_+ and λ_- are real and

$$\lambda_+ > \lambda_- > 1. \tag{14.450}$$

However if $H^2 + k^2 < 1$ then

$$\lambda_+ = \lambda_-^* = \alpha^{-1} e^{-i\psi} \tag{14.451}$$

with

$$\alpha = (\frac{1-k}{1+k})^{1/2} \quad \text{and} \quad \cos\psi = \frac{H}{\sqrt{1-k^2}}. \tag{14.452}$$

Regardless of the sign of $H^2 + k^2 - 1$ the indices are

$$I_x = 0, \quad I_y = -2. \tag{14.453}$$

In the special case $H^2 + k^2 = 1$ where $\lambda_+ = \lambda_-$ we have for the exact result for all R that

$$\langle \sigma_0^x \sigma_R^x \rangle = m_x^2 \tag{14.454}$$
$$\langle \sigma_0^y \sigma_R^y \rangle = 0. \tag{14.455}$$

3) $H = 1$
In this case $\lambda_+ > \lambda_- = 1$ and both $C_x(\theta)$ and $C_y(\theta)$ are discontinuous at one point on the unit circle.
4) $0 \leq H < 0, \; k = 0$
In this case $|\lambda_+| = |\lambda_-| = 1$ and both $C_x(\theta)$ and $C_y(\theta)$ are discontinuous at two points on the unit circle.

The correlations will have a long range order whenever the index $I_j = 0$. This only happens for I_x and only for $0 \leq H < 1$ and $k \neq 0$. In this case Szego's theorem may be applied and we find

$$\lim_{R\to\infty} \langle \sigma_0^x \sigma_R^x \rangle = \frac{2}{1+k}[k^2(1-H^2)]^{1/4} = m_x^2 \tag{14.456}$$

where (14.456)reduces to (14.396) at $H = 0$. There is no long range order for $\langle \sigma_0^y \sigma_R^y \rangle$ for $0 \leq H < 1$, and for $H \geq 1$ neither $\langle \sigma_0^x \sigma_R^x \rangle$ nor $\langle \sigma_0^y \sigma_R^y \rangle$ has long range order.

The approach as $R \to \infty$ to the long range order at $T = 0$ is obtained by the same methods used for the Ising model. For $H \neq 1$ and $k \neq 0$ the decay is exponential. For $H = 1$ or $k = 0, \; 0 \leq H < 1$ the decay is algebraic.

The correlations $\langle\sigma_j^x\sigma_{j+R}^x\rangle$ are treated by exactly the same methods used to study the Ising correlation functions. In particular we see from (14.437) that as $k \to 1$

$$\lambda_+ \to \infty, \quad \lambda_- \to H \tag{14.457}$$

and the correlation $\langle\sigma_0^x\sigma_R^x\rangle$ is identical to the correlation $\langle\sigma_{0,0}\sigma_{N,N}\rangle$ of the Ising model with the identification $\alpha_2 = H$.

In case 2 with $0 \le H < 1$ and $k \neq 0$ the index $I_x = 0$ and thus $\langle\sigma_0^x\sigma_R^x\rangle$ is analogous to the Ising model with $T < T_c$. Thus we find that the leading terms, as $R \to \infty$, are

$$\langle\sigma_0^x\sigma_R^x\rangle \sim m_x^2\{1+\frac{1}{(2\pi)^2}\oint\oint dz_1dz_2\frac{z_1^Rz_2^R}{(1-z_1z_2)^2}\frac{P_2(z_1)P_2(z_1^{-1})}{P_2(z_2)P_2(z_2^{-1})}+O(\lambda_-^{4R})\} \tag{14.458}$$

with $|z_{1,2}| = 1-\epsilon$ where

$$P_2(z) = [(1-\lambda_-^{-1}z)(1-\lambda_+^{-1}z)]^{1/2}. \tag{14.459}$$

By expanding the integrals for large R we see that when $H^2+k^2 > 1$ the approach to m_x^2 is monotonic:

$$\langle\sigma_0^x\sigma_R^x\rangle \sim m_x^2\{1+\frac{\lambda_-^{-2R}}{2\pi(\lambda_- - \lambda_-^{-1})^2R^2}+\cdots\} \tag{14.460}$$

For $H^2+k^2 < 1$ the large R behavior is

$$\begin{aligned}\langle\sigma_0^x\sigma_R^x\rangle \sim m_x^2\{1+\frac{\alpha^{2R}}{\pi R^2}\Bigg[&\mathrm{Re}\frac{e^{2iR\psi}}{(\alpha e^{i\psi}-\alpha^{-1}e^{-i\psi})^2}\\ &+\frac{1}{(\alpha-\alpha^{-1})^2}\mathrm{Re}e^{i(\pi/2-\psi)}[\frac{1-\alpha^2e^{-2i\psi}}{1-\alpha^2e^{2i\psi}}]^{1/2}\Bigg]+\cdots\}\end{aligned} \tag{14.461}$$

where α and ψ are given by (14.452). When $H = 0$ then $\psi = \pi/2$ and (14.461) reduces to

$$\langle\sigma_0^x\sigma_R^x\rangle \sim m_x^2\{1+\frac{\alpha^{2R}[(\alpha+\alpha^{-1})^2-(-1)^R(\alpha-\alpha^{-1})^2]}{\pi R^2(\alpha^2-\alpha^{-2})^2}+\cdots\} \tag{14.462}$$

which agrees with the reduction (14.430)–(14.432). Note that the second term in (14.461) is erroneously omitted in [52].

In case 1 when $H > 1$ where $\lambda_+ > 1 > \lambda_-$ we have to leading order as $R \to \infty$

$$\langle\sigma_0^x\sigma_R^x\rangle \sim \left[\frac{(1-\lambda_-^2)(1-\lambda_+^{-2})}{(1-\lambda_+^{-1}\lambda_-)^2}\right]^{1/4}\frac{1}{2\pi i}\oint dz z^{R-1}P_1(z)P_1(z^{-1})+O(\lambda_-^{4R}) \tag{14.463}$$

with $|z| = 1-\epsilon$ where

$$P_1(z) = \left[\frac{(1-\lambda_+^{-1}z)}{(1-\lambda_-z)}\right]^{1/2}. \tag{14.464}$$

For large R this decays as

$$\langle\sigma_0^x\sigma_R^x\rangle \sim \frac{\lambda_-^R}{(\pi R)^{1/2}}\left[\frac{(1-\lambda_+^{-2})(1-\lambda_+^{-1}\lambda_+^{-1})^2}{(1-\lambda_-^{-2})}\right]^{1/4}. \tag{14.465}$$

For $\langle\sigma_0^y\sigma_R^y\rangle$ when $H>1$ and $0<k\leq 1$ we have $I_y=-1$ and thus we may use the methods used for the Ising model with $T>T_c$:

$$\langle\sigma_0^y\sigma_R^y\rangle \sim \left[\frac{(1-\lambda_-^2)(1-\lambda_+^{-2})}{(1-\lambda_+^{-1}\lambda_-)^2}\right]^{1/4}\frac{1}{2\pi i}\oint dz z^{R-1}P_1^{-1}(z)P_1^{-1}(z^{-1})+O(\lambda_-^{4R}) \tag{14.466}$$

For large R this decays as

$$\langle\sigma_0^y\sigma_R^y\rangle \sim \frac{\lambda_-^R}{2\pi^{1/2}R^{3/2}}\frac{(1-\lambda_-^2)^{3/2}(1-\lambda_+^{-2})^{1/4}}{(1-\lambda_+^{-1}\lambda_-)(1-\lambda_+^{-1}\lambda_-^{-1})^{1/2}}. \tag{14.467}$$

When $0\leq H<1$ and $0<k<1$ we have $I_y=-2$ and we must extend the methods used for the Ising model with $T>T_c$. The result of this extension is that [52]

$$\langle\sigma_0^y\sigma_R^y\rangle \sim \langle\sigma_0^x\sigma_{R+2}^x\rangle\begin{vmatrix}Y_R & Y_{R+1}\\ Y_{R-1} & Y_R\end{vmatrix} \tag{14.468}$$

with

$$Y_R=\frac{1}{2\pi i}\oint dz z^{R-1}P_2^{-1}(z)P_2^{-1}(z^{-1}) \tag{14.469}$$

where $|z|=1-\epsilon$.

When $0\leq H<1$ and $H^2+k^2>1$ the roots $\lambda_\pm$ are real and the leading contribution to Y_R is dominated by $z\sim\lambda_-^{-1}$. Thus we find to leading order

$$Y_R \sim \frac{\lambda_-^{-R}}{(\pi R)^{1/2}}[(1-\lambda_-^{-2})(1-\lambda_+^{-1}\lambda_-^{-1})(1-\lambda_+^{-1}\lambda_-)]^{-1/2}. \tag{14.470}$$

This is sufficient to obtain the leading terms in (14.495) and thus

$$\langle\sigma_0^y\sigma_R^y\rangle \sim -m_x^2\frac{\lambda_-^{-2R}}{2\pi R^3}[(1-\lambda_-^{-2})(1-\lambda_+^{-1}\lambda_-^{-1})(1-\lambda_+^{-1}\lambda_-)]^{-1}. \tag{14.471}$$

When $0<H<1$ and $H^2+k^2<1$ the roots (14.451) $\lambda_\pm$ are complex conjugates. Thus if we write

$$[(1-\lambda_1^{-2})(1-\lambda_+^{-1}\lambda_-^{-1})(1-\lambda_+^{-1}\lambda_-)]^{-1/2}=ce^{-i\phi} \tag{14.472}$$

we have from (14.470)

$$Y_R \sim \frac{2c\cos(R\psi+\phi)\alpha^{-R}}{(\pi R)^{1/2}}. \tag{14.473}$$

This is sufficient to obtain the leading term of (14.468) and thus we obtain

$$\langle\sigma_0^y\sigma_R^y\rangle \sim m_x^2 4c^2\sin^2\psi\frac{\alpha^{-2R}}{\pi R}. \tag{14.474}$$

We thus see that even though the leading term in Y_R oscillates, the leading term in $\langle\sigma_0^y\sigma_R^y\rangle$ is monotonic. Oscillations first occur in the term of order $R^{-3}\alpha^{-2R}$. This behavior of $\langle\sigma_0^y\sigma_R^y\rangle$ is to be contrasted with $\langle\sigma_0^x\sigma_R^x\rangle$ where the oscillations occur in leading order.

The case $H = 1$ is treated by the methods used for Ising correlations at $T = T_c$ and we have

$$\langle\sigma_0^x\sigma_R^x\rangle \sim \frac{2k}{1+k}\frac{e^{1/4}2^{1/12}A^{-3}}{(kR)^{1/4}} + O(R^{-9/4}) \tag{14.475}$$

$$\langle\sigma_0^y\sigma_R^y\rangle \sim -\frac{k}{1+k}\frac{e^{1/4}2^{1/12}A^{-3}}{8(kR)^{9/4}} + O(R^{-17/4}) \tag{14.476}$$

$$\langle\sigma_0^z\sigma_R^z\rangle \sim m_z^2 - \frac{1}{(\pi R)^2} + \cdots \tag{14.477}$$

where $A = 1.282427130\cdots$ is Glaisher's constant [84, page 411].

When $k = 0$ The correlations $\langle\sigma_j^x\sigma_{j+R}^x\rangle$ and $\langle\sigma_j^y\sigma_{j+R}^y\rangle$ are obviously equal. The large R behavior is obtained by an extension of the methods used for the Ising correlations at $T = T_c$. They oscillate as $e^{i\phi R}$ and decay as

$$\langle\sigma_0^x\sigma_R^x\rangle = \langle\sigma_0^y\sigma_R^y\rangle = O(1/R^{1/2}). \tag{14.478}$$

We also note the exact result

$$\langle\sigma_0^z\sigma_R^z\rangle = m_z^2 - \left(\frac{2\sin(R\cos^{-1}H)}{\pi R}\right)^2. \tag{14.479}$$

14.4.4 XYZ correlations

The study of the correlations for the full XYZ model is much more involved than the XY case where $J^z = 0$. The nearest neighbor correlations may be obtained by differentiating the ground state energy but for many years the only additional result known was the fascinating expression obtained by Takahashi [60] in 1977 for the next nearest neighbor correlation for the XXZ model at $\Delta = -1$

$$\langle\sigma_0^z\sigma_2^z\rangle = \frac{1}{3} - \frac{16}{3}\ln 2 + 3\zeta(3) \tag{14.480}$$

where $\zeta(s)$ is the Riemann zeta function

$$\zeta(s) = \sum_{n=1}^{\infty} 1/n^s. \tag{14.481}$$

The systematic study of the correlations of the XXZ and the XYZ model originates with the functional equations for form factors found by Smirnov and the representation theory methods of the Jimbo, Miwa and collaborators. These methods are explained in detail in the books of Smirnov [62] and of Jimbo and Miwa [85]. There is continuing progress being made in this field which is briefly summarized in Table 14.12. These techniques express correlations of the XXZ and the XYZ model as multiple dimensional integrals whose size depends on (and increases with) the separation of the spins being correlated.

The original multiple integral expressions for XXZ and XYZ correlations differ from the correlations of the XY model given above which, because of their determinental form, are all given as sums of products of one-dimensional integrals. However the result of Takahashi [60] for the second neighbor correlation strongly suggests that these multiple dimensional integrals can be further reduced.

The understanding of this reduction was first made by Boos and Korepin [76] in a study of a correlation function $P(n)$ defined as

$$P(n) = \langle \prod_{j=1}^{n} \frac{1+\sigma^z}{2} \rangle. \tag{14.482}$$

This correlation is called the "emptiness probability".

Boos and Korepin studied the cases of the XXZ model with $\Delta = -1$ where $P(n)$ has the integral representation [86]

$$P(n) = \int_C \frac{d\lambda_1}{2\pi i \lambda_1} \int_C \frac{d\lambda_2}{2\pi i \lambda_2} \cdots \int_C \frac{d\lambda_n}{2\pi i \lambda_n} \prod_{a=1}^{n} \left(1 + \frac{i}{\lambda_a}\right)^{n-a} \left(\frac{\pi \lambda_a}{\sinh \pi \lambda}\right)^{n}$$
$$\times \prod_{1 \le j < k \le n} \frac{\sinh \pi(\lambda_k - \lambda_j)}{\pi(\lambda_k - \lambda_j - i)} \tag{14.483}$$

where the contour C in each integral is parallel to the real axis with imaginary part between 0 and $-i$. They found that $P(3)$ and $P(4)$ are expressed in terms of $\zeta(3)$ and $\zeta(5)$. These computations have been extended to $n = 5$ in [77]. The results are conveniently expressed in term of the alternating zeta function

$$\zeta_a(s) = \sum_{n=1}^{\infty} (-1)^{n-1}/n^s = (1 - 2^{1-s})\zeta(s) \tag{14.484}$$

and we note that

$$\zeta_a(1) = \zeta(1) = \ln 2. \tag{14.485}$$

For $1 \le n \le 5$ the results are:

$$P(1) = \frac{1}{2} = 0.5 \tag{14.486}$$

$$P(2) = \frac{1}{3} - \frac{1}{3}\zeta_a(1) = 0.102284273\cdots \tag{14.487}$$

$$P(3) = \frac{1}{4} - \zeta_a(1) + \frac{3}{2}\zeta_a(3) = 0.007628\cdots \tag{14.488}$$

$$P(4) = \frac{1}{5} - 2\zeta_a(1) + \frac{173}{45}\zeta_a(3) - \frac{22}{9}\zeta_a(1)\zeta_a(3) - \frac{17}{15}\zeta_a^2(3) - \frac{22}{9}\zeta_a(5)$$
$$+ \frac{34}{9}\zeta_a(1)\zeta_a(5) = 0.000206270\cdots \tag{14.489}$$

$$P(5) = \frac{1}{6} - \frac{10}{3}\zeta_a(1) + \frac{281}{18}\zeta_a(3) - 30\zeta_a(1)\zeta_a(3) - \frac{163}{3}\zeta_a^2(3)$$
$$- \frac{1355}{36}\zeta_z(5) + \frac{1960}{9}\zeta_a(1)\zeta_a(5) - \frac{85}{9}\zeta_a(3)\zeta_a(5) - \frac{485}{90}\zeta_a^2(5)$$
$$+ \frac{889}{36}\zeta_a(7) - \frac{1645}{9}\zeta_a(1)\zeta_a(7) + \frac{679}{6}\zeta_a(3)\zeta_a(7) = 2.0117259898884\cdots \times 10^{-6} \tag{14.490}$$

The explicit result for $P(6)$ is given in [80] and involves terms such as $\zeta_a^n(5)\zeta_a^{3-n}(3)$ with $0 \leq n \leq 3$. Its numerical value is

$$P(6) = 7.06812753309293\cdots \times 10^{-9}. \tag{14.491}$$

The correlations $\langle \sigma_0^z \sigma_n^z \rangle$ for $\Delta = -1$ may be evaluated by similar methods. For $1 \leq 4$ the results are

$$\langle \sigma_0^z \sigma_1^z \rangle = 2E_{XXX}/3 = \frac{1}{3} - \frac{4}{3}\zeta_a(1) = -0.5908629076\cdots \tag{14.492}$$
$$\langle \sigma_0^z \sigma_2^z \rangle = \frac{1}{3} - \frac{16}{3}\zeta_a(1) + 4\zeta_a(3) = 0.242719798\cdots \tag{14.493}$$
$$\langle \sigma_0^z \sigma_3^z \rangle = \frac{1}{3} - 12\zeta_a(1) + \frac{296}{9}\zeta_a(3) - \frac{224}{9}\zeta_a(1)\zeta_a(3)$$
$$\frac{32}{3}\zeta_a^2(3) - \frac{200}{9}\zeta_a(5) + \frac{320}{9}\zeta_a(1)\zeta_a(5) = -0.2009945090\cdots \tag{14.494}$$
$$\langle \sigma_0^z \sigma_4^z \rangle = \frac{1}{3} - \frac{64}{3}\zeta_a(1) + \frac{1160}{9}\zeta_a(3) - 288\zeta_a(1)\zeta_a(3) - \frac{4688}{9}\zeta_a^2(3) - \frac{2800}{9}\zeta_a(5)$$
$$- \frac{1600}{3}\zeta_a^2(5) + \frac{18560}{9}\zeta_a(1)\zeta_a(5) - \frac{220}{9}\zeta_a(3)\zeta_a(5) + \frac{455}{9}(7)$$
$$- \frac{15680}{9}\zeta_a(1)\zeta_a(7) + 1120\zeta_a(3)\zeta_a(7) = 0.1386111079\cdots \tag{14.495}$$
$$\tag{14.496}$$

where (14.493) is the result of Takahashi [60], the result (14.494) is obtained in [78] and result (14.495) in [79]. The result for $\langle \sigma_0^z \sigma_5^z \rangle$ is also known [80] and the numerical value is

$$\langle \sigma_0^z \sigma_5^z \rangle = -0.12356166\cdots \tag{14.497}$$

14.5 Appendix: Properties of the modified theta functions

We recall from chapter 13 that the functions $H(v)$ and $\Theta(v)$ have the following properties

$$H(-v) = -H(v), \qquad \Theta(-v) = \Theta(v) \tag{14.498}$$
$$H(v + 2nK) = (-)^n H(v), \qquad \Theta(v + 2nK) = \Theta(v) \tag{14.499}$$
$$H(v + 2inK') = (-1)^n q^{-n^2} e^{-n\pi i v/K} H(v) \tag{14.500}$$

$$\Theta(v + 2inK') = (-1)^n q^{-n^2} e^{-n\pi iv/K}\Theta(v) \tag{14.501}$$
$$\Theta(v + iK') = iq^{-1/4}e^{-\frac{\pi iv}{2K}} H(v) \tag{14.502}$$
$$H(v + iK') = iq^{-1/4}e^{-\frac{\pi iv}{2K}} \Theta(v) \tag{14.503}$$

where

$$q = e^{-\pi K'/K}. \tag{14.504}$$

It follows immediately from (14.498)–(14.503) that the modified theta functions $H_m(v)$ and $\Theta_m(v)$ defined by (14.19) have the properties

$$H_m(2K - v) = H_m(v), \qquad \Theta_m(2K - v) = \Theta_m(v) \tag{14.505}$$

$$H_m(-v) = -\exp\left(\frac{i\pi m_2 v}{2L\eta}\right) H_m(v) \quad \Theta_m(-v) = \exp\left(\frac{i\pi m_2 v}{2L\eta}\right)\Theta_m(v) \tag{14.506}$$

$$H_m(u + 2rK + 2isK') = \\ (-1)^r(-1)^{rs}H_m(u)\exp\{(\pi i(rm_2/2 - sm_1)[u + (r-1)K + isK']/(L\eta)\} \tag{14.507}$$

and

$$\Theta_m(u + 2rK + 2isK') \\ = (-1)^{rs}\Theta_m(u)\exp\{(\pi i(rm_2/2 - sm_1)[u + (r-1)K + isK']/(L\eta)\} \tag{14.508}$$

$$\Theta_m(v + iK') = iq^{-1/4}\exp\left(\frac{-i\pi m_1 v}{2L\eta}\right) CH_m(v) \tag{14.509}$$
$$H_m(v + iK') = iq^{-1/4}\exp\left(\frac{-i\pi m_1 v}{2L\eta}\right) C\Theta_m(v) \tag{14.510}$$

where

$$C = \exp\left(\frac{\pi m_2 K'}{8KL\eta}(2K - iK')\right). \tag{14.511}$$

For convenience we note the special cases:

$$H_m(v + 2L\eta) = (-1)^{m_1} i^{m_1 m_2} \begin{cases} H_m(v) & \text{if } m_2\text{=even} \\ \Theta_m(v) & \text{if } m_2\text{=odd} \end{cases} \tag{14.512}$$

$$\Theta_m(v + 2L\eta) = i^{m_1 m_2} \begin{cases} \Theta_m(v) & \text{if } m_2\text{=even} \\ H_m(v) & \text{if } m_2\text{=odd} \end{cases} \tag{14.513}$$

The properties (14.512), (14.513) and (14.22)–(14.25) follow immediately from (14.507)–(14.511).

The functions $H_m(v)$ and $\Theta_m(v)$ satisfy the identities

$$H_m(u)H_m(v)H_m(w)H_m(u+v+w) + \Theta_m(u)\Theta_m(v)\Theta_m(w)\Theta_m(u+v+w)$$
$$= \Theta_m(0)\Theta_m(u+v)\Theta_m(u+w)\Theta_m(v+w) \quad (14.514)$$

$$H_m(u)H_m(v)\Theta_m(w)\Theta_m(u+v+w) + \Theta_m(u)\Theta_m(v)H_m(w)H_m(u+v+w)$$
$$= \Theta_m(0)\Theta_m(u+v)H_m(u+w)H_m(v+w). \quad (14.515)$$

For the computation of the free energy for $m_2 = 0$ we need the properties of the function

$$h(v) = H(v)\Theta(v) = -iq^{1/4}e^{\pi iv/2K}H(v)H(v+iK'). \quad (14.516)$$

From (14.498)–(14.503) we see that

$$h(-v) = -h(v) \quad (14.517)$$
$$h(v+2K) = -h(v) \quad (14.518)$$
$$h(v+iK') = -q^{-1/2}e^{-\pi iv/K}h(v) \quad (14.519)$$

where (14.519) is symmetrically written as

$$h(v+iK'/2) = -e^{-\pi iv/K}h(v-iK'/2). \quad (14.520)$$

It thus follows that $h(v)$ has the same quasiperiodicity properties as $H(v)$ with $q \to q^{1/2}$ and thus, making the q dependence explicit

$$h(v;q) = CH(v;q^{1/2}) \quad (14.521)$$

where C is a constant which may be determined from the product forms of $H(v;q)$, $\Theta(v;q)$ given in (13.381) and (13.382) as

$$C = q^{1/2}\prod_{n=1}^{\infty}(1+q^n)^2(1-q^n) = H(K;q^{1/2})/2. \quad (14.522)$$

In the text we need the Fourier transform of

$$\frac{d}{dy}\ln\left(\frac{h(a-iy)}{h(a+iy)}\right) = \frac{d}{dy}\ln\left(\frac{H(i(ia+y))\Theta(i(ia+y))}{H(i(ia-y))\Theta(i(ia-y))}\right) \quad (14.523)$$

where a is either 2η or $\eta + K$ and y is either x or $x - x'$.

Because of the logarithm in (14.523) we need to use a product form of $H(v)$ to compute the Fourier transform. However, the product form (13.381) may not be directly used because the quasiperiodicity properties in the imaginary direction are not in a useful form. To overcome this we need identities known as the complementary modulus transformations:

$$H(iv;k) = i(K/K')^{1/2}e^{\pi v^2/4KK'}H(v,k') \quad (14.524)$$
$$H(iv+K;k) = (K/K')^{1/2}e^{\pi v^2/4KK'}\Theta(v,k') \quad (14.525)$$
$$\Theta(iv;k) = (K/K')^{1/2}e^{\pi v^2/4KK'}H(v+K',k') \quad (14.526)$$
$$\Theta(iv+K;k) = (K/K')^{1/2}e^{\pi v^2/4KK'}\Theta(v+K',k') \quad (14.527)$$

These identities are proven by first using (14.498) and (14.501) to show that both sides of (14.524) have the same quasiperiodicity properties and the same zeros and thus that they must be proportional. The derivation of the constant is given in many places, for example in [61, chapter 15] but because only ratios appear in (14.523) the proportionality constant is not needed for our purposes and thus the proof will be omitted.

Consider first the case $a = \eta + K$. Then using (14.525), (14.527) and (14.498) we find

$$\ln\left(\frac{h(a-iy)}{h(a+iy)}\right) = \ln\left(\frac{H(i(-i\eta-y)+K;k)\Theta(i(-i\eta-y)+K;k)}{H(i(-i\eta+y)+K;k)\Theta(i(-i\eta+y)+K;k)}\right)$$
$$= \ln\left(e^{2\pi i\eta y/KK'}\frac{\Theta(i\eta+y;k')\Theta(i\eta+y+K';k')}{\Theta(-i\eta+y;k')\Theta(-i\eta+y+K';k')}\right). \qquad (14.528)$$

By use of the product form of $\Theta(v)$ given in (13.382) we may write

$$\ln\frac{\Theta(y+x)}{\Theta(y-x)} = \sum_{m=1}^{\infty}\{\ln(1-q^{2m-1}e^{(y+x)\pi i/K}) + \ln(1-q^{2m-1}e^{-(y+x)\pi i/K})$$
$$-\ln(1-q^{2m-1}e^{(y-x)\pi i/K}) - \ln(1-q^{2m-1}e^{-(y-x)\pi i/K})\}. \qquad (14.529)$$

Thus, using the expansion

$$\ln(1-q^{2m-1}z) = -\sum_{n=1}^{\infty}\frac{1}{n}(q^{2m-1}z)^n \qquad (14.530)$$

in (14.529) and performing the sum over m we obtain

$$\ln\frac{\Theta(y+x)}{\Theta(y-x)} = \sum_{n=1}^{\infty}\frac{2}{n\sinh(n\pi K'/K)}\sin(n\pi y/K)\sin(n\pi x/K). \qquad (14.531)$$

By sending $y \to y+K$ we find the companion formula

$$\ln\frac{\Theta(y+x+K)}{\Theta(y-x+K)} = \sum_{n=1}^{\infty}(-1)^n\frac{2}{n\sinh(n\pi K'/K)}\sin(n\pi y/K)\sin(n\pi x/K) \qquad (14.532)$$

and by combining (14.531) and (14.532) we have

$$\ln\frac{\Theta(y+x)\Theta(y+x+K)}{\Theta(y-x)\Theta(y-x+K)} = \sum_{n=1}^{\infty}\frac{1}{n\sinh(2n\pi K'/K)}\sin(2n\pi y/K)\sin(2n\pi x/K). \qquad (14.533)$$

Therefore by using (14.533) in (14.528) we obtain

$$\ln\left(\frac{h(\eta+K-iy)}{h(\eta+K+iy)}\right) = \frac{2\pi i\eta y}{KK'}$$
$$+i\sum_{n=1}^{\infty}\frac{1}{n\sinh(2n\pi K/K')}\sin(2n\pi y/K')\sinh(2n\pi\eta/K'). \qquad (14.534)$$

By replacing in (14.535) η by $2\eta - K$ we also have

$$\ln\left(\frac{h(2\eta - iy)}{h(2\eta + iy)}\right) = \frac{2\pi i(2\eta - K)y}{KK'}$$
$$+ i\sum_{n=1}^{\infty} \frac{1}{n\sinh(2n\pi K/K')} \sin(2n\pi y/K')\sinh(2n\pi(2\eta - K)/K'). \quad (14.535)$$

The results (14.534) and (14.535) are valid in the region $0 \leq \eta < K$ where the series each converge.

Similarly by use of (14.525), (14.527) and (14.533) we find

$$\ln\frac{h(ix + K - v + 2\eta)}{h(ix + k - v)}$$
$$= \ln\left(\frac{H[i(x + iv - 2i\eta) + K; k]\Theta[i(x + iv - 2i\eta) + K; k]}{H[i(x + iv) + K; k]\Theta[i(x + iv) + K; k]}\right)$$
$$= \frac{2\pi\eta}{KK'}(v - \eta - ix) + i\sum_{n=1}^{\infty} \frac{1}{n\sinh 2n\pi K/K'} \sin 2\pi n(x + iv - i\eta)\sinh 2\pi n\eta/K' \quad (14.536)$$

which is valid for

$$2\eta - K < v < K. \quad (14.537)$$

References

[1] W. Heisenberg, Zur Theorie des Ferromagnetismus, Z. Physik, 29 (1928) 619–636.
[2] H.A. Bethe, Zur Theorie der Metalle: I Eigenwerte und Eigenfunktionen der linearen Atomkette, Z. Physik 71 (1931) 205–226;
[3] A. Sommerfeld and H.A. Bethe, Elektronentheorie der metalle, in *Handbuch der Physik* ed. Geiger and Scheel (Verlag Julius Springer, Berlin, 1933) Vol. 24, Part 2, 604–618.
[4] L. Hulthén, Über das Austauschproblem eines Kristalles, Arkiv för Matematik, Astronomi och Fysik 26A No. 11 (1938) 1–105.
[5] R. Orbach, Linear antiferromagnetic chain with anisotropic coupling, Phys. Rev. 112 (1958) 309–316.
[6] L.R. Walker, Antiferromagnetic linear chain, Phys. Rev. 116 (1959) 1089–1090.
[7] J. des Cloizeaux and M. Gaudin, Anisotropic linear magnetic chain, J. Math. Phys. 7 (1966) 1384–1400.
[8] C.N. Yang and C.P. Yang, One-dimensional chain of anisotropic spin-spin interactions I. Proof of Bethe's hypothesis for ground state in a finite system, Phys. Rev. 150 (1966) 321–327.
[9] C.N. Yang and C.P. Yang, One-dimensional chain of anisotropic spin-spin interactions II. Properties of the ground-state energy per lattice site for an infinite system Phys. Rev. 150 (1966) 327–339.
[10] C.N. Yang and C.P. Yang, One-dimensional chain of anisotropic spin-spin interactions III. Applications, Phys. Rev. 151 (1966) 258–264.
[11] E.H. Lieb, Phys. Rev. Lett., Exact solution of the problem of the entropy of two-dimensional ice, 18 (1967) 692–694.
[12] E.H. Lieb, Exact solution of the F model of an antiferroelectric, Phys. Rev. Lett. 18 (1967) 1046–1048.
[13] E.H. Lieb, Exact solution of the two-dimensional Slater KDP model of a ferroelectric, Phys. Rev. Lett. 19 (1967) 108–110.
[14] E.H. Lieb, Residual entropy of square ice, Phys. Rev. 162 (1967) 162–172.
[15] B. Sutherland, Exact solution of a two-dimensional model for hydrogen-bonded crystals, Phys. Rev. Lett. 19 (1967) 103–104.
[16] C.P. Yang, Exact solution of a model of two-dimensional ferroelectrics in an arbitrary external electric field, Phys. Rev. Letts. 19 (1967) 586–588.
[17] B. Sutherland, C.N. Yang and C.P. Yang, Exact solution of a model of two-dimensional ferroelectrics in an arbitrary external electric field, Phys. Rev. Lett. 19 (1967) 588–591.
[18] B.M. McCoy and T.T. Wu, Hydrogen bonded crystals and the anisotropic Heisenberg chain, Il Nuovo Cimento 56 (1968) 311–315.

[19] M. Takahashi, One dimensional Heisenberg model at finite temperature, Prog. Theo. Phys. 46 (1971) 401–415.

[20] M. Gaudin, Thermodynamics of the Heisenberg-Ising ring for $\Delta \geq 1$. Phys. Rev. Lett. 26 (1971) 1301–1304.

[21] M. Takahashi and M. Suzuki, One-dimensional anisotropic Heisenberg model at finite temperature, Prog. Theo. Phys. 48 (1972) 2187–2209.

[22] L.A. Takhtajan and L.D. Faddeev, The quantum method for the inverse problem and the XYZ Heisenberg model, Uspekhi Mat. Nauk 34(5) (1979) 13–63 (English translation: Russian Math. Surveys 34(5) (1979) 11–68).

[23] T. Deguchi, K. Fabricius and B.M. McCoy, The sl_2 loop algebra symmetry of the six-vertex model at roots of unity, J. Stat. Phys. 102 (2001) 701–736.

[24] K. Fabricius and B.M. McCoy, Evaluation parameters ans Bethe roots for the six-vertex model at roots of unity, in *MathPhys Odyssey 2001, Integrable models and beyond*, (Birkhäuser, Boston, 2002), 119–144.

[25] Yu. Stroganov, The importance of being odd, J. Phys. A 34 (2001) L179–L185.

[26] A.V. Razumov and Yu. G. Stroganov, Spin chains and combinatorics, J. Phys. A 34 (2001) 3185–3190.

[27] J. de Gier, M.T. Batchelor, B. Nienhuis and S. Mitra, The XXZ spin chain at $\Delta = -1/2$; Bethe roots, symmetric functions and determinants, J. Math. Phys. 43 (2002) 4135–4146.

[28] R.J. Baxter, Completeness of the Bethe ansatz for the six and eight vertex models, J. Stat. Phys. 108 (2002) 1–48.

[29] B. Sutherland, Two-dimensional hydrogen bonded crystals without the ice rule, J. Math. Phys. 11 (1970) 3183–3186.

[30] R.J. Baxter, Eight-vertex models in lattice Phys. Rev. Lett. 26 (1971) 832–834.

[31] R.J. Baxter, One-dimensional anisotropic Heisenberg chain, Phys. Rev. Lett. 26 (1971) 834.

[32] R.J. Baxter, Partition function of the eight-vertex model, Ann. Phys. 70 (1972) 193–228.

[33] R.J. Baxter, One-dimensional anisotropic Heisenberg chain, Ann. Phys. 70 (1972) 323–337.

[34] R.J. Baxter, Eight-vertex model in lattice statistics and one dimensional anisotropic Heisenberg chain I. Some fundamental eigenvectors, Ann. Phys. 76 (1973) 1–24.

[35] R.J. Baxter, Eight-vertex model in lattice statistics and one dimensional anisotropic Heisenberg chain II. Equivalence to a generalized ice-type lattice model, Ann. Phys. 76 (1973) 25–47.

[36] R.J. Baxter, Eight-vertex model in lattice statistics and one dimensional anisotropic Heisenberg chain III. Eigenvectors and eigenvalues of the transfer matrix and Hamiltonian, Ann. Phys. 76 (1973) 48–71.

[37] J.D. Johnson, S. Krinsky and B.M. McCoy, Vertical-arrow correlation length in the eight-vertex model and the low-lying excitations of the XYZ Hamiltonian, Phys. Rev. A8 (1973) 2526–2547.

[38] R.J. Baxter, Solving models in statistical mechanics, Adv. Stud. Pure Math. 19 (1989) 95–116.

[39] K. Fabricius and B.M. McCoy, New developments in the eight-vertex model, J. Stat. Phys. 111 (2003) 323–337.
[40] V.V. Bazhanov and V.V. Mangazeev, Eight-vertex model and non-stationary Lamé equation, J. Phys. A38 (2005) L145–L153.
[41] K. Fabricius and B.M. McCoy, New developments in the eight-vertex model II. Chains of odd length, J. Stat. Phys. 120 (2005) 37–70.
[42] K. Fabricius and B.M. McCoy, An elliptic current operator for the eight-vertex model, J. Phys. A39 (2006) 14869–14886.
[43] V .V. Bazhanov and V.V Mangazeev, The eight-vertex model and Painlevé VI. J. Phys. A39 (2006) 12235–12243.
[44] K. Fabricius, A new Q matrix for the eight-vertex model, J.Phys. A 40 (2007) 4075–4086.
[45] S-S. Roan, The Q operator and functional relations for the eightvertex model at root-of-unity $\eta = 2mK/N$ for odd N, J. Phys. A 40 (2007) 11019–11040.
[46] K. Fabricius and B.M. McCoy, The TQ equation of the 8 vertex model for complex elliptic roots of unity, J. Phys. A40 (2007) 14893–14926.
[47] K. Fabricius and B.M. McCoy, New Q matrices and their functional equations for the eight vertex model at elliptic roots of unity, J. Stat. Phys. 134 (2009) 643–668.
[48] V.V.Bazhanov and V.V. Mangazeev, Analytic solution of the eight vertex model, Nucl. Phys. B775 (2007) 225–282.
[49] O.I. Patu, Free energy of the eight vertex model with an odd number of lattice sites, J. Stat. Mech. (2007) P09007.
[50] E.H. Lieb, T. Schultz and D. Mattis, Two soluble models of an antiferromagnetic chain, Ann. Phys. 16 (1961) 407–460.
[51] B.M. McCoy, Spin correlations in the XY model, Phys. Rev. 173 (1968) 531–310.
[52] E. Barouch and B.M. McCoy , Statistical Mechanics of the X-Y Model II, Phys. Rev. A3 (1971) 786–804.
[53] M. Suzuki, Equivalence of the two-dimensional Ising model to the ground state of the linear XY-model, Phys. Lett. A34 (1971); 94–95.
[54] M. Suzuki, Relationship among exactly soluble models of critical phenomena.I 2D Ising model, dimer problem and the generalized XY model, Prog. Theo. Phys. 46 (1971) 1337–1359.
[55] R.J.Baxter, Spontaneous staggered polarization of the F-model, J. Stat. Phys. 9 (1973) 145–182
[56] M.N. Barber and R.J. Baxter, On the spontaneous order of the eight-vertex model, J.Phys. C 6 (1973) 2913–2921.
[57] R.J. Baxter and S.B. Kelland, Spontaneous polarization of the eight-vertex model, J. Phys. C (1974) L403–406.
[58] R.J. Baxter, Corner transfer matrices of the eight-vertex model I. Low temperature expansions and conjectured properties, J. Stat. Phys. 15 (1976) 485–503.
[59] R.J. Baxter, Corner transfer matrices of the eightvertex model II. The Ising model case, J. Stat. Phys. 17 (1977) 1–14.
[60] M. Takahashi, Half-filled Hubbard model at low temperatures, J. Phys. C10 (1977) 1289–1301.

[61] R.J. Baxter, *Exactly solved models in statistical mechanics*, (Academic Press, 1982)
[62] F.A. Smirnov, *Form factors in completely integrable models of quantum field theory* (World Scientific, 1992, Singapore).
[63] M. Jimbo, K. Miki, T. Miwa and A. Nakayashiki, Correlation functions of the XXZ model for $\Delta < -1$, Phys.Letts. A168 (1992) 256–263.
[64] M. Jimbo, T. Miwa, and A. Nakayashiki, Difference equations for the correlation functions of the eight-vertex model, J. Phys. A (1993) 2199–2209.
[65] M. Jimbo and T. Miwa, quantum KZ equation with $|q| = 1$ and correlation functions of the XXZ model in the gapless regime, J. Phys. A29 (1996) 2923–2958.
[66] M. Lashkevich and Y. Pugai, Free field construction for correlation functions of the eight-vertex model, Nucl. Phys. B516 (1998) 623–651.
[67] N. Kitanine, J.M. Maillet, and V. Terras, Form factors of the XXZ Heisenberg spin 1/2 chain, Nucl. Phys. B554 (1999) 647–678.
[68] N. Kitanine, J.M. Maillet, and V. Terras, Correlation functions of the XXZ Heisenberg spin-1/2 chain in a magnetic field, Nucl. Phys. B567 (2000) 554–582.
[69] J.M. Maillet and V. Terras, On the quantum inverse scattering problem, Nucl. Phys. B575 (2000) 627–644.
[70] N. Kitanine, J.M. Maillet, N.A. Slavnov, Spin-spin correlation functions of the XXZ-1/2 Heisenberg chain in an external magnetic field, Nucl. Phys. B641 (2002) 487–518.
[71] N. Kitanine, J.M. Maillet, N.A. Slavnov, Emptiness formation probability of the XXZ spin-1/2 Heisenberg chain at $\Delta = 1/2$, J. Phys. A 35 (2002) L385–L388.
[72] N. Kitanine, J.M. Maillet, N.A. Slavnov, Large distance asymptotic behavior of the emptiness formation probability of the XXZ spin-1/2 Heisenberg chain, J. Phys. A 35 (2002) L753–L758.
[73] N. Kitanine, J.M. Maillet, N.A. Slavnov, Master equation for spin-spin correlation functions of the XXZ spin 1/2 chain, Nucl. Phys. B712 (2005) 600–622.
[74] N. Kitanine, J.M. Maillet, N.A. Slavnov, Dynamical correlation functions of the XXZ spin 1/2 chain, Nucl. Phys. B729 (2005) 558–580.
[75] N. Kitanine, J.M. Maillet, N.A. Slavnov, On the spin-spin correlation functions of the XXZ spin 1/2 chain, J.Phys. A 38 (2005) 7441–7460.
[76] H.E. Boos and V.E. Korepin, Quantum spin chains and Riemann zeta function with odd arguments, J. Phys. A 34 (2001) 5311–5316.
[77] H.E. Boos, V.E. Korepin, Y. Nishiyama and M. Shiroishi, Quantum correlations and number theory, J. Phys. A 35 (2002) 4443–4451.
[78] K.Sakai, M. Shiroishi, Y. Nishiyama, and M. Takahashi, Third neighbor correlators of the spin-1/2 Heisenberg antiferromagnet, Phys. Rev. E67 (2003) 065101-(1-4).
[79] H.E. Boos, M. Shiroishi, and M. Takahashi, First principle approach to correlation functions of the spin-1/2 Heisenberg chain: fourth neighbor correlations, Nucl. Phys. B 712 (2005) 573–599.
[80] J. Sato, M. Shiroishi, and M. Takahashi, Correlation functions of the spin-1/2 antiferromagnetic Heisenberg chain: exact calculation via the generating function,

Nucl. Phys. B729 (2005) 441–466.

[81] H. Boos, M. Jimbo, T. Miwa, F. Smirnov, and Y. Takeyama, Traces of the Sklyanin algebra and correlation functions of the eight-vertex model, J. Phys. A38 (2005) 7629–7660.

[82] A. Erdélyi, W. Magnus, F. Oberhettinger and F.G. Tricomi, *Higher Transcendental Functions* vol. 2 (McGraw-Hill, NY 1953).

[83] A. Erdélyi, W. Magnus, F. Oberhettinger and F.G. Tricomi, *Higher Transcendental Functions* vol. 1 (McGraw-Hill, NY 1953).

[84] B.M. McCoy and T.T. Wu,*The two dimensional Ising model* (Harvard University Press 1973).

[85] M. Jimbo and T. Miwa, *Algebraic analysis of solvable lattice models* (Providence, RI: American Mathematical Society 1995)

[86] V.E. Korepin, A.G. Izergin, F.H.L. Essler and D.B. Uglov, Correlation function of the spin 1/2 XXX antiferromagnet. Phys. Lett. A 190 (1994) 182–184.

15

The hard hexagon, RSOS and chiral Potts models

The final two models which we will discuss whose Boltzmann weights were seen in chapter 13 to satisfy star–triangle equations are the hard hexagon model, which is a special case of the more general RSOS models, and the chiral Potts model. Both of these models were first introduced for physical reasons to model important physics which is different from the Ising and the eight-vertex models. It was only after their introduction as physical models that they were in fact found to be exactly solvable.

We conclude in section 15.3 with a discussion of open questions concerning the SOS chiral Potts, eight- and six-vertex models.

15.1 The hard hexagon and RSOS models

In chapter 7 and 8 we discussed in detail the physics of hard particle systems. Most of these systems could only be studied at low density by means of virial expansions and at high density by means of molecular dynamics computations. Neither of these methods was capable of giving precise analytic information about possible phase transitions. The one exceptional system is particles on a triangular lattice with nearest neighbor exclusion. The Boltzmann weights of this model were shown in chapter 13 to satisfy the face version of the star–triangle equation and from the parametrization in terms of elliptic functions the critical value of the fugacity was found to be (13.281)

$$z_c = [(1+\sqrt{5})/2]^5 = (11+5\sqrt{5})/2 = 11.09017\cdots \tag{15.1}$$

In this section we will present in detail the results for the grand partition function and the phase transition of the hard hexagon model. We begin with a historical overview in 15.1.1. The results in the low density regime are given in 15.1.2 and in the high density regime in 15.1.3. We conclude in 15.1.4 with a discussion of the implication of the hard hexagon results for more general systems.

15.1.1 Historical overview

The hard hexagon model is now recognized as a special case of the SOS and RSOS models discussed in chapter 13 and is exactly solvable. However, the hard hexagon model was never expected to have an exact solution when it was first introduced and, indeed, the history of the study of the phase transition provides a very useful example in the caution which needs to be exercised when approximation methods are used to study phase transitions.

In 1960 Burley [1] computed the virial coefficients of hard hexagons through B_6 and found them to be all positive and from this Burley concludes in the abstract that the gas *"seems to be able to condense into an ordered phase without any transition"*. In a subsequent paper in 1965 Burley [2] used an "approximate technique" from which he concluded that hard hexagons had a first order transition. However, one year later Runnels and Combs [3] did exact computations for strips of finite width and concluded that the transition was of second order. This was confirmed by Gaunt [4] in 1967 who in fact made the (unpublished) conjecture [11] that the critical value of the fugacity is the value in (15.1).

The hard hexagon model was shown in chapter 13 to be a special case of the model of hard squares with diagonal interactions which satisfies a star–triangle equation. The grand partition function of the model of hard squares with diagonal interactions has been computed in several different ways:

The original computation is by Baxter [11–13] in 1980 who used the corner transfer matrix method invented [5] in 1968 for the study of the monomer dimer problem and used in 1976 [10] for the order parameters of the eight-vertex model.

A second method of computation is by Baxter and Pearce [14, 15] in 1982 who showed that the row-to-row transfer matrix for hard squares with diagonal interactions satisfies the functional equation [14, eqn.(3.3)]

$$T(u)T(u+\eta) = [\theta(\eta-u)\theta(\eta+u)/\theta^2(\eta)]^N + [\theta(u)\theta(\eta)]^N T(u-2\eta) \tag{15.2}$$

where $\eta = \pi/5$ and

$$\theta(u) = \sin u \prod_{n=1}^{\infty}(1-q^{2n}e^{2iu})(1-q^{2n}e^{-2iu})(1-q^{2n}). \tag{15.3}$$

The final method is by Andrews, Baxter and Forrester [16] in 1984 who realized that the hard hexagon model is in fact a special case of the SOS models introduced by Baxter [7, 8] for the computation of the eigenvectors of the eight-vertex model. Thus in [16, p.201] it is given that from [7, (1.23)] the eigenvalues $t(u)$ of the transfer matrix $T(u)$ will, for some z, satisfy [16, (1.3.10)]

$$t(u)q(u) = z[\Theta(0)h(u-\eta)]^N q(u+2\eta) + z^{-1}[\Theta(0)h(u+\eta)]^N q(u-2\eta) \tag{15.4}$$

where $q(u)$ is the scalar function

$$q(u) = \prod_{j=1}^{N/2} h(u-u_j) \tag{15.5}$$

with $\eta = K/5$ and from (15.5) it follows that

$$q(u+10\eta) = (-1)^{N/2}q(u). \tag{15.6}$$

We refer to (15.4) as the "scalar tq" equation to distinguish it from the matrix TQ equation established in chapter 14 for the eight-vertex model. When z is chosen to be

$$z^5 = -(-1)^{N/2} \tag{15.7}$$

then it is shown [16, p.202] that (15.4) with (15.7) is automatically satisfied if (15.2) holds. Therefore to quote [16, p.202] "Thus there may be (and almost certainly are) eigenvalues satisfying both (15.2) and (15.4) provided z in the unrestricted SOS model is given by (15.7)."

It should be, of course, possible to obtain the ground state energy directly from the scalar tq equation without recourse to the original computation of the corner transfer matrix [11]. This solution is begun by using the form (15.5) in (15.4) to obtain an equation for the roots u_j just as was done for the eight-vertex model in chapter 14. Thus we find

$$\left[\frac{h(u_k+\eta)}{h(u_k-\eta)}\right]^N = z^2 \prod_{j=1}^{N/2} \frac{h(u_k-u_j+2\eta)}{h(u_k-u_j-2\eta)} \tag{15.8}$$

which is analogous to (14.297) for the eight-vertex model. However, as was the case with the eight-vertex model, in order to use the functional equation (15.8) to obtain an explicit solution for the free energy it is necessary to have independent information about the qualitative location of the roots u_j of $q(u)$. For the eight-vertex model this information was obtained in chapter 14 by directly studying the matrix $Q(u)$. However, for the scalar tq equation where no matrix $Q(u)$ is known this approach is not available. For the critical case of the RSOS models the root patterns have been studied in [20]. However, for the general (noncritical case) such a study does not seem to be in the literature.

In principle, the density of the hard hexagon model can be computed from the grand partition function and from this the equation of state can be obtained. However, this is not what was actually done. Instead Baxter [11–13] made a separate computation of the densities in both the disordered and ordered phases using the corner transfer matrix. Unlike the computation of the free energy this is the only way the density has been obtained. These computations prove that there is a second order phase transition at the value of the fugacity given by (15.1). These original papers on the hard hexagon (and more generally the model of hard squares with diagonal interactions introduced in chapter 13) made no use or contact with the row-to-row transfer matrix methods explained in detail in chapter 14.

Our principal interest will be in comparing the exact results of the hard hexagon model with the series expansions and numerical studies of the equation of state for hard particle systems presented in chapters 7 and 8. Therefore we need to express the fugacity and the partition function per site

$$\kappa = \lim_{N\to\infty} Z_{hh}^{1/N} \tag{15.9}$$

in terms of the density instead of the nome q used in chapter 13 to parametrize the Boltzmann weights. This was studied in the low density regime in 1987 by Richie and Tracy [17] using results obtained in [18]. The corresponding study for high density was obtained also in 1987 by Joyce [19] who also explicitly obtained the virial coefficients which were previously given in Table 7.15. These studies reveal the important fact that in both the low and high density regimes the partition function per site and the fugacity are algebraic functions of the density.

The chronology of the solution of hard hexagons is summarized in Table 15.1.

Table 15.1 Chronology of the hard hexagon model.

Date	Author(s)	Development
1960	Burley [1]	Hard hexagons, no transition
1965	Burley [2]	Hard hexagons, first order
1966	Runnels, Combs [3]	Hard hexagons, second order
1967	Gaunt [4]	Hard hexagons; second order
1968	Baxter [5]	Corner transfer matrices invented
1973	Baxter [6–9]	The SOS models invented scalar tq equation
1980–1982	Baxter [11–13]	Hard hexagon grand partition function and densities using corner transfer matrices
1982–1983	Baxter, Pearce [14, 15]	Partition function from row transfer matrix functional equations
1984	Andrews, Baxter, Forrester [16]	Hard hexagons and the scalar tq equation
1987	Richie, Tracy [17]	Low density equation of state
1987	Tracy, Grove, Newman [18]	
1987	Joyce [19]	Low density virial expansion High density equation of state

15.1.2 Hard hexagons for $0 \le z \le z_c$

We found in (13.302) and (13.303) that in the low density regime the fugacity z of the grand canonical ensemble is expressed parametrically in terms of a nome of the elliptic functions which parametrize the Boltzmann weights as

$$z = -x \left[\prod_{n=1}^{\infty} \frac{(1-x^{5n-4})(1-x^{5n-1})}{(1-x^{5n-3})(1-x^{5n-2})} \right]^5 \tag{15.10}$$

where $-1 \le x \le 0$

$$x = -e^{-\frac{\pi K}{5K'}}, \tag{15.11}$$

the modulus k of $K(k)$ is the same as the parametrization of the Boltzmann weights (13.277) and when $x = -1$ then $z = z_c$.

In [11] Baxter found that the partition function per site is given in terms of x by

$$\kappa = \prod_{n=1}^{\infty} \left(\frac{1-x^{5n}}{1-x^{6n}} \right)^2 \frac{(1-x^{5n-1})^2(1-x^{5n-4})^2(1-x^{6n-3})^2(1-x^{6n-2})(1-x^{6n-4})}{(1-x^{5n-2})^3(1-x^{5n-3})^3(1-x^{6n-1})(1-x^{6n-5})}. \tag{15.12}$$

In principle the density $\rho(z)$ can be found from $\kappa(z)$ by use of the basic relation of the grand canonical ensemble that

$$\rho(z) = z \frac{\partial \ln \kappa}{\partial z}, \tag{15.13}$$

but in actual practice this has not been done. Instead the density was computed by Baxter in 1981 [12] directly from the definition

$$\rho = \langle \sigma_k \rangle \tag{15.14}$$

by the same methods used to compute the order parameter of the eight-vertex model with the result that

$$\rho = -x \prod_{n=1}^{\infty} \frac{(1-x^{6n-3})}{(1-x^{2n-1})(1-x^{5n-1})(1-x^{5n-4})(1-x^{30n-12})(1-x^{30n-18})}, \tag{15.15}$$

The results (15.12) and (15.15) give a parametric representation of the partition function per site κ and the density in terms of the variable x (instead of the fugacity z) and from this the equation of state may be obtained. In particular, Baxter [11, 12] found that at the critical point where $z = z_c$ and $x = -1$ that

$$\rho_c = (5 - \sqrt{5})/10 = 0.27633932\cdots \tag{15.16}$$

$$\kappa_c = [(27/250)(25 + 11\sqrt{5})]^{1/2} = (27 z_c \sqrt{5}/125)^{1/2}. \tag{15.17}$$

It is shown in [19, (12.10)] that z and ρ satisfy an algebraic equation which is quartic in z and order 12 in ρ:

$$\begin{aligned}
&f(\rho, z) \equiv \rho^{11}(\rho-1)z^4 \\
&-\rho^5(22\rho^7 - 77\rho^6 + 165\rho^5 - 220\rho^4 + 165\rho^3 - 66\rho^2 + 13\rho - 1)z^3 \\
&+\rho^2(\rho-1)^2(119\rho^8 - 476\rho^7 + 689\rho^6 - 401\rho^5 - 6\rho^4 + 125\rho^3 - 63\rho^2 + 13\rho - 1)z^2 \\
&+(\rho-1)^5(22\rho^7 - 77\rho^6 + 165\rho^5 - 220\rho^4 + 165\rho^3 - 66\rho^2 + 13\rho - 1)z \\
&+\rho(\rho-1)^{11} = 0
\end{aligned} \tag{15.18}$$

The singular points of ρ as a function of z are found by eliminating ρ between $f(\rho, z) = 0$ and $\partial f(\rho, z)/\partial \rho = 0$. This object is called the resultant. In general for an algebraic equation

$$f(x, y) = \sum_{k=0}^{n} a_n(x) y^k = 0 \tag{15.19}$$

of degree n in y the resultant is a polynomial in x which may be computed by the determinant of the $(2n-1) \times (2n-1)$ Sylvester matrix

$$Res(P_n(x), P_n'(x); x) = \begin{vmatrix}
a_n & a_{n-1} & a_{n-2} & \cdots & a_1 & a_0 & 0\cdots & \cdots & 0 \\
0 & a_n & a_{n-1} & a_{n-2} & \cdots & a_1 & a_0 & 0\cdots & 0 \\
\vdots & \vdots & \vdots & \vdots & \vdots & \vdots & \vdots & \vdots & \vdots \\
0 & 0\cdots & 0 & a_n & a_{n-1} & a_{n-2} & \cdots & a_1 & a_0 \\
a_n' & a_{n-1}' & a_{n-2}' & \cdots & a_1' & 0 & 0\cdots & \cdots & 0 \\
0 & a_n' & a_{n-1}' & a_{n-2}' & \cdots & a_1' & 0 & \cdots & 0 \\
\vdots & \vdots & \vdots & \vdots & \vdots & \vdots & \vdots & \vdots & \vdots \\
0 & 0 & \cdots & 0 & a_n' & a_{n-1}' & a_{n-2}' & \cdots & a_1'
\end{vmatrix} \tag{15.20}$$

where

$$a_n' = n a_n. \tag{15.21}$$

The zeros of this resultant polynomial are the singular points of the equation (15.19). At any nonsingular point in the finite x plane the n branches of the algebraic function

$y(x)$ are analytic functions, whereas at a singular point there is in general at least one branch of $y(x)$ which is not analytic, However there can be exceptional cases where there are zeros of the resultant where all branches of $y(x)$ are analytic. This special singular point is called an apparent singular point.

The resultant $\mathrm{Res}(f, \partial f/\partial\rho; \rho)$ of (15.18) may be computed from (15.20) most easily by using computer algebra and is found to be [19, (12.11)]

$$\mathrm{Res}(f, \partial f/\partial\rho; \rho) = -2^8 \cdot 3^9 z^{22}(1 + 11z - z^2)^{24}. \qquad (15.22)$$

The zeros of (15.22) are at

$$z = 0, \quad z_c, \quad -1/z_c \qquad (15.23)$$

where z_c is the critical fugacity (15.1) and thus $\rho(z)$ is singular at these points. On the physical branch of the curve $\rho(z)$ is analytic at $z = 0$. Thus the singularity closest to the origin is at $z = -1/z_c$ and hence in the cluster expansion of the density

$$\rho(z) = \sum_{l=1}^{\infty} l b_l z^l \qquad (15.24)$$

the cluster integrals b_l alternate in sign as is required by the theorem of Groeneveld proven in chapter 6.

Near the physical singularity at $z = z_c$ the density has the expansion [19, (12.15)]] which is derived from (15.18)

$$\rho(z) = \frac{1}{10}(5 - \sqrt{5}) - \frac{1}{\sqrt{5}}t^{2/3} + \frac{1}{\sqrt{5}}t - \frac{1}{15}(25 + 4\sqrt{5})t^{5/3} + O(t^2) \qquad (15.25)$$

where

$$t = 5^{-3/2}[1 - (z/z_c)]. \qquad (15.26)$$

The corresponding expansion for $P(z)/k_BT$ is derived from (15.25) using the relation

$$\rho(z) = z\frac{\partial}{\partial z}(P(z)/k_BT) \qquad (15.27)$$

as [19, (12.17)]]

$$\frac{P}{k_BT} = \ln\kappa_c - \frac{5}{2}(\sqrt{5} - 1)t + 3t^{5/3} - \frac{5}{4}(27 - 5\sqrt{5})t^2 + \frac{5}{2}(1 + 5\sqrt{5})t^{8/3} + O(t^3) \qquad (15.28)$$

with κ_c given by (15.17). From (15.25) and (15.28) it follows that as $\rho \to \rho_c-$ the singular behaviour of the pressure is

$$\frac{P(\rho)}{k_BT} = \ln\kappa_c - \frac{5^{7/4}}{2}(\sqrt{5} - 1)(\rho_c - \rho)^{3/2} + \cdots \qquad (15.29)$$

Furthermore the reduced isothermal compressibility $\chi(\rho)$ which is defined as [17]

$$\frac{\chi(\rho)}{k_BT} = \frac{\rho}{v}\left(\frac{\partial v}{\partial P}\right)_T \qquad (15.30)$$

is given in terms of κ as

$$\chi(\rho) = \left(\frac{\ln\kappa}{\partial\rho}\right)^{-1} \tag{15.31}$$

or as [17, (5)]

$$\chi(\rho) = \kappa\left(\frac{\partial\Gamma/\kappa}{\partial P/\partial\rho}\right) \tag{15.32}$$

and has the expansion [17]

$$\chi(\rho) = (\rho_c - \rho)^{-1/2}\sum_{n=0}^{\infty} c_n(\rho_c - \rho)^{n/2} \tag{15.33}$$

for which the leading terms are

$$\chi(\rho) = \frac{5+\sqrt{5}}{75}[5^{-1/4}(\rho_c - \rho)^{-1/2} - 1 + O(\rho_c - \rho)^{1/2}]. \tag{15.34}$$

To further study the behavior of the pressure as a function of the density in the low density region we use the expression derived in chapter 6 as part of the derivation of the Mayer virial expansion

$$-\ln[z(\rho)/\rho] = \sum_k \beta_k\rho^k \tag{15.35}$$

and thus we need z as a function of ρ which can be explicitly obtained from (15.18) which is a quartic equation in z as [19, (12.28)]:

$$\begin{aligned}
&4\rho^6(1-\rho)z(\rho) = (1-2\rho)(1-11\rho+44\rho^2-77\rho^3+66\rho^4-33\rho^5+11\rho^6)\\
&+(1-\rho+\rho^2)^{1/2}(1-5\rho+5\rho^2)^{5/2}\\
&-(1-5\rho+5\rho^2)[2(1-16\rho+106\rho^2-378\rho^3+803\rho^4-1080\rho^5\\
&\qquad+962\rho^6-576\rho^7+219\rho^8-50\rho^9+10\rho^{10})\\
&+2(1-2\rho)(1-11\rho+44\rho^2-77\rho^3+66\rho^4-33\rho^5+11\rho^6)\\
&\times(1-\rho+\rho^2)^{1/2}(1-5\rho+5\rho^2)^{1/2}]^{1/2}.
\end{aligned} \tag{15.36}$$

The coefficients β_k are now easily computed by using (15.36) in the left-hand side of (15.35) and expanding the resulting expression in a power series in ρ to obtain β_k. From this the coefficients in the virial expansion

$$\frac{Pv}{k_BT} = 1 + \sum_{k=1}^{\infty} B_{k+1}v^{-k} \tag{15.37}$$

are obtained as

$$(k+1)B_{k+1} = -k\beta_k. \tag{15.38}$$

These virial coefficients have been given in Table 7.15.

The virial coefficients tabulated in chapter 7 oscillate in sign and thus the radius of convergence is determined by a singularity which is not on the real ρ axis. To study these singularities further we compute the resultant polynomial for $f(\rho, z)$ as a function of ρ [19, (12.19) and (12.20)]

$$\begin{aligned}&\mathrm{Res}(f, \partial/\partial z; z) = -\rho^{25}(1-\rho)^{15}(1-\rho+\rho^2)^2(1-5\rho+5\rho^2)^{14}\\&\times(1-10\rho+33\rho^2+33\rho^2-36\rho^3+18\rho^4-70\rho^5+140\rho^6-100\rho^7+25\rho^8).\end{aligned} \tag{15.39}$$

This resultant vanishes at

$$\rho = 0, 1, e^{\pm i\pi/3}, \rho_c, \frac{1}{5\rho_c}, \quad \text{and} \quad \rho_k^{\pm} \tag{15.40}$$

where ρ_c and $(5\rho_c)^{-1}$ with ρ_c given by (15.16) are the roots of

$$1 - 5\rho + 5\rho^2 = 0 \tag{15.41}$$

and $\rho_k^{\pm}$ are the eight roots of the eighth order polynomial in (15.39). If we set

$$\rho = \rho_*/(1+\rho*) \tag{15.42}$$

we see that the eighth order polynomial in (15.39) vanishes when

$$1 - 2\rho_* - 9\rho_*^2 + 8\rho_*^3 + 54\rho_*^4 + 8\rho_*^5 - 9\rho_*^6 - 2\rho_*^7 + \rho_*^8 = 0. \tag{15.43}$$

This polynomial is invariant if $\rho_* \to 1/\rho_*$ and thus setting

$$y = \rho_* + \rho_*^{-1} \tag{15.44}$$

we obtain the quartic equation

$$y^4 - 2y^3 - 13y^2 + 14y + 73 = 0. \tag{15.45}$$

Thus, by explicitly solving the quartic equation (15.45) and then using those solutions in (15.44) and solving for ρ_* we obtain the locations $\rho_k^{\pm}$

$$\begin{aligned}\rho_0^{\pm} &= \frac{1}{2} - \frac{\sqrt{10}}{20}[(4\sqrt{10}-5\sqrt{5}-4\sqrt{2}+7)^{1/2} \pm i(4\sqrt{10}-5\sqrt{5}+4\sqrt{2}-7)^{1/2}]\\&= 0.234862\cdots \pm i0.056041\cdots\\\rho_1^{\pm} &= \frac{1}{2} - \frac{\sqrt{10}}{20}[(4\sqrt{10}+5\sqrt{5}+4\sqrt{2}+7)^{1/2} \pm i(4\sqrt{10}+5\sqrt{5}-4\sqrt{2}-7)^{1/2}]\\&= -0.455069\cdots \pm i0.528502\cdots\\\rho_2^{\pm} &= 1-\rho_0^{\mp} = 0.765137 \pm i0.055041\cdots\\\rho_3^{\pm} &= 1-\rho_1^{\mp} = 1.455069\cdots \pm i0.528502\cdots\end{aligned} \tag{15.46}$$

There are four branches of the algebraic function $z = z(\rho)$ and all of the 14 possible singularities (15.40) occur on at least one branch. The physical branch of $z(\rho)$ is analytic at $\rho = 0$ and has singularities at $\rho_c, 1/(5\rho_c), \exp(\pm i\pi/3), \rho_0^{\pm}$ and $\rho_3^{\pm}$. These singularities are plotted in the complex ρ plane in Fig. 15.1. In this figure we see that the radius of convergence ρ_r of the virial expansion is determined by the singularity at $\rho_0^{\pm}$ and thus from (15.46) we obtain the value

$$\rho_r = |\rho_0^{\pm}| = \frac{\sqrt{5}}{10}[(4\sqrt{10} - 5\sqrt{5} + 5) - \sqrt{10}(4\sqrt{10} - 5\sqrt{5} - 4\sqrt{2} + 7)^{1/2}]^{1/2}$$
$$= 0.2414560 \cdots \qquad (15.47)$$

which is less than the critical value of the density $\rho_c = 0.2763932 \cdots$

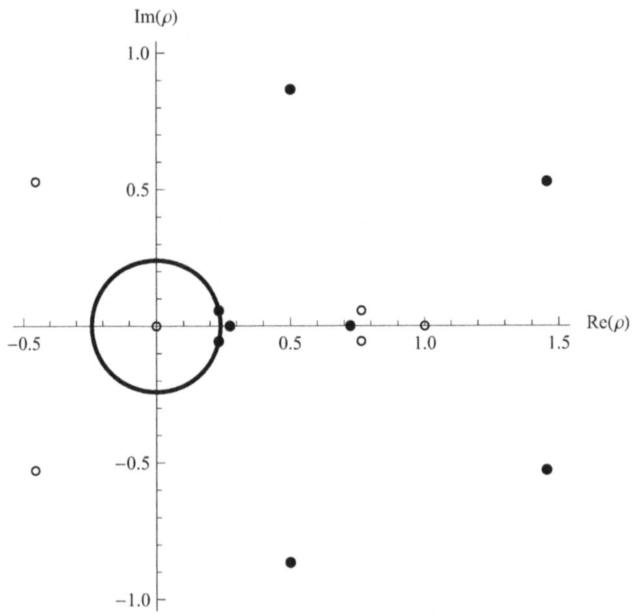

Fig. 15.1 The singularity structure of the algebraic function $z = z(\rho)$ in the complex ρ plane and the circle of convergence of the virial expansion (15.37). The singular points marked by the filled circles are on the physical branch of $z(\rho)$, The unfilled circles are on unphysical branches. The singular points are reflection symmetric about Re(ρ)=1/2.

We conclude by noting that it has been shown in [17] that κ is an algebraic function of ρ which, by setting

$$y = \rho - 1, \qquad (15.48)$$

satisfies

$$F(\kappa, y) = 0 \qquad (15.49)$$

where

$$F(\kappa, y) = \sum_{n=0}^{4} g_n \kappa^{2n} \qquad (15.50)$$

with

$$
\begin{aligned}
g_0 &= 432y^{22}\\
g_1 &= -432y^{10}g_3\\
g_2 &= 16y^4 + 192y^5 + 645y^6 - 516y^7 - 5822y^8 - 4116y^9 + 9349y^{10} - 11400y^{11}\\
&-42672y^{12} - 9800y^{13} + 73y^{14} - 4500y^{15} + 1750y^{16} + 3125y^{22}\\
g_3 &= -1 - 12y - 48y^2 - 56y^3 + 42y^4 + 12y^5 - 100y^6 + 132y^7 + 625y^{12}\\
g_4 &= y^2.
\end{aligned}
\tag{15.51}
$$

15.1.3 Hard hexagons for $z_c \leq z < \infty$

In the high density regime where $z_c < z$ we found in (13.287)–(13.289) that the fugacity z is related to the modulus k of the parametrization of the Boltzmann weights by

$$z' = 1/z = x\left[\prod_{n=1}^{\infty}\frac{(1-x^{5n-3})(1-x^{5n-2})}{(1-x^{5n-4})(1-x^{5n-1})}\right]^5 \tag{15.52}$$

where $0 \leq x \leq 1$

$$x = e^{-\frac{4\pi K}{5K'}}. \tag{15.53}$$

Baxter [11] computed the partition function per site in 1980 in this regime as

$$\kappa = x^{-1/3}\prod_{n=1}^{\infty}\frac{(1-x^{3n-2})(1-x^{3n-1})(1-x^{5n-3})^2(1-x^{5n-2})^2(1-x^{5n})^2}{(1-x^{3n})^2(1-x^{5n-4})^3(1-x^{5n-1})^3}. \tag{15.54}$$

In the high density regime for $z_c < z$ the hard hexagon model is ordered on the triangular lattice. The triangular lattice has three sublattices as is shown in Fig. 15.2. In the fully ordered state the particles occupy only one of the three sublattices. Thus denoting the three sublattices by the indices 1, 2, 3 and by convention defining sublattice 1 to be the lattice of the fully ordered system we need to compute separately the sublattice densities

$$\rho_k = \langle\sigma_k\rangle \tag{15.55}$$

where

$$\rho_1 \neq \rho_2 = \rho_3. \tag{15.56}$$

We define the order parameter R as

$$R = \rho_1 - \rho_2 = \rho_1 - \rho_3 \tag{15.57}$$

and the mean density

$$\rho = \frac{1}{3}(\rho_1 + \rho_2 + \rho_3) \tag{15.58}$$

and note that it is the mean density which appears in the equation of state.

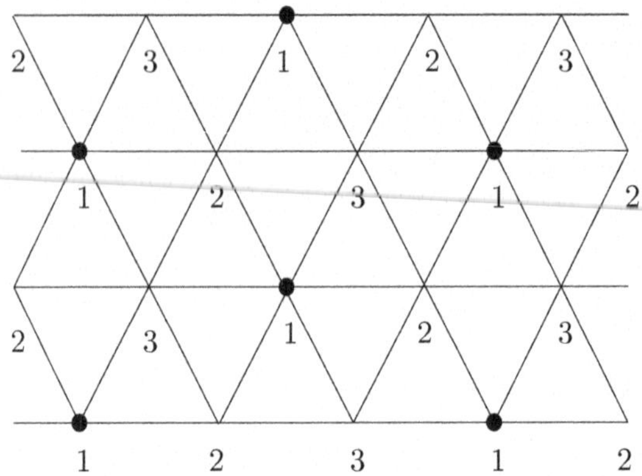

Fig. 15.2 The three sublattices of the triangular lattice. The filled circles represent the lattice sites filled at close packing.

Baxter [12] computed the sublattice densities in 1981 by means of the corner transfer method used to compute the order parameter of the eight-vertex model. Using the notation

$$Q(x) = \prod_{n=1}^{\infty}(1-x^n) \tag{15.59}$$

$$G(x) = \prod_{n=1}^{\infty}\frac{1}{(1-x^{5n-4})(1-x^{5n-1})} \tag{15.60}$$

$$H(x) = \prod_{n=1}^{\infty}\frac{1}{(1-x^{5n-3})(1-x^{5n-2})}. \tag{15.61}$$

Baxter's results are [12, (73) and (69)]

$$\rho_1 = H(x)Q(x)[G(x)Q(x) + x^2H(x^9)Q(x^9)]/Q(x^3)^2 \tag{15.62}$$

$$\rho_2 = \rho_3 = x^2H(x)H(x^9)Q(x)Q(x^9)/Q(x^3)^2 \tag{15.63}$$

and thus

$$R = G(x)H(x)[Q(x)/Q(x^3)]^2 = \prod_{n=1}^{\infty}\frac{(1-x^n)(1-x^{5n})}{(1-x^{3n})^2}. \tag{15.64}$$

Joyce [19] has reduced these high density results of Baxter expressed parametrically in terms of x to algebraic functions. He finds [19, (10.1)] with $z' = 1/z$ that ρ satisfies the equation quadratic in z'

$$z'^2(2-3\rho)(1-\rho)^3 + z'(1-12\rho+45\rho^2-66\rho^3+33\rho^4) + \rho^3(1-3\rho) = 0 \tag{15.65}$$

which is to be compared with the corresponding low density result (15.18) which is a quartic equation in z.

On the physical branch we must have $\rho = 1/3$ (the maximum density at close packing) for $z' = 0$. Solving (15.65) for z' on the physical branch [19, (10.3)]

$$z'(\rho) = -\frac{1}{2}(2-3\rho)^{-1}(1-\rho)^{-3}[(1-12\rho+45\rho^2-66\rho^3+33\rho^4) \\ +(-1+5\rho-5\rho^2)^{3/2}(-1+9\rho-9\rho^2)^{1/2}] \quad (15.66)$$

which is to be compared with the corresponding low density result (15.36). The singularities of (15.66) are at

$$\rho = \frac{1}{10}(5 \pm \sqrt{5}),\ \frac{1}{6}(3 \pm \sqrt{5}),\ \frac{2}{3},\ 1. \quad (15.67)$$

These singularities are all on the real axis in contrast with the corresponding low density result whose singularities are shown in Fig. 15.1.

The reduced isothermal compressibility is given as an explicit algebraic function of the density [19, (10.17)]

$$\chi(\rho) = \frac{1}{15\rho}[(1-2\rho)\left(\frac{-1+9\rho-9\rho^2}{-1+5\rho-5\rho^2}\right)^{1/2} - (1+9\rho-9\rho^2)] \quad (15.68)$$

where $\rho_c < \rho \leq 1/3$ and $\rho_c = (5-\sqrt{5})/10$.

The partition function per site is expressed as an algebraic function of the density [19, (10.44)]

$$\kappa^6(\rho) = \frac{1}{2\cdot 5^5 \rho'^2}\{S(\rho') + [1-5\rho'-5\rho'^2]T(\rho')\sqrt{Q}(\rho')\} \quad (15.69)$$

where

$$\rho' = 1-3\rho \quad (15.70)$$

$$Q(\rho') = [1-\rho'-\rho'^2][1-5\rho'-5\rho'^2] \quad (15.71)$$

and the polynomials $S(\rho')$ and $T(\rho')$ are given by

$$S(\rho') = \sum_{n=0}^{12} S_n \rho'^n \quad (15.72)$$

$$T(\rho') = \sum_{n=0}^{8} T_n \rho'^n \quad (15.73)$$

where S_n and T_n are given in Table 15.2.

Finally we note that Joyce has also found an explicit algebraic expression of the order parameter $R(\rho)$ in terms of the density [19, (10.54)]:

Table 15.2 The coefficients S_n and T_n.

n	S_n	T_n
0	3125	3125
1	2500	50000
2	368394	228481
3	1648220	510878
4	2597775	656219
5	−1194660	510592
6	−11001870	238586
7	−19426812	6176
8	−18739575	6820
9	−11120900	
10	−4063750	
11	−843000	
12	−76250	

$$\begin{aligned}
&R^9(\rho) = \\
&\frac{2^2}{5^3}\frac{(1-\rho')^6[1-5\rho'-5\rho'^2]^{3/2}}{\{[1+8\rho'-3\rho'^2-22\rho'^3-11\rho'^4]+[1-\rho'-\rho'^2]^{1/2}[1-5\rho'-5\rho'^2]^{3/2}\}^2} \\
&\times\{[1-\rho'-\rho'^2]^{1/2}[89+228\rho'+195\rho'^2+55\rho'^3] \\
&-[1-5\rho'-5\rho'^2]^{1/2}[39+106\rho'+91\rho'^2+25\rho'^3]\}^{-1} \\
&\times\{S(\rho')+[1-\rho'-\rho'^2]^{1/2}[1-5\rho'-5\rho'^2]^{3/2}T(\rho')\} \qquad (15.74)
\end{aligned}$$

15.1.4 Discussion

We conclude this sketch and summary of results for the hard hexagon model with some general comments concerning the relation of this exactly solvable model to the many, presumably not exactly solvable, problems of hard particles which were studied in chapter 7 and 8.

1. There is no analytic continuation possible from the high to the low density regime as a function either of fugacity or density. This must mean that the zeros of the partition function on the finite lattice will divide the density plane into two disconnected regions. This is presumably a general feature for transitions in all of the hard particle systems

2. In the hard hexagon model the partition function per site and the order parameter are algebraic functions of the density. There is no reason to expect that such an algebraic relation will hold for the nonintegrable models.

3. The original parametric expressions for the fugacity, density, order parameter and partition function per site are much simpler than the expressions with the density as the independent variable. The implication of this is that there is no reason to expect that simple model equations of state which use the density as the independent variable will capture the true physics of hard particle systems (or indeed of any real system).

15.2 The chiral Potts model

The final model we will discuss is the chiral Potts model. For reasons of space our goal in this section is not to compute the free energy of the chiral Potts model in its full generality as we did for the eight-vertex model in chapter 14. Instead we will content ourselves with a discussion of the physics of the model and a presentation of the solution in the particularly simple superintegrable case where the model is most analogous to the Ising model.

In 15.2.1 we trace the chronology of the chiral Potts model from the initial approximate studies to the discovery of the integrable manifold and we will see that there are two very distinct and different regimes which may be called physical. In one case the Boltzmann weights of the classical two-dimensional statistical model are real and positive. This is the "physical region" discussed in Chapter 13 for the eight-vertex model. In the other case the Hamiltonian of the associated quantum spin chain derived in section 13.7 is Hermitian. This spin chain has the property of a **level crossing transition** which is not seen in any other of the solvable models discussed in this book.

In 15.2.2 we discuss the classical statistical model where the Boltzmann weights are real and positive. In 15.2.3 we discuss the Hermitian quantum spin chain and introduce the very special superintegrable specialization. In 15.2.4 we give the functional equation for the $N = 3$ superintegrable case; in 15.2.5 we use this functional equation to compute the ground state energy of the spin chain for small λ; in 15.2.6 we show for λ near unity that level crossing occurs; in 15.2.7 we give a brief discussion of the order parameter; and we conclude in 15.2.8 with the phase diagram for the spin chain off the superintegrable point and a discussion of open questions and a comparison of the real and the Hermitian manifolds.

15.2.1 Historical overview

The Boltzmann weights of the integrable N-state chiral Potts model were derived in chapter 13 and when $N = 2$ the model reduces to the Ising model. Furthermore the algebra invented and used by Onsager [21] to solve the Ising model extends to a special case of the $N \geq 3$ chiral Potts model. Thus it is fair to say that Onsager's 1944 paper is the first paper on the subject. On the other hand there is no notion of chirality in the Ising model because this feature cannot occur when $N = 2$.

For $N \geq 3$ the most general chiral Potts model is defined as a two-dimensional classical model with interaction energy

$$\mathcal{E}_{cp} = -\sum_{j,k}\sum_{n=1}^{N-1}\{E_n^h(\sigma_{j,k}\sigma_{j,k+1}^*)^n + E_n^v(\sigma_{j,k}\sigma_{j+1,k}^*)^n\} \tag{15.75}$$

where for each site j, k of the lattice

$$\sigma_{j,k} = e^{2\pi i m/N} \quad \text{with} \quad m = 0, 1, \cdots N-1. \tag{15.76}$$

The interaction (15.75) is real (and thus physical) for

$$E_n^{v,h} = E_{N-n}^{v,h*}. \tag{15.77}$$

In the literature the energies $E_n^{v,h}$ are often written as

$$E^{v,h}/k_BT = K_n^{v,h} e^{2\pi i \Delta_n^{v,h}/N} \tag{15.78}$$

where $K_n^{v,h}$ and $\Delta_n^{v,h}$ are real. Thus on the real manifold manifold where

$$K_n^{v,h} = K_{N-n}^{v,h} \quad \text{and} \quad \Delta_n^{v,h} = -\Delta_{N-n}^{v,h} \tag{15.79}$$

the interaction energy is written for N odd as

$$\begin{aligned}\mathcal{E}_{cp}/k_BT = -\sum_{j,k}\Bigg(\sum_{n=1}^{(N-1)/2} &\{2K_n^h \cos[\frac{2\pi}{N}(n(n_{j,k}-n_{j,k+1})+\Delta_n^h)]\\ &+2K_n^v \cos[\frac{2\pi}{N}(n(n_{j,k}-n_{j+1,k})+\Delta_n^v)]\}\Bigg)\end{aligned} \tag{15.80}$$

and for N even as

$$\begin{aligned}\mathcal{E}_{cp}/k_BT = -\sum_{j,k}\Bigg(\sum_{n=1}^{(N-1)/2} &\{2K_n^h \cos[\frac{2\pi}{N}(n(n_{j,k}-n_{j,k+1})+\Delta_n^h)]\\ &+2K_n^v \cos[\frac{2\pi}{N}(n(n_{j,k}-n_{j+1,k})+\Delta_n^v)]\}\\ &+2K_{N/2}^h(-1)^{(n_{j,k}-n_{j,k+1})}+2K_{N/2}^v(-1)^{(n_{j,k}-n_{j+1,k})}\Bigg)\end{aligned} \tag{15.81}$$

where in each site (j,k) the variables $n_{j,k}$ take the values

$$n_{j,k} = 0, 1, \cdots N-1. \tag{15.82}$$

If we only take the term $n = 1$ in (15.80) and (15.81) the model is referred to in the literature as the chiral clock model with the interaction energy

$$\mathcal{E}_{cc}/k_BT = -\sum_{j,k}\{2K_1^h \cos[\frac{2\pi}{N}(n_{j,k}-n_{j,k+1}+\Delta_1^h)]+2K_1^v \cos[\frac{2\pi}{N}(n_{j,k}-n_{j+1,k}+\Delta_1^v)]\}. \tag{15.83}$$

The general chiral Potts model is said to be symmetric when $\Delta_n^v = \Delta_n^h$; to be asymmetric when $\Delta_n^v \neq \Delta_n^h$; and to be fully asymmetric when $\Delta_n^v = 0,\ \Delta_n^h \neq 0$.

The first occurrence of the chiral Potts model in the literature seems to be the 1976 paper of Wu and Wang [22] where it is shown that if we write

$$\mathcal{E}_{cp} = -\sum_{j,k}\{\mathcal{E}^h(n_{j,k}-n_{j,k+1})+\mathcal{E}^v(n_{j,k}-n_{j+1,k})\} \tag{15.84}$$

and if we consider the transposed Fourier transform

$$\tilde{\mathcal{E}}^{h,v}(k) = \sum_{n=1}^{N} e^{2\pi ikn/N}\mathcal{E}^{v,h}(n) \tag{15.85}$$

then the partition function $Z(\mathcal{E}_{cp})$ on a lattice of L sites is related to the partition function $Z(\tilde{\mathcal{E}}_{cp})$ on the L_D faces of the lattice by

$$Z(\mathcal{E}_{cp}) = N^{1-L_D} Z(\tilde{\mathcal{E}}_{cp}) \tag{15.86}$$

The transformation from $\mathcal{E}^{h,v}_{cp}(n)$ to $\tilde{\mathcal{E}}^{h,v}_{cp}(n)$ is called a dual transformation. If $\mathcal{E}^{h,v}_{cp}(n) = \tilde{\mathcal{E}}^{h,v}_{cp}(n)$ the model is called selfdual.

Studies of the phase diagram of the chiral Potts model were initiated for the fully asymmetric case by Ostlund [23] for $N \geq 3$, by Huse [24] for $N = 3$, and also by Yeomans and Fisher [25], and for the symmetric case by Kardar [26]. All of these studies reveal that there are regions in the phase diagram where the correlations oscillate with a period which is temperature dependent. These phases are referred to as incommensurate phases. In Fig 15.3 we reproduce the phase diagram of Ostlund [23] for the fully asymmetric model $\Delta^v = 0$ with $N = 3$ and $K^v = K^h$.

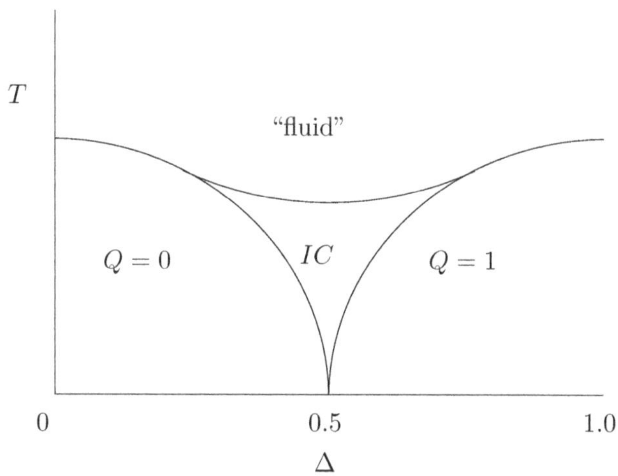

Fig. 15.3 The schematic phase diagram of the fully asymmetric model $\Delta^v = 0$ with $N = 3$ and $K^v = K^h$ suggested in [23].

A prominent feature of the phase diagram of Fig. 15.3 is that there is a phase boundary between the disordered (Li) phase and the ordered phase which ends in a multicritical point where the ordered, disordered and incommensurate phases meet. This point has been called the *Lifshitz* point. However, the existence of this point has been a subject of controversy since 1983 when Haldane, Bak and Bohr [27] published computations which lead to the conjecture that a phase boundary between the ordered and disordered phases does not exist. These studies have been considerably extended and improved by Au-Yang, Perk and Jin [28–30] providing further support for the conclusion that the Lifshitz point does not in fact exist.

This chronology of the classical two-dimensional model is summarized in Table 15.3. There was no indication in the studies done before 1985 of the classical model (15.75) that any exact computations could be done.

Table 15.3 Selected developments in the study of the N-state chiral clock and chiral Potts models.

Date	Author(s)	Development
1976	Wu, Wang [22]	Duality for chiral Potts
1981	Ostlund [23]	N-state asymmetric chiral clock model
1981	Huse [24]	Three-state asymmetric chiral Potts
1981	Yeomans, Fisher [25]	Three-state asymmetric chiral Potts
1982	Kardar [26]	N-state symmetric chiral Potts
1983	Haldane, Bak, Bohr [27]	Lack of Lifshitz point in three-state asymmetric chiral Potts
1995	Au-Yang, Perk [28]	Connection with integrable chiral Potts
1996	Au-Yang, Perk [29]	Phase diagrams for $N = 3, 4, 5$
2002	Jin, Au-Yang, Perk [30]	Three-state asymmetric chiral Potts

The quantum spin chain version of the chiral Potts model for the case $N = 3$ was introduced in 1983 by Howes, Kadanoff and den Nijs [31] as a model of the physical phenomena of level crossing transitions. Just as in the classical case the model was introduced for physical reasons and not because anyone thought that it might be integrable in the sense of solving the star–triangle equation. However, unlike the computations in the classical model (15.75) the computations of [31] lead to some remarkable simple results which demanded some further investigation.

The understanding that the N-state chiral Potts model can actually be solved begins with the 1985 paper of von Gehlen and Rittenberg [32] and the chronology of developments is summarized in Table 15.4. The chiral Potts model is substantially more difficult to investigate than either the eight-vertex or the RSOS models previously studied because, as seen in chapter 13, the spectral variable lies on a curve of genus greater than one and thus the machinery of elliptic functions cannot be used in its solution.

15.2.2 Real and positive Boltzmann weights

In order for the chiral Potts model which satisfies the star–triangle equation to be able to describe the chiral clock models (15.80), (15.81) the Boltzmann weights as given by either of the forms (13.165), (13.166) or (13.181), (13.182) must be real and positive. To examine this positivity requirement we express the Boltzmann weights in terms of the variables x_p, x_q, y_p, y_q which from (13.181)–(13.184) of chapter 13 are written as

$$\frac{W^h_{pq}(n)}{W^h_{pq}(0)} = \left(\frac{y_p^N - x_q^N}{y_q^N - x_p^N}\right)^{n/N} \prod_{j=1}^{n} \left(\frac{y_q - x_p\omega^j}{y_p - x_q\omega^j}\right) \tag{15.87}$$

and

$$\frac{W^v_{pq}(n)}{W^v_{pq}(0)} = \left(\frac{y_q^N - y_p^N}{x_p^N - x_q^N}\right)^{n/N} \prod_{j=1}^{n} \left(\frac{\omega x_p - x_q\omega^j}{y_q - y_p\omega^j}\right) \tag{15.88}$$

with $\omega = e^{2\pi i/N}$ and we recall (13.187) that

Table 15.4 Selected developements in the study of the integrable N-state chiral Potts model.

Date	Author(s)	Development
1944	Onsager [21]	Onsager's algebra for $N = 2$
1983	Howes, Kadanoff, den Nijs [31]	$N = 3$ spin chain investigated
1985	von Gehlen, Rittenberg [32]	Superintegrable case discovered
1987	Au-Yang, McCoy, Perk Tang, Yan [33]	Star triangle for $N = 3$
1987	McCoy, Perk, Tang, Sah [34]	Fermat curve for $N = 4$
1987	Perk [35]	Onsager algebra for general N
1988	Au-Yang, McCoy, Perk, Tang [36]	Star triangle for self-dual $N = 5$
1988	Baxter, Perk, Au-Yang [37] Au-Yang, Perk [38]	Star-triangle for general N
1988	Baxter [39]	Free energy for $P = 0$
1988	Baxter [40]	Superintegrable free energy for $P = 0$
1989	Albertini, McCoy, Perk, Tang [41]	Conjecture of order parameter
1989	Albertini, McCoy, Perk [42]	Level crossing to $P \neq 0$
1989	Albertini, McCoy, Perk [43]	Functional equations
1990	Bazhanov, Stroganov [44]	Six-vertex chiral Potts relation
1990	Baxter, Bazhanov, Perk [45]	Functional equations
1990	McCoy, Roan [46]	Phase diagram for the spin chain
1991	Baxter [47]	Eigenvalues of transfer matrix
1999	Au-Yang, Perk [48]	Limit $N \to \infty$
2005	Baxter [49]	Derivation of the order parameter

$$x_p^N + y_p^N = \lambda'(1 + x_p^N y_p^N). \tag{15.89}$$

To make the reality conditions manifest we follow [28] and write (15.87)–(15.88) in terms of the variables

$$e^{2i\theta} = \frac{x_p}{y_q}, \quad e^{2i\phi} = \frac{x_q}{y_p}, \quad e^{2i\bar{\theta}} = \frac{x_q}{\omega x_p}, \quad e^{2i\bar{\phi}} = \frac{y_p}{y_q} \tag{15.90}$$

as

$$\frac{W^h(n;\theta,\phi)}{W^h(0;\theta,\phi)} = \left(\frac{\sin(N\phi)}{\sin(N\theta)}\right)^{n/N} \prod_{j=1}^{n} \frac{\sin(\theta + \pi j/N)}{\sin(\phi + \pi j/N)} \tag{15.91}$$

and

$$\frac{W^v(n;\bar{\theta},\bar{\phi})}{W^v(0;\bar{\theta},\bar{\phi})} = \left(\frac{\sin(N\bar{\phi})}{\sin(N\bar{\theta})}\right)^{n/N} \prod_{j=1}^{n} \frac{\sin(\bar{\theta} + \pi j/N)}{\sin(\bar{\phi} + \pi j/N)} \tag{15.92}$$

which depend on the four angles $\theta,\ \phi,\ \bar{\theta},\ \bar{\phi}$ instead of the three parameters p, q and λ. The four angles must be real in order that the Boltzmann weights can be real. We regain the Boltzmann weights $W_{pq}^{h,v}(n)$ when the four angles $\theta,\ \phi,\ \bar{\theta},\ \bar{\phi}$ satisfy the one constraint which follows from (15.90)

$$\phi + \bar{\phi} = \frac{\pi}{N} + \theta + \bar{\theta} \tag{15.93}$$

and λ' satisfies

$$\lambda'^2 = \sin^{-2}[N(\theta-\phi)]\{\cos^2[N(\theta+\bar{\theta})]+\cos^2[N(\phi-\bar{\theta})] - 2\cos[N(\theta+\bar{\theta})]\cos[N(\phi-\theta)]\cos[N(\phi-\bar{\theta})]\} \tag{15.94}$$

which may be verified by substituting (15.90) into (15.94) and eliminating y_p^N and y_q^N using (15.89).

The general chiral Potts interaction energies on the real manifold (15.80) and (15.81) have $2(N-1)$ independent parameters and when the angles are not constrained by (15.93) the weights (15.91) and (15.92) give a four-dimensional subspace of this manifold. For $N=3$ this manifold is the entire space of parameters (as $2(N-1)=4$).

To relate parameters $K_n^{v,h}$ and $\Delta_n^{v,h}$ to the four angles θ, ϕ, $\bar{\theta}$, $\bar{\phi}$ we write the chiral Potts interaction energy (15.75) as (15.84) with

$$\frac{\mathcal{E}^{h,v}(n)}{k_BT} = \sum_{j=1}^{N-1} K_j^{h,v}\omega^{\Delta_j^{h,v}}\omega^{jn}. \tag{15.95}$$

By the definition of Boltzmann weights

$$\ln\frac{W^h(n;\theta,\phi)}{W^h(n-1;\theta,\phi)} = -\frac{\mathcal{E}^h(n)-\mathcal{E}^h(n-1)}{k_BT} = \sum_{j=1}^{N-1} r_j^h\omega^{jn}$$

$$\ln\frac{W^v(n;\bar{\theta},\bar{\phi})}{W^v(n-1;\bar{\theta},\bar{\phi})} = -\frac{\mathcal{E}^v(n)-\mathcal{E}^v(n-1)}{k_BT} = \sum_{j=1}^{N-1} r_j^v\omega^{jn} \tag{15.96}$$

with

$$r_j^{h,v} = K_j^{h,v}\omega^{\Delta_j^{h,v}}(1-\omega^{-j}). \tag{15.97}$$

By use of (15.91) and (15.92) we have

$$\ln\frac{W^h(n;\theta,\phi)}{W^h(n-1;\theta,\phi)} = A^h + B_n^h$$

$$\ln\frac{W^v(n;\bar{\theta},\bar{\phi})}{W^v(n-1;\bar{\theta},\bar{\phi})} = A^v + B_n^v \tag{15.98}$$

with

$$A^v = \frac{1}{N}\ln\frac{\sin(N\phi)}{\sin(N\theta)}, \quad B_n^v = \ln\frac{\sin(\theta+\pi n/N)}{\sin(\phi+\pi n/N)}$$

$$A^h = \frac{1}{N}\ln\frac{\sin(N\bar{\phi})}{\sin(N\bar{\theta})}, \quad B_n^h = \ln\frac{\sin(\bar{\theta}+\pi n/N)}{\sin(\bar{\phi}+\pi n/N)}. \tag{15.99}$$

The Fourier transform (15.96) is inverted to find for $1\le j\le N-1$

$$r_j^{h,v} = N^{-1}\sum_{n=1}^{N}\omega^{-nj}B_n^{h,v} \tag{15.100}$$

and hence

$$K_j^{v,h}\omega^{\Delta_j^{v,h}} = \frac{r_j}{1-\omega^{-j}} = -\frac{S_j^{v,h}+iC_j^{v,h}}{2N\sin(\pi j/N)} \tag{15.101}$$

where

$$S_j^{v,h} = S_{N-j}^{v,h} = \sum_{n=1}^{N} B_n^{v,h}\sin[(2n-1)j\pi/N], \tag{15.102}$$

$$C_j^{v,h} = -C_{N-j}^{v,h} = \sum_{n=1}^{N} B_n^{v,h}\cos[(2n-1)j\pi/N] \tag{15.103}$$

and therefore when $K^{v,h}\geq 0$

$$K_j^{v,h} = K_{N-j}^{v,h} = \frac{[(S_j^{v,h})^2+(C_j^{v,h})^2]^{1/2}}{2N\sin(\pi j/N)}, \quad \Delta_j^{v,h} = -\Delta_{N-j}^{v,h} = \frac{N}{2\pi}\arctan\frac{C_j^{v,h}}{S_j^{v.h}} \tag{15.104}$$

which expresses all of the variables $K_j^{v,h}$ and $\Delta_j^{v,h}$ in terms of the four angles θ, ϕ, $\bar{\theta}$, $\bar{\phi}$.

There are several features to be noted about the weights (15.91) and (15.92) which do not depend on the integrability condition (15.93):

1) When θ and ϕ ($\bar{\theta}$ and $\bar{\phi}$) are interchanged $r_j \to -r_j$. We will restrict our considerations to

$$\theta \leq \phi, \qquad \bar{\theta} \leq \bar{\phi} \tag{15.105}$$

2) When

$$\theta \to \phi \;\; (\bar{\theta} \to \bar{\phi}) \;\; \text{then} \;\; K_j^h = 0 \; (\bar{K}_j^v = 0) \tag{15.106}$$

while $\Delta^{h,v}$ are arbitrary. Thus the point $\theta = \phi$, $\bar{\theta} = \bar{\phi}$ is the point $T \to \infty$.

3) When (15.105) holds and

$$\theta, \bar{\theta} \to -\frac{\pi}{N} \tag{15.107}$$

then $B_1^{v,h} \to \infty$ and thus $K_j^{v,h} \to \infty$ for all j which is the point $T \to 0$.

4) When

$$(\theta,\phi,\bar{\theta},\bar{\phi}) \to (\theta,\phi,\bar{\theta},\bar{\phi}) + \frac{\pi}{N} \tag{15.108}$$

we see from (15.99) that $B_n^{v,h} \to B_{n+1}^{v,h}$ and thus from (15.100) and (15.101) K_j is unchanged and

$$\Delta_j \to \Delta_j + j. \tag{15.109}$$

5) When

$$(\theta,\phi,\bar{\theta},\bar{\phi}) \to -(\theta,\phi,\bar{\theta},\bar{\phi}) - \frac{\pi}{2N} \tag{15.110}$$

then $B_n \to -B_{-n}$ and thus K_j is unchanged and

$$\Delta_j \to -\Delta_j. \tag{15.111}$$

We found in (13.192) that when

$$y_{p,q} = \omega^{1/2} x_{p,q} \tag{15.112}$$

the integrable chiral Potts model reduced to the model of Fateev and Zamolodchikov [50] whose Boltzmann weights are given by (13.196) and (13.197). When (15.112) holds we find from (15.90)

$$\theta = -\frac{\pi}{2N} - \theta_{FZ}, \quad \phi = -\frac{\pi}{2N} + \theta_{FZ}, \quad \bar{\theta} = -\frac{\pi}{N} + \theta_{FZ}, \quad \bar{\phi} = -\theta_{FZ} \tag{15.113}$$

where in this limit we have set

$$x_q / x_p = e^{2i\theta_{FZ}} \tag{15.114}$$

to agree with (13.195) and we see from (15.94) that $\lambda' = 0$ as required. Using (15.113) the constraint (15.93) is satisfied and the weights (15.91) and (15.92) reduce to (13.196) and (13.197) as desired. These weights are real and positive when

$$0 \leq \theta_{FZ} \leq \frac{\pi}{2N} \tag{15.115}$$

and thus at the Fateev–Zamolodchikov point

$$-\frac{\pi}{N} \leq \theta, \bar{\theta} \leq -\frac{\pi}{2N}, \quad \phi = -\frac{\pi}{N} - \theta, \quad \bar{\phi} = -\frac{\pi}{N} - \bar{\theta} \tag{15.116}$$

which at the isotropic point $\theta_{FZ} = \pi/4N$ reduces to

$$\theta = \bar{\theta} = -\frac{3\pi}{4N}, \quad \phi = \bar{\phi} = -\frac{\pi}{4N}. \tag{15.117}$$

The isotropic case

In the isotropic (a.k.a. symmetric) case $K_j^v = K_j^h = K_j$ and $\Delta_j^v = \Delta_j^h = \Delta_j$ where $\theta = \bar{\theta}$ and $\phi = \bar{\phi}$ we find from (15.93) that

$$\phi = \frac{\pi}{2N} + \theta \tag{15.118}$$

and thus the Boltzmann weights (15.91), (15.92) are expressed in terms of θ as

$$\frac{W_{pq}^h(n)}{W_{pq}^h(0)} = \frac{W_{pq}^v(n)}{W_{pq}^v(0)} = \left(\frac{\cos N\theta}{\sin N\theta}\right)^{n/N} \prod_{j=1}^{n} \frac{\sin(\theta + \pi j/N)}{\sin(\theta + (\pi(j+1/2)/N)}. \tag{15.119}$$

The weights (15.119) are real and positive for

$$-\frac{\pi}{N} < \theta < -\frac{\pi}{2N} \tag{15.120}$$

We also note that for

$$-\frac{\pi}{2N} < \theta < 0 \pmod{\pi/N} \tag{15.121}$$

the Boltzmann weights are not all positive. Thus (15.120) may be considered as the fundamental region of positivity in the sense that all other regions are obtained by use of the transformations (15.108) and (15.110).

When $\theta = -3\pi/4N$ the model reduces to the isotropic Fateev–Zamolodchikov Z_N symmetric model where

$$\Delta_j = 0 \tag{15.122}$$

and for odd N

$$K_j = \frac{1}{N\sin(\pi j/N)} \sum_{n=1}^{[N/2]} \sin\left((2n-1)\frac{j\pi}{N}\right) \ln \frac{\sin[(n-1/4)\pi/N]}{\sin[(n-3/4)\pi/N]}. \tag{15.123}$$

When $\theta \to -\pi/N$ we see from (15.99) that $B_1^{v,h} \to -\infty$ and thus $K_j \to \infty$ and that $T \to 0$. From (15.100) and (15.101) we see that

$$K_j \omega^{\Delta_j} \to \frac{B_1}{N(\omega^j - 1)} \tag{15.124}$$

and thus we obtain the final result that as $T \to 0$

$$\frac{K_j}{K_1} = \frac{\sin(\pi/N)}{\sin(\pi j/N)}, \quad \Delta_j = \frac{1}{4}(N - 2j). \tag{15.125}$$

The K_j decrease from $K_j = \infty$ at $\theta = -\pi/N$ to (15.123) at $\theta = -3\pi/4$. There are no values of K_j smaller than (15.123) on the integrable manifold.

The general anisotropic case

The considerations of the isotropic case are readily extended to the general anisotropic case where from (15.105), (15.107) and (15.116) we find that the Boltzmann weights (15.91), (15.92) are real and positive for the four dimensional subspace

$$-\frac{\pi}{N} \le \theta \le \phi \le -\frac{\pi}{N} - \theta, \quad -\frac{\pi}{N} \le \bar{\theta} \le \bar{\phi} \le -\frac{\pi}{N} - \bar{\theta}. \tag{15.126}$$

All other real regimes are obtained from (15.126) by the transformations (15.108)–(15.111). The three-dimensional integrable submanifold is obtained when the four angles are restricted by the integrability condition (15.93). We note that the point $T = \infty$ where $\theta = \phi$ and $\bar{\theta} = \bar{\phi}$ is excluded from the integrable subspace by the integrability restriction (15.93).

The case $N = 3$

To explicitly plot phase diagrams we will restrict our consideration to the case $N = 3$ where (15.126) reduces to

$$-\frac{\pi}{3} \le \theta \le \phi \le -\frac{\pi}{3} - \theta, \quad -\frac{\pi}{3} \le \bar{\theta} \le \bar{\phi} \le -\frac{\pi}{3} - \bar{\theta}. \tag{15.127}$$

All other real regimes are obtained from (15.127) by the transformations (15.108)–(15.111). . The cases most studied in the literature are $K_1^v = K_1^h$ with either $\Delta_1^v = \Delta_1^h = \Delta$ (the symmetric case of Kardar [26]) or $\Delta_1^v = 0,\ \Delta_1^h = \Delta$ (the fully asymmetric case of Ostlund [23] and Huse [24]). We will thus restrict our attention to the relation between the phase diagrams in the two-dimensional space

$$K_1^v = K_1^h = K, \quad \Delta_1^v = p\Delta_1^h = p\Delta \tag{15.128}$$

and the one-dimensional integrable submanifold obtained by imposing the two conditions (15.128) on the four equations contained in (15.104). From this the one-dimensional integrable submanifold can be plotted in the two dimensional space of K and Δ.

The endpoints of the integrable submanifold are at the isotropic point of the Fateev–Zamolodchikov model where

$$K = K_c = \frac{2}{3}\ln(1+\sqrt{3}) = 1.4924\cdots, \quad \Delta^v = \Delta^h = 0 \tag{15.129}$$

and at $T = 0$ where $K \to \infty$ and $\Delta = 1/4$. Expansions near these endpoints are derived in [28].

The expansion near the Fateev–Zamolodchikov point (15.129) is uniform in p and is derived in [28] as

$$\frac{K}{K_c} - 1 = C_t(1+p^2)\Delta^2 \tag{15.130}$$

with

$$C_t = \frac{\pi^2}{18}[2 + (7\sqrt{3}+12)K_c] \tag{15.131}$$

which is valid for all p.

For the expansion near $T \to 0$ where $K \to \infty$ the cases of $p = 0,\ 0 < p < 1$, and $p = 1$ must be treated separately [28].

For the symmetric case $p = 1$

$$\Delta = \frac{1}{4} - \frac{\ln 2}{2\pi K}. \tag{15.132}$$

For the asymmetric case $0 < p < 1$

$$\Delta = \frac{1}{4} - \frac{1}{2\pi K}\exp\left(-3K\sin[(1-p)\pi/6]\right). \tag{15.133}$$

For the fully asymmetric case $p = 0$

$$\Delta = \frac{1}{4} - \frac{1}{\pi K}e^{-3K/2} \tag{15.134}$$

where we remark that the result (15.134) fails to agree with the limit $p \to 0$ of (15.133) by a factor of two.

In the isotropic case at $\Delta^h = \Delta^v = 3/2$ the energy (15.80) reduces to the antiferromagnetic (scalar) Potts model

$$\mathcal{E}_{cp}/k_BT = 2K\sum_{j,k}\{\cos[\frac{2\pi}{N}(n_{j,k} - n_{j,k+1})] + \cos[\frac{2\pi}{N}((n_{j,k} - n_{j+1,k})]\}. \tag{15.135}$$

This point is particularly interesting because the ground state of (15.135) is macroscopically degenerate. Furthermore at this point the model is equivalent to a six-vertex model with $\Delta_6 = 2/3$ as first pointed out by Lenard as cited by Lieb [51] in the language of three colorings of the square lattice [52]. From this correspondence it follows that the correlation functions have a power law behavior and thus $T = 0$ can be considered to be a critical point [53, 54]. It is thus a most natural conjecture that the free energy of the isotropic chiral Potts model at $\Delta = 3/2$ is analytic for $T > 0$. There is strong numerical evidence to support this [55] but there appears to be no definitive proof.

Physics on the real manifold

We plot in Fig. 15.4 and Fig. 15.5 the phase diagrams for $p = 0$ and $p = 1$ [28–30] where the phase boundaries are qualitatively drawn with the assumption that there are no Lifshitz points, and we indicate the one-dimensional integrable submanifold computed in the previous subsection. The free energy on the integrable real manifold has been computed and studied in several papers [39, 45, 47, 56]. This free energy lies totally within the ordered phase and is analytic except at the Fateev–Zamolodchikov point where it becomes critical.

The qualitative behavior of the order parameters in the phases marked by $Q = 0, 1, 2$ is determined by the configuration of spins which makes the interaction energy (15.80) the most negative. For example, on the sector marked $Q = 0$ all spins point in the same direction at $T = 0$ and this sector is often referred to as "ferromagnetic". In the other sectors there is an oscillation of the order parameter with a period $e^{2\pi i Q/3}$. In the phase marked "fluid" the system has no long range order and correlations decay exponentially. These features of the phase diagram are identical with the low and high temperature phases of the Ising model.

The new feature of these phase diagrams which is not present in the Ising model is the phase marked IC which stands for incommensurate. In this phase there is no long range order. The correlations decay algebraically and oscillate with a period which depends on both K and Δ. As the phase boundary between the fluid and incommensurate phase is approached from the fluid side the correlation length is believed to diverge exponentially as it does for the classical XY spin model in two dimensions at the Kosterlitz–Thouless temperature discussed in chapter 9.

It is unfortunate that the integrable manifold lies only in the ordered phase and thus we have no exact solutions in the incommensurate phase.

15.2.3 The superintegrable chiral Potts model and Onsager's algebra

We now turn our attention to the chiral Potts spin chain (13.314)

$$H_{cp}(p) = -\sum_{j=1}^{\mathcal{N}} \sum_{n=1}^{N-1} \{\bar{\alpha}_n (X_j)^n + \alpha_n (Z_j Z_{j+1}^\dagger)^n\} \tag{15.136}$$

with

$$\alpha_n = \frac{e^{i(2n-N)\phi/N}}{\sin \pi n/N} \tag{15.137}$$

$$\bar{\alpha}_n = \lambda \frac{e^{i(2n-N)\bar{\phi}/N}}{\sin \pi n/N} \tag{15.138}$$

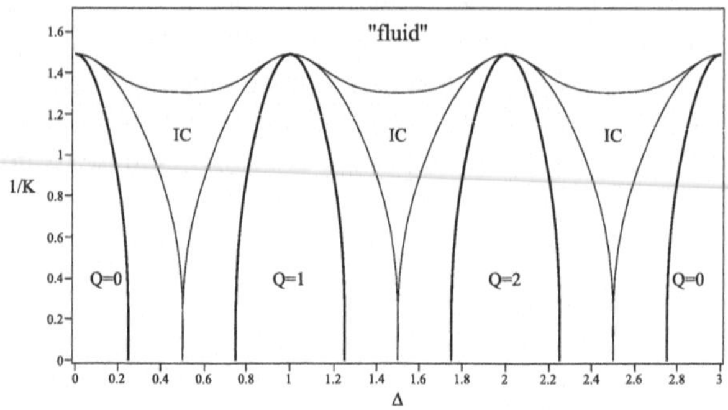

Fig. 15.4 The schematic phase diagram following [28–30] of the fully asymmetric $p = 0$ model $\Delta^v = 0$ with $N = 3, K^v = K^h = K$. The phase boundaries are indicated by the thin lines and the integrable submanifold (15.134) is indicated by the solid lines. We have assumed that there are no Lifshitz points. The integrable submanifold is totally within the ordered phases. The three ordered phases with $Q = 0, 1, 2$ are related by symmetry.

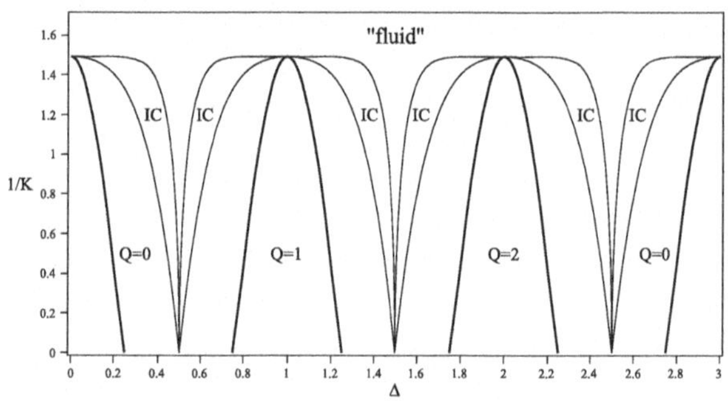

Fig. 15.5 The schematic phase diagram following [28–30] of the fully symmetric $p = 1$ model $\Delta^v = \Delta^h = \Delta$ with $N = 3, K^v = K^h = K$. The phase boundaries are indicated by the thin lines and the integrable submanifold (15.134) is indicated by the solid lines. We have assumed that there are no Lifshitz points. The integrable submanifold is totally within the ordered phases. The three ordered phases with $Q = 0, 1, 2$ are related by symmetry.

where the angles ϕ and $\bar{\phi}$ (which are not the same as the angles in the previous subsection) are expressed in terms of the variables of the chiral Potts curve:

$$a^N + \lambda b^N = \lambda' d^N \tag{15.139}$$

$$\lambda a^N + b^N = \lambda' c^N \tag{15.140}$$

with

$$\lambda' = (1 - \lambda^2)^{1/2} \tag{15.141}$$

by

$$e^{2i\phi/N} = \omega^{1/2} \frac{a_p c_p}{b_p d_p} \tag{15.142}$$

$$e^{2i\bar{\phi}/N} = \omega^{1/2}\frac{a_p d_p}{b_p c_p}. \tag{15.143}$$

From (15.142) and (15.143) and the chiral Potts curve (15.139), (15.140), it follows that ϕ, $\bar{\phi}$ and λ are related by

$$\cos\phi = \lambda\cos\bar{\phi}. \tag{15.144}$$

When

$$\phi = \bar{\phi} = \pi/2 \tag{15.145}$$

the constraint (15.144) is identically satisfied for all λ and the chiral Potts Hamiltonian (15.136) may be written as

$$H_{sicp} = A_0 + \lambda A_1 \tag{15.146}$$

where

$$A_0 = -\sum_{j=1}^{\mathcal{N}}\sum_{n=1}^{N-1}\frac{e^{i\pi(2n-N)/(2N)}}{\sin\pi n/N}(Z_j Z_{j+1}^{\dagger})^n \tag{15.147}$$

$$A_1 = -\sum_{j=1}^{\mathcal{N}}\sum_{n=1}^{N-1}\frac{e^{i\pi(2n-N)/(2N)}}{\sin\pi n/N}(X_j)^n. \tag{15.148}$$

We refer to (15.146) as the superintegrable chiral Potts spin chain. Furthermore we see from (15.142) and (15.143) that the parameters a_p, b_p, c_p, d_p which enter the Boltzmann weights and which are constrained to lie on the curve are given by the point

$$a_p = b_p, \quad c_p = d_p, \quad \frac{c_p}{a_p} = \left(\frac{1+\lambda}{1-\lambda}\right)^{\frac{1}{2N}} \equiv \eta \tag{15.149}$$

where the last line defines η.

The spin chain (15.146)–(15.148) has the remarkable property found in 1985 by von Gehlen and Rittenberg [32] that

$$\begin{aligned}
[A_0,[A_0,[A_0,A_1]]] &= \text{const}[A_0,A_1]\\
[A_1,[A_1,[A_1,A_0]]] &= \text{const}[A_1,A_0].
\end{aligned} \tag{15.150}$$

This relation is more restrictive than the star–triangle equation and for this reason we refer to the chiral Potts model with the restriction (15.145) as the *superintegrable* chiral Potts model. From the equality (15.150) of a triple commutator with a single commutator one can introduce [35, 57] operators A_n and G_n which satisfy the conditions

$$\begin{aligned}
[A_j,A_k] &= 4G_{j-k}\\
[G_m,A_l] &= 2A_{l+m} - 2A_{l-m}\\
[G_j,G_k] &= 0.
\end{aligned} \tag{15.151}$$

The three relations (15.151) define an algebra which is exactly the same as the algebra found and used by Onsager in his original 1944 solution of the Ising model [21] which is the special case $N = 2$ of the chiral Potts model. We therefore refer to the algebra (15.151) as "Onsager's algebra". Thus it is very appropriate to view the superintegrable chiral Potts model as the "most natural" generalization of the Ising model. It is quite surprising that it took 41 years for this generalization to be discovered.

It follows [57–59] from the theory of the representations Onsager's algebra (15.151) that the eigenvalues of any Hamiltonian of the form (15.146) will have the form

$$E(\lambda) = A + \lambda B + 2N\sum_{j=1}^{n} m_j(1+\lambda^2+2\lambda\cos\phi_j)^{1/2} \tag{15.152}$$

where ϕ_j are a set of real numbers and m_j takes on all values in the set

$$m_j = -s_j/2,\ 1-s_j/2,\ \cdots, s_j/2 \tag{15.153}$$

where s_j can be any integer and $s_j/2$ can be thought of as the "spin" of the representation. The numbers ϕ_j are not restricted by the algebra (15.151). For the superintegrable chiral Potts model $s_j/2 = 1/2$ for all j. It is an open question whether or not there are representations of Onsager's algebra for which $s_j/2 \geq 1$.

We note also that (15.152) satisfies the duality relation

$$\lambda E(1/\lambda) = (1-\lambda)(B-A) + E(\lambda). \tag{15.154}$$

15.2.4 The functional equation for the superintegrable case for $N = 3$

The transfer matrix of the chiral Potts model was defined in chapter 13 as

$$T_{p,q} = W^v_{pq}(j_1 - j_1')W^h_{pq}(j_1 - j_2')\cdots W^v_{pq}(j_{\mathcal{N}} - j_{\mathcal{N}}')W^h_{pq}(j_{\mathcal{N}} - j_1') \tag{15.155}$$

where $\mathcal{N}$ (and not N) is the length of the chain, and the weights $W^v_{pq}(n)$ and $W^h_{pq}(n)$ are given by (13.165) and (13.166). We write the transfer matrix as

$$T_{pq} = \frac{(\eta a_q/d_q - 1)^{\mathcal{N}}}{[(\eta a_q/d_q)^N - 1]^{\mathcal{N}}}\mathcal{T}_{p,q} \tag{15.156}$$

and note for $N = 3$ that $\mathcal{T}_{pq}$ satisfies the functional equation which was conjectured in [42] and [43] and proven in [45]

$$\mathcal{T}_{p,q}\mathcal{T}_{p,Rq}\mathcal{T}_{p,R^2q} = \mathcal{K}^{2\mathcal{N}}e^{-iP}\{(\frac{ab}{cd}\eta^2-1)^{\mathcal{N}}(\frac{ab}{cd}\eta^2\omega^2-1)^{\mathcal{N}}\mathcal{T}_{p,q} +$$
$$(\frac{ab}{cd}\eta^2\omega^2-1)^{\mathcal{N}}(\frac{ab}{cd}\eta^2\omega-1)^{\mathcal{N}}\mathcal{T}_{p,R^2q} + (\frac{ab}{cd}\eta^2-1)^{\mathcal{N}}(\frac{ab}{cd}\eta^2\omega-1)^{\mathcal{N}}\mathcal{T}_{p,R^4q}\} \tag{15.157}$$

where R is the automorphism

$$R(a_q, b_q, c_q, d_q) = (b_q, \omega a_q, d_q, c_q) \tag{15.158}$$

with $\omega = e^{2\pi i/N}$, e^{-iP} is the right shift operator where the eigenvalues of the momentum operator P are $2\pi k/\mathcal{N}$ for $0 \leq k \leq \mathcal{N}-1$ and $\mathcal{K}$ is some constant normalizing factor which is irrelevant for our computation. The functional equation (15.157) is the counterpart of the functional equation (14.299) of for $Q_{72}(v)$ and thus the chiral Potts transfer matrix may be thought of as the analogue of the $Q_{72}(v)$ matrix of the eight-vertex model.

The eigenvalues of $t_{p,q}$ of $T_{p,q}$ satisfy (15.157) and are of the following remarkably simple form as a function of q which is in fact valid for all N

$$\begin{aligned} t_{p,q} &= \frac{N^{\mathcal{N}}(\eta\frac{a}{d}-1)^{\mathcal{N}}}{[(\eta\frac{a}{d})^N-1]^{\mathcal{N}}}(\eta\frac{a}{d})^{P_a}(\eta\frac{b}{c})^{P_b}(\frac{c^N}{d^N})^{P_c} \\ &\prod_{l=1}^{m_p}\left(\frac{1+\omega v_l\eta^2\frac{ab}{cd}}{1+\omega v_l}\right)\prod_{l=1}^{m_E}\left(\frac{1+\lambda}{1-\lambda}\right)^{1/2}\left[\frac{a^N+b^N}{2d^N}\pm\frac{w_l(a^N-b^N)}{(1+\lambda)d^N}\right] \end{aligned} \tag{15.159}$$

where we note that, by the use of (15.149), as desired $t_{p,q} \to 1$ as $q \to p$. This representation of the eigenvalues $t_{p,q}$ as a factorization in terms of the coordinates of the curve is a special property of the superintegrable case and is the counterpart of the form (15.152) of the Hamiltonian eigenvalues which followed from Onsager's algebra. In general, a function on a curve can only be represented in terms of its zeros by use of what are called "prime forms" in algebraic geometry. We also note that the separation of the zeros of $t_{p,q}$ into the two classes of v_l and w_l is the analogue of the Bethe roots and complete L-strings of the $Q_{72}(v)$ matrix of the eight-vertex model.

The eigenvalues of the superintegrable chiral Potts Hamiltonian (15.146) are of the form (15.152) mandated by Onsager's algebra of the form

$$\begin{aligned} E &= N(2P_c+m_E)-\mathcal{N}(N-1) \\ &+\lambda[\mathcal{N}(N-1)-N(2P_c+m_E)+2(P_b-P_a)]+2N\sum_{l=1}^{m_E}\pm w_l. \end{aligned} \tag{15.160}$$

Furthermore from (13.378)

$$\lim_{q\to p} T_{p,Rq} = e^{-iP} = \omega^{P_b}\prod_{l=1}^{m_p}\left(\frac{1+\omega^2 v_l}{1+\omega v_l}\right) \tag{15.161}$$

where P is the momentum which has the values

$$P = \frac{2\pi n}{\mathcal{N}} \quad \text{with} \quad n = 0, 1, \cdots \mathcal{N}-1 \tag{15.162}$$

and we note the constraints

$$P_a + 3P_c = 3m_E + m_p \leq 2\mathcal{N} \tag{15.163}$$

$$P_b + m_p \leq 3P_c \tag{15.164}$$

$$3m_e + 2m_p = \text{const.} \tag{15.165}$$

15.2.5 Superintegrable ground state energy for small λ

We consider first the case $m_p = 0$. Then from (15.159) we see that

$$\mathcal{T}_{p,R^2q} = \omega^{P_a+P_b}\mathcal{T}_{p,q} \tag{15.166}$$

and setting $P_b = 0$ as discussed in [43] the functional equation (15.157) reduces to

$$\begin{aligned}\omega^{P_a}\mathcal{T}_{p,q}\mathcal{T}_{p,Rq} &= \mathcal{K}^{2\mathcal{N}}\{(\frac{ab}{cd}\eta^2-1)^{\mathcal{N}}(\frac{ab}{cd}\eta^2\omega^2-1)^{\mathcal{N}}\\ &+(\frac{ab}{cd}\eta^2\omega^2-1)^{\mathcal{N}}(\frac{ab}{cd}\eta^2\omega-1)^{\mathcal{N}}\omega^{P_a}+(\frac{ab}{cd}\eta^2-1)^{\mathcal{N}}(\frac{ab}{cd}\eta^2\omega-1)^{\mathcal{N}}\omega^{2P_a}\}.\end{aligned} \tag{15.167}$$

The action of the automorphism R is

$$R\left(\frac{a^3-b^3}{a^3+b^3}\right) = -\left(\frac{a^3-b^3}{a^3+b^3}\right) \tag{15.168}$$

and thus using (15.159) the functional equation (15.157) reduces to

$$\begin{aligned}&\bar{\mathcal{K}}\left(\eta^2\frac{ab}{cd}\right)^{P_a}\prod_{l=1}^{m_E}\left[\frac{(a^3+b^3)^2}{4c^3d^3}-w_l^2\frac{(a^3-b^3)^2}{(1+\lambda)^2c^3d^3}\right]\\ &=\{(\frac{ab}{cd}\eta^2-1)^{\mathcal{N}}(\frac{ab}{cd}\eta^2\omega^2-1)^{\mathcal{N}}+(\frac{ab}{cd}\eta^2\omega^2-1)^{\mathcal{N}}(\frac{ab}{cd}\eta^2\omega-1)^{\mathcal{N}}\omega^{P_a}\\ &+(\frac{ab}{cd}\eta^2-1)^{\mathcal{N}}(\frac{ab}{cd}\eta^2\omega-1)^{\mathcal{N}}\omega^{2P_a}\}.\end{aligned} \tag{15.169}$$

To solve (15.169) for w_l we note that if we multiply the equations (15.139) and (15.140) together we find for $N = 3$ that

$$\lambda(a^6+b^6)+(1+\lambda^2)a^3b^3 = (1-\lambda^2)c^3d^3 \tag{15.170}$$

and we find

$$\frac{(a^3+b^3)^2}{c^3d^3} = \lambda^{-1}\{(1-\lambda^2)-(1-\lambda)^2(\frac{ab}{cd})^3\} \tag{15.171}$$

$$\frac{(a^3-b^3)^2}{c^3d^3} = \lambda^{-1}\{(1-\lambda^2)-(1+\lambda)^2(\frac{ab}{cd})^3\}. \tag{15.172}$$

Hence both sides of (15.169) are functions of the single variable $ab/(cd)$. Thus, defining

$$t = \eta^2\frac{ab}{cd}, \tag{15.173}$$

we see that the right-hand side of (15.169) vanishes at the roots t_l of the polynomial equation

$$0 = P_Q(t) = t^{-P_a}\{(t-1)^{\mathcal{N}}(t\omega^2-1)^{\mathcal{N}}\omega^{-P_a}$$
$$+(t\omega-1)^{\mathcal{N}}(t\omega^2-1)^{\mathcal{N}} + (t-1)^{\mathcal{N}}(t\omega-1)^{\mathcal{N}}\omega^{P_a}. \qquad (15.174)$$

We note that

$$P_Q(\omega t) = P_Q(t) \qquad (15.175)$$

and thus $P_Q(t)$ is a polynomial in t^3 and the zeros lie on the lines where t^3 is real.

Thus using (15.171) and (15.172) in the left-hand side of (15.169) and setting $t = t_l$ we find the solution

$$w_l^2 = \frac{(1+\lambda)^2}{4}\left(\frac{a^3+b^3}{a^3-b^3}\right)^2 = \frac{(1+\lambda)^2}{4}\left[\frac{1-(\frac{1-\lambda}{1+\lambda})^2 t_l^3}{1-t_l^3}\right]$$
$$= \frac{1}{4}(1-\lambda)^2 + \frac{\lambda}{1-t_l^3}. \qquad (15.176)$$

To complete the computation of the eigenvalue (15.160) of H_{cp} we need to determine P_a and P_c. In [43] these integers are determined as

$$P_a = -Q - \mathcal{N} \pmod 3, \qquad P_c = 0. \qquad (15.177)$$

Thus choosing the minus sign for all w_l in (15.160) we obtain the desired result

$$E^0_{\mathcal{N}}(\lambda; Q) = A^Q_{\mathcal{N}} + B^Q_{\mathcal{N}}\lambda - 3\sum_{l=1}^{m_E^Q}\left[(1-\lambda)^2 + \frac{4\lambda}{1-t_l^3}\right]^{1/2} \qquad (15.178)$$

where

$$A^Q_{\mathcal{N}} + B^Q_{\mathcal{N}}\lambda = -2Q - [2\mathcal{N} - 2Q - 3m_E^Q](1+\lambda) \qquad (15.179)$$

with

$$m_E^Q = \text{integer part of}[\frac{2\mathcal{N}-Q}{3}] \qquad (15.180)$$

and t_l are the real roots of the polynomial $P_Q(t)$.

It remains to consider the limit

$$e_0(\lambda; Q) = \lim_{\mathcal{N}\to\infty} E^0_{\mathcal{N}}(\lambda; Q)/\mathcal{N}. \qquad (15.181)$$

In this limit we see from the definition (15.174) that all real zeros occur for $t < -1$ where the first and third terms are of equal magnitude and oscillate, and the second term is exponentially smaller than the magnitude of terms 1 and 3.

We will use this information on the zeros of $P_Q(t)$ to write $E^0_{\mathcal{N}}(\lambda; Q)$ as a contour integral and to do this it is convenient to first write (15.178) as

$$E^0_{\mathcal{N}}(\lambda; Q) = A^Q_{\mathcal{N}} + B^Q_{\mathcal{N}}\lambda - 3m_E^Q|1-\lambda| - 3\sum_{l=1}^{m_E^Q}\{\left[(1-\lambda)^2 + \frac{4\lambda}{1-t_l^3}\right]^{1/2} - |1-\lambda|\}. \qquad (15.182)$$

Then by use of Cauchy's theorem we find

$$
\begin{aligned}
&E^0_{\mathcal{N}}(\lambda;Q)\\
&= A^Q_{\mathcal{N}} + B^Q_{\mathcal{N}}\lambda - 3m^Q_E|1-\lambda|\\
&-\frac{3}{2\pi i}\int_{c_1} dt\frac{d}{dt}(\ln P_Q(t))\{\left[(1-\lambda)^2+\frac{4\lambda}{1-t_l^3}\right]^{1/2} - |1-\lambda|\}\\
&= A^Q_{\mathcal{N}} + B^Q_{\mathcal{N}}\lambda - 3m^Q_E|1-\lambda|\\
&-\frac{1}{2\pi i}\int_{c_1+c_2+c_3} dt\frac{d}{dt}(\ln P_Q(t))\{\left[(1-\lambda)^2+\frac{4\lambda}{1-t_l^3}\right]^{1/2} - |1-\lambda|\} \qquad (15.183)
\end{aligned}
$$

where the contours c_1, c_2 and c_3 encircle the zeros of $P_Q(t)$ as shown in Fig. 15.6, and in the last line we have summed over the three equivalent contours.

We now obtain a useful form for the $\mathcal{N} \to \infty$ limit. We note that the integrand in (15.182) has branch cuts at $t^3 = 1$ and $t^3 = (\frac{1+\lambda}{1-\lambda})^2$ and we deform the contour from $c_1 + c_2 + c_3$ to $c_1'' + c_2'' + c_3''$ where c_j'' encircles the branch cuts as shown in Fig. 15.6. The contributions from all three contours are equal. On the contour c_1'' the second term in (15.174) is exponentially large compared with terms 1 and 3. Thus, noting from (15.180) that as $\mathcal{N} \to \infty$

$$
m^Q_E = \frac{2}{3}\mathcal{N} + O(1) \qquad (15.184)
$$

we find

$$
e_0(\lambda;Q) = -2|1-\lambda| - \frac{3}{2\pi i}\int_{c_1''} dt\left(\frac{\omega^2}{\omega^2 t-1}+\frac{\omega}{\omega t-1}\right)\left[(1-\lambda)^2+\frac{4\lambda}{1-t^3}\right]^{1/2} \qquad (15.185)
$$

where the right-hand side is independent of Q. Then using the identity

$$
\begin{aligned}
&\frac{1}{2\pi i}\int_{c_1''} dt\left(\frac{\omega}{\omega t-1}+\frac{\omega^2}{\omega^2 t-1}+\frac{1}{t-1}\right)\left[(1-\lambda)^2+\frac{4\lambda}{1-t^3}\right]^{1/2}\\
&= \frac{1}{2\pi i}\int_{c_1''+c_2''+c_3''} dt\frac{1}{t-1}\left[(1-\lambda)^2+\frac{4\lambda}{1-t^3}\right]^{1/2}\\
&= -|1-\lambda| \qquad (15.186)
\end{aligned}
$$

where the last line is obtained by closing the contours on the pole at infinity we find for all three values of Q

$$
e_0(\lambda;Q) = -\frac{1}{2\pi i}\int_{c_1''}\left(\frac{\omega}{\omega t-1}+\frac{\omega^2}{\omega^2 t-1}-\frac{2}{t-1}\right)\left[(1-\lambda)^2+\frac{4\lambda}{1-t^3}\right]^{1/2} \qquad (15.187)
$$

which is simplified as

$$
e_0(\lambda;Q) = \frac{3}{2\pi i}\int_{c_1''} dt\frac{t+1}{t^3-1}\left[(1-\lambda)^2+\frac{4\lambda}{1-t^3}\right]^{1/2}. \qquad (15.188)
$$

The integrand in (15.188) has no poles and is recognized as an Abelian integral of the second kind over a hyperelliptic curve of genus 2.

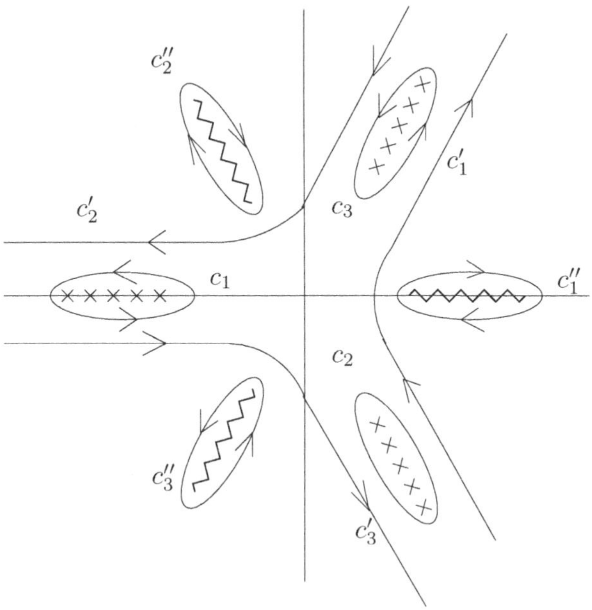

Fig. 15.6 The contours c_i, c_i' and c_i'' in the complex t plane. The zigzag lines denote the branch cuts from $t^3 = 1$ to $t^3 = (\frac{1+\lambda}{1-\lambda})^2$. The crosses indicate the zeros of the polynomial $P(t)$.

To complete the evaluation of $e_0(\lambda; Q)$ deform the contour c_1'' of Fig. 15.6 to the rays

$$t = -\omega^2 z \quad \text{and} \quad t = \omega z \tag{15.189}$$

with $0 \leq z \leq \infty$ shown in Fig. 15.7. On these contours the square root in (15.188) is real and we find

$$e_0(\lambda; Q) = -\frac{3\sqrt{3}}{2\pi} \int_0^\infty dz \frac{z+1}{z^3+1} \left[1 + \lambda^2 - 2\lambda \frac{z^3-1}{z^3+1}\right]^{1/2}. \tag{15.190}$$

In (15.190) the symmetry

$$\lambda e_0(\lambda^{-1}; Q) = e_0(\lambda; Q) \tag{15.191}$$

is obvious and the symmetry

$$e_0(-\lambda; Q) = e_0(\lambda; Q) \tag{15.192}$$

follows by sending $z \to 1/z$.

Setting

$$z = (\tan\theta)^{2/3} \tag{15.193}$$

(15.190) becomes

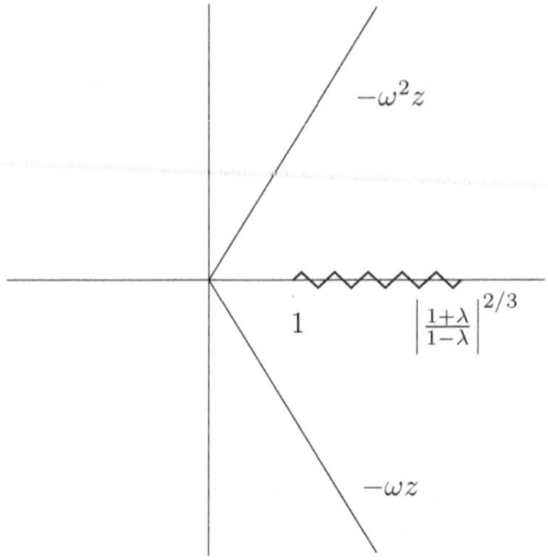

Fig. 15.7 The rays in the t plane $t = -\omega^2 z$ and $t = -\omega z$ for $0 \leq z \leq \infty$. The zigzag lines indicate the branch cut.

$$e_0(\lambda; Q) = \frac{\sqrt{3}(1+\lambda)}{\pi} \int_0^{\pi/2} d\theta \{(\tan\theta)^{1/3} + (\tan\theta)^{-1/3}\} \left[1 - \frac{4\lambda}{(1+\lambda)^2} \sin^2\theta\right]^{1/2} \tag{15.194}$$

from which, by using $\sin^2\theta = u$ and recalling the integral representation of the hypergeometric function

$$F(a, b; c; t) = \frac{\Gamma(c)}{\Gamma(b)\Gamma(c-b)} \int_0^1 du u^{b-1}(1-u)^{c-b-1}(1-tu)^{-a}, \tag{15.195}$$

we obtain

$$e_0(\lambda; Q) = -(1+\lambda)\{F(-\frac{1}{2}, \frac{1}{3}; 1; \frac{4\lambda}{(1+\lambda)^2}) + F(-\frac{1}{2}, \frac{2}{3}; 1; \frac{4\lambda}{(1+\lambda)^2})\}. \tag{15.196}$$

In this form the symmetry (15.191) is obvious but the symmetry (15.192) is no longer manifest.

The result (15.196) is obviously analytic except at $\lambda = 1$. To study the singularly at $\lambda = 1$ we use the identity of hypergeometric functions (1) of [60, 2.10]

$$\begin{aligned} F(a, b; c; \frac{4\lambda}{(1+\lambda)^2}) &= \frac{\Gamma(c)\Gamma(c-a-b)}{\Gamma(c-a)\Gamma(c-b)} F(a, b; a+b-c+1; \left(\frac{1-\lambda}{1+\lambda}\right)^2) \\ &+ \frac{\Gamma(c)\Gamma(a+b-c)}{\Gamma(a)\Gamma(b)} \left(\frac{1-\lambda}{1+\lambda}\right)^{2(c-a-b)} F(c-a, c-b; c-a-b+1; \left(\frac{1-\lambda}{1+\lambda}\right)^2) \end{aligned} \tag{15.197}$$

to find that

$$e_0(1,Q) = -4\pi^{-1/2}\left(\frac{\Gamma(\frac{7}{6})}{\Gamma(\frac{2}{3})} + \frac{\Gamma(\frac{5}{6})}{\Gamma(\frac{1}{3})}\right) \tag{15.198}$$

and that near $\lambda = 1$ we have the expansion

$$\begin{aligned} e_0(\lambda;Q) &= \sum_{n=0}^{\infty}\left(\frac{1-\lambda}{1+\lambda}\right)^{2n} a_n + \left(\frac{1-\lambda}{1+\lambda}\right)^{2-1/3}\sum_{n=0}\left(\frac{1-\lambda}{1+\lambda}\right)^{2n} b_n \\ &+ \left(\frac{1-\lambda}{1+\lambda}\right)^{2+1/3}\sum_{n=0}^{\infty}\left(\frac{1-\lambda}{1+\lambda}\right)^{2n} c_n. \end{aligned} \tag{15.199}$$

We note that in [40] Baxter has obtained $e_0(\lambda; Q, N)$ for the general N-state superintegrable model and that that result may be written as

$$e_0(\lambda; Q, N) = (1+\lambda)\sum_{l=1}^{N-1} F(-\frac{1}{2}, \frac{l}{N}; 1; \frac{4\lambda}{(1+\lambda)^2}). \tag{15.200}$$

We finally note, as discussed in [43], that the dependence on Q of $E^0_{\mathcal{N}}(\lambda, Q)$ for $0 \leq \lambda < 1$ vanishes exponentially in $\mathcal{N}$ as $\mathcal{N} \to \infty$ and thus in this case the ground state energy is three-fold degenerate and corresponds to the $T < T_c$ phase of the Ising model.

However, for $\lambda > 1$ it is shown in [43] that

$$\lim_{\mathcal{N}\to\infty}\{E^0_{\mathcal{N}}(\lambda; Q) - E^0_{\mathcal{N}}(\lambda; 0)\} = 2Q(\lambda - 1) \tag{15.201}$$

and thus in this case the ground state eigenvalue is nondegenerate which corresponds to the $T > T_c$ phase of the Ising model.

15.2.6 Single particle excitations and level crossing

The eigenvalue $E^0(\lambda; Q)$ computed in the previous subsection is the lowest eigenvalue for small (and by the duality (15.154) and (15.201) also for large) λ, and $e_0(\lambda; Q)$ fails to be analytic only at $\lambda^2 = 1$. However, this is no guarantee that $E^0(\lambda; Q)$ will be the lowest eigenvalue for all $0 \leq \lambda \leq 1$. We will, in fact, demonstrate that there is indeed a range of values about $\lambda = 1$ where $E^0(\lambda; Q)$ ceases to be the lowest eigenvalue and thus a level crossing transition occurs.

Consider states where the eigenvalues are given by (15.159) with

$$m_P = 1, \quad P_b = 0. \tag{15.202}$$

These states are shown in [43] to correspond to $Q = 1$ and we denote these eigenvalues as $E^{\text{ex}}(P, \lambda; 1)$. Then if we note that the variables a, b, c, d can be eliminated in favor of the single variable

$$t = ab/cd \tag{15.203}$$

the functional equation (15.157) reduces to

$$
\begin{aligned}
&\mathcal{K}t^{\bar{P}_a}(1+t^3v^3)\prod_{l=1}^{\bar{m}_E}\left[1-\left(\frac{1-\lambda}{1+\lambda}\right)^2 t^3 - w_l^2\frac{4}{(1+\lambda)^2}(1-t)^3\right]\\
&= e^{-iP}\Big[(t-1)^{\mathcal{N}}(t\omega^2-1)^{\mathcal{N}}(1+\omega vt)\omega^{-\bar{P}_a}\\
&(t\omega-1)^{\mathcal{N}}(t\omega^2-1)^{\mathcal{N}}(1+vt)+(t-1)^{\mathcal{N}}(t\omega-1)^{\mathcal{N}}(1+\omega^2 vt)\omega^{\bar{P}_a}\Big].
\end{aligned}
\tag{15.204}
$$

This equation must be used to determine the possible values of v as well as w_l.

We first obtain the values for v by setting $t=-v^{-1}$ where the left-hand side and the second term of the right-hand side of (15.204) vanish. Thus we find that v satisfies the equation

$$
\begin{aligned}
0 &= (-v^{-1}-1)^{\mathcal{N}})(-\omega^2 v^{-1}-1)^{\mathcal{N}}(1-\omega)\omega^{-\bar{P}_a}\\
&+(-v^{-1}-1)^{\mathcal{N}}(-\omega v^{-1}-1)^{\mathcal{N}}(1-\omega^2)\omega^{\bar{P}_a}
\end{aligned}
\tag{15.205}
$$

which we rewrite as

$$
\omega^{\mathcal{N}}\left(\frac{1+\omega v}{1+\omega^2 v}\right)^{\mathcal{N}} = \omega^{2\bar{P}_a-1}. \tag{15.206}
$$

Using the relation (15.161) between P and v we find

$$
\omega^{\mathcal{N}}e^{i\mathcal{N}P} = \omega^{2\bar{P}_a-1} \tag{15.207}
$$

which, using the quantization condition on the momentum (15.162), gives the identity which fixes $\bar{P}_a$

$$
2\bar{P}_a - 1 - \mathcal{N} \equiv 0 \pmod 3. \tag{15.208}
$$

We now determine the w_l as before in terms of the roots of the polynomial

$$
\begin{aligned}
&\bar{P}_1(t) = t^{-\bar{P}_a}\{(\omega^2 t-1)^{\mathcal{N}}(\omega t-1)^{\mathcal{N}}\frac{1+tv}{1+t^3v^3}\omega^{\bar{P}_a}\\
&+(t-1)^{\mathcal{N}}(\omega^2 t-1)^{\mathcal{N}}\frac{1+\omega tv}{1+t^3v^3}+(t-1)^{\mathcal{N}}(\omega t-1)^{\mathcal{N}}\frac{1+\omega^2 tv}{1+t^3v^3}\omega^{-\bar{P}_a}\}.
\end{aligned}
\tag{15.209}
$$

Therefore we may write

$$
\begin{aligned}
&E_{\mathcal{N}}^{\mathrm{ex}}(P,\lambda;1) = \bar{A}_{\mathcal{N}} + \bar{B}_{\mathcal{N}}\lambda - 3\bar{m}_E|1-\lambda|\\
&-\frac{1}{2\pi i}\int_{c_1+c_2+c_3} dt\frac{d}{dt}\ln\bar{P}_1(t)\{\left[(1-\lambda)^2+\frac{4\lambda}{1-t^3}\right]^{1/2} - |1-\lambda|\}
\end{aligned}
\tag{15.210}
$$

where it is shown in [43] that

$$\begin{aligned}\bar{A}_{\mathcal{N}} = 0, \quad & \bar{B}_{\mathcal{N}} = -4 \quad \text{if } \mathcal{N} \equiv 0(\text{mod } 3)\\ \bar{A}_{\mathcal{N}} = 1, \quad & \bar{B}_{\mathcal{N}} = -3 \quad \text{if } \mathcal{N} \equiv 1(\text{mod } 3)\\ \bar{A}_{\mathcal{N}} = 2, \quad & \bar{B}_{\mathcal{N}} = -2 \quad \text{if } \mathcal{N} \equiv 2(\text{mod } 3).\end{aligned} \tag{15.211}$$

Finally we must compute

$$\Delta E(P,\lambda) = \lim_{\mathcal{N}\to\infty} \{E_{\mathcal{N}}^{\text{ex}}(P,\lambda;1) - E_{\mathcal{N}}^{0}(\lambda;Q)\} \tag{15.212}$$

where for $0 \leq \lambda \leq 1$ $E_{\mathcal{N}}^{0}(\lambda;Q)$ is independent of Q to order one but for $\lambda > 1$ we must have $Q = 0$ in order that $E_{\mathcal{N}}^{0}(\lambda;Q)$ is the ground state energy computed in the previous subsection. In (15.212) $\Delta E(P,\lambda)$ is of order one even though each term separately is of order $\mathcal{N}$.

The computation of (15.212) is slightly simplified if we first note for all $\mathcal{N}$ that $\bar{P}_a = P_a(Q = 1)$. Thus setting $Q = 1$ in (15.212) and using (15.179), (15.180) and (15.211), we have for all $\mathcal{N}$

$$\bar{A}_{\mathcal{N}} + \bar{B}_{\mathcal{N}}\lambda - (A_{\mathcal{N}}^{1} + B_{\mathcal{N}}^{1}\lambda) = 3(1-\lambda) \tag{15.213}$$

and

$$\bar{m}_E - m_E^{1} = -1. \tag{15.214}$$

Then subtracting (15.182) from (15.210), using the result that for $1 \leq t$

$$\lim_{\mathcal{N}\to\infty} \frac{d}{dt} \ln \frac{\bar{P}_1(t)}{P_1(t)} = \frac{d}{dt} \ln \frac{1+tv}{1+t^3v^3} = -\left(\frac{\omega v}{1+\omega tv} + \frac{\omega^2 v}{1+\omega^2 tv}\right) \tag{15.215}$$

and deforming the contours c_i to c_i'' and then to the branch cut running from $t = 1$ to $t = (\frac{1+\lambda}{1-\lambda})^{1/2}$, we find for all λ that

$$\begin{aligned}&\lim_{\mathcal{N}\to\infty} \{E_{\mathcal{N}}^{\text{ex}}(P,\lambda;1) - E_{\mathcal{N}}^{0}(\lambda;1)\} = 3(1-\lambda) + 3|1-\lambda|\\ &+\frac{3}{\pi}\int_1^{|\frac{1+\lambda}{1-\lambda}|^{2/3}} dt\{\frac{\omega v}{1+\omega tv} + \frac{\omega^2 v}{1+\omega^2 tv}\}\left[\frac{4\lambda}{t^3-1} - (1-\lambda)^2\right]^{1/2}\end{aligned} \tag{15.216}$$

and thus, using (15.201) we obtain the final result

$$\begin{aligned}&\Delta E(P,\lambda) = 2(1-\lambda) + 4|1-\lambda|\\ &+\frac{3}{\pi}\int_1^{|\frac{1+\lambda}{1-\lambda}|^{2/3}} dt\{\frac{\omega v}{1+\omega tv} + \frac{\omega^2 v}{1+\omega^2 tv}\}\left[\frac{4\lambda}{t^3-1} - (1-\lambda)^2\right]^{1/2}.\end{aligned} \tag{15.217}$$

Now consider the point $\lambda = 1$. Then (15.217) reduces to

$$\Delta E(P,1) = \frac{3v}{\pi}\int_1^{\infty} dt \left(\frac{\omega}{1+\omega tv} + \frac{\omega^2}{1+\omega^2 tv}\right)(t^3-1)^{-1/2}. \tag{15.218}$$

This excitation energy vanishes linearly at $v = 0$ and therefore there is a range of v for which this “excitation” energy is negative and thus there is a range of v for which the

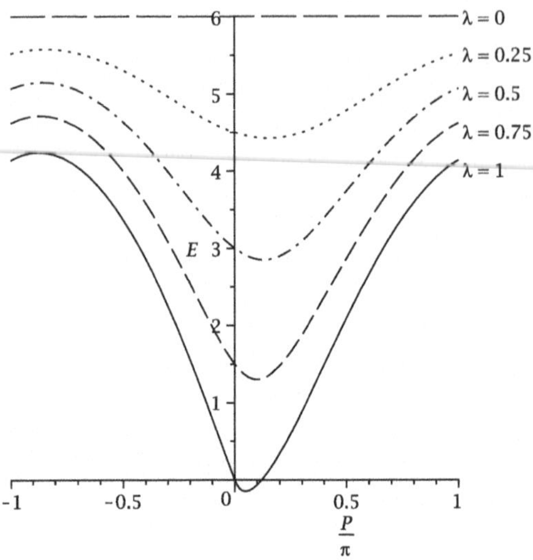

Fig. 15.8 The excitation energy $\Delta E(P,\lambda) = \lim_{\mathcal{N}\to\infty}\{E_{\mathcal{N}}(P,\lambda) - E^0_{\mathcal{N}}(\lambda,0)\}$ for $0 \leq \lambda \leq 1$ as function of P. This excitation curve is tangent to the P axis at $P = 0.1612$ for $\lambda = 0.9735$.

eigenvalue called $e^0(\lambda;0)$ is in fact *not* the ground state eigenvalue of the spin chain. This is the phenomenon of level crossing.

To investigate this level crossing phenomenon in more detail we plot in Fig. 15.8 the excitation energy $\Delta E(P,\lambda)$ for several values for $0 < \lambda \leq 1$ as a function of P. We find numerically that the smallest value of λ for which $\Delta(P,\lambda) = 0$ is $\lambda_L = 0.9735\cdots$ at which point $P = 0.1612\cdots$. Similarly for $\lambda > 1$ the largest value of λ for which $\Delta(P,\lambda) = 0$ is $\lambda_U = 1.109\cdots$ at which point $P = 0.225\cdots$ Thus in the range

$$\lambda_L \leq \lambda \leq \lambda_U \tag{15.219}$$

the energy $E^0(\lambda,0)$ is no longer the ground state. This is the phenomenon of level crossing and is the reason why Howes, Kadanoff and den Nijs [31] conducted their original study of the three-state chiral Potts model.

We finally remark that once it has been shown that "single particle" excitations lie below the "ground state" that it may be expected that multiparticle excitations can lie still lower. This has been investigated in [43] but is beyond the scope of the present discussion. In this region where level crossing has occurred there is no gap in the excitation spectrum and thus the system has massless excitations.

15.2.7 Order parameter

For the Ising model there is only one order parameter $\langle\sigma\rangle$ where $\sigma = \pm 1$ and we take the thermodynamic limit from a finite lattice with all the boundary spins fixed to be $+$. For the chiral Potts model where the "spins" have the values

$$\sigma = e^{2\pi i n/N} \quad \text{with } n = 0,\ 1,\ \cdots, N-1 \tag{15.220}$$

there are $N-1$ order parameters

$$M_k = \langle \sigma^k \rangle \quad \text{with } k = 1,\ 2, \cdots, N-1. \tag{15.221}$$

In [31] this order parameter was perturbatively expanded in a series in λ to order λ^{13} and from this it was conjectured for $N = 3$ that

$$M_k = (1 - \lambda^2)^{1/9}. \tag{15.222}$$

For general N a similar low order expansion was made in 1989 in [41] and from these two results the order parameter was conjectured to be

$$M_k = (1 - \lambda^2)^{k(N-k)/(2N^2)} \tag{15.223}$$

which reduces (as it must) to the spontaneous magnetization of the Ising model for $N = 2$ and $k = 1$. This order parameter is independent of the points p and q and thus is equally valid on both the real and the Hermitian manifolds. This independence of p and q is a necessary property which follows from the star–triangle equation.

For the Ising model the time elapsed between Onsager's announcement [61] in 1949 of the spontaneous magnetization and Yang's published proof [62] in 1952 was only three years. For the chiral Potts model the conjecture (15.223) was proven by Baxter [49] in 2005 which is 16 years after the conjecture was first published by Albertini, McCoy, Perk and Tang [41] in 1989.

We note, however, that for $N \geq 3$ the result (15.223) will not be valid for all $0 \leq \lambda < 1$ but only in the massive phase $0 \leq \lambda \leq \lambda_L(N)$. This is in contrast with the Ising case $N = 2$ where the result (15.223) is valid for all $0 \leq \lambda < 1$.

15.2.8 The phase diagram of the spin chain

The computation of the eigenvalues of the chiral Potts spin chain for the case of general values of ϕ and $\bar{\phi}$ where the Onsager algebra does not hold is substantially more involved than the superintegrable case treated above. For this general case the eigenvalues of the transfer matrix do not have a simple product form like (15.159) and the functional equation (15.157) must be extended to a more elaborate set of equations which, in fact, have a close resemblance to the functional equations of the eight-vertex model. These functional equations are derived in [45] where the ground state for small λ is computed. The work of [45] was used in [46] to compute the region in the λ, $\bar{\phi}$ plane where the ground state of [45] becomes unstable against single particle excitations. The result of this study is graphed in Fig. 15.9.

There is an important difference between the real and the Hermitian cases which must be noted. In the real case the elements of the transfer matrix are real and positive and therefore on a finite lattice level crossing is forbidden by the Perron–Frobenius theorem which states that the maximum eigenvalue of a matrix with positive real elements cannot be degenerate. However, for the Hermitian spin chain the matrix elements of the Hamiltonian are not all real and positive and thus level crossing is not forbidden and indeed we have demonstrated that level crossing does indeed take place. This leads to the question of whether there are properties of the commensurate/incommensurate phase transition that are different in the two different cases.

The most curious of these properties is the question of what happens to the order parameter in the incommensurate phase. The order parameter (15.223) depends

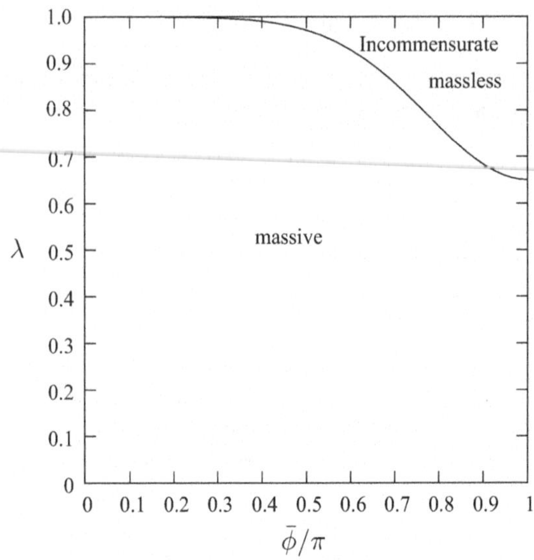

Fig. 15.9 The phase diagram for the $N = 3$ chiral Potts Hamiltonian (15.136) as a function of λ and the angle $\bar{\phi}$. The line $\bar{\phi} = \pi/2$ is the superintegrable case. The value of λ at the phase boundary is always less than unity for $0 < \bar{\phi} < \pi$.

only on λ and not on $\bar{\phi}$, so as a consequence the order parameter of the Hermitian spin chain does not vanish when the phase boundary is approached from the ordered phase. The question then arises whether the order parameter is zero in the massless incommensurate phase, which is what happens in the incommensurate phase on the real manifold, or if there is an order parameter in the incommensurate phase which oscillates with a phase that depends on λ and $\bar{\phi}$ even though the approach to this long range order is algebraic.

This question of the order parameter of the spin chain forces us to return to a question which was not discussed for the real manifold. The order parameter (15.223) is the order parameter on the integrable subspace of the real manifold. However, we have not discussed what the order parameter is off the integrable line on the real manifold. It seems to be assumed in the literature that on the real manifold the order parameter will vanish at the incommensurate phase boundary. If correct this is in striking contrast to the Hermitian spin chain, and the physics of the two cases are quite distinct. Further work is needed to elucidate these differences.

15.3 Open questions

The presentation of the SOS, RSOS, hard hexagon and chiral Potts models of this chapter and the eight- and six-vertex models of chapter 14 have raised several important open questions. Some of these questions have been either explicitly or implicitly presented above. Nevertheless, it is most appropriate to conclude this chapter by bringing these questions together in one place. These questions are listed in table 15.5 and a selection of relevant references is given in Table 15.6.

Table 15.5 Open questions for the eight-vertex, RSOS and chiral Potts models

1. Is there a Q operator for the RSOS models?
2. Characterize the degenerate eigenvectors for eight-vertex model at roots of unity.
3. Find the symmetry algebra for the eight-vertex model at roots of unity.
4. Compute the chiral Potts correlations.

Table 15.6 A selection of references relevant to the open questions of table 15.5

Date	Author(s)	Development
1972	Baxter [63]	Eight-vertex Q matrix
1973	Baxter [6–8]	Eight-vertex Q matrix
		SOS scalar tq equation
		8 vertex eigenvectors
1979	Takhtajan,Faddeev [64]	Algebraic Bethe's ansatz
1984	Andrews, Baxter, Forrester [16]	Scalar tq equation
1989	Bazhanov, Reshetikhin [20]	Roots for critical RSOS
1990	Bazhanov, Stroganov [44]	Chiral Potts six-vertex connection
1990	Baxter, Bazhanov, Perk [45]	Chiral Potts functional equations
2001	Deguchi, Fabricius, McCoy [65]	Six-vertex loop algebra
2001	Fabricius, McCoy [66, 67]	Six-vertex degeneracy
2002	Deguchi [68]	Six-vertex loop algebra
2002	Deguchi [69]	Eight-vertex degenerate eigenvectors
2002	Baxter [9]	Eight-vertex Q matrix
2003	Fabricius, McCoy [70]	Eight-vertex Q matrix
2005	Fabricius, McCoy [71]	Eight-vertex Q matrix
2005	Bazhanov, Mangazeev [72]	Q operators
2006	Bazhanov, Mangazeev [73]	Q operators
2006	Fabricius, McCoy [74]	Eight-vertex degenerate eigenvectors
2007	Bazhanov, Mangazeev [75]	Q operators
2007	Fabricius [76]	Eight-vertex Q matrix
2007	Roan [77]	Eight-vertex Q matrix
2007	Deguchi [78, 79]	Six-vertex loop algebra
2007	Fabricius, McCoy [80]	EIght-vertex Q matrix
2008	Au-Yang, Perk [81, 82]	Chiral Potts eigenvectors
2008	Nishino, Deguchi [83]	Chiral Potts eigenvectors
2008	Baxter [84]	Chiral Potts order parameter
2009	Baxter [85]	Chiral Potts order parameter
2009	Fabricius, McCoy [86]	Eight-vertex Q matrix

15.3.1 Q operators

In the original paper [63] of 1972 a matrix TQ equation was derived for the eight-vertex model with the root of unity condition

$$L\eta = 2m_1K + im_2K' \tag{15.224}$$

At that time it was thought that this Q matrix was valid for all m_1 and m_2 and all lengths N of the chain and that it had the same symmetries as the transfer matrix. However, starting in 2003 subsequent investigations [70,71,76,79,80,86] have discovered that the original Q matrix of [63] does not have the same symmetries as the transfer matrix and does not cover all cases. Different Q matrices are needed for cases not covered in [63], and for the case N odd, with m_1 and m_2 even, there is as yet no Q matrix known. These Q matrices have been discussed in detail in chapter 14. In addition a separate Q matrix was discovered [6] for generic values of η which does not reduce to any of the matrices computed using the procedure of [63].

In the papers of 1973 [6–8] some eigenvectors of the eight-vertex transfer matrix were investigated by relating these eigenvectors to the eigenvectors of an SOS model which was solved by a Bethe's ansatz. The Bethe's ansatz for these SOS models yields a scalar tq equation. This scalar tq equation for SOS eigenvalues is developed in [9, 16, 73, 75] and has been discussed in section 15.1 above. In particular an extensive discussion is given in [75] of the dependence of the eigenvalues q on the phase factor in the tq equation and the relation of the phase factor to the RSOS models. However, as yet, there is no known Q operator for the SOS and RSOS models which, when reduced to an equation for eigenvalues, gives the tq equation of [6–9, 16, 72, 73, 75].

15.3.2 Degenerate subspaces for the eight-vertex model

In [6–8] eigenvectors of the eight-vertex model at roots of unity are constructed by defining a set of vectors in the space of the eight-vertex model which depend on two parameters s and t and by forming linear combinations of these vectors using the solutions of the scalar tq equation of the corresponding SOS model. For nondegenerate eigenvectors of the eight-vertex model there is no problem with this construction. However, for the degenerate eigenvalues a difficulty arises because to one vector in the SOS model there will correspond the entire degenerate subspace in the eight-vertex case.

Some of these degenerate vectors are obtained by variation of the parameters s and t. However, even for the simplest case treated in [6] where there are "no Bethe roots" it was found [70] that the dimension of the degenerate subspace is much larger than what is obtained in [6] by the variation of the parameters s and t. In [9] it is proposed that a complete set of eight-vertex eigenvectors can be obtained by varying the "string" center of the complete L-strings discussed in chapter 14 which "cancel out" of the scalar tq equation. However this suggestion has never been developed to actually compute the dimension of the degenerate subspaces. Thus a complete set of eigenvectors for the 8 vertex model has not yet been constructed using the methods of [6–9].

An alternative construction of the eight-vertex eigenvectors at roots of unity is the algebraic Bethe's ansatz of [64] which also utilizes the parameters s and t of [6–9]. This original construction also fails to compute a complete set of vectors. However, this problem was in large part overcome in [74] where an operator for complete strings with arbitrary string centers was discovered, but even with this operator it remains an open question to compute the dimension of the degenerate subspace.

15.3.3 Symmetry algebra for the eight-vertex model at roots of unity

The underlying problem is that the symmetry algebra responsible for the degeneracies of the eight-vertex model is not known. The corresponding problem for the degeneracies of the six-vertex model has been studied in detail in [65–67, 78, 79] where it is shown that the symmetry group is the loop algebra of sl_2 and that the size of the degenerate multiplets is obtained from a "Drinfeld polynomial" which is explicitly given in terms of the solutions to the Bethe equations. A major open question is to obtain an explicit form for the generators of the analogous symmetry algebra for the eight-vertex model.

15.3.4 Chiral Potts correlations

It remains to discuss the chiral Potts model and strangely enough this is better understood than the eight-vertex model. In [44] it is shown how the chiral Potts model is obtained from the six-vertex model and in [45] a complete set of functional equations is obtained which are analogous not only to the matrix TQ equation of the eight-vertex model but also to the equation involving only Q itself. For the eight-vertex model this equation for Q in terms of itself is needed only to determine the degeneracy of the multiplet but for the chiral Potts model the analogous equation is satisfied by the transfer matrix and is the equation used in section 15.2 to compute the ground state energy of the superintegrable spin chain.

However, it remains a major challenge to compute the eigenvectors of the chiral Potts model. For the superintegrable case which obeys Onsager's algebra it might be thought that the methods of the $N = 2$ Ising model could be applied and generalized. Unfortunately all computations of Ising correlations have been done either by the Pfaffian methods presented in chapters 11 and 12 or by free Fermi operator methods. The relation of the eigenvectors of the Ising and chiral Potts model to Onsager's algebra and loop sl_2 algebra is, at the time of writing, under development by several authors [81–83].

The most important open questions of the superintegrable chiral Potts model concern the order parameter and the correlation functions. The proof in 2005 of Baxter [49] of the order parameter conjectured in 1989 by Albertini, McCoy, Perk and Tang [41] while ingenious and correct is not completely insightful, and it would be far preferable to have a more algebraic proof which would use the Onsager algebra. Such a study has initiated in [84,85] and must be connected with the eigenvector studies of [81–83].

What would be most desirable would be to have an expression for the correlation function $\langle Z_0^n Z_R^{\dagger n}\rangle$ which reduced to M_n^2, the square of the order parameter, when $R \to \infty$ and agrees with the result obtainable from the ground state energy of the superintegrable spin chain when $R = 1$.

For the case $N = 2$, where the superintegrable chiral Potts model reduces to the transverse Ising chain, expressions for the correlation functions which satisfy these two criteria have been given in 14.4.3.

For all N we obtain a sum rule for the nearest neighbor correlations by using what is called "Feynman's theorem" which says that for a Hamiltonian of the form

$$H = A_0 + \lambda A_1 \qquad (15.225)$$

if $e_0(\lambda)$ denotes the ground state energy per site the expectation of the operator A_1 is

$$\langle A_1 \rangle = \frac{\partial e_0(\lambda)}{\partial \lambda}. \tag{15.226}$$

Thus if we apply (15.226) to the Hamiltonian (15.146) of the superintegrable chiral Potts model with A_0 and A_1 given by (15.147) and (15.148) we obtain the sum rules

$$-\sum_{n=1}^{N-1} \frac{e^{i\pi(2n-N)/(2N)}}{\sin \pi n/N} \langle X_0^n \rangle = \frac{\partial e_0(\lambda)}{\partial \lambda} \tag{15.227}$$

and

$$-\sum_{n=1}^{N-1} \frac{e^{i\pi(2n-N)/(2N)}}{\sin \pi n/N} \langle (Z_0 Z_1^{\dagger})^n \rangle = e_0(\lambda) - \lambda \frac{\partial e_0(\lambda)}{\partial \lambda}. \tag{15.228}$$

with $e_0(\lambda)$ given by (15.200) We note that when $N = 2$ these sum rules are verified by (14.417)–(14.422) by use of the relation

$$\frac{\partial}{\partial z} F(a, b; c; z) = \frac{ab}{c} F(a+1, b+1; c+1; z). \tag{15.229}$$

References

[1] D.M. Burley, A lattice model of a classical hard sphere gas, Proc. Phys. Soc. London 75 (1960) 262–274.

[2] D.M. Burley, A first order transition in a plane lattice gas with rigid repulsions, Proc. Phys. Soc. London 85 (1965) 1173–1176.

[3] L.K. Runnels and L.L. Combs, Exact finite methods of lattice statistics I. Square and triangular lattice gases of hard molecules, J. Chem. Phys. 45 (1966) 2482–2492.

[4] D.S. Gaunt, Hard-sphere lattice gases II. Plane triangular and three dimensional lattices, J. Chem. Phys. 46 (1967) 3237–3259.

[5] R.J. Baxter, Dimers on a rectangular lattice, J. Math. Phys. 9 (1968) 650–654.

[6] R.J. Baxter, Eight-vertex model in lattice statistics and one dimensional anisotropic Heisenberg chain I. Some fundamental eigenvectors, Ann. Phys. 76 (1973) 1–24.

[7] R.J. Baxter, Eight-vertex model in lattice statistics and one dimensional anisotropic Heisenberg chain II. Equivalence to a generalized ice-type lattice model, Ann. Phys. 76 (1973) 25–47.

[8] R.J. Baxter, Eight-vertex model in lattice statistics and one dimensional anisotropic Heisenberg chain III. Eigenvectors and eigenvalues of the transfer matrix and Hamiltonian, Ann. Phys. 76 (1973) 48–71.

[9] R.J. Baxter, Completeness of the Bethe ansatz for the six and eight vertex models, J. Stat. Phys. 108 (2002) 1–48.

[10] R.J. Baxter, Corner transfer matrices of the eight-vertex model I. Low-temperature expansions and conjectured properties, J. Stat. Phys. 15 (1976) 485–503.

[11] R.J. Baxter, Hard hexagons: exact solution, J. Phys. A 13 (1980) L61–L70.

[12] R.J. Baxter, Rogers–Ramanujan identities in the hard hexagon model, J. Stat. Phys. 26 (1981) 427–452.

[13] R.J. Baxter, *Exactly solved models in statistical mechanics*, (Academic Press, 1982)

[14] R.J. Baxter and P.A. Pearce, Hard hexagons: interfacial tension and correlation length, J. Phys. A 15 (1982) 897–910.

[15] R.J.Baxter and P.A. Pearce, Hard squares with diagonal interactions, J.Phys. A 16 (1983) 2239–2255.

[16] G.E. Andrews, R.J. Baxter and P.J. Forrester, Eight-vertex SOS model and generalized Rogers–Ramanujan-type identities, J. Stat. Phys. 35 (1984) 193–266.

[17] M.P. Richey and C.A. Tracy, Equation of state and isothermal compressibility for the hard hexagon model in the disordered regime, J. Phys. A20 (1987) L1121–L1126.

[18] C.A. Tracy, L. Grove and M.F. Newman, Modular properties of the hard hexagon model, J. Stat. Phys. 48 (1987) 477–502.

[19] G.S. Joyce, On the hard hexagon model and the theory of modular functions. Phil. Trans. R. Soc. Lond. A 325 (1988) 643–702.
[20] V.V. Bazhanov and N. Yu. Reshetikhin, Critical RSOS models and conformal field theory, Int. J. Mod. Phys. 4A (1989) 115–142.
[21] L. Onsager, Crystal statistics I: A two dimensional model with an order disorder transition, Phys. Rev. 65 (1044) 117–149.
[22] F.Y. Wu and Y.K. Wang, Duality transformation in a many-component spin model, J. Math. Phys. 17 (1976) 439–440.
[23] S. Ostlund, Incommensurate and commensurate phases in asymmetric clock models, Phys. Rev. B 24 (1981) 398–404.
[24] D.A. Huse, Simple three-state model with infinitely many phases, Phys. Rev. B 24 (1981) 5180–5194.
[25] J.M. Yeomans and M.E. Fisher, Many commensurate phases in the chiral Potts or asymmetric clock models, J. Phys. C14 (1981) L835–L839.
[26] M. Kardar, Phase boundaries of the isotropic helical Potts model on a square lattice, Phys. Rev. B26 (1982) 2693–2695,
[27] D. Haldane, Phase diagrams of surface structures from Bethe-ansatz solutions of the quantum sine-Gordon model, Phys. Rev. B28 (1983) 2743–2745.
[28] H. Au-Yang and J.H.H. Perk, The chiral Potts models revisited, J. Stat. Phys. 78 (1995) 17–76.
[29] H. Au-Yang and J.H.H. Perk, Phase diagram in the generalized clock models, Physica A228 (1996) 78–101.
[30] B-Q. Jin, H. Au-Yang and J.H.H. Perk, Int. J. Mod. Phys. B 16 (2002) 1979–1986.
[31] S. Howes, L.P. Kadanoff and M. den Nijs, Quantum model for commensurate-incommensurate transitions, Nucl. Phys. B215 [FS7] (1983) 169–208.
[32] G. von Gehlen and V. Rittenberg , Z_n-symmetric quantum chains with an infinite set of conserved charges and Z_n zero modes, Nucl. Phys. B257[FS14] (1985) 351–370.
[33] H. Au-Yang, B.M. McCoy, J.H.H. Perk, S. Tang and M.L. Yan, Commuting transfer matrices in the chiral Potts models: Solutions of star–triangle equations with genus > 1, Phys. Letts. A123 (1987) 219–223.
[34] B.M. McCoy, J.H.H. Perk, S. Tang and C.H. Sah, Commuting transfer matrices for the four-state self-dual chiral Potts model with a genus-three uniformizing curve, Phys. Letts. A125 (1987) 9–14.
[35] J.H.H. Perk, Star-triangle equations, quantum Lax pairs, and higher genus curves, in *Prod. 1987 Summer Research Institute on theta functions* (Am. Math. Soc. Providence, RI 1989) 341-354; Proceedings of Symposia in Pure Mathematics, 49 (1989) part 1.
[36] H. Au-Yang, B.M. McCoy, J.H.H. Perk and S. Tang, Solvable models in statistical mechanics and Riemann surfaces of genus greater than one, in *Algebraic Analysis* vol.1, eds M. Kashiwara and T. Kawai (Academic Press, 1988) 29–39.
[37] R.J. Baxter, J.H.H. Perk and H. Au-Yang, New solutions of the star–triangle relations for the chiral Potts model, Phys. Letts. A128 (1988) 138–142.
[38] H. Au-Yang and J.H.H. Perk, Onsager's star–triangle equation: master key to integrability, Advanced Studies in Pure Mathematics, vol. 19 (1989) 57–94.

[39] R.J. Baxter, Free energy of the solvable chiral Potts model, J. Stat. Phys, 52 (1988) 639–667.
[40] R.J. Baxter, The superintegrable chiral Potts model, Phys. Letts. A133 (1988) 185–189.
[41] G. Albertini, B.M. McCoy, J.H.H. Perk and S. Tang, Excitation spectrum and order parameter for the integrable N-state chiral Potts model, Nucl. Phys. B 314 (1989) 741–763.
[42] G. Albertini, B.M. McCoy and J.H.H. Perk, Commensurate-incommensurate transition in the ground state of the superintegrable chiral Potts model, Phys. Letts. A 135 (1989) 159–166.
[43] G. Albertini, B.M. McCoy and J.H.H. Perk, Eigenvalue spectrum of the superintegrable chiral Potts model, Adv. Stud. in Pure Math. 19 (1989) 1–55.
[44] V.V. Bazhanov and Yu. G. Stroganov, Chiral Potts model as a descendant of the six vertex model, J. Stat. Phys. 59 (1990) 799–817.
[45] R.J. Baxter, V.V. Bazhanov and J.H.H. Perk, Functional relations for transfer matrices of the chiral Potts model, Int. J. Mod. Phys. B4 (1990) 803–870.
[46] B.M. McCoy and S.-S. Roan, Excitation spectrum and phase structure of the chiral Potts model, Phys. Letts. A 150 (1990) 347–354.
[47] R.J. Baxter, Calculation of the eigenvalues of the transfer matrix of the chiral Potts model, in *Proc. of Fourth Asia Pacific Physics Conference*, Vol. 1, S.H. Ahn, I.-T. Cheon, S.H. Choh and C.Lee, eds (World Scientific, Singapore, 1991) 42–57.
[48] H. Au-Yang and J.H.H. Perk, The large-N limits of the chiral Potts model, Physica A 268 (1999) 175–206.
[49] R.J. Baxter, The order parameter of the chiral Potts model, J. Stat. Phys. 120 (2005) 1–36.
[50] V.A. Fateev and A.B. Zamolodchikov, Self dual solutions of star triangle relations in Z_N models, Phys. Letts. 92A (1982) 37–39.
[51] E.H. Lieb, Residual entropy of ice, Phys. Rev. 162 (1967) 162–172.
[52] R.J. Baxter, Three-colorings of the square lattice: a hard squares model, J. Math. Phys. 11 (1970) 3116–3124.
[53] R.J. Baxter, Critical antiferromagnetic square lattice Potts model, Proc. R. Soc. Lond. A383 (1982) 43–54.
[54] J. Salas and A.D. Sokal, The three state square lattice Potts antiferromagnet at zero temperature, J. Stat. Phys. 92 (1998) 729–753.
[55] S.J. Ferreira and A.D. Sokal, Antiferromagnetic Potts models on the square lattice: a high-precision Monte Carlo study, J. Stat. Phys. 96 (1999) 461–528.
[56] H. Au-Yang, B-Q. Jin and J.H.H. Perk, Baxter's solution for the free energy of the chiral Potts model, J. Stat. Phys. 102 (2001) 471–499.
[57] B. Davies, Onsager's algebra and superintegrability, J. Phys. A 23 (1990) 2245–2261.
[58] B. Davies, Onsager's algebra and the Dolan–Grady condition in the non-self-dual case, J. Math. Phys. 32 (1991) 2945–2950.
[59] E. Date and S-S. Roan, The structure of quotients of the Onsager algebra by closed ideals, J. Phys. A 33 (2000) 3275–3296.

[60] *Higher Transcendental Functions*, vol. 1 ed. A. Erdelyi, McGraw Hill, (New York, 1953).
[61] L. Onsager, discussion, Nuovo Cimento 6 Suppl. (1949) 261.
[62] C.N. Yang, The spontaneous magnetization of the two dimensional Ising model, Phys. Rev. 85 (1952) 808–816.
[63] R.J. Baxter, Partition function of the eight-vertex model, Ann. Phys. 70 (1972) 193–228.
[64] L.A. Takhtajan and L.D. Faddeev, The quantum method for the inverse problem and the XYZ Heisenberg model, Uspekhi Mat. Nauk 34(5) (1979) 13–63 (English translation: Russian Math. Surveys 34(5) (1979) 11–68).
[65] T. Deguchi, K. Fabricius and B.M. McCoy, The sl_2 loop algebra symmetry of the six vertex model at roots of unity, J. Stat. Phys. 102 (2001) 701–736.
[66] K. Fabricius and B.M. McCoy, Bethe's equation is incomplete for the XXZ model at roots of unity, J. Stat. Phys. 103 (2001) 647–678.
[67] K. Fabricius and B.M. McCoy, Completing Bethe's equations at roots of unity, J. Stat. Phys. 104 (2001) 573–587.
[68] T. Deguchi, The 8V CSOS model and the sl_2 loop algebra symmetry of the six vertex model at roots of unity, Int. J. Mod. Phys. B16 (2002) 1899–1905.
[69] T. Deguchi, Construction of some missing eigenvectors of the XYZ spin chain at the discrete coupling constants and the exponentially large spectral degeneracy of the transfer matrix, J. Phys. A35 (2002) 879–895.
[70] K. Fabricius and B.M. McCoy, New developments in the eight vertex model, J. Stat. Phys. 111 (2003) 323–337.
[71] K. Fabricius and B.M. McCoy, New developments in the eight vertex model II. Chains of odd length, J. Stat. Phys. 120 (2005) 37–70.
[72] V.V. Bazhanov and V.V. Mangazeev, Eight-vertex model and non-stationary Lamé equation, J. Phys. A38 (2005) L145–L153.
[73] V.V. Bazhanov and V.V. Mangazeev, The eight-vertex model and Painlevé VI, J. Phys. A39 (2006) 14869–14886.
[74] K. Fabricius and B.M. McCoy, An elliptic current operator for the eight-vertex model, J. Phys. A39 (2006) 14869–14886.
[75] V.V. Bazhanov and V.V. Mangazeev, Analytic theory of the eight-vertex model, Nucl. Phys. B 775 (2007) 225–282.
[76] K. Fabricius, A new Q matrix for the eight-vertex model, J. Phys. A 40 (2007) 4075–4086.
[77] S-S. Roan, The Q operator and functional relations for the eight-vertex model at root-of-unity $\eta = 2mK/N$ for odd N, J. Phys. A 40 (2007) 11019–11040.
[78] T. Deguchi, Regular XXZ Bethe states at roots of unity as highest weight vectors of the sl_2 loop algebra, J. Phys. A 40 (2007) 7473–7508.
[79] T. Deguchi, Irreducible criteria for a finite-dimensional highest weight representation of the sl_2 loop algebra and the dimensions of reducible representations, J. Stat. Mech. (2007) P05007.
[80] K. Fabricius and B.M. McCoy, The TQ equation of the 8 vertex model for complex elliptic roots of unity, J. Phys. A40 (2007) 14893–14926.

[81] H. Au-Yang and J.H.H. Perk, Eigenvectors of the superintegrable model I. sl_2 generators, J. Phys. A 41 (2008) 275201–275210.

[82] H. Au-Yang and J.H.H. Perk, Eigenvectors of the superintegrable model II. Ground state sector, arXiv:0803.3029.

[83] A. Nishino and T. Deguchi, An algebraic derivation of the eigenspaces associated with an Ising-like spectrum of the superintegrable chiral Potts model, arXiv:0806.1268.

[84] R.J. Baxter, A conjecture for the superintegrable chiral Potts model, J. Stat. Phys. 132 (2008) 983–1000.

[85] R.J. Baxter, Some remarks on a generalization of the superintegrable chiral Potts model, J. Stat. Phys (in press).

[86] K. Fabricius and B.M. McCoy, New Q matrices and their functional equations for the eight vertex model at elliptic roots of unity, J. Stat. Phys. 134 (2009) 643–668.

Part IV

Conclusion

However beautiful the strategy, you should occasionally look at the results.

Winston Churchill

16 Reductionism versus complexity

We began this book with a presentation of the general principles of statistical mechanics in chapter 1, and in chapter 2 we outlined the hopes of reductionism to use statistical mechanics to explain the properties of large systems in terms of the underlying microscopic interactions. We have found conditions under which the formalism is consistent and we have used the formalism to discover and prove many theorems about order and critical phenomena. We have presented a variety of computational techniques which are used to study macroscopic systems starting from the underlying interactions and seen that the use of these techniques has resulted in the creation of a great deal of mathematics which was unknown 40 years ago.

However, a comparison of what has been accomplished in chapters 3–15 with the goals set out in chapter 2 immediately reveals that there remains an immense gap in our ability to understand complex systems by means of a reductionist philosophy. At the most mundane level we have not been able to realistically derive or explain many of the phase diagrams presented in chapter 2. At the more refined level we are very far from explaining the relationships in the various levels of description set forth in Table 2.1 of chapter 2.

It must be freely admitted that many of the most interesting properties of large complex systems have not been mentioned at all in the first 15 chapters. Some topics have been omitted because of lack of space; some topics have been omitted because they are well treated in other books and disciplines; and many open questions have been discussed in detail. However, the goal of explaining complex phenomena by a reductionist philosophy is of such fundamental importance in the use of what we call the "scientific method" that it is worthwhile to examine reductionism and complexity from a much broader perspective. This concluding chapter is devoted to such a discussion.

16.1 Does history matter?

The fundamental starting point of the basic principles of thermodynamics presented in chapter 1 is the statement that we are confined to the study of systems in equilibrium, and the fundamental property of equilibrium as abstracted in the laws of thermodynamics is that we are dealing with systems that are described by state functions. To state this a more general language

Equilibrium statistical mechanics deals with phenomena where history does not matter.

It is this assumption of independence of history which allows these phenomena to be studied by the use of ensembles and averages. Conversely in order for the use of averages to be useful in the study of a complex system the phenomena must be such that is it independent of history. This is often encapsulated in the phrase **"all else being equal"**. In statistical mechanics we have implemented this phrase in the assumption of **equal a priori probability** that is embodied in the microcanonical ensemble.

There is a large and important literature which attempts to justify the use of the microcanonical ensemble from classical mechanics. The foremost goal of these studies is the reconciliation of the time reversal invariance of classical equations of motion with the concept of an irreversible approach to the equilibrium state. These studies usually consider systems where it can be proven that the only conserved quantity is the energy and that there is sufficient "chaos" in the system so that an initial volume of points in phase space will spread out and "cover" the surface of constant energy as $t \to \infty$. For such a system, if we wait "long enough" the past history of the system will become irrelevant.

However, it is important to admit that we have actually gone beyond this assumption of independence of history when in we discussed first order phase transitions where the system went from one ergodic component to another. This was seen in the observation that no singularity is seen in the hard sphere equation of state at the freezing density. When we try to study these phenomena by analytic continuation we call the procedure "metastability" but we should admit that in so doing we have actually abandoned the philosophy which leads to the microcanonical ensemble in the first place and have elevated analytic continuation to a law of nature.

The use of the concept of equilibrium and the use of averages is not restricted to statistical mechanics and is often used in economics, sociology, political science, traffic flow and animal behavior where the fundamental degrees of freedom are people or animals [1]. Of particular importance is the widespread occurrence in these systems of phenomena which are very similar to the phase transitions seen in the Ising model and in liquid-solid transitions.

But to use the concept of equilibrium and averages in such areas as the analysis of stock prices, in statistical studies of political systems, or sociology the independence of history assumption must be verified before an analysis based on averages can be trusted to give valid conclusions. It is often not clear how to verify this, and experience shows that there are many situations where the "most probable" result will in fact not occur because of past history or culture. For such phenomena the statistical method of averages can at best be used only with great caution and at worst is useless and can be grossly misleading.

Perhaps the most vivid example of a phenomenon driven by past history is the theory of evolution. It is universally accepted (but by no means mathematically proven) that the existence of life on earth is a very improbable event which cannot possibly be described by equilibrium statistical mechanics. To understand evolution we must be able to develop methods of analysis that will explain how systems that are far from equilibrium can exist for very long periods of time. In other words we must be able to study the physics of rare events which are very far from the average or from what is most probable.

16.2 Size is important

It was Saint Thomas Aquinas [2] who first explained that systems which exist on different length scales can exhibit totally different phenomena. The particular use that Aquinas makes of this principle is to define God as infinitely great as compared to man and that attributes of men such as form and personality cannot be used to describe God (i.e. Michelangelo's painting of God in the Sistine chapel is logically impossible).

In equilibrium statistical mechanics the principle that size is important is vividly seen in the phenomena of phase transitions. The notion of a phase which distinguishes solid from liquid from gas is a notion only applicable to a large number of molecules which has no counterpart in small systems.

The understanding that size is important and that large systems can have properties which small systems cannot have is not limited to the phase transitions observed in equilibrium statistical mechanics and is of vital importance in understanding the relations between the levels of description of Table 2.1.

One of the most important of these size-dependent properties is seen by comparing the phenomena which exist at the size of polymers with phenomena which exist at the size of a cell. At the level of cells it makes abundant sense to discuss the concept of life. It is surely possible to tell if a one-cell organism is alive or dead. However the concept of life certainly does not exist for systems of the size of polymer or small molecules. The very profound question then arises as to how many atoms a system must contain before it can have the property of life. In terms of Table 2.1 the question may be posed as follows

Can viruses be said to be alive?

Stated in this way the question can lead to heated and emotional debates and thus it is perhaps better to rephrase the question in a more quantitative fashion as:

Give a definition of life and, using this definition, compute the minimum size a system must have before life is possible.

A second and equally important size-dependent property is the concept of thought. In terms of computer science thought must be related to information capacity. One water molecule cannot think. Therefore the size-dependent question arises of how far up the scale of sizes in Table 2.1 do you have to go before the concept of thought is applicable? As one well-known example, it can be asked if it is logically possible to describe a gene as being "selfish". To the extent that being "selfish" involves an act of thought it would seem highly unlikely that a single gene could possess sufficient information capacity for the concept of being "selfish" to apply.

Size is also important to the question of the relevance of history discussed in the previous section. Larger systems not only have different properties from small systems but they will develop different time scales in which different properties will dominate. Even for systems where theorems can be proven that show that in the limit $t \to \infty$ the system will approach the equilibrium state of equal a priori probability it is not at all clear that for time scales of minutes, hours or days that this equilibrium state is reached. The fact that this "behavior in the long run" is often irrelevant for phenomena of interest is well expressed in the well-known quote of the economist John Maynard Keynes:

In the long run we are all dead.

In some sense equilibrium statistical mechanics deals with this dead situation.

16.3 The paradox of integrability

Any proof in classical mechanics of the relevance of the microcanonical ensemble relies ultimately on showing that the system has no other exact conservation laws in a finite box other than the conservation of energy. In other words, we use the existence of chaos in the classical equations of motion to justify the use of the microcanonical ensemble.

On the other hand the only problems in classical mechanics that have ever been solved are solved because of the existence of (a large number of) independent algebraic integrals of the motion. In statistical mechanics we have seen in great detail in this book that the only problems that we can solve in the thermodynamic limit are those where there are a large number of conservation laws such as are exhibited in the star–triangle equations of chapter 13.

So there arises the following paradox

If chaos is needed to justify the use of equilibrium statistical mechanics why are we allowed to derive intuition from integrable models which have no chaos at all?

Of course the easy resolution of this is to say that the integrable system is put into contact with a chaotic heat bath which enforces the microcanonical ensemble but has no other effect. But surely this is a situation which is not realized in practice. We discussed magnets in terms of Heisenberg or Ising Hamiltonians without asking how the thermalization actually occurred and we have no shred of evidence that a real system can be integrable and be described by a microcanonical ensemble at the same time.

An alternative statement of the paradox can be framed in terms of scaling theory. In chapter 5, scaling theory is given in terms of length scales: the size of the lattice (or range of the interatomic interaction) and the correlation length which becomes infinite as the temperature approaches the critical temperature. There are no other length scales in the problem, and the scaling function smoothly connects the scales together. This picture was confirmed in great detail in chapters 10–12 on Ising models and is further confirmed by all existing computations on the integrable models introduced in chapter 13. But our ability to do these computations relies on the existence of the conservation laws which follow from the star–triangle equations. A generic system will not obey a star–triangle equation and therefore the question arises of whether there are more length scales in the generic systems than can occur in the integrable systems. The answer to this question is unknown and controversial.

The question of the relation between integrability and chaos can be phrased in a more general framework. For a classical dynamical system with no chaos at all where the entire dynamics is determined by the large number of nongeneric conservation laws the initial conditions will be very important and thus the independence of past history will almost surely be lost. Thus the only problems we can solve are precisely those for which the assumptions needed for the existence of equilibrium fail.

16.4 Conclusion

The foregoing three sections highlight a few of the problems needed to be studied and overcome if a reductionist explanation is to be possible for large complex systems. This brief discussion is by no stretch of the imagination to be considered as complete or definitive. No attempt has been made at giving references for the reason that the literature is vast and ultimately is beyond the scope of this book.

However, it must be kept in mind that there are severe limitations to a reductionist philosophy of science and there are even more severe limitations if we wish to make actual quantitative predictions instead of just giving a qualitative plausibility argument. Eventually scientific thought and method merge into philosophy, and matters of fact merge into matters of belief. To carry out scientific investigation it is important to know where these boundaries lie. The discovery of these boundaries is not a science, it is an art and must be treated as such.

References

[1] P. Ball, *Critical mass: how one thing leads to another* (Farrar, Straus and Giroux 2004).

[2] T. Aquinas, *Expositio supra Librum Boethii de Trinitate* in *Thomas Aquinas Selected Philosophical Writings* (Oxford Univ. Press. 1993) pp 1–50.

Index